Textbook of Wood Technology

THE AMERICAN FORESTRY SERIES

Henry J. Vaux, *Consulting Editor*

ALLEN AND SHARPE An Introduction to American Forestry
AVERY Forest Measurements
BAKER Principles of Silviculture
BOYCE Forest Pathology
BROCKMAN Recreational Use of Wild Lands
BROWN, PANSHIN, AND FORSAITH Textbook of Wood Tèchnology
Volume II—The Physical, Mechanical, and Chemical Properties of the Commercial Woods of the United States
BRUCE AND SCHUMACHER Forest Mensuration
CHAPMAN AND MEYER Forest Mensuration
CHAPMAN AND MEYER Forest Valuation
DANA Forest and Range Policy
DAVIS American Forest Management
DAVIS Forest Fire: Control and Use
DUERR Fundamentals of Forestry Economics
GRAHAM AND KNIGHT Principles of Forest Entomology
GUISE The Management of Farm Woodlands
HARLOW AND HARRAR Textbook of Dendrology
HUNT AND GARRATT Wood Preservation
PANSHIN AND DE ZEEUW Textbook of Wood Technology
Volume I—Structure, Identification, Uses, and Properties of the Commercial Woods of the United States and Canada
PANSHIN, HARRAR, BETHEL, AND BAKER Forest Products
RICH Marketing of Forest Products: Text and Cases
SHIRLEY Forestry and Its Career Opportunities
STODDART AND SMITH Range Management
TRIPPENSEE Wildlife Management
Volume I—Upland Game and General Principles
Volume II—Fur Bearers, Waterfowl, and Fish
WACKERMAN Harvesting Timber Crops
WORRELL Principles of Forest Policy

Walter Mulford Was Consulting Editor of this Series from Its Inception in 1931 until January 1, 1952.

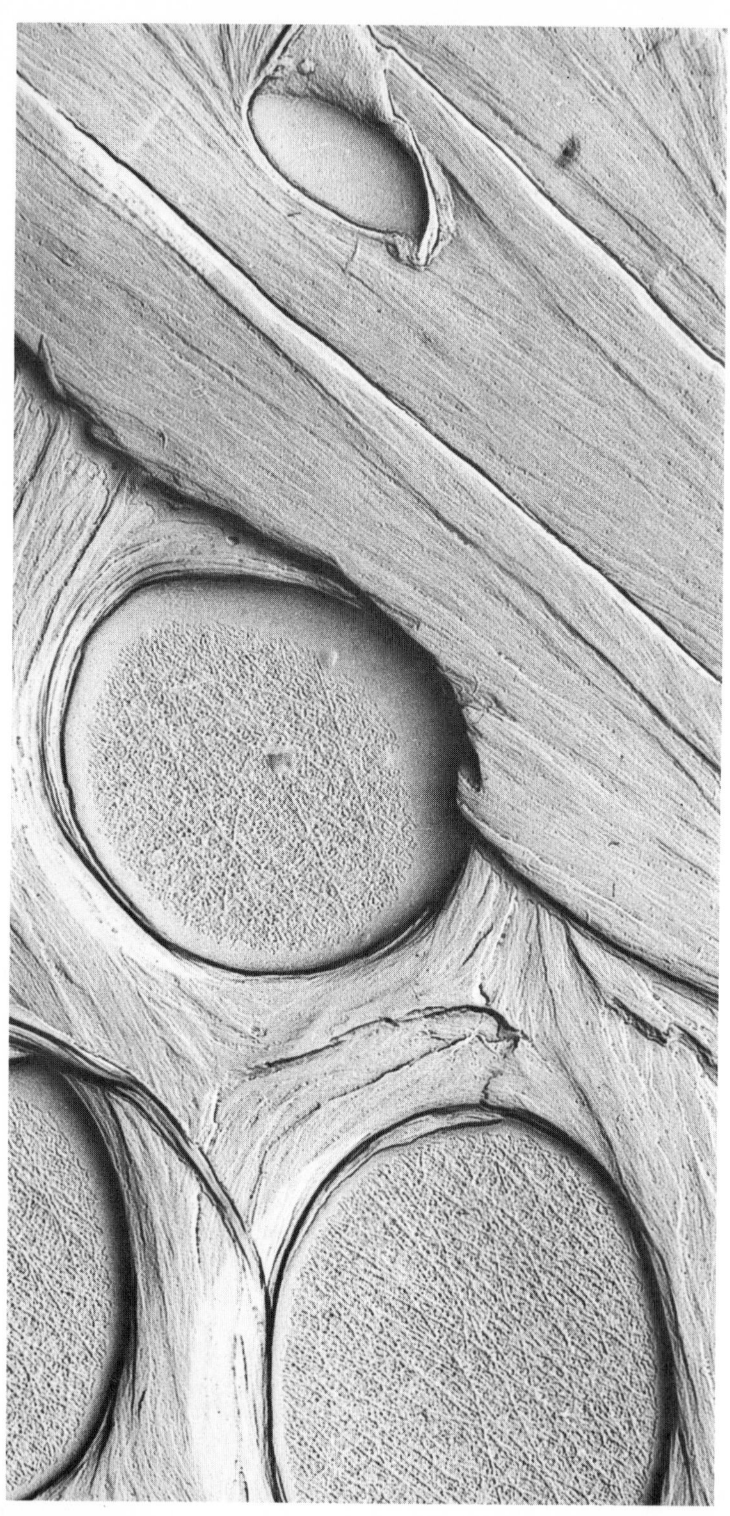

Intervessel pitting in basswood (*Tilia* sp.) revealed by split surface of the secondary wall. The lower left portion shows several pit chambers with the primary texture of their closing membranes and the distortion of the *S*2 layer about the pits. The diagonal part of the photograph at the upper right is the *S*3 layer with several pit apertures whose long axes are parallel to the microfibrillar direction in this layer.

Electron micrograph of a carbon shadowed replica at about 15,000×. (*Courtesy of W. A. Côté, Jr.*)

TEXTBOOK OF WOOD TECHNOLOGY

THIRD EDITION

A. J. Panshin

Professor and Chairman Emeritus of Forest Products
Michigan State University

Carl de Zeeuw

Professor of Wood Products Engineering
State University College of Forestry
at Syracuse University

VOLUME 1

STRUCTURE, IDENTIFICATION, USES,
AND PROPERTIES OF THE
COMMERCIAL WOODS OF THE
UNITED STATES AND CANADA

McGRAW-HILL BOOK COMPANY

New York, St. Louis, San Francisco, Düsseldorf, London, Mexico, Panama, Sydney, Toronto

Textbook of Wood Technology
VOLUME I

Library of Congress Catalog Card Number 73-118779
ISBN 07-048440-6

34567890 MAMM 7987654321

This book was set in Optima by Progressive Typographers, and printed on permanent paper and bound by The Maple Press Company. The designer was Paula Tuerk; the drawings were done by John Cordes, J. & R. Technical Services, Inc. The editors were James L. Smith and Margaret LaMacchia. Stuart Levine supervised the production.

Dedicated to the Late H. P. Brown, and to C. C. Forsaith, Whose Pioneering Work Helped to Elevate Wood Technology to the Status of a Science and Whose Enthusiasm for a Scientific Approach to this Subject Was an Inspiration to Their Students and Associates.

PREFACE

The following statement appeared in the preface to the second edition of this text:

During and following the Second World War the science of wood technology received a tremendous impetus because of the growing recognition of the role forest resources play in the economy of a modern society. This realization brought forth a demand for more accurate, as well as more extensive, basic information about wood as an industrial material.

The need for this information was met by vastly expanding research effort on the part of government-, private-, and university-supported laboratories in the United States and abroad. The very scope of this effort is an indication of the growing awareness by all concerned that utilization of wood must be placed on the same level of technological development as utilization of other major industrial materials. This is true even when the supremacy of wood is not challenged, as, for instance, when it is used in the manufacture of pulp and paper products. Even then competition for the available supply of timber and rising costs of production call for greater refinement of techniques of manufacture and hence for more intimate knowledge of the raw material involved.

It is not surprising, therefore, that a text which attempts to present information in a technical field that currently is undergoing a rapid transition from the traditionally empirical to science-oriented concepts requires an extensive revision to keep it abreast of the latest developments.

This statement is as true today as it was when first written. The material found in the second edition of this text has been further revised, amplified, and reorganized, in an attempt to bring it up to date and to make this volume more useful as a reference book to individuals interested in wood science and as a textbook for classroom purposes.

The book has now been divided into two parts: Part 1—Formation, Anatomy, and Properties of Wood, and Part 2—Wood Identification and Descriptions of Woods by Species. This has been done with the expectation that the separation of the material in this text into two parts will make it better serve the classroom needs, by allowing simultaneous use of this volume as a textbook and as a laboratory manual.

The introductory statement, pages 3 to 12, has been enlarged by the addition of some information contained in Chap. 12 of the second edition, now eliminated.

Chapter 2, Tree Growth, Chaps. 4 and 5, dealing with the structure of conifers and hardwoods, and Chap. 6, The Physical Nature of Wood, have been revised and updated in the areas where pertinent new material has become available.

Chapter 3, The Woody Plant Cell, and Chap. 7, Variability of Wood within a

Species, have undergone thorough revision. It is in these areas, particularly, where most of the active research in wood science has been in evidence in recent years, resulting in large amounts of new information.

The remaining sections of Part 1, Chaps. 8, 9 and 10, dealing with growth-related features and wood defects, have been revised to bring their contents in line with current information on these subjects, for instance in the treatment of reaction wood and spiral grain, and in Chap. 10 in the treatment of wood deterioration, stains, and natural durability of wood.

Part 2 opens with a brief Chap. 11, entitled Features of Wood of Value in Identification. This chapter, in some instances, touches upon subjects discussed in Chap. 3 and elsewhere. This has been done to allow immediate use of the keys and other information dealing with wood identification, without the necessity of first studying several chapters in Part 1 of this text. All the information on wood identification has been rearranged by separating softwoods from hardwoods. Keys I and IV have been considerably simplified. A new chapter, Wood Fibers and Their Identification, has been added to provide additional information on wood as a source of industrial fiber. Keys for Identification and Descriptions of Woods by Species have been critically reviewed, especially the sections dealing with the uses of the various woods.

The authors wish to express appreciation to their colleagues for advice and constructive criticism in the preparation of the manuscript. Thanks are due to P. R. Larson of the U.S. Forest Service, for assistance in providing information for some parts of Chap. 2, and A. C. Van Vliet and M. D. McKimmy of Oregon State University for their review of uses and other information on the western species of wood. The generous help of W. A. Côté, Jr. in opening the files of the electron microscopy laboratory at State University College of Forestry, and of Arnold Day in preparing excellent block photographs, Figs. 4-1 and 5-2, is gratefully acknowledged. The authors are indebted to Christen Skaar for careful analysis and criticism of the sections dealing with physical properties of wood, and to Harold Core for the thoughtful comments which have led to some important changes in presentation of material which, it is hoped, will make this edition a better teaching tool.

The authors are grateful to R. W. Meyer of the Canadian Forest Products Laboratory at Vancouver, Lawrence Leney of the University of Washington, R. J. Thomas of the State University of North Carolina, and J. J. Balatinecz of the University of Toronto for several photographs used in Chaps. 5, 6, and 9. A special note of recognition is due R. W. Wellwood of the University of British Columbia, whose thorough review of the second edition brought to the authors' attention a need for updating certain parts of the text, especially those under the heading of Descriptions of Woods by Species.

A. J. PANSHIN

CARL DE ZEEUW

CONTENTS

Textbook of Wood Technology

Introduction

We may use wood with intelligence
only if we understand wood.

—Frank Lloyd Wright
In the Cause of Architecture: Wood
The Architectural Record,
May, 1928

The progress of mankind from the primitive state to the present-day technologically highly advanced society has always been closely associated with man's dependence on wood. In prehistoric times man relied on wood for survival, for shelter, weapons, and for fire to cook his food and warm himself. Later, when he grasped the significance of its ability to float, and followed this with invention of the wheel, he relied on wood as the means of transportation over the land and across large bodies of water. A good case could be made for the contention that without wood the early migration of men and the later exploration of the world would have been impossible, or at least long delayed by inability to cross large bodies of water.

The greater the technological advances, the more diverse and sophisticated the uses man has found for wood. Today, in spite of the availability of numerous new materials, products of scientific research, even the most highly developed countries would find it difficult to maintain their high standard of living if deprived of access to wood products. It is easy to demonstrate that the higher the level of economic development, the greater is the dependence of men on wood, not only in many of its conventional forms but also in a variety of less readily recognizable items, such as paper and other wood-pulp products, transparent films, filaments, and fabrics, plastics and chemicals, to name but a few.

Since man's dependence on wood is of such antiquity it is only natural that an extensive empirical background on its uses should have been accumulated. But, as is often true when heavy reliance is placed on tradition, it is not surprising to find that even today, though some of the information on wood is factual, much of it is colored with prejudice or is distorted by the rule-of-thumb approach, while some is no more than fantasy. As a result, its exceptional qualities have only too often been taken for granted, or have remained unappreciated, while its faults have been grossly exaggerated. Paradoxically, the reason for this situation is the extreme versatility of wood, coupled with its almost universal availability to men of different cultures and degrees of technical attainment. On the one hand its fabrication and many of its applications are so simple that they require no special skills or technical knowledge by the user; on the other hand wood is a substance of a greater complexity than any other major engineering material, and its utilization under the competitive conditions of modern technology calls for a degree of scientific and technical understanding not possessed by most of its consumers.

A comprehensive knowledge of the characteristics of any material is essential to its correct utilization. In the case of wood the acquisition of basic information is complicated by the fact that wood is a product of the metabolism of a living organism, a tree, and because of that its properties are subject to wide variations, brought about by the external factors affecting the growth of a tree. Furthermore, wood is a product of not one but of many species of trees, and hence each kind of wood exhibits its own anatomical, physical, and chemical characteristics. It is important therefore to keep in mind that in dealing with wood we are confronted not with one standard, man-made material, manufactured to exact and relatively easily reproducible specifications, but with a variable substance whose basic nature is by and large beyond man's control. Finally, even disregarding the problems inherent in the variability of wood due to the influence of external growth factors and the multiplicity of species, wood is a substance of great complexity because of its intricate chemical and anatomical composition. This in turn is responsible for the complex nature of its physical properties.

The fundamental studies of wood must therefore take into account (1) properties that are common to all woods regardless of their botanical origin, (2) properties that are distinctive of certain species only and which frequently dictate the specific uses of these woods, and (3) the degree of variability that may be present within a species, because of environmental growth conditions and perhaps also because of the influence of genetic factors.

A. Properties Common to All Woods

All woods possess certain characteristics in common, for the following reasons:

1. All tree stems have a predominantly vertical arrangement and radial symmetry.
2. Wood, regardless of the source, is cellular in structure, and the chemical composition of the wood cell walls is remarkably similar in that the principal cell

wall constituents of all woods, regardless of their botanical origin, are cellulose, noncellulosic carbohydrates, and lignin.

3. All wood is anisotropic in nature, i.e., it exhibits different physical properties when tested along three major directional axes; this fact arises from the structure and orientation of cellulose in the cell walls, the elongated shape of the wood cells, and the longitudinal-radial arrangement of the cells in respect to the horizontal and vertical axes of the tree stem. This common pattern in the structure and orientation of the woody cells holds for all woods regardless of the great diversity in size, shape, and function of these structures encountered in the various species of trees.

4. Wood is a hygroscopic substance, i.e., it loses and gains moisture as a result of changes in the atmospheric humidity and temperature. Because of its anisotropic nature these moisture variations produce dimensional changes in wood which are unequal in the three axial directions; they are quite small in the longitudinal and appreciable in the radial and tangential directions.

5. Wood is susceptible to attack by fungi and certain microorganisms and insects; it is also flammable, especially when dry.

B. Properties Characteristic of Certain Species Only

It is common knowledge that woods of different species may vary widely in appearance and in physical and chemical properties. The most obvious and readily recognizable are the differences in appearance, i.e., variations in color, texture, and figure; these are traceable to the variability in the cell pattern produced by different combinations in the kinds, size, arrangement, and alignment of the cells found in the various woods and to the different kinds of extraneous substances found in the heartwood. Considerable variations between woods may also exist with respect to weight, hardness, strength properties, degree of dimensional changes, penetrability, insulating properties, machining, gluing and finishing characteristics, ability to hold fasteners, durability, and reaction to chemicals. These differences may be traced to the anatomical structure of the wood, its density, and quite frequently to the type and also to the amounts of the extraneous cell materials.

Existence of almost infinite variations and combinations of properties among the different kinds of woods presents a unique opportunity for matching the particular use requirements with the wood possessing the best combination of features for that specific purpose.

From this it is obvious not only that reliable information should be available on the appearance and the expected behavior of each type of wood in a given situation, but that its positive identification should be assured, as a basis for its marketing and its proper use. Sometimes positive identification of a wood is possible only by means of the botanical features of the tree itself; in most cases, however, the structural diversity of the wood, coupled with the recognizable variations in color, odor, weight, and appearance of grain and figure, affords a reliable means of identification. In some instances this can be done on the basis of the gross features of the wood; in other instances the use of a magnifying glass or of a microscope is essential. For this reason in this text emphasis is placed on the distinguishing characteristics of the more important commercial species of the United States as a means of wood identification.

C. The Extent of Variability within a Species

It is equally important to be able to recognize the extent of the variability that may exist in the wood of the same species. It is frequently overlooked that variability in the properties of wood of the same species may be of even greater significance than the relatively minor discrepancies in the wood characteristics among different species.

Variability in the wood characteristics of the same species may occur from tree to tree, may be evident in the wood taken from various places in the same tree, or may even be found within a given piece of wood. Such diversity in wood properties is largely a result of the growth pattern of trees, as well as of environmental influences affecting the growth of trees, such as climate, soil, moisture, and growing space; undoubtedly genetic factors, at present mostly unrecognized, also play an important role.

Again, of the factors responsible for the variability in the wood characteristics of the same species, appearance, rate of growth, density, and grain alignment are most readily recognizable. On the other hand, differences in the strength properties, dimensional stability, or durability may be related to less-obvious reasons, such as kinds and arrangement of cells within the growth ring, composition and amounts of the extractives, alignment of the microfibrils in the cell walls, or the thickness and the structure of the secondary wall.

The extent of the variability in these more obscure wood characteristics is difficult to recognize without special equipment and techniques, and their precise effect on the physical and chemical properties of wood is even harder to interpret.

The variability may create significant problems in the utilization of a wood. It should be noted, however, that it may also serve to extend the range of usefulness of the species, provided the extent and the effect of departures from the norm are recognized and steps are taken to segregate the material accordingly. Grading of wood is one means for coping with the effects of variability in the wood characteristics of the same species, but considerable equalization of wood properties may also be achieved by design and treatment of the manufactured articles.

What effect heredity may have on the variability of wood quality within a species is as yet largely unknown. That the diversity in wood characteristics in forest trees grown under natural conditions, due to genetic factors, may exist seems quite plausible, even though this possibility has remained largely unexplored. Selective breeding and other forest-tree-improvement procedures, recently instituted as an important tool in forest management, hold great promise for a more scientific approach to the quantity and quality of wood production. For maximum results this approach to selectively controlled wood production calls for a thorough knowledge of the biological nature of wood, as well as for an insight into the specific requirements of its major industrial users.

D. Wood as Industrial Raw Material

Most of the characteristics of wood that make it an outstanding industrial material are traceable to one or more of its basic properties. Likewise its faults are derived from the same basic wood characteristics. A comprehensive knowledge of wood, i.e., of its composition, its chemical and physical behavior, and the causes of its variability as they affect its utility, should therefore be the basis not only of its present and potential utilization but of future forest management policies, as well.

The uses of wood are so many and so diverse that they cannot be enumerated in this text. Broadly speaking, the major uses of wood may be classified as those for (1) fuel; (2) pulp and paper, including a number of derived products, such as cellulose filaments and films, plastics, and explosives; (3) chemicals; and (4) structural and other applications in which it is employed in its solid and largely unmodified state.

Wood is an important source of energy. In technologically less-advanced countries it may still be the exclusive source of domestic as well as industrial heat. In many parts of the world its continuing use as fuel has brought about depletion of forest resources to a dangerously low point. On the other hand in the more advanced countries, including the United States, its role as domestic fuel has largely been relegated to the luxury status as fireplace wood or as charcoal, while coal, oil, gas, and more recently nuclear power have replaced wood as a source of industrial power. There should be no regrets about the declining use of wood as fuel, because it constitutes by far its lowest and least-efficient form of utilization. Nevertheless the ability of wood, and of the wood charcoal and methyl and ethyl alcohols derived from it, to produce energy is another indication of its many-sided versatility.

The supremacy of wood as a raw material for pulp and paper products is unquestioned. There is no other natural substance that can meet the ever-increasing demands of modern society for paper and other pulp products. At present it seems unlikely that a synthetic material could be produced economically to rival wood as a source of pulp.

The list of chemicals that could be obtained from wood might require pages of text. Only a few of these are commercially important in the United States at the present time. What is important, however, is the potential that wood offers as raw material for the highly diversified chemical industry.

But it is when wood is employed in its basically unaltered form, in construction, in furniture, containers, power-transmission lines, transportation, and a host of other industrial applications, that it offers the greatest challenge and opportunity to the designer and the user. It is available to them in a large variety of shapes and sizes, in the form of poles, piling, ties, lumber, shingles, veneer, and plywood, and in various types of wood-composition products.

But it is also in some of these applications that wood meets its stiffest competition from other natural and from many synthetic products developed by modern technology. Some of these materials have rightfully supplanted wood because of their technical superiority in certain properties or the economic advantages they offer. In other instances these new products, though they offer no clear-cut technical and financial advantages, capitalize on the glamour attached to their newness and benefit from superior merchandising. In still other cases wood, though clearly superior to the other known materials, suffers from inability or unwillingness on the part of many of its users to avail themselves of the wealth of technical data available to them.

E. Wood as a Construction Material

By far the most extensive use of wood, especially in its solid form, is in construction. Because of this, in the paragraphs that follow, a brief summary is offered of the characteristics of wood that account for its remarkable versatility and the unique position it occupies among other structural materials.

1. Wood may be cut and worked into various shapes with the aid of simple hand tools or with power-driven machinery. It therefore lends itself well not only to conversion in a plant but also to on-the-site fabrication. It is the latter fact that largely keeps conventional wood-frame construction fully competitive with any method of complete prefabrication of houses yet designed.

2. Wood can be joined with nails, screws, bolts, or connectors, all of which require the simplest kinds of tools and produce strong joints. Wood also can be fastened with adhesives, which when properly used produce a continuous bond over the entire surface to which they are applied. Because of this, glued surfaces can develop the full shear strength possible for wood, in contrast with welding in steel or with mechanical fasteners, all of which develop only limited areas of shear resistance.

Gluing also provides a practical means of fabricating wood members of different shapes and of almost unlimited dimensions. Gluing allows wood to be combined with other materials, such as plastics or metals, to produce composite products with properties that exhibit the best characteristics of the joined materials. In the form of laminated beams, wood structural members are available in sizes capable of supporting roofs over clear spans 100 feet and more in extent. Plywood sheets provide a ready means of stressed-skin panel applications in roofs, floors, and walls. Prefabricated trusses and panelized wall sections offer equal opportunity for partial or complete prefabrication of houses.

3. Although wood undergoes dimensional changes with variations in the moisture content, such changes are relatively insignificant in the direction parallel to the grain, the direction of the grain most important in structures. Also these changes normally require a long time to occur. For example, a moisture-content pickup from 6 to 12 percent in a 60-foot-long wood beam will cause a change in length of only 0.34 inch, and it will call for several weeks of continuing high relative humidity, a condition which seldom exists in practice, to effect it.

The very real problems arising from across-the-grain dimensional changes in wood can be considerably minimized by careful attention to selection of the species and the stock, by proper design, and by following the simple rule that, where dimensional changes may cause a significant problem, wood, before being placed into a structure, must be dried to the average moisture content it is expected to attain in service.

Finally, stabilization of wood is achieved by forming it into plywood, in which the anisotropic dimensional properties of wood are so distributed that shrinkage and swelling of the resulting sheet are at a minimum and are approximately equal in the two principal directions. It is expected that some of the chemical methods of wood stabilization, e.g., treatment with polyethylene glycol, or wood polymer systems using catalyst-heat or radiation techniques, may soon become economical for large-scale applications.

4. Dimensional changes that may take place as a result of rise in temperature are less significant in wood construction than they are in construction utilizing metal structural members. When heated, wood expands across the grain as much as metals or more, but only little in the longitudinal direction, which is of the utmost importance in construction. Moreover, in wood, increase in dimensions with rise in temperature is frequently balanced to a considerable degree by shrinkage caused by drying, with a corresponding increase in strength. There is no such compensating effect in the metal structural members, which expand and lose strength progressively when heated.

For instance, if during a fire room temperature rises from 70°F to 1100°F, or higher, a wood beam of a cross section adequate to span 60 feet would increase in length by about 1.5 inches. In fact, this elongation would be considerably less because of the concurrent shrinkage induced by the moisture loss. Furthermore, the stated amount of elongation is predicated on the assumption that the wood beam would have reached a uniform temperature across the entire section. Because of the low conductivity of wood and the insulating properties of the charred surface, this would not be the case.

Under similar conditions, a steel beam of the same length will elongate approximately 4.95 inches. If by then the side walls of the building have not yet collapsed from the effect of steel elongation, the steel beam will sag because it is unable to support even its own weight when heated to that temperature or beyond.

5. Wood is a combustible material, but it loses its strength gradually under fire conditions, if used in large-enough sizes. Wood structures built with heavy timbers therefore provide an element of safety in fire not possessed by most other construction materials.

When subjected to extreme heat wood decomposes with evolution of combustible gases and tarry substances, leaving behind a charcoal residue. This process is known as pyrolysis. The low thermal conductivity and high specific heat of wood keep the pyrolysis reaction from spreading rapidly into the interior of a piece of wood. Charcoal which forms on the surface, until it reaches the glowing stage, is an excellent heat-insulating material. As it forms on the surface it slows the pyrolysis reaction still further. The charring rate of untreated wood is $^1/_{30}$ to $^1/_{50}$ inch per minute, or about $1^1/_2$ inches per hour. At the same time the temperature $^1/_4$ inch ahead of the charring is only about 360°F, i.e., much lower than is required to sustain pyrolysis.

Because of this self-insulating property of wood, the strength of the burning beam is reduced gradually and only in proportion to the amount converted to charcoal. This is particularly important in heavy solid and laminated structural wood members. In case of fire buildings constructed of such members retain their structural integrity long enough to allow firemen to combat fire in relative safety, without imminent danger of collapse. A wood building is also easier to repair, if it has suffered limited damage, than are structures built of other materials.

Additional fire protection can be provided to a wood structure by the use of wood impregnated with fire-retardant chemicals, or by application of fire-retardant coatings.

6. Wood is surprisingly durable when used under conditions which are not deliberately favorable to the wood-destroying agencies. Many instances are known of wood, protected from dampness and from attacks by insects, lasting centuries and even several millennia. Such extreme examples of the enduring qualities of wood were found, for instance, in the 2700-year-old wood beams in the tomb of King Gordius, near Ankara, Turkey. Sound beams, many centuries old, have recently been removed from the ancient temples in Japan. In this country, perhaps less spectacular but nevertheless impressive are the old colonial homes in New England and in the South, with the original wood exterior and interior in a well-preserved state after 200 years or more.

There is no reason why, if properly used, wood should not last indefinitely. Decay and insect damage can be largely eliminated by following sound methods of design in construction and by using properly seasoned wood. In situations where biological wood-destroying agencies are especially difficult to control, wood can be made to last by impregnation with suitable preservatives.

7. Wood does not corrode. The major constituents of wood are quite inert to the action of a great many chemicals. This recommends it for many industrial applications where resistance to the disintegrating action of chemicals and to corrosion is important.

When unprotected wood is exposed to atmospheric conditions it will weather; i.e., it will darken and its surface will become roughened. A slow erosion, at the rate of about 0.25 inch in a century, will follow. It is mainly to prevent weathering that the exterior surfaces of wood buildings are painted. The other reasons for painting wood are decoration and the reduction of the natural porosity of the wood surface.

8. Because of its fibrous structure and the quantity of entrapped air, wood has excellent insulating properties. With the exception of wood, the common building materials used in house construction are not good insulators (see Table 6-1). In comparison with wood, the heat loss through common brick is 6 times, and through a glass window 8 times as great. Concrete made with sand and gravel aggregate is 15 times as conductive as wood, steel 390 times, and aluminum more than 1700 times. Well-constructed wood windows provide a considerable advantage over highly heat-conductive metal windows by reducing heat transmission in and out of the building and minimizing vapor condensation in cold weather.

Wood provides thermal insulation the year around; i.e., wood insulation is effective not only in the winter against cold but also in the summer against heat. Finally, wood is the only major structural material that combines outstanding structural characteristics with effective thermal insulating properties.

Wood when reasonably dry is also a poor conductor of electricity. This is important in such uses as powerline poles and ladders used in fire fighting, and may be significant in dwellings by minimizing the danger of shock from broken or exposed wires coming into contact with the structural members of a building.

9. Because of the nature of the cell wall substance and its distribution as a system of thin-walled tubes, wood possesses excellent flexural rigidity and outstanding properties in bending. For instance, when Douglas-fir is compared with low-carbon structural steel on a specific strength basis (see Chap. 6), it is found that for *equal weights* of material, even when comparison is made on the basis of reduced design stresses allowable for commercial timbers, Douglas-fir is superior to steel in bending by a ratio of 2.6 to 1.

The high flexural rigidity of wood is most effective in members, such as beams, in which length is far in excess of depth, and for long slender columns. For example, an 8 × 4 inch steel I beam, 20 feet long, used as a column, will support an axial load of 62,150 pounds when designed according to the American Bridge Company's standards. A wood column of comparable weight, in the form of an air-dry tapered Douglas-fir pole, with an effective diameter of 10.5 inches, will carry 82,000 pounds when axially loaded.

Wood is at least nine times as good an energy-absorbing medium as steel. This makes it an excellent material for floors and similar applications where energy absorption is important to the object responsible for the applied load.

Wood structures can be designed to carry impact loads that are twice as great as those they can sustain under static loading. This can be contrasted with steel and concrete, for which no increase in loads is allowed under similar conditions. The exceptional impact strength of wood gives it a considerable mechanical and economic advantage for structures designed to resist earthquakes, or for situations where abrupt loads are imposed, e.g., in aircraft-carrier decking.

Unlike steel, wood possesses excellent vibration-damping characteristics. This property is of utmost importance in bridges and other structures subjected to dynamic loads.

10. Not the least of the superior characteristics of wood is the enduring beauty of its surface pattern, formed by almost infinite variations in grain, texture, and coloration. Wood paneling and wood furnishings are the choice when it is desired to create a feeling of warmth and intimacy in a home or to produce impressively dignified and beautiful surroundings in public buildings. It is no accident that competing materials pay an unconscious tribute to wood by imitating, if not always successfully, its grain and color.

These few examples do not by any means present an exhaustive treatment of all the outstanding characteristics of wood. Rather they serve to illustrate the unlimited possibilities that exist when advantage is taken of the available technical knowledge.

However, in spite of its many proved advantages as a construction material, or as a raw material in its many other industrial applications, adherence to antiquated methods of construction and failure to understand and to make full use of its outstanding technical qualities frequently place wood at a disadvantage in competition with other products. Many of these disadvantages, real or implied, could be overcome by intelligent use of wood, based on comprehensive knowledge of its properties.

Finally, it should never be forgotten that wood is the only major raw material with a wide range of industrial uses, which is renewable, and hence capable of meeting the needs of a nation in perpetuity. It is therefore inconceivable that this priceless material should be either grown or used without due consideration of the known scientific facts concerning its nature.

PART 1

Formation, Anatomy, and Properties of Wood

1

The Plant Origin of Wood

I. TYPES OF PLANTS PRODUCING WOOD

Wood is of plant origin. Not all plants, however, possess woody stems, and not all that do possess woody stems produce timber suitable for use as an industrial material. A brief discussion of the characteristics of woody plants and of plant classification follows in order to trace wood to its botanical source.

A. Characteristics of Woody Plants

The following criteria serve to distinguish woody from nonwoody plants:

1. Woody plants must be *vascular plants*, i.e., must possess specialized conducting tissues consisting of *xylem* (wood) and *phloem* (inner bark). The xylem is lignified (see page 71) and is the wood of the mature plant. Plants devoid of vascular tissue cannot produce wood.

2. They must be *perennial plants*, i.e., must live for a number of years.

3. They must possess a stem that *persists* from year to year. Many perennials fail to be classed as woody plants because their stems die back to the ground each autumn, the roots persisting through the winter and producing a new stem the fol-

lowing spring. Other plants possess persistent creeping stems and hence fall into the category of woody plants, even though they appear to be herbaceous.

4. Typical woody plants exhibit *secondary thickening;* i.e., they have a means of thickening their stems by subsequent growth in diameter, not traceable to terminal growing points. This is achieved through the activities of a growing layer, called *cambium,* which is situated just outside the last-formed layer of wood and beneath the inner bark (phloem); new wood and new phloem are thus produced yearly and are inserted between the previously formed wood and bark. In this manner, in the case of trees, the trunk eventually attains a diameter of sufficient size to be converted into lumber or other useful products.

B. Kinds of Woody Plants

Woody plants are of three types, *trees, shrubs,* and woody *lianas,* among which no hard and fast lines can be drawn. For example, a plant may be shrubby near the limits of its range and arborescent elsewhere. Certain species of *Ficus* (figs) begin life as woody lianas but eventually become arborescent. Many woody plants that are reduced to dwarfed, scraggly shrubs in the boreal zones or at high elevations attain to the dignity of large shrubs or even trees to the southward or at lower altitudes where they are not forced to contend with such a hostile environment. In general, the kinds of woody plants may be defined as follows:

1. A *tree* is a woody plant that attains a height of at least 20 feet at maturity in a given locality and usually (not always) has but a single self-supporting stem or trunk.

2. A *shrub* is a woody plant that seldom exceeds 20 feet in height in a given locality and usually (not always) has a number of stems. Many shrubs have prostrate primary stems embedded in the soil or leaf mold; these send up persisting secondary branches at intervals (ex., *Gaultheria procumbens* L.—aromatic wintergreen), which appear as separate individuals.

3. A woody *liana* is a climbing woody vine. Woody lianas climb by twining, clambering, aerial roots, tendrils, etc., and are characteristic features of tropical rain forests in many parts of the world (ex., *Calamus* spp.—rattans).

II. PLANT CLASSIFICATION

In classification, plants are first divided into large divisions which have certain gross features in common, then successively into smaller divisions. The characters enumerated under the successive divisions are increasingly specific and of narrower latitude.

There are four main divisions in the vegetable kingdom (each divided further into classes, orders, families, genera, and species, as necessity demands), viz.,

Thallophytes... Algae, fungi, bacteria, etc.
Bryophytes... Liverworts and mosses

Pteridophytes.. Ferns, scouring rushes, horsetails, club mosses, and quillworts
Spermatophytes... All seed plants

Gymnosperms
Cycadales Coniferales
Ginkgoales Gnetales

Angiosperms
Monocotyledons Dicotyledons

Thallophytes constitute the lowest division of the vegetable kingdom and include the simplest forms of plants. The plant body, or thallus, exhibits little variation or specialization in structure (though often a wide range of form) and usually carries on its life activities either in water or on a moist substratum. Included in this group are the algae (pond scums, seaweeds, etc.) and the fungi (mushrooms, bracket fungi, etc.), both of which exhibit a remarkable variation in form and size of the thallus, but extreme simplicity in its structure. Many of the simplest Thallophytes are unicellular, and some are free-swimming and resemble minute animals. Sexuality has become well developed in many forms, whereas in others it is totally lacking, possibly through degeneration occasioned by a parasitic or saprophytic habit.

Bryophytes are best represented by the mosses, although a second group, the liverworts, is also included. The Bryophytes show a distinct advance in specialization over the Thallophytes. This is evinced through the definite establishment of a sexual stage in which the sexes may be distinguished, and an "alternation of generations," whereby a sexual stage or generation is followed by a semidependent, asexual stage, which in turn again gives rise to sexual forms. Though more specialized than Thallophytes, Bryophytes are, relatively speaking, simple plants. The plant body is an elementary structure, which possesses chlorophyll and is in some cases thalloid, whereas in others it develops a primitive stem and leaves. True vascular tissue (vascular bundles) is entirely lacking.

Vascular plants made their appearance for the first time in the *Pteridophytes*, a group that includes the true ferns and the forms that are recognized as fern allies, the horsetails, scouring rushes, club mosses, and quillworts. True roots, stems, and leaves equipped with special conducting or vascular tissue, composed of xylem (wood) and phloem, have become established as definite structures and function as in the seed plants. As in the Bryophytes, there is a sexual stage in which the sexes may be distinguished, but the sexual organs have become increasingly specialized. A fertilized egg develops without pause into an asexual stage in which sexless individuals, through spore formation, again give rise to sexual forms. In the higher Pteridophytes, it is the asexual or sporophytic stage that has become dominant, and the sexual generation has been relegated to an obscure, independent existence or has become actually dependent on the asexual generation. Pteridophytes were formerly represented by a vast assemblage of plants, many of which were arborescent and flourished during the Carboniferous period, contributing largely to the formation of our coal deposits of today. Owing to an altered environment and the development of seed plants, which are better adjusted to withstand present-day conditions, the group is now on the wane and is represented by only some 4000 species.

The dominant plants of today are the *Spermatophytes,* or seed plants. They are vascular plants, which represent the highest type of specialization to date, though not necessarily the final type. As in the Pteridophytes, the asexual, or sporophytic, stage bears true roots, stems, and leaves and is independent; the sexual, or gametophytic, stage has undergone further reduction and is wholly dependent on this. The most striking difference between Spermatophytes and Pteridophytes lies in the formation of seeds in the first-mentioned group; these are dormant structures entailing a pause in the development of the young sporophyte that has developed from a fertilized egg nucleus. Seeds permit of the wider and more rapid dissemination of the plants possessing them and tide the plants over unfavorable periods. Sexuality is a necessary part of the life cycle of Spermatophytes and is brought about through the transfer of male nuclei to the proximity of the female nuclei by means of pollen grains. Following the fusion of the sex nuclei, made possible by the formation of a pollen tube serving as a siphon, a young sporophyte (embryo) is formed within the ovule (developing seed), which, as the latter matures, passes into a dormant condition. Upon subsequent germination of the seed, the young sporophyte again assumes an active existence.

The Spermatophytes in turn are divided into two subdivisions: the *Gymnosperms* and the *Angiosperms,* distinguished by the manner in which the seeds are borne. The word "gymnosperm" is derived from the Greek *γυμνός*, meaning naked, and *σπέρμα*, seed, and includes those Spermatophytes in which the seeds are not enclosed in an ovary but are borne naked, subtended by scales or fleshy structures. "Angiosperm" comes from the Greek *ἀγγεῖον*, meaning vessel, and *σπέρμα*, seed, and embraces those forms in which the seeds are enclosed in an ovary, which may or may not dehisce at maturity. The boundary between the two groups is sufficiently clear to serve the purposes of classification, although it in no way indicates the disparity in numbers and size.

Gymnosperms are very ancient and, in terms of numbers, form but a small part of the present seed-plant vegetation. Some 650 living forms exist today, grouped in four orders, which are to be regarded as the surviving remnant of a vast phylum that had its genesis in the Carboniferous period and flourished during the Triassic. Angiosperms were evolved comparatively recently (lower Cretaceous period) in a geological sense and now are represented by a vast assemblage of at least 250,000 species, which comprise the bulk of the seed-plant vegetation of the present day.* The most obvious characters that set off this group from the Gymnosperms are the presence of the flower with its generally showy perianth, stamens, and pistil and the manner in which the ovules, or immature seeds, are borne enclosed in an ovary.

* One school of thought holds that the Angiosperms have been able to attain and hold the ascendancy over other groups because of adaptive features which they have developed to meet the environmental conditions in force at the present time.

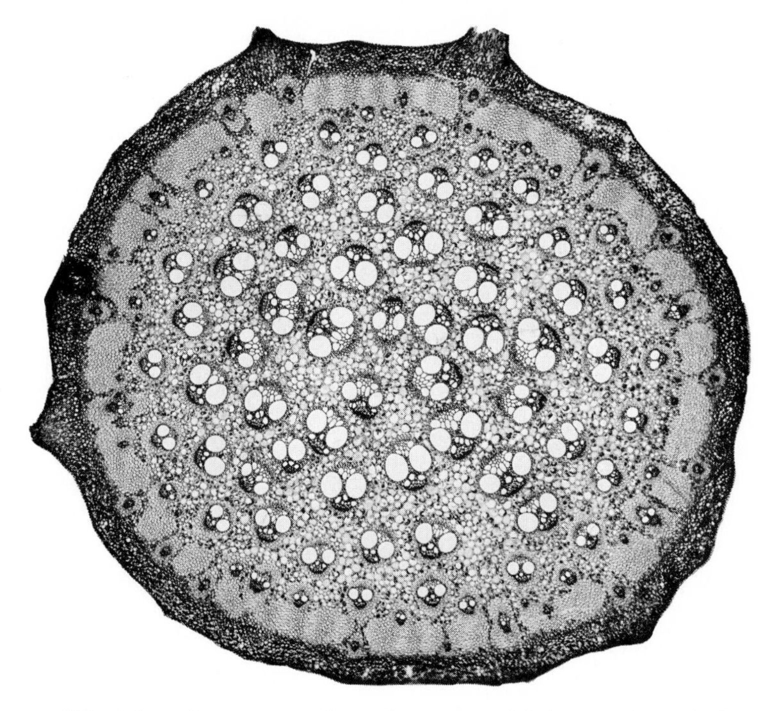

FIG. 1-1 Transverse section of a monocotyledonous stem, *Smilax hispida* Muhlb., showing scattered vascular bundles. (15×.)

Two classes of Angiosperms are recognized, the ***Monocotyledons*** and the ***Dicotyledons***. As the terms imply, the Monocotyledons are featured by one seed leaf (cotyledon), which is terminal on the axis, the Dicotyledons by two seed leaves, which are lateral. Another striking difference between these plants, and an important one in plant anatomy, rests in the fact that the vascular bundles of Monocotyledons are scattered in the stem (Fig. 1-1), whereas those of the Dicotyledons are arranged in a ring, or the stem contains a vascular cylinder enclosing a pith (Fig. 1-2). Monocotyledons are now held to be the more modern group that arose originally from the Dicotyledons. They include at least 30,000 species and are arranged in 7 orders embracing 48 families. The larger division, the Dicotyledons, embraces 30 orders, and over 200,000 species; some of these orders are represented wholly by herbaceous forms whereas others consist wholly of woody plants or contain both woody and herbaceous species.

III. POSITION OF TIMBER-PRODUCING TREES IN PLANT CLASSIFICATION

The vascular origin of wood has already been indicated. Thallophytes and Bryophytes are ruled out as potential sources of wood because they are devoid of vascu-

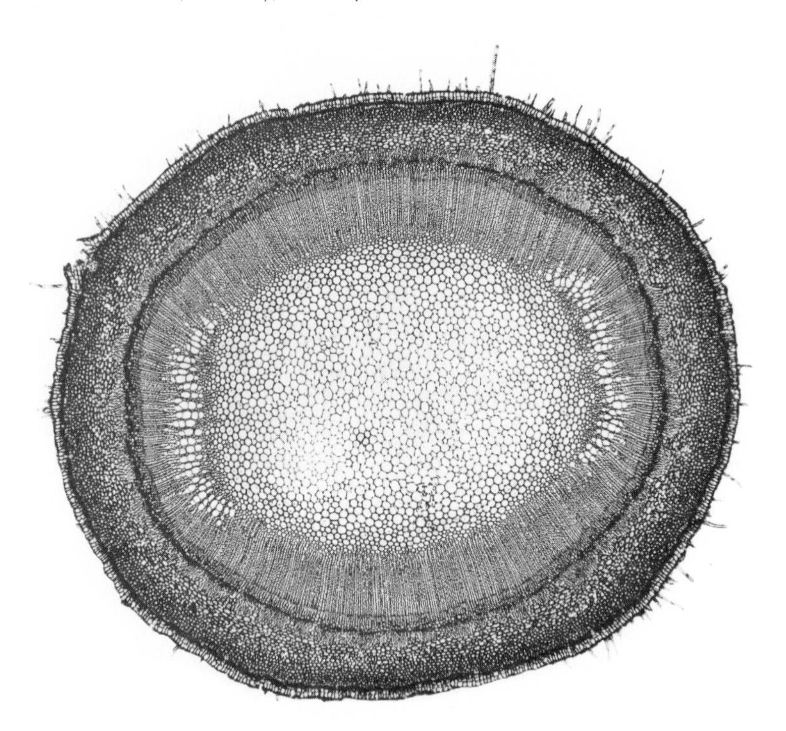

FIG. 1-2 Transverse section of a 1-year dicotyledonous stem, *Fraxinus pennsylvanica* Marsh., showing a vascular cylinder (xylem, cambium, and phloem) surrounding the pith. The cylinder is bounded by cortex, periderm, and epidermis, in the sequence stated. (25×.)

lar (specialized conducting) tissue. The vascular nature of Pteridophytes was pointed out on page 15; they are all woody in the sense of possessing persistent stems, but they have not been a source of wood commercially in modern times. The stems of the great majority are too small and short-lived for conversion into lumber; further, the structure of the stem (arrangement of the vascular tissue) is not of the type that characterizes the higher plants, a situation that holds for the arborescent ferns even though they attain the stature of trees.

Lumber-producing trees are restricted to the Gymnosperms and the Angiosperms of the Spermatophytes; but, here again, certain reservations must be made. The Gymnosperms, like the Pteridophytes, are all woody, and four living orders are recognized: *Cycadales, Ginkgoales, Coniferales,* and *Gnetales,* listed in the probable order of their evolution. The Cycads consist of nine genera and about eighty species and are woody plants of the tropics that resemble tree ferns and palms; the stem is unbranched, attains a height of 60 feet in some species, and bears a cluster of large pinnate leaves at the top. Although the stature of these trees is such as to permit of their conversion into lumber, the structure of the stem is not of the normal type, precluding this possibility.

The next order, the Ginkgoales, is restricted to a single species, *Ginkgo biloba* L., of China and Japan.* This is a deciduous tree with the habit of a conifer and fan-shaped leaves like the pinnae of the maidenhair fern; hence the name "maidenhair tree." The growth is of the normal type, and the wood is quite suitable for commercial use; but the tree is too limited in its range to make it a factor in the production of wood.

Nor are the Gnetales, probably the most recent order of Gymnosperms in a phylogenetic sense, a source of wood: some sixty-odd species are recognized, grouped in three genera. One, *Welwitschia mirabilis* Hook. f., is a monotype, found in the deserts of western South Africa, with the habit of a large turnip from the top of which two broad, flat, strap-shaped leaves are borne; the others (species of *Ephedra* and *Gnetum*) are vines, shrubs, or small trees.

The Coniferales (Coniferae) alone among Gymnosperms are productive of timber on a commercial scale; the various members of this group are known as conifers, evergreens, or softwoods and are the source of the softwood lumber of the trade. Forty-six genera† and about five hundred species are recognized, many of which are important timber trees; in fact, the chief economic product of this order is wood, and it occupies a unique position in this respect which is out of all proportion to its size botanically. The situation holds for the following reasons:

1. Coniferous trees are frequently gregarious and often cover wide tracts with almost pure stands, or stands consisting of relatively few species. The economics involved in the logging of such stands are less complex than of stands composed of a greater number of species.
2. Coniferous forests attain their maximum development and area in the temperate zones where, to date, industry, with its many demands for wood, has attained its greatest impetus.
3. Coniferous trees exhibit excurrent (monopodial) growth, i.e., possess stems that continue through the crown and end in a terminal leader, frequently with little taper below; such stems convert well into lumber, with the minimum amount of wastage.
4. Coniferous woods are of a type entirely different from those produced by broad-leaved trees; they lend themselves to many uses for which the woods of broad-leaved trees are less suited.

As has been previously pointed out, the Angiosperms, in comparison with the Gymnosperms, comprise a very large and dominant group of at least 250,000 herbaceous and woody plants, which in turn are divided into Monocotyledons and Dicotyledons. Many Monocotyledons are arborescent (palms, yuccas), and some attain

* Widely grown as an ornamental tree in the United States.

† Forty-seven genera if the monotypic genus *Sequoiadendron* proposed by Bucholz is accepted. For further information, see J. T. Bucholz, The Generic Segregation of the Sequoias, *Am. J. Botany*, **26**(7):535–538 (1939). In 1940 still another monotypic genus, *Metasequoia*, was described from China, increasing the total to 48 genera.

to the dignity of large trees. Their trunks are not infrequently used "in the round" for house posts, poles, etc., but cannot be sawn into lumber of the normal type because the vascular bundles retain their individuality as such and are scattered in the stem.

Dicotyledons are the source of the hardwood lumber of the trade. This heterogeneous group includes herbaceous and woody plants, the latter consisting of shrubs, trees, and lianas. Hardwoods, as such, are readily distinguished from softwoods by certain easily recognized features, to be explained subsequently. Since the botanical group producing hardwoods is much larger than the group containing softwoods, greater diversification in wood structure may be anticipated in the former. This is the case and it explains why the identification of hardwood timbers is less difficult than that of softwoods.

By way of summation, it is well to make note again of the fact that two types of woods are recognized: softwoods and hardwoods. Softwoods are produced by various genera and species, mostly evergreen, belonging to the Order Coniferales of the Gymnosperms; hardwood trees are Angiosperms of the subgroup known as Dicotyledons and are distributed in many orders and families. These terms are sufficiently convenient to meet the needs of the trade, indicating as they do the general relationship that holds between these two classes of wood. But they are not specific to species; hard pine is a "softwood" in this sense, and basswood is a "hardwood"; yet the former is much harder than the latter.

IV. PLANT FAMILIES CONTAINING TREES IN THE UNITED STATES

Experience has taught that it is convenient to think of woods in terms of plant families; following this procedure, the families containing trees in the United States are listed in botanical sequence in Table 1-1; those in boldface are the sources of one or more important woods of the United States.

V. FACTORS CONTROLLING THE DESIGNATION OF A WOOD AS COMMERCIALLY IMPORTANT

In different parts of the world, the richness of the arborescent flora fluctuates between wide extremes and is probably at a peak in the watersheds of the Amazon River and in the Malay peninsula. A conservative estimate places the number of tree species in the first-mentioned area as at least 5000, and fully 2500 trees are known from the Malay region. Nearer home, the "Check List of Native and Naturalized Trees"* lists 865 species, 61 varieties, and 101 hybrids, for a total of 1027 kinds of trees, native and naturalized in the United States, exclusive of the Hawaiian

* E. L. Little, "Check List of Native and Naturalized Trees of the United States (Including Alaska)," Agriculture Handbook 41, Forest Service, Washington, 1953.

TABLE 1-1 Plant Families Containing Trees in the United States*

Gymnosperms	**Hamamelidaceae**	Koeberliniaceae
Pinaceae	**Platanaceae**	Caricaceae
Cupressaceae	**Rosaceae**	Punicaceae
Taxaceae	**Leguminosae**	Rhizophoraceae
Taxodiaceae	Zygophyllaceae	Myrtaceae
Angiosperms	Malpighiaceae	Melastomaceae
Palmae	Rutaceae	Combretaceae
Liliaceae	Simaroubaceae	Araliaceae
Casuarinaceae	Burseraceae	**Cornaceae**
Juglandaceae	Meliaceae	Clethraceae
Myricaceae	Euphorbiaceae	**Ericaceae**
Leitneriaceae	Anacardiaceae	Myrsinaceae
Salicaceae	Cyrillaceae	Theophrastaceae
Betulaceae	**Aquifoliaceae**	Sapotaceae
Fagaceae	Celastraceae	**Ebenaceae**
Ulmaceae	Staphyleaceae	Symplocaceae
Moraceae	**Aceraceae**	Styracaceae
Proteaceae	**Hippocastanaceae**	**Oleaceae**
Olacaceae	Sapindaceae	Apocynaceae
Polygonaceae	**Rhamnaceae**	Boraginaceae
Nyctaginaceae	**Tiliaceae**	Verbenaceae
Magnoliaceae	Malvaceae	Solanaceae
Annonaceae	Sterculiaceae	**Bignoniaceae**
Lauraceae	Theaceae	Rubiaceae
Capparidaceae	Canellaceae	Caprifoliaceae
Moringaceae	Tamaricaceae	Compositae

* Classification according to the Engler and Prantl system.

Islands. But there is no such number of commercial timber species in this country. In fact, only about 80 species in the United States are recognized as being commercially important—30 softwoods and 50 hardwoods, or a ratio of about 1 to 12.

This disparity between the total number of tree species and those which are commercially important may be due to several factors which are discussed briefly in the following paragraphs.

Many arborescent species seldom attain a size that would make them of merchantable value in terms of saw logs; however, as, for instance, in the case of aspen, they still may be of considerable commercial importance for pulpwood or for uses other than lumber. The quality of the wood is also of prime importance; trees that may have the necessary size are often ruled out because the wood is not of sufficient value under the prevailing economic conditions.

The question of accessibility is likewise of great importance, though less so in the United States than in the technologically less-developed countries. But even in the United States there are still extensive stands of Engelmann spruce in the West, for example, that have not been fully exploited, in part because it is a mountain species that grows in locations that are not readily accessible, and in part because that region of the country lacks industries which could utilize large quantities of this wood. There are many potentially valuable timbers in the tropics, the exploitation of which has been retarded by lack of adequate transportation facilities.

Closely coupled to accessibility is the matter of quantity obtainable. Adequate supplies must be available at a price not prohibitive to the consumer. It would prove futile to expend large sums of money in advertising and in working up a market if a constant supply of that wood is not available.

The number of commercial timbers in a country or area is also never so large as the number of tree species because woods frequently cannot be separated with certainty "to species." The commercial timbers in the United States run to about 80, but many more than 80 tree species contribute to the production of these. Six oaks, for example, produce the bulk of the white oak lumber of the trade, and the same holds for red oak; white ash lumber in the East may be almost any ash but black ash (*Fraxinus nigra* Marsh.). This is because wood of a number of botanically closely related species cannot be distinguished with any certainty after a tree is cut and a log converted into lumber; hence the product is sold under a single commercial designation.

It must not be inferred from the preceding paragraphs that the number of commercially important woods by a country, or a region, is fixed for all time. To the contrary, the quota of usable woods in the area in question depends on such considerations as local scarcity of better woods, improvements in manufacture, development of new products, and a better knowledge of the technical properties of the wood. In the United States some species that were considered "weed trees" at the turn of the century, or even much later, are important today. Sweetgum (*Liquidambar styraciflua* L.) and western hemlock [*Tsuga heterophylla* (Raf.) Sarg.] may be cited as cases in point. The former quickly attained commercial significance when methods were devised to kiln-dry it properly. The intrinsic worth of western hemlock was not fully recognized at first because it inherited the name of its less-desirable eastern counterpart. Aspen was held in low regard until its importance as a pulp-wood species was firmly established.

From this discussion it should be evident that woods become commercially important through rigorous selection based on the size of the tree species producing them, the quality of the wood, the accessibility and volume of the stands of a given kind of timber, the status of technological development of consuming industries, and the currently prevailing economic conditions.

2

Tree Growth

The body of all vascular plants consists of a cylindrical *axis* bearing lateral *appendages*. The axis, in turn, is made up of two structurally and functionally distinct parts: the stem and the root.

The *stem,* also called *trunk* or *bole,* is an aerial portion of the axis; it supports, successively, limbs, branches, and branchlets, often in a manner characteristic for the given tree species. The stem provides mechanical support for the crown, serves as an avenue for conduction between the crown and the roots, and on occasion stores appreciable amounts of reserve food materials.

The *root* likewise is subdivided into the lateral branch roots and rootlets; these, in aggregate, form the *root system.* Roots are organs of anchorage and support; in addition they perform the functions of absorption, by means of root hairs or mycorrhiza, and of conduction and storage.

The *appendages* of the axis are of three kinds: (1) those which contain vascular tissue (*leaves, thorns*), (2) those devoid of vascular tissue and usually formed only by the outermost layers of the stem, i.e., by the epidermis and cortex (*emergences*—e.g., prickles of the rose), and (3) those which arise as extensions of epidermis only (*hairs*). The last two categories, i.e., emergences and hairs, may occur on both stems and root parts.

In addition to the appendages described in the preceding paragraph, a tree in season may bear flowers, some of which through the process of pollination and fertilization will produce fruit and seed. There is no agreement among botanists on the nature of the flowers and their phylogenetic relationship to the other parts of the plant. The majority, however, subscribes to the idea that flowers are modified forms of shoots or leaves.

This text deals mainly with the part of the tree that produces commercial wood, i.e., with the stem. More specifically, further discussion will emphasize primarily the structure and properties of that part of the stem which contains wood (*xylem*). A brief discussion of how the stem develops is included in this text only for the purpose of rounding out the information on the origin and development of xylem.

I. THE STEM

Wood on a commercial scale is obtained only from trees of some maturity. The axis in such trees consists of a central core containing the conductive (*vascular*) tissues on the inside and the protective outer-bark layers (*epidermis, cortex*) on the outside. The central core, also called the *stele,* is composed mainly of the wood (*xylem*) toward the inside and the inner bark (*phloem*) toward the outside, with a *cambium layer* separating the two. The innermost part of the stele is occupied by the pith, with a few cells of *the primary xylem* adjacent to it (see page 30).

The axis is formed through the process of elongation and through growth in thickness. Elongation of tree stems is traceable to the *primary growth,* which takes place at or near the *apical growing points.* The primary growth is responsible for the elongation not only of the main stem but also of its branches, and hence it controls the ultimate height of the tree and also to a large extent the form assumed by the mature plant. The plant tissues arising from the apical growing points are called the *primary tissues.*

Growth in diameter is due mainly to the activities of the *vascular wood cambium,* a growing layer situated between the *phloem* and the *xylem.* Growth produced by the cambium is designated as the *secondary growth* or the *secondary thickening,* to differentiate it from growth in length traceable to the apical growing points. Plant tissues originating through cell formation in the vascular cambium are known as *secondary tissues.* Such tissues add to the bulk of the tree stem.

A. Meristematic and Permanent Tissues

A basic structural and physiological unit of wood, as of all plant material, is a cell.* The higher plants are characterized by a complexity of types of cells that go into

* Few people have any conception of the cell count in a relatively small volume of wood. It is decidedly worthwhile in teaching wood anatomy to require students to make certain computations that will clarify their ideas on this subject. Coniferous woods serve best for this purpose since their structure is relatively simple. The volume of an average tracheid can be computed, and, after allowing for ray volume (2 to 11 percent), the number of tracheids per unit of volume can be determined approximately. According to C. C. Forsaith, 1 cubic foot of spruce wood contains roughly 10 to 14 billion tracheids; a cord, 1000 to 1200 billion; 1 ton of pulp, 1400 to 1800 billion. If the tracheids in 1 cubic foot could be placed end to end, they would more than encircle the earth at the equator.

their makeup, varying in origin, form, and function. A mass of cells of similar origin, or with similar function, is called *tissue*. Various tissues may be classified on several bases. For instance, wood (xylem) is a tissue, because of its location in the stem and its common origin from the wood cambium. However, a number of tissues may be recognized also within the xylem on a physiological basis; i.e., cells involved in conduction, such as vessels in hardwoods, may be called a *conducting tissue,* those involved primarily in support (fibers) are considered to be a *supporting* or a *mechanical tissue,* while parenchyma cells are a *storage tissue*. The same cells, e.g., tracheids in conifers, may be considered as both the conductive and the supporting tissues. Xylem tissues may also be designated on the basis of alignment of wood cells in respect to the axis of the stem as longitudinal or axial, and as horizontal or ray tissues.

Finally, tissues are also distinguished as *meristematic,* or those involved in the new-cell formation, and *permanent,* which are regions within a plant where growth has ceased at least temporarily and in which cells and tissues have become fully differentiated and mature. Parts of the permanent tissue, and sometimes the entire permanent tissue, may again become meristematic and then be involved in further cell formation. An example of this is formation of the *phellogen* or *cork cambium* (see page 60), which arises in the permanent bark tissues. Fully differentiated xylem and phloem are permanent tissues.

The meristematic tissues, or *meristems,* in which plant tissues are formed, are classified as *primary* and *secondary*. The *primary meristems* consist of cells which are direct descendants of the embryonic cells and which never cease to be involved with the formation of the fundamental parts of the plant. In trees the principal primary meristems are those involved in elongation of stem, shoots, and roots; they are also called *apical meristems,* or *growing points*. The *secondary,* or *lateral, meristems* arise in permanent tissues; i.e., they originate from the cells which first differentiate and function as a part of some mature tissue, and then again become meristematically active. Such meristems are involved in the lateral, i.e., in the diameter growth of the stem. *Phellogen* or *cork cambium* is a secondary meristem. The vascular or *wood cambium,* which is responsible for the formation of xylem and phloem, may also be regarded as a secondary meristem, because tissues formed by it are permanent in nature, even though the cambium itself, in part at least, arises from the procambium, a primary meristematic tissue.

B. Apical Stem Meristem

As stated before, height growth in trees is brought about by the activity of the apical meristems, also called the *growing points*. These meristems occur at the tips of the stem and shoots. Similar meristems also occur at the tips of the roots.

Active cell formation in the apical meristems during the growing season is confined to the very tips of these regions. These regions within the meristems are sometimes designated as *promeristems*. In the lower plants cell division in the promeristems is confined to a single cell, which is therefore responsible for the elongation of that plant. In the higher plants the initiating layer may consist of a group or groups of

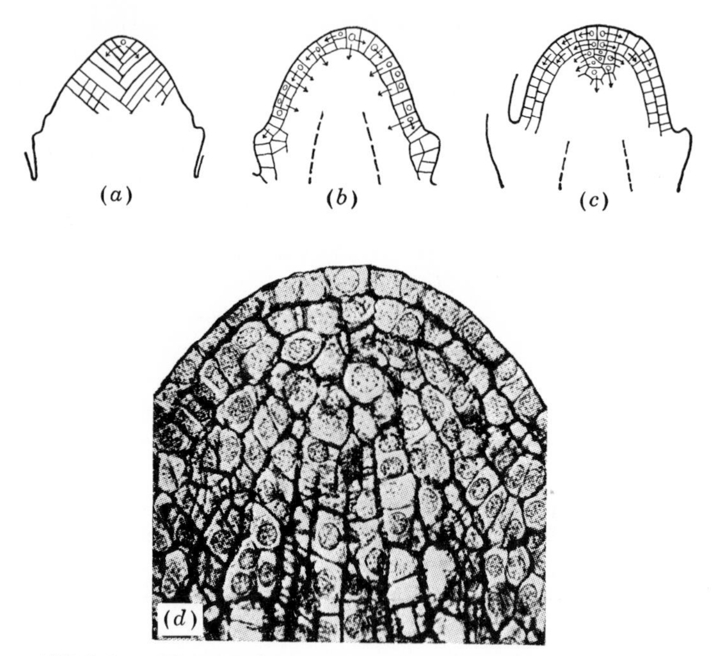

FIG. 2-1 Diagrams showing position and planes of division of stem-apex initials.

(*a*) Single initial with oblique anticlinal division only.

(*b*) Initials many, with planes of division both anticlinal and periclinal.

(*c*) Initials in three layers; the two outer tiers divide anticlinally only, forming a two-layered tunica; the innermost layer, with division in all planes, forms the corpus.

(*d*) *Sequoia sempervirens*, initials of the surface layer divide chiefly anticlinally, subapical initials divide in all planes.

[(*a*), (*b*), *and* (*c*) *after A. J. Eames and L. H. MacDaniels, "An Introduction to Plant Anatomy," 2d ed., McGraw-Hill Book Company, New York, 1947;* (*d*) *after G. L. Cross, Comparison of the Shoot Apices of the Sequoias, Am. J. Botany,* **30:**130–142 (1943).]

such cells. In the angiosperms there is a recognizable stratification of cells in the promeristem, giving rise to the *tunica-corpus* theory of the primary-cell formation.[41]* According to this theory the initiating cells in the apex are grouped into a central core (the *corpus*), which is enveloped by one or more layers or cells forming the *tunica* (Fig. 2-1). The tunica-corpus arrangements have many variations. In the dicotyledons, which include all the hardwood species, the tunica commonly consists of two layers of cells, though tunica layers of up to five cells in thickness have been reported.[23,41] The mantle-like arrangement of the tunica can be achieved only by the cell division limited to the plane at right angles to the surface (anticlinal division). This type of division brings about an enlargement in surface area. Cells

* Superscript numbers pertain to references at the end of the chapter.

below the tunica are less regularly arranged, indicating that in the corpus region cell division is in all directions, resulting in volume growth of the meristematic tissue.[17,19]

In most of the coniferous species cell division in the surface layer of the apex is both at right angles to the surface (anticlinal) and parallel to the surface (periclinal); therefore in gymnosperms the tunica-corpus arrangement of apical initials is either wanting or is much less definite than in the hardwoods. When evident it is confined to a single layer of tunica-like cells and weakly differentiated corpus-like grouping of cells.[28]

Below the region where active cell formation takes place (promeristem), the cells formed in the apical growing points undergo changes in size, shape, and function, leading finally to differentiation into permanent tissues. This partially differentiated meristematic zone consists of (1) *protoderm,* from which the epidermal system develops, (2) the *procambium,* from which the vascular tissues evolve, and (3) the *ground meristem,* which gives rise to pith.

When a growing shoot is split longitudinally, the sequence of these changes can be readily seen. Figure 2-2 is a longitudinal section of eastern white pine (*Pinus strobus* L.) cut through an apical growing point, subtended by several bud scales. It depicts the transition from the apical promeristem to permanent primary tissue. The promeristem of the growing point is a triangle designated *a-a-a*. The outermost layer of the promeristem, as well as the group of cells (central mother cells) directly below the topmost letter *a*, may be considered as the initiating meristematic cells. Below the promeristem is pith, designated as *b-b*. The initial stages in the develop-

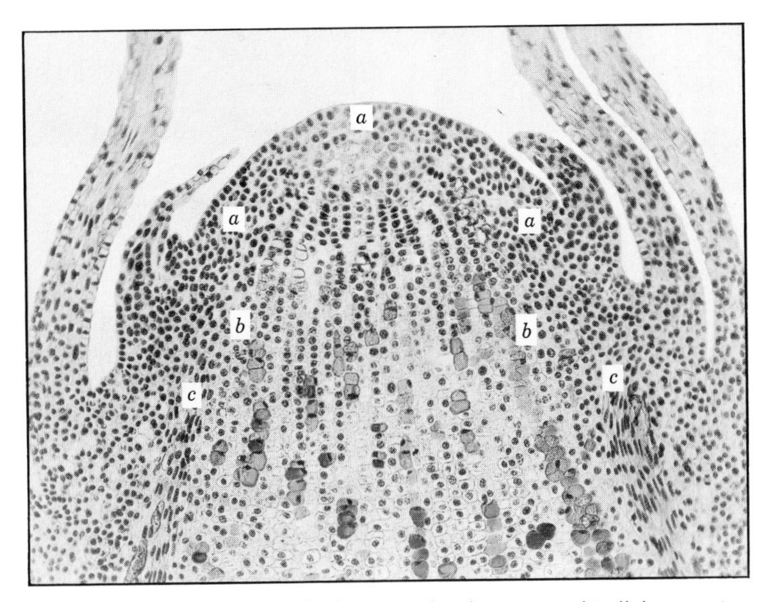

FIG. 2-2 Photomicrograph depicting the dormant, multicellular growing point at the tip of a white pine shoot, lateral sectional view. Promeristem, *a-a-a*; pith, *b-b*; procambium, *c-c*. (50×.) (*Photograph by W. M. Harlow.*)

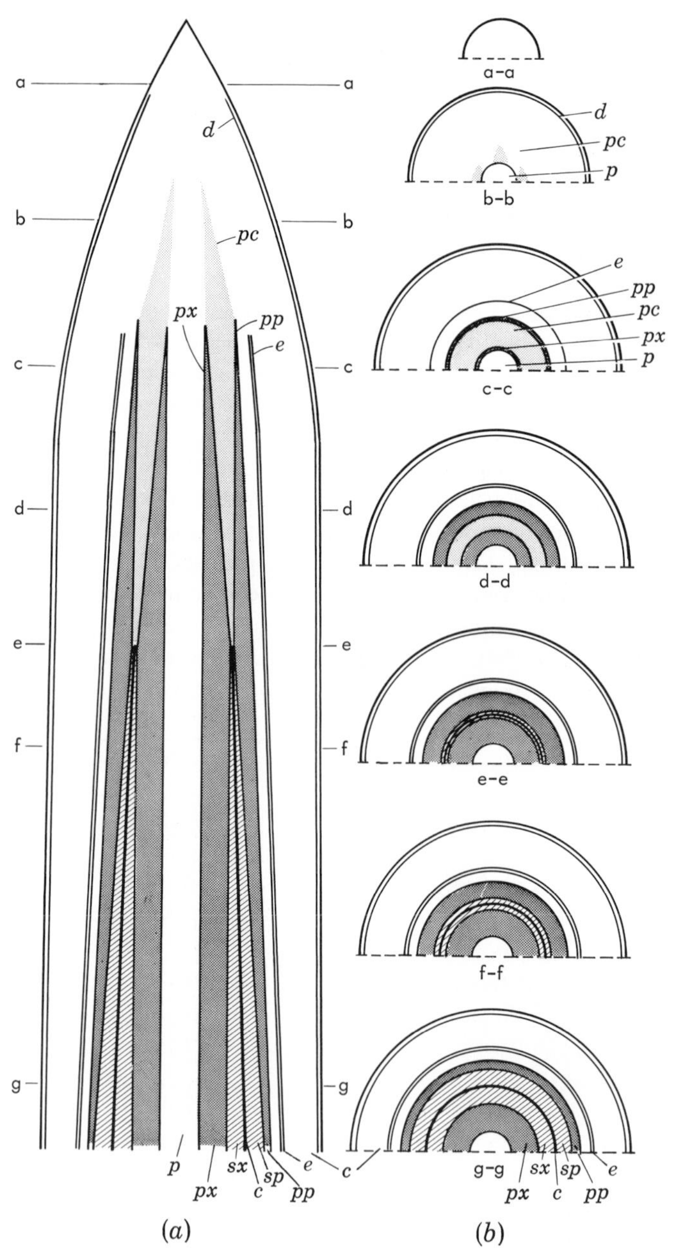

FIG. 2-3 Schematic drawing portraying the ontogeny of a young tree stem (twig). The explanation is given in the text. (*After Eames and MacDaniels.*)

ment of the primary vascular system can be seen at *c-c*. This zone is procambium; most of it will differentiate into the primary vascular tissues, called *primary phloem* and *primary xylem*. However, the groups of cells in the center of the procambium, instead of being transformed into permanent primary xylem and phloem, will remain meristematic, giving rise to strands of *lateral (vascular) cambium*. These strands of lateral cambium become interconnected through addition of meristematic cells originating from the surrounding parenchyma. This gives rise to the complete cylinder of the lateral cambium and leads to formation of the permanent secondary tissues, i.e., the secondary xylem and phloem.

Figure 2-3 depicts schematically, and in greatly simplified form, the composite sequence of the development of the primary and secondary tissues originating from the division of cells in the apical growing point. This drawing is only approximate, since it does not show the variations in the developmental sequence of the primary and secondary tissues found among the hardwood and coniferous species. It does, however, help to visualize the general path of transition from the apical meristem to the permanent secondary tissues.

The promeristem is at *a-a*. Below this point cells formed in the promeristem undergo changes in size and shape indicative of the cell types into which they are destined to develop. At *b-b* three different layers of cells are distinct: *d*, a uniseriate layer of cells which later on will be transformed into the outermost protective layer or *epidermis; pc*, a few strands of the *procambium;* and *p, pith*.

At *c-c* the procambium strands (*pc*) have united to form a complete cylinder, and the outermost and innermost procambium cells have begun to differentiate into the *primary phloem* (*pp*) and *primary xylem* (*px*). A new layer—*endodermis* (*e*)—is now in evidence. This is a uniseriate row of cells forming a protective sheath around the vascular region (*stele*). The layer (*d*) now becomes the *epidermis*, or the outer covering of the stem. The space between the endodermis and epidermis is occupied by the *cortex* (*c*), a layer of primary tissue several to many cells wide.

At *d-d* no new tissues are in evidence but more of the procambium layer has changed to the primary xylem and phloem. Somewhere between the transverse planes of *d-d* and *e-e* all the procambium has been transformed into the primary xylem and phloem except for cells in the middle, which become the *cambium*. This layer, designated as *c*, is evident at *e-e* as a dark line. The cambium has begun to form the *secondary phloem* on the outside and the *secondary xylem* on the inside. Up to this point, i.e., before the lateral (wood) cambium has been formed and become active, all the tissues in the young stem consist of the primary tissue. Above *e-e* the complete sequence of tissues from the outside are epidermis, cortex, endodermis, primary phloem, procambium, primary xylem, and pith. Below section *e-e*, in addition to these, the lateral cambium, and the cambium-formed secondary phloem and secondary xylem, make their appearance.

From there on, as shown in *f-f* and *g-g*, the amount of the secondary phloem and xylem increases while that of the primary phloem is reduced by crushing. The primary xylem is buried inside, between the secondary xylem and the pith, and occupies the same space as in *e-e*.

The condition portrayed at *g-g* will be found to exist at the end of the first year's growth (see Fig. 2-4*b*). The following year the cambium will produce another layer of secondary xylem on the inside and phloem on the outside, respectively. No new primary tissues will be formed at this level in the stem, and in fact, the remaining primary tissues except the pith and primary xylem will be gradually crushed and finally sloughed off in the process of formation of the outer layers of bark (see page 61).

C. Vascular Cambium

In all typical woody plants the root and the stem axes continue to grow not only in length but also in diameter. The diameter growth is accomplished mainly by cell division in the vascular cambium, also called ***lateral*** or ***wood cambium***. In the three-dimensional view the vascular cambium is a sheath of meristematic cells

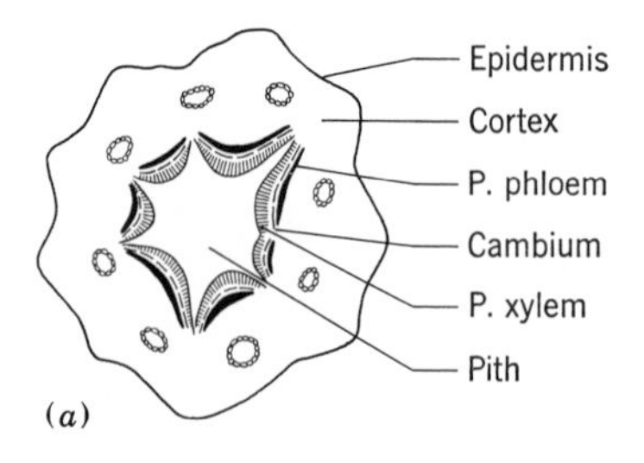

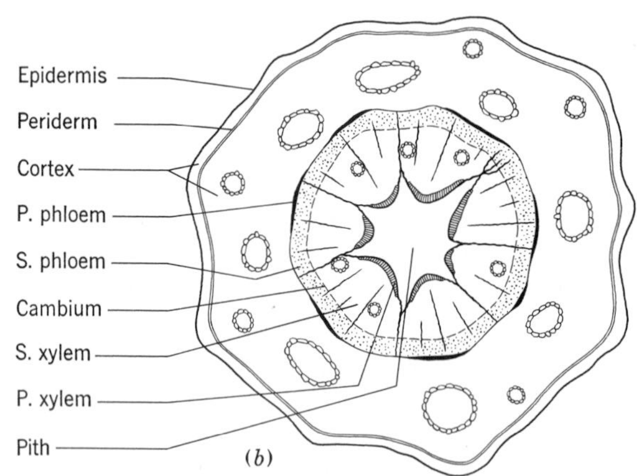

FIG. 2-4 Schematic drawing of transverse sections of a young pine twig (***Pinus strobus*** L.), showing the arrangement of the tissues prior to and after secondary thickening.

(*a*) About 1 inch from the apical growing point; the primary tissues are complete, but secondary thickening has not begun.

(*b*) At the end of the growing season; secondary thickening is well advanced.

situated between the secondary xylem and the secondary phloem; it extends from the growing tips in the stem and the branches into the corresponding regions in the roots. It is through the activities of this meristematic layer that the tree continues to increase in diameter year after year by addition of successive layers of secondary xylem and phloem.

The vascular cambium consists of two kinds of cells (Fig. 2-5): the *fusiform initials,* which give rise to all the longitudinally oriented cells in the wood and the phloem, and the *ray initials,* which produce rays in both these tissues. The addition of new cells in xylem and phloem is accomplished by the tangential division of the cambial initials, in the manner shown in Fig. 2-6. Two apparently identical cells are formed by such division. One of these becomes either a xylem mother cell or a phloem mother cell, while the other increases to the original size of the cambial initial and continues to function as a meristematic cell. The cambial initials continue to divide in this manner throughout the growing season, with one cell always remaining a cambial initial and the other giving rise to new phloem or xylem cells.

The newly formed xylem or phloem mother cells either mature directly into permanent wood or phloem cells or more frequently they, in turn, divide one or more times before all the derived cells become transformed into permanent types (Figs. 2-6 and 2-7).

When viewed in a transverse section it is frequently difficult to ascertain which cells are the true cambial initials and which are the daughter cells in the various stages of division and maturation. For this reason careful distinction should be made between the *cambial zone,* also called the *cambial region,* a term used to describe the entire region consisting of the meristematic and partially differentiated cells, and the *cambium,* which is a single row of permanent initiating cells capable of repeated division for as long as a tree remains alive.

1. CELL STRUCTURE IN THE VASCULAR CAMBIUM

As stated in the preceding paragraphs, the vascular cambium consists of two types of cells: the *fusiform initials* and the *ray initials*. The fusiform initials are elongated tapering cells. In coniferous woods and in the less highly specialized hardwoods the initials are quite variable in length and their gradually tapering ends overlap extensively, as seen in the tangential section. In conifers these cells range in length from less than 2000 to more than 9000 microns, and are 30 or more microns in diameter. In the less-specialized hardwoods, such as birch or yellow-poplar, the fusiform initials range from 1000 to 2000 microns in length.[3]

The highly specialized hardwoods have relatively short and nearly uniform fusiform cambial initials ranging from 300 to 600 microns in length. In some woods, such as black locust and persimmon, they are more or less symmetrically grouped in parallel, horizontal rows (Fig. 2-9). Such cambia are called *stratified* or *storied*. Ripple marks (see Chap. 5) are traceable to cambia with the storied initials.

Fusiform initials increase, in all trees, in length with the age of the tree, until it reaches maturity (usually 30 to 60 years), after which the length of initials remains

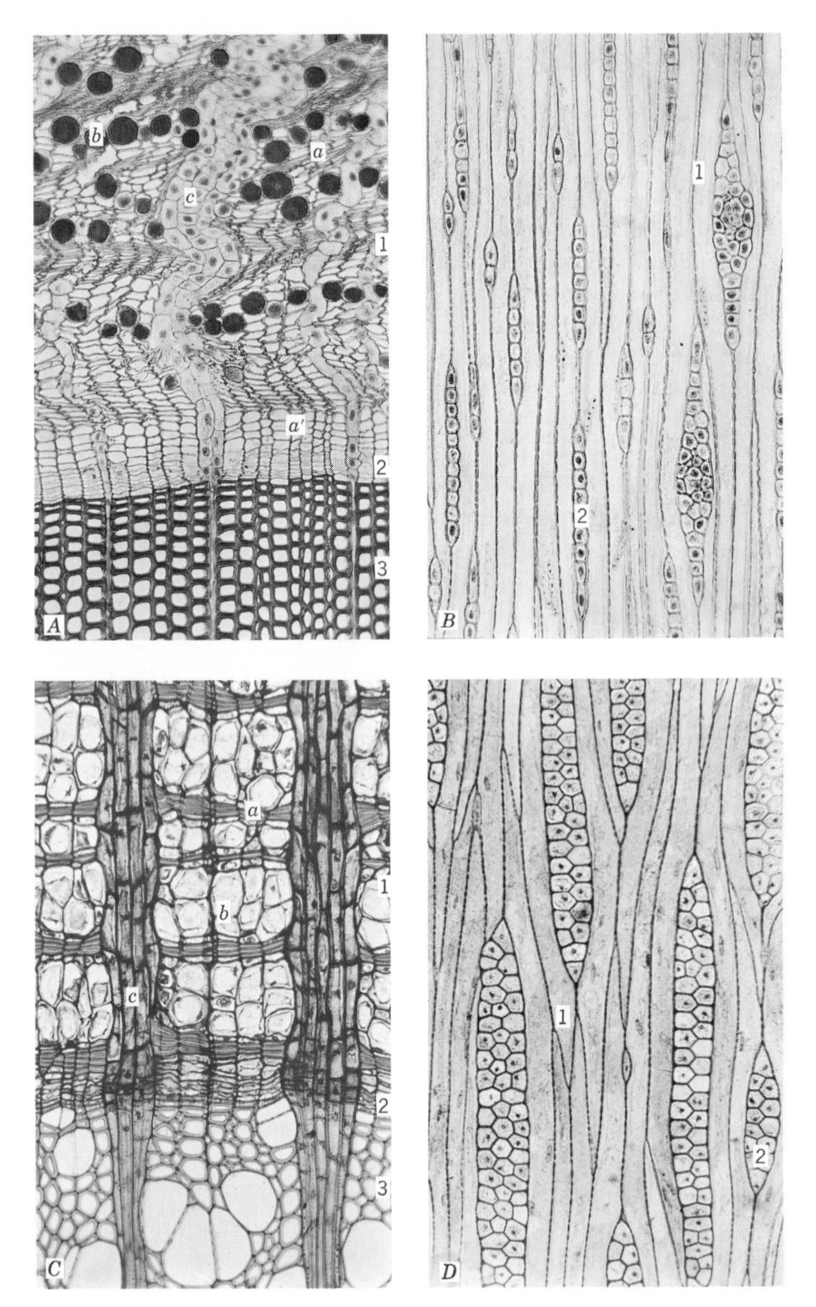

FIG. 2-5 Cambia of trees in the dormant (winter) conditions.

A. Inner bark (1), dormant cambial region (2), and portion of the last annual ring (3) of eastern white pine (***Pinus strobus*** L.). The sieve tubes function only for one year; they have collapsed in the older living bark (*a*) but have yet to function in the

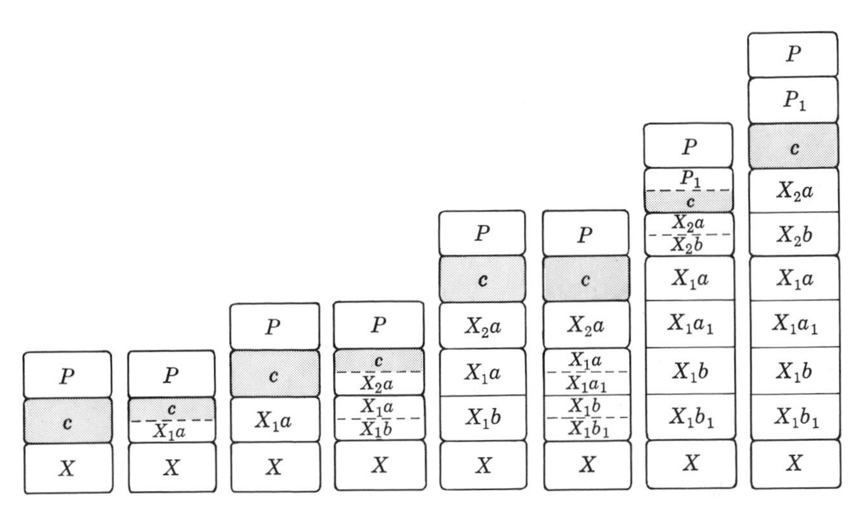

FIG. 2-6 Schematic drawing showing division of the cambial longitudinal initials; *c*—cambial initials; *x*—a mature xylem cell; x_1 (a, a_1, b, b_1), x_2 (a, b)—xylem mother cells and their derivatives; *p* and p_1—phloem mother cells.

---

vicinity of the cambium (*a'*); the enlarged cells with dark contents (*b*) compose the phloem parenchyma; the phloem portion of a ray is visible (*c*); the cambial region (2) consists of 6–10 rows of tabular cells (in this view), one row of which, though indistinguishable, is the true cambium. The transition from the thin-walled cells of the cambial region (2) to the mature tracheids of the wood (3) is very abrupt. (100×.)

B. Cambium of eastern white pine (***Pinus strobus*** L.) in tangential view. The cambium is composed of very long (several millimeters) longitudinal mother cells (1) with tapering ends, and short-ray mother cells (2), grouped in cambial rays. Some of the longitudinal cells extend completely across the photograph, lengthwise; the tapering ends only of others are visible. Uniseriate and multiseriate fusiform cambial rays can be distinguished; the transverse resin canals which are a feature of the fusiform xylary and phloem rays of this species do not show since these structures are of postcambial origin. (100×.)

C. Inner bark (1), dormant cambial region (2), and portion of the last annual ring (3) of yellow-poplar (***Liriodendron tulipifera*** L.). The phloem is stratified in this species and consists of alternating layers of bast fibers (*a*) and soft bast (*b*) between the rays (*c*); the soft bast is composed of sieve tubes (large quadrangular cells), companion cells, and phloem parenchyma. The cambial region (2) consists of about six rows of tabular cells (in this view), one row of which, though indistinguishable, is the true cambium. The transition from thin-walled cambial cells (2) to the mature elements of the wood (3) is abrupt; vessels, fiber tracheids, and the xylary portions of rays are evident in the latter. (100×.)

D. Cambium of yellow-poplar (***Liriodendron tulipifera*** L.) in tangential view. The cambium is composed of long longitudinal mother cells (1) with tapering ends, and short-ray mother cells (2), grouped in cambial rays. The longitudinal cells are not sufficiently long to extend completely across the photograph lengthwise, a situation which held for the comparable units in white pine. The cambial rays are 2- to 3-seriate except for one lone ray which consists of a single cell. (100×.)

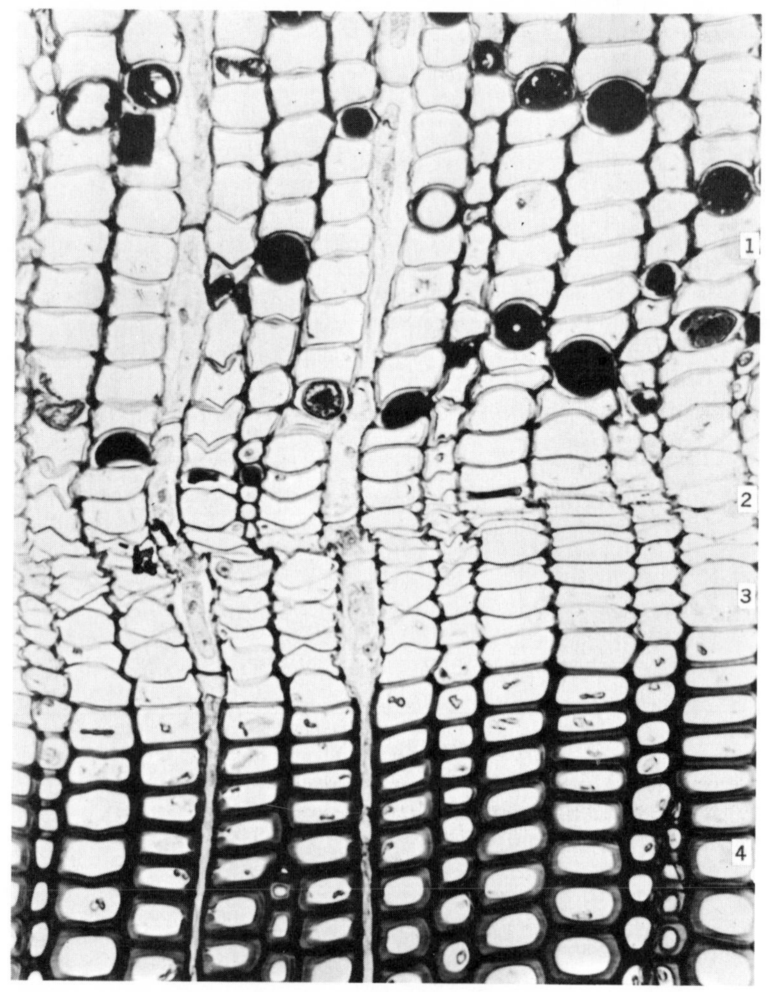

FIG. 2-7 Active cambium of pine (*Pinus* sp.), in transverse section. 1. Inner living bark consisting of sieve tubes, phloem parenchyma (cells with dark contents), and phloem rays. 2. Cambium. 3. Immature longitudinal tracheids with thin walls. 4. Mature longitudinal tracheids with thick walls. (*Courtesy of the Forest Research Institute, Dehra Dun, India.*)

Note the apparent rapidity with which the xylary cells mature behind the cambium.

fairly constant.[3,5] This increase in length of the initials is more pronounced in the coniferous species, in which the fusiform initials may increase in length from 100 to 400 percent in the first 40 to 60 years. In hardwoods the increase in length of the fusiform initials is seldom more than 100 percent and usually only half that much, and in some species even less. The size of the cambial initials may also vary somewhat with the position in the tree, and in response to ecological factors.

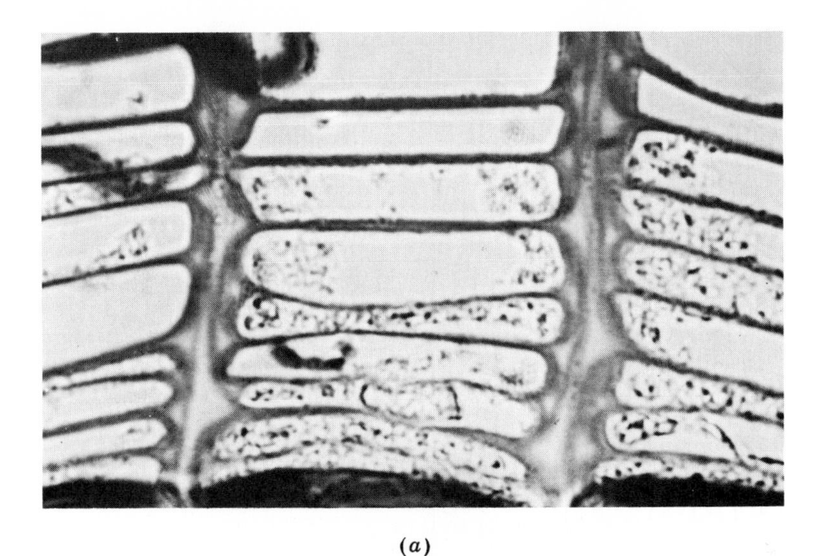

(*a*)

(*b*)

FIG. 2-8 *Pinus strobus* L.

(*a*) Transverse section of the cambial zone from a large stem in the resting winter condition. Preparation made and photographed after treatment with chloriodide of zinc by Kerr and Bailey for class demonstration. The cellulosic layers of the thick radial walls were stained blue and appear dark in the photograph, in contrast to the colorless intercellular substance. Each protoplast is enclosed in a cellulosic wall layer. In places pairs and groups of cells are still enclosed within the wall of the mother cell. In other places this wall has been ruptured and continuity of the intercellular substance has been established. (1400×.)

(*b*) The same, more highly magnified. On the left the cells are still enclosed within the wall of the mother cell and there is discontinuity in the intercellular substance. (2000×.)

(*Hitherto unpublished photographs and legends courtesy of I. W. Bailey.*)

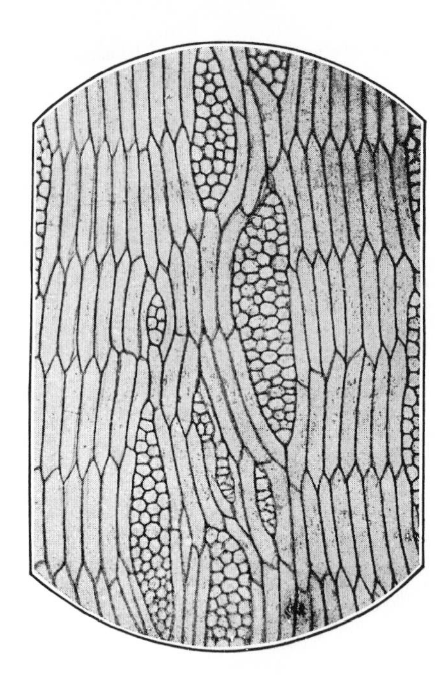

FIG. 2-9 The dormant cambium (*Robinia pseudoacacia* L.); the longitudinal cambial initials are stratified. (*From Eames and MacDaniels.*)

The ray initials, when viewed tangentially (Figs. 2-5, 2-9), are rounded cells, with little variation in size or shape. These cells are grouped in the form of a ray, the number of cells within each ray varying greatly in different species and often in the same species and even in individual plants.

Each cambial initial is enclosed on all sides by a wall (primary wall) of its own. The walls of the adjacent cambial cells are separated by an amorphous intercellular layer (*true middle lamella*). Because of the difficulties involved in isolation of the cambial initials from the adjacent partially differentiated cells there is no agreement on the chemical composition of the cambial walls and the intervening intercellular material. The preponderance of the available information favors the concept of the cambial wall as a truly anisotropic structure, composed of a relatively low amount of loosely dispersed cellulose microfibrils interspersed in a large volume of polyuronides and other noncellulosic polysaccharides and a smaller amount of pectin. The intercellular layer is an amorphous material, composed principally of calcium pectate; later this layer becomes lignified in the xylem cells but remains largely unlignified in the phloem cells.

When viewed under a light or an electron microscope the cambial primary walls give a false impression of considerable concentration of microfibrils. This, however, is due to a large extent to the techniques employed in preparation of such material for observation with a microscope, resulting in drying and removal of noncellulosic wall constituents. As a result the cellulosic cell wall material is brought closer together by lateral contraction of the walls.

According to Bailey and Kerr,[6] the intercellular material between the cambial initials "is a plastic colloidal substance that passes readily into a semi-liquid phase, thus facilitating those movements and adjustments of cells that are such characteristic features of the actively growing cambium." In their view the cambial wall "is a discrete morphological structure which maintains its identity under all conditions of growth and development, whereas the intercellular material is passively molded into various forms and possesses few of the attributes of a true membrane."

2. CELL DIVISION IN THE VASCULAR CAMBIUM

The cambial fusiform initials divide tangentially (periclinally), thus providing new xylem and phloem cells. They also divide radially or pseudotransversely (anticlinally), thus accounting for the increase in the circumference of the cambium as a tree grows in diameter.

a. Tangential Division of the Cambial Fusiform Initials. All the longitudinal xylary and phloem cells arise from the division of the fusiform initials in the tangential-longitudinal plane. The cells so formed are aligned in radial rows; all the cells in a row, regardless of whether phloem or xylem, are descendants of the same initial in the cambium. The radial alignment of the elements is largely retained in coniferous woods; porous woods, on the other hand, show less evidence of it because of the postcambial enlargement of certain cells, particularly of the vessel segments, producing cell distortion in the tissue. It also follows that the same cambial fusiform initial is capable of forming all the types of longitudinal cells found in a given wood (Fig. 5-1).

Cell division in the cambium was described in detail by Bailey[2,4] for *Pinus strobus* L. It is believed that his observations apply equally to most of the coniferous species and perhaps to some of the hardwoods.

In the resting state each cambial initial contains cytoplasm with a single nucleus centrally located and placed with its longest axis approximately parallel to the long axis of the cell. During mitosis, the polar axis of the division figure shifts in such a manner that it is placed diagonally across the cell (Fig. 2-10). The cell plate is laid down between the two daughter nuclei, and it is gradually extended across the cell from one radial wall to the other and then up and down the cell until it reaches the end of the cell (Fig. 2-10). This divides the protoplast into two parts, each containing one of the daughter nuclei.

On the basis of evidence obtained with the electron microscope, Frey-Wyssling and others[20,22,37] present the following sequence of the cell plate formation and its transformations into the intercellular layer (true middle lamella) and the two primary walls of the newly formed adjacent cells. According to these investigators the first evidence of cell plate formation is accumulation of small, dark-stained vesicles* in the equatorial plane. These saclike bodies, extruded by the Golgi apparatus,*

* Vesicles are saclike structures, containing isotropic carbohydrates; they are presumed to be formed in the Golgi apparatus. The Golgi apparatus are structures in the cytoplasm, composed of double membrane and containing vesicles, which are extruded during cell division.

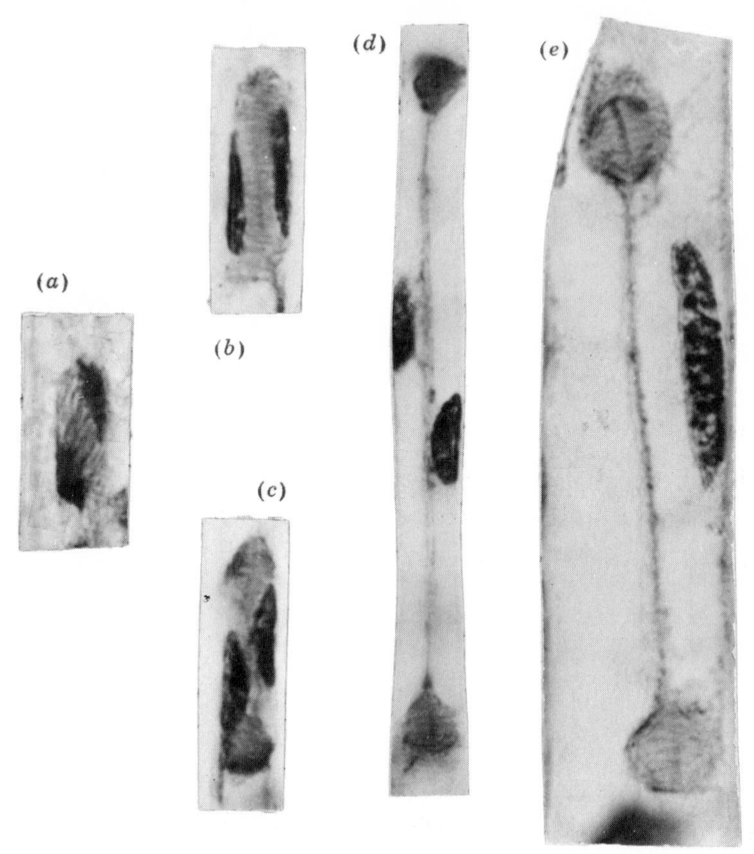

FIG. 2-10 Cell division of the fusiform cambial initials in ***Pinus strobus*** L.
(*a*) Early telophase from longitudinally dividing fusiform initials, seen in radial, longitudinal extension. (710×.)
(*b*) The same, widening of central spindle and beginning of cell plate formation. (710×.)
(*c*) The same, later stage than in (*b*), showing disappearance of connecting fiber and movement of nuclei toward the cell plate. (710×.)
(*d*) The same, showing later stage of plate formation, daughter nuclei, and kinoplasmoses. (710×.)
(*e*) Obliquely dividing fusiform initial, as seen in tangential longitudinal extension, one daughter nucleus out of plane of section. (710×.)
(*Photograph by I. W. Bailey, by permission, from Ref.* 4.)

fuse together, forming a plate (Figs. 2-11 and 2-12) which extends, usually in a pseudotransverse direction, until it reaches the longitudinal walls of the mother cell and fuses with it.

Since the vesicles consist of isotropic carbohydrates (hemicelluloses and pectic substances), the first stage of the new cell wall formation is the deposition of a plastic, isotropic gel (cell wall matrix). But even before the cell plate joins the walls of the dividing mother cell, two narrow bright layers can be observed, under po-

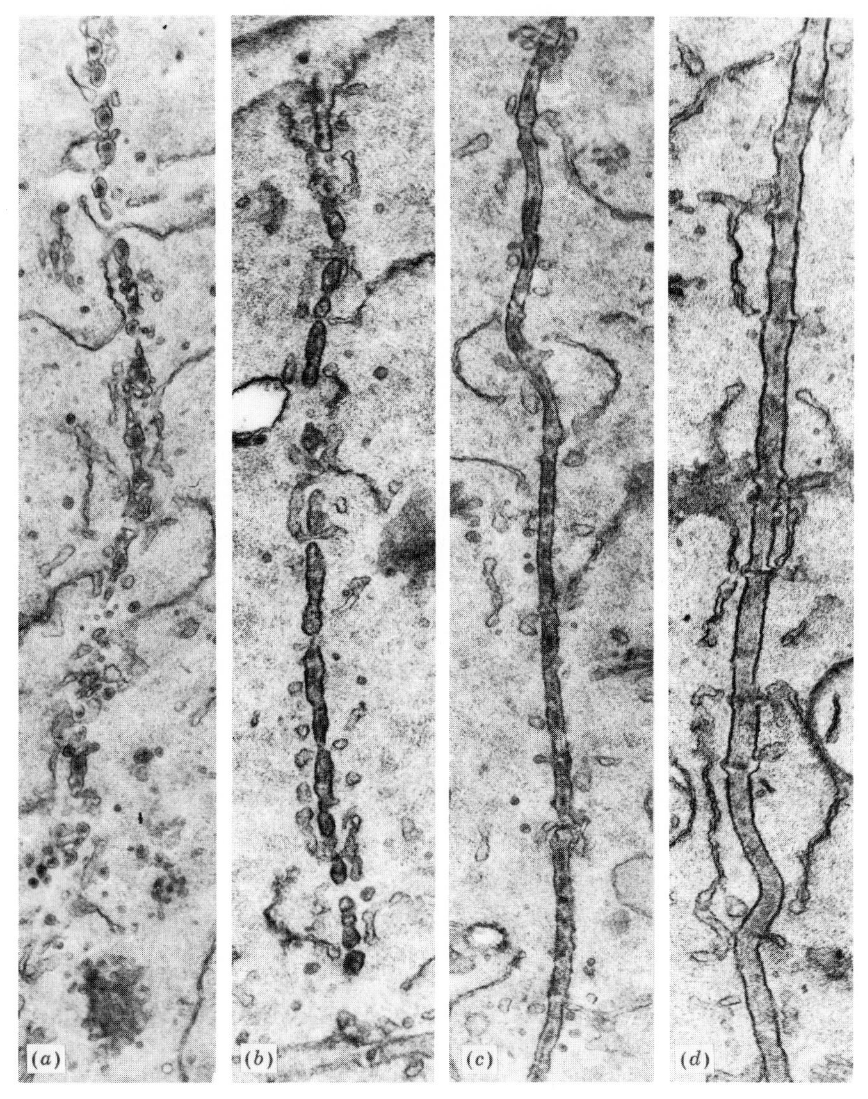

FIG. 2-11 Successive stages of matrix deposition during the formation of the cell plate in mitosis.

(*a*) Golgi-produced vesicles. (25,000×.)

(*b*) Golgi vesicles line up and fuse. (29,000×.)

(*c*) Lateral incorporation of additional vesicles. (26,000×.)

(*d*) Young wall consisting of matrix (middle lamella and two adjacent primary walls). (32,000×.)

(*Courtesy of A. Frey-Wyssling, Ref.* 20.)

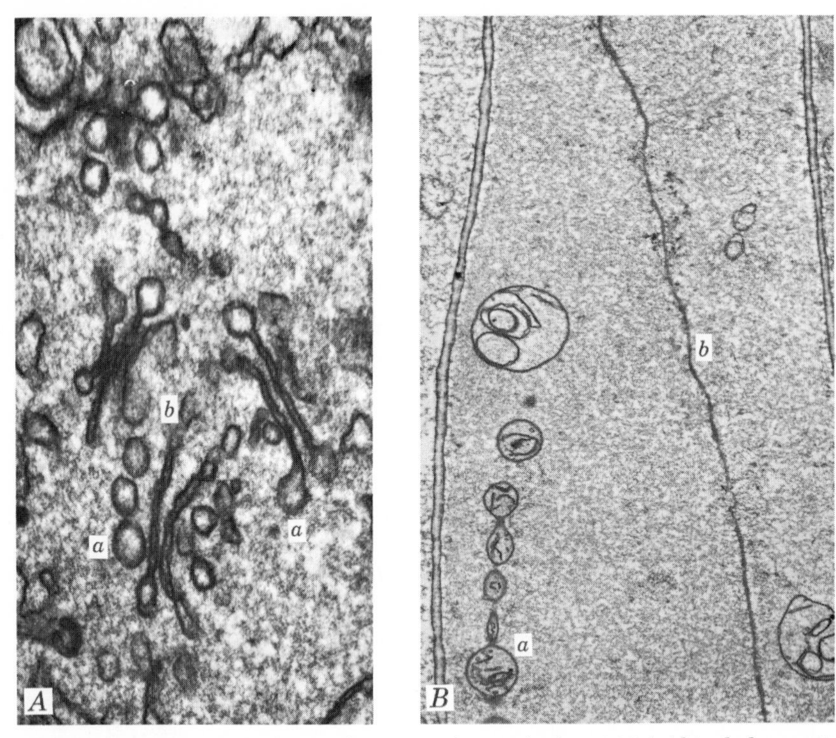

FIG. 2-12 *A*. Golgi bodies producing vesicles in a fusiform initial, ***Abies balsamea*** (L.) Mill., cross section; *a*. vesicle, *b*. Golgi body. (44,300×.)

B. Pseudotransverse orientation of cell plate in the dividing fusiform initial, in compression wood of *A*. ***balsamea*** (L.) Mill.; *a*. plastids, *b*. cell plate. (5900×.)

(*Photographs by N. P. Kutscha*.)

larized light, on either side of the central, isotropic layer, indicating the presence of small amounts of cellulose microfibrils in these newly differentiated layers. In the mature cell these layers become the primary walls of the daughter cells, while the central isotropic layer becomes the true middle lamella (intercellular layer). This represents the second stage of the new cell wall formation. In the third stage of the cell wall ontogeny the monomeric precursors of lignin diffuse into the matrix and polymerize there to form lignin "by pushing aside the matrix which in a dehydrated state is reduced to a minimum volume."[20]

The resulting mother cell generally subdivides once or more than once before it and its progeny mature into the xylem elements. This results in the formation of a group of cells (Fig. 2-8), all enclosed in the common wall belonging to the originating meristematic cambial cell. When the new cells begin to enlarge, the original parent wall is ruptured and fragments of it become embedded in the intercellular layer. This breakup of the parent wall establishes continuity of the intercellular substance between the adjacent cells.

b. Increase in Girth of the Vascular Cambium. Growth in diameter of trees brings about circumferential increase in the vascular cambium. Bailey[5] postulates that this may involve one or more of the following factors:

1. An increase in the tangential diameter of the already existing fusiform initials
2. An increase in the length of fusiform initials
3. An increase in the number of fusiform initials
4. An increase in the diameter of the ray initials
5. An increase in the number of ray initials

His investigations have shown conclusively that, although all these five factors are involved in the lateral expansion of the cambium, most of the increase in girth is due to the addition of new initials. For instance, in comparing a 1-year-old stem of white pine (*Pinus strobus* L.) with a 60-year-old stem of the same species, Bailey[5] found that in 59 years the average length of the fusiform initials increased from 870 to 4000 microns and the average diameter from 16 to 42 microns. At the same time the average diameter of the ray initials increased from 14 to 17 microns. This enlargement of the cambial initials in length and in diameter, considerable though it was, accounts for only a small percentage of the total diameter expansion. Most of it was found to be due to the increase in the number of initials, as shown by increase in a 59-year period in the number of the fusiform initials from 724 to 23,100 and that of the ray initials from 70 to 8796.

The manner in which the number of the fusiform initials in the vascular cambium is increased is illustrated in Fig. 2-13. In the coniferous species and in the less-specialized hardwoods with relatively long and unstratified initials, new initials are formed mostly by anticlinal division. The anticlinal divisions are pseudotransverse, resulting in formation of an oblique wall between the two new cells (Fig. 2-13*a*). The sloping ends of these cells then grow past each other, the two initials coming to be parallel and in contact with each other, though they are not necessarily of the same length. In addition to pseudotransverse division, new cambial initials may be formed (at least in conifers) by lateral division, in which the cell plate curves, intersecting only one side of the parent cell (Fig. 2-13*d* and *e*). According to Bannan,[8,10] lateral divisions are uncommon, representing only about 1 percent of all the anticlinal divisions recorded in his investigations. The girth of the cambium is thus increased by the formation of new cells, by increase in their length, and to some extent by increase in the lateral (tangential) diameter of the cambial initials.

The highly specialized dicotyledons (e.g., *Diospyros virginiana* L., *Robinia pseudoacacia* L.) have stratified cambia, with relatively short initials of nearly uniform length, arranged in a more or less horizontal series. In such species new fusiform initials are formed by radial-longitudinal division (Fig. 2-13*f* to *h*) so that the newly produced cambial cells are of the same length and lie parallel to each other. The tangential diameter of the new cells then increases to approximately the original diameter of the mature cells before division. In this instance the girth of the cambium increases because of the lateral expansion of cells formed by the division, rather than through elongation of the newly formed initials. In the species with storied cambium, the fusiform initials show practically no increase in length with increase in age of the tree.

In the evolutionary development of hardwoods the fusiform initials become pro-

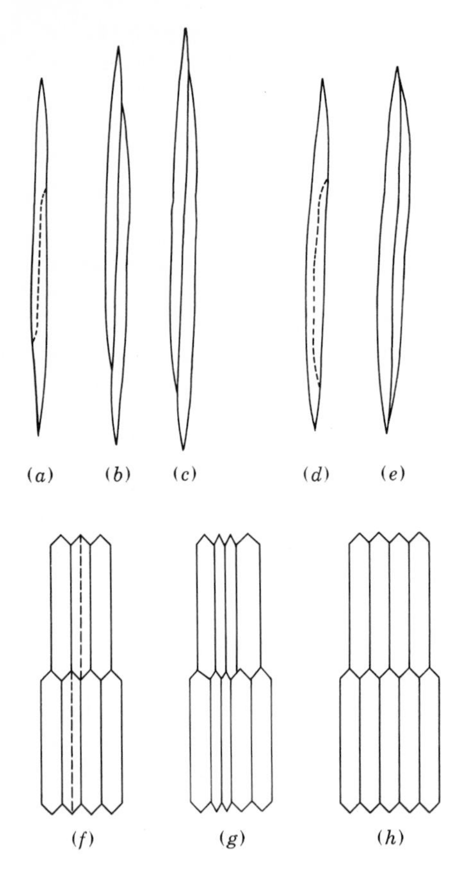

FIG. 2-13 Diagrams illustrating the manner in which the increase in the girth of the cambium proceeds in nonstratified and in stratified cambia.

(*a*) to (*c*) Fusiform initial from a nonstratified cambium, dividing pseudotransversely; the daughter cells resulting from this division elongate and slide by one another in the tangential plane.

(*d*) and (*e*) Fusiform initial formed by lateral division, in which the cell plate curves, intersecting only one side of the parent cell.

(*f*) to (*h*) Fusiform initials from a stratified cambium; the cells resulting from the radial-longitudinal division expand laterally but do not elongate.

gressively shorter and gradually approach the storied condition. This shortening of the cambial initials is accompanied by a shift in the position of the newly formed wall from the pseudotransverse to the more nearly radial-longitudinal plane. This is accompanied by reduction in the extent of the longitudinal sliding growth.

Bailey* also reports that in trees with unstratified cambia the mean length of fusiform initials in different parts of the plant varied within wide limits. For instance, in *Pseudotsuga menziesii* (Mirb.) Franco the shortest fusiform initials were 280 microns and the longest 8600 microns, the mean length varying from 900 to 6000 microns, while those in *Robinia pseudoacacia* L. (stratified cambium) ranged in length from 70 to 320 microns and the mean of their length varied between 150 and 170 microns.

c. Frequency of Fusiform-initial Formation in the Vascular Cambium. The frequency of formation of the new fusiform initials varies considerably in different trees and in different parts of the same tree as well as at different ages of the cam-

* I. W. Bailey, The Potentialities and Limitations of Wood Anatomy in the Study of the Phylogeny and Classification of Angiosperms, *J. Arnold Arb.*, **38**:243–254 (1957).

bium. Usually a greater number of initials is produced than is necessary for circumferential expansion of the tree. This leads to elimination of the excess number of newly formed initials.

According to Bannan and Bayly[11] the survival of the newly formed fusiform initials is dependent largely on the length of the new initial and the extent of contact with the rays. The longest initials formed by pseudotransverse, anticlinal division usually continue to function as meristematic cells. The survival of initials of intermediate length depends to a large degree on the extent of ray contact, while the shortest initials almost always fail even when bordered by rays. They lose their capacity for further division and either mature into malformed permanent cells in the xylem or phloem, or are gradually reduced to ray initials (Fig. 2-14). The survival of the longest initials and elimination of the shortest effectively maintains or gradually increases the cell length, which otherwise would decline or at least vary over wide limits.

New initials formed by lateral, anticlinal division may be lost by maturation into

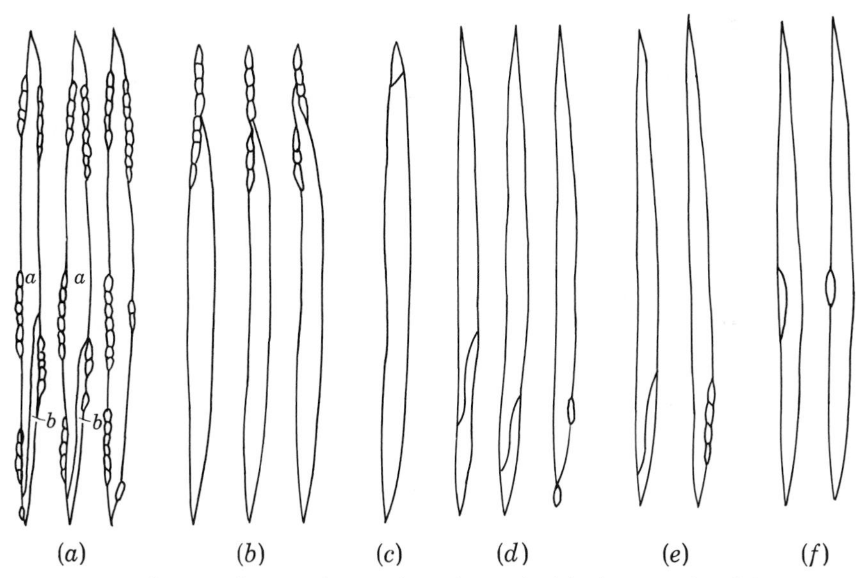

FIG. 2-14 Schematic drawing showing loss of a cambial fusiform initial and formation of new ray initials in the cambium.

(*a*) Initial *a* with extensive ray contact survives, while initial *b* with sparse ray contact matures into a deformed cell and disappears.

(*b*) A ray is split by intrusive growth of a fusiform initial.

(*c*)A new ray initial arising from pinching off the top of a fusiform initial.

(*d*) Two single ray cells are formed through reduction of a short fusiform initial; either or both of these cells may survive and later develop into rays consisting of a number of cells formed by subsequent division of these initials, or they may be eliminated.

(*e*) A new ray is formed by septation of the entire short fusiform initial.

(*f*) A new ray initial formed on the side of a fusiform initial, which will continue to function as such.

[(*a*) *and* (*d*) *after M. W. Bannan;* (*b*), (*c*), *and* (*f*) *after E. S. Barghoorn.*]

the xylem or phloem cells, transformed into and continued to function as ray initials, or enlarged and become functional fusiform initials.

d. Effect of Frequency of Fusiform-initial Formation on Cell Length. A negative relationship between frequency of anticlinal division of the fusiform initials and their length has been noted by several investigators.[10] Since frequency of anticlinal division is greater near the pith, and gradually declines outward in the stem, this fact may explain, in part at least, the gradual outward increase in the length of the fusiform initials and hence in the cells derived from them. Considerable evidence exists that in the adult wood the ultimate length of cells is genetically controlled. However considerable fluctuation in the rate of anticlinal division of the fusiform initials, and hence in the length of cells, may occur as a result of fluctuating environmental conditions which govern the seasonal growth. Bannan[9,10] concludes that any factors favoring uniform ring width through the middle to late years of tree growth tend to favor a constant rate of anticlinal division and cell size. Conversely, any fluctuation in the rate of seasonal formation of wood will be reflected in the varying rate of anticlinal division of the cambial fusiform initials and hence in the fluctuating length of the cells derived from them. When division of the cambial fusiform initials is rapid, a greater number of shorter initials is formed and is retained, resulting in formation of a comparable number of shorter mother cells. When, in turn, these divide rapidly, as they may during early-wood formation, a greater number of shorter wood cells is produced for two reasons: (1) there are more shorter mother cells present and (2) there is less elongation in the cells formed by the rapidly dividing mother cells. The final effect of this is that on the average early-wood cells tend to be shorter than comparable cells formed later in the season.

e. Ray-initial Formation in the Vascular Cambium. Nearly all new ray initials arise by the method just described; i.e., they are formed through reduction of short fusiform initials to a single ray cell, or by septation of the entire such initial into a number of ray cells[7,8] (Fig. 2-14*d* and *e*). Barghoorn,[12,13,14] however, reports that new ray initials are also formed at the side of a fusiform initial or arise by pinching off the tips of the fusiform initials (Fig. 2-14*c* and *f*). In these cases the remainder of the fusiform initial continues to function as such. Finally, rays may be formed by splitting of an existing ray by apical growth of a fusiform initial (Fig. 2-14*b*).

In hardwoods with multiseriate rays there is usually a gradual increase in the width of rays. This is accomplished either (1) by addition of more ray initials at the side of the existing initials, (2) by division of ray initials themselves, or (3) by fusion of two or more proximate rays. Likewise, increase in height of rays is occasioned by division of the ray initials or by the vertical fusion of one or more of the cambial rays.

3. SEASONAL ACTIVITY OF THE VASCULAR CAMBIUM

Under normal climatic conditions the cambium of temperate-zone trees is inactive in the winter. While in the dormant state the walls of the cambial initials become considerably thickened and the protoplast passes into what Priestley[40] has termed the *gel state*. On the reawakening of growth in the spring the cambial initials extend radially and the cell walls become thinner and more plastic. The cytoplasm becomes semifluid (*sol state*) and assumes a parietal position around a large central vacuole. This is the stage when the bark will slip easily on the wood. Shortly after the establishment of this condition the cambial initials and frequently the last-formed daughter cells of the previous season begin to divide.

Wilcox[50] reports that in upper New York cambial cells begin to enlarge radially when the mean temperature rises above 40°F for about a week. This generally occurs 3 weeks before the buds open and 4 weeks prior to initiation of the cambial cell division. From this it may be concluded that the initial enlargement of cambial cells depends on sufficiently high temperature to permit metabolic activities. It also appears that the required auxins must have been formed in the preceding season, since no new auxin could have been produced in the buds, which are still dormant.

It is generally accepted that the renewal of cell division in the cambial zone in the spring is closely dependent on the resumption of growth in the buds and on the appearance of *auxins* (plant growth hormones) in the expanding buds and later in the actively growing terminal points and the young leaves. The currently prevailing point of view among plant physiologists is that auxins produced in these young expanding parts of the plant are reawakening cell division in the cambium.[32,34,48]

Although initiation of cambial activity in the spring is dependent on the stimulus provided by the expanding buds, there are important differences in the spread of cambial cell division throughout the stem among ring-porous, diffuse-porous, and coniferous species.[48] In the ring-porous species the resumption of cambial activity becomes evident at an early stage of the bud swelling and extends rapidly down the branches and trunk, giving the appearance of a simultaneous initiation of cambial cell division throughout the stem. In the diffuse-porous species the spread of cambial activity down the stem is rather slow, requiring in some instances, several weeks before it reaches the base of the trunk. Lobjanidze,[35] on the basis of a 2-year study in the Georgian Soviet Republic, reports that in the ring-porous species some cambial activity in the branches and the stem is evident even prior to the development of buds, but that the cambium becomes more active after the leaves have developed. In the diffuse-porous trees, on the other hand, the cambium becomes active, both in the branches and in the stem, only after the buds have opened.

In the coniferous species cambial activity begins also at the base of the breaking buds, but the exact pattern of initiation of cell division in the cambium in the branches and in the main stem is still uncertain. The discrepancies in reports on the resumption of cambial activity in the coniferous trees may be because in conifers some auxin synthesis may occur in the old foliage before or simultaneously with the development of new foliage. Therefore the pattern of initiation of cell division in the branches and along the stem may vary in different trees depending on the vigor and distribution of the old foliage.

In any case considerable variations exist in the spread of cambial activity throughout the trunk, depending on the age of the tree, its vigor, and the exposure. In old trees the cambium may become active earlier in the twigs than in the main stem, while in young trees of the same species cell division in the cambium may commence simultaneously throughout the plant. In the Northern Hemisphere the cambium may become active earlier on the southwest side when exposed to the heat of the sun. If resumption of cambial activity in the spring is followed by a sudden fall of temperature the active cambium and the newly formed, partially differentiated xylem and phloem cells may be killed, resulting in the frost injury known as *sun scald.*

There is no agreement on whether the cambial initials produce first the phloem or the xylem cells. Bannan[7] found that in white-cedar phloem develops later than xylem. Artschwager,[1] on the other hand, reports that in pecan the formation of new xylem and phloem begins simultaneously. Both investigators agree that formation of the new phloem is slow in comparison with that of xylem, and that a broad band of xylem is usually formed before there is any appreciable addition to the phloem.

The final xylem and phloem increment ratio is dependent on the environmental conditions as they are reflected in the rate of seasonal growth. The rate of phloem formation is rather stable throughout the growing season and is little affected by changes in growing conditions. The xylem increment, on the other hand, is subject to wide fluctuations depending on the environmental and climatic conditions affecting seasonal growth. Therefore, while under favorable growth conditions four to six times as many xylem as phloem cells are formed by a given cambial initial, the ratio becomes progressively smaller as conditions for growth become less favorable.

As stated before, the initiation of cambial activity appears to be closely correlated with the growth of buds and with formation of growth hormones (*auxins*). The duration of cambial activity, on the other hand, is not always correlated with the continuation of active primary growth. There is an indication that in some diffuse-porous species cambial activity is correlated with the duration of extension growth, whereas in the ring-porous species and in conifers the cambium continues to divide for some time after the terminal shoot growth has ceased. This suggests that while inception of cambial activity and early-wood formation is closely correlated with opening of buds and shoot growth, further wood formation may be attributed to more limited meristematic activities of developing needles or leaves, and perhaps to auxin production within the cambium itself.

The duration of cambial activity depends also on the age of the plant and on the plant part; it also varies with latitude and environment. For instance, the cambium remains active longer in the rapidly grown twigs, in which apical growth is still active. Abundance of nutrients and water may account for the prolongation of cambial activity. Experimental evidence tends also to indicate that cessation of cambial activity is photoperiodic; i.e., it is related to day-length conditions, formation of new cells slowing down or stopping altogether as daylight decreases.[49]

D. Growth Increment

In trees growing in the temperate zones periodicity of the vascular cambium activity is registered in formation of the growth increments, representing the accumulation of

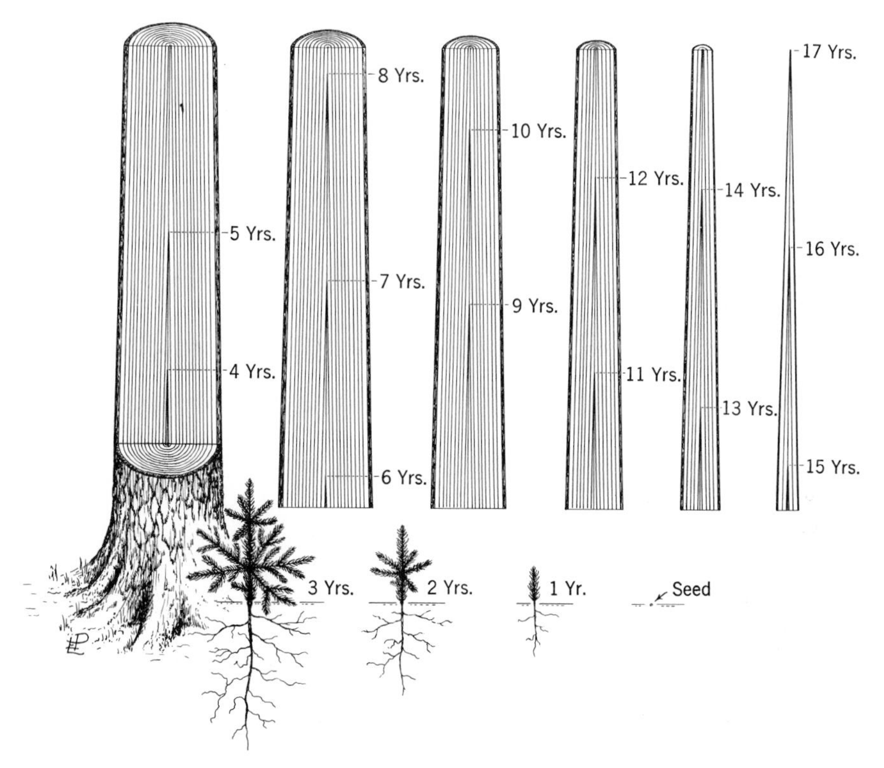

FIG. 2-15 Schematic drawing of a 17-year-old coniferous tree showing the manner in which the trunk increases in thickness and in height through the addition of growth (annual) increments.

xylem for any given year. Figure 2-15 depicts diagrammatically the manner in which the main stem (trunk) of a tree increases in circumference over a period of years.* The growth increments appear as concentric rings when cut transversely, i.e., as seen on the end of a log, and as superposed, cone-shaped, or paraboloid zones of tissue, when viewed in longitudinal section. They approach in shape a hollow cone or a hollow paraboloid, the end of each cone showing where the tip of the terminal leader was the preceding season.

The tree shown in Fig. 2-15 is 17 years old. Only 14 growth rings show on the stump because it took the tree 3 years to attain stump height. The growth rings formed during that period will not show on the stump since the terminal leader has not yet grown past that plane. Ring counts of the stump therefore may not indicate the actual age of the tree, since the seedling stages are not recorded at that height. In longleaf pine (***Pinus palustris*** Mill.) during the "grass stage," which may last 4 to 12 years or more, no distinct growth rings are formed. Therefore, considerable

* In this sketch the effect of branching on the axis formation is ignored.

error may result if the age of the tree is estimated by counting growth rings at the stump.

Growth increments, as they appear on the end of a log, show as a succession of rings around the pith. In the literature these frequently are called ***annual rings***. However, since under certain conditions more than one ring of wood may be formed in the same year (***false rings***), or because of sluggish growth the increment may form on only one side of the trunk (***discontinuous rings***), ***growth rings*** is a more acceptable term than annual rings.

On the faces of boards the figure resulting from growth increments varies according to the plane of the wood that is exposed. On the radial surface (edge-grained or quarter-sawn stock) the increments assume the form of parallel striae. Concentric and nested U- and V-shaped patterns are characteristic of the tangential (plain or flat-sawn stock) face.

Growth increments stand out in wood to varying degree because the growth intensity, and consequently the cell size and arrangement and the density of the wood produced, are not uniform throughout the growing period. This holds for trees grown in temperate regions, where growing seasons are followed by a period of complete dormancy. In tropical woods there may be more than one period of active growth in a year and in the majority of species there is no demarcation to indicate the beginning or the end of the successive growth increments. As a consequence most tropical trees do not exhibit well-defined growth rings.

1. EARLY AND LATE WOOD

In temperate-zone trees usually there is rapid growth in diameter early in the season; the rate of growth slows down appreciably as the season advances. The portion of the growth formed in the spring abounds in conductive tissue and is often quite porous, i.e., the cells are large and the wood relatively low in density; this part of the ring is commonly designated as ***early wood*** (also called ***springwood***). Denser and hence frequently darker wood produced later in the season is termed ***late wood*** (or ***summerwood***). The transition between the early and late wood may be gradual or abrupt, giving rise to differentiations between certain softwoods (e.g., hard pines, transition from early to late wood abrupt, vs. soft pines, transition gradual), and between ***ring-porous*** and ***diffuse-porous*** hardwoods.

On the basis of experimental evidence Larson[30] concludes that in conifers:

> Large-diameter, springwood cells will be produced during the period of active elongation growth and high auxin synthesis. Narrow-diameter, summerwood cells will be produced following the cessation of terminal growth and the consequent reduction in auxin synthesis; a growth-inhibiting system may also become more prominently active at this time. Any factor that causes terminal elongation growth to begin prematurely or to cease will bring about a respective increase or decrease in cell diameter.

Further experiments with the effect of the length of daily illumination (photoperiod) on tracheid diameter in ***Pinus resinosa*** Ait. led Larson[31] to observe: "Long days promoted needle elongation and large-diameter tracheid production, whereas short days brought about cessation of needle elongation and the transition to narrow-diameter tracheids." This response to photoperiod varied with the stage of seasonal growth. During active terminal growth the developing bud, i.e., the growing shoot, exerted a decisive influence on tracheid diameter, while the needles became the principal factor in controlling tracheid diameter when terminal growth ceased. Larson suggests that the influence of photoperiod on tracheid diameter is indirect and is due to its effect on production and distribution of auxin in the terminal shoots. He assumes that insofar as auxin provides the stimulus to the cell development, the size of any tracheid is determined by the amount of this substance reaching that cell at the time of its differentiation.

In addition to the effect of the growth-regulating substance (auxins), Zahner[51] lists four other factors that may influence tracheid development in conifers:

> . . . (2) amount of available carbohydrates from stored and currently synthesized sources; (3) level of metabolism and rate of new cell division in the cambial region; (4) length of time new tracheids remain alive in assimilation; and (5) rate of upward flow of transpirational water and tension forces in the functioning xylem adjacent to differentiating tracheids.

In turn, all these factors, including auxin production, are controlled, to some extent, by atmospheric conditions governing transpiration and by the availability of soil moisture. According to this hypothesis, when the soil moisture is depleted the forces of transpiration cause removal of moisture stored within the tree. This creates an internal moisture deficit. The indirect effect of this deficit is reduction in the amount of auxin reaching the cambium region, due to the smaller quantities of it produced in the terminal shoots, and the lower rate of translocation in the phloem. The direct effect of the water deficiency in the stem is on the overall level of metabolism, expressed in reduction in the rate of cell division in the cambium and decrease in the radial enlargement of newly formed xylem cells.

Normally late-wood formation in conifers, characterized by decrease in the radial diameter of tracheids, is also accompanied by a simultaneous increase in thickness of the secondary wall. In any given year the reduction in the tracheid diameter and the wall thickening typical of late wood usually commences at the base of the stem and progresses up the trunk. It is believed that changes in the tracheid diameter and cell wall thickness are a result of different physiological influences.[18,43,44,48] According to Larson[34] the ultimate radial diameter attained by a tracheid depends "primarily upon the quantity of auxin reaching that tracheid during its critical stage of enlargement." Therefore cessation of terminal and foliar growth, which leads to reduction in auxin synthesis, is reflected in reduction in the radial tracheid diameter, i.e., in formation of cell sizes typical of late wood.*

As the current-year foliage approaches maturity its contribution to the production of photosynthate rises. But since the mature foliage no longer requires large amounts of such products for its own development the excess amounts of photosynthetic products become available for formation of wood, in this instance for formation of thicker cell walls, especially in the late early-wood cells. Larson[34] concludes: "any

* For a more comprehensive discussion of the physiology of wood formation, see P. R. Larson, Wood Formation and the Concept of Wood Quality, *Yale Univ. School of Forestry Bull.* 74, 1969.

condition promoting maturation of the terminal meristems will result in a concomitant increase in wall thickness, providing photosynthesis remains at an optimum level.'' In short, evidence indicates that while cell diameter is regulated largely by plant growth hormones, cell wall thickness is determined mainly by the availability of the photosynthesis products.

If the hormone theory of wood formation is accepted as valid, the influence of environment, i.e., nutrients, water temperature, and light, must be indirect, through its effect on the vegetative growth of the crown, which is the principal site for formation of the growth regulators responsible for tracheid size and for the production of the photosynthates involved in the cell wall development.

It is also possible that the higher rate of cell formation and the larger size of cells produced in the spring and early summer bring about rapid exhaustion of the available cytoplasm, and with it cessation of the wall thickening. Later in the season, because of the reduced cambial activity and hence lesser competition for the available carbohydrates, coupled with the smaller overall size of the individual cells, the cytoplasm remains alive for a longer time, resulting in formation of thicker walls in the late-wood cells.

In the same tree, once initiated, the extent of normal formation of the late wood (see false rings, page 52) is closely correlated with the amount of summer precipitation and with the length of the growing season. Between trees of the same species grown under comparable conditions the difference in the late-wood formation may also be due to heredity.

2. VARIATION IN THE WIDTH OF GROWTH RINGS

Growth increments differ greatly in width and in wood density in different tree species, in individual trees of the same species, and at different heights of a given tree. Certain trees, for instance, poplars (*Populus* spp.) and the catalpas (*Catalpa* spp.), are naturally fast-growing and under normal conditions of growth develop wide rings. Slash pine (*Pinus elliottii* Engelm.) is a more vigorous tree and almost invariably forms wider rings than longleaf pine (*Pinus palustris* Mill.) growing on the same or on slightly better-drained sites. On the other hand, 30 or 40 rings per inch are not at all unusual, especially in broad-leaved trees and also in the trees of such coniferous genera as *Taxus, Thuja,* and *Juniperus*.

As a rule the rings of shade-enduring trees vary more in width than those of light-demanding species. Alternating zones of narrow and broad increments are frequently visible in samples of hemlock (*Tsuga* spp.), shade-tolerant species; the growth increments of true firs (*Abies* spp.), rated as trees of lesser tolerance than hemlocks, are usually fairly wide and uniform in width.

Site likewise has a direct bearing on ring width. For instance, northern white-cedar (*Thuja occidentalis* L.) occurs in the Northeast on two very distinct sites, in deep swamps, where the water gives an acid reaction, and on comparatively dry limestone outcrops. Northern white-cedar has wider rings, and seeds more prolifically, when grown on the latter.

On the same site and at the same age individuals of a given species may exhibit an appreciable difference in ring width depending on the crown class. A moisture-loving species grown on a dry site may survive, but the adverse conditions may be reflected in reduced growth increments.

Finally, in a given tree the ring width varies at different heights and distance from the pith. Lengthwise, in a given tree grown under optimum conditions, the growth rings are widest near the lower part of the crown and decrease in width toward both the tip and the base of the trunk. At any given level along the stem the widest rings normally occur nearer the pith, the narrowest at the periphery of the trunk. The latter condition is particularly in evidence in overmature individuals.

Fluctuations in the growth-increment width in different individuals of the same species and in a given tree have a profound effect on the wood density and on the physical properties of wood. This question is discussed in Chap. 7.

3. DISCONTINUOUS AND FALSE RINGS

Normally growth rings completely encircle the pith. Not infrequently, however, particularly in very old or in suppressed trees, some of the rings are interrupted; i.e., they do not completely surround the stem but are present only in part of it. Such increments are called ***discontinuous rings*** (Figs. 2-16, 2-17*b*). The discontinuous rings result when the cambium remains dormant at one or more places throughout its circumference during a current growth season, meanwhile functioning more or less normally elsewhere. As a result the new growth layer appears to run into the late wood of the preceding ring.

Hartig[24] has advanced a "nutritional theory" to explain this phenomenon. He conjectures that food materials, elaborated in the crown, would be utilized in the upper stem parts first and therefore would be available to the lower parts in progressively decreasing amounts. If the crown is strongly suppressed, ring may decrease in width or be absent entirely in some parts of the stem. The nutritional theory fails to explain the frequently haphazard year-to-year distribution of discontinuous rings. Also the fact that late-wood bands in such increments are commonly of approximately normal width tends to discredit the nutritional theory. More recently it was proposed[29] that discontinuous rings in suppressed and overmature trees may be due to the delayed arrival or to the deficiency of the growth hormones during the early part of the growing season. This local deficiency results either in complete dormancy of a portion of the cambial layer, where therefore no cell division occurs during that growing season, or in a delay in the reactivation of cambial activity. When this happens a portion of a growth ring is formed, consisting either of an underdeveloped early-wood zone and fairly normal late wood, or of only the late wood.

Whatever the causes, the conditions which bring about the formation of discontinuous rings may hold for only one season or may prevail in a given vicinity for two or more seasons; in such case a number of rings may be discontinuous (Fig. 2-16).

Since there is no interruption in the continuity of wood tissue, the quality of the wood is little, if at all, impaired by the presence of discontinuous rings. But the determination of tree age based on growth-ring count, especially when increment cores are used, may be in serious error, resulting in the underestimate of tree age.

False rings often lead to an overestimate of tree age. False rings are wholly included within the boundaries of true rings. A band of what appears to be late wood is formed which simulates normal late wood in appearance and density; this is followed by tissue resembling early wood, after which true late wood is produced (Fig. 2-17*a*).

False rings can be distinguished from true rings in that the cells composing the false late wood grade to the inside and to the outside into more porous tissue, whereas in true rings the transition from the late wood of one year to the early wood of the succeeding year is rather abrupt.

If only one false ring is included in the true ring, the latter is designated as a *double ring;* if more than one of such false rings occur within the boundaries of the true ring, the latter is said to be a *multiple ring.*

False rings arise for various reasons. Late frosts resulting in temporary defoliation, destruction of foliage through insect infestation, periods of drought followed by

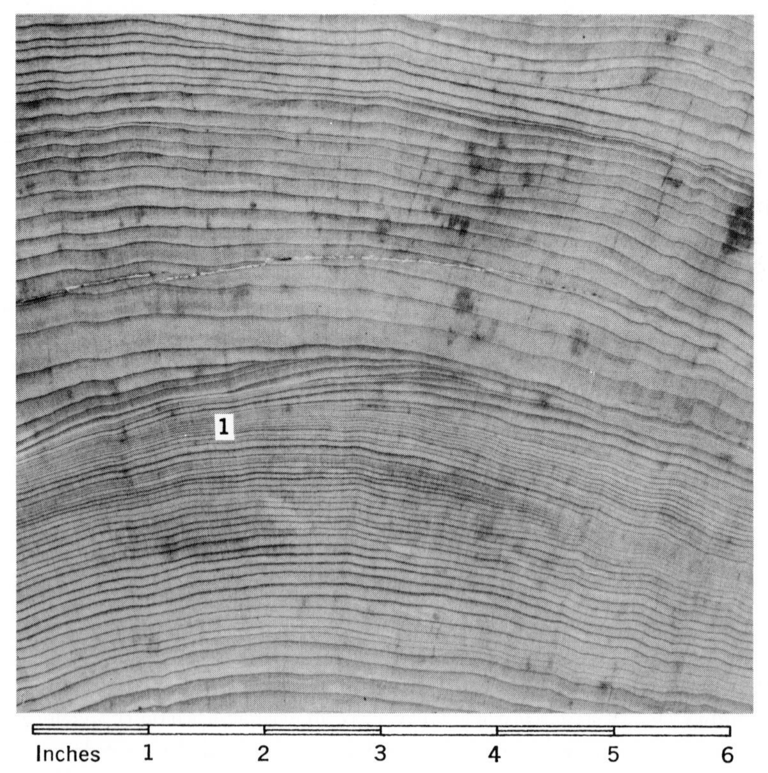

FIG. 2-16 Discontinuous rings (above 1) in redwood [*Sequoia sempervirens* (D. Don) Endl.]. (*Photograph by E. Fritz.*)

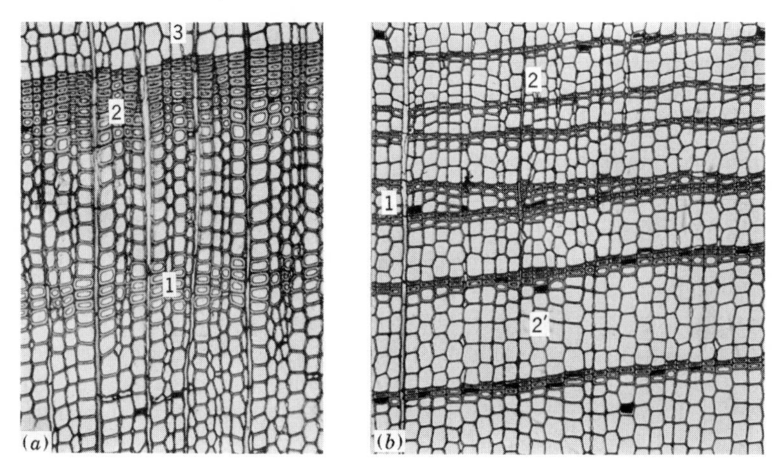

FIG. 2-17 Discontinuous and false rings.
(*a*) False rings in baldcypress [*Taxodium distichum* (L.) Rich.]. 1. Flattened, thicker-walled tracheids denoting outer boundary of a false ring; the cells grade into wider-lumened tracheids above and below. (55×.) 2. Dense late wood marking the true boundary of the ring. 3. Early wood of the succeeding ring.
(*b*) Discontinuous rings in redwood [*Sequoia sempervirens* (D. Don) Endl.]. (70×.) 1. A discontinuous ring. 2, 2′. Normal rings.

heavy precipitation and temperatures suitable for growth, and resumption of growth in the late summer or early autumn occasioned by unseasonably warm weather, all may be cited as contributory causes. In every case formation of a false ring is caused by a temporary cessation of terminal activity, accompanied by decrease in auxin production, and followed by conditions favorable to resumption of growth in the apical regions and hence to renewed production of growth hormones. The cessation of terminal growth must be of sufficient duration to produce cells with narrow diameters resembling those of typical late wood and growth resumption intense enough to produce an auxin gradient favorable to formation of the early-wood type of cells, with large radial diameters.

If resumption of terminal growth is slight, early-wood cells, characteristic of false rings, may form only in the upper reaches of the crown, where they may be indistinguishable from cells typical of juvenile wood.[32] It follows also that in any given growth increment the false ring may not extend to the base of the tree. A failure to recognize this may account for the apparent absence of false rings, following a new flush of growth, reported by some investigators.

E. Sapwood and Heartwood

For a period of years the newly formed xylem not only provides mechanical support to the tree but participates in its conduction functions, and to some extent serves as a place for the storage of reserve food as well. These physiological activities in the xylem are performed by the living cells, mainly the parenchyma cells. That part of the woody core in the tree in which some xylem cells are living and hence physiologically active is called *sapwood.* After an indefinite length of time, which varies greatly in different kinds of trees and with the condition of their growth, the protoplasm of the living cells in the xylem dies. Secondary changes that take place as a result of this lead to formation of a physiologically dead part of the xylem, called *heartwood.* In some instances an intermediate zone between the sapwood and the heartwood has been reported. This intermediate zone, though it still contains the living cells, may have coloration similar to that of heartwood, and is therefore frequently mistaken for the latter.

The transition of sapwood into heartwood is accompanied by the formation of various organic substances, known collectively as *extractives, extraneous materials,* or *infiltrations,* and in hardwoods frequently also by the accelerated formation of tyloses in the vessels. The development of these extraneous chemical compounds in the xylem is usually marked by an appreciable darkening of the tissue, resulting in a coloration of the heartwood which contrasts with the light-colored sapwood. The darker color is not, however, always a necessary condition for the existence of heartwood, since the inner core of xylem in *Picea, Abies, Tsuga, Populus,* and others is not marked with a perceptible color change; nevertheless, since it consists of physiologically dead tissue it is technically heartwood.

It has been a common practice, especially in Europe, to classify all trees on the basis of distinctiveness of heartwood as:[15,47]

1. *Sapwood trees,* in which wood of the sapwood type, i.e., wood containing some living cells, is found in the vicinity of the pith (ex., *Alnus* spp.). However it is believed that in such species, changes characteristic of the heartwood formation are considerably delayed and, hence, are evident only in trees of advanced age.[15]

2. *Ripewood trees,* in which the heartwood remains light-colored, because extraneous materials remain unpigmented, but all of the parenchyma cells are dead (ex., *Abies* spp.)

3. *Trees with regularly formed heartwood,* in which pigmented heartwood materials are always formed in the parenchyma cells. These pigmented materials are found not only in the cell cavities but in the cell walls as well (exs., oak, walnut, cherry).

4. *Trees with irregular heartwood,* in which colored heartwood may be present in all the specimens or may be formed only on one side of the stem. In these woods the pigmented substances are retained as inclusions in the storage (parenchyma) cells, while the cell walls of these and other xylary cells remain unpigmented (ex., *Fraxinus* spp.).

However, Bosshard[16] points out that this classification, implying absence of heartwood in the so-called sapwood trees, is misleading because though they are delayed, typical signs of the biological alterations characteristic of heartwood formation also occur in such species. On this basis he has proposed a new terminology for classification of heartwood formation, shown in the following table.*

* From Ref. 16, with slight rearrangement and modifications in terminology and examples.

Old terminology	*Example*	*New terminology*
Sapwood trees	*Alnus* spp.	Trees with retarded formation of heartwood
Ripewood trees	*Abies* spp.	Trees with light heartwood
Trees with regularly formed heartwood	*Quercus* spp.	Trees with obligatory colored heartwood
Trees with irregular heartwood formation	*Fraxinus* spp.	Trees with facultatively colored heartwood

Bosshard[16] states further that in the species with obligatory colored heartwood the polyphenolic infiltrations penetrate the cell wall and coat the cellulosic and hemicellulosic materials present in them; this causes decrease in the shrinking and swelling capacity of the wood, progressively from the cambium to the pith. In the species with facultatively colored heartwood no penetration of the polyphenolic substances occurs, and therefore there is no corresponding effect on the dimensional changes that can be attributed to these infiltration substances.

There are two hypotheses of heartwood formation. The older advocates that heartwood formation is caused by the accumulation of air in the closed cell system; this brings about secondary changes in the protoplasm of the parenchyma cells, leading to formation of extraneous substances and causing the death of the parenchyma cells.

More recently Rudman[45] suggested that heartwood forms (1) when water requirements of the crown reduce the moisture content of the inner part of the stem of the tree at certain periods of the year, and (2) when reserve food materials accumulate in quantities exceeding those required by the photosynthetic activities of the tree. Under these conditions reserve starch in the parenchyma cells of the sapwood will be hydrolized into soluble carbohydrates, which are eventually transformed, by an irreversible reaction, into extractive substances. A species which is not efficient in utilizing food products photosynthetically will start forming heartwood while the tree is still young. Mature trees of such species will tend to have narrow sapwood and wide heartwood. A species which is efficient in the use of food materials will delay heartwood formation or may never form it. In between these two extremes are species with wide sapwood and narrow heartwood.

Transformation of sapwood into heartwood is accompanied by a number of well-identifiable biochemical changes in the storage (parenchyma) cells, apparently brought about by progressive depletion of oxygen, culminating in complete breakdown of the normal respiratory activities in the living cells in the inner sapwood. The most noticeable of these are:[16,21,25,26]

1. Changes in size and shape of the nuclei from the cambium zone toward the heartwood. The shape of nuclei changes rapidly from typically round in the cambium, and the cells immediately adjacent to it, to radially elongated. This change occurs in the first 5 to 10 growth rings from the cambium. After that the nuclei undergo further changes in size and shape and finally disappear altogether. The

disappearance of nuclei in the storage cells is considered to be one reliable indication of the complete transformation of sapwood into heartwood.

2. The amount of nitrogenous materials, which is high in the cambium zone, decreases sharply through the sapwood toward the heartwood boundary.

3. The storage materials—starch and sugars—decrease from the outer sapwood toward the heartwood. This loss of storage materials alters the enzyme system of the cells in such a way as to allow the residual oxygen to oxidize and to polymerize phenolic materials present in the parenchyma cells, leading to formation of pigments characteristic of colored heartwood.

4. Since the outermost heartwood cells provide a barrier to the inward movement of the extraneous materials formed in the parenchyma cells at the sapwood-heartwood boundary, these accumulate in lethal quantities, causing the death of parenchyma cells and formation of additional heartwood.

The organic compounds found in the heartwood are chemically extremely variable and complex. The origin of the heartwood extractives has not been adequately explained. Since a variety of colorless phenolic compounds has been found in the sapwood of many species, it is possible that some of these products, or their precursors, are synthesized in leaves or needles, or even in the cambium, and then transported downward in the phloem and transversely, via the rays, through the sapwood to the boundary of the heartwood. In the process of translocation these products of biosynthesis may undergo considerable further chemical alterations before they are finally infiltrated as extractives in the heartwood.[21,46] More recent evidence indicates that heartwood extractives may arise *in situ* in the parenchyma cells from translocated or stored carbohydrates before these cells die. After extractives are formed they may be deposited as amorphous materials in the cells, partially or completely filling the cell cavities, or they may be infiltrated in the cell walls of the storage and the adjacent xylary cells.

The characteristic color of normal heartwood, usually in shades of yellow, orange, red, and brown, is due to deposition of the extractives. In a number of cases the distinctive color of the heartwood, e.g., in black cherry, walnut, or mahogany, explains why these woods are preferred for furniture, fixtures, and wood paneling.

The presence of extractives in heartwood, and in some hardwoods the formation of tyloses (see page 105) as well, materially reduces the permeability of the heartwood. This loss of permeability makes heartwood more resistant to impregnation with preservatives and fire-retardant chemicals, and causes difficulties in drying and pulping. On the other hand, the occlusion of the vessels in such wood as white oak, makes the wood suitable for uses where permeability must be low, e.g., tight cooperage.

Heartwood is often more durable, i.e., it is more resistant to fungal and insect attacks than sapwood. The decrease in permeability is a factor in the increased durability of heartwood because of the reduction in the amount of available air and moisture for fungal growth. The main reason, however, for the increased durability of heartwood of some species of wood is the presence of the extractives in heartwood that are toxic, to some degree, to the wood-deteriorating organisms. The presence of infiltrations that inhibit fungal and insect attack is not closely related to color of the heartwood, as is sometimes supposed. There are a number of kinds of timbers with dark-colored heartwoods which are quite durable; examples of such woods are redwood [*Sequoia sempervirens* (D. Don) Endl.] and Osage-orange [*Maclura pomifera* (Raf.) Schneid.]. On the other hand, the durable heartwood of northern white-cedar (*Thuja occidentalis* L.) and Atlantic white-cedar [*Chamaecyparis thyoides* (L.) B.S.P.] is only a pale straw to reddish brown, at best, while the dark reddish-brown heartwood of sweetgum (*Liquidambar styraciflua* L.) is nondurable. The reason for this is that durability of wood depends on the amount and the toxicity of extractives, which may be light-colored, as are the oils in cedars, or dark, as are tannins in redwood.

Because of the extractives, heartwood is generally somewhat heavier than sapwood at the same moisture content. On the other hand, green sapwood, especially in softwoods, may be heavier because of much higher moisture content (often up to 200 percent, based on the ovendry weight). However, water distribution in the sapwood of living trees, especially in hardwoods, varies with species, parts of the tree, season, and the site conditions. The heartwood of some hardwoods, such as *Carya, Fraxinus, Ulmus, Quercus, Nyssa, Juglans, Populus,* and *Betula,* may contain a higher moisture content than the corresponding sapwood. In these woods the moisture content rises abruptly from the inner sapwood to the heartwood boundary. The moisture content of the heartwood apparently does not fluctuate because there is no water movement in this region, comparable to that of the sapwood.[27]

In pulping, large quantities of extractives in the heartwood interfere with bleaching of the pulp to the desired whiteness; additional treatments are required for the removal of the extraneous materials. When acid in nature (tannins), extractives may cause corrosion in metals; they may interfere with setting of paints and varnishes and have an adverse effect on the adhesion of surface films and on the formation of strong glue joints.

Heartwood forms a core with a roughly conical shape in the tree trunk. Once its formation is initiated, the diameter and height of the core continue to increase throughout the life of a tree. It is not known whether heartwood formation is a continuous process that goes on year after year, or whether periods of active heartwood development are followed by periods of relative inactivity when no new heartwood is added. The boundary of the heartwood bears no direct relationship to growth increments; it may include a given ring in one part of a cross section and not at another point at the same height in the tree.

The width of the sapwood expressed as a linear measure or in terms of numbers of growth increments varies in different trees of the same species and at different heights in the same tree. Within a species the width of sapwood is directly related to dominance of the tree in the stand, with the more vigorous trees having the wider sapwood. Within the tree the sapwood is widest in the upper trunk toward the crown, and decreases in width toward the base. This difference in sapwood width at different heights in the trunk may be a reflection of the fact that the tree requires a fairly constant volume of sapwood at all levels in the stem. On geometrical considerations it is evident that the sapwood shell in a tree can decrease in width as the tree diameter increases, and still maintain a constant cross-sectional area.

There are also wide variations in sapwood width between species of trees. For example, in the pines, longleaf pine (*Pinus palustris* Mill.) has a narrow sapwood, while in slash (*Pinus elliottii* Engelm.) and loblolly pine (*Pinus taeda* L.) the sapwood zone is wide. In terms of numbers of growth rings in the sapwood, the variation is also large, as is shown in Table 2-1 for hardwoods native to the United States. Similar variation occurs in the conifers, with as many as 150 to 200 rings in sapwood of ponderosa pine (*Pinus ponderosa* Laws.) and with very few rings in some other species.

Wood resembling heartwood in appearance may form as a result of pathological disturbances which kill the parenchymatous cells in the sapwood prematurely. The tissue containing these dead cells darkens and simulates normal heartwood. One of the most common causes of such disturbances is wounding of the tree; for this reason such abnormal tissue is called *wound heartwood.*

False heartwood often forms in trees which do not normally have colored heartwood. It is believed that the cause of this condition is fungal infection of the tree spreading from dead branches and branch stubs into the trunk and killing the living parenchyma. The changes associated with the death of these living cells cause a browning of the cell contents and stimulate the formation of tyloses. In those cases in which the incipient decay is arrested, the affected wood retains its normal strength

TABLE 2-1 Number of Rings in the Sapwood of Hardwood Trees*

Scientific name of tree	*Number of rings in the sapwood*
Catalpa speciosa Warder	1–2
Robinia pseudoacacia L.	2–3
Castanea dentata (Marsh.) Borkh.	3–4
Prunus serotina Ehrh.	10–12
Gleditsia triacanthos L.	10–12
Juglans nigra L.	10–20
Fagus grandifolia Ehrh.	20–30
Magnolia acuminata L.	25–30
Acer saccharum Marsh.	30–40
Cornus florida L.	30–40
Cornus nuttallii Audubon	30–40
Acer saccharinum L.	40–50
Betula lenta L.	60–80
Magnolia grandiflora L.	60–80
Nyssa sylvatica Marsh.	80–100
Nyssa aquatica L.	100

* This information was taken from C. S. Sargent, "Manual of the Trees of North America," Houghton Mifflin Company, Boston, 1926.

and durability and is apparent only by its changed color. The colors produced in false heartwood are a grayish green, brown, or reddish brown. False heartwood is known to occur in beech, birch, maple, poplar, aspen, and pine, among others.

Occasionally, streaks of light-colored wood that have the appearance and properties of normal sapwood are found embedded in heartwood; such tissue is described as ***included sapwood***. This is a misnomer, however, since such light-colored areas contain no living cells. There is no exact information relative to the cause of the inclusion of such light-colored tracts of tissue in the heartwood. Included sapwood may occur in any tree species; it is especially common in western redcedar (***Thuja plicata*** Donn) and eastern redcedar (***Juniperus virginiana*** L.).

Lighter zones or streaks are often present in heartwood, simulating included sapwood; they form as a result of incipient decay. In such instances, portions of the darker heartwood are bleached as a result of fungous action.

F. Stem Form*

Stem form, as interpreted in this text, refers to the taper of the bole occasioned by the rate of change in stem diameter from the base to the top of the tree.

The basic form of the tree stem is presumed to be genetically controlled. However, it can be strongly modified by the environmental influences and cultural practices which affect the vigor, size, and shape of the tree crown. In addition to inheritance several other theories have been advanced to account for the variations in stem form. Of these, the mechanistic and hormonal theories appear to provide the most plausible explanation of the reason for stem-form variations.

The mechanistic theory assumes that the tree stem must be strong enough not only to support the weight of the stem and crown of the tree and the additional weight of snow and ice, but also to be able to resist the forces of wind exerted on the crown. As a consequence of these requirements a cylindrical stem would require a higher proportion of denser and stronger late wood in the lower part of the bole than a tapering stem.

The hormonal theory advocates that auxin gradient throughout the stem governs the distribution of radial growth of the stem and hence determines the stem form; therefore the proximity, shape, and vigor of the crown are the determining factors in development of the shape of the stem. Perhaps the two theories can be reconciled by assuming that while the hormonal activities of a tree govern its stem formation, the mechanical requirements for stress resistance exert considerable influence on the stem form of the individual tree that actually develops.

The stem within the crown is generally strongly tapered because of the downward increase in the number and cumulative effect of branches on formation of wood in the upper stem. The so-called crown-formed wood consists mostly of juvenile wood composed mainly of the early-wood type of cells. In trees with long and vigorous crowns, stem diameter, below the crown level, continues to increase appreciably down the branch-free bole, forming a decidedly tapering stem. Because of high

* For a comprehensive discussion of stem-form development, see Ref. 33.

auxin production and its favorable distribution along the stem, such trees produce higher overall quantities of early wood in the lower part of the trunk. When the crown recedes because of the tree's senescence, increased competition, or artificial pruning, the stem becomes more cylindrical because of a concentration of growth in the vicinity of the crown base. The decreased synthesis of auxin and the greater distances involved in its translocation down the stem result in formation of lesser amounts of early wood and proportionately greater quantities of late wood.

The overall effect of the tree crown on the shape of the stem is that in long-crowned trees maximum ring width is generally found in the lower part of the trunk; as the crown recedes, the maximum width of the growth ring shifts upward, resulting in a more cylindrical bole. Favorable seasonal growth conditions tend to shift the greater width of the growth increments down the stem and unfavorable conditions shift them upward.

These observations on stem form apply only to trees grown under normal conditions prevailing in forest stands. They do not apply to hardwoods with strongly deliquescent habits or to aberrant forms, such as wolf trees or trees grown in the open.

II. CORK CAMBIUM

The young stems of trees are protected from desiccation, and to some extent from mechanical injury, by the *epidermis*. The outer surface of cells forming this layer is more or less cutinized to prevent undue loss of moisture. The epidermis is generally pierced by stomatal openings which ensure the proper aeration of tissue lying beneath.

In the great majority of woody plants no provision exists for the increase in the diameter of the epidermis as the stem enlarges in girth through secondary thickening. As a consequence it soon ruptures, usually during the first year, and is sloughed off. However, before the epidermis is ruptured a new protective layer, called the *periderm,* is formed (Figs. 1-2 and 2-18).

The periderm consists of three layers: (1) the *phellogen* (cork cambium), a uniseriate layer of meristematic initials, (2) the *phellem,* the layer of cork cells formed by the phellogen to the outside, and (3) the *phelloderm,* a layer consisting of one or more rows of thin-walled cells formed by the cork cambium to the inside.

The cork cambium is a secondary lateral meristem. It arises from the permanent, i.e., completely differentiated living cells in the tissue outside the phloem, mostly from the mature parenchyma cells of the epidermis and cortex, or from those of the secondary phloem. These cells become meristematic and form a continuous layer of cork-cambium initials (*phellogen*) around the stem.

In trees with smooth bark, such as beech and white birch, the first-formed periderm may persist for many years. In these species the phellogen continues activity for a number of years and forms a *ring-periderm,* which is marked by a distinct zonation in cell sizes. In most woody plants, however, the first layer of periderm is replaced in a few years by the formation of other layers of periderm, deeper in

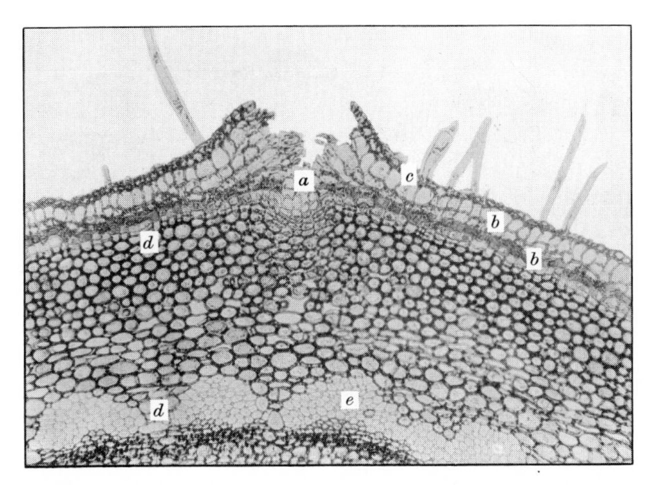

FIG. 2-18 Outer portion of green ash (*Fraxinus pennsylvanica* Marsh.), in transverse section, showing a lenticel (*a*), the first periderm (*b-b*), the epidermis with epidermal hairs (*c*), the collenchymatous cortex (*d-d*), and masses of pericyclic fibers (*e*). (100×.)

the cortex and finally in the secondary phloem. These later-formed layers of periderm are not continuous but are in the shape of overlapping short arcs or lunes (Fig. 2-19), which are active only for a short time. Meanwhile the bark outside these layers of phellogen cracks and eventually is sloughed off. Roughening of the bark indicates the inception of deep phellogen formation and coincides with phellogen formation in the secondary phloem.

From this point on the bark consists of two parts: dark *outer dead bark,* which in transverse and radial sections consists of dead phloem (and possibly of cortical tissue) bounded by anastomosing lines of periderm (1); light-colored *inner living bark,* composed of accumulated annual layers of phloem, in which the sieve tubes no longer function except in the layers immediately adjacent to the vascular (wood) cambium (2), where some sieve tubes and parenchyma cells are still living.

The width of these two layers and their composition vary greatly in different species and in individuals of the same species grown under different conditions* as well as at different heights of the same tree. For example, in *Picea* spp. the layer of brown bark is comparatively thin because it weathers fairly rapidly; as a result the outer surface of the tree trunk is scaly but relatively smooth. In giant sequoia [*Sequoia gigantea* (Lindl.) Decne.] and Douglas-fir [*Pseudotsuga menziesii* (Mirb.) Franco], on the other hand, brown bark 1 to 2 feet in thickness is occasionally encountered. This may represent the accumulation of several to many hundreds of years of old phloem cut off by lunes of periderm but not cast off. The surface of the bark of such trees is deeply fissured.

* For details on bark structure of the coniferous species, see T. P. Chang, Bark Structure of the North American Conifers, *U.S. Dept. Agr. Tech. Bull.* 1095, 1954.

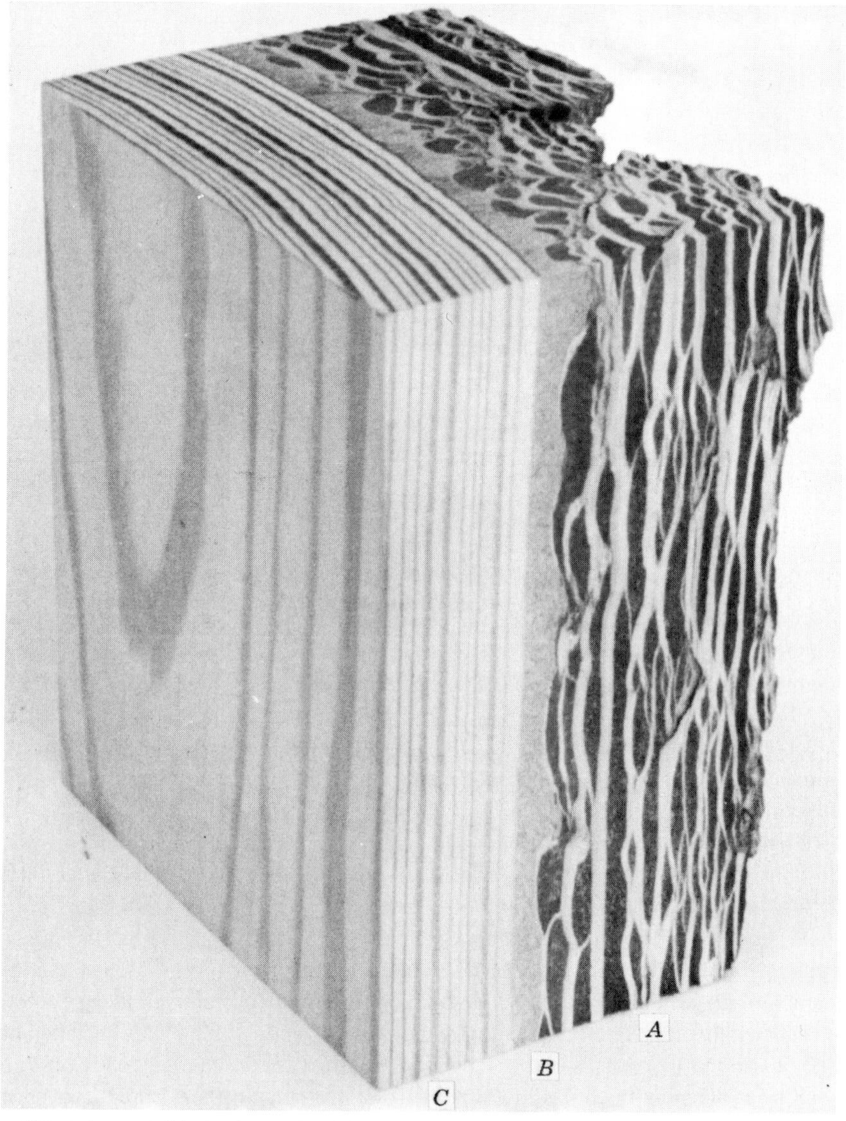

FIG. 2-19 A block of Douglas-fir [*Pseudotsuga menziesii* (Mirb.) Franco].
A. The outer dead bark.
B. The inner living bark.
C. Wood.

In the cork oaks the first-formed periderm arises in the epidermis. It forms a large number of cork cells on the outside and a few cells of phelloderm cells internally. This phellogen layer persists indefinitely, unless disturbed. In commercial cork production, the first or the virgin layer of cork is stripped when the trees reach a diameter of about 40 centimeters. The cork strip separates at the phellogen layer, which dies, but a new layer of the cork cambium is formed a few millimeters deeper in the cortex. New phellogen forms cork more rapidly than before, and cork of sufficient thickness for stripping is accumulated in 10 years or less. When this is stripped the new phellogen layer is formed and the process is repeated as long as the tree is alive.

The number of successive periderms persisting outside the living bark is variable and is not necessarily indicative of the thickness of the dead bark. In Fig. 2-20*a*, which is a photomicrograph of red spruce, only one layer of periderm is in evidence. In Fig. 2-20*b*, four periderm layers in ***Tilia americana*** L. are visible over a shorter distance than in *a*. In both *a* and *b*, the strips of bark shown in the photographs are too short to indicate that the periderm anastomosed to the right or left.

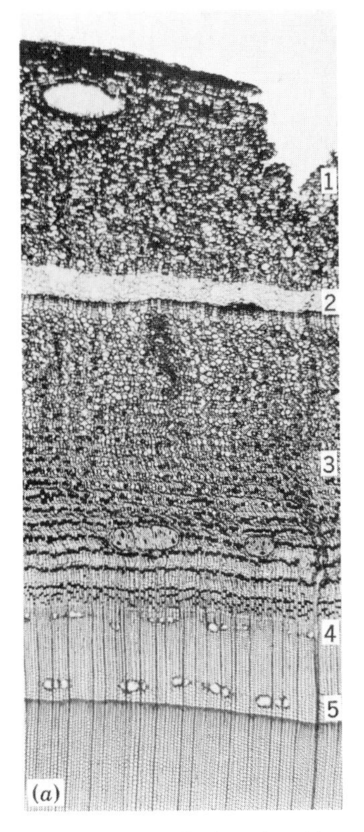

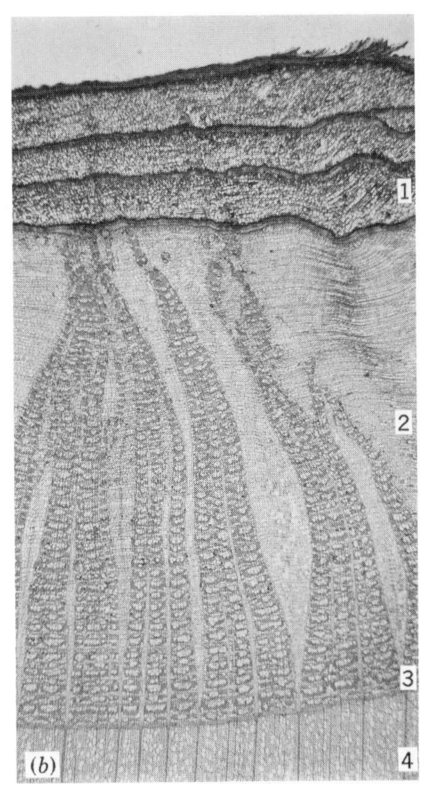

FIG. 2-20 Bark and periderm formation in trees.

(*a*) Bark and last-formed wood of ***Picea rubens*** Sarg., showing the outer dead brown bark (1), the last periderm (2), the inner bark (secondary phloem) (3), location of the cambium (4), and the wood (5). (12×.)

(*b*) Bark and portion of the last annual ring of ***Tilia americana*** L., showing the outer dead brown bark composed of four successive periderms alternating with bands of dead phloem tissue (1), a wide band of inner living bark (secondary phloem) in which some of the rays have flared (2), position of the cambium (3), and portion of the last annual ring (4). (12×.)

Even though the bark on a mature tree is formed through the activity of the vascular cambium (inner bark, secondary phloem) and the bark cambium (phellem and phelloderm), it is never so thick as the wood. There are three reasons for this:

1. The layer of xylem formed during a given year usually consists of many more rows of cells and therefore is much thicker than the corresponding layer of phloem; for this reason wood in a tree stem accumulates faster than the phloem.

2. In the phloem only bast fibers, when they are present, and occasionally phloem parenchyma, become lignified. Sieve tubes and companion cells (in hardwoods) remain unlignified and usually collapse the second year after they are formed, because of bark pressure (Fig. 2-5*A*). Therefore, the older phloem tissue is compressed radially and hence no longer occupies so much space as formerly.

3. Outside layers of bark formed in the later years consist of old phloem tissues; they are shed from time to time through the process of periderm formation already described.

SELECTED REFERENCES

1. Artschwager, Ernst: The Time Factor in the Differentiation of Secondary Xylem and Phloem in Pecan, *Am. J. Botany,* **37:**15–24 (1950).
2. Bailey, I. W.: Phenomena of Cell Division in the Cambium of Arborescent Gymnosperms and Their Cytological Significance, *Proc. Natl. Acad. Sci. U.S.,* **5:**283–285 (1919).
3. ———: The Cambium and Its Derivative Tissues, II, Size Variations of Cambial Initials in Gymnosperms and Angiosperms, *Am. J. Botany,* **7:**355–367 (1920).
4. ———: The Cambium and Its Derivative Tissues, III, A Reconnaissance of Cytological Phenomena in the Cambium, *Am. J. Botany,* **7:**417–434 (1920).
5. ———: The Cambium and Its Derivative Tissues, IV, The Increase in Girth of the Cambium, *Am. J. Botany,* **10:**499–509 (1923).
6. ——— and Thomas Kerr: The Cambium and Its Derivative Tissues, X, Structure, Optical Properties and Chemical Composition of the So-called Middle Lamella, *J. Arnold Arb.,* **15:**327–349 (1934).
7. Bannan, M. W.: The Vascular Cambium in Radial Growth in *Thuja occidentalis* L., *Can. J. Botany,* **33:**113–138 (1955).
8. ———: The Relative Frequency of the Different Types of Anticlinal Division in Conifer Cambium, *Can. J. Botany,* **35:**875–884 (1957).
9. ———: Sequential Change in Rate of Anticlinal Division, Cambial Cell Length and Ring Width in the Growth of Coniferous Trees, *Can. J. Botany,* **45**(9):1359–1367 (1967).
10. ———: Anticlinal Divisions and Cell Length in Conifer Cambium, *Forest Prod. J.,* **17**(6): 63–69 (1967).
11. —— and I. L. Bayly: Cell Size and Survival in Conifer Cambium, *Can. J. Botany,* **34:**769–776 (1956).
12. Barghoorn, E. S.: The Ontogenetic Development and Phylogenetic Specialization of Rays in the Xylem of Dicotyledons, I, The Primitive Ray Structure, *Am. J. Botany,* **27:**918–928 (1940).
13. ———: The Ontogenetic Development and Phylogenetic Specialization of Rays in the Xylem of Dicotyledons, II, Modification of the Multiseriate and Uniseriate Rays, *Am. J. Botany,* **28:**273–282 (1941).
14. ———: The Ontogenetic Development and Phylogenetic Specialization of Rays in the Xylem of Dicotyledons, III, The Elimination of Rays, *Bull. Torrey Botan. Club,* **68:**317–327 (1941).

15. Bosshard, H. H.: Notes on the Biology of Heartwood Formation, *News Bull., Intern. Assoc. Wood Anatomists,* **1**:11–14 (1966).
16. ———: On the Formation of Facultatively Colored Heartwood in *Beilschmiedia tawa, Wood Sci. & Technol.,* **2**(1):1–12 (1968).
17. Clowes, F. A. L.: Apical Meristems, *Botan. Monogr.,* vol. 2, Blackwell Scientific Publications, Ltd., Oxford (1961a).
18. Dinwoodie, J. M., and S. D. Richardson: Studies on the Physiology of Xylem Development. Part II: Some Effects of Light Intensity, Daylength and Provenance on Wood Density and Tracheid Length in *Picea sitchensis, J. Inst. Wood Sci.,* **7**:34–47 (1961).
19. Esau, K.: "Plant Anatomy," 2d ed., John Wiley & Sons, Inc., New York, 1964.
20. Frey-Wyssling, A.: The Ultrastructure of Wood, *Wood Sci. & Technol.,* **2**(3):78–83 (1968).
21. ——— and H. H. Bosshard: Cytologie of the Ray Cells in Sapwood and Heartwood, *Holzforschung,* **13**(5):129–137 (1959).
22. ———, J. F. Lopez-Saez, and K. Mühlethaler: Formation and Development of the Cell Plate, *News Bull., Intern. Assoc. Wood Anatomists,* **1**:4–12 (1965).
23. Gifford, E. M., Jr.: The Shoot Apex in Angiosperms, *Botan. Rev.,* **20**:477–529 (1954).
24. Hartig, R.: Ein Ringlungversuch, *Allgem. Forst-u. Jagdztg.,* **65**:365–373, 401–410 (1889).
25. Higushi, T., K. Fukazawa, and M. Shimado: "Biochemical Studies of Heartwood Formation," Proc. 14 Congress IUFRO, pt. XI, sec. 22/41, Munich (1967).
26. Hillis, W. E. (ed.): "Wood Extractives and Their Significance to the Pulp and Paper Industry," Academic Press, Inc., New York, 1962.
27. ———: "Chemical Aspects of Heartwood Formation," Proc. 14 Congress IUFRO, pt. XI, sec. 22/41, Munich (1967).
28. Johnson, M. N.: The Shoot Apex in Gymnosperms, *Phytomorphology,* **1**:188–204 (1951).
29. Larson, P. R.: Discontinuous Growth Rings in Suppressed Slash Pine, *Trop. Woods,* **104**:80–99 (1956).
30. ———: A Physiological Consideration of the Springwood Summerwood Transition in Red Pine, *Forest Sci.,* **6**:110–122 (1960).
31. ———: The Indirect Effect of Photoperiod on Tracheid Diameter, *Am. J. Botany,* **49**(2): 132–137 (1962).
32. ———: Auxin Gradients and the Regulation of Cambial Activity, pp. 97–117 in Th. T. Kozlowski (ed.), "Tree Growth," The Ronald Press Company, New York, 1962.
33. ———: Stem Form Development of Forest Trees, *Forest Sci.,* Monograph 5 (1963).
34. ———: Some Indirect Effects of Environment on Wood Formation, in M. H. Zimmermann (ed.), "The Formation of Wood in Forest Trees," Academic Press, Inc., New York, 1964.
35. Lobjanidze, E. D.: "Kambyi i Formirovaneye Goditchnich Koletz Drevesini," publication Academy of Science, Georgian SSR, Tbilici, 1961.
36. Mühlethaler, K.: Growth Theories and Development of the Cell Wall, in W. A. Côté, Jr. (ed.), "Cellular Ultrastructure of Woody Plants," Syracuse University Press, Syracuse, N.Y., 1965.
37. ———: Ultrastructure and Formation of Plant Cell Wall, *Ann. Rev. Plant Physiol.,* **18**:1–24 (1967).
38. Priestley, J. H.: Studies in the Physiology of Cambial Activity, I, Contrasted Types of Cambial Activity, *New Phytologist,* **29**(1):56–73 (1930).
39. ———: Studies in the Physiology of Cambial Activity, II, The Concept of Sliding Growth, *New Phytologist,* **29**(2):96–140 (1930).
40. ———: Studies in the Physiology of Cambial Activity, III, The Seasonal Activity of the Cambium, *New Phytologist,* **29**(5):316–354 (1930).
41. Reeve, R. M.: The Tunica-corpus Concept and Development of Shoot Apices in Certain Dicotyledons, *Am. J. Botany,* **35**:65–75 (1948).
42. Reinders-Gouentak, C. A.: Physiology of the Cambium and Other Secondary Meristems of the Shoot, *Handb. Pflanzenphysiologie,* **15**:1077–1105 (1965).

43. Richardson, S. D.: Studies on the Physiology of Xylem Development, III, Effect of Temperature, Defoliation, and Stem Girdling on Tracheid Size in Conifer Seedling, *J. Inst. Wood Sci.,* **12:**3–11 (1964).
44. ——— and J. M. Dinwoodie: Studies on the Physiology of Xylem Development. Part 1: The Effect of Night Temperature on Tracheid Size and Wood Density in Conifers, *J. Inst. Wood Sci.,* **6:**3–13 (1960).
45. Rudman, P.: Heartwood Formation in Trees, *Nature,* **210**(5036):608–610 (1966).
46. Stewart, C. M.: Excretion and Heartwood in Living Trees, *Science,* **153:**1068 (1966).
47. Trendelenburg, R., and H. Mayer-Wegelin: "Das Holz als Rohstaff," 2 überarbeitete Auflage, Carl Hanser Verlag, Munich, 1955.
48. Wareing, P. F.: The Physiology of Cambial Activity, *J. Inst. Wood Sci.,* **1:**34–42 (1958).
49. ——— and D. L. Roberts: Photoperiodic Control of Cambial Activity in *Robinia pseudoacacia* L., *New Phytologist,* **55:**356–366 (1956).
50. Wilcox, H.: In Th. T. Kozlowski (ed.), "Tree Growth," The Ronald Press Company, New York, 1962.
51. Zahner, R.: Internal Moisture Stresses and Wood Formation in Conifers, *Forest Prod. J.,* **13**(6):240–247 (1963).

3

The Woody Plant Cell

The characteristic feature of woody cells is the special nature of their cell walls. The gross structure and properties of the mature woody cell arise from the nature and organization of the chemical compounds which make up the wall and the systematic way in which the cell wall is constructed by the cytoplasm. The discussion in this chapter will be concerned first with the characteristic chemical compounds of the cell wall, then with the organization of this wall, its relation to postcambial cell development, and the modifications that are common in normal cells.

I. CHEMICAL COMPONENTS OF THE PLANT CELL WALL

Wood tissue is composed principally of a group of naturally formed organic polymer substances that are remarkably uniform regardless of origin in the plant kingdom; these polymers make up the bulk of the cell walls. In contrast, there is considerable diversity in the minor organic and inorganic materials which are found in admixture with the cell wall polymers or in the cell lumina; it is these minor substances, called extractives, or minor components, that impart special properties to many kinds of wood. Wood also contains varying amounts of water

TABLE 3-1 Schematic Classification of the Chemical Components of Cell Wall Substance in Normal Wood

I. Primary components
  A. Total polysaccharide fractions, expressed as holocellulose—60 to 70 percent
    1. Cellulose—40 to 50 percent
       Long-chain polymer with low solubility
    2. Hemicellulose—20 to 35 percent
       Noncellulosic polysaccharides; these are readily soluble in dilute alkali and hydrolyzable by dilute acids to component sugars and uronic acids
  B. Lignin—15 to 35 percent
II. Secondary components
  A. Tannins
  B. Volatile oils and resins
  C. Gums, latex, alkaloids, and other complex organic compounds including dyes and coloring materials
  D. Ash—usually less than 1 percent

in liquid and vapor form (see pages 201 to 205). In this chapter all references to the percentages of the various chemical components of the cell wall and extraneous materials are based on the ovendry weight of wood.

The chemical components of the cell wall can be classified under an arbitrary scheme which is shown as Table 3-1. The first group basically establishes the chemical and physical nature of the cell wall; it also constitutes the bulk of material in the wall. The most important single component of this group is the cellulose, which has a primary relationship to the physical behavior of the wood as a whole. The hemicelluloses and lignin present in the wall also exert important influences on behavior of the wood through their volume and their characteristics but are distinctly secondary in importance to the cellulose. The materials in the second group are not essential components in the organization of wood, but their presence profoundly influences much of the physical and some of the chemical behavior of the wood by modification of the properties of the principal elements, even when the modifiers are present only in small amounts.

A. Polysaccharide Fractions of the Cell Wall

Holocellulose is produced by first removing the materials from the wood that are soluble in ether alcohol and water, then removing the lignin.[64] This residue of the wood substance contains the entire polysaccharide material from the original wood. Treatment of the holocellulose with dilute alkali resolves the polysaccharides into two fractions: (1) the insoluble *cellulose*, and (2) a complex mixture of alkali-soluble compounds collectively known as *hemicelluloses*.

1. CELLULOSE

Cellulose is the most important single component in the woody cell wall in terms of its volume and its effect on the characteristics of wood. As can be seen from

Table 3-1, roughly half the dry weight of wood is cellulose. Furthermore, it can be estimated that the cellulose constitutes one-third of the total material produced by all plants collectively. On this basis alone, it is the most important raw material of botanical origin available to man.[48] It has been shown to be present in all the higher plants, many of the algae, and some of the fungi. In addition, cellulose has been demonstrated in the Tunicate group of the animal kingdom. However, there is still some doubt whether this animal cellulose is precisely the same as that found in plants.

Cellulose has been characterized as a stable, fibrous residue of woody tissue, markedly resistant to the attack of nitric acid, caustic soda, or potash, and most other acids, bases, and solvents. Cellulose can be dispersed only with cuprammonium hydroxide, some salt solutions, such as concentrated aqueous zinc chloride, 70 percent sulfuric acid, 40 percent hydrochloric acid, and a few other reagents.[64] This stability of cellulose is of considerable importance in the use of wood and cellulose fibers but makes the modification of the natural properties of cellulose difficult.

The cellulose molecule itself is generally believed to consist of β-D-glycopyranose residues linked by glucosidic bonds between the one and four atoms of adjacent residues to form continuous linear chains.[49,54,64] This definition was written on the basis of studies of cotton cellulose, but there is every indication that the cellulose molecules from all sources, i.e., wood, cotton, flax, ramie, straws, bamboo, etc., are chemically identical. Most sources of cellulose, such as wood, have small amounts of other simple sugar chain compounds so closely associated physically with the actual cellulose that separation of cellulose conforming completely to the definition cannot be attained.

A portion of the linear chain molecule structure of cellulose is shown in Fig. 3-1. The anhydro-D-glucopyranose residues are indicated as glucose units in the diagram for simplicity. These units (*monomers*) are the anhydro form of D-glucose.[64] The units are joined end to end, i.e., *polymerized* through a beta (β) oxygen linkage between number one carbon atom of one unit and number four atom of the unit next in line. The intervalence angles of the carbon atoms cause the configurations of the chain in one plane of the molecule to appear as a series of hexagonal links with alternating side groups, and in a plane at right angles to the above, to appear as a flat ribbon with a characteristic sawtooth folding about a central axis.

The basic repeating element of the cellulose chain is the *cellobiose* unit, indicated in Fig. 3-1 as consisting of two of the anhydro-D-glucopyranose units. X-ray diffraction studies have shown cellobiose units to have a length of 10.3 angstrom units (Å).*[64] Cellobiose is the shortest polymer of anhydro-D-glucopyranose to exhibit any of the characteristics of cellulose.

The *cellulose polymer* as it exists in the cell wall consists of macromolecules whose lengths are stated in terms of the numbers of monomers joined to form the polymer, i.e., *degree of polymerization*, or D.P. Estimates of the D.P. of cellu-

* One angstrom unit = Å = 1/10,000 micron = 1/100,000,000 centimeter. In other notation: $Å = 10^{-4}\ \mu = 10^{-8}$ centimeter $= 10^{-10}$ meter
$= 3.937 \times 10^{-9}$ inches

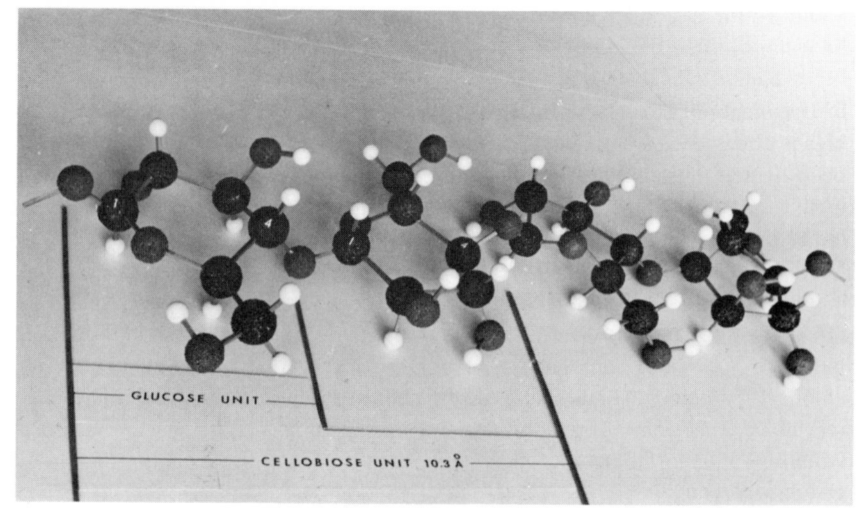

FIG. 3-1 Model of a portion of a cellulose molecule consisting of four glucose units joined through their 1–4 carbon atoms. The black spheres represent carbon atoms, the gray spheres oxygen, and the light-colored balls hydrogen atoms. The bars for the valence linkages are extended to allow a better spatial view of the structure. The cellobiose unit indicated could include any two adjacent glucose units.

lose macromolecules range from 5000 to 10,000 at a maximum,[47,48] with about 500 as a minimum.[52] In terms of molecular weight the native cellulose, i.e., in place in the cell wall, is in the order of 600,000. If the length of the macromolecule is calculated from the basic monomer length of 5.15 Å, it can be shown that they are from 2.5 to 5 microns long (2.5 to 5 $\times$ 10^{-3} millimeter). Single macromolecules of cellulose, in spite of their large size, molecularly speaking, are still too small to be observed by any means yet known.

The mechanism for polymerization in higher plants has been studied by extracellular synthesis methods. It is believed that the synthesis in plant cells is carried out first through the production of an activated monomer from D-glucose-1-phosphate and triphosphate compounds and is then catalyzed by a phosphorylase enzyme to yield either uridine diphosphate-D-glucose (UDGP) or guanosine diphosphate-D-glucose (GDPG). These activated monomers are added to the ends of existing chains of cellulose with loss of the phosphate portion.[49,52]

2. HEMICELLULOSE

Hemicelluloses constitute from 35 to 50 percent of the polysaccharides in the cell wall substance and from 20 to 35 percent of the total dry weight of the wall.[64] The definition of hemicelluloses is even less precise than that of most other plant constituents and is based mainly on chemical behavior. *Hemicelluloses* constitute that portion of the total polysaccharides in wood which for the most part is soluble in dilute alkali and hydrolyzes readily in dilute acid to form sugars and sugar acids. This hemicellulose fraction is composed of two general classes of substances: (1) those collectively called *xylans*, whose molecules are formed by polymerization of the anhydro forms of pentose sugars, and (2) hemicelluloses commonly called *glucomannans*, whose molecules are formed by polymerization of anhydro forms of hexose sugars, mainly glucose and mannose.

The hemicelluloses are polymeric in structure but have rather short chain lengths, as is indicated by their solubility and relative ease of removal from wood. The presence of this class of polysaccharides in the cell wall has a tremendous influence on certain of the physical properties of wood, and for this reason it is an important class of compounds. However, the nature of the mixture of hemicelluloses that are present in the different kinds of wood and their association with the long-chain cellulosic polysaccharides is still incompletely known. Reference should be made to Timell[54] for a discussion of the current status of knowledge of hemicelluloses.

B. Lignin

Lignin is the fraction of the cell wall material which is characteristic of woody plants. In fact, its name derives from *lignum*, the Latin name for wood. Lignin is formed only in the walls of living plants in the Spermatophytes, the Pteridophytes, and the mosses.[48]

The woody cell wall contains from 15 to 35 percent lignin. Determination of the lignin content depends on the action of strong acids which break down the polysaccharide fractions of the wood and leave an insoluble, chemically stable lignin residue. The residue is changed so much by this drastic chemical treatment that its properties give very little indication of the nature of the lignin as it exists in the cell wall (*native lignin* or *protolignin*). Progress has been made in recent years toward understanding the organization of native lignin, but only a few facts are consistently shown by different studies.[19,44,64] It is generally agreed that the basic structural unit of lignin is the phenyl propane "skeleton." This is a phenol ring that may be substituted in two or three positions. The addition of one methoxyl group to the phenol ring produces a *guaiacyl unit* which is characteristically present in all conifers. The presence of two methoxyl groups on the phenol ring results in a *syringyl unit*. Angiosperms exhibit both guaiacyl and syringyl units in their lignins, while gymnosperms contain only guaiacyl. These two forms of phenyl propane units yield different colors with the *Maüle reaction*. The Maüle test consists of treatment with chlorine, followed by a weak base: a red color is produced in angiosperms and a yellow or pale brown color in gymnosperms.[64]

About one in fifty of the phenyl propane units contains a chemical group which is responsible for the *phloroglucinol reaction*. This test yields a red color which is taken as an indication of the presence of lignin in the wood to which the reagent has been applied.

Protolignin is known to be a three-dimensional condensation polymer with an amorphous structure of high molecular weight. The amorphous nature of this polymer has been demonstrated by the fact that the presence of 15 to 35 percent lignin in the woody cell wall does not alter the x-ray diffraction pattern of the cellulose component in any way from the diagrams that are identified with pure cellulose.

Protolignin is markedly thermoplastic; this property is used to advantage in the bonding of some types of fiber and particle board. After removal from wood, lignin loses this thermoplastic character. Protolignin is quite insoluble and much lower in hygroscopicity than cellulose. It is present in the fine cavities within the mature cell wall, where it acts as a bulking agent and reduces the dimensional changes in the cell wall with moisture changes because of its low hygroscopicity. However, the most important physical property of this polymer is its rigidity and the increased stiffness it imparts to cell walls in which it is located.

C. Secondary Cell Wall Components

This class of compounds is noted here even though they are not an essential part of the structure of the wall, as can be shown by their presence in both the cell wall and the cell lumina and by their virtual absence in many kinds of wood.

1. EXTRANEOUS MATERIALS

These substances, also called *extractives,* may be infiltrated completely into the cell walls or they may occur as surface deposits or plugs in cell lumina. Wood extractives represent an extremely wide range of classes of organic compounds. The most important groups in terms of amounts that occur naturally and in economic importance are the polyphenols and wood resins. The latter are the source of steam-distilled wood turpentine, tall oil, and rosin. Polyphenols occur in both the angiosperms and the gymnosperms; they include a large number of important chemicals such as tannins, anthocyanins, flavones, catechins, kinos, and lignans. Other types of organic materials that commonly occur are gums, tropolones, fats, fatty acids, waxes, and volatile hydrocarbons.[8,64]

Typically these organic materials occur as a complex mixture of related compounds. For example, it has been reported that volatile oils extracted from wood occur in mixtures of from 3 to 35 separate compounds.[64] The phenolic types of extractives have also been shown to be mixtures. Incense-cedar (*Libocedrus decurrens* Torrey) wood, in a study by Anderson and Zavarin,[1] is reported to have a total extractive content amounting to 19.18 percent of the dry-wood weight. More than 80 percent of these extractive materials are phenolic in character and have been shown to consist of four identified, oily phenolic compounds and a large fraction of nonvolatile, phenolic-based materials classified as phenolic ethers, phlobaphenes, and tannins.

In general, the extractives constitute only a few percent of the ovendry weight of wood. However, the amounts may be considerable, as in the reported 20 to 25 percent (dry-weight basis) content of tannins in quebracho wood (*Schinopsis lorentzii* Engl.). Extractives in excess of 40 percent have been reported.

The extractives provide a major contribution to many of the properties of wood, in spite of their usually limited volume. The odors and colors typical of most woods derive from extractives. Without extractives there would be fewer obvious external characteristics to distinguish between species and this would leave anatomical structures as the only means for differentiation. Fungal and insect resistance of wood are closely linked to the presence of specific forms of toxic or inhibitory organic chemicals. In addition, their presence in wood is directly related to permeability and physical properties such as specific gravity, hardness, and compressive strength.

2. ASH CONTENT

The ash content is normally 0.1 to 0.5 percent of the ovendry weight of wood for domestic woods;[17] but it may be much greater. The alkaline earths, i.e., calcium, potassium, and magnesium, usually account for 70 percent of the total ash present. Occasionally manganese is present in appreciable amounts, as in the mineral streaks of hard maple. Silicon occurs commonly in certain groups of trees and may amount to more than 2 percent of the dry weight of wood. The presence of silica in amounts less than 0.5 percent dry weight of wood will cause marked dulling of machine tools. The rate of loss of sharpness of cutting edges in tools employed in working wood with excessive amounts of silica is inversely related to moisture content of the wood and directly related to the rate of motion of the cutting edge. These inorganic elements may be present in wood as components of the extractives or in the form of crystals. The crystals in wood most commonly are diamond or square in shape, but druses, raphides, and "crystal sand" are also encountered.[27] Single crystals are most commonly calcium oxalate, although relatively few analyses of single crystals from wood have been made.

D. Analytical Data

In general terms, the chemical composition of North American timbers on an ovendry basis can be summarized as cellulose 40 to 50 percent; hemicelluloses 20 to 35 percent; lignin 15 to 35 percent; ash less than 1 percent; and a remaining fraction consisting of resins, tannins, and miscellaneous compounds usually amounting to only a few percent.

Chemical analysis of wood according to standard methods yields results which conform to the generalized scheme as outlined. However, some consistent differences are apparent in proximate analysis data of wood. Angiosperms generally have lignin contents which represent the low end of the range given, and hemicelluloses are at the high end of the scale of percentages. In the gymnosperms the reverse holds true, as can be seen by reference to Table 3-2. In addition to the above differences between angiosperms and gymnosperms, experimental data commonly show considerable variation within the general ranges quoted. A part of this arises from species or generic differences, but a much greater cause for variability in data is traceable to the fact that the chemical composition of wood from the different parts

TABLE 3-2 Chemical Analysis of Certain Softwoods and Hardwoods as Determined at Forest Products Laboratory, Madison*

Species	Shipment No.	Holocellulose, %	Alpha cellulose, %	Lignin, %	Total pentosan, %	Solubility in: Alcohol benzene, %	Solubility in: Ethyl ether, %	Solubility in: 1% NaOH, %	Solubility in: Hot water, %	Ash, %
Noble fir (*Abies procera*)	2657	61.3	42.8	29.3	9.0	2.7	0.6	9.6	2.3	0.4
Western larch (*Larix occidentalis*)	2815	66.5	50.0	26.8	7.8	1.4	0.4	13.4	4.9	0.4
Engelmann spruce (*Picea engelmannii*)	2659	67.9	44.3	26.3	9.2	2.8	1.4	12.2	3.7	0.2
Slash pine (*Pinus elliottii*)	2404	68.5	46.1	28.0	8.6	2.6	2.0	9.9	2.5	0.2
Western white pine (*Pinus monticola*)	2660	64.3	42.3	25.4	7.9	8.3	5.6	15.6	3.7	0.3
Red pine (*Pinus resinosa*)	2607	71.2	46.8	26.2	10.0	3.5	2.5	13.4	4.4	
Douglas-fir (*Pseudotsuga menziesii*)	2655	67.0	50.4	27.2	6.8	4.4	1.2	15.1	5.6	0.2
Western hemlock (*Tsuga heterophylla*)	2765	74.0	52.5	27.8	9.2	1.6	0.8	9.2	0.4	0.3
Red maple (*Acer rubrum*)	2731	71.0	44.5	22.8	17.1	2.5	0.8	17.9	4.4	0.7
Yellow birch (*Betula alleghaniensis*)	2732	72.5	51.0	22.7	22.6	2.6	0.8	15.4	2.7	0.8
Beech (*Fagus grandifolia*)	2733	75.7	51.2	21.0	20.2	1.8	0.7	14.7	1.5	0.5
Quaking aspen (*Populus tremuloides*)	2861	78.5	48.8	19.3	18.8	2.9	1.0	18.7	2.8	
Chestnut oak (*Quercus prinus*)	2741	75.7	46.8	24.3	19.2	4.7	0.6	21.1	7.2	0.4
Basswood (*Tilia americana*)	2735	76.7	48.2	20.0	16.6	4.1	2.1	19.9	2.4	0.7
American elm (*Ulmus americana*)	2734	72.9	55.2	20.5	16.2	2.0	0.5	14.3	1.6	0.4

* All percentages based on moisture-free wood.

of the same tree may be quite dissimilar (see Chap. 7). The presence of reaction wood in the analysis sample can also be a major cause for variable results.

Another cause for differences in data on chemical composition of wood that is botanically similar consists of differences in analytical methods. As was discussed earlier, the boundaries separating the classes of compounds are not distinct; even minor changes in procedures can alter percentages of the compounds. Therefore when considering data from several sources, the methods of the various investigators should be taken into account.

II. BASIC STRUCTURAL UNITS OF THE CELL WALL

The cellulose in plant cell walls occurs in the form of long slender strands or fibrils. These consist of aggregations of the cellulose molecules large enough to be studied with the electron microscope and well enough ordered to be analyzed by x-ray crystallographic methods. The aggregation arises from characteristics of the cellulose molecule:[64] (1) the molecule itself is a uniform ribbon-shaped structure, (2) the carbon-to-carbon bonding gives the chain high rigidity and coherence, and (3) numerous hydroxyl (OH) groups are available along the length of the molecule for lateral bonding to other chains.

A. Crystalline Nature of Cellulose

The parallel arrangement of molecules of cellulose has been shown to give a crystalline structure which has a characteristic organization as shown by x-ray analysis. The smallest grouping that is associated with crystallinity is the *unit cell*. This has been shown to be a parallelopiped in the monoclinic system which consists of portions of five cellulose molecules arranged in an X pattern in cross section. The wide faces of this unit are paired chains held together on their edges by hydroxyl (OH) bonds and having a distance of 8.35 angstroms from center to center of the chains. The smaller dimension of the cross section is 7.9 angstroms, with three layers of cellulose molecules held together flatwise by weak Van der Waals forces. The unit cell length is 10.3 angstroms and is determined by the length of the cellobiose unit.[20,24,42,43] The unit cell of crystalline organization is a hypothetical structure which does not exist as an independent entity. It is simply a definition of the basic pattern of intermolecular orientation and spacing.

B. Elementary Fibrils

The smallest aggregations of cellulose molecules that are found in plant materials are known as *elementary fibrils*.[40] These are lateral extensions of the unit cell order with average diameters of about 35 angstroms, and they contain from 37 to 42 parallel cellulose molecules in the model. Careful studies using negative staining methods[25] have shown that groupings with these dimensions exist in the cell wall of the living plant. Sullivan, as quoted by Rollins,[49] has shown that the observed sizes range from 20 angstroms in slash pine to 55 angstroms in hardwoods. The elementary fibrils are believed to be crystalline in structure throughout their length, i.e., bands consisting of cellulose chains that are closely and regularly spaced.

C. Microfibrils

In the cell wall the elementary fibrils are characteristically bundled to form larger strands that are known as *microfibrils*. These are large enough to be easily observed with an electron microscope (Fig. 3-3). The elementary fibrils are believed to aggregate into ribbon-like bands with more or less rectangular cross sections having average widths from 100 to 300 angstroms, thicknesses ranging from 50 to 100 angstroms, and lengths of several microns.[40,47,48,49] The organization of the elementary fibrils within the microfibril is not completely crystalline. The exact system for the internal arrangement of the microfibril has been the subject of much study and speculation. The system which is most in agreement with the observed physical behavior is essentially that proposed by Rånby.[47] This organization is diagrammed in Fig. 3-2. The microfibril is highly ordered for portions of its length that vary in size from 50 to 600 angstroms. These ordered portions are called *crystallites* and are alternated with short portions of the microfibril length that have a different crystalline order because of the introduction of nonglucose sugar in monomer or short-chain form; these parts of the microfibril exhibit greater porosity.[49] The presence of crystallites has been demonstrated by the use of electron-dense stains and by the shape and dimensions of the fragments produced from drastic chemical action which cleaves the microfibrils at the regions of different molecular order. It is also agreed that there are spaces up to 10 plus angstroms in diameter between some of the elementary fibrils[20] which may be open capillaries and are the sites for lignin deposition within the microfibril.

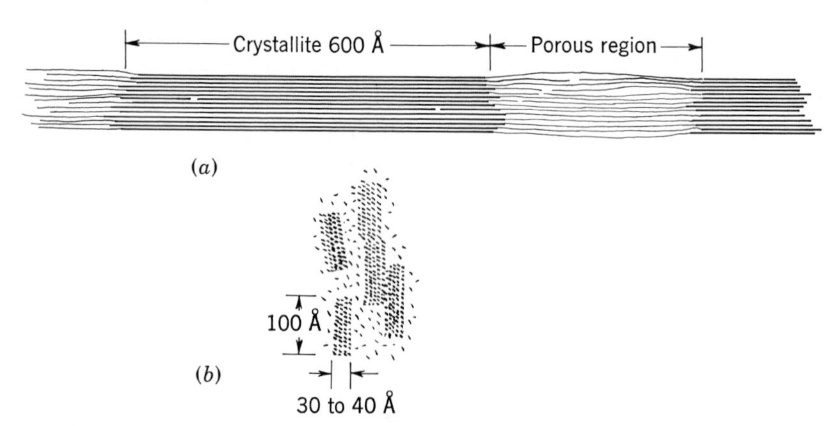

FIG. 3-2 Diagrammatic representation of microfibrillar organization in the woody cell wall. A portion of one microfibril in longitudinal view is shown at (*a*). Cross sections of five adjacent microfibrils are indicated at (*b*), with three of these joined laterally by co-crystallization.

D. Higher-order Aggregations in the Cell Wall

1. REINFORCED-MATRIX CONCEPT

A higher order of aggregation of the microfibrils also occurs to form the primary framework of the stress-resisting system of the cell wall. Groups of microfibrils are apparently organized into bundles, with some points along their lengths bonded laterally by co-crystallization[47] (Fig. 3-2*b*). However, for the most part the microfibrils are embedded in a matrix which consists of hemicelluloses. This *paracrystalline sheath* generally has a very low order of orientation and is the *amorphous region* of the cell wall. The ratio of crystalline to paracrystalline material is approximately 2 to 1,[24,45] with some evidence that the crystalline portion represents a larger part of the total. There are some spaces in the matrix of paracrystalline material which are estimated to be 100 angstroms or more in diameter and which are coated or filled with encrusting substances such as lignin, polyphenols, resins, and other infiltrates. This organization of wall components has been generalized into the *reinforced-matrix theory* to explain many of the physical properties of the cell wall.[20] Strands of microfibrils, with high tensile strength parallel to their length, are embedded in a plastic matrix which consists of the hemicelluloses, pectic materials, lignin, and infiltrates. The reinforcing effect of the microfibrils in the matrix depends on the square of their length, their diameters, the friction coefficient with the matrix, and the rigidity of the matrix. Such a system accounts for physical behavior which is in agreement with the viscoelastic mechanical properties of wood under stress and also explains the dimensional changes wood exhibits under changing moisture conditions (see Chap. 6).

2. LAMELLATION WITHIN THE CELL WALL

A further gross aggregation of microfibrillar bundles into thin sheets approximately 300 to 600 angstroms thick is evident in the cell wall under electron microscopy (Fig. 3-4). These sheets tend to separate when the cell wall has been subjected to heavy beating in the preparation of paper. Such sheets are layered to form *lamellae*. These are clearly visible with the electron microscope and have been distinguished by optical means.[6] The successive lamellae are set off by slight changes in the direction of the microfibrils and by concentration of lignin-rich material between the lamellae.[7]

3. MICROCAPILLARY SYSTEMS

Optical microscopy shows the presence of striations which are long slender void spaces in the cell wall that are large enough to contain crystals of iodine or gold. These cavities are parallel to the general microfibrillar direction in the secondary cell wall and have been widely used as a means for measuring microfibrillar orienta-

FIG. 3-3 Bundles of microfibrils in the secondary wall of Douglas-fir [*Pseudotsuga menziesii* (Mirb.) Franco]. (35,800×.) (*Electron micrograph courtesy of W. A. Côté, Jr.*)

FIG. 3-4 Secondary wall of basswood (*Tilia* sp.). Electron micrograph of a thin layer which has been peeled off and shows lamellation of the microfibrils. (14,500×.) (*Courtesy of W. A. Côté, Jr.*)

tion.[5,45] These cavities are the largest in a system of *microcapillaries,* most of which have diameters beyond the optical range. This system of microcapillaries is also called the *second-order space* in the cell wall; it forms a network which penetrates between the lamellae, within microfibrillar bundles, and within the microfibrils. It is believed that the volume of these microcapillaries is at a maximum in the fully swollen wall and reduces to a minimum in the ovendry cell wall. There is considerable controversy over the porosity of the dry cell wall. Some workers believe that there is no void space in the dry cell wall, but recent evidence, as summarized by Wangaard,[55] indicates the presence of approximately 2 percent void volume in the dry cell wall. The presence of these spaces within the cell wall causes the actual density of the cell wall to be less than that for solid cell wall substance. This actual cell wall density is termed *packing density* by some authors.[16] Reported values for the density of cell walls range from 0.7 to 1.5 grams per cubic centimeter.[16,55] The ratio of actual cell wall density to density of solid cell wall substance is called the *packing fraction.* The value of the packing fraction is at a maximum of 1.0 for the condition when the dry cell wall has a density equal to that of solid cell wall substance. Therefore, if the hypothesis is accepted that the actual cell wall in the ovendry condition contains a small volume of voids, the values of the packing fraction will always be less than one.

E. Folded-chain Theory for Cellulose Organization

The theory of molecular organization of the cell wall which has been presented so far represents a synthesis of the majority of the investigations and theories. There are numerous disagreements between workers in the field and many points at which theory and observations do not completely agree. Recently a distinctly different proposal for the basic organization of cellulose has been advanced which merits separate discussion.

Manley[35] proposes that the cellulose molecule is a folded ribbon structure which is again helically folded. According to his theory each of the cellobiose units is doubled back on the previous cellobiose unit, in a series of accordion pleats, to form a flat ribbon. This ribbon is in turn wound into a low-angle helix so that the cellobiose units in the helix lie with their axes parallel to the axis of the helical structure. This helix is proposed to have a cross section of 35 angstroms wide and 10 angstroms thick. These dimensions agree with the elementary fibril in width at least and would be readily aggregated to the dimensions usually observed for the microfibril. This idea is in agreement with the known organization of certain other polymer macromolecules, such as starch, and has been demonstrated by Manley for cellulose synthesized in solution. However, it is believed that cellulose as laid down in the cell wall is formed under constraint and is not free to form a helix system. Furthermore an even more important deterrent to the general acceptance of this theory is that it does not provide a structure with high tensile strength along the chain axis, and this is a known and important characteristic of cellulose.

F. Summary of Organization of the Cell Wall Materials

The constituent components of the woody cell wall are oriented in a regular manner. The linear cellulose molecules are closely and regularly spaced with a crystalline order whose smallest characteristic grouping is known as the *unit cell.* This crystalline unit consists of five parallel cellobiose units arranged in the monoclinic system. The crystalline order, as described by the "unit cell," is continuous throughout the bundles of cellulose microfibrils which are called *elementary fibrils* and which are approximately 35 angstroms in diameter and are believed to be made up of 37 to 42 molecules in parallel. These elementary fibrils are aggregated into ribbonlike bands called *microfibrils* that are more or less rectangular in cross section. Microfibrils are easily distinguished with an electron microscope since they are 100 to 300 angstroms wide, 50 to 100 angstroms thick, and several microns long.

The microfibrillar strands are embedded in a matrix consisting of hemicelluloses, lignin, and extractives. The ribbon-like microfibrils are aligned edge to edge in the matrix so that they form thin, discontinuous sheets of cellulose strands which are in turn layered to build up the cell wall structure. The whole framework of cellulosic strands and layers, as well as the matrix material, is penetrated by a system of very fine capillary cavities that change volume with the changing moisture content of the cell wall.

III. FORMATION OF THE CELL WALL

A. Primary Cell Wall Development

At the onset of the differentiation of the woody cell the living protoplasm is enclosed in a primary wall which is capable of being extensively increased in surface area to keep pace with the enlarging volume of the cell as the latter matures. Between adjacent cells is an isotropic layer called the *true middle lamella.** In the cambial stage of cell formation and during the enlargement of the cell this true middle lamella consists mostly of pectic material and is very plastic in nature, so that slippage between the surfaces of the enlarging cells is possible. Both the true middle lamella and the primary wall are thin and difficult to differentiate with optical microscopy. In practice the complex formed by the intercellular material and the two primary walls on either side is called the *compound middle lamella.* This term corresponds to Kerr and Bailey's *cambial wall*[29] and to Roelofsen's *meristematic wall.*[48] The use of this terminology for the combined three layers is a convenience, since only very special optical techniques or electron microscopy can separate these regions.

1. THE PRIMARY CELL WALL

The primary cell wall is associated with the meristematic and enlarging cells. Its characteristic structure and composition alter only slightly from the cambium to that found in the fully enlarged cell. It is estimated that the thickness of the primary cell wall in the living tree is about 0.1 micron. In dry wood, shrinkage reduces the thickness to 0.03 micron.[33] This primary wall in the living tree is estimated to consist of only about 9 percent cellulose[48] embedded in an amorphous plastic matrix of hemicelluloses, pectic materials, and lignin, with some 70 percent or more water. In the early postcambial stages more pectin is present than lignin. This ratio is reversed in the primary wall of the mature cell.

The cellulose occurs in the form of a stress-resisting framework of microfibrils very much dispersed in a loose, irregular interwoven pattern that is known as the *primary-wall texture* (see Fig. 3-11*F* and Fig. 3-13). The orientation that is present is transverse and is about 15° from the direction normal to the longitudinal cell axis. Part of the microfibrils slope in one direction and part in the opposite direction to form the interwoven texture or *tube structure.*

The primary wall is in close contact with the limiting membrane or *plasmalemma* of the cytoplasm of the cell. At numerous locations over the wall surface the loose texture of the microfibrillar framework is penetrated by groups of strands of cytoplasm each enclosed in a tubular extension of the plasmalemma. These strands, called *plasmodesmata,* join the cytoplasm of adjacent cells into a single interconnecting system. The areas in which the plasmodesmata are clustered are known as *primary pit fields.*

* Equivalent to the term intercellular substance of many authors.

2. CELL ENLARGEMENT

Increase in the overall dimensions of the cell is the characteristic feature of primary growth. The first phase of enlargement in the postcambial development of the cell is an increase in diameter. This is particularly evident in the early-wood tracheids of the gymnosperms and the vessel elements of the angiosperms. Diameter increase in the gymnosperms is almost entirely in the radial direction, and almost no tangential enlargement takes place. In contrast, the vessel elements of hardwoods, especially those in the early wood of ring-porous woods, exhibit major increase in tangential diameters; followed by that in the radial direction. Other cells, such as the late-wood tracheids of conifers, hardwood fibers, and parenchyma cells, undergo little of this diameter increase, as can be seen from the cross-sectional dimensions of the mature cells.

The second phase of cell enlargement is an increase in length along the longitudinal axis. Length increase is typical of the gymnosperm tracheids and fibers in the angiosperms. Other types of cells elongate to a lesser extent, or not at all, as in the case of early-wood vessel elements in the more highly specialized angiosperms. The extent of the elongation varies between angiosperms and gymnosperms in general. Conifer tracheids increase from 10 to 15 percent in length from that of the cambial initials, at a maximum, and in some species show little or no elongation.[3] In contrast, the hardwood fibrous elements, which are much shorter at all stages of development than conifer tracheids, will more than double their lengths during maturation from the cambial initials. The average elongation of fibers in the dicotyledons is 140 percent calculated from 54 species, with a range from 20 to 460 percent. The hardwoods with unspecialized or primitive vessel elements, such as *Alnus, Liquidambar,* and *Nyssa,* show average increases of about 120 percent for the fibers and only 15 percent for the vessel elements. The hardwoods with highly specialized vessel elements, such as *Quercus* and *Robinia,* show fiber elongations of almost 200 percent, but a 25 percent decrease in vessel-element lengths.[3]

Tangential enlargement in most growing cells, except vessel elements in hardwoods, shows little increase over the tangential diameter of the cambial initial. This can be understood by recognizing that the circumferential increases that must occur as a result of diameter growth in the tree stem are accomplished mainly by the production of new cambial initials through anticlinal division, not through increase in the tangential diameter of the cambial initials. An exception to the usual lack of tangential increase in the growing cell is the massive enlargement of early-wood vessel elements in ring-porous hardwoods. In these cases it is evident by making counts of the numbers of cells in the radial direction within the ring that the enlargement of vessel elements is accompanied by a suppression of the rate of division in the cambial daughter cells to provide the additional space required for these specialized cells. The early-wood vessels of the oaks, for example, are from ten to fifty times greater in the tangential diameter than the width of their cambial initials.

Longitudinal increase in cell dimensions appears to be a matter of fairly general surface area increase, with somewhat more increase occurring at the ends than at the middle. The most widely accepted theory for the mechanism to account for the increase in individual cell lengths and the corresponding adjustments with neighboring cells is known as *intrusive growth.* According to this theory, the tips of the elongating cell force their way between the other developing cells above and below. The plastic nature of the middle lamella allows considerable adjustment, but it seems likely that many of the cytoplasmic connections through the plasmodesmata must be broken and re-formed in this process. The intrusive growth theory is also known as *tip growth, interposition growth,* or *apical growth.* This last name is confusing because of its usual application to the apical meristematic growth of trees.

An explanation for the greater length than diameter increase in growing cells is based on the stress behavior of the wall. In the primary wall the largest component of resistance from the microfibrillar framework is in the transverse direction of the cell. This results in restraint in the primary-wall expansion transversely and in greater growth parallel to the cell axis.[46]

3. MECHANISM FOR PRIMARY-WALL FORMATION

Studies of the walls of growing cells by means of autoradiography show that the increase in surface is fairly uniform except for slightly greater elongation at the ends of the cell.[61] This conforms to the *multinet theory* of wall enlargement in which the network of microfibrils in the wall is expanded by the cell growth and new microfibrils are inserted into the spaces formed. This insertion process is known as *intussusception* and continues as long as enlargement of the cell occurs. As the growth rate slows, more transversely oriented microfibrils are deposited on the surface of the first-formed primary wall by *apposition.* In the final stages of the enlarged cell the primary wall is slightly thicker than that found in the cambial stages; it is also somewhat zonate, with a more regular transverse microfibril orientation on its inner surface.

The actual mechanism by which the microfibrils of the wall are formed and laid down in the cell wall is still not understood. Studies of the cytoplasm of growing cells indicate that certain of its organelles may be associated with the formation and transfer of molecules of wall substance. It appears that the new molecular material for microfibril formation may be brought into close proximity with the plasma membrane. At this point a new plasma membrane is formed to the inside and the old one disintegrates, leaving the new microfibrillar material in the wall matrix.[14] The means by which the direction of microfibrillar orientation is maintained is even more obscure. It has been suggested on the basis of tissue-culture studies that the chemical growth hormones, or auxins, control the orientation of the cellulose molecules in the wall.[46] Another theory for orientation control has been advanced by Preston[45] on the basis of algal cell studies. He found a system of granules 300 to 500 angstroms in diameter in the matrix just outside the plasmalemma. The granules are arranged in several layers in close cubic packing and are the sites of the enzymes which synthesize cellulose and add units onto the ends of existing molecules in a direction determined by the granule arrangement.

B. Normal Secondary-Wall Formation

Growth of the cell wall, after the enlargement phase is complete, consists of the formation by apposition of additional wall material to the inside of the primary wall. These wall layers are incapable of further surface enlargement and are classified as *secondary-wall layers*. They are relatively dense and contain high proportions of cellulose. Because of this structure they constitute the major source of strength for the woody cell.

The first material in the secondary layers is laid down near the central part of the cell, with deposition moving toward the ends of the cell.[55,61] The microfibrillar orientation of this first lamella is very little different from that in the primary wall. Successive lamellae are laid down in waves from the middle to the end of the cell, each one with slightly different orientation of the microfibrils. The apposition of cell wall material continues for some time, with the production of a wall whose thickness depends partly upon the type of cell and partly upon the growth conditions of that particular season. The deposition of material is apparently a cyclic process, with the lamellae of cellulosic components deposited during the daylight hours, and lignin produced and infiltrated into the newly formed wall during the middle of the night.[9] The process of lignification parallels the wall thickening; it begins at the cell corners, then modifies the remainder of the middle lamella by replacement or alteration of the pectic materials with lignin so that the intercellular material becomes quite rigid. As the secondary thickening of the wall occurs, lignin deposition progresses inward with the wall development.

During the period of secondary-wall formation a number of specialized local developments of the wall take place. The most typical of these modifications are the pits, which are essentially gaps in the secondary-wall material, with shapes typical of both the cell types and the kind of wood.

In prosenchyma cells, the final stage in the development of the woody cell occurs at the time of the death of the protoplasm. The residues of the cell contents are deposited on the inner face of the cell either as an amorphous layer or in the form of a warty deposit. The parenchyma cells, of course, continue to function as living cells for some time after the formation of the secondary wall is complete, but they too die eventually, as heartwood is formed. Except for this difference and the greater amount of infiltrates in the parenchyma of the heartwood, the wall structure of these two types of cells is essentially similar.

The mature secondary wall is extremely variable in thickness and in the complexity of its internal structure. However, all forms of secondary wall have certain characteristics in common: no further enlargement in area is possible, and the organization of the material in these walls is relatively dense and very rigid. The maturation of the cell, with its increase in rigidity, is called *lignification* by botanists. The preferred use of the term by wood anatomists and chemists is to indicate the deposition of the chemical substance lignin.

1. STRUCTURE OF THE SECONDARY CELL WALL

In the normal development of a tracheid or fiber type of cell the secondary wall consists of three layers: a thin outer layer with a nearly horizontal helix of microfibrils, a thick central layer, with the microfibrils nearly parallel to the cell axis, and an inner, thin layer made up of microfibrils roughly parallel to those in the outer layer. These three layers have come to be designated by letters corresponding to the succession of their formation. Under this terminology the outer layer is labeled as $S1$, the central layer as $S2$, and the inner one as $S3$. This model for the secondary wall is based on the normal conifer tracheid, and while it is appli-

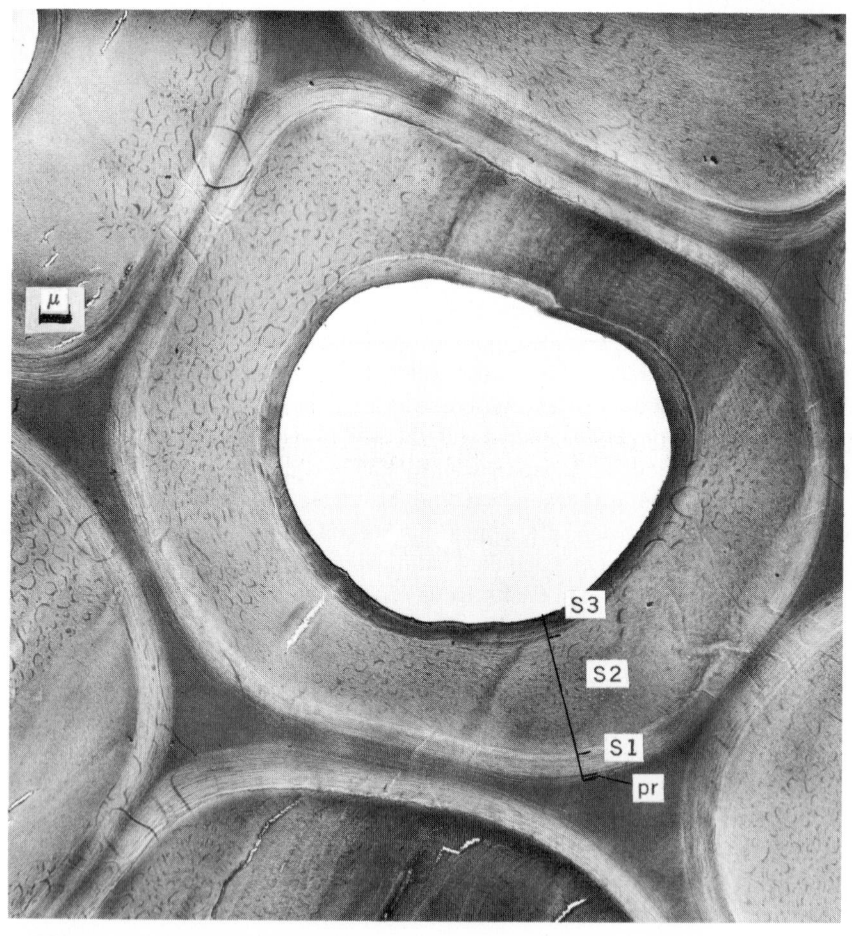

FIG. 3-5 Cross section of late-wood tracheid of longleaf pine (***Pinus palustris*** Mill.) showing the layers of the secondary wall (S1, S2, and S3), the thin primary wall (*pr*), and the true middle lamella, which is markedly thickened at the cell corners. (4200×.) The small dark areas in the wall are artifacts resulting from fractures in the embedding medium. (*Electron micrograph courtesy of W. A. Côté, Jr.*)

FIG. 3-6 Model of a portion of a conifer tracheid with the walls cut away to show the helical organization of the microfibrils in the secondary wall. The relation to adjacent tracheids, the thickness of the true middle lamella, and the appearance of layers in cross section are shown by the pieces of tracheids attached to the base. The warty layer is visible as a lumen lining.

cable to many other cell types in hardwoods as well, it must be realized that it is a generalization and that other arrangements of the secondary wall do exist in plants. As Bailey has noted,[5] it is not possible to construct a single model that represents all the structural variations that exist in secondary walls of plants. However, the three-layered model does represent the usual development of the secondary wall of the normal, fiber type of cell, which constitutes the bulk of the mechanical tissue in both hardwoods and softwoods. The important differences associated with reaction wood are discussed on pages 288 to 308.

The most obvious difference in the layers of the secondary wall is in the direction and angle of the helix of the microfibrillar orientation. In general, this difference has been determined by the direction of the striations on the surface of the cell wall, the orientation of the slitlike openings of the pit apertures, deposition of minute crystals of metals and iodine in the fine cavities of the wall, preferential attack of fungal hyphae, special optical and x-ray methods, and electron microscopy. The angular orientation, in degrees, is usually given with respect to the longitudinal axis of the cell, while the helix directions are described according to convention as *Z* or *S* helices. If the helix direction on the side of the cell next to the observer is parallel to the middle bar of the letter *Z*, then the cell has a *Z helix,* which corresponds to a right-hand thread. The opposite direction of helical orientation is known as an *S helix,* with the near side of the cell having microfibrils parallel to the middle bar of the letter *S*. The *S* helix is threaded in a left-hand manner.

The *S*1, or outer layer of the secondary cell wall, is only 0.1 to 0.2 micron thick,[20] but in spite of this, it has been proved to be made up of a complex of several thin sublayers, differing in the orientation of their microfibrillar bundles and in their thickness. Wardrop[56] has defined the system as it exists in tracheids. He describes the *S*1 as a wall layer consisting of a few lamellae of alternating *S* and *Z* helical orientation. These lamellae are symmetrically arranged with respect to the longitudinal axis of the cell. The successive lamellae from the outside contain wider bands of microfibrils. On the inner surface of the *S*1 layer, one or more lamellae, with nearly continuous microfibrillar structure and orientation more nearly parallel to the cell axis, may exist.[15,61] The inner lamellae with their *S* helix and average microfibril angle of 50 to 70°, measured from the cell axis, define the general character of the *S*1. The network of the crossed structure associated with the *S*1 region accounts for the high tensile strength of this cell wall zone in the transverse direction; this network also is responsible for the fact that moisture-dimensional changes in the *S*1 layer occur almost entirely in the longitudinal direction of the cell. These properties are related to the well-known *ballooning* of isolated tracheids, swollen by reagents other than water (Fig. 3-7). The organization of the *S*1 layer is apparently transitional from the structure of the primary wall to that shown in the dominant, central region of the secondary wall. For this reason some authors have labeled the *S*1 portion of the secondary wall the *transition layer.*

The central, or *S*2, zone of the secondary wall consists of a dense organization

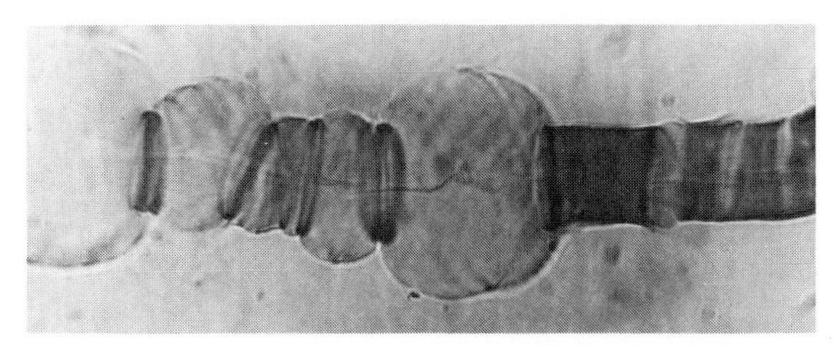

FIG. 3-7 Ballooning in a portion of a white pine tracheid delignified to 6.5 percent by the Kraft process and swollen in 3-methyl-benzylammoniumhydroxide. (*Courtesy of W. M. Harlow.*)

of microfibrillar lamellae that has a nearly parallel alignment of microfibrils (Fig. 3-8). The microfibrillar orientation in the *S*2 layer is about 10 to 30° from the cell axis. This *S*2 region is the thickest of the three secondary-wall zones and constitutes the bulk of the cell wall. Consequently the orientation of the wall material in the *S*2 layer is dominant in the cell wall, and its mass largely controls the physical behavior of the plant cell wall. The *S*2 layer of the secondary cell wall is the one which shows most variation in thickness and reflects the major differences between early and late wood.

The *S*3 layer of the secondary wall appears to be thinner than the *S*1. Most investigators indicate that the microfibrillar orientation in this layer is predominantly a very flat helix that lies at an angle of 60 to 90° to the cell axis (Fig. 3-8). However, major deviations from this pattern are known, and others may be revealed with additional study. Several lamellae have been observed within the *S*3 layer, but their organization is less precise than in the *S*1. The *S*3 layer differs from the remainder of the secondary wall in its chemical and staining reactions.[48] Because of these differentiating reactions, certain authors have called the *S*3 layer of the secondary wall a ***tertiary layer*** or ***tertiary wall***. On the whole, this terminology is not being accepted by anatomists.

As has been noted, the wall thickness may be variable, because of differences largely in the *S*2 layer. Some work has been done to define the relative percentages of the various cell wall layers[32,36,38] using tracheids of *Picea abies*, *Pinus sylvestris*, and *Abies balsamea*. It has been shown that the true middle lamella, plus the primary wall, is remarkably uniform, constituting about 2 percent of the total volume of the cell wall. *S*1 is more variable, with an average of 16 percent, but ranging from 10 percent in pine to 16 percent in true fir and 22 percent in spruce. The *S*2 averages 74 percent, with a range from 68 in spruce to 78 percent in the pine. The *S*3 layer was shown to be smaller than the *S*1 and averaged 8 percent of the total cell wall volume.

Variations in the angular direction of microfibrils within the secondary wall, from the pattern that has been described, are quite common. There are considerable differences recorded in the literature for different kinds of wood. Furthermore, it must also be realized that the angle which the microfibrils exhibit, in a

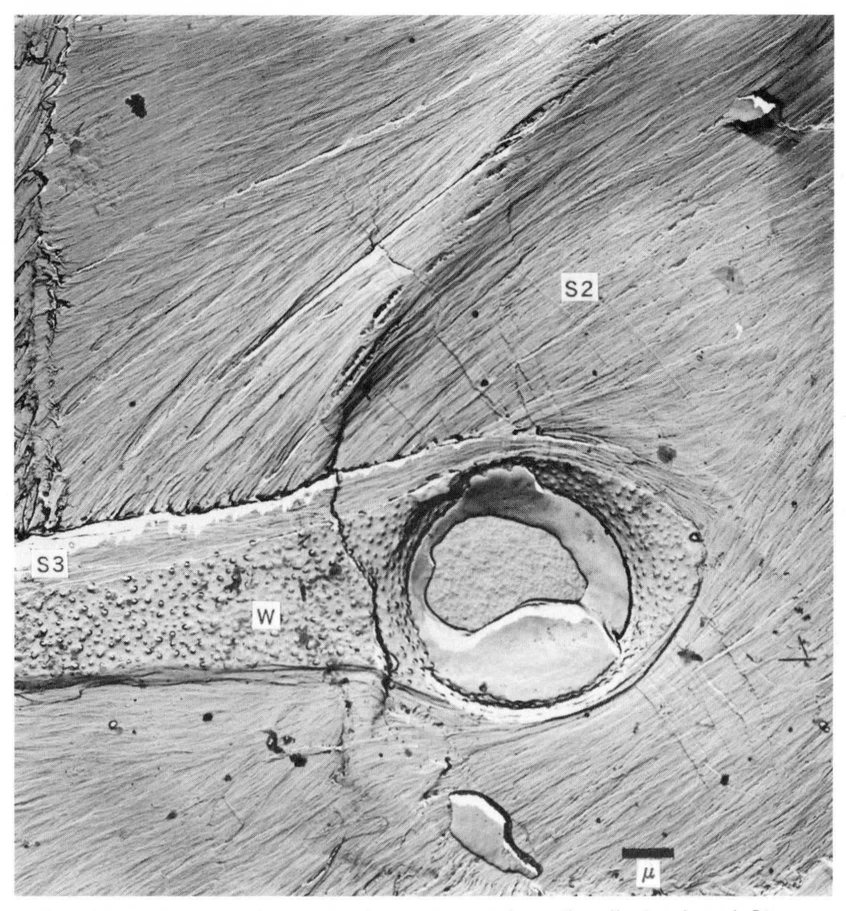

FIG. 3-8 Electron micrograph of a replica of the cell wall of redwood [*Sequoia sempervirens* (D. Don) Endl.]. The warty layer (*W*) and the S3 layer of the secondary wall have been stripped away from the S2 layer except in the vicinity of the pit. (5250×.) (*Courtesy of W. A. Côté, Jr.*)

given layer of the wall, with respect to the cell axis is not uniform throughout a single cell. For example, Preston[45] has shown that the helix angle on the radial face of a tracheid is larger, as measured from the cell axis, than it is on the tangential face of the cell. The angle appears to be a function of the width of the cell in the transverse direction. Early-wood tracheids in conifers exhibit larger microfibrillar angles on their radial faces than are evident on the radial faces of the late-wood cells in the same row. In addition, the helix angle has been shown to be a function of elongation; the angle decreases with respect to the cell axis, as the length of the cell increases.[45] The increase in angular deviation of the microfibrils from the transverse axis of the cell with increasing age of the cambium can be explained on this basis, since there is a gradual increase in the length of the cambial initials, and cells derived from them, with age.

2. VARIATIONS IN HELICAL ORIENTATION

The helical orientation of the microfibrils within the secondary wall is still incompletely known. Most of the work has been done with conifer tracheids, and the consensus is that the majority of the genera and species will show an *S* helix in the *S*1, with a *Z* helix in the *S*2, and a reversal to the *S* helix in the *S*3 layer (Fig. 3-6). However, other work[60] has shown that, for ***Pinus radiata*** D. Don at least, the helix is reversed with a *Z-S-Z* organization in the successive layers. Others indicate that there may be no essential differences in the orientation for the helical structure of the three layers.

3. WALL STRUCTURE OF PARENCHYMA CELLS AND VESSEL ELEMENTS

The wall structure of parenchyma cells and vessel elements conforms to that described previously for tracheids and fibers. The primary wall is similar in all cell types. The variations which exist are confined to the secondary wall and consist of modifications in the thickness of the wall layers or in the number of layers formed.

In general, the walls of parenchyma cells are thinner than those of fibers and tracheids. This makes it difficult to separate the layering which is present in the secondary walls of parenchyma cells. There is agreement that the inner and outer layers of the secondary wall are oriented in flat helices with an angle of 30 to 60°

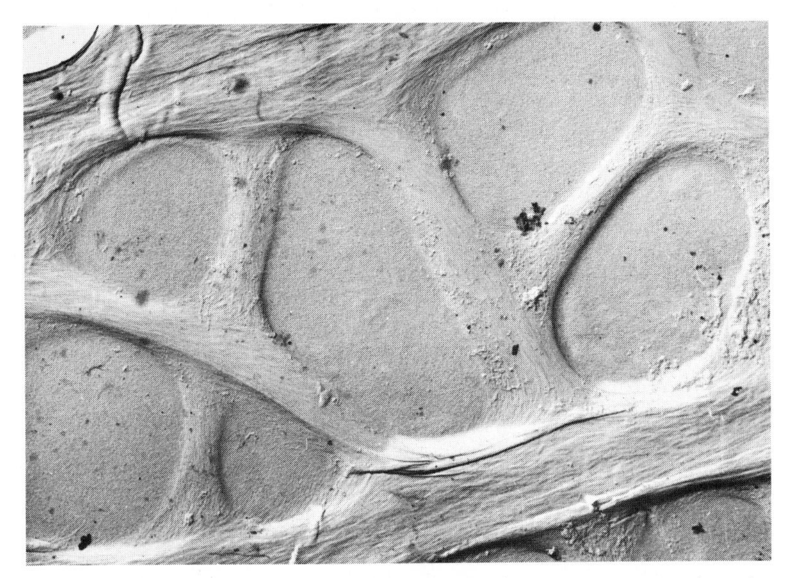

FIG. 3-9 Electron micrograph of vessel-parenchyma pitting as seen from the vessel side in white oak (*Quercus alba* L.). Note the modification of the microfibril orientation in the secondary wall from the close spacing of the pitting. (4000×.) (*Electron micrograph courtesy of R. W. Meyer.*)

from the long axis of the cell.[23,57,59] The central wall layer is relatively thin and may not be as thick as either of the other layers. Orientation in the central part of the wall is approximately axial in the usual pattern. No differences have been shown between ray and longitudinal parenchyma cell walls, but interspecies differences apparently do occur.

The primary-wall structure of vessel elements is similar to that found in other cell types.[48,57] The structure of the secondary wall in the less-specialized dicotyledons is similar to that found in the thin-walled tracheids. In the vessel elements of the more specialized hardwoods the layering is not very marked and at best consists of small differences in orientation between lamellae.[57] In those species in which the intervessel pitting is crowded, the microfibrils are reduced to a network of strands intertwining between the pits and the usual patterns of microfibril orientation in the secondary wall are not evident (Fig. 3-9).

C. Distribution of the Chemical Constituents in the Cell Wall

The chemical components of the cell wall are not uniformly distributed throughout the cell wall. This has been known in a qualitative way for many years, but quantitative data have been lacking because of the difficulty of analysis. The first direct analysis of the compound middle lamella, for example, was made by A. J. Bailey in 1936.[2] He used a microdissection technique on Douglas-fir cross sections and determined that for wood with a total lignin content of 30 to 40 percent on a dry-weight basis, 71 percent of the lignin and 14 percent of the pentosans were associated with the compound middle lamella. Later work with ultraviolet spectroscopy[33,50] has confirmed that the greatest concentration of lignin is in the compound middle lamella and that it decreases in the secondary-wall layers in definite ratios of concentration. When these concentration ratios are combined with Bailey's quantitative information, the $S2$ is calculated to contain from 18 to 25 percent lignin and the $S3$ from 11 to 18 percent. These distributions are shown diagrammatically in Fig. 3-10. Consideration of this diagram shows that if the area delineated by the lignin curve is measured for each of the cell wall layers, their areas are proportional to the quantity of lignin in the layer. On this basis it is evident that well over half the lignin occurs in the $S2$ layer where the concentration is lowest but the cell wall volume is greatest. In spite of the high concentration of lignin in the compound middle lamella, the volume of this wall layer is so small that only about 10 percent of the total lignin can be contained in this region. This suggested mass distribution of lignin is contrary to the earlier views expressed in the literature.

Limited information on specific distribution of polysaccharides in the cell wall is available. Only recently fibers and tracheids have been analyzed by separating the postcambial xylem tissue into four stages of cell wall development.[32,36,38] The percentages of cellulose and other polysaccharides in the wall layers have been determined for *Abies balsamea*, *Picea abies*, *Pinus sylvestris*, and *Betula*

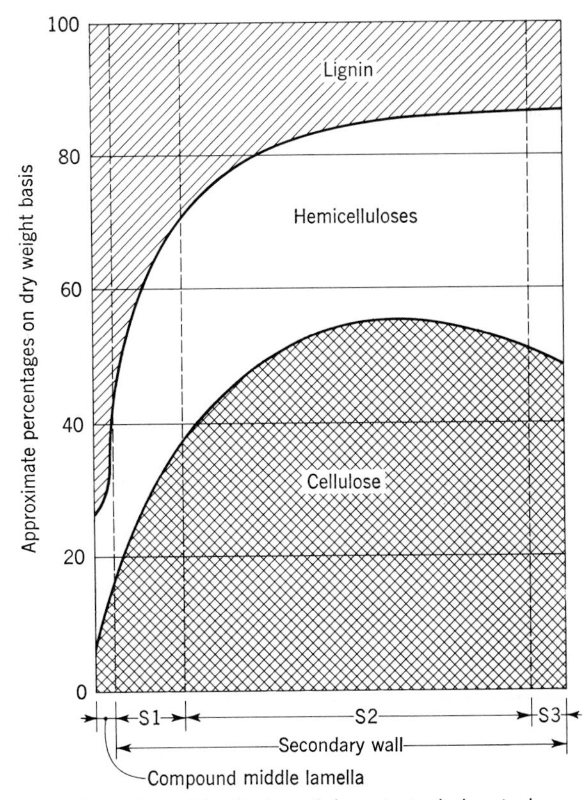

FIG. 3-10 Distribution of the principal chemical constituents within the various layers of the cell wall in conifers.

verrucosa. These data have been used to construct the curves for the polysaccharide components in Fig. 3-10. It is apparent that less than 10 percent of the primary wall is cellulosic; in the *S*2 layer the cellulose increases to more than 50 percent of the material in that cell wall layer; a reduction in amount of cellulose is evident in the *S*3. In general it can be seen that the lignin content is inversely related to that of the total polysaccharides. The two components of the latter, i.e., cellulose and hemicellulose, tend to vary more or less directly.

D. Modifications of the Cell Wall

The general nature of the structure of the cell wall which has been discussed up to this point is further complicated by modifications arising as normal features of the cell wall. These can take the form of gaps, called *pits,* in the secondary wall, perforations leaving a visibly open passage between cells, thickening on the inner surface of the cell wall, or material deposited on the wall surface after the death of the cell.

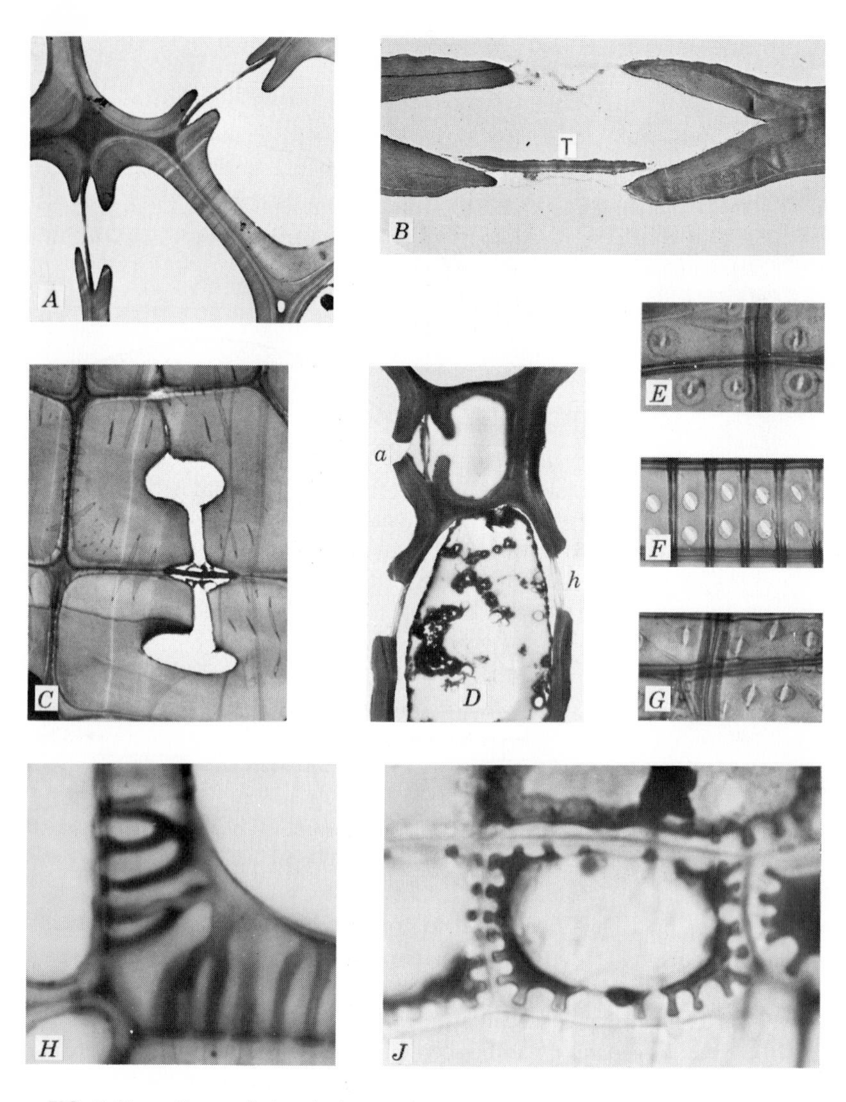

FIG. 3-11 Types of pit pairs in wood.

A. Bordered pit pairs in relatively thin-walled fibers of red oak (***Quercus rubra*** L.). Electron micrograph of a thin section at 2500×. (*Courtesy of W. A. Côté, Jr.*)

B. Bordered pit pair between tracheids in longleaf pine (***Pinus palustris*** Mill.). The torus (*T*) is aspirated but does not completely close the pit aperture in this instance. The supporting strands of the pit membrane are too thin to show as a continuous line. (*Electron micrograph of a thin section at* 2350×, *courtesy of W. A. Côté, Jr.*)

C. Bordered pit pair with extended pit canals and small cavities in thick-walled fibers of southern blue gum (***Eucalyptus globulus*** Labill.). (*Electron micrograph of thin section at* 2500×, *courtesy of W. A. Côté, Jr.*)

D. Ray cells from Douglas-fir [***Pseudotsuga menziesii*** (Mirb.) Franco] in tangential section showing a small bordered pit pair (*a*) between the ray tracheid cell at the top and a longitudinal tracheid; half-bordered pit pairs (*h*) between the ray parenchyma

1. PITTING OF THE CELL WALL

The surface of all woody cells is marked with normal structures, called pits, that are evident at low magnifications. A pit is defined as a recess in the secondary wall of the cell, open to the lumen on one side and including the membrane closing the recess on the other side. Normally two complementary pits in adjacent cells occur as a *pit pair*. Single pits are seen in wood occasionally as *blind pits*. When this occurs, one of the normal pair is lacking or a pit leads to an intercellular space. Single pits are also present in cells separated from each other, as in pulping. The primary parts of the pit are the *pit cavity* and the *pit membrane*. The pit cavity is the entire space within the wall recess, between the pit membrane and the lumen. The pit membrane is that section of primary wall and true middle lamella which limits the outer end of the pit cavity (Fig. 3-11).

Cell wall pitting can assume a wide variety of forms. In general, however, these can be reduced to two types on the basis of the shape of the pit cavity. A *simple pit* is formed if the diameter of the pit cavity is uniform or increases slightly toward the lumen. On the other hand, *bordered pits* are characterized by a constriction of the pit cavity toward the cell lumen (Fig. 3-11). Pit pairs may be made up from similar pits to form bordered pit pairs or simple pit pairs, or they may consist of dissimilar halves to form a *half-bordered pit pair*.

The shape of the pit cavity formed in the secondary wall is the result of both the thickness of the wall and the diameter change in the pit cavity. In conifers the usual simple pit forms in relatively thin walls and the pit cavity increases in diameter toward the lumen. A typical example of such a pit is shown in Fig. 3-11. In hardwoods, where thicker secondary walls are common, the simple pitting shows a more striking form with elongated canal-like cavities of uniform diameter passing through the secondary wall (Fig. 3-11). In such structures the opening of the pit leading into the lumen is called the *pit aperture*. Sclerosed tyloses, and some cells with very thick walls, such as phloem sclerids and vessel elements in *Diospyros*, show tubular cavities of several pits that coalesce to form branching or *ramiform pits*.

There are numerous variations in the size and shapes of simple pits, the larger pits usually occurring in prosenchyma cells. In the conifers, the simple pits are

---

cell and longitudinal tracheids. (*Electron micrograph of thin section at* 2500×, *courtesy of* W. A. Côté, *Jr.*)

E. Surface view of small bordered pits with included inner apertures.

F. Surface view of bordered pits in which the inner apertures extend to the outer edge of the pit border.

G. Surface view of bordered pits in which the inner apertures of most of the pits are extended beyond the pit borders.

H. Ramiform pitting in the cross-sectional view of two adjacent vessel elements of common persimmon (*Diospyros virginiana* L.). (820×.)

J. Simple pit pairs in walls of ray parenchyma cells of true fir (*Abies* sp.). Note the nodular appearance characteristic of the end walls in the genus. The pitting is accentuated by the deposits of material on the lumen surface. (470×.)

confined to parenchyma. On the other hand, in hardwoods, while parenchyma cells are usually simply pitted also, simple pits are not limited to this type of cell; in general the variation in simple pitting in hardwoods is greater than in conifers. The common feature of all simple pits is that the surface view of the structure shows only a single, circular outline.

Bordered pits are characterized by a constriction of part of the pit cavity to form a dome-shaped recess called the *pit chamber*. The overarching part of the secondary wall, which defines the pit chamber, is called the *pit border*. A projection of this type of pitting, in surface view, appears as two or three concentric rings. The name bordered pit is derived from this appearance. The remaining portion of the cavity, between the chamber and the lumen, may be quite shallow, as in the bordered pits of most conifers. In this case the opening is known simply as an *aperture*. However, if the wall is thick the passage from the pit chamber into the lumen may be elongated to form a *pit canal*. In such cases the chamber end of the canal is known as the *outer aperture* and the lumen end as the *inner aperture* (Fig. 3-11).

The pit canal may appear in a variety of forms; it may be a simple tube with uniform diameter, or the inner end may flare with a flattened elliptical section. In this case the long axis of this flattened canal lies parallel to the mean microfibrillar direction of the $S2$ layer of the secondary wall and can be used as an indicator of the orientation of the microfibrils in the $S2$ layer of the secondary wall. If the surface view of the pit, having a flattened aperture of this sort, shows an inner opening whose length is less than the membrane diameter, the aperture is said to be *included;* on the contrary, if the aperture is longer than the diameter of the pit membrane, then it is said to be *extended* (Fig. 3-11). In extreme cases of such extension, in hardwoods, the inner apertures may be united to form grooves on the inner surface of the secondary wall; these are called *coalescent apertures*. In face view the bordered pits under a microscope will appear as concentric circles, with or without a diagonal ellipse arising from the inner aperture. In the large bordered pitting between conifer tracheids it is a simple matter to distinguish such pitting. However, in certain of the hardwoods the separation of bordered and simple pitting is not an easy matter and requires close observation.

The development of a pit can be typified by the process of formation of the conifer bordered pit pair. The location of a pit in any cell is marked in the cambial stages by a slightly thinner region of the primary wall and a clustering of plasmodesmata to form a primary pit field. At the end of primary growth a ring of circularly oriented microfibrils is laid down enclosing the pit field and outlining the wall area which will be the closing membrane of the pit. This bordering ring constitutes the *pit annulus*. A similar structure forms in the matching cell where the other half of the pit pair is located, and subsequent development continues in a parallel fashion. The initial ring develops into an arched shell structure with a circular opening at the top on the lumen side. The opening defines the pit aperture, and the shape of the shell structure defines the convex pit cavity. The shell of the pit border appears to be poorly joined to the later-formed wall layers.[57] The secondary wall layers are laid down enclosing the pit-cavity shell, with the microfibrils deviating from their normal path in the same pattern as is seen with wood grain around a knot (Fig. 3-8). The *S*1 is thicker than normal and forms over most of the shell defining the cavity but thins out toward the aperture.[22] The transition between *S*1 and *S*2 is marked by a zone of greater density. The *S*2 decreases in thickness over the bulge which marks the pit-cavity location and tapers out near the aperture (Fig. 3-11*B*). While the wall thickening is progressing, several changes occur in the closing membrane. An interconnecting net of more or less radially oriented microfibril strands is formed over the existing primary wall, and a ring of microfibrils is laid down over the central portion of the membrane (Fig. 3-12*b*). This ring defines the area called the *torus*, which is slightly larger in diameter than the pit aperture. In the last stages the torus is thickened by deposition of encrusting materials and may become almost double convex in cross section (Fig. 3-11*B*). At this time also the membrane between the torus and the pit border is further differentiated. The region between the torus and the pit border is called the *margo;* it consists of an open net of radially oriented supporting strands superimposed on an unoriented primary-wall network, and it extends from the torus to the pit border (Fig. 3-12*f*).[53] It is believed that the intercellular material is removed by enzymatic action, leaving openings between the microfibrillar bundles which are from 0.1 to 1 micron wide and which allow passage of liquids and materials in a particulate form up to 0.2 micron in diameter.[20]

The tori of conifer pits are usually flattened in cross section and fairly circular in outline. The torus may be thickened by deposits of encrusting materials, and the outline may be irregularly extended over the margo, as in *Tsuga canadensis* (L.) Carr. On the other hand the torus and thickening may be absent in exceptional cases, as in *Thuja plicata* Donn (Fig. 3-12*c*). The torus is generally conceded to be impermeable, at least when thickened, but minute holes have been reported in the tori of some of the European pines.[48]

The supporting strands in the margo of the closing membrane are flexible and allow the rigid torus to be displaced laterally by surface tension[34] against one or the other of the apertures, to form an *aspirated pit* (Fig. 3-11*B*). Such a condition closes the pit to the normal passage of liquids and has been used as an explanation for the lack of permeability in such woods as *Pseudotsuga menziesii* (Mirb.) Franco. It is not understood, as yet, whether the pits can be aspirated during the period of the functioning of the tracheids. Flat tori, in combination with thin supporting strands as in the early wood, are associated with aspiration in pits. Heavy, lens-shaped tori and thick supporting strands, as are commonly found in the late-wood zones, are too rigid to become aspirated.[34]

Two other mechanisms for reduction of flow through bordered pits of conifers have been proposed,[30] in addition to pit aspiration: (1) occlusion of the pit with heartwood extractives; e.g., extraction of incense-cedar wood (*Libocedrus decurrens* Torr.) with ethanol can increase permeability of the heartwood as much as ten thousand times; (2) incrustation of the pit with insoluble lignin-like materials (Fig. 3-13).

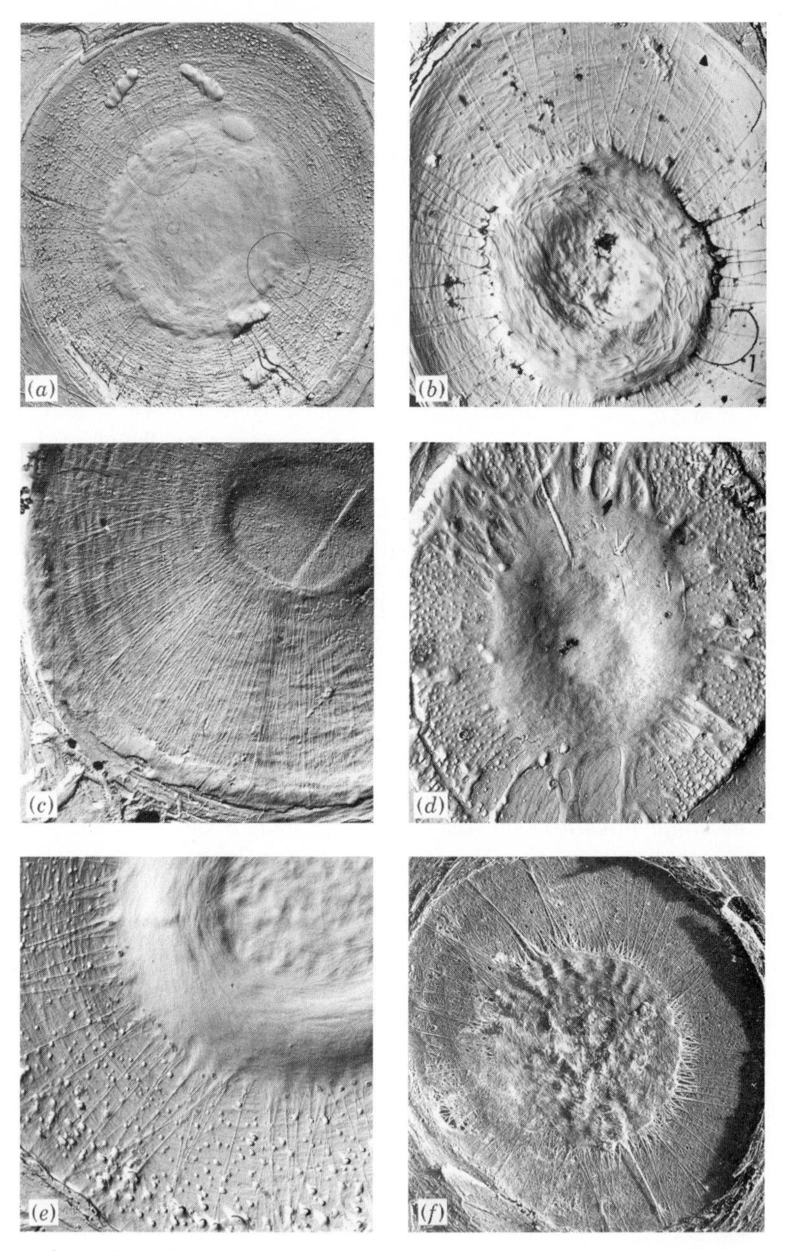

Fig. 3-12 Pit membranes.

(*a*) Bordered pit in ponderosa pine (***Pinus ponderosa*** Laws.). The thin torus at the center is not encrusted. The inner face of the pit cavity can be seen between the radial supporting strands of the membrane. The circular orientation of the microfibrils on the face of the pit cavity and the warty deposits on its surface are evident. (2200×.)

Recent studies of bordered pits in conifers make it plainly evident that, in the sapwood at least, actual openings in the margo of the pit membrane exist. Therefore, the conduction from cell to cell in conifers is through an open system as long as the pits are not aspirated, encrusted, or occluded in heartwood formation.

The specialized pit membrane as described for the bordered pit pairs of conifers differentiates these pits from all others. The pit membrane in simple and half-bordered pit pairs in both conifers and hardwoods, and in the bordered pit pairs of hardwoods, is a uniform structure exhibiting the primary wall texture with no evidence of a torus or radial supporting strands (Figs. 3-9, 3-13, and frontispiece). Electron micrographs taken at magnifications up to 80,000 times also show no openings in bordered pit membranes of the hardwoods or simple and semibordered pitting in conifers. Frequently the primary wall texture is overlaid with a continuous smooth deposit, as in the window pitting of ***Pinus strobus*** L. (Fig. 3-13*d*).

As can be seen from the previous discussion, the lateral shape of pits can vary a great deal. The same can be said of the surface appearance of the pits under a microscope. Where the pits are not crowded, the normal surface outline is circular. However, where the pitting is crowded, the outlines of the pits assume a hexagonal shape, as seen in surface view. In hardwoods, a number of other forms of pit outline occur in addition to the normal circular or hexagonal appearance. For example, there may be a lateral elongation of the pit structure in such woods as *Magnolia* to form *linear pitting.* A regular horizontal pattern made up of a series of linear pits is called *scalariform pitting.* Another pit arrangement that is found on the surfaces of cells is *opposite pitting;* it occurs when the pit centers are aligned both horizontally and vertically. In contrast, when the pit centers are arranged diagonally, this formation is called *alternate pitting.* Examples of these forms of pitting are shown in Figs. 12-141 to 12-160.

(*b*) Bordered pit in bigcone Douglas-fir [*Pseudotsuga macrocarpa* (Vasey) Mayr]. The microfibrillar orientation in the torus and its central depression at the pit aperture are particularly evident. One of the supporting strands can be seen at 1, curled away from its original position. (4500×.)

(*c*) Bordered pit in the sapwood of western redcedar (*Thuja plicata* Donn). The central region consists of crossed microfibrillar strands depressed into the pit aperture but lacks any thickening on the primary wall and cannot be classed as a torus. (4600×.) (*Electron micrograph courtesy of* W. A. Côté, Jr., *and* R. L. Kramer.)

(*d*) Bordered pit in eastern hemlock [*Tsuga canadensis* (L.) Carr.]. The torus is irregularly extended by thickening of groups of the radial supporting strands. Details are obscured by a warty deposit. (3700×.)

(*e*) Bordered pit in pond pine (*Pinus serotina* Michx.). The circular orientation of the microfibrils in the torus and the supporting strands in the region of the margo are particularly evident. (4500×.)

(*f*) Bordered pit from a ray tracheid in loblolly pine (*Pinus taeda* L.). The wood was pentane-dried to prevent aspiration of the torus. The section was extracted to show the fine detail of the microfibrillar structure. (6500×.) (*All electron micrographs courtesy of* W. A. Côté, Jr., *except as noted.*)

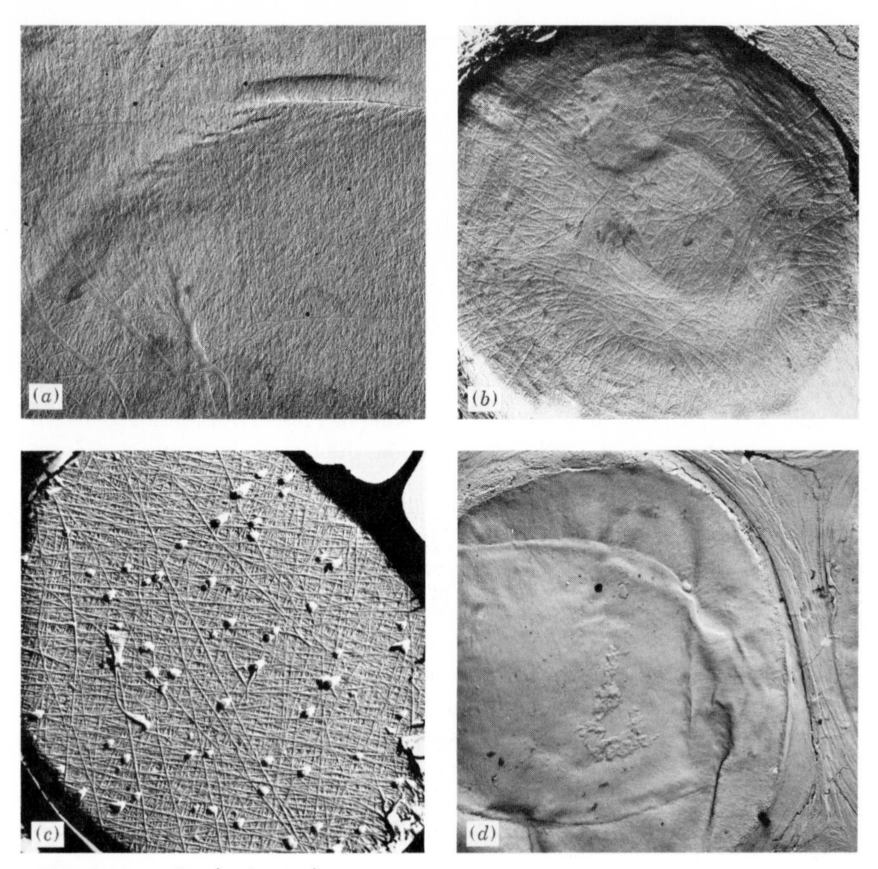

FIG. 3-13 Simple pit membranes.

(*a*) Half-bordered pit membrane in sand pine [*Pinus clausa* (Engelm.) Vasey]. The texture of the primary wall is apparent, with the addition of some aggregated strands on the surface. (6000×.)

(*b*) Pit membrane surface in sand pine [*Pinus clausa* (Engelm.) Vasey]. Microfibrillar strands are evident on the surface of this half-bordered pit membrane, but no torus has formed. (3200×.)

(*c*) Microfibrillar network in the primary wall of an intervessel membrane of a bordered pit in basswood (*Tilia americana* L.). The globules on the membrane are extraneous materials. (10,700×.)

(*d*) Pit membrane of a half-bordered pit in eastern white pine (*Pinus strobus* L.) at 3600×. The pattern of the microfibrils is visible around the edge up to the depressed border of the pit aperture. The central portion of the membrane is encrusted.

(*All electron micrographs courtesy of* W. A. Côté, Jr.)

2. PERFORATIONS OF THE CELL WALL

The major liquid conduction in hardwoods is through the tubular vessels, which arise by formation of openings, called *perforations*, in the common end walls of adjacent vessel elements.

The primary wall at the ends of the vessel elements, where the perforations will eventually appear, consists of an open mesh similar to that described for the simple pit membrane. At the completion of enlargement of the vessel element, the microfibrils of the primary wall and the intercellular substance are dissolved by an enzyme called *cellulase*.[31] This action leaves one or more holes in the end walls of the vessel elements.

The area of the common end walls of two adjacent vessel elements, which is involved in their coalescence, is known as a *perforation plate*. The actual opening from one vessel to another is called a *vessel perforation*. These perforations may occur singly or in multiples within the perforation plate. A single, usually large and rounded opening is called a *simple perforation*. *Multiple perforations* are formed by two or more openings in the perforation plate. The common form of multiple perforation is the *scalariform perforation*, which is formed by a series of elongated and parallel openings separated by narrow strips of the thickened perforation plate. These strips between the scalariform perforations are called *bars*. The bars are formed with the usual compound middle lamella and double secondary walls (Fig. 5-5).

3. SPIRAL THICKENING

The interior surface of the secondary wall may develop localized ridges of microfibrils that are helically oriented about the cell axis. This type of thickening of the cell wall is known as *spiral thickening*. The ridges are formed by the deposition of parallel bundles of microfibrils, either as a single helix or as more than one set of coils wound one within the other about the cell axis, or in the form of branching ridges. Figure 3-14 illustrates the deposition of microfibrils in the ridges and the branching of the spirals in Douglas-fir.

The spiral thickenings are usually oriented as an *S helix* in the same direction as the microfibrils in the *S*3 layer of the secondary wall. The angle of the helix pitch varies inversely with the lumen diameter. Large-lumened cells will show nearly horizontal helices, and the late-wood cells, with narrow lumens, will have steeply pitched spiral thickenings. Wardrop and Dadswell[58] note that there is a direct increase in steepness of pitch of the spiral thickenings with increase in length of mature cells, but that this relationship is not very close. Spiral thickening normally occurs over the entire length of those cells in which it occurs, but some cells, such as the vessel elements of *Liquidambar* and *Nyssa*, show spirals only in the cell ends. The coarseness and spacing of the helical ridges have some value in separating different kinds of wood.

Spiral thickening occurs in the tracheids of only a few of the native gymnosperms. It is found in *Pseudotsuga*, *Taxus*, and *Torreya* as a regular feature and sporadically in *Larix laricina* (Du Roi) K. Koch and the hard pines. In the angiosperms native to North America, spiral thickening is relatively common and may occur in the vessel elements, fiber tracheids, vascular tracheids, and even in ray cells.

As can be noted in Fig. 3-14*b*, the thickenings consist of aggregates of micro-

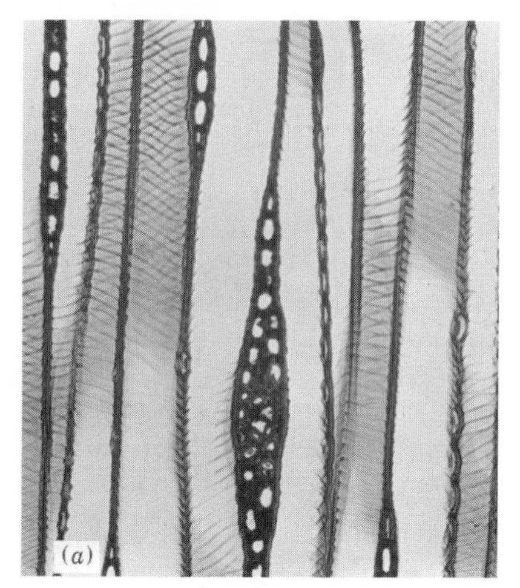

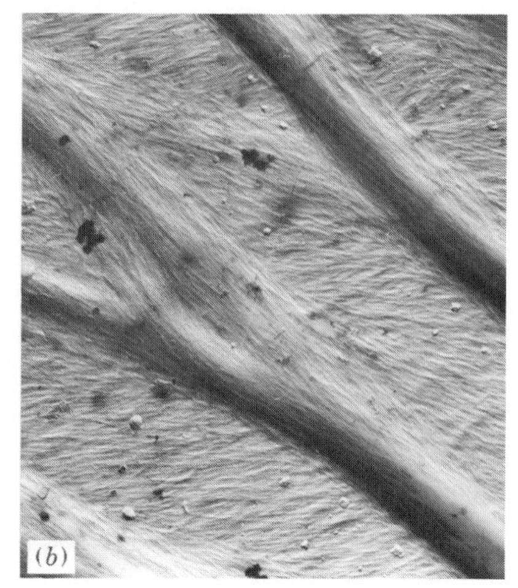

FIG. 3-14 Spiral thickening in Douglas-fir [*Pseudotsuga menziesii* (Mirb.) Franco]. (*a*) Appearance of spiral thickenings under a light microscope. (110×.)
(*b*) Electron micrograph of the surface of a tracheid showing the microfibrillar structure of the spiral thickenings. (6800×.)
(*Courtesy of W. A. Côté, Jr.*)

fibrils formed on top of the *S*3 layer and differing from it in direction. Investigation of this type of thickening[57] has shown that they are a specialized noncontinuous layer on the inner surface of the *S*3.

4. DENTATIONS

The ray-tracheid cells in the hard pines are characterized by the presence of localized wall thickenings which appear toothlike (dentate) in the usual microscope preparation. In reality when the cell is considered as a three-dimensional structure, these are irregular ringlike structures formed as extensions of the ray cell wall and are analogous to the spiral thickenings except for their irregularity.[12,26] The magnitude of these irregular rings varies among species; they may be only slightly raised, as in *Pinus resinosa* Ait., or at times form a closed bulkhead in the ray cells of some of the southern hard pine species (Fig. 3-15).

5. CALLITROID THICKENINGS

These structures are local thickenings of the cell wall that are typical of the genus *Callitris*, a conifer which occurs mostly in Australia. However, recent investigations of the southern hard pines[12,26] have shown that they occur also in this group, but as a fairly uncommon structure. These structures appear in surface view as a pair of slightly curved parenthesis-shaped bars positioned above and

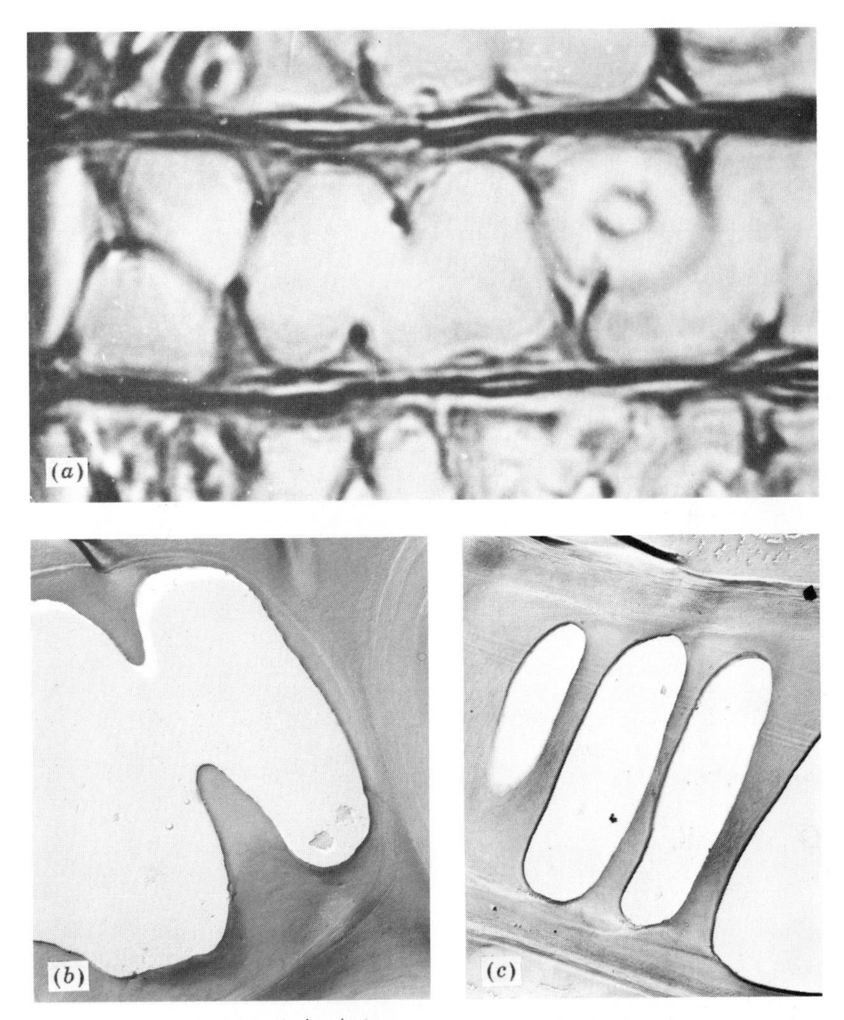

FIG. 3-15 Dentations in hard pines.

(*a*) Dentations in ray tracheids of longleaf pine (*Pinus palustris* Mill.). The dentations are formed by sectioning the ridges on the cell walls. In some cases these ridges are shallow, and in others the ridges are deep and meet to close off a section of the cell.

(*b*) Electron micrograph of dentation ridges in spruce pine (*Pinus glabra* Walt.). (7000×.)

(*c*) Electron micrograph of coalesced dentation ridges in sand pine [*Pinus clausa* (Engelm.) Vasey] forming locules in the ray tracheid cells. (7000×.)

[*Electron micrographs* (*b*) *and* (*c*) *courtesy of* W. A. Côté, *Jr.*]

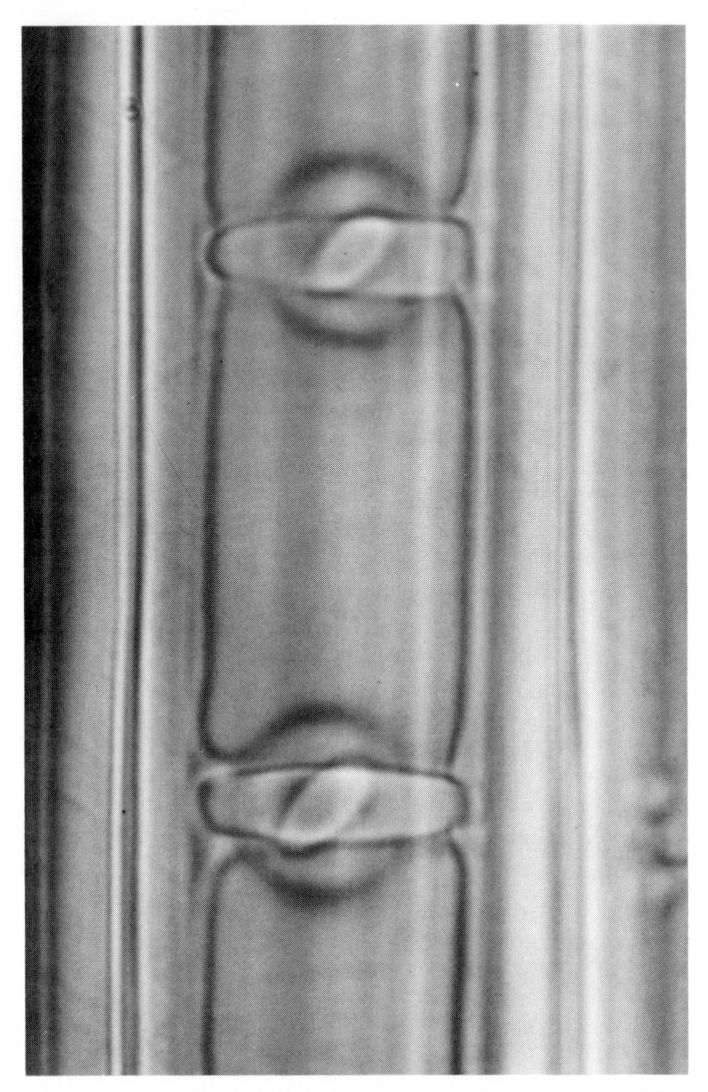

FIG. 3-16 Callitroid-like thickenings of the radial intertracheid pitting of the transition zone between early and late wood in pond pine (*Pinus serotina* Michx.). Note the ridges that continue onto the tangential walls. (2000×.) (*Courtesy of E. T. Howard and F. G. Manwiller.*[26])

below the pit aperture normal to the axis of the cell (Fig. 3-16). Cronshaw[13] has shown these to be ridges on the inner wall surface and probably elaborations of the *S*3 layer.

6. CRASSULAE

Surface view of bordered pits in conifer tracheids frequently reveals a pair of dark, curved zones lying above and below the pits themselves. These structures are the ***crassulae*** (see Fig. 4-4*a* and *c*). Not too much is known about them, but it appears that they are a slight localized thickening of the primary wall which may serve as stiffening around the primary pit fields. The secondary wall in these areas shows no special modifications.

7. TRABECULAE AND SEPTATIONS

Trabeculae are rodlike extensions across the cell lumen from wall to wall in conifer tracheids (Fig. 4-5). It is apparent that the secondary wall is continuous and forms a shell structure enclosing material of a different organization than that usually found in the cell wall. These are rather uncommon structures and are believed to be associated with fungal activity.

Septations are transverse partitions in hardwood fibers and are formed entirely of secondary cell wall layers. Septate fibers are uncommon in temperate-zone woods, but are found in the woods of a number of tropical families as a normal feature.

8. WART STRUCTURES

The last aspect of cell formation is the death of the cytoplasm. This results in the deposition of the remains of the membranous portions of the cytoplasm over the inner face of the cell wall. This denatured material consists of two layers representing the remains of the plasmalemma and the tonoplast with fragments of proteinaceous material enclosed. The layer is quite distinct and can be isolated by mechanical and chemical means. It has chemical reactions similar to those for lignin, but in addition shows evidence of proteins that are normally not found to any extent in the secondary wall.[63]

Small wartlike protuberances are often evident and closely associated with the cytoplasmic remains; the formation is known as the ***warty layer*** (Figs. 3-8 and 3-12). The warts appear in a large number of conifers and hardwoods. However, they possess little taxonomic significance other than to separate the hard pines, which have them, from the soft pines, which do not show them.

The largest of these warts can be observed with an optical microscope, but critical study requires electron microscopy. These structures measure from 0.05 to 0.5 micron.[57] Their shape is frequently that of a rounded cone, but extreme variations from this are known to occur. Critical studies[57] have shown that the wart itself is a localized outgrowth of the S3 layer whose chemical reactions indicate a higher proportion of nonglucose sugars than is usually found in the secondary-wall layers. In addition, a small nodule of proteinaceous material is found in the remains of the cytoplasm directly in association with the wart; the combination is considered to be the wart structure. The nodules of cytoplasmic material are believed to arise from the remains of the organelles which were involved in the localized deposition of the wart itself.

Electron microscope studies have shown that the warty layer is continuous over the inner surface of the cell wall and the interior surface of the pit chamber.

9. VESTURED PITTING

In the angiosperms a development of the warty structure into large simple or branched forms associated with the pit chamber is known as ***vestured pitting***. The structures have been known for a long time under the name of ***cribriform pitting***, but the structures causing their characteristic appearance in the microscope were not described prior to I. W. Bailey's studies with the light microscope.[4] His descriptions have been corroborated and extended by work with electron

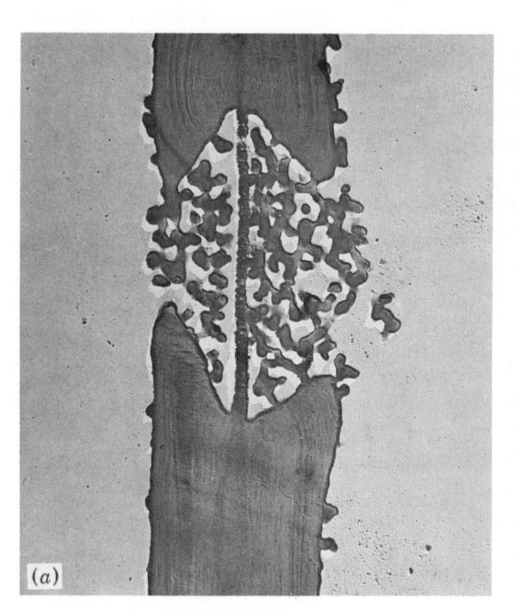

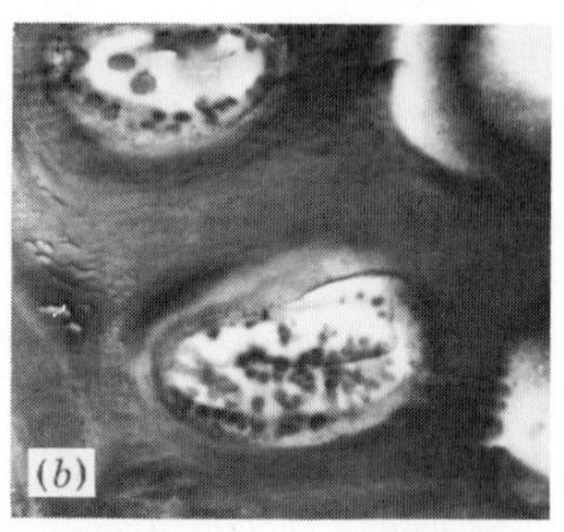

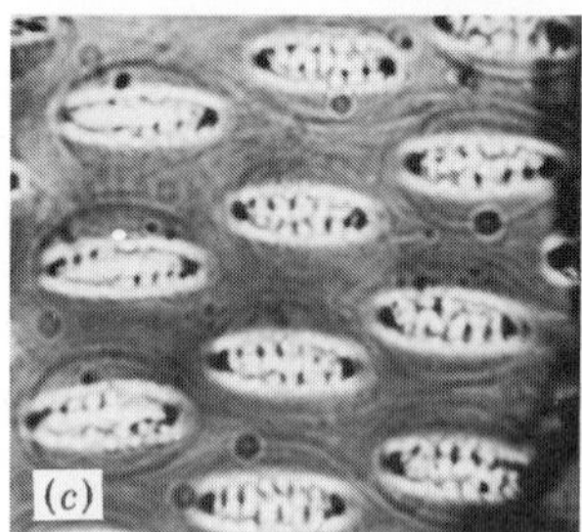

FIG. 3-17 Vestured pitting.

(*a*) Section view of a bordered pit pair in bagtikan (*Parashorea plicata* Brandis). Note the continuity of the warts on the lumen side of the wall and the vesturing in the pit cavities. (7100×.) (*Electron micrograph courtesy of W. A. Côté, Jr.*)

(*b*) Surface view of vestured pits in the intervessel pitting of black locust (*Robinia pseudoacacia* L.). Electron micrograph at 3300×. (*Courtesy of W. A Côté, Jr.*)

(*c*) Surface view of vestured pits on the intervessel walls in black locust (*Robinia pseudoacacia* L.) as seen with a light microscope. (1250×.)

microscopy to show that vestured pitting consists of either branched or unbranched processes, resembling wart structures, that arise from either the chamber wall or the pit membrane[11,62] (Fig. 3-17*a*).

Extensive studies, principally by I. W. Bailey, have shown that vestured pits are very consistent in their distribution. They may be present in all the woods of a family, or in certain genera, or in some cases only in single species within a genus. In our native timbers the woods of the Leguminosae can be separated from all other families by vestured pitting, which is present in all the woody native genera of that family except *Cercis* (Fig. 3-17). However, the presence of vestured pits must be verified with care, when using light microscopy, to be certain that there is no confusion with artifacts in the section, such as can be seen in *Carya ovata* (Mill.) K. Koch (Fig. 12-143).

10. TYLOSES

In the angiosperms, specialized structures may be produced in the vessels in the process of heartwood formation. These are bubble-like outgrowths of the parenchyma cells adjacent to the vessels and are called *tyloses* (singular *tylosis*). They may partially or completely fill the vessels.

Tyloses are formed as a part of the process of transformation of sapwood to heartwood in many kinds of trees. The causative mechanism for tylosis formation may be related to low water content in the vessels, or to injury which may be either mechanical[28] or the result of fungal or virus infection. The tyloses arise from activity of parenchyma cells in contact with the vessel. Chattaway[10] believes that tyloses arise from ray cells alone and that tyloses formation as the result of activity of longitudinal parenchyma cells in contact with the vessels is very rare. Formation of tyloses can be induced artificially in wood from newly felled trees by maintenance of temperatures at about 25°C and relative humidity at 98 percent. Experiments by Jurášek[28] with *Fagus sylvatica* L. led him to conclude that tyloses may arise from any parenchyma cell adjacent to a vessel as long as the pit pairs have diameters 10 to 15 microns or larger.

The production of a tylosis is preceded by the formation of a special wall layer in the parenchyma cell from which the tylosis will arise.[39,51] This wall layer has been termed the *protective layer* by Schmidt[51] because of its assumed function, but in reality the wall is meristematic in nature with the usual ability of such walls to allow expansion of the wall surface. This layer first forms over the pit membrane region on the inside of the secondary cell wall and later extends until the entire cytoplasm of the cell is enclosed. The protective layer has a primary texture, is heavier in the regions near the pits, and shows evidence of lignification (Fig. 3-18).[39] Studies of tyloses in the Myrtaceae[18] indicate that the protective layer may be two-layered. This multiple layering is also evident in some ray cells in the heartwood that exhibit thick protective layers formed by the successive formation of additional protective layers.[39]

At the time of formation of the tylosis, enzymatic action partly removes the

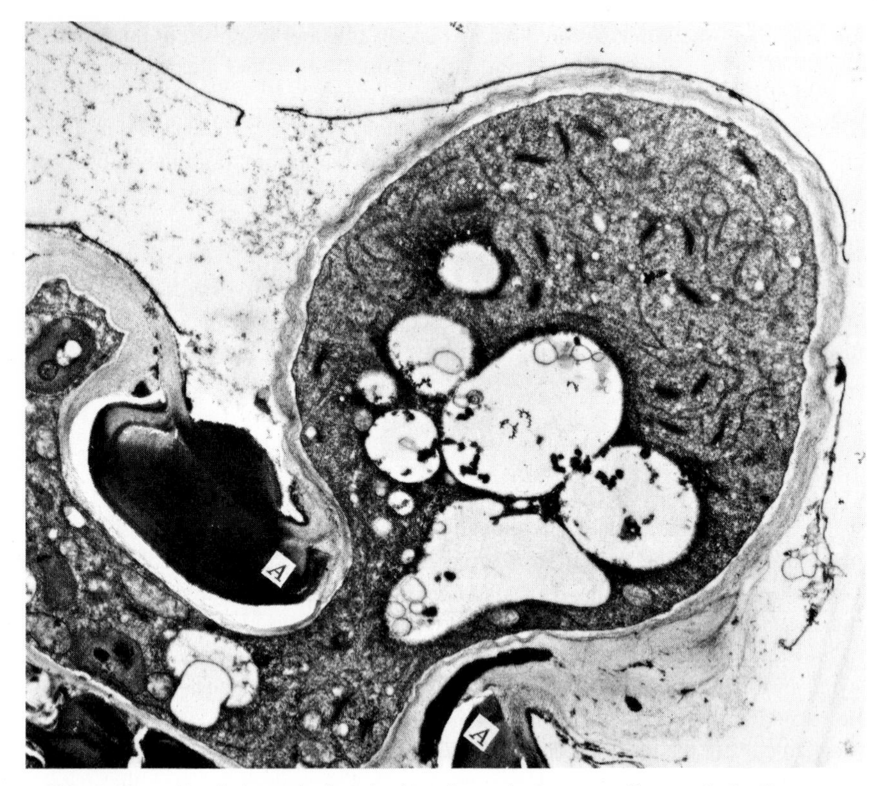

FIG. 3-18 Newly formed tylosis bud in white oak (*Quercus alba* L.). The bud has grown from the pit (*A-A*) in the ray cell at the lower left, into the vessel lumen at upper right. Fragments of the ruptured pit membrane are visible in the vessel lumen near the pit. (5900×.) (*Electron micrograph courtesy of R. W. Meyer.*)

pit membrane and any vesturing or encrustations that are present. The ray cell cytoplasm then increases in volume, as a budlike protrusion of its enlarging protective layer develops into the vessel lumen (Fig. 3-18). The fragmentary remains of the pit membrane are evident in the early stages of the tylosis formation.[39] It is evident from the electron micrograph (Fig. 3-18) that the cytoplasm of the developing tylosis is continuous with that of the parenchyma cell from which it arises. The wall of the tylosis develops by intussusception, in the usual manner for primary wall growth, from that portion of the protective layer protruding into the vessel lumen. The fully developed tylosis wall in *Eucalyptus* is described as two-layered, the outer layer with random orientation of microfibrils and the inner one consisting of numerous lamellae of more regular orientation.[18] The ultimate size and shape of a tylosis are determined by the available space in the vessel lumen. When individual tyloses contact each other, the common walls become lignified and develop intertylosic pitting. The walls of the tyloses in many kinds of wood are quite thin with little evident pitting, but these walls may become heavily thickened (*sclerotic*) with a layered structure and distinct simple pits (Fig. 5-6*j*).[18]

E. Intercellular Spaces

A normal feature associated with the ray parenchyma is a system of radial intercellular spaces which are believed to constitute a gas canal system.* These spaces are small with cross-sectional dimensions in the range from 1 to 15 square microns, but essential for transport of the oxygen needed in metabolism. The connections between the spaces and ray parenchyma cells are through slitlike, blind pits at the ray cell corners.

* E. L. Back: Intercellular Spaces along the Ray Parenchyma—The Gas Canal System of Living Wood? *Wood Sci.*, **2**(1):31–34 (1969).

SELECTED REFERENCES

1. Anderson, A. B., and E. Zavarin: The Influence of Extractives on Tree Properties, III, Incense Cedar (*Libocedrus decurrens* Torr.), *J. Inst. Wood Sci.*, **15:**3–24 (1965).
2. Bailey, A. J.: Lignin in Douglas Fir—Composition of the Middle Lamella, *Ind. Eng. Chem., Anal. Ed.* **8:**52–55 (1936).
3. Bailey, I. W.: The Cambium and Its Derivative Tissues, II, Size Variations of Cambial Initials in Gymnosperms and Angiosperms, *Am. J. Botany*, **7**(9):355–367 (1920).
4. ———: The Cambium and Its Derivative Tissues, VIII, Structure, Distribution and Diagnostic Significance of Vestured Pits in Dicotyledons, *J. Arnold Arb.*, **14:**259–273 (1933).
5. ———: Aggregations of Microfibrils and Their Orientation in the Secondary Wall of Coniferous Tracheids, *Am. J. Botany*, **44**(5):415–418 (1957).
6. ——— and T. Kerr: The Visible Structure of the Secondary Wall and Its Significance in Physical and Chemical Investigations of Tracheary Cells and Fibers, *J. Arnold Arb.*, **16**(3):276–300 (1935).
7. ——— and ———: The Structural Variability of the Secondary Wall as Revealed by "Lignin" Residues, *J. Arnold Arb.*, **18**(4):261–272 (1937).
8. Berlyn, G. P.: Recent Advances in Wood Anatomy—the Cell Walls in Secondary Xylem, *Forest Prod. J.*, **14**(10):467–576 (1964).
9. Bobak, M., and V. Nečesaný: Changes in the Formation of the Lignified Cell Wall within a Twenty-four Hour Period, *Biol. Plant., Praha*, **9**(3):195–201 (1967).
10. Chattaway, M. M.: The Development of Tyloses and Secretion of Gum in the Heartwood Formation, *Australian J. Sci. Res.*, **B2**(3):227–240 (1949).
11. Côté, W. A., Jr., and A. C. Day: Vestured Pits—Fine Structure and Apparent Relationship to Warts, *Tappi*, **45**(12):906–910 (1962).
12. ——— and ———: "Wood Ultrastructure of the Southern Yellow Pines," State University College of Forestry, Syracuse, N.Y., tech. publ. in press, 1969.
13. Cronshaw, J.: The Nature of Callitroid Thickenings, *J. Inst. Wood Sci.*, **8:**12 (1961).
14. ———: Cytoplasmic Fine Structure and Cell Wall Development in Differentiating Xylem Elements, in W. A. Côté, Jr. (ed.), "Cellular Ultrastructure," Syracuse University Press, Syracuse, N.Y., 1965.
15. Dunning, C. E.: Cell-wall Morphology of Longleaf Pine Latewood, *Wood Sci.*, **1**(2):65–76 (1968).
16. Elliott, G. K., and S. E. G. Brook: Microphotometric Technique for Growth Ring Analysis, *J. Inst. Wood Sci.*, **18:**24–43 (1967).
17. Ellis, E. L.: Inorganic Elements in Wood, from W. A. Côté, Jr. (ed.), "Cellular Ultrastructure," Syracuse University Press, Syracuse, N.Y., 1965.
18. Foster, R. C.: Fine Structure of Tyloses in Three Species of the Myrtaceae, *Australian J. Bot.*, **15**(1):25–34 (1967).
19. Freudenberg, K.: The Formation of Lignin in the Tissue and *in vitro*, in M. H. Zimmerman

(ed.), "The Formation of Wood in Forest Trees," Academic Press, Inc., New York, 1964.
20. Frey-Wyssling, A.: The Ultrastructure of Wood, *Wood Sci. & Technol.*, **2**(2):73–83 (1968).
21. ———, A. K. Mühlethaler, and R. Muggli: Elementarfibrillen als Grundbausteine der nativen Cellulose, *Holz Roh-Werkstoff*, **24**(10):443–444 (1966).
22. Harada, H., and W. A. Côté, Jr.: Cell Wall Organization in the Pit Border Region of Softwood Tracheids, *Holzforschung*, **21**(3):81–85 (1967).
23. ———, Y. Miyazaki, and T. Wakashima: Electronmicroscope Investigations on the Cell Wall Structure of Wood, *Govt. Forestry Expt. Sta., Bull.* 104, Tokyo, Japan, 1958.
24. Hermans, P. H.: "Physics and Chemistry of Cellulose Fibers," Elsevier Publishing Company, Amsterdam, 1949.
25. Heyn, A. N. J.: The Microcrystalline Structure of Cellulose in Cell Walls of Cotton, Ramie and Jute Fibers as Revealed by Negative Staining of Sections, *J. Cell. Biol.*, **29**(2):181–197 (1966).
26. Howard, E. T., and F. G. Manwiller: Anatomical Characteristics of Southern Pine Stemwood, *Wood Sci.*, **2**(2):77–86 (1969).
27. Jongebloed, W. L., and S. M. Jutte: X-ray Projection Microscopy of Crystals in Tropical Hardwoods, *Holzforschung*, **19**(2):36–42 (1965).
28. Jurášek, L.: Vznik Thyl v Bukovèm Drěvě (The Origin of Tyloses in Beech Wood), *Drev. Vyskum*, **1**(1/2):7–15 (1956).
29. Kerr, T., and I. W. Bailey: The Cambium and Its Derivative Tissues, X, Structure, Optical Properties and Chemical Composition of the so-called Middle Lamella, *J. Arnold Arb.*, **15**:327–349 (1934).
30. Krahmer, R. L., and W. A. Côté, Jr.: Changes in Coniferous Wood Cells Associated with Heartwood Formation, *Tappi*, **46**(1):42–49 (1963).
31. Küster, E.: "Die Pflanzenzelle," G. Fischer, Jena, 1956.
32. Kutscha, N. P.: Cell Wall Development in Normal and Compression Wood of Balsam Fir, *Abies balsamea* (L.) Mill., Ph. D. dissertation, State University College of Forestry at Syracuse University, Syracuse, N.Y., 1968.
33. Lange, P. W.: Molecular Organization of the Cell Wall, 4th Cellulose Conference, State University College of Forestry at Syracuse University, Syracuse, N.Y., 1962.
34. Liese, W., and J. Bauch: On the Closure of Bordered Pits in Conifers, *Wood Sci. & Technol.*, **1**(1):1–13 (1967).
35. Manley, R. S. J.: Fine Structure of Native Cellulose Microfibrils, *Nature*, **204**(4964): 1155–1157 (1964).
36. Meier, H.: The Distribution of Polysaccharides in Wood Fibers, *J. Polymer Sci.*, **51**(155): 11–18 (1961).
37. ———: General Chemistry of Cell Walls and Distribution of the Chemical Constituents across the Wall, in M. H. Zimmerman (ed.), "The Formation of Wood in Forest Trees," Academic Press, Inc., New York, 1964.
38. ——— and K. C. Wilkie: The Distribution of Polysaccharides in the Cell-wall of Tracheids of Pine (*Pinus sylvestris* L.), *Holzforschung*, **13**(6):177–182 (1959).
39. Meyer, R. W., and W. A. Côté, Jr.: Formation of the Protective Layer and Its Role in Tylosis Development, *Wood Sci. & Technol.*, **2**(2):84–94 (1968).
40. Mühlethaler, K.: Die Feinstruktur der Zellulosemikrofibrillen, *Beih. Z. Schweiz. Forstver.*, **30**:55 (1960).
41. ———: The Fine Structure of the Cellulose Microfibril, in W. A. Côté, Jr. (ed.), "Cellular Ultrastructure," Syracuse University Press, Syracuse, N.Y., 1965.
42. Northcote, D. H.: The Cell Walls of Higher Plants; Their Composition, Structure and Growth, *Biol. Rev., Cambridge Phil. Soc.*, **33**(1):53–102 (1958).
43. Ott, E., et al.: "High Polymers," vol. V, "Cellulose and Cellulose Derivatives," 2d ed., Interscience Publishers, Inc., New York, 1954.
44. Pearl, I. A.: Annual Review of Lignin Chemistry, *Forest Prod. J.*, **17**(2):23–32 and 58–68 (1967).

45. Preston, R. D.: Structural and Mechanical Aspects of Plant Cell Walls with Particular Reference to Synthesis and Growth, in M. H. Zimmerman (ed.), "Formation of Wood in Forest Trees," Academic Press, Inc., New York, 1964.
46. Probine, M. C.: Chemical Control of Plant Cell Wall Structure and of Cell Shape, *Proc. Roy. Soc., London,* **B161:**526–537 (1965).
47. Rånby, B. G.: The Fine Structure of Cellulose Fibrils, in F. Bolam (ed.), "Fundamentals of Papermaking Fibers," First Tech. Section, British Paper and Board Makers Association, London, 1961.
48. Roelofsen, P. A.: "The Plant Cell Wall," vol. III, part 4, "Encyclopedia of Plant Anatomy," Gebrüder Borntraeger, Berlin-Nikolassee, 1959.
49. Rollins, M. L.: Cell Wall Structure and Cellulose Synthesis, *Forest Prod. J.,* **18**(2):91–100 (1968).
50. Ruch, F., and H. Hengartner: Quantitative Bestimmung der Ligninverteilung in der pflanzlichen Zellwand, *Beih. Z. Schweiz. Forstver.,* **30:**75–90 (1960).
51. Schmidt, R.: The Fine Structure of Pits in Hardwoods, in W. A. Côté, Jr. (ed.), "Cellular Ultrastructure of Woody Plants," Syracuse University Press, Syracuse, N.Y., 1965.
52. Stanley, R. G.: Biosynthesis of Cellulose Outside of the Living Tree, *Forest Prod. J.,* **16**(11):62–68 (1966).
53. Thomas, R. J.: The Development and Ultra Structure of Bordered Pit Membranes in the Southern Yellow Pines, *Holzforschung,* **22**(2):38–44 (1968).
54. Timell, T. E.: Recent Progress in the Chemistry of Wood Hemicelluloses, *Wood Sci. & Technol.,* **1**(1):45–70 (1967).
55. Wangaard, F. F.: Cell-wall Density of Wood with Particular Reference to the Southern Pines, *Wood Sci.,* **1**(4):222–226 (1969).
56. Wardrop, A. B.: The Organization and Properties of the Outer Layer of the Secondary Wall in Conifer Tracheids, *Holzforschung,* **11**(4):102–110 (1957).
57. ———: The Structure and Formation of the Cell Wall in Xylem, in M. H. Zimmerman (ed.), "Formation of Wood in Forest Trees," Academic Press, Inc., New York, 1964.
58. ——— and H. E. Dadswell: Helical Thickenings and Micellar Orientation in the Secondary Wall of Conifer Tracheids, *Nature,* **168**(4275):610–612 (1951).
59. ——— and ———: The Cell Wall Structure of Xylem Parenchyma, *Australian J. Sci. Res. Ser. B,* **5**(2):223–236 (1952).
60. ——— and ———: Variations in the Cell-wall Organization of Tracheids and Fibers, *Holzforschung,* **11**(2):33–41 (1957).
61. ——— and H. Harada: The Formation and Structure of the Cell Wall in Fibers and Tracheids, *J. Exp. Botany,* **16**(47):356–371 (1965).
62. ———, H. D. Ingle, and G. W. Davies: Nature of Vestured Pits in Angiosperms, *Nature,* **197**(4863):202–203 (1963).
63. ———, W. Liese, and G. W. Davies: The Nature of the Wart Structure in Conifer Tracheids, *Holzforschung,* **13**(4):115–120 (1959).
64. Wise, L. E., and E. C. Jahn: "Wood Chemistry," 2d ed., Reinhold Publishing Corporation, New York, 1952.

4

The Minute Structure of Coniferous Woods

This chapter deals with the wood anatomy of coniferous species. A detailed description of the minute anatomical features of white pine (***Pinus strobus*** L.) wood is presented first. This is followed by a summary of the histological departures that occur in the North American coniferous woods as a group.

I. THE WOOD ANATOMY OF EASTERN WHITE PINE
(*Pinus strobus* L.)

Different kinds of elements (cells) found in white pine wood are listed in Table 4-1. All these elements need not necessarily show in a given section of wood, as viewed under the microscope. Their presence or absence depends on whether cells of the type in question happened to project into the plane under examination. It follows, however, that all the cell types found in a wood will appear sooner or later if serial sections are cut in any plane.

A. Longitudinal Cells

1. LONGITUDINAL TRACHEIDS

More than 90 percent of the volume of white pine wood consists of one kind of cells, viz., the longitudinal tracheids (see Table 4-3). These cells are relatively long (3.0 to 4.0 millimeters), four- to six-sided, prismatic elements, with closed ends.

TABLE 4-1 Elements of White Pine Wood

Longitudinal	*Transverse*			
1. Prosenchymatous	1. Prosenchymatous	Homocellular (uniseriate) rays	Heterocellular (uniseriate) rays	Fusiform rays
a. Tracheids	*a*. Ray tracheids			
2. Parenchymatous	2. Parenchymatous			
a. Epithelial cells:* excreting cells of longitudinal resin canals	*a*. Cells of ray parenchyma			
	b. Epithelial cells:* excreting cells of transverse resin canals			

* Longitudinal and transverse resin canals are not listed, since they are intercellular spaces encircled by excreting cells (epithelium) and are hence not true elements. Longitudinal parenchyma is not present in white pine wood.

The ends taper sharply to a point in the tangential plane (Fig. 4-3*C*,*m*) but are blunt as seen in the radial view (Fig. 4-3*B*,4). When cut transversely the longitudinal tracheids in normal white pine wood are frequently hexagonal in the early wood, and rectangular and somewhat tabular in the late wood.

The radial diameter of white pine tracheids is not constant but is contingent on the position of the cell in the annual increment; the tracheids are widest in the radial plane at the inception of the early-wood cells and the narrowest in the last-formed late wood (see Fig. 4-1). The tangential diameter of longitudinal tracheids is much less variable than the radial. In fact, in coniferous woods the tangential diameter of longitudinal tracheids is used as a measure of texture (Table 4-4). This relative consistency in width of the longitudinal tracheids is explained by the fact that the fusiform initials in the cambium of a coniferous tree gradually enlarge over a period of 20 to 40 years, after which the size remains relatively constant. In maturing, tracheids increase considerably in length and to some extent in radial direction, especially in the early wood, while in width, i.e., in the tangential direction, they remain practically unchanged. It should be noted that since the longitudinal tracheids are tapered cells their diameters, as viewed in the transverse section, depend also on the place along their length at which they were cut transversely (compare in Fig. 4-3, 1^a with a^1).

TABLE 4-2 Volumetric Composition of White Pine Wood

	Percent
Longitudinal tracheids	93.00
Longitudinal resin canals	1.00
Wood rays	6.00
Total	100.00

FIG. 4-1 A cube of eastern white pine (*Pinus strobus* L.). (75×.) (*Photograph by Arnold Day.*)

According to data compiled at the State University of New York College of Forestry, the average length of the longitudinal tracheids of mature white pine wood ranges from 3 to 4 millimeters and the tangential diameter reaches a maximum of 45 microns (average 25 to 35 microns); i.e., they are ninety to one hundred times as long as they are wide.

Reference to Fig. 4-2 reveals that the longitudinal tracheids are equipped with three kinds of pits, viz., (*a*) those which compose pit pairs that ensure communication laterally between a longitudinal tracheid and a ray tracheid, (*b*) those belonging to pit pairs in the walls between longitudinal tracheids and a ray-parenchyma cell, and (*c*) those leading laterally from one longitudinal tracheid to an adjacent tracheid.

Pits of types *a* and *b* (Fig. 4-2*B*) occur at points where wood rays were in contact with the longitudinal tracheids, as seen in the radial section. These rectangular areas, formed by the walls of a ray cell and a longitudinal tracheid, are called *cross fields* (formerly known as *ray crossings*). When a ray tracheid is involved, the resulting pits are bordered (type *a*); they are similar to intertracheid pits, only smaller. When a ray parenchyma crosses a longitudinal tracheid, the pits (type *b*) are large, round or square in shape, with a slight border which is almost indistinguishable except under very high magnification. The complementary pits in the ray parenchyma are of the same size, but simple, and the resulting pit pairs of type *b* are therefore half-bordered. It is these pits of type *b* that are of considerable value in identification of the various kinds of coniferous woods (see page 134). In white pine, because of their shape and size, they are described as *window-like*, or *fenestriform*.

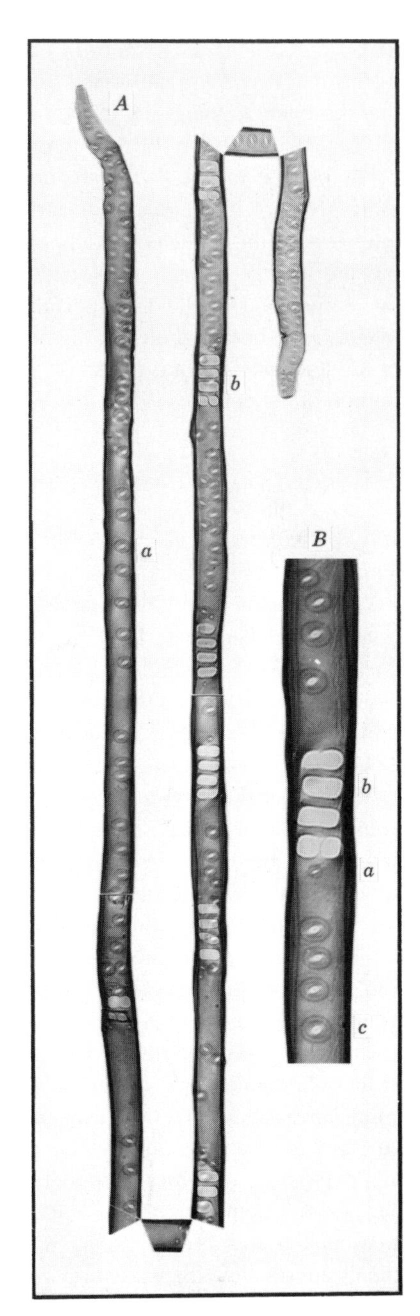

FIG. 4-2 *A*. A longitudinal tracheid of eastern white pine (***Pinus strobus*** L.) in radial surface aspect. (65×.) *a*, bordered pits leading laterally to a contiguous longitudinal tracheid; *b*, cross field.

B. Portion of the same, at a higher magnification, depicting a cross field. (130×.) *a*, a small bordered pit leading laterally to a ray tracheid; *b*, windowlike pits leading laterally to cells of ray parenchyma; *c*, bordered pits leading laterally to a contiguous longitudinal tracheid.

(*Photograph by C. H. Carpenter.*)

Seven cross fields are visible on one radial wall of the tracheid (Fig. 4-2*A*). This means that it was crossed seven times by the ray on one side. It may be assumed that the count is the same on the opposite radial wall. This tracheid was therefore in contact with 10 to 15 rays throughout its length, on one side or the other.

Pits of type *c* (Fig. 4-2*B*) are bordered. They are large and numerous on the radial faces of longitudinal tracheids, and much smaller and restricted to the last few rows of the late-wood tracheids on the tangential walls. On the radial walls pits are arranged mostly in a single row; occasionally they are found in pairs in the early-wood tracheids, especially when the texture is coarser than normal. Intertracheid bordered pits on the radial walls are most numerous toward the ends of the tracheids and are sparse or absent over extensive areas through the central portion of the cell.

2. LONGITUDINAL EPITHELIAL CELLS

The epithelial cells are excreting, thin-walled parenchyma cells which surround longitudinal resin canals. The cavity of a canal, as depicted on surface *A* in Fig. 4-3, is in reality an intercellular space, i.e., a place where immature longitudinal xylary cells pulled apart, leaving a tubular cavity. The resin canal as such is therefore not a wood element but a cavity surrounded by thin-walled, parenchymatous cells (***epithelial cells***). In eastern white pine the longitudinal resin canals attain a maximum diameter of 200 microns, measured to include the surrounding epithelial cells; the average ranges from 135 to 150 microns.

The epithelial cells are traceable to daughter cells arising from fusiform initials in the cambium. As seen in the transverse section clusters of longitudinal daughter cells fail to mature into longitudinal tracheids in the normal way. The true middle lamella between cells near the center of the cluster splits and an intercellular cavity is formed, presumably by the fluid pressure from the epithelial cells. Meanwhile each of the daughter cells adjacent to the cavity divides by transverse walls into a number of smaller units aligned in a longitudinal row. The epithelial cells in white pine are generally devoid of pits, and are apparently not lignified.

The entire layer of epithelial cells surrounding a resin canal is called ***epithelium***. In the white pine wood it is usually several layers of cells in thickness. The cells immediately adjacent to the canal retain their shape, they also remain functional, i.e., they remain alive and excrete resin as long as they are a part of the sapwood in a living tree. Some of the epithelial cells behind the first row frequently collapse (Fig. 4-14*a*), permitting the cells in the first layer to enlarge peripherally, thus increasing the size of the tubular intercellular space. In the heartwood the resin canals are usually plugged with tylosoids (see page 144 and Fig. 4-14*h*) and hence may be considered vestigial structures.

From this discussion it is evident that the longitudinal resin canals in the wood of white pine, and other conifers possessing resin canals, are intercellular spaces and are of postcambial development. The continuity of the cambium is not interrupted prior to their formation, nor is there any evident change in the nature of the cambial cells. The tubular intercellular spaces, as noted before, are formed by splitting of the true middle lamella in the center of a cluster of longitudinal cambial daughter cells which then become epithelium, i.e., layers of thin-walled secreting cells.

B. Transverse Cells

All the transverse cells found in any given wood are included in the wood rays, a ribbon-like aggregate of cells formed by the cambium and extending radially in the xylem, i.e., radially from the cambium toward the pith.

In white pine there are two kinds of rays: *uniseriate* and *fusiform*. The first type of rays consists of one row of cells as seen in the tangential and transverse sections. These rays are usually *heterocellular,* i.e., they are composed of two kinds of cells: *ray tracheids* and *ray parenchyma*. This feature is seen best in the radial section. Occasionally uniseriate rays may consist entirely of ray tracheids, in which case they are termed *homocellular*. Usually, however, the uniseriate rays are composed predominantly of ray parenchyma, with ray tracheids forming one or more marginal rows of cells and an occasional row of ray tracheids in the body of a ray. In the tangential plane the ray cells are exposed in cross section. The last cell (a ray tracheid) at each end of the ray is gable-shaped, to permit the longitudinal tracheids to close in above and below the ray without leaving a large intercellular space. The height of the uniseriate rays in white pine varies from one to a dozen, or more cells.

The fusiform rays, so called because of their spindle-like shape, as seen in the tangential section, consist of the same elements as the uniseriate rays, i.e., of ray tracheids and ray parenchyma, occupying the same relative position, but in addition they include one or rarely two resin canals, surrounded by epithelial cells (Figs. 4-14*c* and *d*). No exact information is available on the ratio of fusiform to uniseriate rays, but it is believed to be at least 1 to 20, or higher.

All the ray cells have their long axis in the radial direction, and all these develop from ray initials in the cambium. The appearance of a ray in the transverse section is that of a uniseriate ribbon, except for the fusiform rays which, if cut through the middle part of the ray, may consist of several rows of ray parenchyma cells, or ray parenchyma cells and epithelial cells, with the resin canal cavity in the center. In the radial section the rays appear as ribbons composed of a number of rows of more or less brick-shaped ray tracheid and ray parenchyma cells. In the case of fusiform rays these are further supplemented with the epithelial cells and the resin canal cavity in the central portion of the ray.

1. RAY TRACHEIDS

The ray tracheids are more or less brick-shaped cells, 0.1 to 0.2 millimeter in length, often tapered toward one edge. They are equipped with bordered pits, of the same type found in the longitudinal tracheids, only smaller. The resulting pit pairs are bordered when found between the adjacent ray tracheid cells and between ray tracheids and longitudinal tracheids. Pit pairs connecting the ray tracheids with the ray parenchyma are half-bordered; i.e., the pit is bordered in the ray tracheid cell, while the complementary pit in the adjacent ray parenchyma cell is simple.

In white pine the ray tracheids may occasionally compose the entire ray, if it consists of a few rows of cells. More commonly they are restricted to one row on the upper and lower margins of the ray, or several rows may be present on one or both margins. Not infrequently one or more rows of ray tracheids are also found in the body of the ray.

2. RAY PARENCHYMA

The cells of the ray parenchyma are comparable with ray tracheids but are usually somewhat longer; the longest ray parenchyma cells occur in the early-wood zone. They can be readily distinguished from ray tracheids by their simple pits. In white pine simple pits on the side walls, leading to the longitudinal tracheids, are large, one or two per cross field (see Fig. 12-34). Such pits are described as *fenestriform*. The complementary pits on the longitudinal tracheid side are narrowly bordered, and hence the resulting pit pairs are half-bordered. As is evident on surface *A* in Fig. 4-3, the imperforate membranes of these pits are frequently arched into lumina of the longitudinal tracheids. The pit pairs on the end walls of adjacent parenchyma cells in the same longitudinal row and on the horizontal walls in the vertically adjacent rows of cells, are simple; these pits are much smaller than the window-like pits on the side walls.

3. RAY EPITHELIAL CELLS

These cells are similar to those of the longitudinal type. They surround the horizontal (transverse) resin canals, which, like the longitudinal resin canals, form through splitting of the true middle lamella of a group of cells, in this case ray parenchyma daughter cells. The transverse resin canals in white pine, and in other conifers with resin canals, are therefore always included within the ray. The resulting rays, as already stated, are called *fusiform*. In white pine the transverse resin canals are much smaller than the longitudinal; usually they are less than 60 microns in diameter (compare Figs. 4-14*a* and *c*).

C. Description of a Block of White Pine Wood

By way of summation of information presented on the structure of white pine wood, attention is directed to Fig. 4-3. This figure is a schematic drawing of a small block of white pine at high magnification (400X), showing the wood in three planes: *A*—transverse, *B*—radial, and *C*—tangential.

The sense of proportion is frequently lost in the study of wood in shifting from a hand lens to a compound microscope. For this reason it may be helpful to indicate that the area of the faces of the cube is in the neighborhood of 9/100 square millimeter, or roughly 1/7000 square inch, and the volume of the cube approximately 27/1000 cubic millimeter, i.e., less than 1/600,000 cubic inch.

It is quite apparent from this figure that white pine wood consists of several different kinds of cells, held together by intercellular substance.* Cells of similar shape and function collectively comprise a *type of wood element*. Those which have their long axis directed vertically in the standing tree are the longitudinal elements (cells), traceable to fusiform initials in the cambium. The cells which are elongated transversely and radially in the tree are the transverse elements; collectively the transverse elements comprise the wood (xylary) rays. They result from the division of ray initials in the cambium.

1. DESCRIPTION OF SURFACE A—TRANSVERSE SURFACE

On surface *A*, a segment of an annual ring (1-1^a) is shown in which growth proceeded from right to left when this layer was formed. Two types of longitudinal elements, viz., longitudinal tracheids (*a*, *a'*) and epithelial cells (*b*) belonging to a longitudinal resin canal (2), and one type of transverse elements [ray cells (*c*) forming part of a wood ray (3-3^a)] are shown. The longitudinal tracheids are arranged in radial rows, those in the early wood (1) being thinner-walled and tending toward the hexagonal in comparison with the thick-walled, rectangular (tabular) tracheids of the outer late wood (1^a). The orderly arrangement of the cells of this type is traceable to the fact that those of a given row are all descendants of the same fusiform initial in the cambium; as this divides again and again, the daughter cells thus produced to the inside of the cambium mature into longitudinal tracheids.

Bordered pits are visible both on the radial walls and on the tangential walls of the tracheids. If surface A could be viewed frontally (as a section of wood is seen under the microscope) instead of obliquely, as in the drawing, only pit pairs in sectional view would be visible (*d*, *f*, *g*); as it is, some bordered pits show on back walls in oblique view, appearing as bosses perforated at the center (*e*).

Attention should also be called to the varying appearance of coniferous pit pairs as seen in sectional view on the transverse surface. If the section is median, i.e., cut through the middle of pit canals, then the two pit apertures are visible as in *d*. If the cut is made to one side of the pit canals, the pit chamber appears as a biconvex cavity, spanned by a pit membrane and without visible means of communication, with the cell lumina lying to either side of the wall. The pit membrane may be either thickened or unthickened, depending on whether or not the plane of the section includes the edge of the *torus* (compare *f* and *g*).

The uniseriate portion of a fusiform wood tray (3-3^a) crosses surface A diagonally beyond the fifth row of tracheids. Since the ray cells are seen in longitudinal section view, they are longer than the cross section of the longitudinal tracheids

* The dark medial layer in the transverse section is the compound middle lamella (see page 80).

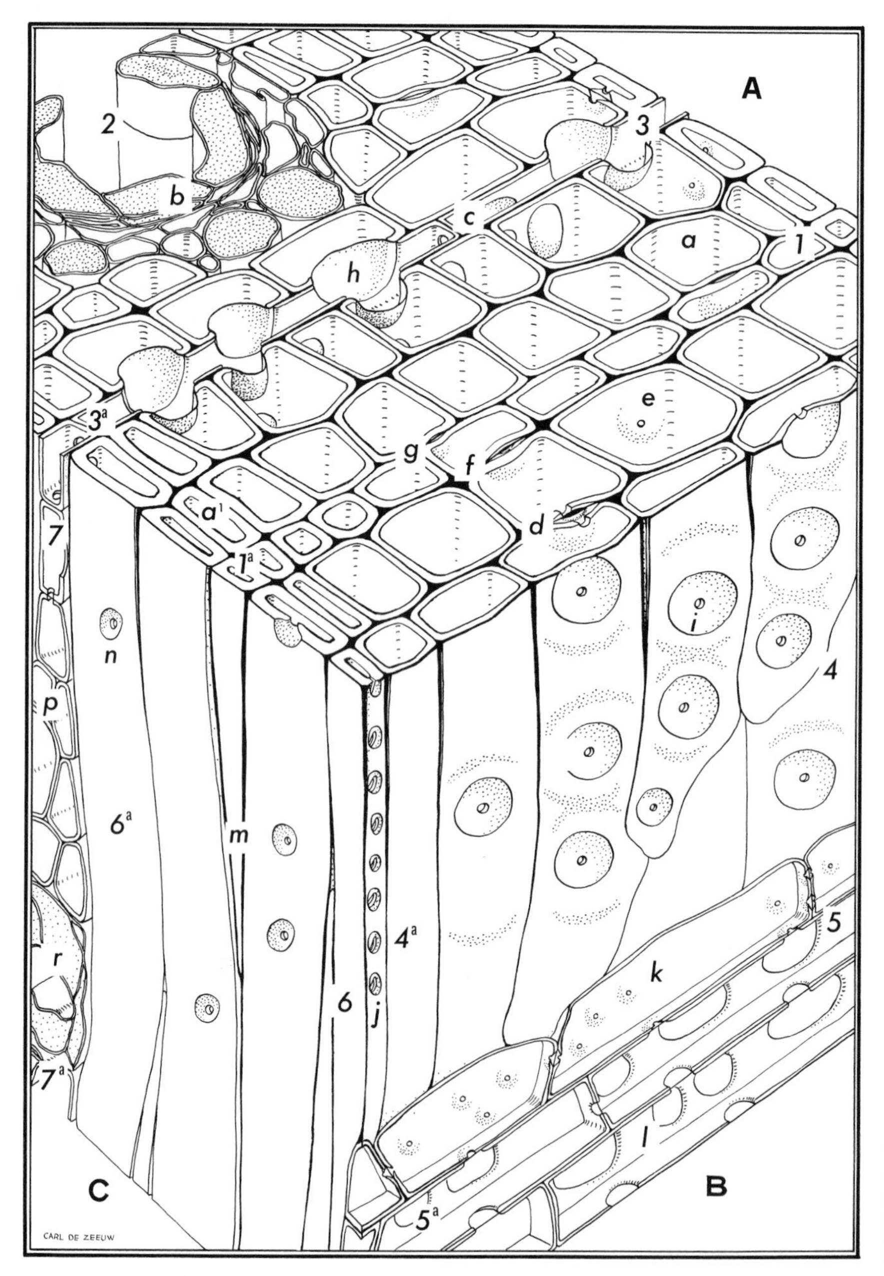

FIG. 4-3 Schematic three-plane drawing of the wood of eastern white pine (*Pinus strobus* L.). (400×.)

Surface A. 1-1^a, portion of an annual ring; 2, resin canal; 3-3^a, wood ray; *a*-*a'*, longitudinal tracheids; *b*, epithelial cells; *c*, ray cells; *d*, pit pair in median sectional view; *e*, bordered pits in the back walls of longitudinal tracheids, in surface view; *f*, pit pair in sectional view, showing the margin of the torus but so cut that the pit apertures are not included in the plane

arranged along the flanks of the ray. Pit pairs in the walls between the ray parenchyma cells and the longitudinal tracheids are half-bordered. The pit on the tracheid side has a wide aperture and a very narrow border; the complementary pit in the contiguous ray cell is simple. Because of the wide aperture, the unthickened compound pit membrane (*h*) often arches outward into the lumen of a longitudinal tracheid.

The final feature of surface *A* is the longitudinal resin canal (2) in the upper left-hand portion of the figure. It consists of a tubular cavity extending along the grain, sheathed by thin-walled epithelial cells (*b*).

2. DESCRIPTION OF SURFACE B—RADIAL SURFACE

Surface *B* pictures longitudinal tracheids (4-4^a) and the upper part of a uniseriate wood ray (5-5^a). The schematic conception of the surface pictured in *B* was obtained by splitting the wood radially in such a manner that all the compound middle lamella was removed, exposing the secondary wall of the longitudinal tracheids.

The blunt tracheid ends visible on surface *B* do not terminate at the same height, even though these tracheids originated from the same cambial fusiform initial. The reason for this is that when newly formed cells enlarge they do not grow in length exactly the same amount. Were it not so, all the ends of the tracheids in the upper row would be even. It also means that the tracheids in the upper row and in the row immediately below (not shown in the drawing) grow past each other so that only one series of tracheids is visible in any given place on the radial surface. This condition may be seen at the bottom of the radial surface in Fig. 4-1. Neither do tracheids in the row immediately behind, i.e., in the second row of tracheids, end at the same level as those in the first. This is indicated by the parts of the two tracheids shown immediately above the right-hand end of the ray in *B*-4, Fig. 4-3. This is because the fusiform initials in the cambium are staggered, as well as because of the difference in the postcambial elongation of the individual tracheids.

Since the compound middle lamella was removed, the large bordered pits in the walls of longitudinal tracheids show not as bosses but as bowl-shaped recesses in the wall. ***Crassulae***, too (see page 125), are not evident, since in reality they are

---

of section; *g*, pit pair in which neither the pit aperture nor the torus shows; *h*, window-like pit pairs between longitudinal tracheids and ray parenchyma.

Surface *B*. 4-4^a, portions of longitudinal tracheids in radial aspect (the ends are blunt); 5-5^a, upper part of a uniseriate ray; *i*, bordered pits on the radial walls of longitudinal, early-wood tracheids (the base of the pit is toward the observer); *j*, small bordered pits on the radial walls of longitudinal late-wood tracheids, in the same view as in *i*; *k*, ray tracheids; *l*, cells of ray parenchyma.

Surface *C*. 6-6^a, portions of longitudinal tracheids in tangential aspect; 7-7^a, portion of a xylary ray; *m*, tapering ends of longitudinal tracheids; *n*, a small bordered pit on the tangential wall of a longitudinal late-wood tracheid; *p*, cells of ray parenchyma; *r*, transverse resin canal.

thickenings in the intercellular material, which is not shown in the drawing. Their position, however, is indicated by dumbbell-shaped or lunate stippled depressions in the secondary wall.

The bordered pits on the radial walls of the longitudinal tracheids of white pine are usually in one row, although paired pits occur occasionally in the early-wood tracheids or when the wood is coarser textured than usual. They are largest in the early-wood portion of an annual increment (*i*) and become successively smaller toward the outer margin [outer late wood, see (*j*)]. The shape of the apertures changes as the pits are reduced in size from rounded to flattened and their axes then are set at an angle to vertical.

The upper three rows of a uniseriate ray shown in 5-5*a* of Fig. 4-3 on surface *B* consist of two kinds of elements; ray tracheids (*k*) and cells of ray parenchyma (*l*). One row of the former is present on the upper margin; this is underlaid by two rows of ray parenchyma.

The ray tracheids are provided with bordered pits, and the cells are seldom more than 0.1 to 0.2 millimeter in length. Pit pairs between ray tracheids and longitudinal tracheids, and between adjacent ray tracheid cells are bordered; those between ray tracheids and ray parenchyma are half-bordered.

The cells of ray parenchyma are somewhat longer than those of ray tracheids. They possess simple pits, of which those on the side wall are fenestriform.

3. DESCRIPTION OF SURFACE C—TANGENTIAL SURFACE

Surface *C* shows portions of seven longitudinal tracheids ($6\text{-}6^a$); two cells of a uniseriate ray at the right, and a part of a fusiform ray ($7\text{-}7^a$) at the left.

The ends of the longitudinal tracheids taper to a point (*m*), in contrast with blunt rounded contours in the radial plane. Several small bordered pits (*n*) are shown. These pits are comparable with those on the radial face, except that they are smaller and less numerous. In white pine the tangential pits are confined to the last few layers of late-wood tracheids.

Both rays are shown cut transversely. Two ray cells of a uniseriate ray are visible; the top pointed cell is a ray tracheid and the bottom cell a ray parenchyma. A portion of the fusiform ray at $7\text{-}7^a$ shown in the drawing consists of a uniseriate row of ray parenchyma at the top, broadening into a wider portion of the ray, consisting of ray parenchyma (at *p*) and epithelial cells (at *r*) which surround the transverse resin canal. If the entire fusiform ray were shown it would have terminated with one or more rows of ray tracheids, with the topmost cell gable-shaped (see Fig. 4-14).

It should be pointed out that, unlike the cells on section *B*, each element shown on section *C* is derived from a different mother cell in the cambium, and because these cambial initials are staggered, each mature tracheid ends at a different level (see also tangential surface in Fig. 4-1).

II. COMPARATIVE ANATOMY OF THE CONIFEROUS WOODS OF THE UNITED STATES

The elements of coniferous woods native to North America are listed in Table 4-3; these should be compared with those in Table 4-1, and the departures noted. The wood elements that do not occur in white pine are set in Table 4-3 in boldface type. Examination of this table reveals that all coniferous woods possess few types of cells and are therefore relatively simple in structure.

Not all the cell types listed in Table 4-3 are found in every coniferous wood. Longitudinal tracheids are always present and constitute fully 90 percent of the volume of the wood. Strand tracheids, which are a transitional element between longitudinal parenchyma and longitudinal tracheids, may occur in coniferous woods with resin canals or longitudinal parenchyma. Longitudinal parenchyma may or may not be present, and therefore its presence or absence, relative abundance, and its position in the growth ring are of diagnostic significance. Epithelial cells are a constant feature in ***Pinus***, ***Picea***, ***Larix***, and ***Pseudotsuga***, the four genera of conifers in which both longitudinal and transverse resin canals are always present: they may also occur in other coniferous woods when traumatic resin canals, formed presumably as a result of injury, are present. Different ray combinations, indicated in Table 4-3, are traceable to varying cell composition.

A. The Longitudinal Coniferous Tracheids

In their arrangement and shape the tracheids of white pine may be considered as representative for coniferous woods as a group. Neither in their grouping nor in their shape do these cells exhibit sufficient variation to render this feature of any value diagnostically.

TABLE 4-3 Elements of Coniferous Woods

Longitudinal	*Transverse*			
1. Prosenchymatous	1. Prosenchymatous	Homocellular (uniseriate)† rays	Heterocellular (uniseriate)† rays	Fusiform rays
a. Tracheids	*a*. Ray tracheids			
a′. Resinous tracheids				
b. **Strand tracheids**				
2. Parenchymatous	2. Parenchymatous	Homocellular (uniseriate)† rays		
a. **Cells of longitudinal parenchyma**	*a*. Cells of ray parenchyma			
b. Epithelial cells:* excreting cells of longitudinal resin canals	*b*. Epithelial cells:* excreting cells of transverse resin canals			

* Longitudinal and transverse resin canals are not listed, since they are intercellular spaces encircled by excreting cells (epithelium) and are hence not true elements.

† Rarely biseriate and then usually only through the middle portion of the ray, as seen in the tangential section.

In contrast, longitudinal tracheids manifest considerable fluctuation in size and markings, to permit separation of the wood to genus and even occasionally to species.

1. SIZE OF LONGITUDINAL TRACHEIDS

The radial diameter of a tracheid varies according to its position in the growth increment; this dimensional fluctuation is not of generic or specific significance. The tracheids of greatest radial diameter are produced soon after growth starts in the spring; the early-wood tracheids are relatively thin-walled. As the season progresses the tracheids become narrower and thicker-walled. The transition from the early-wood to the late-wood tracheids may be gradual (ex., eastern white pine, balsam fir) or abrupt (ex., hard pines, western larch). The narrowest tracheids with the thickest wall are produced at the outer margin of the growth increment, i.e., just before the end of the growing season.

The tangential diameter of a longitudinal tracheid varies according to the place within its length where the measurement is taken, the position in the tree, and finally, according to the kind of wood. The tangential diameter remains about the same through the length of a tracheid except at the tapering ends where it is smaller. The relationship which exists between tracheid size and position in the tree, as affected by the age of the tree and the height in the stem, is discussed at length on pages 239 to 246. The variation in the tangential diameter of tracheids, in the coniferous woods discussed in this text, is indicated in Table 4-4. *Sequoia sempervirens* and *Taxodium distichum* head the list with tracheids that attain a maximum diameter of 80 and 70 microns, respectively. The other extreme is represented by *Taxus brevifolia*, with a maximum tracheid breadth of 25 microns and the average of 15 to 20 microns. On the basis of these figures, a medium-textured coniferous wood is defined arbitrarily in this text as one with the average tangential diameter of the tracheids somewhere between 30 to 45 microns. Above these figures, softwoods are considered to be coarse-textured; below these, as fine-textured.

The tracheid lengths of the conifers included in this text are given in Table 4-5. The average length ranges from 1.18 millimeters in *Juniperus osteosperma* to 7.39 millimeters in *Sequoia sempervirens*. Comparison of information in Tables 4-4 and 4-5 indicates a correlation between tracheid breadth, i.e., tangential diameter, and tracheid length; those whose tracheids are widest also possess the longest.* Tracheids measuring 3.0 to 5.0 millimeters may be considered as average for coniferous woods. Above or below these figures, the wood in question will be considered as unusually long- or short-fibered, respectively.

* The same correlation also holds within the same species and even in the same tree (see pages 239 to 249).

TABLE 4-4 Tangential Diameter (Breadth) of Longitudinal Tracheids of the Coniferous Woods of the United States
Arranged in order of averages

Scientific name	*Diameter, μ*	*Scientific name*	*Diameter, μ*
Sequoia sempervirens	80 (avg 50–65)	*Libocedrus decurrens*	50 (avg 35–40)
Taxodium distichum	70 (avg 45–60)	*Tsuga heterophylla*	50 (avg 30–40)
Pinus lambertiana	65 (avg 40–50)	*Abies balsamea*	50 (avg 30–40)
Larix occidentalis	60 (avg 38–50)	*Abies fraseri*	50 (avg 30–40)
Pinus monticola	60 (avg 35–45)	*Pinus resinosa*	45 (avg 30–40)
Pinus palustris	60 (avg 35–45)	*Thuja plicata*	45 (avg 30–40)
Pinus echinata	60 (avg 35–45)	*Tsuga canadensis*	45 (avg 28–40)
Pinus taeda	60 (avg 35–45)	*Larix laricina*	45 (avg 28–35)
Pinus elliottii	60 (avg 35–45)	*Torreya californica*	55 (avg 25–35)
Pinus ponderosa	60 (avg 35–45)	*Torreya taxifolia*	55 (avg 25–35)
Pinus jeffreyi	60 (avg 35–45)	*Pinus strobus*	45 (avg 25–35)
Abies concolor	60 (avg 35–45)	*Chamaecyparis nootkatensis*	40 (avg 25–35)
Abies grandis	60 (avg 35–45)		
Abies magnifica	60 (avg 35–45)	*Chamaecyparis thyoides*	40 (avg 25–35)
Abies procera	60 (avg 35–45)	*Picea glauca*	35 (avg 25–30)
Pinus contorta	55 (avg 35–45)	*Picea mariana*	35 (avg 25–30)
Picea sitchensis	55 (avg 35–45)	*Picea rubens*	35 (avg 25–30)
Pseudotsuga menziesii	55 (avg 35–45)	*Thuja occidentalis*	35 (avg 20–30)
Chamaecyparis lawsoniana	50 (avg 35–40)	*Juniperus virginiana*	35 (avg 20–30)
		Taxus brevifolia	25 (avg 15–20)

2. MARKINGS OF LONGITUDINAL TRACHEIDS

Variations in the nature of markings on the lateral walls of coniferous tracheids are traceable to pits and in some species to spiral thickenings in the walls of the tracheids.

a. Pits. The pits in the longitudinal tracheids fall into three categories: (1) intertracheal, i.e., those belonging to pit pairs between the adjacent longitudinal tracheids, (2) those of pit pairs providing communication between the longitudinal tracheid and ray parenchyma in contact with it, and (3) those of pit pairs that connect the longitudinal tracheid with the ray tracheid, when these are present in the ray.

The intertracheal bordered pits are always numerous and conspicuous on the radial walls and are either absent or confined mainly to the last few rows of the late-wood tracheids on the tangential walls. In size or shape they do not present enough variation from one wood to another to be of diagnostic value. Considerable variation in pit size and in the shape of pit apertures, however, can be observed on the radial walls of the early-wood and late-wood tracheids. In the former the pits reach their maximum size and generally have a rounded aperture. In the latter the pits are smaller, because of the narrowing of the radial diameters of the longitudinal tracheids toward the outer margin of the growth rings, and may assume

TABLE 4-5 Average Length of Longitudinal Tracheids of the Coniferous Woods of the United States

Determined from specific samples

Scientific name	Avg length, mm	Standard deviation*	Scientific name	Avg length, mm	Standard deviation*
Pinus lambertiana	5.14	0.94	*Pinus resinosa*	2.51	0.54
	5.40	1.03		2.63	0.45
	5.24	0.96		2.67	0.27
Pinus monticola	2.83	0.59		2.70	0.89
	2.97	0.58	*Larix laricina*	2.86	0.39
	3.79	0.69		3.00	0.46
Pinus strobus	3.00	0.55		3.68	0.65
	3.70	0.66	*Larix occidentalis*	2.82	0.48
	3.99	0.69		2.97	0.55
	4.00	0.91		4.09	0.71
Pinus contorta	3.19	0.44	*Picea glauca*	2.92	0.41
	3.26	0.44		3.76	0.79
Pinus palustris	4.90	0.83		3.24	0.51
Pinus echinata	4.46	0.91	*Picea mariana*	3.25	0.40
	4.64	0.92		3.81	0.52
	4.85	0.76		3.60	0.72
Pinus taeda	4.33	0.91	*Picea rubens*	3.01	0.49
Pinus elliottii	4.58	0.87		3.17	0.60
Pinus rigida	3.57	0.74		3.01	0.61
	3.75	0.83	*Picea engelmannii*	2.49	0.48
Pinus serotina	2.73	0.36		2.75	0.66
Pinus ponderosa	3.53	0.75		3.63	0.55
	3.71	0.86	*Picea sitchensis*	5.22	0.85
	4.08	0.98		5.37	1.06
Pinus jeffreyi	3.20	0.61		5.45	0.98

an oval or even elliptical shape; the pit apertures, too, lose their rounded contours and become oval, elliptical, or slitlike.

The bordered pits on the radial walls of the tracheids may be crowded or distant; the first condition is more frequent toward the ends of the tracheid, the second in places throughout the median portion. Crosswise of the cell there may be one, two, or more pits, arranged in a transverse row.* This last condition is found in the early wood but does not necessarily hold throughout the whole length of a tracheid.

Transverse rows of pits on the radial walls of early-wood tracheids, consisting of two, three, or more pits, are indicative of coarseness of texture. The biseriate pits are common in western larch and sugar pine, while in the early wood of ***Sequoia sempervirens*** and ***Taxodium distichum*** rows of three to four pits are not uncommon (Fig. 4-7*c*, 3).

With one reported exception, that of western redcedar (***Thuja plicata*** Donn), intertracheal pits of all native conifers are equipped with well-defined tori. The

* In one family of the Coniferales, the Araucariaceae, the pits are in oblique series and where crowded are hexagonal. This family is largely confined to the Southern Hemisphere.

TABLE 4-5 Average Length of Longitudinal Tracheids of the Coniferous Woods of the United States (*Continued*)
Determined from specific samples

Scientific name	Avg length, mm	Standard deviation*	Scientific name	Avg length, mm	Standard deviation*
Pseudotsuga menziesii	3.00	0.31	*Taxodium distichum*	3.14	0.58
	3.32	0.39		5.23	1.42
	3.88	1.41		5.79	1.10
Pseudotsuga menziesii var. *glauca*	2.85	0.45	*Libocedrus decurrens*	3.60	0.59
			Thuja occidentalis	2.17	0.47
Tsuga canadensis	3.37	0.57		2.16	0.43
	3.80	0.82	*Thuja plicata*	3.00	0.45
	4.24	0.99		3.18	0.48
Tsuga heterophylla	2.87	0.40	*Chamaecyparis lawsoniana*	3.18	0.47
	2.91	0.58			
	3.10	0.59	*Chamaecyparis nootkatensis*	2.24	0.39
Abies balsamea	3.33	0.43			
	3.37	0.50	*Chamaecyparis thyoides*	3.20	0.45
	3.53	0.61		3.34	0.42
Abies concolor	3.79	0.63	*Juniperus virginiana*	2.15	0.50
Abies grandis	3.05	0.47	*Juniperus silicicola*	2.38	0.62
	3.35	0.43	*Juniperus osteosperma*	1.18	0.29
	3.53	0.65	*Taxus brevifolia*	2.31	0.34
Abies magnifica	3.27	0.52		2.32	0.45
Abies procera	3.60	0.58	*Torreya californica*	3.23	0.67
Sequoia sempervirens	5.79	1.03	*Torreya taxifolia*	2.76	0.39
	7.39	1.31			

* The standard deviation is a length in millimeters that, added to and subtracted from the tracheid average length, establishes limits of length within which approximately two-thirds of the tracheids will fall. Limits obtained as above by adding and subtracting three times the standard deviation will include over 99 percent of the tracheids.

absence of these structures in western redcedar can therefore serve as a means of identification of this species. Eastern and western hemlocks are characterized by the torus extensions in the form of broad supporting strands and a wartlike layer, extending from the lumen into the pits. In a number of species, e.g., incense-cedar, western redcedar, and hemlocks, the intertracheal pits are heavily encrusted with extractives. Such pit encrustations are either absent or sparse in Douglas-fir.* Further studies of the details of pit structures are needed. But since many of these features require high magnification obtainable only with an electron microscope, their usefulness in everyday identification of wood is limited.

The bordered pits on the radial wall of tracheids are bounded above and below by zones of darker substances. These structures, formerly called *bars of Sanio*, are now known as *crassulae* (Fig. 4-4*a* and *c*). They are traceable to a concen-

* R. L. Krahmer and W. A. Côté, Jr., Changes in Coniferous Wood Cells Associated with Heartwood Formation, *Tappi*, **46**(1):42–49 (1963).

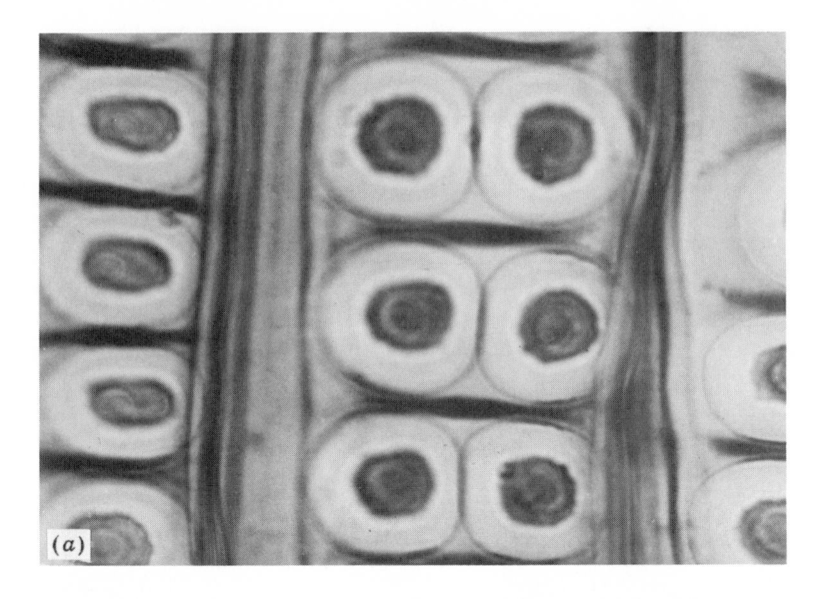

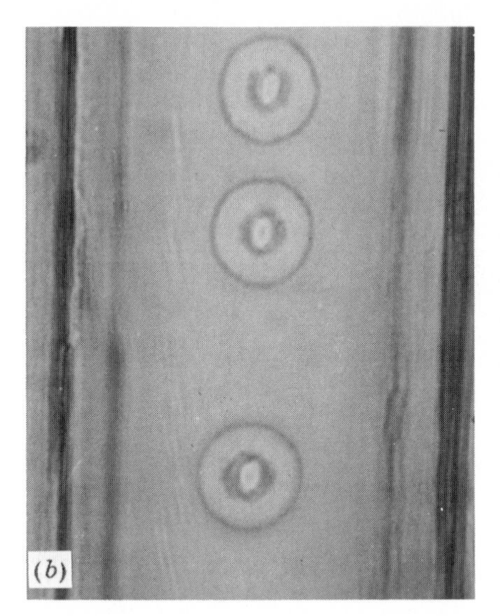

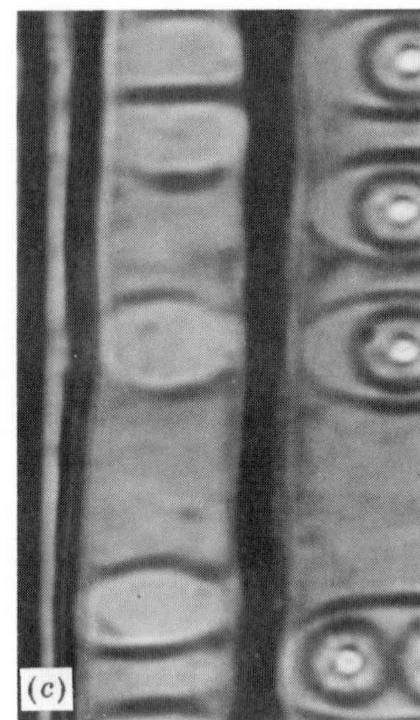

FIG. 4-4 Nature of the pits on the lateral walls of coniferous tracheids.

(*a*) Paired bordered pits on the radial walls of the longitudinal early-wood tracheids of sugar pine (*Pinus lambertiana* Dougl.). The outer circle of the pit is the pit annulus; the dark irregular central area is the torus; the smaller rounded spot within the boundary of the torus is the pit aperture; the dark bars above and below the pits are crassulae. (875 ×.)

(*b*) Bordered pits on the tangential wall of a late-wood tracheid of sugar pine (*Pinus lambertiana* Dougl.) taken at the same magnification as *A*; the pits are much smaller and crassulae are wanting.

(*c*) Pit fields on the radial walls of the longitudinal tracheids of baldcypress [*Taxodium distichum* (L.) Rich.], bounded above and below by crassulae. Five

tration of intercellular substance at these points, this layer being overlaid by the primary and secondary walls of the cell.

Crassulae are confined to the radial walls of tracheids. They are common throughout the coniferous woods with the exception of the Araucariaceae and hence are of no diagnostic value in wood identification. They are also visible in the vessels and tracheids of certain hardwoods (Fig. 5-7*a*).

In most coniferous woods, the bordered pits on the radial walls of the longitudinal tracheids develop in places where the true middle lamella is thinner than elsewhere. Such a thin place, bounded above and below by crassulae, is called a ***primary pit field***. It may contain one or more pits. Not infrequently, however, pits fail to develop, resulting in an empty pit field (Fig. 4-4*c*).

The bordered pits on the tangential walls of longitudinal tracheids are always smaller than those on the radial walls; this fact is evident when (*a*) and (*b*) of Fig. 4-4, taken at the same magnification, are compared. They occur throughout coniferous woods with the exception of the hard pines, in which species the pits on the tangential walls of longitudinal tracheids are either absent or found only occasionally. Where present, as stated before, tangential pitting is confined mainly to the last few rows of late-wood tracheids and the inner tangential walls of the early-wood tracheids of the succeeding growth increment.

b. Spiral Thickenings on the Wall of Longitudinal Tracheids. Spiral thickenings (Fig. 3-14) are present as a constant feature in the longitudinal and ray tracheids in Douglas-fir, Pacific yew, and *Torreya* spp. They may also occasionally occur in other woods, notably in the ray tracheids and in the longitudinal tracheids in the juvenile wood of the spruces and in the late-wood tracheids in tamarack and in some hard pines; in these woods, however, presence of the spirals is of no diagnostic value.

In Douglas-fir the spirals are most conspicuous in the early-wood tracheids; they are less well developed and sometimes absent in the late-wood tracheids. In the tracheids of *Taxus brevifolia* and *Torreya* spp. spirals occur throughout the growth increment; Pacific yew differs from *Torreya* in possessing steeper-angled and more compact spirals. These features are of some generic significance but difficult to use since the angle of the spirals is also contingent on the width of the cells and the thickness of the cell wall. Cells that are narrow-lumened for either of the above reasons possess steeper spirals than those with wider cell cavities. The angle of the spirals therefore varies according to the position of the tracheid in the growth increment.

c. Trabeculae in Longitudinal Coniferous Tracheids. A *trabecula* is a cylindrical, barlike, or spool-shaped structure extending across the lumen of a tracheid from one tangential wall to another (Fig. 4-5). The origin of trabeculae is traceable to the cambium where they form as a delicate filament of wall substance across a fusiform initial. A secondary wall is laid down on this filament after the cell division, at the same time and in the same manner as wall thickening is progressing

---

of the pit fields are empty, three contain one pit each, and one has two bordered pits. (230×.) Three and four pits, aligned in a transverse row, frequently occur in a pit field in this species (see Fig. 4-7).

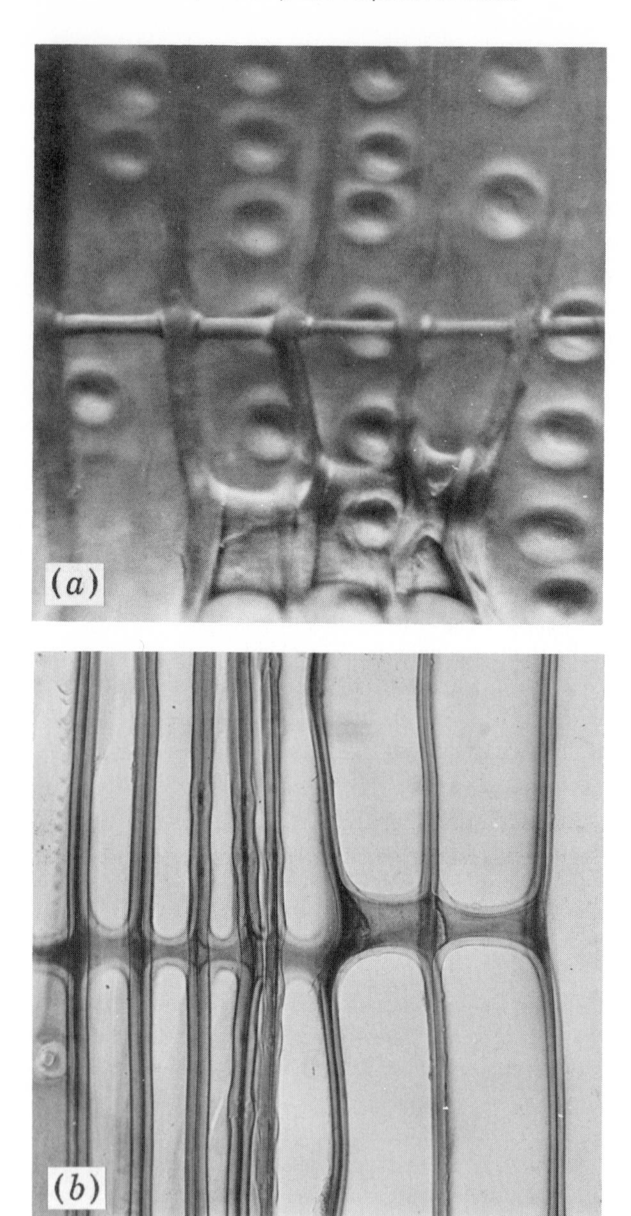

FIG. 4-5 (*a*) Trabeculae in western white pine (*Pinus monticola* Dougl.), photographed with incident light. (430×.) (*Photograph by R. E. Pentoney.*)
(*b*) Trabeculae in Alaska-cedar [*Chamaecyparis nootkatensis* (D. Don) Spach]. (360×.)

in the cell. T. A. McElhanney and his associates[11] at the Forest Products Laboratory of Canada ascribe the formation of the initial thin filament to fungal hyphae in the cambium. They state that "the rods seem to be due to the results of the deposition of cell-wall material about a fungus filament in the cambium, with the result that the structure is perpetuated in the series of cells developed from the cambial cell." This contention finds support in the fact that trabeculae are likely to be present in regions where the cambium has been exposed to fungal infection, notably in wood formed after wounding and in close proximity to fungous cankers.

Trabeculae are best seen in radial and transverse sections of wood. As a rule they are formed at the same height in a number of tracheids in the same tier, i.e., in a radial row of tracheids that arose from the same mother cell in the cambium.

3. LONGITUDINAL RESINOUS TRACHEIDS

As longitudinal coniferous tracheids of the normal type pass from sapwood into heartwood, resinous material sometimes accumulates in these cells at or near the places where they are in contact with wood rays. Such deposits are usually reddish brown or nearly black and are amorphous in nature. In transverse sections, if the plane of the cut happens to include them, they appear to fill the cells either completely or partly. Along the grain, they assume the form of transverse plates extending across the cell and simulating cross walls, or of lumps on the walls on one or both sides of the tracheid. Tracheids with such inclusions are designated as *resinous tracheids*.

Resinous tracheids are comparatively rare in coniferous woods. They are most frequent in *Agathis* and *Araucaria* of the Araucariaceae. Penhallow[12] reported them as occasional in *Abies fraseri* and *A. grandis;* they also occur sporadically in some of the soft pines and possibly elsewhere.

4. LONGITUDINAL STRAND TRACHEIDS

Strand tracheids differ from longitudinal tracheids in being shorter and in possessing end walls, one or both of which are at right angles to the longitudinal walls; when the second condition holds, the cells are narrowly rectangular along the grain. Both the end and the radial walls are provided with bordered pits of the normal type.

Strand tracheids arise from a single cambial cell through further division of a daughter cell, which otherwise would have developed into a longitudinal tracheid. All the segments thus produced through formation of cross walls may become prosenchymatous, or some may be parenchymatous, resulting then in a "mixed strand" (Fig. 4-6).

Strand tracheids may be regarded as transitional elements between longitudinal tracheids and epithelial or longitudinal parenchyma. When present in a wood, they are found in the vicinity of the longitudinal resin canals, on the outer face of the annual increment, i.e., terminally, or in or near traumatic tissues. Terminal

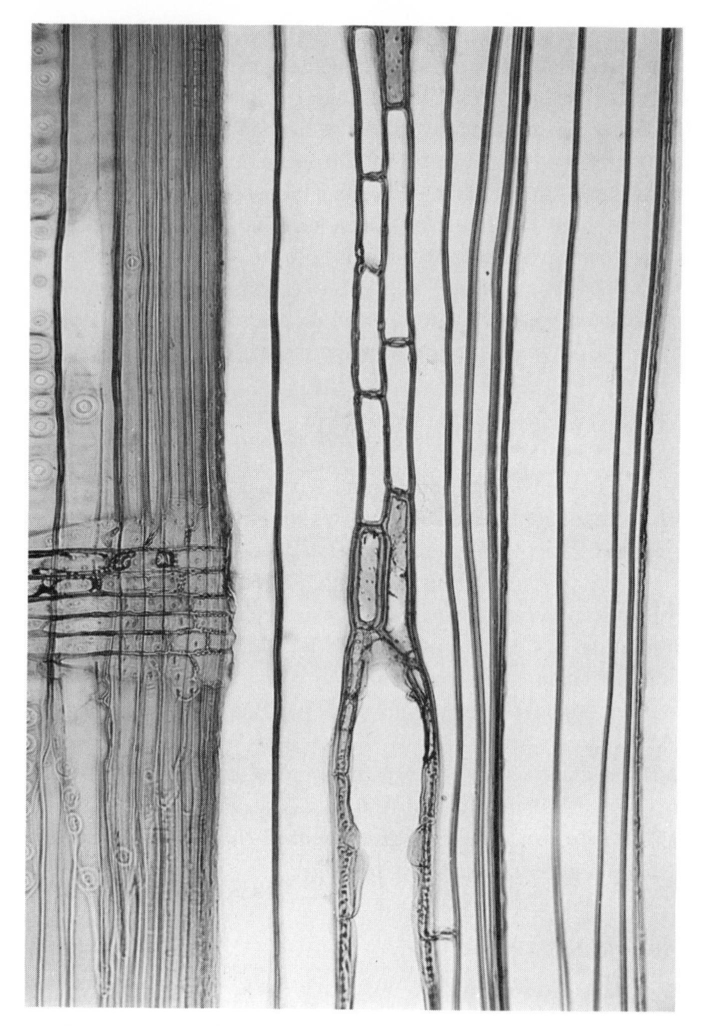

FIG. 4-6 Strand tracheids in western larch (*Larix occidentalis* Nutt.). The strands are mixed passing above and below into epithelial cells. (120×.)

strand tracheids are sometimes present in *Larix* spp. and in Douglas-fir, where they apparently replace the strands of longitudinal parenchyma that occupy the same position and are sporadic in their distribution in these genera.

B. Longitudinal (Axial) Parenchyma in Coniferous Wood

There are three types of parenchyma in coniferous woods, viz., *longitudinal parenchyma, ray parenchyma,* and *epithelial parenchyma*. The last two types are discussed in conjunction with wood rays and resin canals, respectively. The following paragraphs are devoted to a discussion of longitudinal parenchyma.

The longitudinal parenchyma in softwoods occurs in the form of strands extending along the grain (Fig. 4-7*c*, 1). Each strand arises by the further division of a daughter cell formed by division of a fusiform initial in the cambium. When cut transversely, such a strand appears as a cell which is usually thinner-walled than the neighboring tracheids and frequently contains extraneous materials (Fig. 4-7).

The number of cells composing a strand is undoubtedly variable in different coniferous woods, in different samples of a given species, and even in the same sample. No definite figure can be cited because of the variations in length of the strands, which precludes accurate count. In some woods (***Larix*** spp., ***Pseudotsuga*** spp.), the strands consist in part of prosenchymatous cells (strand tracheids).

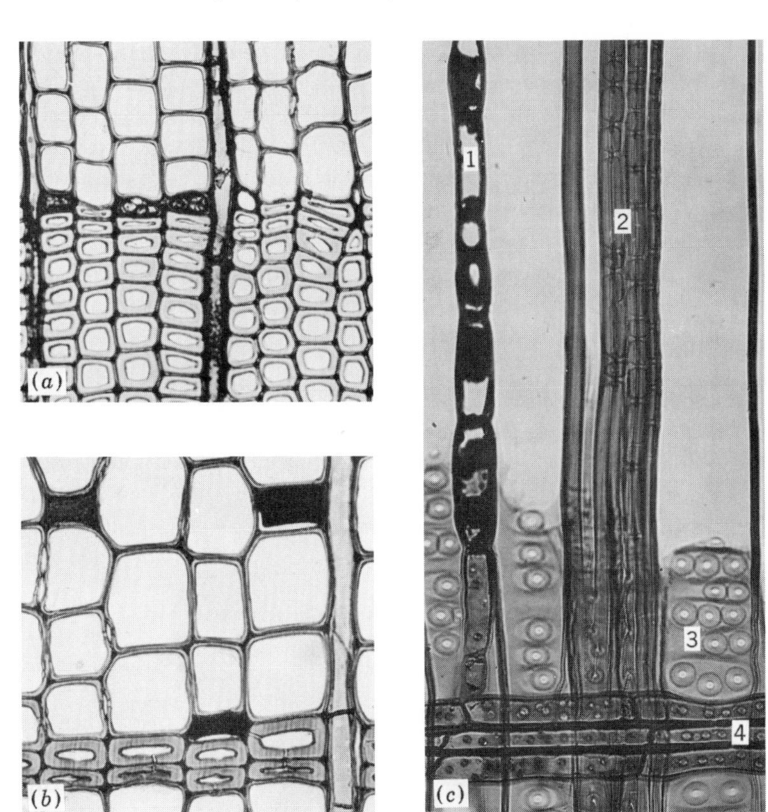

FIG. 4-7 Longitudinal parenchyma in coniferous woods.

(*a*) Marginal parenchyma in western hemlock [***Tsuga heterophylla*** (Raf.) Sarg.] (three cells on the outer margin of an annual ring). (200×.)

(*b*) Diffuse parenchyma in redwood [***Sequoia sempervirens*** (D. Don) Endl.] (three dark cells in the first-formed early wood). (200×.)

(*c*) Portion of a strand of longitudinal parenchyma of baldcypress [***Taxodium distichum*** (L.) Rich.], radial view. (165×.) (1) strand; (2) bordered pit pairs in sectional view, the pits with pit canals; (3) bordered pits in transverse rows of three on the radial walls of early-wood tracheids; (4) wood ray in radial aspect.

The units composing a strand vary in shape according to their position. Those at the ends are square or rectangular at the base and taper abruptly or gradually to a point; the remainder of the strand consists of cells that approach the rectangular as viewed in the longitudinal section.

Pits in the cross walls of axial parenchyma when present are invariably simple. When viewed in the longitudinal sections (i.e., radial or tangential) simple pits give the appearance of beadlike (*nodular*) thickenings on the transverse walls of parenchyma cells (Fig. 12-43). Simple pits are absent or sparse and inconspicuous in *Chamaecyparis thyoides* and *Sequoia*. In the case of *Sequoia* the absence of simple pits (and hence nodular thickenings) in the transverse walls of the axial parenchyma serves as a ready means for separating it from *Taxodium*.

As viewed in the transverse section the axial parenchyma cells may be widely scattered among the tracheids. In that case the parenchyma is said to be *diffuse*. If the cells are sufficiently numerous to be loosely grouped in a row or a band extending concentrically in a growth ring (i.e., tangentially), the parenchyma is termed *banded* (formerly called *metatracheal*). These two types of arrangement intergrade. If the cells are confined to the outer margins of the growth ring, i.e., to the last-formed or first-formed rows of cells, they are called *terminal* or *marginal*.*

Axial parenchyma does not occur in the wood of *Pinus*, *Taxus*, and *Torreya*, among the genera represented in the United States.† It is present in the root wood of *Picea* but not in the trunk wood. It is sporadic and sparse in *Larix*, *Pseudotsuga*, *Tsuga*, and *Abies* and is either terminal, diffuse, or both. In the Cupressaceae (*Thuja*, *Chamaecyparis*, *Cupressus*, *Libocedrus*, and *Juniperus*) it is relatively abundant, banded or diffuse, but unevenly distributed; i.e., parenchyma cells may be found in appreciable numbers in one ring or one sample of wood, and be sparse or lacking in the others. Axial parenchyma is best developed in the Taxodiaceae (*Sequoia* and *Taxodium*), in which woods it ranges from diffuse to banded but is always abundant enough to be found in any given section of the wood under examination.

C. Rays of Coniferous Woods

A coniferous wood may have either one or two types of rays. When normal resin canals are absent the rays are usually uniseriate, as viewed in the tangential plane. The presence of transverse resin canals (which are invariably accompanied by longitudinal canals) results in formation of fusiform rays, so called because of their spindle shape when cut transversely (Fig. 4-14*c*, *d*, and *e*).

* In the strict sense, parenchyma cells formed at the beginning of a season's growth should be called *initial*, and those formed at the close of a season, *terminal*. But because it is difficult to differentiate between the two the term terminal or marginal is generally used to include both forms.

† Also in *Araucaria* and *Agathis* of the Southern Hemisphere.

1. UNISERIATE RAYS

The rays of coniferous woods, without resin canals, are almost exclusively uniseriate and will be so called throughout this discussion. Occasionally, in some species, rays with one or more pairs of cells in the central portion of the ray (*t*) may be found. Rays completely biseriate, except for the end cells, are sometimes encountered in redwood. The biseriate rays are of no diagnostic significance, since of the 15 genera of coniferous woods covered in this text, sporadic biseriation of rays occurs in 11.

The uniseriate rays may be composed of ray parenchyma exclusively, or of only ray tracheids, as in the low rays of *Chamaecyparis nootkatensis*, or of both these kinds of cells. When the uniseriate rays are composed of only one type of cells, whether ray parenchyma or ray tracheids, they are said to be *homocellular;* when both kinds of cells are found in the rays, such rays are called *heterocellular.*

The uniseriate rays vary in height, as viewed in tangential section, both in number of cells and in linear measurements. This holds not only for different woods but in the same wood. Some of the variations in the number of cells and in the overall height of the rays is genetically or ecologically controlled, and some is due to the manner in which new ray initials are formed in the cambium. As the diameter of a tree increases, the rays diverge outward into the cambium. Since, in a given wood, the spacing of rays, as they appear on the transverse section, remains approximately constant, additional rays are formed to maintain the spacing characteristic of the species. The different ways in which new ray initials may arise in the cambium are discussed on page 44 and portrayed in Fig. 2-14. Once a new cambial ray is formed, its increase in height and width is effected by transverse anticlinal division.*

The upper limit in the height of coniferous xylary rays varies according to the kind of wood and the sample thereof. The highest figures are reached in *Taxodium distichum* and *Sequoia sempervirens*, in which rays 40 to 60 plus cells and 500 to 1000 plus microns in height are occasionally found, although the averages are much lower. *Juniperus virginiana* and *Chamaecyparis lawsoniana* are examples of woods in which the tallest rays are seldom over 6 cells and 300 microns in height. The mean maximum height of the narrow rays of conifers, taken as a group, is somewhere in the neighborhood of 10 to 15 cells.

The range of ray heights and the maximum height are occasionally of sufficient significance to be of some value, as a secondary characteristic, in differentiating between various kinds of coniferous woods. For instance, the firs (*Abies* spp.) have taller rays than the hemlocks (*Tsuga* spp.), and the same holds as between *Taxodium distichum* and *Sequoia sempervirens*.

a. Ray Parenchyma. The cells of ray parenchyma are thin-walled and are equipped with simple pits. The shape, size, and arrangement of pits in the cross

* For more detailed discussion of the original development of rays in conifers, see Ref. 3. Development of new rays in hardwoods is similar in nature.

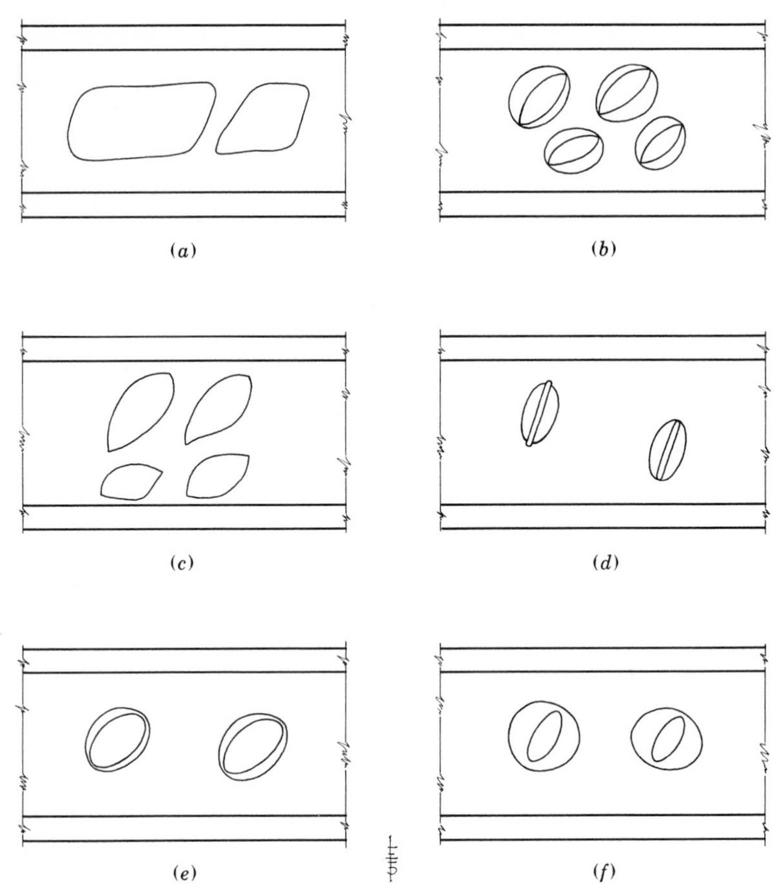

FIG. 4-8 Types of pit pairs occurring in the cross fields of coniferous woods. (*After Phillips.*[13]) (For corresponding photographs see Fig. 4-9.)

(*a*) Window-like pit pairs of the type that characterize the three commercial soft pines, and red pine (*Pinus resinosa* Ait.).

(*b*) and (*c*) Pinoid pit pairs of the type occurring in the hard pines other than red pine (*P. resinosa* Ait.). A border may (*b*) or may not (*c*) be present.

(*d*) Piceoid pit pairs of the type encountered in *Picea, Larix,* and *Pseudotsuga*.

(*e*) Taxodioid pit pairs of the type that feature *Sequoia* and *Taxodium* of the Taxodiaceae, *Abies* of the Pinaceae, and *Thuja* of the Cupressaceae.

(*f*) Cupressoid pit pairs of the type that occur in *Chamaecyparis, Libocedrus,* and *Juniperus* (but not in *Thuja*) of the Cupressaceae, and in *Tsuga* of the Pinaceae.

fields vary in different coniferous woods and are therefore of considerable importance in the identification of softwoods.

Phillips[13] has described five fairly well-defined types of cross-field pits in conifers, viz.,

1. Large pits with broad apertures and an overhanging border on the tracheid

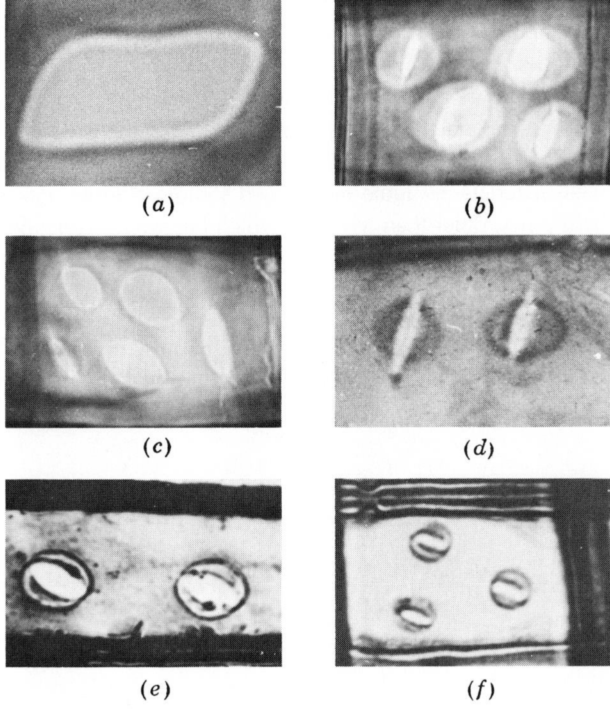

FIG. 4-9 Types of pit pairs occurring in the cross fields of coniferous woods.

(*a*) Window-like pit pairs of the type that characterize the three commercial soft pines, and red pine (*P. resinosa* Ait.).

(*b*) and (*c*) Pinoid pit pairs of the type occurring in the hard pines other than red pine (*P. resinosa* Ait.). A border may (*b*) or may not (*c*) be present.

(*d*) Piceoid pit pairs of the type encountered in *Picea*, *Larix*, and *Pseudotsuga*.

(*e*) Taxodioid pit pairs of the type that feature *Sequoia* and *Taxodium* of the Taxodiaceae, *Abies* of the Pinaceae, and *Thuja* of the Cupressaceae.

(*f*) Cupressoid pit pairs of the type that occur in *Chamaecyparis*, *Libocedrus*, and *Juniperus* (but not in *Thuja*) of the Cupressaceae, and in *Tsuga* of the Pinaceae.

side so narrow that careful focusing is required to determine its presence. Such pits are termed *window-like*, or *fenestriform* (Fig. 4-8*a*). They are found in the three commercial soft pines and in red pine, among native woods; also in Scotch pine and in a number of other exotic species of conifers.

2. *Pinoid* pits differ from the window-like pits by being smaller, more variable in size, and more numerous per cross field (Figs. 12-32 and 12-33). According to Thomas and Nicholas,[18] no secondary thickenings has been detected in the ray parenchyma cell walls. Therefore the ray side of the pinoid pit membrane is ap-

parently continuous with the rest of the ray walls (Fig. 4-10). A distinctive border was found on the longitudinal tracheid side of the pit, and thus the pinoid pit pair can be classified as half-bordered; with the lower magnifications (below 1000X) this border may or may not be evident (Fig. 4-9*b* and *c*). The same investigators found that the pinoid pit membranes are void of detectable openings (Figs. 3-13 and 4-11). In the heartwood the pit membranes are heavily encrusted with materials insoluble in acidified sodium chlorite. Pinoid pits feature all native hard pines, other than red pine, which has window-like pits.

3. ***Piceoid*** pits, shown in Fig. 4-8*d*, are small bordered pits, generally elliptical

FIG. 4-10 Cross section of a pinoid pit membrane from the outer sapwood of loblolly pine. Note the pit-border formation on the longitudinal tracheid side of the pit membrane and the continuous nature of the ray parenchyma cell wall and the pit membrane. Electron micrograph; direct carbon replica preshadowed with platinum. (5800×.) (*Courtesy of R. J. Thomas and D. D. Nicholas, Ref.* 18.)

FIG. 4-11 Pinoid pits from untreated, air-dried, outer sapwood of pond pine, as viewed from the lumen of a longitudinal tracheid. Electron micrograph; direct carbon replica preshadowed with platinum. (3900×.) (*Courtesy of R. J. Thomas and D. D. Nicholas, Ref.* 18.)

in shape, with a narrow (linear) and frequently slightly extended aperture. Pits of piceoid type are found in *Picea*, *Larix*, and *Pseudotsuga*.

4. *Taxodioid* pits (Fig. 4-8*e*) possess apertures which range from oval to circular, are included, and are much wider than the narrow, fairly even border. Pits of this type feature the Taxodiaceae, i.e., *Sequoia* and *Taxodium*, although pits of the cupressoid type may also be present in the latter. Taxodioid pits also are present in *Abies* and *Thuja*.

5. *Cupressoid* pits (Fig. 4-8*f*) resemble piceoid but differ from them in that the aperture is included and elliptical, rather than linear as in the piceoid type, and the border remains wide. Cupressoid pits are found in *Chamaecyparis*,

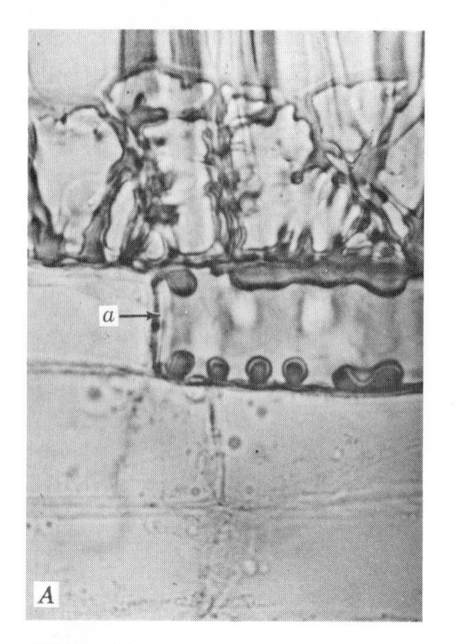

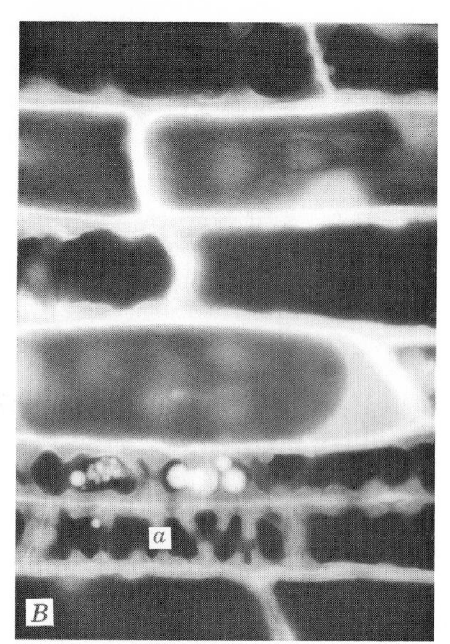

(FIG. 4-12)

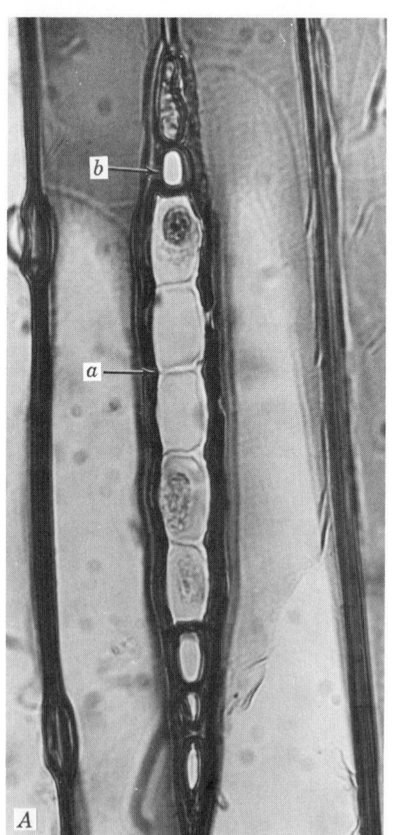

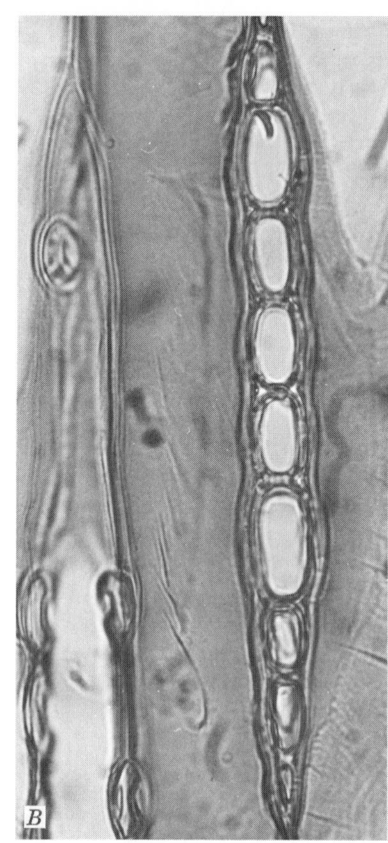

(FIG. 4-13)

Juniperus, *Libocedrus*, and *Taxus*, and are occasionally present in *Tsuga*. They are also present in a number of foreign conifers.

Pits in the transverse (end) walls of ray parenchyma, when present, are simple. As viewed in the radial section they impart a beadlike (*nodular*) appearance to the transverse walls. Their presence or absence may be of diagnostic value in the case of species which are otherwise difficult to separate. For instance, *Libocedrus*, in which the transverse (end) walls of ray parenchyma exhibit conspicuous nodules, can be readily separated from *Thuja* and *Chamaecyparis*, which do not.

Thomas and Nicholas[18] report that in the four species of yellow pine studied by them the end walls of the ray parenchyma cells were void of simple pits, presumably because of absence of the secondary-wall thickening. On the other hand Balatinecz and Kennedy[2] present evidence that in hard pines, unlike what happens in all other coniferous species, the secondary-wall thickening and apparently the lignification* of ray parenchyma cells and hence development of pits is a gradual process, related to aging of the sapwood. They have found that most ray parenchyma cells in the sapwood adjacent to the cambium are thin and apparently unlignified, and possess smooth walls. The number of lignified cells increases toward the heartwood, until in the heartwood itself all but about 10 percent of the ray parenchyma cells are completely lignified (Figs. 4-12 and 4-13). Some of the lignified cells possess thin and smooth walls, while others exhibit knoblike projections, with the gaps between them representing variable forms of simple pits. From this it may be concluded that absence or presence of simple pits in the ray parenchyma of hard pines depends on location of a given cell in the stem, in relation to sapwood and heartwood. No such lagging in maturation of ray parenchyma cells was observed by the authors in the soft pines and other coniferous species.

b. Ray Tracheids. Ray tracheids are provided with bordered pits of the same general type as those found in the longitudinal tracheids, only smaller (Figs. 12-31 to 12-36).

* The extent of lignification was determined by histological staining techniques, a method of determining the chemical composition of the cell wall that is not regarded as absolutely reliable. However, the authors point out that they employed four staining methods and fluorescence microscoping technique on the same material, and since the results were in agreement, the conclusions drawn from them may be considered valid.

FIG. 4-12 *A*. A ray in sapwood of *Pinus contorta* Dougl., (*r*), stained with phloroglucinol-HCl. Only a single parenchyma cell (*a*) has thickened and lignified.

B. Fluorescence micrograph of *Pinus banksiana* Lamb., indicating general wall thickening and lignification of ray cells typical of heartwood of the hard pines. Note interspersed ray tracheids at *a*.

(*Photographs courtesy of J. J. Balatinecz and R. V. Kennedy, Ref.* 2.)

FIG. 4-13 Tangential sections of *Pinus banksiana* Lamb., stained by the Coppick and Fowler technique.

A. Tangential section from sapwood. The almost transparent ray parenchyma cell wall (*a*), contrasted with thick, lignified walls of ray tracheid (*b*).

B. Tangential section from heartwood. Ray parenchyma walls are thickened and lignified and stain identically to ray tracheids.

(*Photographs courtesy of J. J. Balatinecz and R. V. Kennedy, Ref.* 2.)

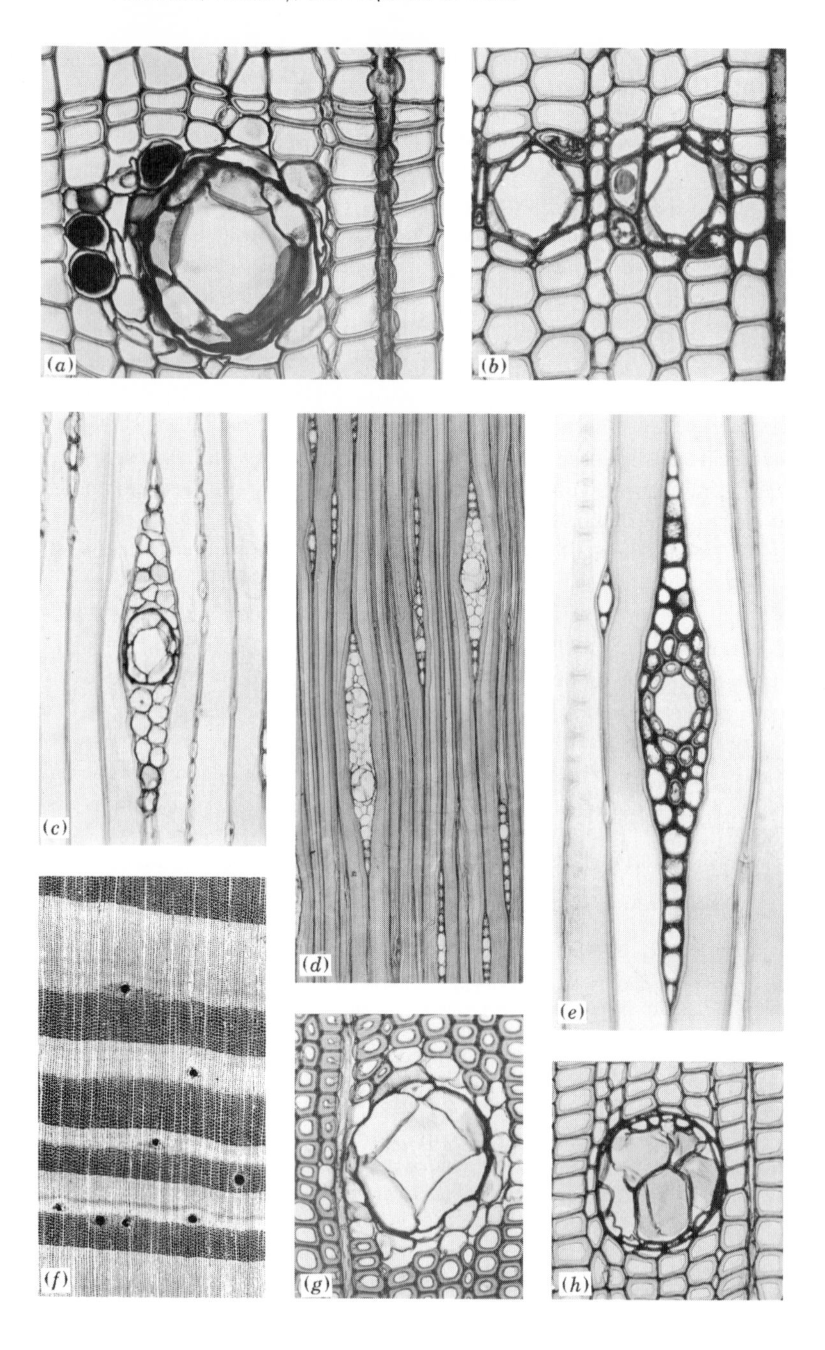

FIG. 4-14 Normal resin canals in coniferous woods.

(*a*) Longitudinal resin canal in transverse section, with thin-walled epithelial cells. Eastern white pine (*Pinus strobus* L.). (150×.) (*Photograph by W. M. Harlow.*)

Ray tracheids occur normally in the Pinaceae and are constant features of the woods of *Pinus, Picea, Larix, Pseudotsuga,* and *Tsuga*. It should be noted that the first four genera also possess resin canals and hence fusiform rays. Ray tracheids also occur sporadically in certain species of *Abies* and quite erratically in the Taxodiaceae (*Taxodium, Sequoia*) and the Cupressaceae (*Thuja, Chamaecyparis, Libocedrus, Juniperus*). They are invariably present in the wood of Alaska-cedar [*Chamaecyparis nootkatensis* (D. Don) Spach.], in which species ray tracheids and ray parenchyma seldom occur in the same ray; i.e., a ray is composed wholly of either the one or the other type of cell. (Compare Figs. 12-41 and 12-42.)

Ray tracheids attain their best development in the genus *Pinus*, especially in the hard pines. In this genus, one or more rows of ray tracheids are present on the upper and lower margins of the ray, and not infrequently in the median portion as well. The low rays of hard pines frequently consist wholly of ray tracheids.

The inner wall of ray tracheids in hard pines possesses irregular tooth-like projections, or is reticulated (Figs. 3-15, 12-31, and 12-33). Ray tracheids with these types of wall irregularities are said to be ***dentate***. They serve as an easy and positive means of identification of hard pines. A number of investigators,[6] using the electron microscope, have found that pines with dentate ray tracheids also have "warts" on the inner walls of the bordered pits of the longitudinal tracheids, and that these are absent in pines with smooth-walled ray tracheids (soft pines).

The ray tracheids of Douglas-fir possess spiral thickening comparable to that on the longitudinal tracheids.

2. FUSIFORM RAYS

Fusiform rays occur in the four genera which possess normal resin canals, viz., *Pinus, Picea, Pseudotsuga,* and *Larix*. As previously pointed out, these four genera are also among the conifers in which the rays always contain ray tracheids.

(*b*) Two longitudinal resin canals in transverse section, with thick-walled epithelial cells. Sitka spruce [*Picea sitchensis* (Bong.) Carr.]. (150×.)

(*c*) Transverse resin canal in a fusiform ray, cut at right angles, with thin-walled epithelial cells. Eastern white pine (*Pinus strobus* L.). The magnification is the same as in (*a*).

(*d*) Transverse resin canals in fusiform rays, cut at right angles; the canals are occluded with tylosoids; rays with two canals are relatively rare. Longleaf pine (*Pinus palustris* Mill.). (75×.)

(*e*) Transverse resin canal in a fusiform ray, cut at right angles, with thick-walled epithelial cells. Sitka spruce [*Picea sitchensis* (Bong.) Carr.]. The magnification is the same as in (*a*).

(*f*) Distribution of longitudinal resin canals in longleaf pine (*Pinus palustris* Mill.). (8×.)

(*g*) Longitudinal resin canal of longleaf pine (*Pinus palustris* Mill.) in transverse section; the canal is bounded on the left by a ray and elsewhere by thick-walled late-wood tracheids. (125×.)

(*h*) Longitudinal resin canal of red spruce (*Picea rubens* Sarg.) in transverse section; the orifice of the canal is occluded with tylosoids. (150×.)

Fusiform rays are composed of ray tracheids, on the upper and lower ends of a ray, ray parenchyma, and epithelial cells surrounding a resin canal. Occasionally, especially in pines, two resin canals may be found in the same ray (Fig. 4-14*d*). Ray tracheids and ray parenchyma in the fusiform rays are identical with those found in the narrow rays. The epithelial cells are similar to those found around the longitudinal resin canals. They are thin-walled in the genus *Pinus* and thick-walled in *Picea, Larix,* and *Pseudotsuga.* (Compare Fig. 4-14*c* and *e*.)

Fusiform rays are always decidedly in the minority, the ratio being in the neighborhood of 1 fusiform ray to 20 narrow rays. Insight relative to their size and abundance can be gained by reference to Figs. 12-45 to 12-58. Fusiform rays attain their maximum diameter in pines. The height in cells is variable and apparently of no diagnostic value; the upper limit may be set at about 30 cells.

3. AVERAGE RAY VOLUME OF THE CONIFEROUS WOODS OF THE UNITED STATES

In their spacing, the wood rays of coniferous woods exhibit no marked departures that are of diagnostic significance. Across the grain they range from six to nine per millimeter. Considerable difference exists, however, in ray volume, as indicated in Table 4-6.

The average ray volume for all coniferous woods discussed in this text is rela-

TABLE 4-6 Average Ray Volumes of the Coniferous Woods of the United States
Arranged in botanical sequence, by genera

Scientific name	Avg ray volume, %	Variation in samples examined, %	Scientific name	Avg ray volume, %	Variation in samples examined, %
Pinus lambertiana	5.7	0.7	*Picea sitchensis*	7.2	3.1
Pinus monticola	6.5	2.6	*Pseudotsuga menziesii*	7.3	2.1
Pinus strobus	5.3	0.4	*Tsuga canadensis*	5.9	0.7
Pinus contorta	5.7	0.9	*Tsuga heterophylla*	8.0	0.4
Pinus palustris	8.3	3.7	*Abies balsamea*	5.6	2.3
Pinus echinata	8.0	3.3	*Abies concolor*	9.4	1.8
Pinus taeda	7.6	1.6	*Abies grandis*	6.6	5.2
Pinus elliottii	11.7	1.8	*Abies procera*	6.5	2.1
Pinus rigida	7.2	2.7	*Sequoia sempervirens*	7.8	2.5
Pinus ponderosa	6.7	2.4	*Taxodium distichum*	6.6	2.6
Pinus jeffreyi	8.1		*Libocedrus decurrens*	8.9	0.9
Pinus resinosa	7.0	1.4	*Thuja occidentalis*	3.4	0.6
Larix laricina	11.0	4.2	*Thuja plicata*	6.9	1.4
Larix occidentalis	10.0	1.1	*Chamaecyparis lawsoniana*	5.5	1.7
Picea glauca	7.0	0.4			
Picea rubens	4.9	2.7	*Chamaecyparis thyoides*	5.1	2.4
Picea engelmannii	5.9	2.5	*Juniperus virginiana*	6.2	1.3

tively low (7.08 percent), compared with that of porous woods (17.0 percent). The lowest figure of 3.4 percent was recorded for *Thuja occidentalis*, the highest of 11.7 percent for slash pine (*Pinus elliottii* Engelm), closely followed by larches with volumes of 11.0 and 10.0 percent.

D. Resin Canals in Coniferous Woods

Resin canals in coniferous woods are of two kinds, those normal to the wood and those that presumably arise as a result of wounding and hence are termed *traumatic*.

1. NORMAL RESIN CANALS

Normal canals are a constant feature in *Pinus*, *Picea*, *Larix*, and *Pseudotsuga* (Figs. 14-1 to 14-14). They extend both longitudinally and transversely, the latter always included in the rays (fusiform rays).

The manner of their origin has already been described (page 114). Since the origin of normal resin canals is identical throughout coniferous woods, such departures as occur in these structures are traceable to variations in the relative number and grouping of resin canals and in the size and the nature of the epithelial cells. Some of these are of diagnostic significance.

Normal resin canals are more abundant and more evenly distributed in pines than in the other three genera. A trend is evident to greater concentration of longitudinal resin canals in the late wood, particularly in hard pines.

The distribution of normal longitudinal canals is far less regular in the spruces, larches, and Douglas-fir. Spruces and Douglas-fir are peculiar in that the normal longitudinal canals are solitary in some samples and fairly evenly distributed in the outer portion of the growth ring; in others, a strong tendency is manifested toward the grouping of the canals in tangential rows of 5 to 30 or more (see Figs. 12-11 and 12-22). In *Larix* they occur singly or in groups of several.

Normal resin canals vary in size not only according to whether they are longitudinal or transverse but also according to the genus and sometimes the species. Variation in size of canals also depends on the age of the tree and the rate of growth, insofar as it affects texture. In a given wood, the longitudinal resin canals are invariably larger than those of the transverse type [compare (*a*) to (*c*), (*b*) to (*e*) in Fig. 4-14]. For example, in eastern white pine, the longitudinal canals attain a maximum of 200 microns, and the average diameter ranges from 135 to 150 microns, while the transverse canals are usually less than 80 microns in diameter.

Normal longitudinal and transverse resin canals are largest in *Pinus*. In some species, for instance, *P. lambertiana*, the longitudinal canals (average 175 to 225) reach a maximum diameter of more than 300 microns; on the other hand, in *P. banksiana* the longitudinal resin canals seldom exceed 100 microns in diameter and usually are much smaller.

The resin canals in *Picea*, *Larix*, and *Pseudotsuga* are considerably smaller;

the longitudinal type occasionally reaches a maximum diameter of 150 microns, averaging 50 to 90 microns. Among *Picea*, Sitka spruce stands out because of the somewhat greater size of its canals.

The nature of the epithelial cells offers a means of separating pines from the other three genera. Pines are characterized by thin-walled epithelial cells (Fig. 4-14*a* and *c*). They are generally devoid of pits and apparently are unlignified. In *Picea*, *Larix*, and *Pseudotsuga* the epithelial cells are mostly thick-walled, are generally provided with pits, and appear to be lignified (Fig. 4-14*b*, *e*, and *h*).

Repeated wounding, as in tapping of longleaf and slash pines for naval stores, results in production of an abnormal number of canals in the wood that forms in the vicinity of the wound.

As normal resin canals pass from sapwood into heartwood, they cease to function and are frequently occluded with ***tylosoids*** (Fig. 4-14*h*). These result from proliferation of unlignified epithelial cells, until the canal is entirely plugged throughout or at various points along its course.

2. TRAUMATIC RESIN CANALS

Traumatic resin canals may accompany transverse canals of the normal type (ex., Sitka spruce) or may occur in a wood devoid of such canals [ex., deodar (*Cedrus deodara* Loudon), hemlocks, and true firs]. They may be longitudinal or transverse, and they result from the cleavage or separation of cells, in a manner entirely comparable with that occurring in the formation of normal canals.

Longitudinal and transverse traumatic canals seldom occur in the same sample, although deodar is an exception to this rule. The longitudinal traumatic resin canals are generally arranged in a tangential row (Fig. 4-15*a*) that may extend for an inch or two along a ring. Usually they are restricted to the early-wood portion of the growth increments.

The traumatic transverse canals, like the normal transverse kind, are confined to wood rays. They are much larger than the normal canals occurring in this position, and the wood rays containing traumatic canals are greatly enlarged as a result. When present in a coniferous wood, transverse traumatic canals frequently show as radial streaks extending across the grain in edge-grained stock.[4]

The epithelial cells of traumatic resin canals as a rule are thick-walled and pitted, and have every appearance of being lignified.

E. Crystalliferous Wood Elements in Coniferous Woods

Crystal-bearing cells are rarely encountered in coniferous woods and are of such sporadic occurrence that they are of no diagnostic significance. They are apparently restricted to the Pinaceae. When present, crystals occur in the form of exceedingly small cubes, octahedra, or rectangular prisms and are ordinarily associated with parenchyma cells. Pierce[14] has reported crystals in the longitudinal tracheids of

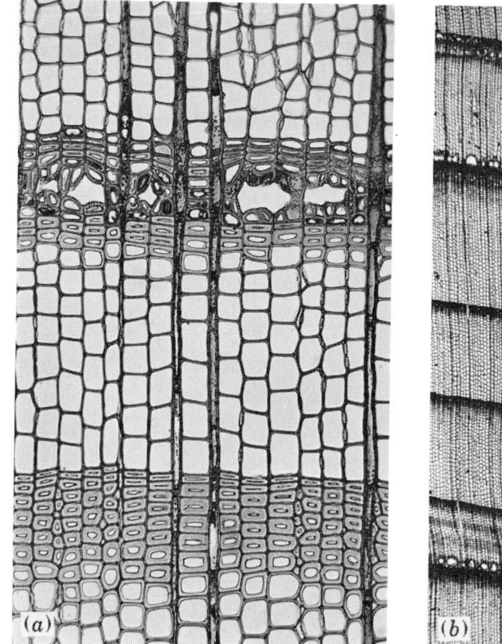

FIG. 4-15 Traumatic resin canals in conifers.
(*a*) Traumatic schizogenous resin canals (*x*) in western hemlock [*Tsuga heterophylla* (Raf.) Sarg.]. Part of a longer row extending tangentially. (95×.)
(*b*) Three rows of traumatic schizogenous resin canals in redwood [*Sequoia sempervirens* (D. Don) Endl.]. The upper row also exhibits a small resin cyst in cross section. (6×.)

the monotypic genus *Pseudolarix*. Kennedy et al.* report that crystals of rhomboidal and elongated forms, mainly in the marginal ray parenchyma cells, are regularly found in *Abies bracteata, A. procera, A. grandis, A. magnifica,* and *A. concolor;* they are absent in *A. amabilis, A. balsamea,* and *A. lasiocarpa.* Occasional crystals of calcium oxalate in the ray parenchyma cells were also reported in *Picea sitchensis.*[17]

F. Summary of Anatomical Information

Anatomical information on the wood of coniferous trees is summarized in Table 4-7. The data presented in this table can be used directly for identification of softwoods or can be transferred to the multiple-entry punch cards used as a means for keying out woods.† A type card is shown in Fig. 4-16.

* R. W. Kennedy, E. B. R. Sastry, G. M. Barton, and E. L. Ellis, Crystals in the Wood of the Genus *Abies* Indigenous to Canada and the United States, *Can. J. Botany,* **46**(10):1221–1228 (1968).

† Details for construction and use of card keys can be obtained from the following references: E. W. J. Phillips, Identification of Softwoods, *Forest Prod. Res. Bull.* 22, London, 1948; J. D. Brazier and G. L. Franklin, Identification of Hardwoods, *Forest Prod. Res. Bull.* 46, London, 1961.

TABLE 4-7 Anatomical Data of Coniferous Woods*

Species	General						Tracheids			Canals		Rays						Cross-field pits					Parenchyma		
	Heartwood distinct† (1)	Heartwood odor‡ (2)	Heartwood greasy (3)	Tangential surface dimpled (4)	Abrupt early-late wood transition (5)	Gradual early-late wood transition (6)	Radial pits 2-more (7)	Pits on tangential walls (8)	Spirals (9)	Epithelium walls thin (10)	Epithelium walls thick (11)	Fusiform epithelium walls thin (12)	Fusiform epithelium walls thick (13)	Tracheids present (14)	Tracheids dentate (15)	End walls nodular (16)	Indentures (17)	Fenestriform (18)	Pinoid (19)	Piceoid (20)	Taxodioid (21)	Cupressoid (22)	Present (23)	End walls smooth (24)	End walls nodular (25)
Abies spp		*s*				+	*s*	+						*s*		+	+				+		*s*		+
Chamaecyparis lawsoniana	+	+				+		+						*s*								+	+		+
Chamaecyparis nootkatensis	+	+				+		+						+§		+	+					+	+		+
Chamaecyparis thyoides	+	+				+		+						*s*								+	±	+	
Juniperus virginiana	+	+				+		+								+	+					+	+		+
Larix spp	+				+		±	+	*s*		+		+	+		+	+			+			*s*		+
Libocedrus decurrens	+	+				+	*s*	+								+						+	+		+
Picea sitchensis	+			±	*s*	+	*s*	+			+		+	+		+	+			+					
Picea spp					*s*	+					+		+	+		+	+			+					
Pinus lambertiana	+	*s*				+	±	+		+		+		+		+		+							

Species	(1)	(2)	(3)	(4)	(5)	(6)	(7)	(8)	(9)	(10)	(11)	(12)	(13)	(14)	(15)	(16)	(17)	(18)	(19)	(20)	(21)	(22)	(23)	(24)	(25)
Pinus monticola	+	*s*				+	±	+		+		+		+		+		+							
Pinus ponderosa	+	*s*		±	+			*s*		+		+		+	+				+						
Pinus resinosa	+	*s*			+			*s*		+		+		+	+			+							
Pinus spp., hard pines	+	*s*		*s*	+				*s*	+		+		+	+				+						
Pinus strobus	+	*s*				+	*s*	+		+		+		+		+		+							
Pseudotsuga menziesii	+	+			+	+	*s*	+	+		+		+	+		+	+			+			*s*		+
Sequoia sempervirens	+	+	+		+		+	+													+		+		*s*
Taxodium distichum	+	+	+		+		+	+								*s*					+	*s*	+		+
Taxus brevifolia	+					+		+	+													+		+	
Thuja occidentalis	+	+			+	+	*s*	+									+				+		±		+
Thuja plicata	+	+			+	+	*s*	+									+				+		±		+
Tsuga canadensis		*s*			+		*s*	+						+		+	+			+		+	*s*		+
Tsuga heterophylla		*s*				+	*s*	+						+		+	+			+		+	*s*		+
Torreya spp	*s*	+				+	*s*	+	+													+			

* In this table the *plus* signs (+) indicate that a given feature is commonly present; the *letter s* indicates that a feature may be present but only sporadically, and hence is of little diagnostic value; when both a *plus* and a *minus* sign appear in the same space, this indicates that the particular feature is abundant in some pieces and sparse in others; blank spaces indicate the absence of the feature.

† See Chap. 12 for specific information on color of the heartwood.

‡ See Chap. 12 for more information on odor.

§ In *Chamaecyparis nootkatensis* ray tracheids generally are confined to low rays, which frequently consist entirely of ray tracheids.

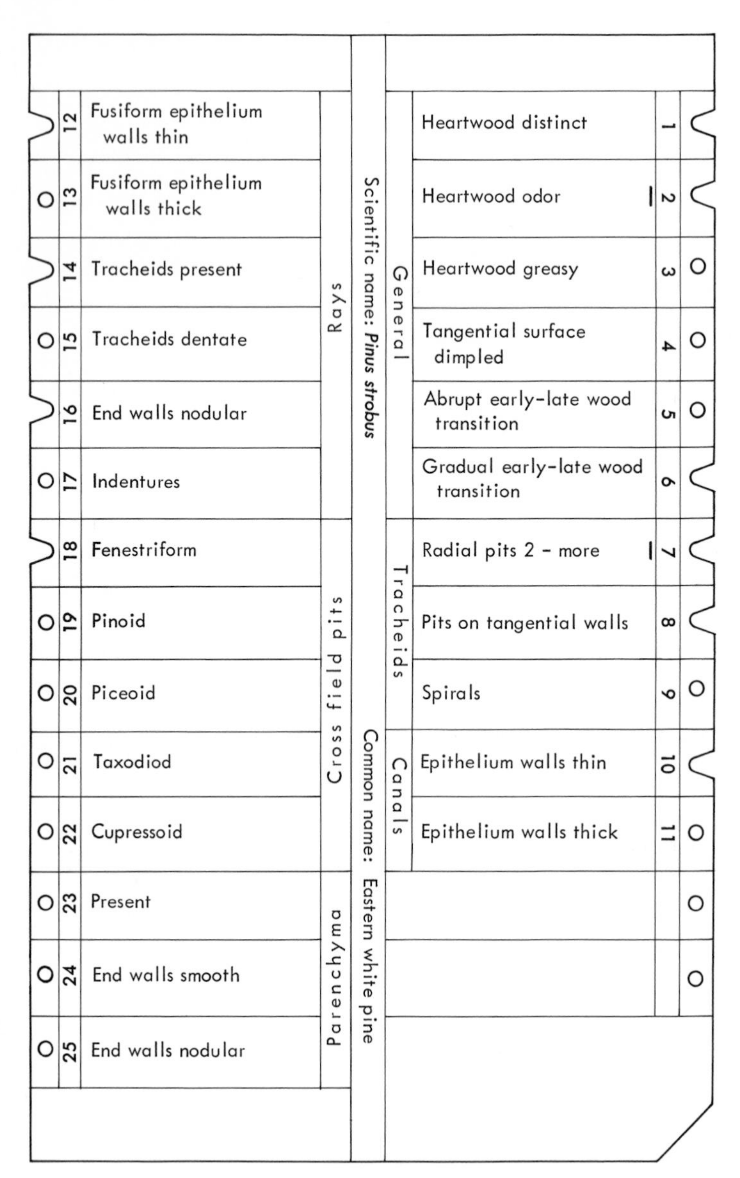

FIG. 4-16 A sample punch card. Eastern white pine (*Pinus strobus* L.). Further information can be included by adding additional titles or using the ends on the reverse side of the card. Punched perforations signify that the feature is present; a black line under the number means that this feature is present but not well developed or requires further amplification, for which see Table 4-7 or Chap. 12.

SELECTED REFERENCES

1. Bailey, I. W., and Anna F. Faull: The Cambium and Its Derivative Tissues, IX, Structural Variability in the Redwood, *Sequoia sempervirens,* and Its Significance in the Identification of Fossil Woods, *J. Arnold Arb.*, **15:**233–253 (1934).
2. Balatinecz, J. J., and R. V. Kennedy: Ray Parenchyma Cells in Pine, *Forest Prod. J.,* **17**(10): 57–64 (1967).
3. Barghoorn, E. S.: Origin and Development of the Uniseriate Ray in Coniferae, *Bull. Torrey Botan. Club,* **67:**303–328 (1940).
4. Gerry, E.: Radial Streak (Red) and Giant Resin Ducts in Spruce, *Forest Prod. Lab., Mimeo. Rept.* 1391, 1942.
5. Hudson, R. H.: The Value of the Fusiform Ray in Separating the Genera *Picea* and *Larix, J. Inst. Wood Sci.,* no. 2, November, 1958.
6. ———: The Anatomy of the Genus *Pinus* in Relation to Its Classification, *J. Inst. Wood Sci.,* no. 6, December, 1960.
7. Jane, F. W.: "The Structure of Wood," The Macmillan Company, New York, 1956.
8. Jeffrey, E. C.: "The Anatomy of Woody Plants," chap. V, The University of Chicago Press, Chicago, 1917.
9. Kukachka, B. F.: Identification of Coniferous Woods, *Tappi,* **43**(11):887–896 (1960).
10. Lewis, F. T.: The Shape of Tracheids in the Pine, *Am. J. Botany,* **22:**741–776 (1935).
11. McElhanney, T. A., et al.: "Canadian Woods, Their Properties and Uses," Forest Products Laboratories of Canada, Ottawa, 1935.
12. Penhallow, D. P.: "North American Gymnosperms," Ginn and Company, Boston, 1907.
13. Phillips, E. W. J.: The Identification of Coniferous Woods by Their Microscopic Structure, *Linn. Soc. London J.,* **52:**259–320 (1941); *Forest Prod. Res. Bull.* 22, London, 1948.
14. Pierce, A. A.: Anatomy of the Xylem of *Pseudolarix, Botan. Gaz.,* **95:**667–677 (1934).
15. ———: Types of Pitting in Conifers, *Trans. Illinois State Acad. Sci.,* **28:**101–104 (1936).
16. Reid, R. W., and J. A. Watson: Sizes, Distribution and Numbers of Vertical Resin Ducts in Lodgepole Pine, *Can. J. Botany,* **44:**519–525 (1966).
17. Sudo, S.: Anatomical Studies on the Wood of Species of *Picea,* with Some Consideration on Their Geographic Distribution and Taxonomy, *The Govt. Forestry Expt. Sta., Bull.* 215, Tokyo, Japan, 1968.
18. Thomas, R. J., and D. D. Nicholas: The Ultrastructure of the Pinoid Pits in So. Yellow Pines, *Tappi,* **51**(2):85–88 (1968).

5

The Minute Structure of Hardwoods (Porous Woods)

The anatomical features that distinguish hardwoods from coniferous woods may be summarized as follows:

1. Hardwoods differ from softwoods in possessing *vessel elements* (also called *vessel members and vessel segments*), which when viewed in the transverse section are called *pores*, hence the name *porous woods*. Softwoods are said to be *nonporous* in the sense that they do not contain vessels.

2. The radial alignment of the longitudinal cells that characterizes softwoods, when viewed in transverse section, is wanting or more or less obscured in hardwoods. This is occasioned in part by the fact that the vessels (pores) of hardwoods usually enlarge appreciably in diameter, following their formation from the cambial initials, crowding some of the longitudinal cells out of their normal alignment and forcing the rays, especially if they are narrow, to bend around the large pores; and in part by the failure of the meristematic cells on the flanks of the cambial initial that is to form a vessel element to divide, while the developing vessel element expands in the radial direction.*

* This phenomenon can be demonstrated most readily in ring-porous woods by counting the number of cells radially, i.e., from the boundary of one ring to that of the next, selecting first a place where an early-wood vessel is flanked on each side by a ray and then a place between the two rays where no large early-wood vessels have been formed. Invariably, it will

3. Hardwoods are much more complex in structure than softwoods because more cell types (elements) enter their composition.

4. The rays of hardwoods are more variable in width than those of conifers. In the latter rays are mostly uniseriate (*t*); uniseriate rays also characterize certain hardwoods, but the great majority of them possess rays two or more cells wide. Ray width, in many instances, may be used as a diagnostic feature in the separation of hardwoods; this does not hold for softwoods.

I. DEVELOPMENT OF THE ELEMENTS OF POROUS WOODS FROM CAMBIAL INITIALS

Figure 5-1 serves to emphasize the fact that the longitudinal elements of wood are all derived from the same type of cambial initial. A cambial fusiform initial is shown in the center at *a*. In a ring-porous wood, daughter cells cut off from this cell mature without or with little elongation into *early-wood vessel members* of the type depicted at *b*, or into *late-wood vessel members* of the kind shown at *c* and *d*, or into other types of cells described in the succeeding paragraphs.

A *vascular tracheid* is illustrated at *g* and a *vasicentric tracheid* at *h*. In contrast to the vessel elements, both types of these cells have imperforate ends. The vasicentric tracheid is the longer of the two, but both grew somewhat in length as they matured, for they are obviously longer than cambial cell *a*.

A fibrous cell is portrayed at *i*. Its exact nature cannot be determined because the type of pitting is not indicated. The long fibrous elements in hardwoods arise from the same type of fusiform cambial initial that produces the vessel elements and tracheids. As the fibrous cells mature they increase in length several hundred percent, meanwhile becoming narrower and hence more fiber-like in nature.

The longitudinal elements *b*, *c*, *d*, *g*, *h*, *i* in Fig. 5-1 are all prosenchymatous; the remaining types, *e* and *f*, are parenchymatous. The cells of type *e* are designated as *strand parenchyma*. Such a strand results through the partition of a longitudinal fusiform cell (*a*) into separate cells by the formation of cross walls. The number of cells in a strand that may arise this way depends on the number of cross walls that are laid down. Evidence of the manner of the formation of such strands is permanently recorded in the wood in that the cells at each end of it taper to a point. The strand parenchyma *e* is somewhat longer than the fusiform initial *a*. This is indicative of the fact that a certain amount of elongation occurs in the parent longitudinal cell before the cross walls are formed. The extent of this elongation, and the number of units composing such a strand, vary in different hardwoods.

Cell type *f* is comparatively rare in commercial hardwoods but abounds in the woods of certain shrubs. It is known as a *fusiform parenchyma cell*. Such an element is parenchymatous in nature in that it retains a living protoplast for some

---

be found that fewer cells have been formed throughout the growing season by the cambial initials in the former than in the latter instance, thus indicating that the initials divided fewer times opposite the vessel.

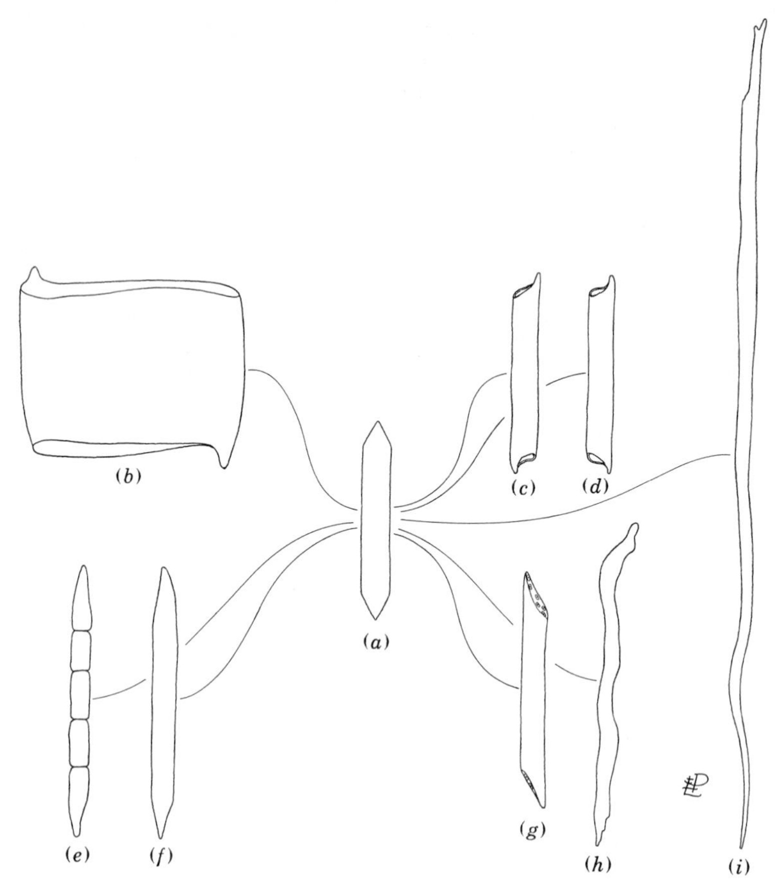

FIG. 5-1 Schematic drawings indicating the manner in which the longitudinal elements of porous wood are derived from fusiform cambial initials. The cells are illustrated in lateral aspect.

(*a*) Fusiform cambial initial.

(*b*) Early-wood vessel element.

(*c*) and (*d*) Late-wood vessel elements with caudate ends, on the opposite and on the same side respectively.

(*e*) A strand of longitudinal parenchyma.

(*f*) A fusiform parenchyma cell.

(*g*) A vascular tracheid (in shape, it is similar to a late-wood vessel element but is imperforate).

(*h*) A vasicentric tracheid.

(*i*) A fibrous cell.

time, but it has the shape of a short fiber. It obviously arises from a longitudinal fusiform initial (*a*), without partition, but usually after some elongation.

By way of summation it should be noted that differentiation of the various types of cells derived from the fusiform initials in the cambium proceeds along three principal lines, with considerable overlapping between them.

1. Elongation, i.e., the mature cell becomes considerably longer than the fusiform initial. This type of differentiation is most characteristic of the fibers (libriform fibers and fiber tracheids). Some elongation, however, occurs also in the tracheids (vasicentric and vascular), in the total length of the axial parenchyma strand, and in the vessel elements of more primitive hardwoods.
2. Increase in diameter, associated with little or no elongation. This is characteristic of the vessel elements (members) and to some extent of the vascular tracheids. The mature early-wood vessel elements in some ring-porous woods, in fact, may become shorter than the cambial fusiform initials.
3. Division into a vertical row of cells by means of transverse walls. This is the manner by which strands of the axial parenchyma are formed from a single fusiform initial.

All the xylary ray cells originate from the ray initials in the cambium. These are not shown in Fig. 5-1, because except for elongation in the horizontal-radial direction, mature xylary ray cells are largely replicas of the cambial ray initials, which through division in the tangential plane give rise to them.

II. THE WOOD OF SWEETGUM (REDGUM)
(*Liquidambar styraciflua* L.)

Following the procedure adopted in the preceding chapter, a detailed description of the wood anatomy of sweetgum, serving as an example of the wood anatomy of hardwoods, will be presented first. This will be followed by a discussion of the anatomical features characteristic of hardwoods as a group.

Wood elements found in sweetgum are listed in Table 5-1, and the dimensional and numerical data on them are compiled in Table 5-2. It should be noted that the epithelial cells of longitudinal resin canals are not a constant feature of this wood, except in the outer portion of the pith; however, they occasionally occur in the xylem (Fig. 5-15), presumably as a result of injury.

A. Longitudinal Cells

1. VESSEL ELEMENTS

Porous woods are characterized by the presence of vessels, composite tubelike structures of indeterminate length. Vessels arise through the fusion of the cells in a longitudinal row by partial or complete disappearance of the cross walls. The component cells of a vessel are called ***vessel elements, vessel members***, or ***vessel segments***. Each vessel element develops from a longitudinal cell that has arisen through division of a fusiform cambial initial. In the early stages of development, a longitudinal cell destined to become a vessel member is imperforate. It enlarges rapidly transversely, but there is no or little elongation along the grain, except in some diffuse-porous hardwoods, of which sweetgum is one. After the

TABLE 5-1 Elements of Sweetgum

Longitudinal	*Transverse*
1. Vessel elements 2. Cells of longitudinal parenchyma 3. Fiber tracheids 4. Epithelial cells of longitudinal traumatic gum canals*	1. Cells of ray parenchyma *a.* Procumbent cells, homocellular rays *b.* Upright cells, homocellular rays (*a* and *b*: heterocellular rays)

* Such longitudinal gum canals are always present in the outer portion of the pith in sweetgum; they may occur in wood (Fig. 4-15*b*), apparently because of injury. The exudation from these gum canals constitutes the American storax of the trade.

vessel element has attained its maximum size, openings called ***perforations*** are formed at or near each end of the member, and the secondary wall is laid down. Similar perforations are formed in the adjoining vessel elements in such a manner that the opening in one member corresponds with that in the other. In this way communication, endwise, is ensured between contiguous vessel elements. In

TABLE 5-2 Dimensional and Numerical Data on Sweetgum Wood*

1. Longitudinal elements

Vessel elements: Length along the grain, 1.32 ± 0.30 mm; diameter in the radial plane, 68 ± 15 μ; diameter in the tangential plane, 58 ± 7.7 μ; thickness of the vessel wall, approx 3 μ; number of vessels (pores) per square millimeter on the transverse surface, 125–145

Fiber tracheids: Length along the grain, 1.82 ± 0.16 mm; diameter in the radial plane, 22 ± 5.7 μ; diameter in the tangential plane, 34 ± 6.2 μ; thickness of the wall, 7 ± 1.2 μ

Cells of longitudinal parenchyma: Length (height) along the grain, 156 ± 38 μ; diameter in the radial plane, 22 ± 6.7 μ; diameter in the tangential plane, 26 ± 7.6 μ; thickness of the wall, approx 2.5 μ

2. Transverse elements

Ray parenchyma

Upright cells: Length (height) along the grain, 54 ± 11 μ; diameter in the radial plane, 54 ± 9.4 μ; diameter in the tangential plane, 14 ± 2.9 μ; thickness of the wall, approx 3 μ

Procumbent cells: Length (height) along the grain, 25 ± 4.5 μ; diameter in the radial plane, 120 ± 42 μ; diameter in the tangential plane, 12 ± 3.7 μ, thickness of the wall, approx 3 μ

3. Height of wood rays in cells and microns

1-seriate rays		*2–3-seriate rays*	
In cells	*In microns*	*In cells*	*In microns*
2–19	30–585	8–49	150–1300

* For interpretation of the figures in Table 5-2, see footnote at bottom of Table 4-5.

TABLE 5-3 Volumetric Composition of Sweet gum Wood*

	Percent
Vessels	54.9
Fiber tracheids	26.3
Longitudinal parenchyma	00.5
Wood rays	18.3
Total	100.0

* Data taken, with one slight change, from G. E. French, The Effect of the Internal Organization of the North American Hardwoods upon Their More Important Mechanical Properties, thesis submitted in partial fulfillment for the degree of Master of Forestry, The New York State College of Forestry, Syracuse, N.Y., 1923.

sweetgum, vessel elements are equipped with ladder-like, *scalariform* perforation plates,* with numerous (15 plus) bars, and spiral thickenings restricted to the tapering ends of the vessel elements (Fig. 5-3*B*, *f*).

Sweetgum is a diffuse-porous wood, i.e., the pores (vessels as seen in the transverse section) in this wood exhibit little or no variation in size or shape throughout the growth rings, indicative of seasonal growth. They are relatively thin-walled and more or less angled, to conform to the contours of the surrounding thick-walled fibers (Fig. 5-3*A*). Angling of the vessels to the extent found in sweetgum is the exception rather than the rule and hence constitutes an anatomical feature of diagnostic importance.

Vessel elements in sweetgum are 1.32 ± 0.30 millimeters in length; this is longer than the average for the hardwoods considered as a group.

Where one vessel is in contact with another, the pit pairs are bordered; intervessel pits range from oval to linear and are 6 to 30 microns in diameter; when they are oval, their arrangement is opposite. Pits leading to fiber tracheids are bordered and round, conspicuous but smaller than the intervessel kinds (Fig. 5-3*B* and *C*).

2. FIBER TRACHEIDS

Among hardwoods the term *fiber tracheid* is applied to longitudinally elongated cells, with pointed, closed ends and bordered pits, having lenticular or slitlike apertures. Fiber tracheids commonly are thick-walled, with small lumina.

In sweetgum the fiber tracheids, as seen in the transverse surface, are strongly angled (Fig. 5-3*A*). Near the outer surface they are rectangular and somewhat flattened in the tangential plane.

* *Perforation plate* is a term of convenience for the area of the wall (originally imperforate) involved in the coalescence of two members of a vessel.

FIG. 5-2 A cube of sweetgum (*Liquidambar styraciflua* L.). (75×.) (*Photograph by Arnold Day.*)

3. LONGITUDINAL PARENCHYMA

The cells of longitudinal parenchyma in sweetgum have the angular shape and dimensions of fiber tracheids but are much thinner-walled. Four such cells are visible on surface A, Fig. 5-3 (c, c^1, c^2, c^3).

The longitudinal parenchyma cells in sweetgum are relatively sparse and ***diffuse***, i.e., distributed irregularly among the fiber tracheids. As viewed in the longitudinal sections, the longitudinal parenchyma is found to be arranged in strands consisting of a number of cells. The end cells of a strand, which are not shown in Fig. 5-3, are pointed, whereas the middle cells are prismatic, with the long axis in the longitudinal direction.

The longitudinal parenchyma of sweetgum is provided with bordered and simple pits. The pits are bordered when the contiguous cells are vessel elements and fiber tracheids (Fig. 5-3A, c). Where longitudinal parenchyma abuts on ray parenchyma the pits are simple; they are relatively large where the longitudinal (axial)

FIG. 5-3 Schematic three-plane drawing of the wood of sweetgum (*Liquidambar styraciflua* L.). (330×.)

Surface A. 1-1^a, boundary between two annual rings (growth proceeded from right to left); 2-2^a, wood ray consisting of procumbent cells; 2^b-2^c, wood ray consisting of upright cells; a-a^6, inclusive, pores (vessels in transverse section); b-b^4, inclusive fiber tracheids; c-c^3, inclusive, cells of longitudinal parenchyma; e, procumbent ray cell.

Surface B f, f^1, portions of vessel elements; g, g^1, portions of fiber tracheids in lateral surface aspect; h, a strand of longitudinal parenchyma in lateral surface aspect; 3-3^a, upper portion of a

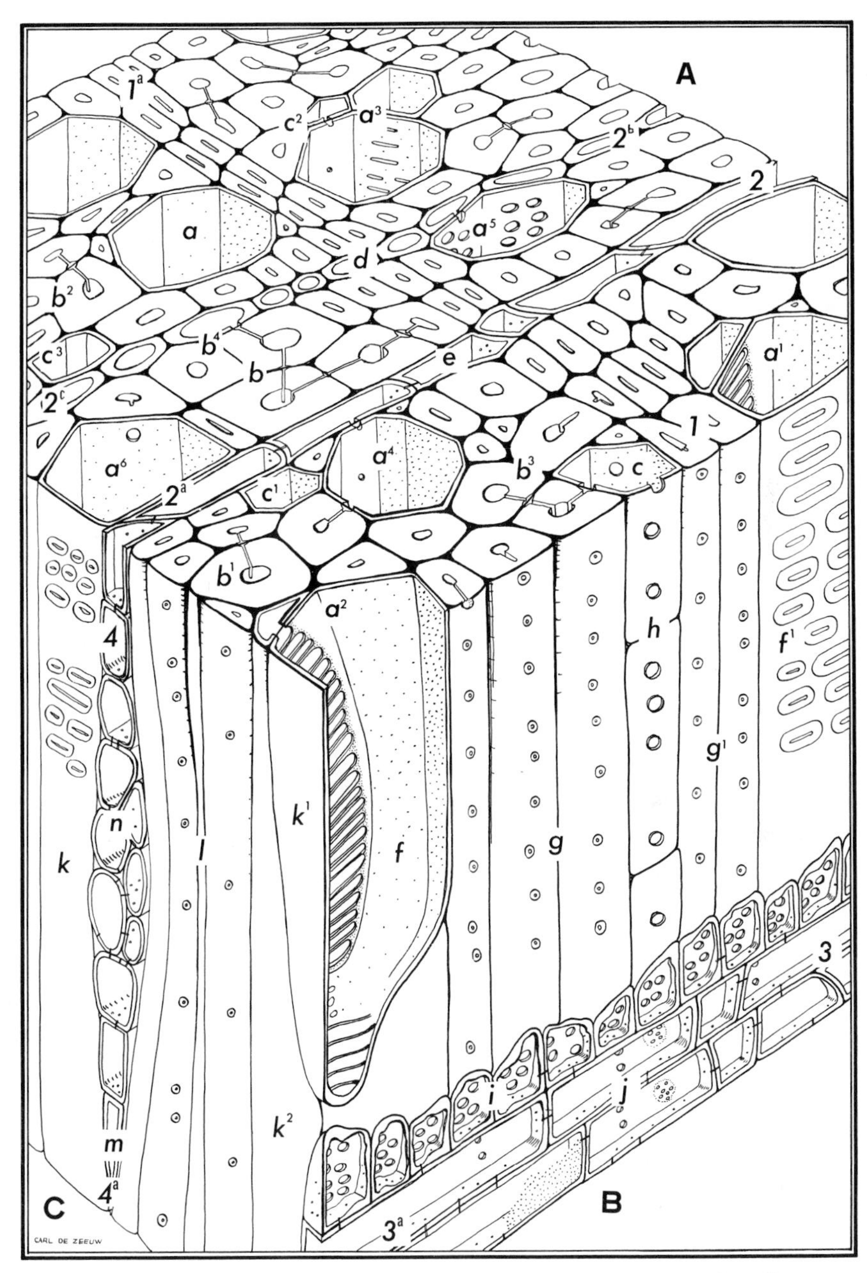

heterocellular wood ray in lateral sectional aspect; *i*, a marginal row of upright cells; *j*, two rows of procumbent cells.

Surface *C*. *k*, portion of a vessel element in tangential surface aspect; k^1, k^2, overlapping vessel elements in tangential surface aspect; *l*, fiber tracheids in tangential surface aspect; 4-4^a, portion of a wood ray in tangential sectional view; *m*, an upright cell in the lower margin; *n*, procumbent cells in the body of the ray.

parenchyma cells are in contact with upright cells, and small and clustered where the axial cells are contiguous with procumbent ray cells (Fig. 5-3*B*).

B. Transverse Cells

As in conifers, all the transverse cells in hardwoods are included in the wood rays. Hardwoods possess no ray tracheids, but parenchyma cells may be of two kinds, *upright* and *procumbent*. The upright ray cell has its longest dimension in the longitudinal direction, while the longest axis of a procumbent ray cell is radial. When both kinds of ray parenchyma cells are present in the same ray, it may be described as *heterocellular*; if only one type is found, a ray is said to be *homocellular*.

In sweetgum, rays are either homocellular or heterocellular (Fig. 5-2). A more detailed description of ray cells in sweetgum follows.

1. UPRIGHT CELLS (Fig. 5-3*B, i*)

The upright cells in sweetgum can be readily distinguished from the procumbent because they differ in shape and pitting. As seen on the radial surface they are higher along the grain and shorter across the grain radially; wedge-shaped, if in the marginal rows of the ray, and prismatic if such cells are included in the body of the ray.

The most striking feature of the upright cells of sweetgum is the pits. In the back wall they are either simple or bordered, depending on the type of contiguous cells; pits are bordered when adjacent cells are fiber tracheids or vessel elements, and simple if the cells are axial parenchyma.

Pits between the upright cells in the same row, and between the cells of adjacent rows of ray cells are simple, but smaller than those in the back walls.

2. PROCUMBENT CELLS (Fig. 5-3*B, j*)

The procumbent cells are longer horizontally, i.e., in the radial direction, than they are high vertically. Pits leading to the adjacent cells are either simple or bordered, the same as in upright cells. They are small and simple, and frequently in clusters, when they provide communication with the other ray cells and the longitudinal parenchyma. Pits leading to the contiguous fiber tracheids and vessel elements are bordered, but somewhat smaller than those in the upright cells.

C. Description of a Block of Sweetgum (Redgum) Wood

The structural details of sweetgum wood are shown schematically in Fig. 5-3. A brief description of each of the three surfaces follows.

1. DESCRIPTION OF SURFACE A—TRANSVERSE SURFACE

Portions of two growth rings are shown; the late-wood portion of one is at the top and the early wood of the succeeding ring is toward the bottom. The line of demarcation ($1\text{-}1^a$) separating the two rings is caused by the fact that the fibrous cells of sweetgum become somewhat flattened tangentially as the outer margin of the ring is approached. These are followed by cells with greater radial diameters in the early wood of the succeeding ring, the discrepancy in size being sufficient to delimit the boundaries of the two rings.

Three kinds of longitudinal elements, viz., vessels (a to a^6), fiber tracheids (b to b^4), and longitudinal parenchyma (c to c^3), are visible. Two kinds of transverse elements, procumbent cells (e) composing the ray ($2\text{-}2^a$) and upright ray cells (d) composing the ray ($2^b\text{-}2^c$), are also shown on the transverse surface. The cells of longitudinal parenchyma are thinner-walled than the fibers and are exceedingly sparse in their distribution. Two cells (c^1 and c^2) are shown in contact with vessels (a^3 and a^4). When longitudinal parenchyma is associated with the vessels, it is termed *paratracheal*. The longitudinal parenchyma cell c^3 is bounded on three sides by fiber tracheids and belongs to the *apotracheal-diffuse parenchyma* type, which is a term applied to single parenchyma cells distributed irregularly among fibers.

The pores in sweetgum either occur singly (a^4) or are occasionally in multiples of two to several (a^3). It should be noted that the condition existing at a^3 is different from that found in a^1 and a^2. At a^3 two vessels are parallel to each other along the grain. At a^1 and a^2, two pores, as they appear in the transverse section, belong in reality to two vessel elements of the same vessel, through overlapping of their ends as shown at a^2.

The perforation plates which are depicted at a^1 and a^2 are of scalariform type because of their ladder-like appearance. The nature of the pits showing on the back walls of the vessels on surface A is contingent on the type of cell that happens to be in contact with the vessel at that point.

2. DESCRIPTION OF SURFACE B—RADIAL SURFACE

Portions of five elements are visible on surface B. Three of these are longitudinal kinds: (1) vessel elements at f and f^1, (2) five fiber tracheids ($g\text{-}g^1$), and (3) a strand of axial parenchyma (h), with three cells showing extending along the grain. The two other kinds of cells are ray cells. The upper row consists of the upright cells and the two lower rows of the procumbent ray cells.

A portion of vessel element f projects downward and ends in a blunt tip. This vessel segment is split longitudinally in such a way as to show a scalariform plate between this segment and adjacent segment k^2. It should be noted that spiral thickening is present on the inner wall at the tip. Thickenings of this sort are not unusual in vessel walls of a number of species, e.g., cherry, basswood, and maple, where they occur through the length of the element. The restriction of spiral thickenings to the tapering tips of the vessel elements in sweetgum may be considered as unusual and hence of considerable diagnostic value.

The vessel element at f^1 illustrates the type of intervessel bordered pits found in sweetgum. In this case they are rather large and variable in shape. These shapes are sufficiently characteristic to be of assistance in identification of sweetgum with a microscope.

3. DESCRIPTION OF SURFACE C—TANGENTIAL SURFACE

A portion of surface *C* included in the drawing contains a portion of vessel element *k*, the overlapping ends of two vessel segments (k^1, k^2), lengths of four fiber tracheids (*l*), a part of a ray (4-4^a), cut transversely, and two cells of a ray shown on surface B.

The structural characteristics of the vessel and fiber tracheid walls are identical with those of the radial surface, except that intervessel pitting is usually more abundant and conspicuous on the tangential walls.

Ray 4-4^a is biseriate through the central portion and possesses uniseriate extensions above and below. Its lower end terminates in an upright cell (*m*) of the kind shown in the upper row of the ray 3-3^a on surface *B*. The remaining cells are of the procumbent type, similar to those of the two bottom rows of the ray 3-3^a.

TABLE 5-4 Elements of Porous Wood

Longitudinal	*Transverse*
A. Prosenchymatous	*A*. Prosenchymatous
1. Vessel elements	1. None†
2. Tracheids	
a. Vasicentric tracheids	
b. Vascular tracheids	
3. Fibers	
a. Fiber tracheids	
b. Libriform fibers	
B. Parenchymatous	*B*. Parenchymatous
1. Cells of axial (strand) parenchyma	1. Cells of ray parenchyma
2. Fusiform parenchyma cells	*a*. Procumbent cells, homocellular rays } Heterocellular rays
3. Epithelial cells; excreting cells encircling the cavities of longitudinal gum canals*	*b*. Upright cells, homocellular rays } Heterocellular rays
	2. Epithelial cells; excreting cells encircling the cavities of transverse gum canals*

* Longitudinal gum canals, apparently of traumatic origin, are occasionally present in sweetgum. They are not found elsewhere in the woods covered by this text. Longitudinal and transverse canals are normal features of certain tropical timbers, but the two types seldom occur in the same wood.

† None in domestic hardwoods.

III. COMPARATIVE ANATOMY OF THE HARDWOODS

The elements that occur in the native species of porous woods are listed in Table 5-4. When compared with Table 4-3, the more complex structure of hardwoods is quite apparent.

A. The Longitudinal Prosenchymatous Elements of Porous Woods

Unlike some coniferous woods, such as pines and spruces, in which both longitudinal and transverse prosenchymatous cells occur, all cells of this type in hardwoods are arranged vertically.

The longitudinal prosenchymatous elements of porous woods, as indicated in Table 5-4, consist of vessel elements, various types of tracheids, and fibers. Such cells are primarily concerned with conduction or the mechanical functions, or both. They are elongated in the direction of the grain of the wood. They differ from the axial (longitudinal) parenchymatous cells, with which they are associated, in that their protoplast disappears as soon as or shortly after these cells attain full maturity.

1. VESSEL ELEMENTS

Vessel elements are units of an articulate tube-like structure of indeterminate length, known as a vessel. Vessel elements vary greatly in shape and size (Fig. 5-4), in their distribution within a growth ring, in their sculpturing, and in the nature of their inclusions.

The length of vessels and the course they follow through the stem have not been studied in detail, and therefore this subject represents a gap in the knowledge of wood structure. Zimmermann and Tomlinson[29] studied the axial course followed by vessels in *Acer rubrum* L., *Populus* spp., *Quercus* spp., and *Fraxinus* spp. by cutting sequential sections from a block of wood and photographing them through a microscope with a modified motion picture camera. When the resulting film is projected, the changes in the relative position of vessels can be observed. The film shows that in maple and poplar, individual vessels move from cluster to cluster of vessels and that they never end in isolation but begin and end within a vessel cluster. Because of this and the fact that in many cases two or more vessels run side by side over considerable distances, the water-conducting units do not have dead ends and water can pass through pits and perforations from one vessel system into several others. Oak vessels were found to be very long and rather straight, while in ash, "Vessels do not run parallel but describe what appears to be random deviation from their path." This tangential deviation was calculated to amount to more than 30 centimeters over an axial distance of 5 meters. Even more extensive spread was found in maple and poplar. This was demonstrated by the upward tangential spread of dye introduced at some point in the stem. The authors suggest that the physiological implication of this is that "water must spread tangentially

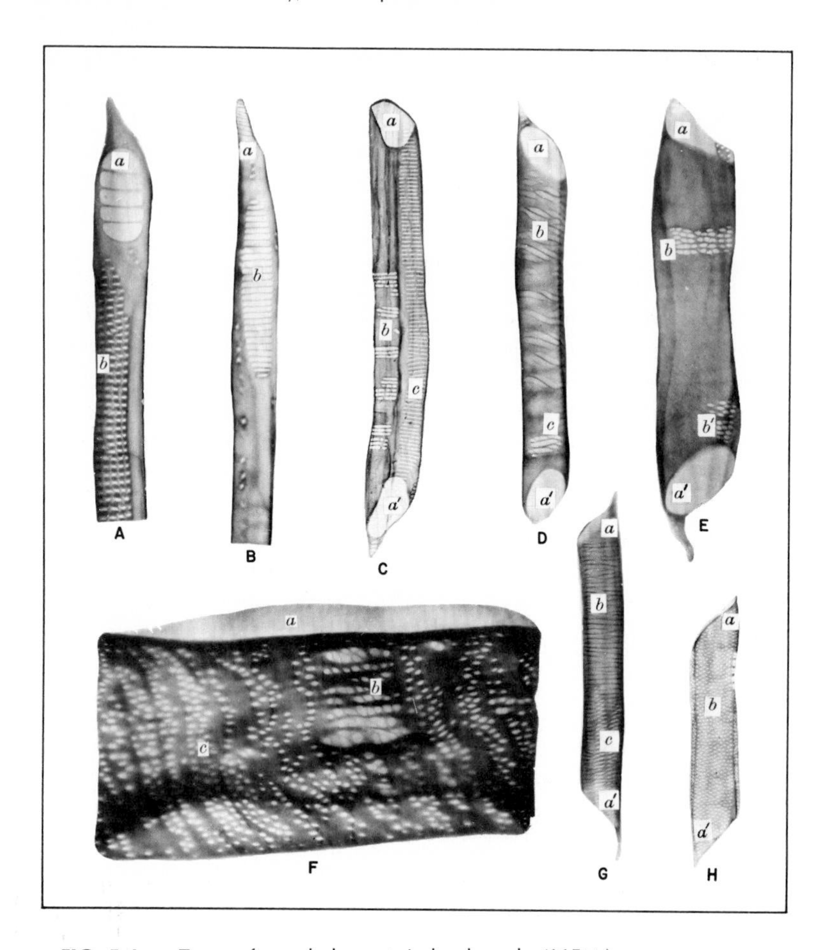

FIG. 5-4 Types of vessel elements in hardwoods. (115×.)

A. Portion of a vessel element of yellow-poplar (***Liriodendron tulipifera*** L.) showing a scalariform perforation plate with few bars (*a*), and opposite pitting (*b*).

B. Portion of a vessel element of sweetgum (***Liquidambar styraciflua*** L.) showing spiral thickening at tip (*a*), and scalariform perforation plate with many bars (*b*).

C. A vessel element of cucumbertree (***Magnolia acuminata*** L.) showing simple perforations at the ends (*a*, *a'*), several cross fields on the radial face (*b*), and scalariform pitting composed of linear pits on the tangential wall (*c*).

D. A vessel element of yellow buckeye (***Aesculus octandra*** Marsh.) showing simple perforations at the ends (*a*, *a'*), spiral thickening (*b*), and a cross field (*c*).

E. A vessel element of black willow (***Salix nigra*** Marsh.) showing simple perforations at the ends (*a-a'*), and two cross fields (*b-b'*).

F. Annular (ring-shaped) vessel element from the early wood of chestnut [***Castanea dentata*** (Marsh.) Borkh.] showing simple perforations at the upper end (*a*), a cross field (*b*), and strips of pits leading to vasicentric tracheids (*c*).

G. A vessel element of silver maple (***Acer saccharinum*** L.) showing caudate ends (*a-a'*) on the same side, spiral thickening (*b*), and pits leading to longitudinal parenchyma (*c*).

during its ascent in the stem"[28] and that "any one point at the base of a tree stem is in direct axial contact with a good part (or all) of the circumference of the top of the tree."[29]

Similar findings were reported from a study of *Eucalyptus maculata*, in which, in one sample, 56 out of 149 vessels were found to be in direct contact with one another for an average of 28 percent of their length and many other vessels were indirectly connected through other wood elements.*

a. The Shape, Size, and Distribution of Vessel Elements. A vessel element when presented in transverse aspect, i.e., when it is cut at a right angle to the grain, appears as a *pore*. The term *pore* includes not only the actual orifice but the encircling wall as well.

Hardwoods can be divided into two groups on the basis of pore size. If the pores formed in the spring are much larger than those formed later in the season, the wood is *ring-porous* (ex., chestnut, oak, elm, black locust, ash); on the other hand, if the pores are fairly uniform in size and quite evenly distributed throughout the ring, the wood is said to be *diffuse-porous* (ex., birch, maple, basswood, gums, yellow-poplar). Certain woods, such as the *Juglans* spp. and persimmon, are intermediate in this respect; they are classed as *semi-ring-porous* or *semi-diffuse-porous*.

Woods of the same genus, and even the same species, may be ring-porous in one region of the North Temperate Zone but exclusively diffuse-porous farther south. The first few rings of ring-porous wood, next to the pith, retain a certain degree of diffuse porousness because the early-wood pores do not attain maximum size in successive rings for a number of years.

Pores may be solitary or grouped in various ways. A *solitary* pore is generally rounded; the oval shape predominates, with the long axis directed radially. In woods with thick-walled fibrous elements and thin-walled vessels, as in sweetgum, the pores are angular. Not infrequently several pores are contiguous in a radial direction. Such a pore cluster appears as one pore consisting of several divisions. This type of pore arrangement is known as a *pore multiple*; it is common in maple, birch, poplar, and many other woods. Radial series of solitary pores are called *pore chains*; this kind of pore grouping is exemplified by holly. Nested late-wood pores (*pore clusters*) are an important diagnostic feature of coffeetree. Tangential grouping of late-wood pores into concentric *wavy bands* is characteristic of elms, hackberry, and a few other woods.

As viewed laterally the vessel elements range from drum-shaped (Fig. 5-4*F*) and barrel-shaped to oblong and linear (*C*, *D*, *E*, *G*, and *H*), with or without tapering or ligulate extensions (tails) at one or both ends.

* Anon., Penetration of Liquids into Wood, *Forest Prod. Newsletter* 356, Division of Forest Products, CSIRO, So. Melbourne, Australia, January-February, 1969.

H. A vessel element of butternut (*Juglans cinerea* L.) showing caudate ends (*a-a'*) on opposite sides, and intervessel pitting on the tangential wall (*b*).

(*Photographs A, B, D, E, F, G, and H, inclusive, by C. H. Carpenter; photograph C by W. M. Harlow.*)

TABLE 5-5 Average Length of the Vessel Elements and Fibers (Fiber Tracheids and Libriform Fibers) of Some of the Hardwoods of the United States, Determined from Specific Samples

All figures are expressed in millimeters. For significance of "standard deviation," see footnote to Table 4-5. The data are presented in botanical sequence by genera.

Scientific name	Vessel elements: Avg length	Vessel elements: Standard deviation	Fibers: Avg length	Fibers: Standard deviation	Scientific name	Vessel elements: Avg length	Vessel elements: Standard deviation	Fibers: Avg length	Fibers: Standard deviation
Juglans cinerea	0.36	0.14	1.13	0.17	*Celtis occidentalis*	0.26	0.03	1.13	0.17
Juglans nigra	0.51	0.08	1.21	0.14	*Morus rubra*	0.21	0.03	0.91	0.12
Carya tomentosa	0.44	0.06	1.62	0.26	*Maclura pomifera*	0.18	0.02	1.14	0.16
Carya cordiformis	0.44	0.11	1.38	0.22	*Magnolia acuminata*	0.72	0.15	1.39	0.28
Carya ovata	0.47	0.09	1.34	0.28	*Magnolia grandiflora*	0.99	0.14	1.81	0.29
Carya illinoënsis	0.41	0.08	1.28	0.20	*Liriodendron tulipifera*	0.89	0.13	1.74	0.29
Populus grandidentata	0.64	0.09	1.33	0.17	*Umbellularia californica*	0.37	0.06	0.94	0.14
Populus tremuloides	0.67	0.18	1.32	0.22	*Sassafras albidum*	0.39	0.06	1.02	0.14
Populus trichocarpa	0.58	0.09	1.38	0.19	*Liquidambar styraciflua*	1.32	0.30	1.82	0.16
Salix nigra	0.42	0.09	0.85	0.17	*Platanus occidentalis*	0.63	0.12	1.08	0.17
Carpinus caroliniana	0.42	0.10	1.17	0.18	*Prunus serotina*	0.39	0.06	1.21	0.18
Ostrya virginiana	0.68	0.11	1.23	0.18	*Gymnocladus dioicus*	0.27	0.06	1.12	0.15
Betula lenta	0.91	0.12	1.52	0.22	*Gleditsia triacanthos*	0.19	0.03	1.24	0.11
Betula alleghaniensis	0.84	0.16	1.38	0.17	*Robinia pseudoacacia*	0.18	0.02	1.13	0.16
Betula papyrifera	1.00	0.26	1.35	0.15	*Cladrastis lutea*	0.23	0.03	0.61	0.14
Betula populifolia	0.74	0.15	1.26	0.14	*Ilex opaca*	0.88	0.24	1.74	0.27
Alnus rubra	0.85	0.14	1.19	0.18	*Acer macrophyllum*	0.33	0.05	0.77	0.15
Fagus grandifolia	0.61	0.11	1.28	0.21	*Acer rubrum*	0.42	0.05	0.92	0.12
Castanea dentata	0.58	0.12	1.22	0.16	*Acer saccharinum*	0.41	0.07	0.76	0.13
Castanopsis chrysophylla	0.67	0.11	0.87	0.11	*Acer saccharum*	0.41	0.09	0.92	0.13
Quercus rubra	0.42	0.09	1.32	0.29	*Aesculus octandra*	0.44	0.08	1.00	0.13
Quercus coccinea	0.43	0.09	1.61	0.26	*Tilia americana*	0.43	0.09	1.21	0.17
Quercus palustris	0.46	0.09	1.30	0.20	*Tilia heterophylla*	0.48	0.04	1.34	0.18
Quercus phellos	0.45	0.08	1.38	0.25	*Nyssa aquatica*	1.11	0.28	1.89	0.33
Quercus shumardii	0.45	0.08	1.44	0.20	*Nyssa sylvatica*	1.33	0.34	2.30	0.36
Quercus velutina	0.43	0.08	1.44	0.29	*Cornus florida*	1.04	0.17	1.74	0.29
Quercus alba	0.40	0.09	1.39	0.20	*Cornus nuttallii*	1.13	0.20	1.64	0.27
Quercus bicolor	0.41	0.07	1.19	0.17	*Oxydendrum arboreum*	0.47	0.09	1.05	0.16
Quercus lyrata	0.42	0.08	1.35	0.20	*Arbutus menziesii*	0.53	0.10	0.79	0.14
Quercus macrocarpa	0.35	0.07	1.20	0.19	*Diospyros virginiana*	0.36	0.04	1.39	0.20
Quercus prinus	0.40	0.09	1.45	0.22	*Fraxinus americana*	0.29	0.03	1.26	0.17
Quercus stellata	0.43	0.09	1.35	0.21	*Fraxinus nigra*	0.27	0.04	1.27	0.17
Lithocarpus densiflorus	0.57	0.11	1.10	0.15	*Fraxinus latifolia*	0.23	0.05	1.20	0.16
Ulmus americana	0.22	0.04	1.55	0.20	*Fraxinus pennsylvanica*	0.26	0.03	1.27	0.17
Ulmus rubra	0.22	0.03	1.30	0.15	*Fraxinus quadrangulata*	0.23	0.04	1.03	0.15
Ulmus thomasii	0.25	0.03	1.21	0.16	*Catalpa speciosa*	0.28	0.09	0.64	0.11

Vessel elements vary considerably in length in different woods (Table 5-5), even though little or no elongation occurs along the grain as they enlarge in diameter. This variation in length of vessel elements is due to the fact that the longitudinal cambial initials are of different length in different species of trees.

Vessel elements may also vary in length to a lesser extent in the same tree, because of increase in size of the cambial initials as a tree matures. Considerable variation in length of the elements may be also found in the same ring, particularly in the ring-porous woods such as oak. This is because, while the vessel members in the late wood remain about the same length as the cambial initials, those in the early wood shorten appreciably as they increase in diameter and become annular or barrel-shaped.

Vessel elements, i.e., the wood pores, vary greatly in diameter. The tangential diameter is the more conservative and hence is the dimension usually measured. The smallest pores have a tangential diameter in the neighborhood of 20 microns; the other extreme is found in the oaks and chestnut, where pores at the beginning of the ring are frequently 300 plus microns in width. Lianas, such as the grape and woodbine, have even larger pores.

The spacing and number of pores per square millimeter of surface are also variable. In some woods tracts of tissue can be found that are entirely devoid of pores, whereas in neighboring tracts of the same portion of the ring numerous pores can be counted. In boxwood (***Buxus sempervirens*** L.) the pores average about 180 per square millimeter. Pore counts between 6 and 20 per square millimeter may be considered as average.

b. The Sculpturing of Vessel Elements. The sculpturing of vessel elements is occasioned in part by the nature of the openings that form at or near their ends, in part by the nature of their pitting, and by thickenings, provided these are present, on the inner surface of the secondary wall.

1. *Nature of the opening between vessel elements* (Fig. 5-5). Communication between elements of a vessel is ensured through the formation of an aperture or a series of parallel, transversely oriented apertures through the walls common to two adjacent vessel elements. The portion of the wall involved in the coalescence of two members of a vessel is known as a *perforation plate,* and the apertures are termed *perforations.*

In reality a perforation plate consists of two *half plates* that are reversed as to direction, each half belonging to different vessel elements meeting at that point. Ordinarily a given vessel element possesses two such half perforation plates, one at each end; but occasionally more than two are present, indicating that the vessel element in question connects with as many vessel elements.

When but one aperture (perforation) is in evidence, the perforation plate is *simple* (Fig. 5-5*a*, *f*). When a number of parallel, transversely oriented apertures are present, it is called a *scalariform perforation plate* (Fig. 5-5*c*, *e*). In either case, that portion of the margin of the plate remaining after perforation is complete is called a *perforation rim* (Fig. 5-5*a*).

2. *Nature and extent of pitting on the walls of vessel elements.* The nature and the extent of the pitting on the walls of a vessel element vary greatly depending upon the kinds of cells that happen to be in contact with them. Pit pairs between vessel elements and another prosenchymatous cell are usually bordered. Where they lead to parenchymatous elements they may be bordered, half-bordered, or simple.

Points of contact of vessel elements and rays usually stand out, because the pits are different in size, in nature, and in arrangement from those leading to longitudinal elements (Fig. 5-4*C*, *D*, *E*, and *F*). Pits leading to tracheids and fibers are usually in vertical or nearly vertical rows, conforming to the configuration of these

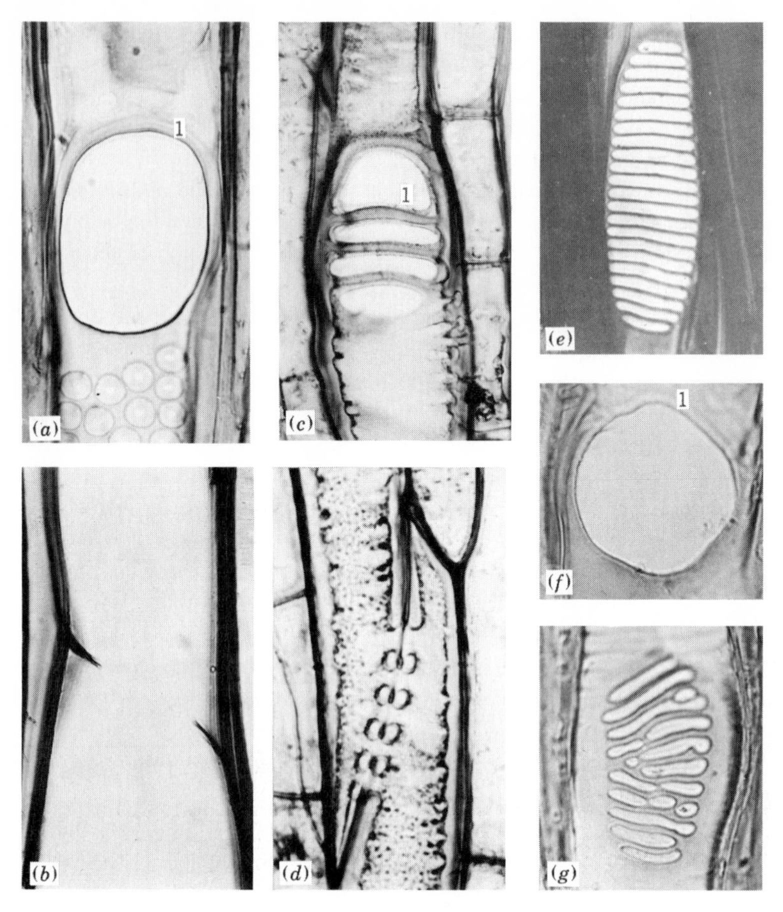

FIG. 5-5 Types of perforations between vessel elements.

(*a*) Simple perforation plate in cottonwood (*Populus deltoides* Bartr.) in oblique surface (radial) view; the perforation rim (1) is visible on the upper margin of the orifice. (550×.)

(*b*) A similar perforation plate in sectional (tangential) view, at the same magnification as (*a*).

(*c*) Scalariform perforation plate with three thick bars (1) in a late-wood vessel in sassafras [*Sassafras albidum* (Nutt.) Nees] in oblique surface (radial) view. (550×.)

(*d*) A similar plate in sectional (tangential) view, at the same magnification as *C*.

(*e*) Scalariform perforation plate with many thin bars, in sweetgum (*Liquidambar styraciflua* L.) in oblique surface (radial) view. (170×.)

(*f*) Simple perforation plate in sparkleberry (*Vaccinium arboreum* Marsh.) in oblique surface (radial) view; the perforation rim (1) is visible at the upper margin of the orifice. (1240×.)

(*g*) Foraminate perforation plate in sparkleberry (*Vaccinium arboreum* Marsh.) in oblique surface (radial) view. (1240×.) Both simple [see (*f*)] and foraminate perforations occur in this wood.

elements (Fig. 5-4*F*). In some woods pit pairs are wanting where libriform fibers are in contact with vessel elements.

Intervessel pitting is generally the most conspicuous, and it appears to best advantage on the tangential faces of a vessel element. It is frequently of diagnostic significance. Three arrangements of intervessel pit grouping are recognized: *alternate pitting* (Figs. 12-148 and 12-153), *opposite pitting* (Figs. 12-159 and 12-160), and *scalariform pitting* (Fig. 12-152).

The pits in the alternate pitting arrangement are in diagnonal rows. They range from circular to oval, if they are uncrowded; if crowded they are polygonal and frequently hexagonal. Where the opposite arrangement prevails, i.e., when the pits are in horizontal pairs or in short horizontal rows, the pits are often rectangular. The pitting is described as scalariform when the pits are linear, with the long axis across the vessel, and are arranged in a ladder-like series.

The variations in shape and size of intervessel pits may best be observed by reference to Figs. 12-141 to Fig. 12-160. Those in birch or persimmon, ranging from 2 to 4 microns in diameter, may be considered as unusually small; pits in maple, ranging from 5 to 10 microns in diameter, are of average size; linear pits in magnolia, varying from 12 to 50 microns in size, are exceptionally large.

3. *Spiral thickening in vessel elements.* Spiral thickening is localized thickening in the form of ridges on the inner surface of the secondary wall and is believed to be an integral part of it. Spiral thickening characterizes the vessel elements of some hardwoods (Figs. 12-141, 12-151, and 12-156), but in some woods it occurs also in the fibers and the tracheids (Fig. 5-7*b*).

In diffuse-porous hardwoods the vessel elements throughout the ring may have spiral thickenings; in ring-porous woods, on the other hand, the spirals, when present, are usually confined to the smaller vessel elements in the late wood. Sweetgum and tupelos are peculiar in that the spiral thickening is restricted to the ligulate tips of the vessel elements.

Spiral thickening is an important diagnostic feature. It occurs, for example, in *Acer* spp. but is lacking in ***Betula*** spp. Tropical hardwoods rarely show thickenings of this sort. The small vessels and vascular tracheids of the elms and hackberry are always characterized by its presence.

c. The Inclusions in Vessel Elements. The inclusions that are most frequent in vessel elements are tyloses and various amorphous exudations that are gummy, resinous, or chalky in nature. Rarely, starch grains and crystals are also in evidence.

A *tylosis* (pl. *tyloses*) may be defined as an outgrowth of protoplasm from an adjacent parenchyma cell through a pit cavity in the cell wall into the cavity of a vessel element, forming a saclike structure. When numerous, tyloses may fill the vessel cavity completely (Fig. 5-6*b*, *e*, *f*).

Tyloses are usually formed in the inner sapwood, just prior to its transformation into heartwood. This is a normal physiological process that occurs in many species of hardwoods. Gerry[12], however, reports that normal tyloses sometimes form in the outer sapwood also, particularly in regions where, for one reason or another, the water content falls below normal. Tyloses may also form as a result of me-

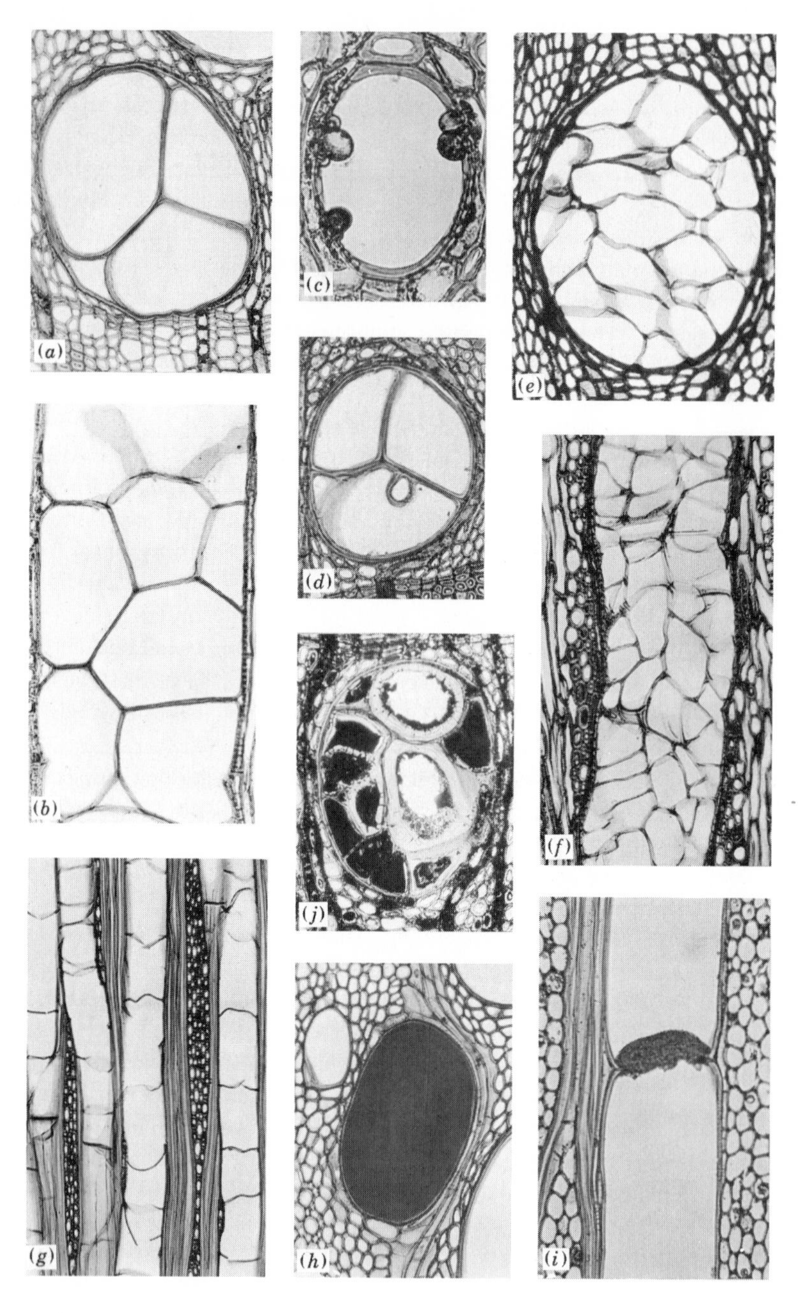

FIG. 5-6 Inclusions in the vessels of hardwoods.

(*a*) Tyloses in an early-wood pore of post oak (*Quercus stellata* Wangenh.); walls somewhat thickened. (115×.)

(*b*) Portion of an early-wood vessel segment (*t*) of post oak (*Q. stellata* Wangenh.) showing contiguous tyloses in lateral view. (115×.)

chanical injury[17], fungous growth, and virus infection. Such tyloses are considered to be traumatic. Likewise, cases are known where tyloses formed in trees after they had been felled, plugging all the sapwood vessels.[23] For further information on tylosis and its formation, see page 105.

Chattaway,[4] states that when the pit apertures in the pit pairs leading from the ray cells into the vessels are small, the activity of ray cells results in secretion of gummy substances, instead of tylosis formation. When these substances come in contact with the air in the vessels, they solidify and form gums and resins of various types and chemical compositions.

Chattaway[4] suggests further that, since the ray cells play an important role in blocking of vessels at heartwood formation, one of the clues to the development of heartwood may be found in the length of life of the ray cells.

Profuse tylosis and gum formation inhibits the penetrability of a wood but does not necessarily increase its durability. The white oaks normally have a high tylosic content, and the wood is used for tight cooperage. The red oaks, in contrast, have so few tyloses that it is possible to blow through a piece of wood for some distance along the grain; red oaks hence are not suited for tight cooperage unless specially treated. On the other hand, woods with open pores treat more easily with preservatives.

Gummy materials in the vessel elements occur in irregular lumps on the inner walls. They may completely occlude the cell (Fig. 5-6*h*) or form false partitions (Fig. 5-6*i*) across the vessel at the juncture where two vessel members meet, thus effectively blocking the vessel at these points. In domestic timbers the color of the gum is usually some shade of red or brown. In pale-yellow woods, of the type included in the Rutaceae, it appears yellow by transmitted light. The jet black of ebony (*Diospyros ebenum* Koenig) is due to copious deposits of gummy infiltration of the same color, not only in the vessel members but in the lumina of the other elements of the wood as well.

The presence or absence of gummy deposits in the vessels is frequently of value in identification. Honeylocust has abundant reddish deposits of this sort in the

(*c*) Tyloses in the bud stage in a late-wood pore of live oak (*Q. virginiana* Mill.). (225×.)

(*d*) Secondary tylosis bud forming on the wall of a tylosis in blue oak (*Q. douglasii* Hook. & Arn.). (115×.)

(*e*) Thin-walled tyloses in an early-wood pore of black oak (*Q. velutina* Lam.). (115×.)

(*f*) Portion of an early-wood vessel (*t*) of black locust (*Robinia pseudoacacia* L.) showing thin-walled tyloses in lateral view. (115×.)

(*g*) Uniseriate rows of tyloses in the vessels of beech (*Fagus grandifolia* Ehrh.); the upper and lower walls (which are in contact) appear as nearly horizontal transverse partitions arranged in a ladder-like series. (115×.)

(*h*) A pore of honeylocust (*Gleditsia triacanthos* L.) occluded with gum. (115×.)

(*i*) A gum plug at the juncture of two vessel elements in large-leaved mahogany (*Swietenia macrophylla* King). (115×.)

(*j*) Sclerosed, pitted tyloses, with dark contents, in a pore in Emory oak (*Quercus emoryi* Torr.). (115×.)

[*Photographs (a)–(f), inclusive, by S. Williams.*]

heartwood; they are wanting or scanty in the wood of the botanically related Kentucky coffeetree, and hence this feature can assist in the separation of these woods.

Chalky deposits occur in the vessels of some woods, although their distribution is too sporadic to serve as a positive aid in identification. They are usually present in mahogany (*Swietenia* spp.) and in teak (*Tectona grandis* L. f.) and occasionally in mulberry. Little is known about the chemistry of these deposits; those of teak are said to consist of calcium phosphate.

2. HARDWOOD TRACHEIDS

Two distinct types of hardwood tracheids are recognized; these are *vascular tracheids* and *vasicentric tracheids*.

a. Vascular Tracheids. These are cells very similar in size, shape, and position to small late-wood vessel elements, except that they are imperforate at the ends. They are arranged in vertical series, like the small vessels with which they are associated (Fig. 5-7*b*); a given series may consist entirely of vascular tracheids, or of a mixture of these and vessel members, without any regular sequence.

The lateral walls of vascular tracheids contain numerous bordered pits, of the same general type as intervessel pits, and frequently possess conspicuous spiral thickenings. Vascular tracheids with spirals are found in *Ulmus* and *Celtis*. The wavy bands of porous tissue that characterize the late wood of these woods are largely composed of small vessel elements and vascular tracheids. Since vascular tracheids, when cut transversely, have the same appearance as pores, they cannot be distinguished from true pores in the cross section of wood.

b. Vasicentric Tracheids. These are short, irregularly shaped cells, with closed ends, that abound in the proximity of the large early-wood vessels of such ring-porous woods as oaks and chestnut; they are also found in lesser numbers in the flame-shaped areas of porous tissue extending into the late wood in these woods. In both instances they are associated with axial parenchyma, which they resemble in cross section (Fig. 5-7*a*). The lateral walls of vasicentric tracheids are copiously pitted with conspicuous bordered pits.

Vasicentric tracheids differ from vascular tracheids in having tapering or rounded ends and in not being arranged in definite longitudinal rows.

The transition from large early-wood vessel elements to neighboring vasicentric tracheids is abrupt; that from the latter to wood fibers, in contrast, is more gradual, often through intermediate cell types.

3. FIBERS

The term *fiber* is frequently used loosely for any kind of wood cells in general. Specifically this term applies to long, narrow cells, with closed ends, other than tracheids. Two types of fibers are recognized, *fiber tracheids* and *libriform fibers*.

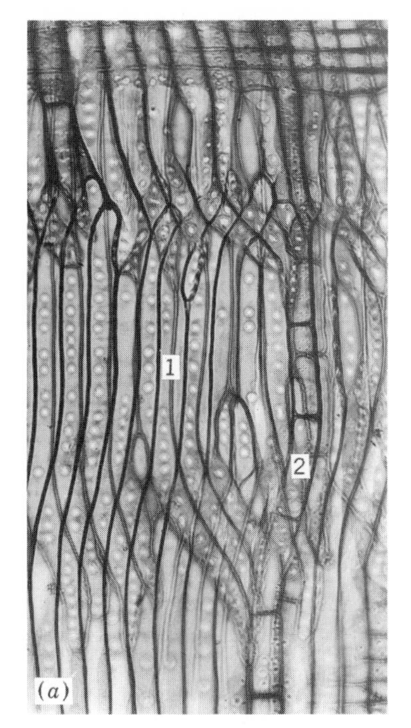

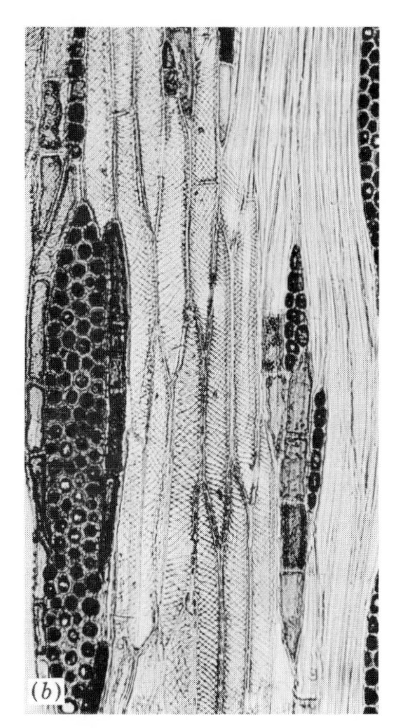

FIG. 5-7 Vasicentric and vascular tracheids in hardwoods.
(*a*) Overlapping vasicentric tracheids with bordered pits (1) and a strand of longitudinal parenchyma (2), in chestnut [*Castanea dentata* (Marsh.) Borkh.] (*r*). (240×.)
(*b*) Vascular tracheids with spiral thickening, which grade into small vessel members elsewhere in the wood. Slippery elm (*Ulmus rubra* Mühl.) (*t*). (160×.)

a. Fiber Tracheids. These are generally relatively thick-walled fibrous cells with pointed ends and bordered pits.*

b. Libriform Fibers. These are elongated fibrous cells which differ from fiber tracheids in possessing simple pits.

Both types of fibers exhibit a number of modifications, some of which are of diagnostic value. As a rule the inner surface of the secondary wall of fibers is smooth; in some woods, such as holly, spiral thickenings may be present. In other woods, for instance, California-laurel [*Umbellularia californica* (Hook. & Arn.) Nutt.], thin transverse walls (*septa*) form in the fibers, resulting in *septate* libriform fibers or fiber tracheids. Such transverse walls extend only to the inner surface of the lateral secondary walls of the fiber, dividing the fiber into two or more compartments. Not infrequently the fibers are falsely septated. This feature is rare in temperate-zone woods but common in tropical timbers. This results from the deposition of transverse plates of gummy or resinous materials, simulating true septa.

* This definition is equally applicable to the late-wood tracheids in coniferous woods. However, most authors restrict this term to woody angiosperms only.

Another modification of fibers consists of the *gelatinous fibers*, encountered mostly in tension wood of hardwoods (see page 303). These fibers differ from the normal kind in possessing an innermost cell wall layer which differs in physical and chemical properties from the usual secondary cell wall layer (Figs. 5-8 and 8-12). This layer, generally described as *gelatinous*, is highly refractory to light and usually gives a cellulose reaction with various staining reagents, indicating absence or a low degree of lignification. The so-called gelatinous layer may be present in addition to layers *S*1, *S*2, and *S*3, or may replace layer *S*3 or layers *S*2 and *S*3.

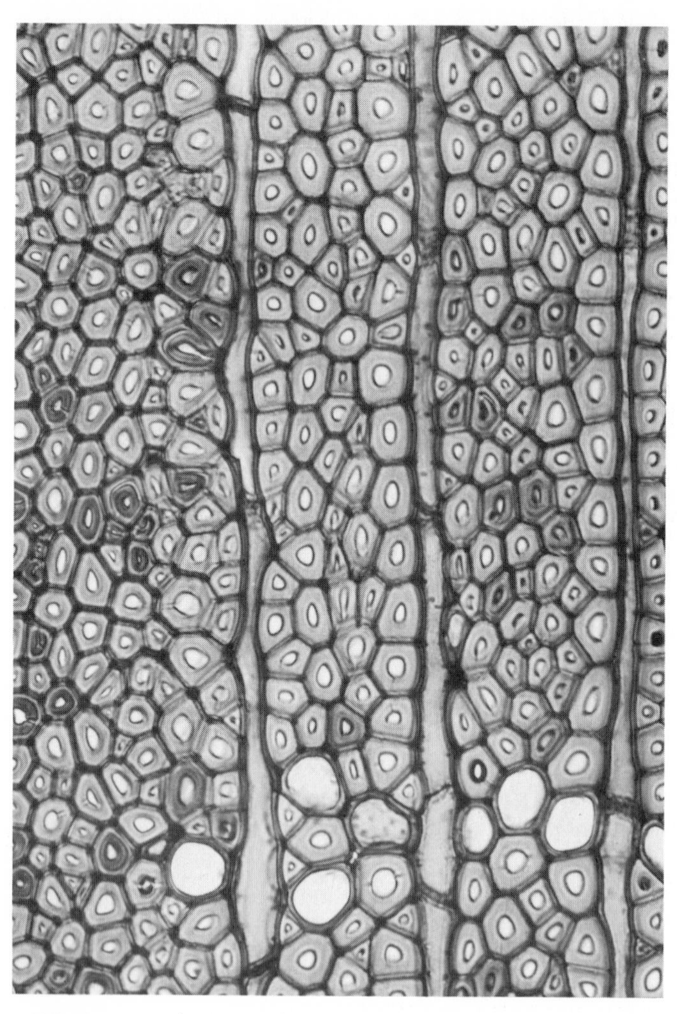

FIG. 5-8 Gelatinous and normal fibers in white oak (*Quercus alba* L.). (450×.)
(*Photomicrograph by S. Williams.*)

TABLE 5-6 Mean Weighted Specific Gravity, Fiber Length in the Thirtieth Ring, and Their Ranges Sampled from Cross Sections 4.5 ft From Tree Base*

Species	*Avg weighted sp gr*	*Range*	*Avg fiber length 30th year, mm*	*Range, mm*
Yellow poplar	0.440	0.380–0.520	2.0	1.40–2.40
Sweetgum	0.496	0.434–0.580	1.93	1.65–2.14
Water tupelo	0.474	0.433–0.550	1.85	1.44–2.13
Sycamore	0.476	0.404–0.533	1.83	1.49–2.16
Swamp blackgum	0.479	0.381–0.587	1.78	1.46–2.21
Water oak	0.599	0.538–0.643	1.54	1.32–1.76
Willow oak	0.587	0.522–0.640	1.54	1.33–1.81
White ash	0.546	0.480–0.593	1.14	0.99–1.42
Red maple	0.496	0.429–0.540	0.97	0.89–1.06

* R. L. McElwee, Wood and Fiber Characteristics of Selected Southern Hardwoods, presented at the Carolinas-Chesapeake Section Meeting, Forest Prod. Res. Soc., Clemson, S.C., Oct. 20, 1967.

Fiber tracheids grade into libriform fibers on the one hand, and, when septate, into strand parenchyma on the other; cell types intermediate between these two kinds occur in many woods. Fiber tracheids and libriform fibers may occur in the same wood, and the transition between them is often so gradual that it may be an arbitrary matter in assigning a given cell to one or the other type.

Fiber tracheids and libriform fibers, either separately or in admixture, often compose 50 percent or more of a given wood by volume; these cells make up the bulk of the fibrous mass obtained when a hardwood is pulped.

Fiber tracheids and libriform fibers vary greatly in diameter, in length (Tables 5-6 and 5-7), in the thickness of walls, and in the volume entering into the structure of a wood. This holds not only for different species but in different individuals of the same species, and even at different places in a tree.

TABLE 5-7 Average Fiber Length at Varying Heights in Six Yellow-Poplar (*Liriodendron tulipifera* L.) **Trees***

Height above ground, ft	*Weighted avg length, mm* (1)	*Avg length in 20th ring, mm* (2)
4.5	1.91	2.02
20	1.77	
40	1.61	1.86
60	1.45	1.40

* From F. W. Taylor, Variations in the Size and Proportion of Wood Elements in Yellow-Poplar Trees, *Wood Sci. & Technol.*, **2**(3):153–165 (1968).

For instance, Taylor,* on the basis of the analysis of six yellow-poplar trees, reports that the weighted average of fiber length in this species decreased steadily with increase in height from the base of the trees, as shown in column 1 of Table 5-7.

Taylor observes further that this decrease in fiber length cannot be attributed to increased proportion of juvenile wood alone, since this same trend was evident in the fibers from the mature wood as well. This is illustrated in column 2, Table 5-7. Discussion of the possible reasons for these observed differences can be found in Chap. 7. The variations in the amount and quality of fibrous tissue have a profound effect on wood density, strength, shrinkage characteristics, and other factors affecting utilization of wood.

B. The Parenchymatous Elements of Porous Woods

As previously defined, wood parenchyma is a tissue composed of comparatively short, brick-shaped, or isodiametric cells equipped in most cases with simple pits only. These cells are involved primarily in storage and to a lesser extent in the conduction of carbohydrates; they remain functional for a much longer period (supposedly as long as they are a part of the sapwood) than the conductive and mechanical prosenchymatous cells.

The wood parenchyma of hardwoods, like that of conifers, is of two types: (1) *longitudinal* (also called *vertical* and *axial*), i.e., derived from fusiform cambial initials and hence directed along the grain, and (2) *transverse*, which is derived from ray initials and extends across the grain, forming wood rays; this type of parenchyma can be properly designated as *ray parenchyma*.

Hardwoods differ from softwoods, as a group, in possessing much more wood parenchyma of both kinds. There are exceptions, however, e.g., the poplars (*Populus* spp.) and willow (*Salix* spp.), in which rays are uniseriate and longitudinal parenchyma is sparsely developed, or cherry (*Prunus* spp.), in which longitudinal parenchyma is either absent or extremely rare.

1. LONGITUDINAL PARENCHYMA

As indicated in Table 5-4, there are three kinds of longitudinal parenchyma: (1) *strand parenchyma*, formed through transverse division of the fusiform cambial daughter cell, with the entire strand of approximately the same shape as the original fusiform mother cell (Fig. 5-9*g*1, *h*2), (2) *fusiform parenchyma*, which are longitudinal parenchyma cells derived from a fusiform cambial initial without subdivision (Fig. 5-9*h*3), and (3) *epithelial parenchyma*, i.e., cells which encircle the cavities of longitudinal canals. Longitudinal resin canals are not present as normal structures in any of the woods covered in this text; they occur sporadically

* F. W. Taylor, Variations in the Size and Proportion of Wood Elements in Yellow-Poplar Trees, *Wood Sci. & Technol.*, **2**(3):153–165 (1968).

in sweetgum, presumably as a result of injury. Consequently, longitudinal epithelial cells are wanting in domestic woods of normal structure. Certain tropical hardwoods, among others those belonging to the Dipterocarpaceae, some species of which produce the Philippine mahogany of commerce wood, possess normal gum canals. As a rule, in the hardwoods, in which canals occur, they are of either the longitudinal or the transverse type only, the two kinds rarely occurring in the same wood.

Strand parenchyma is by far the most common type of longitudinal parenchyma in hardwoods. In some tropical tree species it may comprise an appreciable volume of the wood, at times more than 50 percent; in domestic woods it ranges from less than 1 to about 18 percent in volume. Longitudinal strand parenchyma, when massed in various ways, is often visible on transverse surfaces at low magnification, appearing lighter or darker than the surrounding tissue. Various arrangements of parenchyma are recognized, and these are sufficiently uniform to permit their use in the identification of the various porous woods. The descriptive terms used in this text in defining different arrangements of longitudinal parenchyma, as visible in transverse sections, are as follows:

1. *Apotracheal parenchyma*, longitudinal parenchyma arranged independent of the pores.

a. Diffuse apotracheal, single parenchyma cells distributed irregularly among fibers (ex., sweetgum and tupelo).

b. Diffuse-in-aggregates, apotracheal parenchyma cells that tend to be grouped in short tangential lines, sometimes extending from ray to ray (Fig. 5-9*b*) (ex., yellow birch).

2. *Paratracheal parenchyma*, longitudinal parenchyma associated with the vessels of vascular tracheids.

a. Scanty paratracheal, paratracheal parenchyma confined to a few cells around the vessels (ex., birch, maple).

b. Vasicentric paratracheal (Fig. 5-9*d*), paratracheal parenchyma forming a more or less complete sheath, one or more cells wide, around a vessel (ex., ash, persimmon, California-laurel).

c. Aliform paratracheal, paratracheal parenchyma that extends from the flanks of the pore, forming winglike lateral extensions. This type of parenchyma is more common in the tropical species; it is present occasionally in California-laurel and in the outer late wood in some ash species (*Fraxinus* spp.).

d. Confluent paratracheal (Fig. 5-9*e*), aliform parenchyma forming irregular tangential or diagonal bands that coalesce (ex., honeylocust).

3. *Banded parenchyma*, forming concentric lines or bands. It could be further differentiated into *apotracheal banded*, if mainly independent of the pores (ex., basswood, Fig. 5-9*a*); *paratracheal banded*, if definitely associated with the pores. Since, however, both types of banded parenchyma generally occur in the same wood (ex., hickory, persimmon, beech), the use of the term *banded*, without further qualification, is quite appropriate.

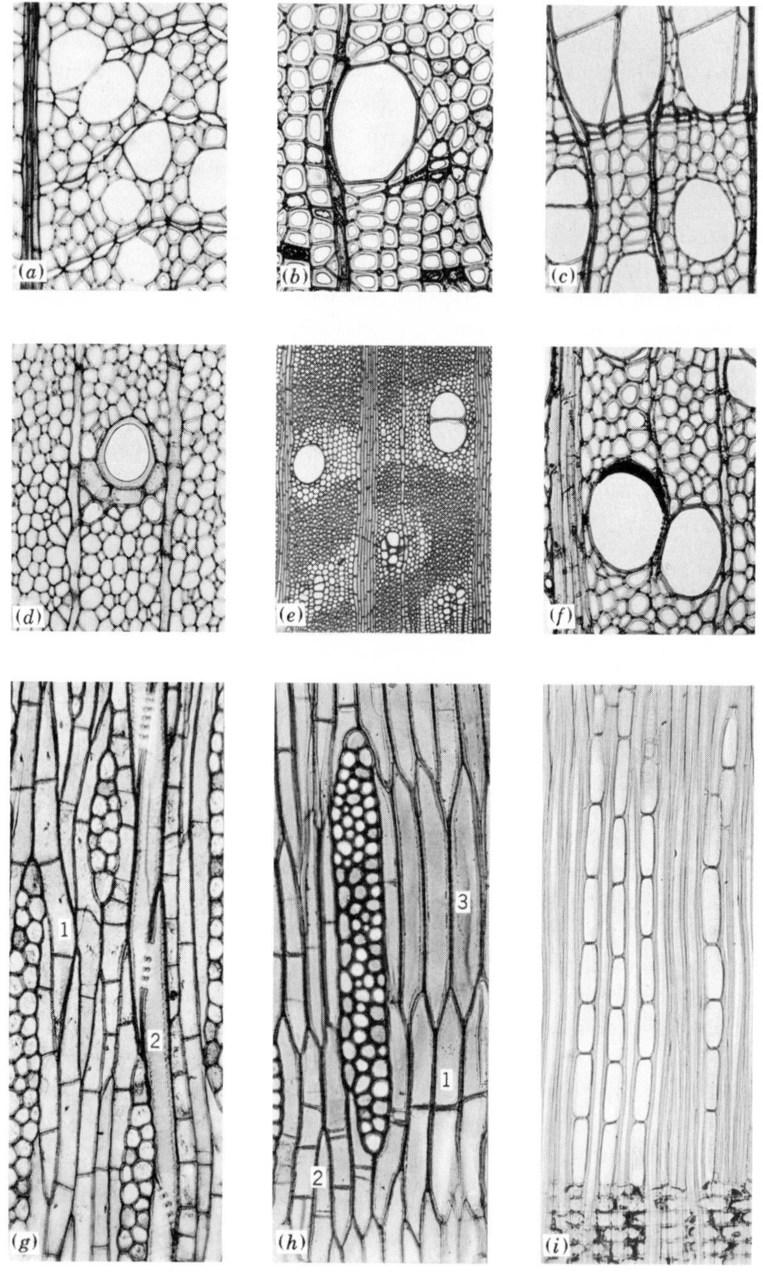

FIG. 5-9 Longitudinal (axial) parenchyma in hardwoods.

(*a*) Longitudinal apotracheal banded parenchyma (*x*) in basswood (***Tilia americana*** L.). (150×.)

(*b*) Longitudinal diffuse-in-aggregates parenchyma (dark cells) (*x*) in yellow birch (***Betula alleghaniensis*** Britton). (150×.)

4. ***Marginal parenchyma***, longitudinal parenchyma occurring either as occasional cells, or forming a more or less continuous layer of one to more cells in width, at the margin of a growth ring (Fig. 5-9*c*). Formerly this type of parenchyma was described as ***terminal***. However, Chowdhury[7] has shown that longitudinal parenchyma can occur either at the end of the season's growth or at its beginning. He has therefore proposed the designation of ***terminal*** for the former and ***initial*** for the latter. But, since in actual practice it is frequently difficult to distinguish between these two types, a single term ***marginal*** is preferred. ***Marginal parenchyma*** occurs as occasional cells in the maples (*Acer* spp.) and birches (***Betula*** spp.), associated with other types of parenchyma, or in more or less continuous bands as in the willows, poplars, yellow-poplar, and magnolias.

More commonly than not several types of parenchyma are present in the same wood. For instance, in red oaks apotracheal parenchyma, ranging from diffuse to banded, occurs together with several different forms of paratracheal parenchyma. Paratracheal parenchyma ranging from scanty to confluent, together with terminal parenchyma, features honeylocust, black locust, and other native species belonging to the family Leguminosae, while hickories are characterized by apotracheal and paratracheal banded and marginal forms of parenchyma.

2. RAY PARENCHYMA (Fig. 5-10)

For emphasis a brief review of the nature, and origin of rays is presented again in this section. A ray, as previously defined, is a ribbon-like aggregate of cells formed by the cambium and extending radially in the stem of a tree. A ***wood*** or ***xylary*** ray is only a part of a larger structure which, for lack of a better term, may be called a ***complete ray***. Such a complete ray consists of a ***xylary ray***, i.e., the portion of the ray in the wood, and a ***phloem ray*** which is the portion of the same ray in the phloem. Both of these originated from a ***cambial ray***, i.e., ray initials in the cambium.

(*c*) Longitudinal marginal parenchyma (*x*) on the outer margin of an annual ring in eastern cottonwood (*Populus deltoides* Bartr.). (150×.)

(*d*) Longitudinal vasicentric paratracheal (*x*) nearly encircling a late-wood pore in white ash (*Fraxinus americana* L.). (150×.)

(*e*) Longitudinal confluent paratracheal parenchyma (*x*) in honeylocust (*Gleditsia triacanthos* L.). (40×.)

(*f*) Xylary tissue (*x*) devoid of longitudinal parenchyma in black cherry (*Prunus serotina* Ehrh.). Longitudinal parenchyma is extremely sparse in this wood. (150×.)

(*g*) Strands of longitudinal parenchyma (1), and a late-wood vessel (2), with scalariform perforation plates, in lateral, tangential view. [*Sassafras albidum* (Nutt.) Ness.] (150×.)

(*h*) Strands of longitudinal parenchyma composed of two (1) and four (2) units, and fusiform longitudinal parenchyma cells (3), in lateral, tangential view. Black locust (*Robinia pseudoacacia* L.). (150×.)

(*i*) Portions of a number of strands of longitudinal parenchyma, separated by fibers, in lateral, radial view. Scarlet oak (*Quercus coccinea* Muenchh.). (150×.)

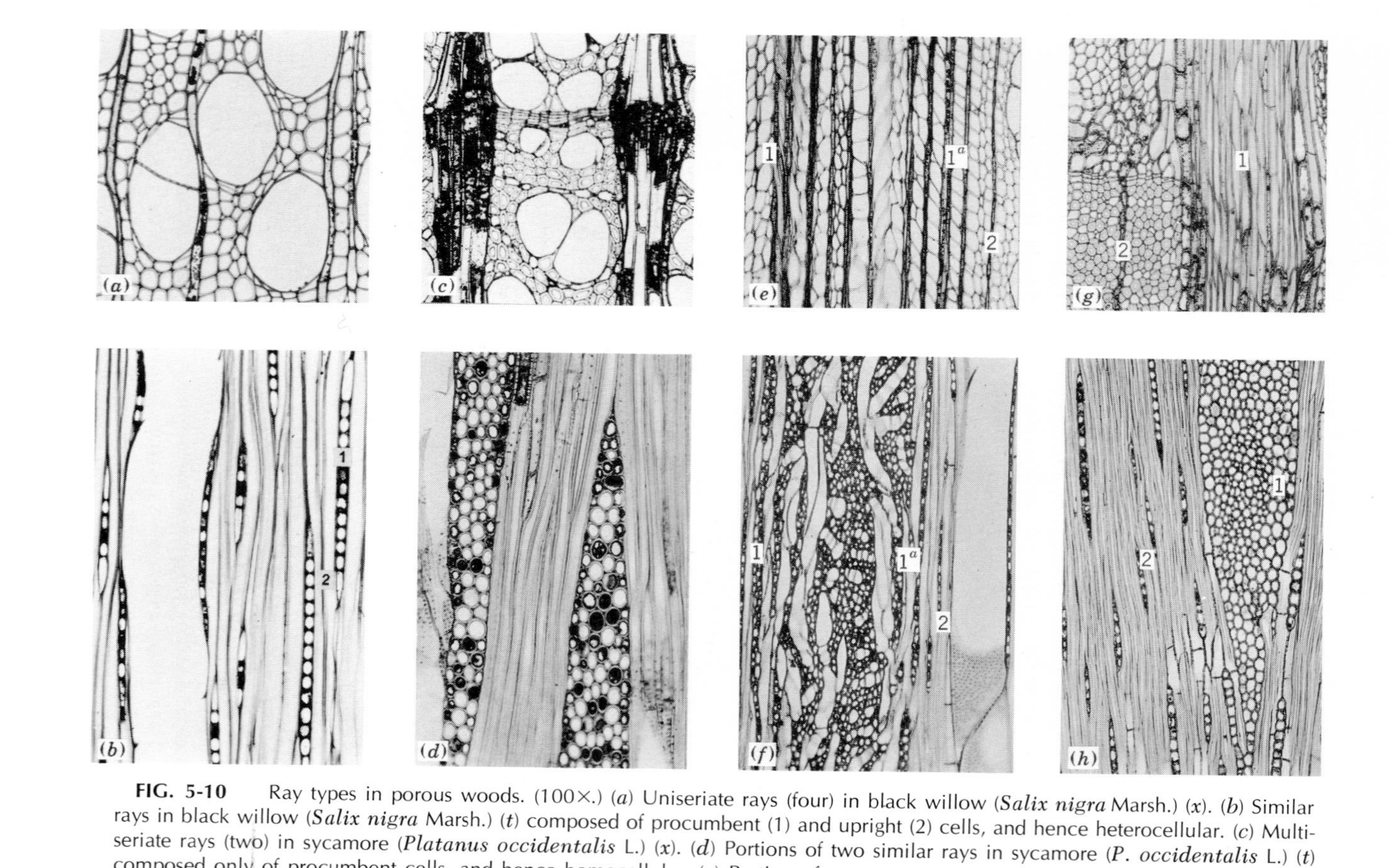

FIG. 5-10 Ray types in porous woods. (100×.) (*a*) Uniseriate rays (four) in black willow (*Salix nigra* Marsh.) (*x*). (*b*) Similar rays in black willow (*Salix nigra* Marsh.) (*t*) composed of procumbent (1) and upright (2) cells, and hence heterocellular. (*c*) Multiseriate rays (two) in sycamore (*Platanus occidentalis* L.) (*x*). (*d*) Portions of two similar rays in sycamore (*P. occidentalis* L.) (*t*) composed only of procumbent cells, and hence homocellular. (*e*) Portion of an aggregate ray ($1\text{-}1^a$), and one uniseriate ray (2) in red alder (*Alnus rubra* Bong.) (*x*). (*f*) Portions of a similar aggregate ray ($1\text{-}1^a$) and uniseriate rays (2) in red alder

The rays in a tree diverge as they extend outward in the wood. Hence the spacing between any two rays increases as the stem increases in diameter. When the distance between two such rays has become sufficiently great, a new ray is formed in the cambium, so that spacing between the rays remains fairly constant in any given wood. Once initiated a new cambial ray may increase in size through the addition of further cells, resulting in an increase in the size of the ray in succeeding growth rings; barring injury to the cambium they are continued as long as the tree lives. The first-formed wood rays extend completely across the secondary xylem and they may initiate opposite the primary xylem. But since all wood rays originate in the cambium, such terms as *pith rays* and *medullary rays*, frequently encountered in literature on wood, should not be used, because they imply that wood rays consist of the same kind of tissue as the pith or medulla and originate from it.

The wood rays of dicotyledonous woods are much more variable in width and height than those of softwoods. Exclusively or predominantly uniseriate rays of the type that characterize coniferous woods are seldom found in domestic hardwoods. Notable exceptions are *Salix* spp. (Fig. 5-10*a* and *b*), *Populus* spp., *Aesculus* spp., and *Castanea dentata* (Marsh.) Borkh. Most hardwoods, however, possess rays ranging from two to many seriate (Fig. 5-10*c* and *d*). In timbers covered by this text, maximum seriation is reached in oaks, where the large rays are often 30 or more cells wide (Fig. 5-10g and *h*) and 300 plus microns in width. In a number of woods the rays are of distinctly two sizes, the smaller frequently being uniseriate. Such ray combinations are found in sugar maple (*Acer saccharum* Marsh.), American beech (*Fagus grandifolia* Ehrh.), and other species but have their optimum development in oaks (Fig. 5-10g and *h*). Multiseriate rays frequently flare where they cross the boundaries of growth rings (Fig. 5-10*c*); this condition is described as *noded*.

Rays may or may not be visible with the naked eye on the transverse and tangential surfaces. When they are in evidence, this feature can be used to separate certain woods. For example, rays are barely visible on the transverse section and indistinct on the tangential in birch, but are distinct on both surfaces in maple. Beech can be separated from maple because in the former, rays are more conspicuous and varied in size, as seen in the flat-sawn lumber (see Figs. 12-79 and 12-100).

In height hardwood rays range from one cell and a few microns (about 20) to the upper limit of 2 inches or more along the grain (50,000 plus microns), found in the oaks.

As viewed in tangential section, the rays in hardwoods are generally quite variable in height, and usually intergrade in size, possibly indicating that those which arose last in the cambium have not yet attained their maximum height and width. Such variously sized rays are commonly staggered in the tangential plane. In persimmon (*Diospyros virginiana* L.) and in yellow buckeye (*Aesculus octandra* Marsh.), as well as in many tropical timbers, rays and sometimes other longitudinal elements are arranged in regular horizontal rows (Fig. 12-219). Such arrangement of wood elements is called *storied*; when viewed with the naked eye or a hand lens, woods with storied cells exhibit ripple marks on the tangential surface. Storied

TABLE 5-8 Spacing of Wood Rays in Porous Woods

Rays per mm	*Spacing*
5 or less	Widely spaced
6–9	Normally spaced
10–13	Fairly close
14–20	Close
21 or more	Extremely close

rays are quite uniform in height, provided that they are restricted to one story. However, in some woods with tiered rays some of the rays extend through several stories. In such instances ray height is variable, even though the wood exhibits ripple marks. If the rays alone are storied, then the cambial rays are likewise storied; when all the longitudinal cells are stratified, then both the fusiform initials, as well as rays in the cambium, are storied.

The rays of porous woods differ not only in size (width and height) but also in their spacing. Ray spacing in hardwoods can be studied to best advantage in transverse sections. It is computed as the number of rays per millimeter that cross the boundary of a ring; the upper and the lower figures, rather than the average, are recorded, e.g., six to nine per millimeter.

Table 5-8 indicates the range of ray spacing in the native species and gives convenient terms to record the degree of spacing. In most domestic woods the spacing is normal, i.e., the rays average six to nine per millimeter. In sweetgum, the species of *Nyssa*, and a few others, rays are so close together (21 or more per millimeter) that often half the area of a transverse surface appears to consist of ray tissue.

It should be pointed out that the figure of wood, especially that on the radial surface, may be affected by the size, and to a lesser degree by spacing, of the rays; this figure is described as ***ray fleck***. In most native woods ray flecks range from inconspicuous (ex., willow, poplar, buckeye, birch) to relatively conspicuous (ex., tulip-poplar, magnolia, maple). The most noticeable ray-fleck pattern occurs in the quarter-sawn stock in the woods with exceptionally high rays, such as beech, sycamore, and especially in oaks.

The ray volume of porous woods exerts an important effect on their physical properties, especially on dimensional changes and formation of checks, and to some extent on strength and penetrability.

Data on the volume of wood rays in the hardwoods native to the United States were compiled by J. E. Myer;[25] some of these data are presented in Table 5-9.*

Ray volume is contingent upon the size attained by the rays and on their number (spacing). It varies not only in different kinds of wood, but also within a species and at different places in the same tree. According to Myer, the larger variations

* Also see Chalk.[3]

TABLE 5-9 Average Ray Volumes of Hardwoods of the United States
Arranged in botanical sequence, by genera

Scientific name	*Avg ray volume, %*	*Variation in samples examined, %*
Juglans cinerea	8.6	2.0
Juglans nigra	16.8	3.7
Carya cordiformis	16.8	7.8
Carya ovata	19.9	9.4
Populus deltoides	13.7	5.8
Populus grandidentata	11.0	5.6
Populus tremuloides	9.6	4.4
Betula lenta	16.6	6.4
Betula alleghaniensis	10.7	0.9
Betula nigra	15.8	3.3
Betula papyrifera	11.0	5.6
Betula populifolia	9.3	2.2
Fagus grandifolia	20.4	5.3
Castanea dentata	11.9	3.3
Quercus rubra	21.2	
Quercus velutina	31.3	
Quercus alba	27.9	
Quercus bicolor	29.7	
Quercus virginiana	32.2	
Ulmus americana	11.4	7.2
Ulmus rubra	13.0	4.2
Ulmus thomasii	18.6	3.8
Celtis occidentalis	13.3	3.0
Magnolia acuminata	13.8	3.5
Liriodendron tulipifera	14.2	2.5
Liquidambar styraciflua	18.3	1.8
Platanus occidentalis	19.2	3.7
Prunus serotina	17.2	3.4
Gleditsia triacanthos	18.4	3.3
Robinia pseudoacacia	20.9	3.1
Acer macrophyllum	18.4	3.8
Acer rubrum	13.3	2.2
Acer saccharinum	11.9	5.0
Acer saccharum	17.9	5.2
Tilia americana	6.0	3.8
Tilia heterophylla	5.3	2.0
Nyssa sylvatica	17.3	9.7
Fraxinus americana	11.9	4.9
Fraxinus nigra	12.0	7.1
Fraxinus latifolia	14.5	4.4
Catalpa speciosa	13.3	2.4

in the volume of the rays in wood are due to inheritance; the smaller variations are the result of differences in age and in climatic and site conditions.

Of the hardwoods studied by Myer, the two *Tilia* species had the lowest ray volume; the highest, in commercial species, was recorded in oaks. The greatest number of species possessed ray volumes between 10 and 20 percent. The average ray volume of all the porous woods examined by Myer was 17 percent; of the conifers he studied, 7.8 percent.

However, White and Robards[27] point out that the number of rays and their width, and hence the ray volume, are affected by the growth rate, as measured by the width of growth rings. They report that in the three species they studied—sweet chestnut, *Castanea sativa* Mill; ash, *Fraxinas excelsa* L.; and sassafras, *Sassafras officinale* Nees and Eberm. [*Sassafras albidum* (Nutt.) Nees]—there were significantly more rays in the samples with wide growth rings than in the slower-grown wood. For instance, in sassafras there was an increase from 41±3 rays per square inch in the ring measuring 3.6 millimeters in width to 65±3.9 rays per square inch in the growth ring 7.2 millimeters wide. In ash and sassafras the extra ray width was due to the increased size and greater number of cells; in chestnut it was due to the presence of a large number of biseriate rays, which are seldom found in chestnut wood of slower growth. No consistent relationship between the radial growth and ray height was found. These findings point out the uncertainties surrounding the use of numerical data involving sizes and frequency of anatomical features, for purposes of wood identification.

a. The Composition of Wood Rays in Porous Woods. With the exception of the specialized type of ray, called *aggregate*, the rays of hardwoods consist entirely of parenchyma cells; the parenchymatous ray cells may, however, vary considerably in size and shape. When ray cells are all of approximately the same size and shape, the rays are termed *homocellular* (formerly called *homogeneous*); the rays are said to be *heterocellular* (formerly termed *heterogeneous*) if containing more than one type of ray-parenchyma cells (see Fig. 5-11*a* and *b*).

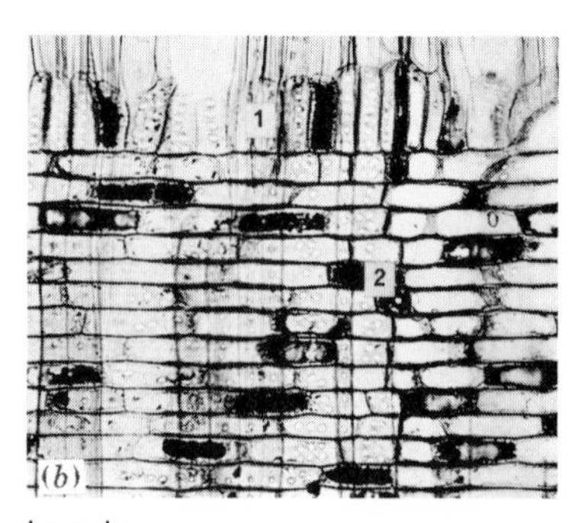

FIG. 5-11 Composition of rays in hardwoods.
(*a*) Homocellular ray (*r*) in sycamore (*Platanus occidentalis* L.) consisting entirely of procumbent cells. (125×.)
(*b*) Heterocellular ray (*r*) in sweetgum (*Liquidambar styraciflua* L.) consisting of marginal upright (1) and of procumbent (2) cells. (125×.)

The radially elongated ray cells, of the type pictured in Fig. 5-11*a* and *b*, are called ***procumbent cells***, while those vertically oriented, as in upper marginal row of 5-11*b*, are called ***upright cells***. Homocellular rays may be composed entirely of either type of cells. In the native species upright cells in the heterocellular rays are generally confined to one or more cells on the upper and lower margins of the ray. In some woods, for instance, in dogwood (*Cornus* spp.), there may be several rows of such marginal upright cells, forming uniseriate, tail-like projections of the rays (Fig. 12-215). In the tropical woods upright cells sometimes form a sheath around the central portion of the ray made up of the procumbent ray cells; in this case the upright cells are called ***sheath cells***.

Ray composition is an important and frequently easily determinable characteristic for separation of woods that are closely related or otherwise easily confused. For instance, *Salix* spp. are characterized by the uniseriate, heterocellular rays, while *Populus* spp. have the uniseriate but essentially homocellular type. Likewise, hackberry (*Celtis occidentalis* L.) is easily separated from the elms (*Ulmus* spp.) by the multiseriate heterocellular rays, in contrast to the homocellular kinds found in elms. Sassafras, frequently mistaken for black ash, can be separated from it by rays ranging from the homocellular to the heterocellular type, while rays in ash are strictly homocellular.

Some woods containing multiseriate homocellular and heterocellular rays may also have strictly uniseriate rays consisting either entirely of procumbent cells (oaks, hard maple) or entirely of upright cells (holly), or of both types (dogwood).*

Red alder, American hornbeam, tanoak, and some *Quercus* spp. are characterized by rays of two kinds; narrow and usually uniseriate, which to all appearances are spaced normally as viewed on the transverse and tangential surfaces, and clusters of such narrow rays which are closely spaced but between which there is tracheary tissue (fibers and vessels, see Fig. 5-10*e* and *f*). To the unaided eye or at a low magnification such formations appear as a single ray and are termed ***aggregate rays***. Such rays may be relatively abundant (American hornbeam) or relatively infrequent and then at wide and irregular intervals (red alder).

Transverse resin canals are not found in any domestic woods; they occur in the rays of some tropical hardwoods. When present they are of the same general nature as those in conifers; i.e., they are surrounded by the epithelial cells, which are parenchymatous in nature.

The pits of ray parenchyma range from simple and minute to bordered and relatively large. The pitting may be equally abundant on all the walls, and their nature is often strongly influenced by the nature of the complementary pit on the wall of the adjacent element. Not infrequently the lateral walls of the upright cells are copiously and conspicuously pitted (ex., sweetgum, willow), whereas pits are sparse or wanting altogether on the lateral walls of the procumbent cells.

The nature of pitting between rays and vessel elements is frequently of some

* Kribs[20] has proposed a classification of rays based on their appearance in radial and tangential sections. This classification is outside the scope of this text but may be found useful when dealing with a large number of woods from tropical regions.

diagnostic value. All the native species of wood can be grouped on the basis of ray-vessel pitting under three headings:

1. Ray-vessel pitting simple and elongated (ex., yellow buckeye, hackberry).
2. Ray-vessel pitting ranging from simple to bordered and quite variable in shape and size (ex., oak, American basswood).
3. Ray-vessel pitting similar to intervessel type (ex., maple, birch, ash).

3. INCLUSIONS IN PARENCHYMA CELLS

The cells of parenchyma frequently contain inclusions; these materials may be crystals, silica, and numerous amorphous materials of complex chemical nature, including gums, resins, tannins, oils, latex, coloring matter, and nitrogenous materials, such as alkaloids. Carbohydrates, usually in the form of starch grains, are also common, particularly in the sapwood portion of the stem. Little exact information is available on the chemical nature of many of these infiltrations. Frequently the color, odor, and taste of wood are traceable to these products. For instance, the rays of sassafras are characterized by oil cysts (Fig. 5-12); the persistent aromatic odor of this wood is due in part to the contents of such cells.

Silica, though it usually occurs in the cell walls, may also be present in the form of inclusions of various size in the cell cavities.

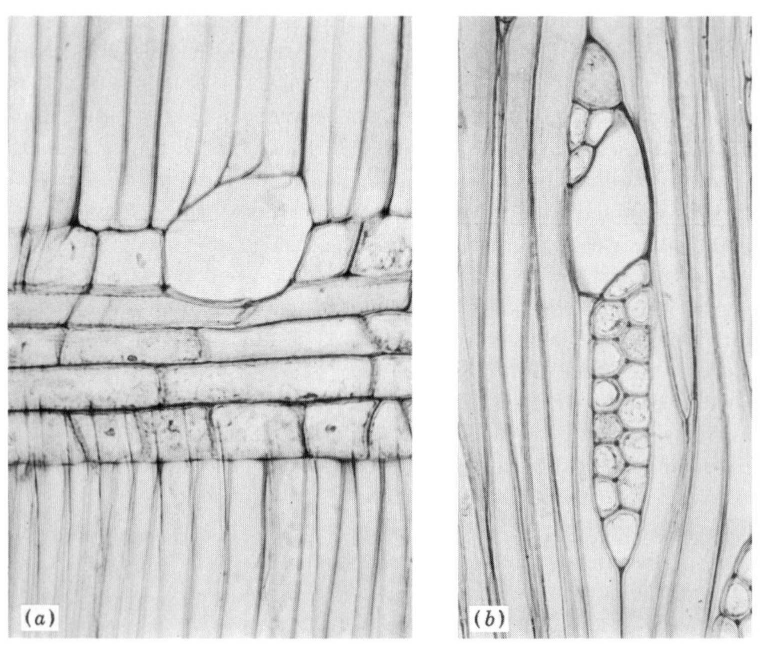

FIG. 5-12 Oil cells in the wood rays of sassafras [*Sassafras albidum* (Nutt.) Nees].

(*a*) Wood ray in a radial section, showing one oil cell. (240×.)

(*b*) Wood ray in a tangential section, showing one oil cell. (240×.)

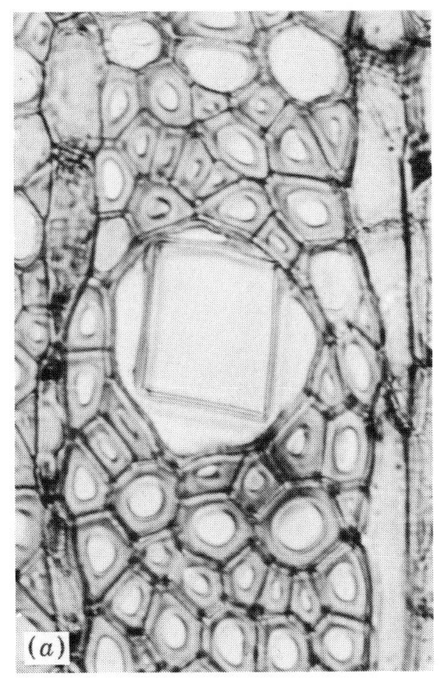

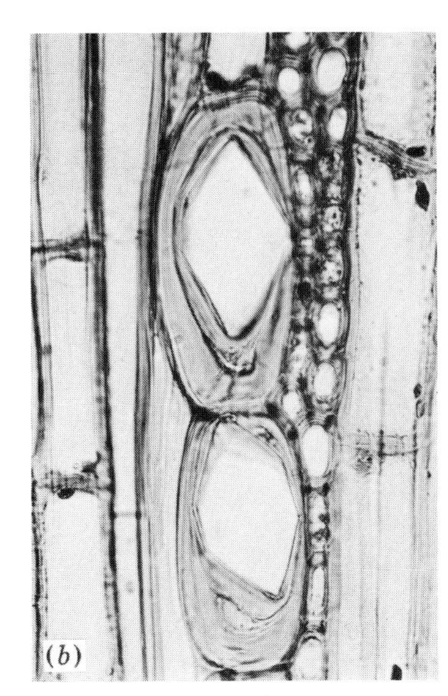

FIG. 5-13 Crystalliferous xylary parenchyma.

(*a*) Crystal locule in an enlarged cell of longitudinal (strand) parenchyma (*x*), in bitternut hickory [*Carya cordiformis* (Wangenh.) K. Koch]. (550×.) The crystal has been dissolved by the acid used in softening the wood.

(*b*) Crystal locules in two sclerosed cells of longitudinal (strand) parenchyma (*t*), in black walnut (*Juglans nigra* L.). (385×.) As in (*a*), the crystals have been dissolved.

Crystals, mostly of calcium oxalate, are quite common, both in the longitudinal parenchyma and in the ray cells. The most common shape of such crystals is rhomboidal (Figs. 5-13, 5-14). However, they also occur in the form of spherical clusters (druses), as separate needle-shaped crystals (acicular), as bundles of long needle-shaped crystals (raphides), and as a granular mass of very fine crystals ("crystal sand").[5,6]

When crystals form in strand parenchyma, a single cell unit of the strand may be transformed into a crystal-bearing loculus, or two or more may serve as crystal repositories (Fig. 5-13*b*). A strand may be in part crystalliferous and in part devoid of crystals. Since the crystal-bearing cells are usually somewhat enlarged, they can be readily recognized. Sometimes one or more units of the strand subdivide, so that the resulting cells are just large enough to accommodate a crystal. Such crystal-bearing parenchyma is designated as *chambered*.

In the rays, if heterocellular, crystals occur more frequently in the upright cells. However, they are not uncommon in the procumbent cells of homocellular rays.

4. NORMAL AND TRAUMATIC GUM CANALS IN POROUS WOODS

Gum canals in porous woods, when present, may be normal or of traumatic (wound) origin. They are not found as normal structures in any of the domestic woods covered by this text.

FIG. 5-14 Crystalliferous xylary parenchyma. Solitary crystal, *in situ*, in the ray parenchyma of yellow birch (*Betula alleghaniensis* Britton). (590×.) (*Photomicrograph by Bror L. Grondal.*)

Normal gum canals occupy the same position in porous woods as the resin canals of conifers; i.e., they extend with the grain, embedded in longitudinal elements, or across the grain, included in the wood rays. The two types, however, are seldom present in the same wood; i.e., the canals in a given hardwood are usually either longitudinal or transverse. Normal longitudinal canals, for instance, feature the dipterocarp woods of the Indo-Malayan region. These include the Philippine mahogany of commerce, and this feature provides a ready separation of these woods from American mahogany (*Swietenia* spp.) and African mahogany (*Khaya* and *Entandrophragma* spp.).

Normal transverse canals are found in the woods of some genera of the Anacardiaceae and Burseraceae. When present, such canals are embedded in the fusiform rays, in the same way as in the coniferous woods.

Traumatic canals in hardwoods may form in one of two ways, or by a combination of these, and they are restricted to the longitudinal kind only. One kind of traumatic canals in hardwoods is of a *schizogenous* nature, of the same kind as in conifers; they result from the separation of cells at the middle lamella. Usually they are arranged in tangential rows, as viewed in cross section (Fig. 5-15*a*). Others are formed through the actual disintegration (gummosis) of cell walls, in which case they are called *lysigenous* canals (Fig. 5-15*b*). In some cases traumatic canals are formed as a result of both these processes; i.e., the cavity first forms through cell fusion and separation of cells at the middle lamella, followed by the enlargement of the cavity through gummosis of the cells immediately surrounding it. This kind of canal formation is described as *schizolysigenous.*

The traumatic schizogenous canals of hardwoods are provided with epithelium, the same as the normal canals; those of lysigenous origin are devoid of it. Since the lysigenous and schizolysigenous canals result from cell disintegration, they are generally much larger (often 1/8 inch or more in diameter) than the schizogenous canals. Lysigenous canals may be scattered singly, or more commonly they are arranged in tangential rows as seen across the grain, and are often conspicuous to the naked eye because of their size and gummy contents. Along the grain the cavity may be continuous, or appear as a series of cysts separated by wound tissue.

Traumatic schizogenous gum canals are occasionally found in sweetgum (Fig. 5-15*a*). They do not occur in any other woods covered by this text. Traumatic lysigenous canals are common in cherry (Fig. 5-15*b*), in which wood they sometimes form as a result of the work of cambial miners.

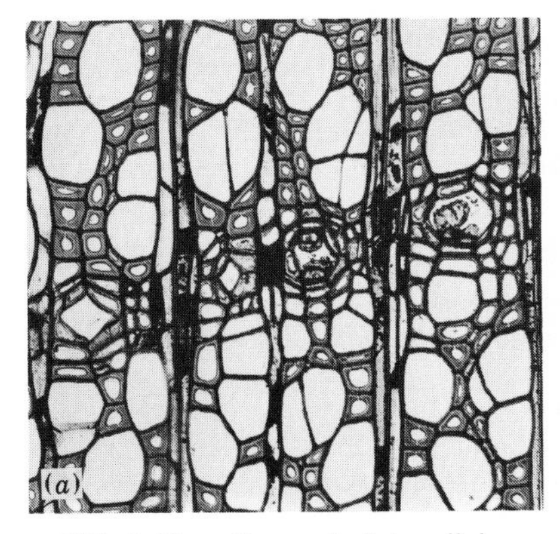

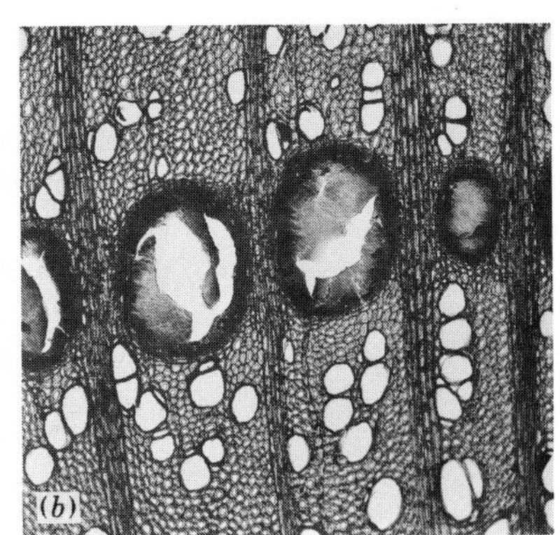

FIG. 5-15 Traumatic intercellular spaces in hardwoods.

(*a*) Traumatic schizogenous gum canals (*x*) in sweetgum (*Liquidambar styraciflua* L.). A row may extend some distance in the tangential direction. (95×.)

(*b*) Traumatic lysigenous gum cavities (*x*) in black cherry (*Prunus serotina* Ehrh.). (50×.)

IV. SUMMATION OF ANATOMICAL CHARACTERISTICS OF XYLEM USEFUL IN IDENTIFICATION OF POROUS WOODS

When confronted with a specimen of an unknown wood it is helpful to bear in mind those gross and minute structural features which may bring about positive identification in the simplest and most dependable way. In some cases the gross features, such as color of the wood, smell, and grain appearance, provide all the identification needed. In other instances these observations on the gross characteristics of wood must be supplemented with information that can be obtained with a hand lens, and in still other cases no positive identification is possible without resorting to a microscope. It is therefore advantageous to proceed in examining a piece of wood from the gross to the minute anatomical features and, when confronted with the latter, to select those which are more easily observable and most reliable. Knowledge of wood anatomy should suggest which structural characteristics are most important and on which of the three surfaces, i.e., transverse, radial, or tangential, they can be seen to best advantage.

A brief summary of the minute features of the xylem, from the standpoint of their usefulness in wood identification, is included in the following paragraphs. Wood sections in which these can be best observed are indicated by letters x for the transverse surface, t for the tangential, and r for the radial. Wood elements, and the features of particular importance under each, are arranged in descending order of their usefulness in identification.

Since all cell dimensions and frequency of cell distribution are subject to variation within a species and in different parts of the same tree, these data should be used as a final resort in the identification of wood. Furthermore, length of longitudinal elements can seldom be accurately ascertained from the wood sections; it requires macerated material for exact measurements (see also page 164).

1. *Vessels and vessel elements*

a. Distribution and shape of pores and pattern formed by them (x)

b. Type of perforations (r, t)

c. Intervessel pitting; size, shape of pits, and their arrangement (r, t)

d. Occurrence of spiral thickenings (r, t)

e. Tyloses and inclusions (x, r, t)

f. Diameter and the number of pores per square millimeter (x)

g. Thickness of walls (x)

h. Length of vessel members (r, t); best determined from macerated material

2. *Rays*

a. Number of cells wide and different combinations of ray widths, e.g., strictly uniseriate rays accompanied by the multiseriate type (t)

b. Composition of rays, i.e., strictly homocellular, heterocellular, or a combination of the two (t, r)

c. Occurrence of specialized cells, e.g., oil cells in sassafras (t, r)

d. Number of rays per linear millimeter (x, t)

e. Height of rays in the longitudinal direction in the number of cells and in microns (*t*). Unlike measurements of individual cells, information on the height of an entire ray is useful and can be easily and accurately determined from the tangential surfaces.

f. Type of pits leading to the contiguous cells (*r*)

g. Occurrence of gum and crystals (*r*, *t*). (Best seen in untreated wood.)

3. *Longitudinal parenchyma*

a. Distribution and arrangement of parenchyma, i.e., apotracheal-diffuse, paratracheal etc. (*x*)

b. Occurrence of inclusions, i.e., gums, crystals, and others (*t*, *r*)

c. Number of cells in a parenchyma strand (*t*, *r*)

4. *Tracheids*

a. If present the type of tracheids, i.e., whether vascular or vasicentric (*r*, *t*)

b. Pattern of distribution (*r*, *t*)

c. Occurrence of spiral thickenings (*r*, *t*)

d. Type of pitting (*r*, *t*)

5. *Fibers*

a. Whether libriform or fiber tracheids (*r*, *t*)

b. Occurrence of septate fibers (*r*, *t*)

c. Occurrence of inclusions (*x*, *t*, *r*)

d. Fiber length; it can be accurately determined only from macerated material

e. Thickness of walls (*x*)

f. Gelatinous fibers

6. ***Storied arrangement of cells*** (*t*). Generally this condition can be determined with the naked eye (ripple marks) or with a hand lens. However, further information on what elements might be storied, in addition to the rays, may be useful and can be obtained by examining a tangential surface of the wood under a microscope.

Anatomical information on the hardwoods is summarized in Table 5-10. The data presented in this table may be used directly for identification of hardwoods, or may be transferred to the multiple-entry punch cards used as a means for keying out woods.* A sample card is shown in Fig. 5-16.

* Details for construction and use of card keys can be obtained from J. D. Brazier and G. L. Franklin, Identification of Hardwoods, *Forest Prod. Res. Bull.* 46, London, 1961.

TABLE 5-10 **Anatomical Data for Hardwoods**[a]

Species	General					Vessels																				
						Pore arrangement[b]							Perforations			Pits										
	Heartwood distinct[c] (1)	Heartwood odor distinctive[d] (2)	Ring-porous (3)	Semi-ring-porous (4)	Diffuse-porous (5)	Early-wood pores one row (6)	Predominantly solitary (7)	Multiples (8)	Chains (9)	Wavy bands (10)	Nested (11)	Spirals present (12)	Simple (13)	Scalariform (14)	More than 10 bars (15)	Opposite (16)	Alternate (17)	Scalariform (18)	Less than 5 μ (19)	Less than 12 μ (20)	More than 12 μ (21)	Linear (22)	Vestured (23)	Miscellaneous (24)	Tracheids, vascular (25)	Tracheids, vasicentric (26)
Acer spp., hard maple	+				+		+	+				+	+				+			+						
Acer spp., soft maple	+				+		+	+				+	+				+			+						
Aesculus spp.	*s*	[e]			+			+				±	+				+			+						
Alnus rubra					+		*s*	+			*s*			+	+	*s*	+		+	*s*						
Arbutus menziesii	+				+	+	*s*	+				+		+[g]			+		+							
Betula spp.	+				+			+						+	+		+		+							
Carpinus caroliniana	+				+			+	*s*			+	+	*s*			+			+	*s*					
Carya spp.	+		+	+				+					+				+			+						
Castanea dentata	+		+					[f]					+	*s*			+			+	+					+

	(1)	(2)	(3)	(4)	(5)	(6)	(7)	(8)	(9)	(10)	(11)	(12)	(13)	(14)	(15)	(16)	(17)	(18)	(19)	(20)	(21)	(22)	(23)	(24)	(25)	(26)
Castanopsis chrysophylla	+		+			+		*f*					+	*s*			+			+	+					+
Catalpa spp.	+	+	+	+						*s*	+	±	+				+			+						
Celtis spp.	+		+							+		+	+				+			+					+	
Cladrastis lutea	+			+				+			+	+	+				+			+	+	*s*	+			
Cornus spp.	+				+		+	+						+	+	+		+		+	+	+				
Diospyros virginiana	*s*			+				+					+				+		+							
Fagus grandifolia	*s*				+		+	+					+	*s*			+			+	+					
Fraxinus spp., white ash	+		+				*s*	+					+				+		+	*s*						+
Fraxinus nigra	+		+				*s*	+					+				+		+	*s*						+
Gleditsia triacanthos	+		+					*s*			+	+	+				+			+			+	*h*		
Gymnocladus dioicus	+		+								+	+	+				+			+			+			
Ilex opaca	*s*				+				+			+		+	+		+		+	*s*						
Juglans cinerea	+			+				+					+				+			+	+					
Juglans nigra	+	+		+				+					+				+			*s*	+					
Liquidambar styraciflua	+				+			+	+			+[k]		+	+	+		*s*		+	+	+				
Liriodendron tulipifera	+				+		*s*	+						+	+	+		*s*		+	*s*	*s*				

TABLE 5-10 Anatomical Data for Hardwoods (*Continued*)

Species	General					Vessels																				
						Pore arrangement[b]							Perforations			Pits										
	Heartwood distinct[c] (1)	Heartwood odor distinctive[d] (2)	Ring-porous (3)	Semi-ring-porous (4)	Diffuse-porous (5)	Early-wood pores one row (6)	Predominantly solitary (7)	Multiples (8)	Chains (9)	Wavy bands (10)	Nested (11)	Spirals present (12)	Simple (13)	Scalariform (14)	More than 10 bars (15)	Opposite (16)	Alternate (17)	Scalariform (18)	Less than 5 μ (19)	Less than 12 μ (20)	More than 12 μ (21)	Linear (22)	Vestured (23)	Miscellaneous (24)	Tracheids, vascular (25)	Tracheids, vasicentric (26)
Lithocarpus densiflorus	+				+		+						+				+			+						+
Maclura pomifera	+		+							*s*	+	+	+				+			+				*m,h*		
Magnolia spp.	+				+		*s*	+				+[l]	+	+[l]				+			+	+				
Morus rubra	+		+							*s*	+	+	+				+			+				*m*		
Nyssa spp.	*s*				+		*s*	+				*s*[k]		+	+	+				+	*s*	*s*				
Ostrya virginiana	+				+			+	*s*			+	+	*s*			+			+	*s*					
Oxydendrum arboreum	+				+		+					+	+	+[g]			+			+						
Platanus occidentalis	+				+		+	+					+	*s*			+		+	*s*						

	(1)	(2)	(3)	(4)	(5)	(6)	(7)	(8)	(9)	(10)	(11)	(12)	(13)	(14)	(15)	(16)	(17)	(18)	(19)	(20)	(21)	(22)	(23)	(24)	(25)	(26)
Populus spp., aspen	*s*	[e]		*s*	+			+					+				+			+	*s*					
Populus spp., cottonwood	*s*	[e]			+			+					+				+			+	*s*					
Prunus serotina	+		*s*	*s*	+	*s*		+			+	+	+				+			+				[h]		
Quercus spp., red oaks	+		+				+[j]						+				+			+						+
Quercus spp., white oaks	+		+				+[j]				+		+				+			+				[m]		+
Quercus virginiana	+				+		+						+				+			+						+
Rhamus purshiana	+				+			+				+	+				+			+						
Robinia pseudoacacia	+		+							*s*	+	+	+				+			+			+	[m]		
Salix nigra	+			*s*	+			+					+				+			+	*s*					
Sassafras albidum	+	+	+				*s*	+					+	*s*			+			+						
Tilia spp.	*s*	[e]			+			+				+	+				+			+						
Ulmus americana	+		+			+[i]				+		+	+				+			+					+	
Ulmus rubra	+	*s*	+							+		+	+				+			+					+	
Ulmus tomasii	+		+			+				+		+	+				+			+					+	
Umbellularia californica	+	+			+		*s*	+				*s*	+				+			+						

TABLE 5-10 Anatomical Data for Hardwoods (*Continued*)

Species	Fibers				Parenchyma									Rays												
	Libriform (27)	*Fiber tracheids* (28)	*Septate* (29)	*Spirals* (30)	*Rare or absent* (31)	*Marginal* (32)	*Paratracheal-scanty* (33)	*Paratracheal-vasicentric* (34)	*Paratracheal-aliform* (35)	*Paratracheal-confluent* (36)	*Apotracheal-diffuse* (37)	*Apotracheal-diffuse-in-aggregate* (38)	*Banded* (39)	*Uniseriate* (40)	*Up to 4 seriate* (41)	*Widest 5 to 10 seriate* (42)	*Widest more than 10 seriate* (43)	*Aggregate* (44)	*Homocellular* (45)	*Heterocellular* (46)	*Storied* (47)	*Sheath cells* (48)	*Oil cells* (49)	*Ray-vessel pitting I*[n] (50)	*Ray-vessel pitting II*[n] (51)	*Ray-vessel pitting III*[n] (52)
Acer spp., hard maple	+	+				+	+				+			+		+			+							+
Acer spp., soft maple	+	+				+	+				+			±	+	*s*			+							+
Aesculus spp.	+	*s*				+	+							+					+	+	±			+		
Alnus rubra		+				*s*	+				+			+				+	+							+
Arbutus menziesii		+	±	±			*s*							+[p]	+	*s*				+					+	
Betula spp.		+				*s*					+	+				+			+							+
Carpinus caroliniana		+				*s*						+	+		+				+	*s*					+	
Carya spp.	+	+				+	+	+			*s*	+	+		+	*s*			+	*s*						+
Castanea dentata	+	+					+				+			+					+						+	
Castanopsis chrysophylla	+	+					+				+		+	+					+						+	
Catalpa spp.	+					+		+	+				+		+	*s*			+	*s*					+	

Cladrastis lutea	+					+		+			+			+[p]		+				+		+				
Cornus spp.		+					+					+	+	+[p]		+				+						+
Diospyros virginiana	+					+		+				+	+		+				+	+	+					+
Fagus grandifolia		+									+	+	+				+		+						+	
Fraxinus spp., white ash	+					+	+	+	+						+				+							+
Fraxinus nigra	+					+	+	+							+				+							+
Gleditsia triacanthos	+					+			+	+	+					+	*s*		+							+
Gymnocladus dioicus	+					+		+	+	+					+	+			+							+
Ilex opaca		+		+			+				+			+[p]		+				+						+
Juglans cinerea		+				+	+	+			+	+	+		+				+	*s*						+
Juglans nigra		+				+	+	+			+	+	+		+	*s*			+	*s*						+
Liquidambar styraciflua		+					+				+			+[p]	+				+	+					+	
Liriodendron tulipifera	+	+				+									+	*s*			+	+					+	
Lithocarpus densiflorus	+	+					+				+	+	+	+			+	+	+						+	
Maclura pomifera	+					*s*		+	+	+					+	*s*			+						+	
Magnolia spp.		+				+									+	*s*			+	+					+	
Morus rubra	+							+	+		+					+			+	+					+	
Nyssa spp.		+					+				+				+					+						+
Ostrya virginiana	+					*s*						+	+		+			+	+	*s*					+	

TABLE 5-10 **Anatomical Data for Hardwoods** (*Continued*)

Species	Fibers				Parenchyma								Rays													
	Libriform (27)	*Fiber tracheids* (28)	*Septate* (29)	*Spirals* (30)	*Rare or absent* (31)	*Marginal* (32)	*Paratracheal-scanty* (33)	*Paratracheal-vasicentric* (34)	*Paratracheal-aliform* (35)	*Paratracheal-confluent* (36)	*Apotracheal-diffuse* (37)	*Apotracheal-diffuse-in-aggregate* (38)	*Banded* (39)	*Uniseriate* (40)	*Up to 4 seriate* (41)	*Widest 5 to 10 seriate* (42)	*Widest more than 10 seriate* (43)	*Aggregate* (44)	*Homocellular* (45)	*Heterocellular* (46)	*Storied* (47)	*Sheath cells* (48)	*Oil cells* (49)	*Ray-vessel pitting I*[n] (50)	*Ray-vessel pitting II*[n] (51)	*Ray-vessel pitting III*[n] (52)
Oxydendrum arboreum		+					*s*				+			+[p]		+				+						
Platanus occidentalis		+					+				+	+					+		+							+
Populus spp., aspen	+					+								+					+						+	
Populus spp., cottonwood	+					+								+					+						+	
Prunus serotina		+			+		*s*				*s*				+	*s*			+	*s*						+
Quercus spp., red oaks	+	+					+				+	+	+	+			+		+						+	
Quercus spp., white oaks	+	*s*					+				+	+	+	+			+		+						+	
Quercus virginiana	+	+					+				+	+	+	+	+		+	+	+						+	
Rhamus purshiana	+	*s*						+			+				+	*s*			+							+
Robinia pseudoacacia	+					+		+		+						+			+	*s*					+	

	(27)	(28)	(29)	(30)	(31)	(32)	(33)	(34)	(35)	(36)	(37)	(38)	(39)	(40)	(41)	(42)	(43)	(44)	(45)	(46)	(47)	(48)	(49)	(50)	(51)	(52)
Sassafras albidum	+	+					*s*	+	+	+					+				+	+			+		+	
Tilia spp.		+				+						+	+	+		+			+		*s*				+	
Ulmus americana	+							+[o]			+					+			+						+	
Ulmus rubra	+							+[o]			+					+	*s*		+						+	
Ulmus tomasii	+							+[o]			+						+		+						+	
Umbellularia californica	+		±					+							+					+			+	+		

[a] In this table the *plus* (+) signs indicate a given feature is commonly present; the *letter s* indicates that a feature may be present but only sporadically, and hence is of little diagnostic value; when both a *plus* and *minus* signs appear in the same space, this indicates that this particular feature is abundant in some pieces and sparse in others; blank spaces indicate the absence of the feature.

[b] In the case of ring-porous and semi-ring-porous woods, pore arrangement refers to the late wood, for diffuse-porous woods to the entire growth ring.

[c] See Chap. 12 for specific information on color of the heartwood.

[d] See Chap. 12 for more specific information on odor.

[e] Odorless or nearly so when dry; with a disagreeable odor when wet.

[f] Arranged in flame-shaped patches of light tissue.

[g] Also occasionally reticulate (foraminate).

[h] Colored deposits common.

[i] Usually in one row; sometimes in several rows, especially in juvenile wood.

[j] In red oak late-wood pores round, thick-walled, distinct with a hand lens; in white oak late-wood pores, more numerous, thin-walled, angular, not sharply defined with a hand lens.

[k] Restricted to the tapering ends of the vessel members.

[l] Spirals rare in ***M. acuminata,*** present in ***M. grandiflora;*** scalariform perforations occasional in ***M. acuminata,*** common in ***M. grandiflora.***

[m] Tyloses abundant in heartwood.

[n] Ray-vessel pitting, class I—pits simple and elongated; class II—pits ranging from simple to bordered; class III—pits same as intervessel.

[o] Paratracheal parenchyma abundant but not forming complete sheaths around the early-wood pores and included in the wavy concentric bands of late-wood vessels and vascular tracheids.

[p] Uniseriate rays consist entirely of upright cells.

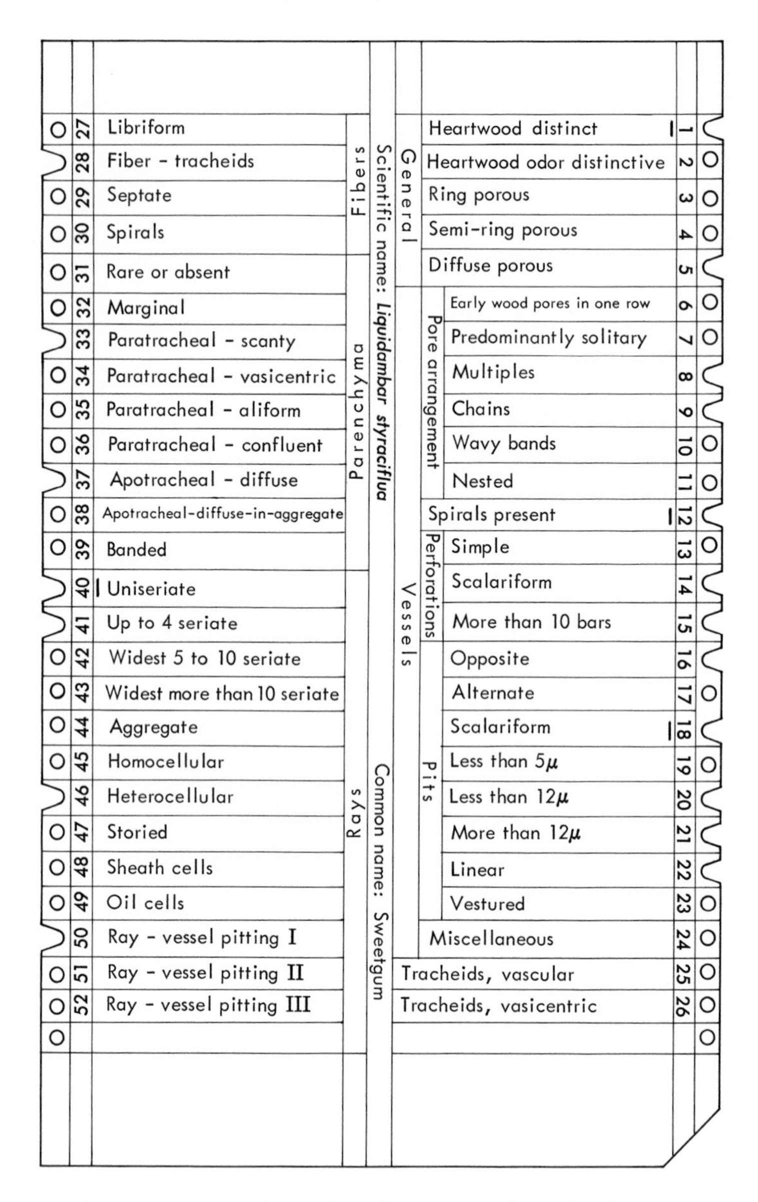

FIG. 5-16 A sample punch card. Sweetgum (*Liquidambar styraciflua* L.). Further information can be included by adding new titles or by using the ends on the reverse side of the card. Punched perforations signify that the feature is present; a black line under the number means that this feature is present but not well developed or requires further amplification, for which see Table 5-10 or Chap. 12.

SELECTED REFERENCES

1. Bailey, I. W.: The Development of Vessels in Angiosperms and Its Significance in Morphological Research, *Am. J. Botany,* **31:**421–428 (1944).
2. Barghoorn, E. S.: The Ontogenetic Development and Phylogenetic Specialization of Rays in the Xylem of Dicotyledons, *Am. J. Botany,* **27:**918–928 (1940); **28:**273–282 (1941).
3. Chalk, L.: Ray Volumes in Hardwoods, *Trop. Woods,* **101:**1–9 (1955).
4. Chattaway, M. M.: The Development of Tyloses and Secretion of Gum in the Heartwood Formation, *Australian J. Sci. Res.,* **B2:**227–240 (1949).
5. ———: Crystals in Woody Tissues, I, *Trop. Woods,* **102:**55–74 (1955).
6. ———: Crystals in Woody Tissues, II, *Trop. Woods,* **104:**100–124 (1956).
7. Chowdbury, K. A.: Terminal and Initial Parenchyma in the Wood of *Terminalia tomentosa* W. & A., *New Phytologist,* **35:**351–358 (1936).
8. Desch, H. E.: "Timber, Its Structure and Properties," The Macmillan Company, New York, 1953.
9. Esau, K.: "Anatomy of Seed Plants," John Wiley & Sons, Inc., New York, 1960.
10. French, E. E.: The Effect of the Internal Organization of the North American Hardwoods upon Their More Important Mechanical Properties, unpublished thesis, The New York State College of Forestry, Syracuse, N.Y., 1923.
11. Frost, F. H.: Histology of the Wood of Angiosperms, 1, The Nature of the Pitting between Tracheary and Parenchymatous Elements, *Bull. Torrey Botan. Club,* **56:**259–264 (1929).
12. Gerry, E.: Tyloses: Their Occurrence and Practical Significance in Some American Woods, *J. Agr. Res.,* **1:**445–469 (1914).
13. Greenidge, K. N. H.: An Approach to the Study of Vessel Length in Hardwood Species, *Am. J. Botany,* **39**(8):570–574 (1952).
14. Hess, R. W.: Classification of Wood Parenchyma in Dicotyledons, *Trop. Woods,* **96:** 1–20 (1950).
15. Isenberg, I.: Microchemical Studies of Tyloses, *J. Forestry,* **31:**961–967 (1933).
16. Jane, F. W.: "The Structure of Wood," The Macmillan Company, New York, 1956.
17. Jurášek, L.: Vznik Thyl v Bukovèm Drěvě (The Origin of Tyloses in Beech Wood), *Drev. Výskum,* **1:**7–15 (1956).
18. Kórán, A., and W. A. Côté, Jr.: Ultrastructure of Tyloses and a Theory of Their Growth Mechanism, *News Bull., Int. Ass. Wood Anatomists,* **2:**3–15 (1967).
19. ——— and ———: The Ultrastructure of Tyloses, pp. 319, 334 in W. A. Côté, Jr. (ed.), "Cellular Ultrastructure of Woody Plants," Syracuse University Press, Syracuse, N.Y., 1965.
20. Kribs, D. A.: Salient Lines of Structural Specialization in the Wood Rays of Dicotyledons, *Botan. Gaz.,* **94:**547–557 (1935).
21. ———: Salient Lines of Structural Specialization in the Wood Parenchyma of Dicotyledons, *Bull. Torrey Botan. Club,* **64:**177–186 (1937).
22. Metcalfe, C. R., and L. Chalk: "Anatomy of the Dicotyledons," vols. I, II, Oxford University Press, Amen House, London, 1950.
23. Meyer, R. W.: Tyloses Development in White Oak, *J. Forest Prod.* **17**(12):50–56 (1967).
24. ——— and W. A. Côté, Jr.: Formation of the Protective Layer and Its Role in Tyloses Development, *Wood. Sci. & Technol.,* **2**(2):84–94 (1968).
25. Myer, J. E.: Ray Volumes of the Commercial Woods of the United States and Their Significance, *J. Forestry,* **20:**337–351 (1922).
26. Weber, I. E.: Intercellular Cavities in the Rays of Dicotyledenous Woods, *De Lilloa (Argentina),* **2:**465–469 (1938).
27. White, D. J. B., and A. W. Robards: Some Effects of Radial Growth Rate upon the Rays of Certain Ring-porous Hardwoods, *J. Inst. Wood Sci.,* **17:**45–52 (1966).
28. Zimmermann, M. H.: Physiological Aspects of Wood Anatomy, *Bull. Intern. Assoc. Wood Anatomists,* **2:**11–14 (1968).
29. ——— and P. B. Tomlinson: A Method for the Analysis of the Course of Vessels, *Bull. Intern. Assoc. Wood Anatomists,* **1:**2–6 (1967).

6

The Physical Nature of Wood

Basically all the physical properties of wood are determined by the factors inherent in its structural organization. These may be summarized under five headings:

1. The amount of cell wall substance present in a given volume of wood
2. The amount of water present in the cell wall
3. The proportionate composition of the primary components of the cell wall and the quantity as well as the nature of the extraneous substances present
4. The arrangement and orientation of the wall materials in the cells and in the different tissues
5. The kind, size, proportions, and arrangement of the cells making up the woody tissue

The first of these factors is measured by the specific gravity of the wood and furnishes the most useful index to the predicted physical behavior of wood. The second factor profoundly affects the total physical behavior of wood, not only because the addition of water to the cell wall changes its density and dimensions but because of its effect on plasticity and transfer of energy within a piece of wood. The third factor is related to many of the special properties of certain kinds of wood as well as to the deviations from expected quantitative behavior. The last two items are the cause for the large differences which are found in the physical responses of wood with respect to the grain direction.

I. NONMECHANICAL PROPERTIES OF WOOD

A. Moisture Content of Wood

Wood is a hygroscopic substance; i.e., it has an affinity for water in both liquid and vapor form. This ability of wood to absorb or to lose water is dependent on the temperature and the humidity of the surrounding atmosphere. As a consequence, the amount of moisture in wood fluctuates with changes in the atmospheric conditions around it.

All the physical properties of wood, mechanical as well as nonmechanical, are greatly affected by the fluctuations in the quantity of water present. In using wood as a raw material it is therefore essential to be able to evaluate its moisture content, and to understand where the moisture is located and how it moves through the wood.

1. DETERMINATION OF THE MOISTURE CONTENT

The total amount of water in a given piece of wood is called the ***moisture content*** (abbreviated M.C.); it is expressed as a percentage of the ovendry weight of the wood. The ovendry weight is used as a basis because it is an indication of the amount of solid substance present. The moisture content is stated as

$$\text{M.C., \%} = \frac{\text{weight of water in wood}}{\text{OD weight of wood}} \times 100 \tag{1}$$

The standard method for determining the amount of water is to dry the wood in an oven to constant weight at 103 ± 2°C. The weight obtained under such conditions is known as the ***ovendry weight*** (abbreviated OD weight). The weight of water present is therefore the difference in weights before and after drying, and Eq. (1) may be rewritten as

$$\text{M.C., \%} = \frac{\text{weight of wood with moisture} - \text{OD weight of wood}}{\text{OD weight of wood}} \times 100 \tag{2}$$

As an example of the application of Eq. (2), consider a block of wood fresh from the tree which weighs 415.0 grams. If after oven-drying the weight is 280.4 grams, substitution in Eq. (2) yields a moisture content of 48 percent.

When wood contains appreciable amounts of extractives such as resins or creosote which are volatilized by heating, the standard methods for moisture determination yield false values. In these cases solvent extraction in special apparatus must be employed.

2. LOCATION OF WATER IN WOOD

Water is taken up by wood in two different ways: (1) in the cell wall materials as ***bound water,*** and (2) as ***free water*** in a liquid form in the cell lumina.

Polar liquids unite with dry cell wall material by means of hydrogen bonding. Formation of these bonds releases energy which can be measured as the ***heat of wetting*** for dry wood. Conversely, energy must be supplied to wet wood to remove any water that is present. The forces of attraction between dry wood and water are so large that it is impossible to prevent the gain of moisture, and in consequence wood is a ***hygroscopic material.*** Molecules of polar liquids diffuse into the spaces in the cell wall structure, the dislocations in the microfibrils, the amorphous regions between the microfibrils, and the interlamellar spaces. As the molecules of liquids enter the cell wall, the microfibrillar framework expands laterally in proportion to the size and quantity of the molecules of liquids which have been introduced. With continued expansion, the framework of the wall substance becomes distorted and builds up a resisting force to the introduction of further liquid. Ultimately the forces tending to attract the molecules of liquid into the wall are balanced by the resistance of the strained lattice to further expansion. As a result a balance point, called the ***fiber saturation point*** (F.S.P.), is reached. By definition, the fiber saturation point represents that moisture content at which the cell wall is completely saturated with water, but no moisture is present in the cell lumen. This condition is associated with maximum swollen volume of the cell wall and with major changes in the physical behavior of wood, and hence is of primary importance. The fiber saturation point is not completely uniform between different kinds of native woods because of variations in their chemical organization, but generally falls between 25 and 30 percent moisture content.

3. MOISTURE EQUILIBRIUM IN WOOD

Wood exposed to an atmosphere containing moisture in the form of water vapor will come, in time, to a steady moisture-content condition, called the ***equilibrium moisture content*** (E.M.C.). This steady moisture state depends on the relative humidity,* the temperature of the surrounding air, and the drying conditions which it has previously undergone; it fluctuates with changes in one or both of these atmospheric conditions, as shown in Fig. 6-1. Woods that are seemingly alike may not conform to the same E.M.C. under identical atmospheric conditions because they contain varying amounts of extractives, or the previous drying conditions may have caused the formation of permanent hydroxyl bonds that decrease the ability of the wood to absorb moisture.

The equilibrium moisture content of wood exposed to normal conditions outdoors, but under cover, is about 12 to 15 percent over much of the United States. This range of moisture content is usually taken as the ***air-dry moisture content of lumber.*** Under heated conditions inside buildings, the equilibrium moisture content will normally be much lower and will range from 8 percent to as low as 4 percent in mid-winter. Closer estimates may be made of the equilibrium moisture contents for wood in service, if atmospheric conditions are known, by referring to Fig. 6-1

* Relative humidity is the ratio of the actual vapor pressure of the air to the pressure of saturated vapor at the prevailing dry-bulb temperature, expressed as a percentage.

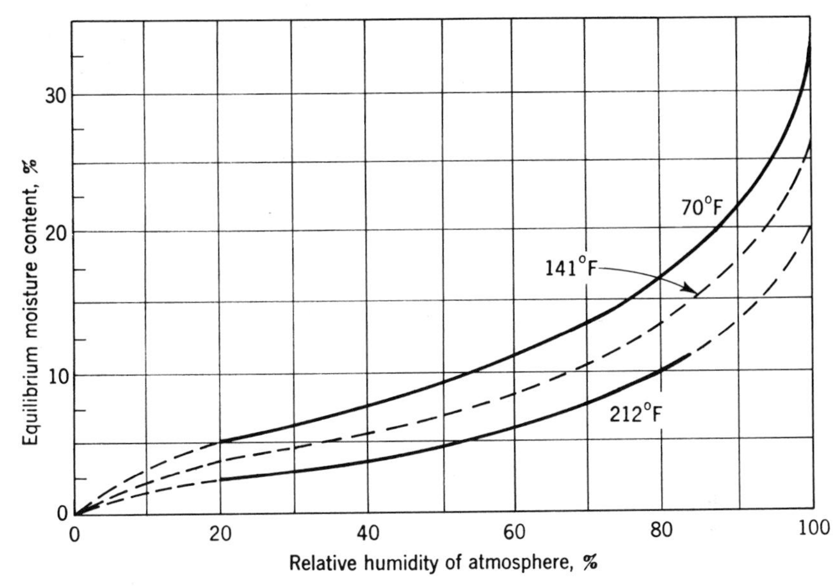

FIG. 6-1 Sorption isotherms for wood at three temperatures. (*Adapted from U.S. Dept. of Agriculture Handbook No.* 188.)

or to tables such as those given in the "Wood Handbook."[10] The changes in the moisture content of wood in response to changes in relative humidity and temperature of the air are a function of time, with rapid adjustments taking place at first and very slow changes occurring as the moisture content approaches the state of equilibrium. These fluctuations in the moisture content may be fairly extensive at the surface, where moisture loss or gain is rapid. They are much less extensive in the interior of the piece, where changes in moisture content proceed at a slower rate, because of the longer time necessary for diffusion of water through the wood to take place.

Where moisture-content differences exist within a piece of wood, there is a gradual transition in amount of moisture between the two extremes. When plotted, the differences in magnitude of moisture content determine the rate or slope of the transition from one moisture condition to another within the piece of wood. This transition rate is known as the *moisture gradient* of the wood.

4. MOISTURE IN WOOD ABOVE THE FIBER SATURATION POINT

The liquid phase of water in contact with wood, whether in the living tree or under service conditions, will cause the cell lumina to accumulate free water. *Maximum moisture content* is reached when all spaces in the wall and the lumens are filled. The amount of water held at this point of total saturation is limited by the void volume of the wood, i.e., by the space not occupied by cell wall substance and extraneous materials. A value for maximum moisture content in wood can be calculated if the weight of water which occupies the void volume is expressed as a percentage of the ovendry weight of the wood. The commercial woods native to the United States have a range of maximum moisture contents from 60 percent for the heavy woods, to 200 percent for the lighter timbers. An extremely low level for maximum moisture content is exhibited by black ironwood [*Krugiodendron ferreum.* (Vahl) Urban], which grows to a limited extent in Florida. This wood has a specific gravity of approximately 1.18 (ovendry weight and volume) and contains only 26 percent moisture at saturation. In contrast, the lightest timbers such as the South American balsa wood (*Ochroma lagopus* Sw.), with specific gravity of 0.2 or less, can hold more than 400 percent moisture when saturated. Normally wood in living trees will not contain more than half to two-thirds as much moisture as is theoretically possible. However, there is no exact value for a species, and variations occur from tree to tree, as well as within single trees. The sapwood of conifers is usually much wetter than the heartwood, and its moisture content may exceed 200 percent. In hardwoods there is commonly no such difference in the moisture content between sapwood and heartwood or with location in the stem. The total water content of the wood of a tree does not appear to fluctuate widely at different times during the year. The distribution of the water within the stem, however, may change from month to month, or from one season to another.

5. MOISTURE MOVEMENT IN WOOD

The equilibrium moisture content conditions that have been described are an expression of the total effect of the moisture on wood. However, within the piece the rates of movement are not the same in all directions with respect to the major axes of the wood. In the longitudinal direction, the movement of water in the vapor form is greatly expedited by the tubular structure of the cells. As a consequence water moves twelve to fifteen times as rapidly along the grain as it does across it, and hence in a cube-shaped piece of wood much more water will evaporate through the ends of the piece. In practice, however, since the length of the pieces generally considerably exceeds their thickness, most water is lost through the sides. For instance, in a piece 12 inches long and 1 inch thick, water will have to travel only a maximum of ½ inch across the grain to reach a surface and twelve times as far from the center of the piece to evaporate from end grain. This situation is exaggerated to an even greater extent in thin veneers.

B. Dimensional Changes in Wood

Addition of water or other polar liquids to the cell wall substance causes the microfibrillar net to expand in proportion to the amount of liquid which has been added. This continues until the fiber saturation point has been reached. Further addition of water to the wood produces no change in the volume of the wall substance, because additional water, above this level, is concentrated in the lumen. Conversely, the removal of moisture from the cell wall below the fiber saturation point causes the wall to shrink. Such dimensional changes are traditionally expressed as a percentage of the maximum dimension of the wood, and since the green size is a condition at which no reduction in dimension has yet occurred, the shrinkage is expressed as a percentage of the green volume or size.

$$\text{Shrinkage, \%} = \frac{\text{change in dimension from swollen size}}{\text{swollen dimension}} \times 100 \qquad (3)$$

The shrinkage curves shown in Fig. 6-2 are typical of wood in general, although they were constructed from data on a limited number of woods and levels of moisture content. It will be observed from these curves that the tangential shrinkage for air-dried wood is about twice as large as the radial, at the same moisture content. The volumetric shrinkage is roughly the sum of the two, since the longitudinal shrinkage of normal wood from the green to ovendry condition is almost negligible, being in the order of 0.1 to 0.2 percent. It will be noted that the portions of the curves between 6 and 18 percent moisture content are approximately straight lines, and for this reason dimensional changes in wood, for any moisture conditions between these points, are assumed to conform to a straight-line relationship.

Dimensional changes that occur in wood are a function not only of the quantity of moisture in the wood, but also of the amount of the cell wall substance. The greater the amount of material present, the larger the dimensional changes that are possible for the same percent moisture-content change. This must be considered as a rough indicator only, since the correlation does not hold well for all woods, as is evident from an examination of the values in Tables 12-1 and 12-2. As an extreme example of such a departure, Honduras mahogany (*Swietenia macrophylla* King),

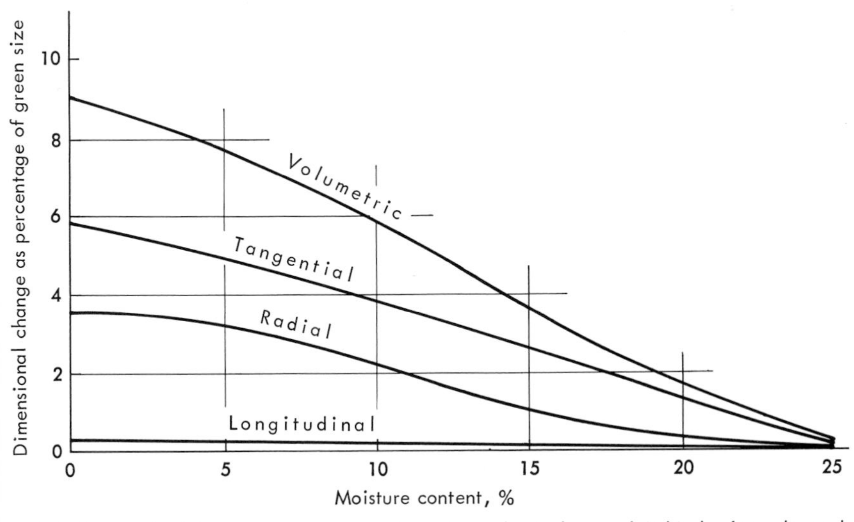

FIG. 6-2 Shrinkage curves for wood. Data from physical tests of six kinds of wood tested at State University of New York, College of Forestry, for the United Fruit Company.

with approximately the same amount of material in the cell wall as sweetgum (*Liquidambar styraciflua* L.), is nearly twice as stable dimensionally. The volumetric shrinkage from green to ovendry condition is 7.7 percent for the mahogany, as against 15.0 percent for the sweetgum.

Collapse (see page 327), a defect which may develop in the drying of some woods, is an exception to the rule that green wood does not shrink above the fiber saturation point. Collapse is characterized by irregular and abnormal shrinkage which occurs above the fiber saturation point. It is due to buckling and crushing of the cell walls; for this reason the amount of shrinkage resulting from collapse cannot be predicted on the basis of the normal dimensional changes in wood. Collapse is common in western redcedar (*Thuja plicata* Donn), redwood [*Sequoia sempervirens* (D. Don.) Endl.], and a number of hardwoods, such as oaks (*Quercus* spp.), black walnut (*Juglans nigra* L.), and most species of the genus Eucalyptus.

Swelling of wood is dimensional change expressed as a percentage of the dry dimensions of wood.

$$\text{Swelling, \%} = \frac{\text{change in dimension from dry size}}{\text{dry dimension}} \times 100 \qquad (4)$$

Swelling in wood cannot be taken as a reciprocal of the shrinkage values since two different reference bases are used in their calculation. Furthermore an irreversible change occurs in the dimensional response of wood to moisture changes when it is first dried from the green condition. As a result, all the subsequent dimensional changes with moisture content are smaller than those which occur during the first drying cycle. This permanent change in dimensional response to moisture varies markedly from species to species.

1. DIMENSIONAL CHANGES OF WOOD IN SERVICE

The total dimensional changes that may occur in a piece of wood, in drying from the green condition, are an important consideration in manufacture. Even more important, however, are the changes in dimensions that accompany the usual fluctuations of relative humidity after wood has been placed in service. This behavior is known as *movement* and cannot be predicted from the dimensional changes based on shrinkage from the green condition. It is a useful index to the employment of wood under the range of atmospheric conditions that affect wood in service. As an example, basswood (*Tilia* sp.) exhibits very high shrinkage from the green to ovendry conditions for its specific gravity, and on the basis of these values it would be unsuitable for use. However, the dimensional changes this same wood undergoes, in the air-dry condition, from winter to summer in heated buildings are from one-third to one-half as great as would be expected on the basis of the shrinkage data from green conditions. As a result basswood is suitable for general use.

2. DIMENSIONAL CHANGES AND SPECIFIC GRAVITY

The magnitude of dimensional changes in wood is directly related to the amount of cell wall material that is present. In general, shrinking and swelling in wood increase with increasing specific gravity. This is true both between pieces of wood of different kinds and within single pieces in which there are local differences in specific gravity, such as those between early-wood and late-wood bands in the hard pines. However, this relationship may be masked by the presence of large amounts of extractives which serve as bulking agents in the cell wall spaces and reduce dimensional changes below that expected for the apparent specific gravity of the wood. Also, abnormally high lignin fractions in wood may reduce dimensional changes, at least in the longitudinal axis of wood.[33]

3. ANISOTROPIC DIMENSIONAL CHANGES IN WOOD

The observed dimensional changes in wood are unequal along the three structural directions; i.e., wood exhibits anisotropy. These dimensional changes include both the swelling, associated with the addition of water to the wood, and shrinking which results from the removal of liquid from the wood. For normal wood the dimensional changes, as indicated by shrinkage data from the green to the ovendry condition, amount to only 0.1 to 0.3 percent in the longitudinal direction.[10] In the radial direction, the shrinkage from the fiber saturation point to the air-dry condition is from 2 to 3 percent. In the tangential direction shrinkage to the air-dry condition is about twice as great as in the radial direction.[10] The ratio of tangential to radial shrinkage (T/R) ranges from 1.40 to more than 2 in our native woods (Tables 12-1 and 12-2) and, along with the shrinkage percentage itself, constitutes a means of assessing the dimensional stability of any given wood. The wood best suited for uses involving critical dimensional stability is one with a low T/R ratio and with low absolute transverse dimensional changes.

The large differential changes between the directions parallel and perpendicular to the grain can be explained by considering the following ideas:

1. The cell walls of wood consist of reinforcing microfibrils of high tensile strength along their axes, embedded in an essentially amorphous matrix of lignin and hemicellulosic material.[1,2,11,30] With the addition or removal of water the microfibrils change very little in length and only very little laterally; at the same time the matrix tends to change dimensions more or less equally in all directions and to a much larger extent than the microfibrils. The strong reinforcement of the microfibrils deforms the matrix and produces unequal strains in the length, width, and thickness of the wall, causing small longitudinal and considerable lateral dimensional changes. The magnitudes of the longitudinal and transverse components of dimensional changes in the cell walls are functions of (*a*) the mechanical properties of the composite wall, i.e., the ratio between the effective modulus of elasticity of the microfibrils and the shear modulus of the matrix for a given moisture condition of

the wood, the square of the microfibril length, friction coefficient of the matrix, and (*b*) the mean angle of the microfibrils with the cell axis. Increasing microfibrillar angle in the cell wall causes increase in longitudinal changes and decrease in transverse changes. The magnitude of these rates is increased by increasing moisture content up to the fiber saturation point. This reinforced matrix theory has been very nicely demonstrated with pine wood but has not been applied to hardwoods as yet.

2. The majority of cells in wood are arranged with their long axes in the longitudinal direction, i.e., parallel to the grain, and because of the effect of the microfibrils discussed in the preceding paragraph, they change little in the longitudinal dimension and considerably in the lateral directions with fluctuations in the moisture content of the wood.

There is fairly general agreement concerning the reason for the small dimensional changes in the longitudinal direction in wood, in comparison with the relatively large lateral dimensional changes. However, the differential between the transverse shrinkage and swelling in the tangential and radial directions is still not understood. A number of theories have been advanced in an attempt to explain the dimensional anisotropy between the lateral axes of wood.[14,27,43] Most of these theories have assumed that an explanation can be given on the basis of one or two factors.

One theory which appears to explain a considerable part of the differences between the tangential and radial dimensional changes is founded on the dissimilarity in wall thickness and cell wall layer orientation in early and in late wood.[14] The large percentage of $S2$ layer in the late-wood cell walls causes a large lateral shrinkage in this portion of the increment. The early-wood cells, with their thinner walls, have a relatively small percentage of $S2$ layer, and the lateral dimensional changes are considerably affected by the low-angle microfibrillar orientation of the $S1$ and $S3$ layers. As a result the early wood has much less transverse shrinkage than the late wood. The transverse dimensional changes in the two axial directions are the result of the differences in the amount and structure of the cell walls in the early and late wood. Tangential shrinkage and swelling are largely controlled by the changes in the late wood, since this part of the growth increment is strong enough to force the early wood to comply with it. The radial dimensional changes, on the other hand, are a summation of the weighted contributions of each part of the annual increment. It is smaller than that in the tangential direction because of the presence of the low-shrinking early-wood component in the total. This lamellar theory for explaining radial and tangential shrinkage differences has been presented in mathematical terms and demonstrated with reasonable precision for pine wood.[43]

Not all the difference between tangential and radial shrinkage and swelling in wood can be accounted for by the above theory. For example, the T/R ratio of red oak late wood is about 1, but the early-wood T/R ratio is almost 2.[27] The only factor which seems applicable in explaining such a discrepancy is the presence of rays which restrain the relatively weak early wood and reduce its radial dimensional change. The low radial shrinkage of ray tissue has been demonstrated[28] and can contribute materially to the T/R differential in woods which have large rays and relatively thin walls in the longitudinal tissue. In the conifers, where rays are small, and in those hardwoods with heavy-walled cells, the influence of the rays is necessarily smaller, but nevertheless still significant.

Other factors undoubtedly contribute to the lateral anisotropy of wood. In fact, the real explanation of this phenomenon probably depends on a set of simultaneously varying factors which may not be entirely recognized as yet. For example, Hittmeier[16] has demonstrated that at low moisture contents the tangential to radial (T/R) ratio for swelling in wood increases as moisture content increases, but at high moisture contents the T/R ratio does not increase. This indicates that the transverse changes at high moisture contents may be largely tangential. This effect can be noted by examining the ratio of radial to tangential shrinkage for the regions near 10 and 20 percent moisture contents in Fig. 6-2.

C. Specific Gravity and Density of Wood

1. SPECIFIC GRAVITY

Specific gravity is usually expressed as a ratio of weight of the substance to the weight of an equal volume of water and is abbreviated as sp gr or G.* In the case of wood the ovendry weight is used as a basis,† and comparison is made with the weight of the displaced volume of water. Stated as an equation this becomes

$$\text{sp gr} = \frac{\text{ovendry weight of wood}}{\text{weight of the displaced volume of water}} \tag{5}$$

In this equation the *ovendry weight* of the wood is *always* used as the numerator. The value of the denominator, which depends on the volume of the wood, varies with the moisture content of the test block, because of the dimensional changes that occur in wood below the fiber saturation point. For this reason it is necessary to specify the moisture content of the wood at which the volume was determined, when stating the specific gravity. As the volume becomes smaller, with decrease in the moisture content, the denominator of the ratio becomes smaller, and the specific-gravity value correspondingly larger. The reverse is true as the moisture content of the test block increases. As a result, the minimum value of the specific gravity is obtained when the green volume is used, and the maximum when the volume of wood is taken at the ovendry condition in determining the weight of the displaced volume of water.

Specific gravity of wood based on green volume, or ***basic specific gravity,*** is one of the most useful and commonly cited values. The term ***basic*** is applied since both green volume and ovendry weight are as nearly constant and reproducible measurements as can be obtained with wood.

*The term *specific gravity* is very misleading since it has nothing to do with gravity. A more suitable name for this concept is *density index*.

† The use of ovendry weight as a reference was developed because this value can be reproduced experimentally.

In general terms, the specific gravity of wood depends upon (1) the size of the cells, (2) the thickness of the cell walls, and (3) the interrelationship between the number of cells of various kinds in terms of (1) and (2). Fibers are particularly important in the determination of specific gravity, since their small cross sections allow a great number of them to be massed in a small place. If the fibers are thick-walled and show small lumina, then the total air space is relatively small, and the specific gravity tends to be high. On the other hand, if they are thin-walled or wide-lumened, or both, the specific gravity will be low. Unusually light woods, such as balsa (*Ochroma lagopus* Sw.), illustrate this last condition with high concentrations of thin-walled large-lumened fibers and low vessel volume. Low specific gravity may also be the result of unusually high vessel volume in wood.

The above discussion assumes that all the solid material in a piece of wood is a part of the cell wall substance, which swells and shrinks with changes in the moisture content. However, the total material in the woody cell can consist not only of the cell wall substance, but also of considerable amounts of infiltrated materials located in the lumen; these generally play no part in the dimensional changes caused by changes in moisture content, but in some cases may reduce the swelling and shrinking. Therefore, the usual specific-gravity measurements are only a relative index of the cell wall materials.

The basic specific gravities (calculated on ovendry weight and volume in the green condition) of the commercial woods included in this text range from 0.29 (*Thuja occidentalis* L.) to 0.81 (*Quercus virginiana* Mill.), most of them falling between 0.35 and 0.65 (see Table 12-1). The extreme ranges of basic specific gravities in our native woods are 0.21 for corkwood (*Leitneria floridana* Chapm.) and 1.04 in black ironwood [*Krugiodendron ferreum* (Vahl) Urban]. However, woods from other parts of the world may exceed the range of these values, with specific gravities reported as low as 0.04 and as high as 1.40.

Woods with basic specific gravities of 0.36 or less are considered to be light; 0.36 to 0.50, moderately light to moderately heavy; above 0.50, heavy. The groupings of specific-gravity classes apply to woods from both the temperate and tropical zones; groups of timbers from either zone show about the same specific-gravity distribution among the three classes.

2. DENSITY AND WEIGHT DENSITY OF WOOD

Density of homogeneous material is defined as its mass per unit volume: $\rho =$ mass/volume. Fundamentally density differs from specific gravity in that the latter is a pure number indicating relative density. Density is expressed as slugs per cubic foot, grams per cubic centimeter, or kilograms per cubic meter. Values for density are commonly cited in the metric system, but in the English system true density values are not often given because the units are unfamiliar to people other than physicists.

Density of dry solid cell wall material in the ovendry condition has been deermined to be about 1.5 grams per cubic centimeter.* Little variation exists in this value for different kinds of wood, as long as the same experimental procedures are employed. This figure of 1.5 agrees well with calculated densities for the entire cell wall determined as the sum of weighted percentages of the densities of the individual cell wall components.[40] However, there is little agreement with the computed densities for each cell wall layer determined separately, because the varying amounts of the components present in each layer are still too little known.

The density of the actual cell wall is believed to be similar to that of solid cell wall substance when both are measured in the ovendry condition. When moisture is present in the wall of the cell below the fiber saturation point, the system of microcavities in the wall increases in volume with increasing moisture content of the wood; this results in a reduction of the cell wall density below that for solid cell wall substance. The ratio of cell wall density to density of dry solid cell wall substance is known as the *packing fraction* (see Chap. 3).

In English-speaking countries, the term ***density*** is employed for ***weight*** per unit volume. This quantity should be distinguished from true density by calling it ***weight density***. As an example, the density of water is 62.4/32.2 = 1.94 slugs per cubic foot; its weight density is 62.4 pounds per cubic foot.

Weight density, or weight of wood per unit volume, is customarily calculated on the basis of both the weight and the volume of the piece taken at the same moisture content. The equation for weight density per unit volume then becomes

$$\text{Weight density of wood} = \frac{\text{weight of wood with moisture}}{\text{volume of wood with moisture}} \tag{6}$$

Weight density of wood, because of the manner of calculation, bears a direct relationship to the moisture content of the piece. Thus the minimum values for weight density occur at the ovendry condition and the maximum when the wood is fully saturated. These relationships can be illustrated by an example. Sugar maple has a specific gravity of 0.656 based on ovendry weight and volume; the volumetric shrinkage from the green size to ovendry condition is 15 percent, and the fiber saturation point (F.S.P.) for this wood is taken as 30 percent moisture content. By reference to Eq. (6), the weight density of 1 cubic foot of ovendry sugar maple is

$$\text{Weight density ovendry} = \frac{0.656 \times 62.4}{1.00} = 40.9 \text{ lb/cu ft}$$

When the dry wood is soaked and brought to the fiber saturation point the increased volume can be determined by proportion from the known shrinkage data:

$$\text{Maximum volume of swollen wood} = \frac{1 \text{ cu ft}}{85} \times 100 = 1.177 \text{ cu ft}$$

* Dry solid cell wall substance has been demonstrated to have a density of 1.53 grams per cubic centimeter in water and 1.46 grams per cubic centimeter in helium gas.

The weight of the wood at fiber saturation then becomes the sum of the ovendry weight of wood (40.9 pounds) plus the weight of water added and

$$\text{Weight density at F.S.P.} = \frac{40.9 + (40.9 \times 0.30)}{1.177} = 45.2 \text{ lb/cu ft}$$

Above the fiber saturation point both the ovendry weight of the cell wall material and the volume of the wood are constant, but the total weight increases because of the addition of water. Thus at the theoretical maximum moisture content of 86 percent (for the maple in this example),

$$\text{Weight density at 86\% M.C.} = \frac{40.9 + (40.9 \times 0.86)}{1.177} = 64.6 \text{ lb/cu ft}$$

3. BUOYANCY OF WOOD

The ability of a piece of wood to float is due to the buoyant force which develops as a result of the difference between the density of the wood and that of the water displaced by the fully submerged piece. Dry cell wall substance has a density of approximately 1.5, and any piece of wood consisting of at least two-thirds cell wall substance would sink in water. However, most woods at usual moisture contents have densities less than 1.00 gram per cubic centimeter and consequently float in water. As these woods are soaked, the air spaces fill with water and the density of the wood increases until it equals or exceeds that of the displaced water, and the block sinks.

D. Thermal Properties of Wood

1. THERMAL CONDUCTIVITY OF WOOD

In the English system thermal conductivity of wood (K) is measured as the amount of heat, in British thermal units (Btu), that will flow in 1 hour through a homogeneous material 1 inch thick and 1 foot square, when a 1°F temperature difference is maintained between the surfaces. The thermal conductivity of wood is a measure of the heat flow and is dependent on three factors: (*a*) direction of heat flow with respect to the axis of grain orientation in the wood, (*b*) moisture content of the wood, and (*c*) specific gravity of the wood. It is nearly the same in the two transverse directions, but in the longitudinal axis it is 2¼ to 2¾ times greater. This relationship is easy to check experimentally by the application of a source of heat concentrated at a point, such as a small soldering iron, to the wet or paraffin-coated surface of a board. Measurement of the elliptical pattern formed by the heat flow on the surface of the board will show the directional difference in thermal conductivity in a qualitative manner.

Thermal conductivity of wood for any direction is approximately one-third greater for moisture contents above 40 percent than it is for wood that is drier. Specific gravity of the wood causes the thermal conductivity to vary directly in a straight-line relationship. According to MacLean,[28] the three factors on which heat conductivity of the wood depends may be combined in a single expression to show the transverse thermal conductivity K:

$$K = G[1.39 + C(\text{M.C.})] + 0.165 \tag{7}$$

where G is specific gravity at a given moisture content, M.C. is moisture content in percent, and C is a constant depending on the moisture content, with a value of 0.028 below 40 percent M.C., and 0.038 above.

The values of the thermal conductivity for wood are quite low compared with the metals (see Table 6-1), and rate favorably with those of other insulating materials. These low transmission rates for heat in wood explain why wood furniture seems so "warm" to the touch and point out one of the reasons for the continued preference for wood in chairs and tables. It also explains why papers stored in thick-walled, tight wooden boxes can sustain the temperatures and fire durations such as occur in residence fires without any damage to the papers.

TABLE 6-1 Thermal Conductivity (K) for Some Kinds of Wood and Certain Other Materials (K in units Btu in./ft^2/hr/°F)

Material	*Weight density, lb/cu ft*	*Temp, °F*	*K*
Wood, ovendry, transverse to grain:			
Balsa (*Ochroma* sp.)	10	85	0.41
Spruce (*Picea* sp.)	21	85	0.62
Basswood (*Tilia* sp.)	24	85	0.69
Aspen (*Populus tremuloides* Michx.)	26	85	0.71
Pine, white (*Pinus strobus* L.)	25	85	0.72
Pine, southern (*Pinus* sp.)	35	85	0.94
Maple, sugar (*Acer saccharum* Marsh.)	43	85	1.13
Elm, rock (*Ulmus thomasii* Sarg.)	48	85	1.16
Insulating materials:			
Styrofoam	1.7	. . .	0.25
Cork, granular	5.4	23	0.34
Asbestos wool	25	212	0.70
Gypsum board	51	99	0.74
Metals:			
Steel (1% carbon)	487	64	314
Lead	710	64	241
Aluminum	165	64	1400
Copper	556	64	2690

Values for wood taken from MacLean[28] and those for other materials from T. Baumeister (ed.), "Marks' Mechanical Engineers' Handbook," 7th ed., McGraw-Hill Book Company, New York, 1958.

2. THERMAL INSULATING VALUE OF WOOD (*R*)

This value is the reciprocal of the conductivity. It is therefore apparent that the insulating value of wood is inversely proportional to the specific gravity and moisture content. This relationship explains the use of low-density dry balsa wood (*Ochroma lagopus* Sw.) for insulating purposes.

3. THERMAL EXPANSION OF WOOD

The measurement of the dimensional changes of wood caused by temperature differences is called the ***coefficient of thermal expansion*** (α). These coefficients vary in wood in inverse order as do those for thermal conductivity. According to Weatherwax and Stamm,[42] the coefficient of thermal expansion for the longitudinal direction (α_L) in the temperature range from -50 to $+50$°C averages 3.39×10^{-6} per degree Celsius, regardless of kind of wood and its specific gravity. In the transverse direction the expansion is about 10 times that in the longitudinal direction. For an average specific gravity of 0.46, the coefficient of radial expansion (α_r) is 25.7×10^{-6} per degree Celsius, while that for tangential expansion (α_t) is 34.8×10^{-6} per degree Celsius. The values of the thermal coefficients in the radial and tangential directions vary directly in a straight-line relationship to the specific gravity of the wood.

Longitudinal thermal expansion in wood is small in comparison with that of other common solid materials. However, the transverse thermal expansion in wood is greater than it is for any of the metals and other common materials. For example, the coefficient of thermal expansion per degree Celsius for steel is 10×10^{-6}, for aluminum 24×10^{-6}, and for flint glass 7.9×10^{-6}. The reason that the thermal changes are not more commonly recognized is that wood is usually used within a narrow range of temperatures and the dimensional changes caused by moisture fluctuations are large enough and usually in an opposite sense so that the thermal effects are masked.

4. IGNITION OF WOOD

It is usually accepted that the ignition of wood begins when the temperature of the wood is approximately 273°C. At this temperature the wood ignites if the supply of oxygen is unlimited; or if the oxygen is controlled, it is the temperature at which gases begin to evolve from ***destructive distillation*** of the wood. The speed with which combustion is initiated is dependent upon the rate of accumulation of heat at the surface of the wood. Several factors influence this rate: the size of the piece of wood, the rate of heat loss from the surface to the interior, the presence of thin outstanding edges, and the rate at which heat is supplied to the surface of the wood. Small pieces with sharp projecting edges, such as square-splint match sticks, ignite easily because a relatively small quantity of heat is necessary to raise the temperature of the whole stick, and especially the thin edges, to the ignition point. Large pieces, with rounded edges, are much slower to catch fire because the conduction of heat into the interior of the piece keeps the surface below ignition temperature for some time. Combustion in a given piece can occur rapidly when large quantities of heat are supplied to the surfaces. However, ignition can occur also if much smaller amounts of heat are supplied to the wood over a long period of time, since the entire piece will eventually be raised to the ignition temperature.

5. FUEL VALUE OF WOOD

Fuel value of wood is primarily determined by the density of the wood and its moisture content. It is modified by variations in lignin content and to a much greater extent by the presence of extractives such as resins and tannins. The heat of combustion (H), i.e., the heat in Btu produced by burning 1 pound of ovendry wood, averages about 8500 Btu for hardwoods and 9000 Btu for conifers. These values for heat of combustion bear little relationship to a particular kind of wood and vary only from 5 to 8 percent at a maximum.

Actual heat produced by burning wood containing some moisture is lower than the value of H quoted above, because part of the heat is lost in removing the water and vaporizing it. An approximation of the actual fuel value for wood is given by the equation

$$\text{Btu per pound of wood} = H \times \frac{100 - (\text{M.C.}/7)}{100 + \text{M.C.}} \tag{8}$$

where H is the heat of combustion of the wood and M.C. is the moisture content of the wood in percentage.

E. Electrical Properties of Wood

1. DIRECT-CURRENT (dc) ELECTRICAL PROPERTIES OF WOOD

The electrical properties of wood are measured by its *resistivity* or *specific resistance* or by its reciprocal, *conductivity*. Dry wood is an excellent electrical insulator, with dc resistivity in the order of 3×10^{17} to 3×10^{18} ohm-centimeters at room temperature.[5] Moisture affects the electrical conductivity of wood as it does other hygroscopic materials. The resistivity decreases rapidly by an approximate factor of three for each percentage moisture content increase up to the fiber saturation point. At 16 percent moisture content, the value of resistivity for wood at room temperatures decreases to 10^8 ohm-centimeters, and at the fiber saturation point it becomes approximately that of water alone, i.e., 10^5 to 10^6 ohm-centimeters.

Conductivity in wood is principally affected by the moisture content, but temperature is also an inseparably related influence. The structural orientation in wood modifies dc resistivity independently of moisture content. In general, resistivity across the grain is from 2.3 to 4.5 times greater than that along the grain for conifers and from 2.5 to 8.0 times greater for hardwoods.[26]

Electrical conductivity in wood is believed to occur primarily by migration of metallic ions which are held in the wood as impurities. This theory at least partly explains the fact that the presence of water-soluble electrolytes in heartwood or the introduction of ionic salts increases conductivity. It also explains the decrease in resistivity of wood with increase of temperature, such as characterizes electrical conduction in ionic solutions.

The electrical-resistance moisture meter is a simple instrument which expresses moisture content in wood as a function of dc resistivity. Since resistivity decreases strongly with increasing temperature, and is affected by the presence of electrolytes and moisture gradients over the depths penetrated by the electrodes of the instrument, the apparatus must be calibrated for a given kind of wood and temperature range.[9] Furthermore these instruments are useful only over a total range from 7 to 25 percent moisture content in the outer shell of the wood.

2. ALTERNATING-CURRENT (ac) CHARACTERISTICS OF WOOD

One measure of the insulating capacity of a material under alternating current is its *dielectric constant* (ϵ). This is expressed as the ratio of the charge held by a condenser in which the electrodes are separated by a dielectric material such as wood, to the charge held by the condenser with the electrodes separated by a vacuum at a given voltage. In wood the dielectric constant varies directly with specific gravity and moisture content, and decreases in a complex manner with increasing frequency of the alternating current.[25,32,35] In dry wood the dielectric constant is from thirty to fifty times greater in the longitudinal direction than it is in the transverse direction. Along the grain, it is uniform when the cellular structure shows no major differences in various parts of the cross section, but it exhibits much greater values for the large-pored early-wood zones in ring-porous woods.[3]

Basically, dry-wood substance is a nonconductor, with a dielectric constant of about 2. It is known that cellulose, lignin, and probably other wall constituents, as well as the contained water, are composed of dipolar molecules, which orient themselves with an applied electrical field. When the direction of the field is reversed, the dipoles change their orientation also. The mobility and volume concentration of the polar groups are directly related to the dielectric constant. Water by itself allows considerable mobility and has a large concentration of polar groups; it thus produces a higher dielectric constant than does dry wood. In addition, the presence of water in the cell wall increases the concentration of polar groups in the cell wall material; thus the dielectric constant increases with increasing moisture content. Above the fiber saturation point, the dielectric constant of fully saturated wood approaches that of water, which has a value of 81.[5]

At low frequencies of alternating current, such as are employed in normal power, the oscillation of the dipolar molecules is slow. When high frequencies of alternating current are used, the oscillation of the molecules is rapid and the movement of the dipolar molecules against each other produces frictional heat, in proportion to the amount of molecular activity. This principle is used in the radio-frequency heating of wood for setting glue lines in the manufacture of furniture. For best performance, some water should be present in the wood, because the water is heated by high-frequency alternating current much more rapidly than is the wood.

Moisture content of wood can be measured with moisture meters which determine dielectric constant rather than electrical resistance. This method has the advantage over dc resistivity meters in that the readings are not affected by ash or mineral content. However, readings must be corrected for density of the wood.

F. Permeability of Wood

Permeability is the term used to indicate the rate of flow of gases and fluids in wood. This property is of great importance for wood preservation and pulping where fluids must be introduced into wood, or for drying where the removal of water vapor and capillary water are important factors.

Permeability in wood is related to the sizes of the passages that are available for flow of liquids or gases. The passage of liquids and gases through openings that are a micron or more in diameter follows the same laws that explain flow in pipes (viscous flow). Movement of vapor through passages smaller than 1 micron is affected by the mean free path between collisions of the vapor molecules being transported. This type of flow may be much less than ordinary viscous flow and controls the longitudinal permeability of conifers and the lateral permeability of both hardwoods and conifers.

In the conifers the openings in the margo of the pit membranes are usually the limiting structures for permeability.[34] The small sizes of these openings are reflected in the reported values of longitudinal permeability in the conifers; these values range from less than 0.001 to 450 cc/sec cm^2 atm/cm.[36] In contrast, the hardwoods with their wide variations in mean vessel diameters and the equally wide variations in extent of plugging with tyloses and gums have a wider range of longitudinal permeability. Reported values in the hardwoods range from less than 0.001 to more than 4500 cc/sec cm^2 atm/cm.[36]

Lateral permeability is much smaller than that in the longitudinal direction because of the large numbers of cell walls which must be traversed. Longitudinal permeabilities may be from 1000 to 100,000 times greater than the transverse values.[7] There is a slight difference between permeability in the radial and tangential directions, with the tangential usually smaller than the radial.

In the hardwoods, resistance to fungal attack is usually greater for the less-permeable woods. In these the growth of fungi is apparently inhibited by reduction in the amount of oxygen available for metabolic activity. In addition, heartwood formation, which is related to much of the low permeability in hardwoods, is also associated with infiltration of growth-inhibiting substances that are effective in reduction of fungal activity.

II. MECHANICAL PROPERTIES OF WOOD

A. Definition of Terms

The mechanical properties of wood are an expression of its behavior under applied forces. This behavior is modified in a number of ways, depending upon the kinds of force exerted on the wood and the basic differences in the organization of wood that have already been noted. Force, expressed on the basis of unit area or volume, is known as a stress (σ). There are three kinds of primary stresses that can act on a body. The force may be acting in compression, if it shortens a dimension or reduces the volume of the body; in this case there is said to be a *compressive stress,* which is defined as total compression force divided by the cross-sectional area of the piece being stressed. If the force tends to increase the dimension, or volume, then it is a tension force, and a *tensile stress* is exerted on the body. *Shear stresses* result from forces which tend to cause one portion of the body to move with respect to another in a direction parallel to their plane of contact. *Bending stresses* result from a combination of all three primary stresses; they cause flexure or bending in the body. The resistance of the body to the applied stress is known as the strength of the material. Since there are a number of different kinds of stresses, the strength of the material must be stated in terms of its compressive, tensile, shear, or bending strength.

In all real materials the stresses which act on a body produce a change in shape and size. The distortion resulting from applied stress is known as *strain* (ϵ); this value is expressed in terms of the deformation per unit area or volume. For example, compressive strain is the reduction in length of a member under compression divided by the length of the member before loading in compression. Each different type of stress produces a corresponding strain; so the kind of strain produced, i.e., compressive, tensile, shear, or bending, must be stated, as is the case with stresses.

The strain induced in a piece of wood is proportional to the applied stress, when the strain is small. It is also fully recoverable if the time of application of the stress is short and the strain remains small. This behavior was described by Robert Hooke in 1678 and is stated in the form of an equation as $\sigma = K\epsilon$. The proportionality constant (K) relating these two terms is a measure or *modulus of elasticity*; it indicates the ability of the material to recover its original shape and size after the stress is removed. The modulus of elasticity for compressive and tensile stresses is known as *Young's modulus* (Y), and the modulus for bending elasticity is commonly indicated as E.

The elastic behavior of wood is illustrated by the straight-line portion of the curve for load and deformation, as shown in Fig. 6-3. The area under the straight-line portion of the curve represents the potential energy, or recoverable work, and is a measure of the resilience of the material. The steepness of the slope of the elastic line is a measure of the magnitude of the elastic modulus, i.e., the steeper the slope, the greater the modulus.

For any given piece of wood subjected to stress, the load-deformation curve reaches a *proportional limit,* beyond which the total deformation is nonrecover-

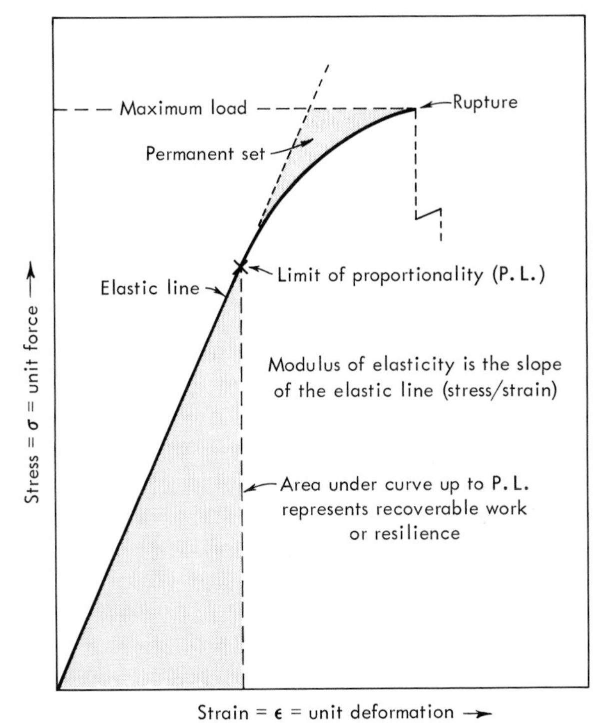

FIG. 6-3 Idealized load-deformation diagram for static loading to failure.

able and some ***permanent set*** is imposed on the specimen. In some kinds of wood there is almost no demarcation of the end of the elastic portion of the curve, and the proportional limit can scarcely be defined. The set is attributed to plastic deformation of the wood and is represented by the area between the projection of the elastic line beyond the proportional limit and the actual curve. This plastic deformation increases with the applied load above the proportional limit until the piece breaks or fails in some manner. The total amount of nonrecoverable strain that a piece can absorb up to the point of complete failure is a reflection of its ***toughness***. Those samples which bend a great deal and break gradually with the absorption of much energy are tough. On the other hand, wood which breaks abruptly and completely with relatively small bending is ***brash*** or ***brittle***.

Failure occurs when the strain limit has been reached, as shown by the upper limit of the load-deformation curve. The ***maximum crushing strength*** (C_{max}) is a measure of the ability of a piece to withstand loads in compression parallel to the grain up to the point of failure. In bending, the magnitude of the load required to cause failure is expressed by the ***modulus of rupture*** (R). The shape of the curve for the bending load-deformation relationship in wood beyond the maximum load is determined by the toughness of the wood in bending; the curve will be abruptly terminated for brash wood but will decrease stepwise for tough material.

Resistance to ***impact bending*** is another type of strength of wood; it is essentially a measurement of the energy absorption or work properties and toughness. ***Shear*** in wood is determined only in a direction parallel to the grain, since wood is weak in shear in this axis. Resistance to shear across the grain is so much greater than other mechanical properties that it is of no practical significance.

B. Effect of Specific Gravity on Strength of Wood

The specific gravity of wood, because it is a measure of the relative amount of solid cell wall material, is the best index that exists for predicting the strength properties of wood. In general terms, without regard to the kind of wood, the relationship between specific gravity and strength can be expressed by the equation[10]

$$S = K(G)^n \tag{9}$$

where S is any one of the strength properties, K is a proportionality constant differing for each strength property, G is the specific gravity, and n is an exponent that defines the shape of the curve representing the relationship. Figure 6-4 presents curves for three important strength properties at two levels of moisture content. It is clear from the curves that the strength in compression parallel to the grain varies in a direct linear relationship to the specific gravity. In other words, doubling the specific gravity will double the compressive strength parallel to the grain. In contrast, the modulus of rupture, whose relationship to the specific gravity is in the order of the 1.25 power, will increase approximately two and a half times when the specific gravity is doubled, while the compression perpendicular to the grain, in which the exponent of the specific gravity is 2.25, will increase more than four times. It should be borne in mind furthermore that the prediction equations given here, and in further detail in the "Wood Handbook,"[10] will not conform precisely to actual mean test values in individual species. This is illustrated in Table 6-2, in which a few kinds of wood are shown with actual and predicted values. The scatter represents differences in the composition and the cellular arrangements of the different woods. A closer prediction can be made by the use of equations derived from test data for the individual species, such as are found in ref. 29.

The effectiveness of wood in resisting any particular form of applied force is a function not only of the total amount of the wall material, but of the proportions of the cell wall components found in a given piece, and also of the amount of extractives in the cell lumen. A measure of the efficiency of the wood to resist stress is given by an index called the ***specific strength***,* which is the ratio of strength to specific gravity. This index is often referred to in general terms as the ***weight-strength ratio***. For example, basswood and sugar maple, with roughly similar percentages of extractives and cellulose, have widely divergent specific gravity

* Specific strengths in compression and tension are strength divided by specific gravity. The specific strength in bending is bending strength divided by specific gravity to the 1.5 power, and the specific index of deflection, or rigidity in bending, is the modulus of elasticity in bending E divided by specific gravity, squared.

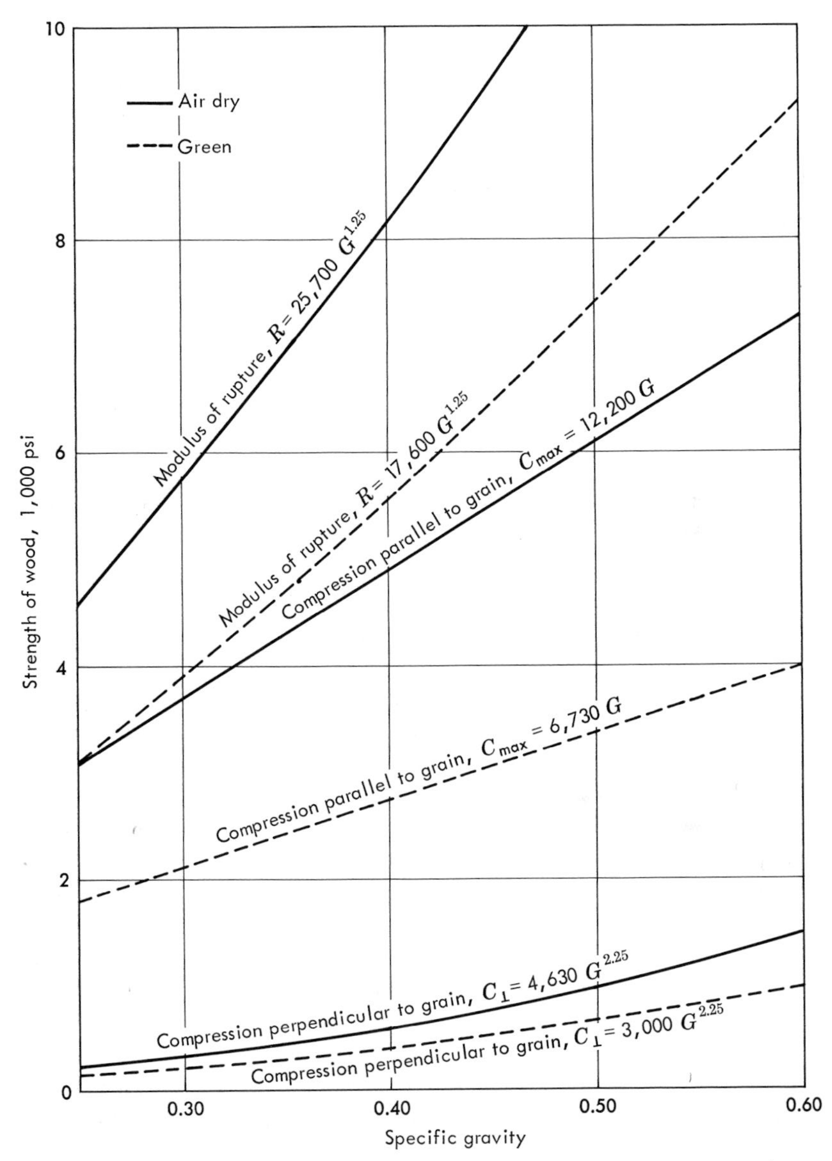

FIG. 6-4 Prediction curves for three strength properties of wood. (*Curves drawn from equations in "Wood Handbook."*[10])

values, but show only 5 percent difference in specific compression strength parallel to the grain for the 75 percent increase in specific gravity (Table 6-2). The remaining three kinds of wood show larger compressive strengths in column 4 than predicted in column 5, and also high specific strengths. In the case of western larch and long-leaf pine, the increases can be explained on the basis of the high extractive contents

TABLE 6-2 Specific Strength of Some Woods Native to the United States

Kind of wood (1)	sp gr green volume (2)	Extractive content, % of OD weight (3)	Comp. parallel to grain ($C_{\parallel}$), green, psi: Actual test (4)	Comp. parallel to grain ($C_{\parallel}$), green, psi: Predicted (5)	Specific strength ($C_{\parallel}$/sp gr) (6)
Basswood (*Tilia* sp.)	0.32	8	2200	2150	6,850
Redwood [*Sequoia sempervirens* (D. Don) Endl.]	0.38	18	4200	2560	11,000
Western larch (*Larix occidentalis* Nutt.)	0.51	24	3990	3440	7,800
Longleaf pine (*Pinus palustris* Mill.)	0.54	20	4300	3640	7,950
Sugar maple (*Acer saccharum* Marsh.)	0.56	7	4020	3770	7,200

Columns 2 and 4 taken from the "Wood Handbook."[10]

Column 3, approximate cold and hot water plus ether extractives. Data from L. E. Wise and E. C. Jahn, "Wood Chemistry," Reinhold Publishing Corporation, New York, 1946.

Column 5 predicted from the equation $C = 6730\ (G)$.

which tend to stiffen and harden the individual cells, increasing their efficiency in compression. In the case of the redwood, the additional strength arises from an unusually high proportion of lignin, as well as the high tannin content in the heartwood. These two components act to cause a dense mass of cell wall material in a form well suited to resist compression. From this discussion it is evident that the specific strength is a useful index for the comparison of samples with differing specific gravities, employed to determine whether the differences are due to factors other than the specific gravity.

In comparison with other structural materials the weight-strength ratio for wood is very favorable for some applications. The dispersal of the cell wall material as thin shells has an important effect on the flexural rigidity of wood. The moment of inertia of a bending member is vastly increased if a given amount of material is arranged as a tubular structure rather than a solid rod. For this reason wood has a high index of rigidity in comparison with solid structural materials and is well suited for use in situations that require elastic stability, such as long beams and columns or stressed skin construction. On the other hand, wood suffers in comparison with metals for uses that require high shear and compression resistance, because the very distribution of wood substance that increases the rigidity in bending reduces the shear and compression efficiency.

C. Effect of Moisture Content on Strength of Wood

Most of the strength properties and elastic characteristics of wood vary inversely with the moisture content of the wood below the fiber saturation point. This behavior is understandable when one considers the dispersal, or concentration, of the solid wood substance which occurs when the wood gains or loses moisture. In effect, the changes in moisture produce the differences in specific gravity, as has been pointed out earlier, and result in changes in strength. The variations in strength of wood, due to the changes in moisture content, are expressed as percentages of change of a given strength property for each 1 percent change in moisture content. Plots of the curves of these relationships for four different properties are shown in Fig. 6-5. These

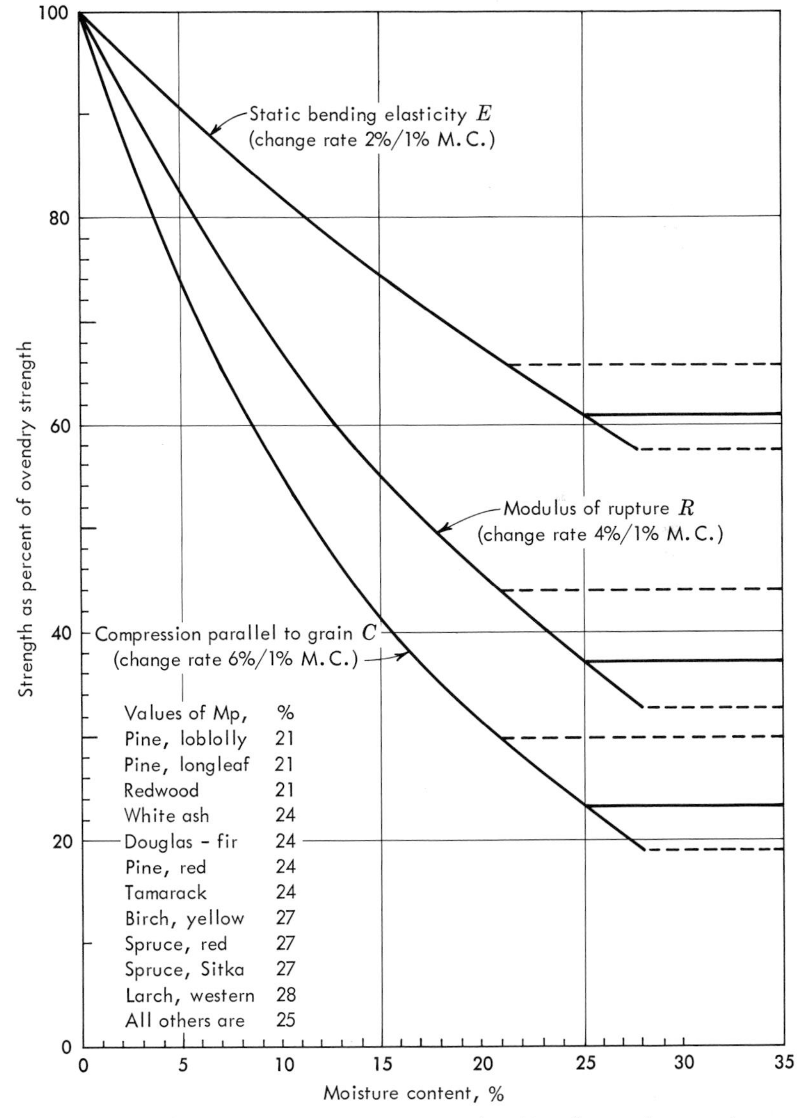

FIG. 6-5 Change of moisture content in wood and its effect on the strength properties. (*Drawn from data in "Wood Handbook."*[10])

curves may be used directly to estimate proportionate changes in the strength properties, or the percentage value shown on each curve may be applied successively as in compound interest calculations to determine strength changes below the fiber saturation point. For example, a piece of wood which has a modulus of rupture R of 10,000 psi at 12 percent moisture content will increase to 11,700 psi at 8 percent moisture content and decrease to 8550 psi at 16 percent moisture content. Either of two methods of determination will produce similar estimates:

1. Reading total changes of percentage directly from the curve in Fig. 6-5, the R value at 12 percent moisture content represents 62.5 percent of the strength of ovendry wood, and R at 8 percent M.C. is 73 percent. Therefore, by proportion,

$$R \text{ at 8 percent M.C.} = 10{,}000 \, \frac{73}{62.5} = 11{,}700 \text{ psi}$$

Similarly, reading from the chart for 16 percent M.C.,

$$R \text{ at 16 percent M.C.} = 10{,}000 \, \frac{53.5}{62.5} = 8550 \text{ psi}$$

2. Using the rate of change with moisture-content changes as the exponent n in the compound-interest calculation $R_2 = R_1(1 + r)^n$,

$$R \text{ at 8 percent M.C.} = 10{,}000\ (1.04)^4 = 11{,}700 \text{ psi}$$
$$R \text{ at 16 percent M.C.} = 10{,}000/(1.04)^4 = 8{,}550 \text{ psi}$$

The general curves for strength changes with moisture changes, below the fiber saturation point, shown in Fig. 6-5, apply to any kind of wood without regard to the species. If a more precise adjustment of strength values for a particular kind of wood is desired, the Forest Products Laboratory "exponential formula" as given in the "Wood Handbook"[10] should be used. However, for the most part, the general relationships obtained from the curves in Fig. 6-5 are sufficiently accurate for any use, except the strength-moisture adjustments made for small clear samples in standard laboratory testing.

Above the fiber saturation point the strength properties are constant with changes in moisture content. This is apparent from the horizontal lines which intersect the curves at the higher-moisture-content levels. Because of limitations in the procedures used in testing, the intersection point for the strength of saturated wood and the curve representing the effect of drying do not meet exactly at the fiber saturation point. The actual intersection point is at a slightly lower moisture content, which is designated as the M_p. Ordinarily this is located at 25 percent moisture content, but certain kinds of woods, as presented in the table in Fig. 6-5, will have other values of M_p. For these 11 kinds of wood, which have M_p values other than 25 percent, the magnitudes of the various strength properties of green wood must be established in Fig. 6-5 according to the point marked by the intersection of the curve and the vertical line for the M_p that is appropriate. Three horizontal lines for each curve are drawn through the upper and lower limits (dashed lines) and the normal location of the M_p (solid line) intersection with the curve. These horizontal lines indicate the range of strength values that apply to the green wood. For example, the modulus of rupture for longleaf pine at all moisture contents above the saturation point (M_p 21 percent) is equivalent to 44 percent of the modulus of rupture for oven-dry wood. At the other extreme value for M_p (28 percent) the modulus of rupture for western larch is only 33 percent as large for green wood as it is for ovendry wood. The normal level of M_p at 25 percent moisture content of the wood indicates that the modulus of rupture for green wood is 37 percent of that for ovendry wood.

A reversal of the usual rule for changes in strength with moisture content is found for shock resistance and toughness. Both these strength properties are work values; they measure the product of applied force times the deformation of the specimen, and are a combined measurement of both bending strength and plasticity of the wood. The latter increases directly with the moisture content up to the fiber saturation point and controls the toughness behavior of wood. Estimates of the rate of change of impact-bending toughness with moisture content indicate that it is in the order of ½ to 1 percent, per 1 percent moisture content difference. Static-bending toughness, as measured by the total work expended, also varies directly with changes in the moisture content of wood.

D. Duration of Stress

The duration of the stress, or in other words the time during which the force acts on a member, has an important effect on the magnitude of the load which that member can sustain. This is true for all forms of stress, but is especially important for bending strength. The maximum bending strength, or modulus of rupture, will decrease in proportion to the logarithm of the time over which the loading is applied.[24,30] Figure 6-6 shows this relationship for times of load application which vary from those required for testing small clear specimens of wood. The reason for this behavior of wood lies in the fact that the cell wall reacts to stress simultaneously in an elastic and a plastic manner. For very short periods of time the deformation is entirely elastic in nature;[18] however, even for increase in time of loading as short as 1 second, as indicated in Fig. 6-6, the plasticity of wood modifies this behavior. Over longer periods of time, the plastic flow of the wood becomes more important in governing the deformation, so that the maximum strain which the wood can withstand is reached at a lower level of load. In other words, the total strain to failure can be governed by either the elastic or the plastic system, with the latter allowing the maximum strain to be reached for a smaller load. For example, a piece of wood which will carry a load to failure of 8000 pounds, for a time of load application of approximately 5 minutes (standard test), will support an 8500 pound load applied for only 1 minute. But when the time of load duration is extended to 1 year, the piece can only be expected to carry a breaking load of 5350 pounds.

The viscous-flow behavior of wood, or its *rheological properties*, is the cause of

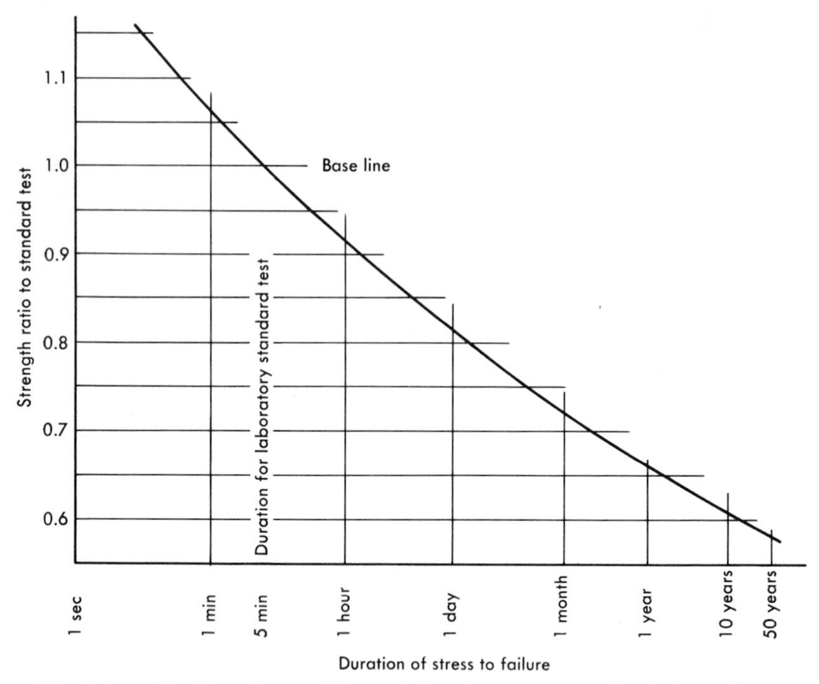

FIG. 6-6 The dependence of the modulus of rupture in Douglas-fir [*Pseudotsuga menziesii* (Mirb.) Franco] on the duration of stress application. (*Adapted from U.S. Forest Prod. Lab. Rept. No.* 1916.)

the normal sag of floor timbers and similar structures. Such deformation under constant stress is called *creep*. This time-dependent behavior of wood has some advantages in that increased levels of stress may be employed in designing wood structures that are to be loaded for very short periods of time. Increase of allowable stress for timber under impact conditions is in direct contrast to the required practice with metals and concrete, which have a limited dependence on time of load application. The rheological behavior of wood is representative of a large group of materials, such as many of the plastics, which are composed of polymerized macromolecules. All members of this class of materials have similar rheological behavior resulting from the chain molecules from which they are composed.

The ability of wood to stand repeated, reversed, or short-duration cyclic loads without failure is a special form of time-dependent behavior called *fatigue resistance*. The chain polymer form of the cellulose in the cell walls and its microfibrillar aggregation are especially well suited to resist this type of stressing. In contrast to crystalline metals, in which fatigue failures are very troublesome, wood seldom fails from this cause, even at high stress levels. Tests on air-dry wood in tension, in reversed bending, or as a rotating beam, at stress levels about 30 percent of the maximum static breaking loads, have shown an endurance limit of approximately 30 million cycles of stress.[10] The above level of fatigue resistance holds true for the normal range of specific gravity in our native woods.

E. Anisotropic Behavior of Wood

A material which has different physical properties in the directions of the various structural axes is said to be *anisotropic*. The cell wall exhibits definite anisotropy because of the structural organization of the materials composing it. In addition to the wall anisotropy, the nature of the thin-walled tubular cells in wood and their arrangement in respect to the axes in the stem reinforce this nonuniformity. The shape of the cells and the orientation of the cellulose in the walls account for a very high ratio of tensile and compressive strength of wood along the longitudinal axis. When subjected to forces in the lateral direction, the same thin-walled tubes can be easily distorted. As a consequence, compressive, tensile, and shear strengths vary widely between the longitudinal and the lateral directions in wood. This basically affects the ways in which wood can be employed. For example, the ratio of compression strength parallel to the grain ($C_{\parallel}$) to the compression perpendicular to the grain ($C_{\perp}$) varies from a minimum of 4, in hardwoods containing small thick-walled fibers, to a maximum of 12 in thin-walled tracheids of conifers. This means that wood is four to twelve times stronger in compression parallel to the grain than it is perpendicular to the grain. Finally, mechanical properties of wood also vary somewhat between the radial (R) and tangential (T) axes in the lateral directions because of the anatomical differences in wood structure.

The changes in magnitude of the compressive and tensile strengths parallel and perpendicular to the grain, as well as of the moduli of elasticity in these directions, are indicated by an empirical equation known as *Hankinson's formula*. This is expressed as

$$N = \frac{PQ}{P \sin^2 \theta + Q \cos^2 \theta} \tag{10}$$

with P and Q being the strengths or elasticity parallel and perpendicular to the grain, respectively; θ is the angle between the applied load direction and the longitudinal direction of the grain in the piece of wood; N is the magnitude of the allowable load or elasticity. The formula is widely employed in the design of wood structures for determining bearing strengths at an angle to the grain and for calculating carrying capacities of connectors in joints. For example, a piece of wood which has compressive strengths of 7250 psi parallel to the grain, and 600 psi perpendicular to the grain, will be expected to carry only 1920 psi when the angle between the direction of the grain and the direction of load application is 30°. The calculation according to Hankinson's formula is

$$\frac{7250 \times 600}{7250(0.5)^2 + 600(0.866)^2} = 1920 \text{ psi}$$

Additional calculations for other angles yield values which define a sigmoid curve approximating the shape of the upper edge of the left-hand face (yz plane) for the block diagram in Fig. 6-7.

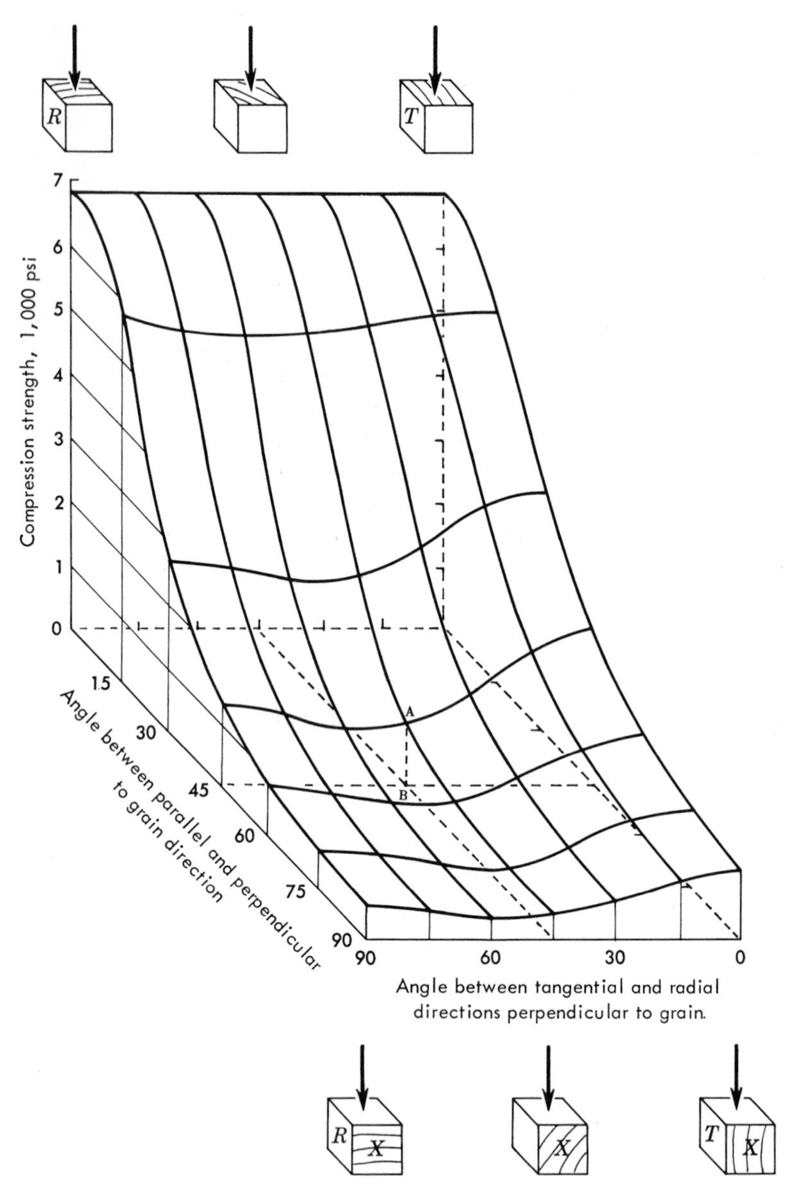

FIG. 6-7 Change in compression strength for Scotch pine [*Pinus sylvestris* L. (from Sweden)] at 12 percent moisture content, as related to the angle between direction of load application, grain angle, and growth-ring orientation. (*Adapted from B. Thunell, TRÄ dess Byggnad och Felaktigheter, Stockholm,* 1955.) For example: the length of the vertical line *A-B* drawn from the center of the lower plane of the diagram indicates that a short column with the grain angle at 45° to the direction of compression load and the rings oriented at 45° to the edges of the cross section will support a stress of 1000 psi. This value is found by measuring *AB* with the scale on the vertical axis.

Figure 6-7 is a graphic representation of the actual variation for compression strengths as the angle between grain orientation and direction of load application varies (1) from parallel to perpendicular to the grain (yz planes), and (2) between the radial and tangential directions with respect to the growth rings (xy planes). This block diagram is an adaptation of experimental data from Sweden for ***Pinus sylvestris*** L. The values plotted on the yz planes are somewhat higher than would be predicted by Hankinson's formula, but this discrepancy is to be expected since the latter equation is an approximation.

F. Cross Grain

The effects of structural anisotropy in wood are also evident when the direction of the grain is not parallel to the loaded axis of the member. This deviation is usually expressed as slope of grain and is measured as unit deviation from a reference edge for x units of length along that edge. Thus for a 1-inch deviation in 12 inches along the reference line (Fig. 6-8*A*) the slope of grain would be 1 in 12, or in alternative notations 1:12 or $^1/_{12}$. For pieces which are neither flat-sawn nor radially sawn,

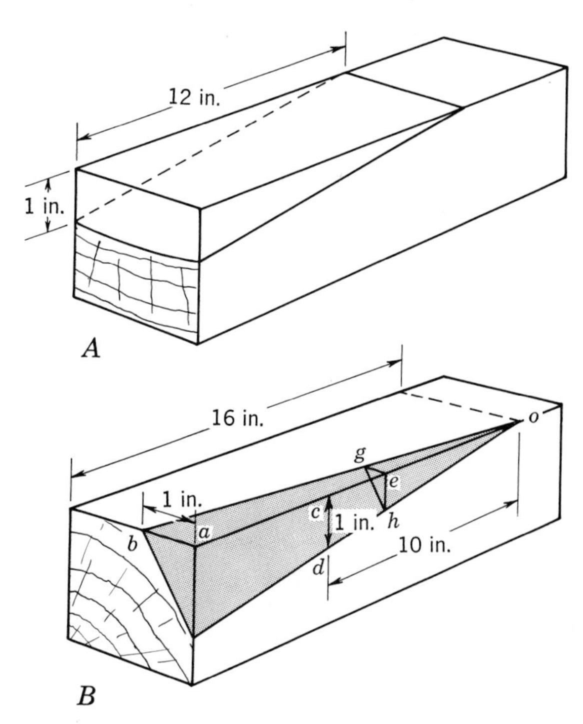

FIG. 6-8 Diagrams of slope of grain in wood. *A*. Simple slope of grain confined to one face. *B*. Combined slope of grain with deviations shown on two adjacent faces.

cross grain may appear on both faces adjacent to the reference edge. In this case the true slope of grain for the combined slopes is greater than that for either face and is the root mean square of the two slope measures. For example, in Fig. 6-8*B* the top face of the piece shows a 1-inch deviation at *a-b* for 16 inches from *o* to *a*. The side face measures 10 inches from *o* to *c* for a 1-inch deviation at *c-d*. The combined slope is the length along the edge *o-e* corresponding to a 1-inch width of the face at *g-h*. The combined slope is calculated as

$$1/x = \sqrt{(1/16)^2 + (1/10)^2} = 1/8.5$$

In another form this equation is stated as $1/x = \sqrt{A^2 + B^2}/AB$ where A and B are the slope distances on the two adjacent faces, or denominators of the slopes.

The tensile behavior of wood is the most sensitive to cross grain and decreases for any slope of grain greater than 1 in 25, i.e., deviations of grain whose *fractional* representation has a larger value, 1 in 20, 1 in 10. This means that as the denominator decreases the slope of grain increases. Bending strength is not reduced until the grain slope is greater than 1 in 20. Compression strengths are the least sensitive and show no change until the grain slopes become larger than 1 in 15. The loss in strength becomes large for cross grain which exceeds the critical limits given above; a slope of 1 in 8 reduces the bending strength approximately one-half and the compressive strength one-third.

In addition to the loss of strength, cross grain increases the tendency toward checking and warp in boards, and can cause very serious twisting in wood poles.

G. Nondestructive Stress Determination in Lumber

In wood both strength and the corresponding elasticity are dependent upon specific gravity and the presence of defects such as reaction wood and slope of grain. It has been demonstrated that there are very strong correlations between modulus of rupture and elasticity in bending, maximum compressive strength parallel to the grain and Young's modulus in compression, and also between maximum tensile strength and modulus of elasticity in tension.[17] The determination of elasticity in structural sized pieces of wood is quite simple and can be performed without damage to the wood using either static or vibrational methods.[15] These facts have been used in recent years as the basis for several related systems of determining stress grades in dimension lumber. The most common methods employ equipment which pass the piece through a series of rollers that apply a small bend as the lumber passes through the rolls. The load or deflection at the rolls is measured electrically; a small computer calculates the elasticity and converts it to bending strength. The effects of all the factors influencing the allowable strength and elasticity are automatically integrated and accounted for in the stress value which the machine stamps on the piece.

The systems are being used or tested quite extensively and hold promise of increasing the reliability of stress rating of construction lumber, over that from the old visual methods, at very low cost.

H. Degradative Changes in Strength of Wood

Wood in service can be subjected to a wide range of conditions which may result in degradative chemical changes in the wood. The most important of the degradative reactions affect the cellulose and depend on a number of interrelated factors, such as (1) temperature, (2) time of exposure to the temperature, (3) moisture content of the wood, and (4) pH of the system in which wood is maintained.

1. IMMEDIATE TEMPERATURE RESPONSE OF WOOD

The short-term response in strength to changes in temperature is degradative only for extremely low temperatures at high moisture contents and for temperatures at or above the ignition point of wood. Both the strength and elastic properties of wood vary inversely with temperature at a given moisture content and directly with specific gravity.[6,13,22,23,37,44] The various mechanical properties are affected to different degrees. The order of decreasing influence of temperature is compression parallel to the grain, modulus of rupture, shear, bending elasticity, tensile strength.[6]

The importance of this type of temperature response becomes most evident at the extremes of the range. Thoroughly frozen wood with high moisture content is much more subject to longitudinal splits and brittle fracture than the same kind of wood at room temperatures. On the other extreme of the temperature scale, *steam bending of wood* employs the principle that wet wood, at elevated temperatures, has lower strength and modulus of elasticity in bending than at room temperatures, because of the greater plasticity. The heated wood can be bent to shape while still wet, and if cooled and dried under restraint will retain most of the induced bend with little loss of strength.

When the temperature of wood has been raised to the point at which ignition occurs, the burned or charred surface of course loses almost all its ability to resist applied forces. The residual strength of the member, in such a case, is a function of the net unburned size and the magnitude of the internal temperature of the remaining unignited wood. Because the conduction of heat is low in wood, and because the charcoal on the surface, if thick enough, also acts as a form of insulation, the internal temperature will not rise materially during the course of a fire in reasonably thick members. Furthermore, since the rate of charring on the surface is relatively slow, wooden members retain a large part of their initial strength during fires. This factor is of practical importance and indicates the reason why roof and floor structures built with heavy wood members will not collapse during the course of an ordinary fire, in contrast to the behavior of unprotected steel exposed to similar temperatures.

2. PERMANENT CHANGES IN WOOD WITH TEMPERATURE AND TIME

Wood at any level of moisture content, when subjected to temperatures lower than 150°F for relatively short periods of time and *then returned to normal room tem-*

peratures, shows no loss of strength or change in elastic properties. This also applies to wood exposed to repeated cycles of temperatures below freezing.[38] In contrast, wood which is heated in the temperature range between 150°F and the ignition point for any appreciable length of time, and subsequently tested at room temperatures, will show a permanent loss of strength and elastic properties. These changes in the mechanical properties of wood so treated are the result of hydrolysis of the cellulose.

The common practice of heating wood in water or in water solutions induces changes in the wood which are related to many of the known chemical reactions of the constituents of the wood. Water which is at or near the boiling point promotes hydrolysis of the chemical compounds in wood.[10,31] As a result of this hydrolysis, wood loses bending strength according to the acidity of the water and the length of time of heating. Curves illustrating this loss of bending strength of wood, when heated in water at 200°F, are shown in Fig. 6-9. The greater degradation of the hardwoods than that shown for softwoods under the same conditions confirms the trend exhibited by the curves in Fig. 6-10.

Exposure of untreated wood at normal temperatures for a sufficiently long period results in characteristic degration of the strength properties. Studies of old timbers from ancient buildings[19,20] free from the effects of decay and not heated above normal temperate zone atmospheric conditions indicate that strength in old timbers is directly related to the hydrolytic changes in the cellulose fraction. The marked loss in strength for the hardwood (*Zelkova*) shown in Fig. 6-10 results from a rela-

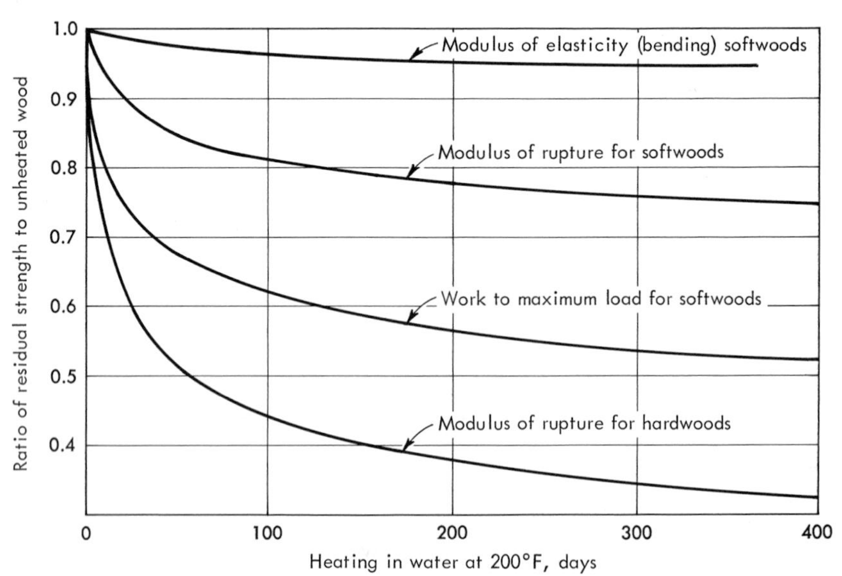

FIG. 6-9 Effect of heating in water at 200°F on bending strength of wood. Samples tested after cooling to room temperature and conditioning to 12 percent moisture content. (*Adapted from "Wood Handbook."*[10])

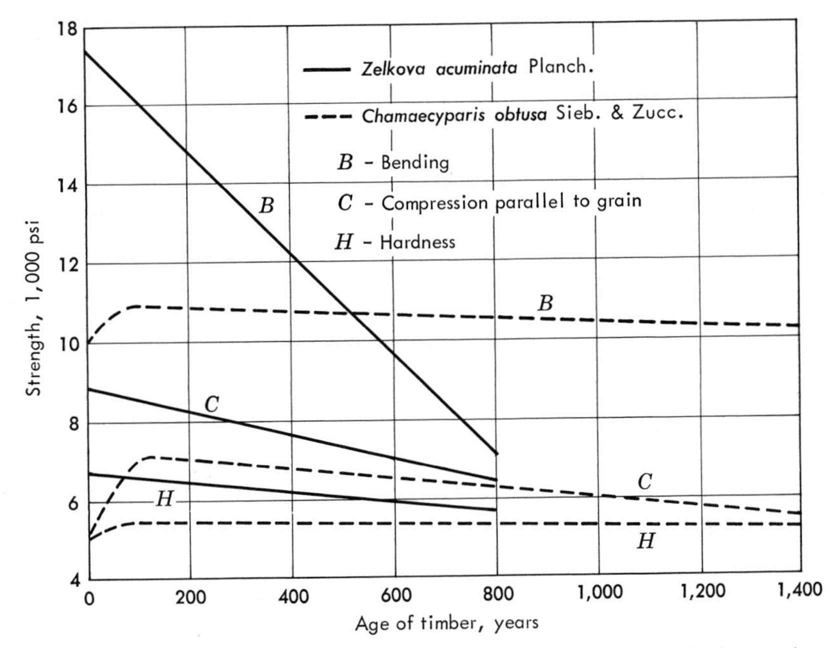

FIG. 6-10 Reduction of strength in wood with time. Old wood which was taken from ancient temples and had been protected from decay. (*Adapted from Kohara et al.*[20])

tively rapid decrease in the percentage of the cellulosic fraction. The strength curves shown for the softwood (*Chamaecyparis*) are significantly correlated with the percentage changes in the crystalline cellulose content. This fraction increases for the first 100 to 200 years, then steadily decreases to an ultimate level about equal to the original condition. All strength properties in Fig. 6-10 follow the same pattern. The pentosans and lignin percentages in wood remain about the same regardless of time.[20] Comparable changes in chemical composition and strength can be induced in wood by heating at temperatures in excess of 100°C, in sealed tubes, for periods as short as a few days. Kohara's experiments[21] on new wood of *Chamaecyparis obtusa* Sieb. & Zucc. and *Zelkova acuminata* Planch., heated at 130°C for periods of 8, 16, and 24 hours, showed that the changes in composition were analogous to those in ancient timbers.

3. ANAEROBIC DECOMPOSITION OF WOOD

Wood that is completely buried under conditions which prevent free interchange of atmospheric gases will hydrolyze slowly. The rate appears to be affected by the continuity of its submersion and the pH of the water. There is a regular decrease of weight with time which results from degradation of the cellulose. Carbon dating of buried samples from 2000 to 47,000 years old[4] has shown that loss of the poorly oriented cellulose fraction occurs first, followed by that of the well-oriented portion of the cellulose. Burial for an estimated 150,000 years has been reported to reduce the material almost wholly to lignin.[19] However in another study of fossil conifers from lignite beds, examination of samples 30 million years old showed well-organized cellulose still remaining. The tracheids in this fossil material were entire and could be formed into reasonably good paper.[8] Estimates of the time required to produce a 50 percent reduction in the cellulose content in wood completely submerged are 1500 to 2000 years for gymnosperms, and 200 to 420 years for angiosperms.[19] These rates for anaerobic deterioration are similar to those given previously for aerobic chemical changes and strength losses, except that the deterioration appears to be somewhat faster under water than in air.

4. DEGRADATION OF WOOD BY CHEMICALS

Wood is remarkably resistant to degradation when used in contact with acids. For example, a group of hardwoods and conifers tested by Wangaard[41] after they had been soaking in 2 percent HCl for 32 days showed only minor loss in modulus of rupture. More drastic treatment, e.g., soaking in 10 percent HCl for 4 days at 50°C, may cause reductions up to 70 percent in bending strength. Strongly basic solutions promote oxidation of the polysaccharides and result in strength loss which may be more serious than that from exposure to acids. Soaking in 2 percent NaOH for 32 days at 20°C results in 50 percent or more loss in bending strength. Wood treated with low concentrations of water-soluble salts as protection against biological attack exhibits little decrease in strength properties under normal conditions of service. In hot dry regions, however, the salts may migrate to the surface, thus increasing the salt concentration to a level high enough to deteriorate the surface. Fire-retardant treatments of wood require heavy concentrations of salts applied at elevated temperatures and high pressures. The net result of this sort of treatment is that fire-retardant treatment of wood can cause loss of strength, especially in tension and bending.

The hemicelluloses of wood appear to be the most susceptible fraction to attack by either acids or alkali. In the study conducted by Wangaard,[41] 70 percent of the variation in strength retention was associated with loss in hemicelluloses. Other effects of treatment that influence strength are acid-induced hydrolysis of cellulose and alkali-induced swelling. Resistance to acid attack is influenced by the degree of crystallinity of the cellulose, while low pentosan content in conifers increases resistance to both acids and alkali.

SELECTED REFERENCES

1. Barber, N. F.: A Theoretical Model of Shrinking Wood, *Holzforschung*, **22**(4):97–103 (1968).
2. ——— and B. A. Meylan: The Anisotropic Shrinkage of Wood, a Theoretical Model, *Holzforschung*, **18**(5):146–156 (1964).
3. Bosshard, H. H.: Some Aspects of the Application of High-frequency Heating in Wood Biology, *News Bull., Intern. Assoc. Wood Anatomists*, 1, 1958.

4. Chowdhury, K. A., R. D. Preston, and R. K. White: Structural Changes in Some Ancient Indian Timbers, *Proc. Roy. Soc. London*, **168B**(1011):148–157 (1967).
5. Clark, J. D., and J. W. Williams: The Electrical Conductivity of Commercial Dielectrics and Its Variation with Temperature, *J. Phys. Chem.*, **37**(1):119–131 (1933).
6. Comben, A. J.: The Effect of Low Temperature on the Strength and Elastic Properties of Timber, *J. Inst. Wood Sci.*, **13**:42–55 (1964).
7. Comstock, G. L.: Physical and Structural Aspects of the Longitudinal Permeability of Wood, Ph.D. dissertation, State University College of Forestry at Syracuse University, Syracuse, N.Y., 1968.
8. Crook, F. M., P. F. Nelson, and D. W. Sharp: An Examination of Ancient Victorian Woods, *Holzforschung*, **18**(5):153–156 (1965).
9. Davidson, R. W.: The Effect of Temperature on the Electrical Resistance of Wood, *Forest Prod. J.*, **8**(5):160–164 (1958).
10. Forest Products Laboratory: "Wood Handbook," U.S. Department of Agriculture Handbook 72, Washington, 1955.
11. Frey-Wyssling, A.: The Ultrastructure of Wood, *Wood Sci. & Technol.*, **2**(2):73–83 (1968)
12. Garlick, G.: Structure and Its Relationship to Utilization, *J. Inst. Wood Sci.*, **14**:3–17 (1965).
13. Goulet, M.: Die Abhängigkeit der Querzugfestigkeit von Eichen-, Buchen- und Fichtenholz von Feuchtigkeit und Temperatur im Bereich von 0° bis 100°C., *Holz Roh-Werkstoff*, **18**(9):325–331 (1960).
14. Hale, J. D.: The Anatomical Basis of Dimensional Changes of Wood in Response to Changes in Moisture Content, *Forest Prod. J.*, **7**(4):140–144 (1957).
15. Hillbrand, H. C., and D. G. Miller: Machine Grading—Theory and Practice, part 2, *Forest Prod. J.*, **16**(12):36–40 (1966).
16. Hittmeier, M. E.: Effect of Structural Direction and Initial Moisture Content on Swelling Rate of Wood, *Wood Sci. & Technol.*, **1**(2):109–121 (1967).
17. Hoyle, R. J.: Background to Machine Stress Grading, *Forest Prod. J.*, **18**(4):87–97 (1968).
18. Kitazawa, G.: Young's Modulus of Elasticity of Small Wood Beams by Dynamic Measurements, *J. Forest Prod. Res. Soc.*, **2**(5):228–231 (1952).
19. Kohara, J.: Permanence of Wood, *Mokuzai Gakkaishi*, **19**(2):191–195 (1956); **20**(2): 195–200 (1956).
20. ——— and H. Okamoto: Studies of Japanese Old Timbers, *Sci. Rept. Saikyo Univ.*, **7**:9–20 (1955).
21. ——— and ———: Permanence of Wood, part XI, *J. Japan. Forestry Soc.*, **37**:392–395 (1955).
22. Kollmann, F.: Die Abhängigkeit der elastischen Eigenschaften von Holz von der Temperatur, *Holz Roh-Werkstoff*, **18**(8):308–314 (1960).
23. ———: Die mechanischen Eigenschaften verschieden feuchter Holzer im Temperaturbereich von −200 bis +200°C, *VDI-Forschungsh.*, **B11**(403):1–18 (1940).
24. Krech, H.: Untersuchungen über den Zusammenhang zwischen Biegefestigkeit des Holzes und Biegegeschwindigkeit, *Holz Roh-Werkstoff*, **18**(7):233–236 (1960).
25. Lin, R. T.: Review of the Dielectric Properties of Wood and Cellulose, *Forest Prod. J.*, **17**(7):61–66 (1967).
26. ———: Review of the Electrical Properties of Wood and Cellulose, *Forest Prod. J.*, **17**(7):54–61 (1967).
27. McIntosh, D. C.: Transverse Shrinkage of Red Oak and Beech, *Forest Prod. J.*, **7**(3): 114–120 (1957).
28. MacLean, J. D.: Thermal Conductivity of Wood, *Am. Soc. Heating Refrig. Air-Cond. Engrs. J. (Heating, Piping and Air Conditioning)*, **13**:380–391 (1941).
29. Markwardt, L. J., and T. R. C. Wilson: Strength and Related Properties of Wood Grown in United States, *U. S. Dept. Agr. Tech. Bull.* 479, 1935.

30. Meylan, B. A.: Cause of High Longitudinal Shrinkage in Wood, *Forest Prod. J.*, **18**(4): 75–78 (1968).
31. Mithel, B. B., G. H. Webster, and W. H. Rapson: The Action of Water on Cellulose between 100 and 225°C, *Tappi*, **40**(1): 1–4 (1957).
32. Peterson, R. W.: The Dielectric Properties of Wood, *Can. Forest Prod. Lab. Tech. Note* 16, 1960.
33. Sadoh, T., and G. N. Christensen: Longitudinal Shrinkage of Wood. Part I: Longitudinal Shrinkage of Thin Sections, *Wood Sci. & Technol.*, **1**(1):25–44 (1967).
34. Sebastian, L. P., W. A. Côté, Jr., and C. Skaar: Relationship of Gas Phase Permeability to Ultrastructure of White Spruce Wood, *Forest Prod. J.*, **15**(9):394–404 (1965).
35. Skaar, C.: The Dielectric Properties of Wood at Several Radio Frequencies, *N.Y. State College of Forestry Tech. Publ.* 69, 1948.
36. Smith, D. N., and E. Lee: The Longitudinal Permeability of Some Hardwoods and Softwoods, *Forest Prod. Res. Spec. Rept.* 13, Dept. of Scientific and Industrial Research, London, 1958.
37. Sulzberger, P. H.: The Effect of Temperature on the Strength of Wood, Plywood and Glued Joints, Commonwealth of Australia Department of Supply, *Aeronautical Research Consultative Committee Rept.*, ACA-46, 1953.
38. Thunell, B.: Inverkan av upprepad frysning och upptining i vattenmättat tillstånd på furuvirkes hållasthetsegenskaper, *Statens Provingsanstalt, Stockholm, Medd.* 82, 1940.
39. Varossieu, W. W.: Buried and Decayed Wood from a Biological Aspect, from W. B. Beekman (ed.), "Hout in Alle Tijden," Central Institute voor Materiaalonderzoek, Delft, 1949.
40. Vorreiter, L.: Rechnunsmassige Bestimmung der Zellwanddichte aus den Holzkonstituenten, *Holz Roh-Werkstoff*, **13**(5):185–187 (1955).
41. Wangaard, F. F.: Resistance of Wood to Chemical Degradation, *Forest Prod. J.*, **16**(2): 53–64 (1966).
42. Weatherwax, R. C., and A. J. Stamm: Coefficients of Thermal Expansion of Wood and Wood Products, *Am. Soc. Mech. Engrs. Trans.*, **69**(44):421–432 (1947).
43. Ylinen, A., and P. Jumppanen: Theory of the Shrinkage of Wood, *Wood Sci. & Technol.*, **1**(4):241–252 (1967).
44. Youngs, R. L.: Mechanical Properties of Red Oak Related to Drying, *Forest Prod. J.*, **7**(10):315–324 (1957).

7

Variability of Wood within a Species

Variability between different kinds of wood arising from differences in anatomical structure and associated physical properties is obvious enough to be accepted without question. However, much of the variability within a species is more subtle and is not immediately evident. Obvious differences from tree to tree and within single trees may appear as fast or slow growth or as variations in weight between pieces of lumber. On the other hand, variables of equal or greater importance such as fiber length are not immediately evident and can be determined only after tedious analysis.

Variation within a species is a product of a complex system of interacting factors which modify the physiological processes involved in formation of wood. Measurements of any of the properties of wood will show that there is a range of values which conforms to statistical laws and which will approximate the normal curve of statistical distribution, provided enough samples are measured.[53] An example of this type of curve is the distribution of 460 measurements of specific gravity for the merchantable trunks of 19 trees of basswood (*Tilia americana* L.) representing a major part of the geographical range for this species (Fig. 7-1). It is apparent that the solid line for curve *A* is a reasonable approximation of the predicted normal curve based on the experimental parameters. The central part of these plotted data having a mean ($\bar{X}$) of 0.349 and values lying between 0.325 and 0.373

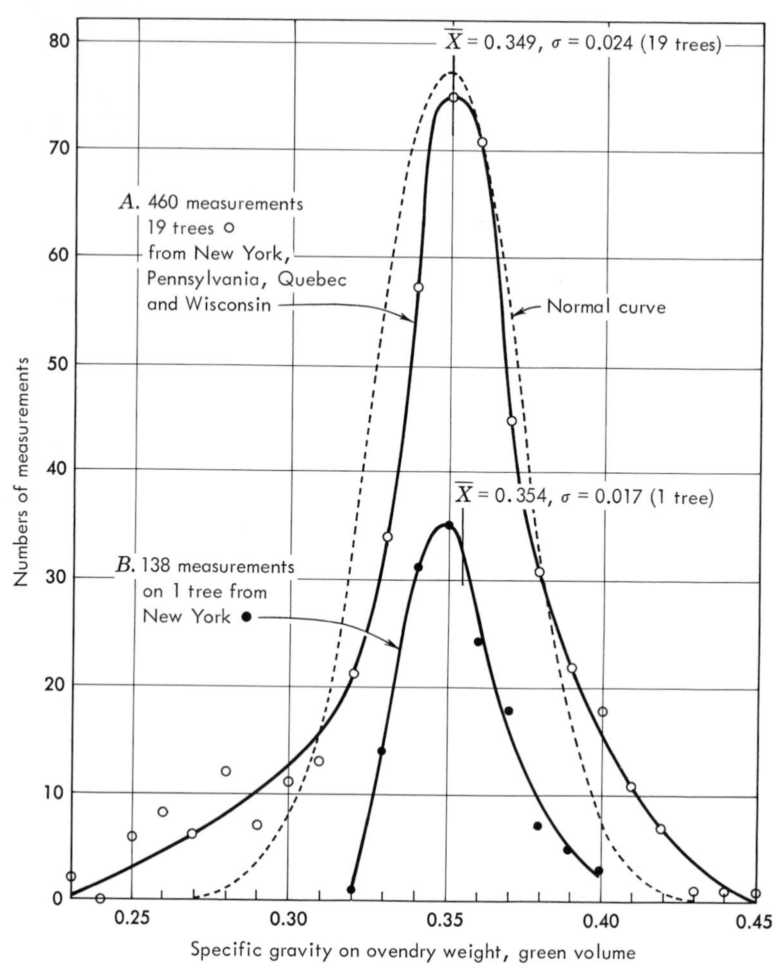

FIG. 7-1 Specific-gravity distribution curves for basswood (*Tilia americana* L.). (*Data for eight trees courtesy of U.S. Forest Products Laboratory; four trees courtesy of Canadian Forest Products Laboratory; remainder of data compiled by Carl de Zeeuw.*)

($\pm 1\sigma$) includes two-thirds of all the measurements and typifies the expected specific-gravity range in basswood. The data do show greater numbers of values at both the high and low extremes than are indicated for a curve of normal distribution. It is believed that this unbalance arises from the inclusion of some tension-wood specimens which introduce values both lower and higher than expected for normal wood.

Sets of measurements from single trees of a species, as in Fig. 7-1 (curve *B*), usually exhibit less variability than is found between trees in the whole population. This variability between trees may be up to ten times greater than the within-tree variance, or it may be only slightly larger, or on occasion smaller. The differences in magnitude arise from the fact that separate trees in the species not only may show genetical differences but are also subject to varying environmental conditions. However, regardless of the differences in the magnitude of distribution of values, measurements from single trees generally conform, more or less, to the normal curve of distribution. The skewness which is evident for curve *B* arose from the fact that relatively few low-specific-gravity measurements could be obtained from the central core of this particular tree because of the presence of rot near the pith.

The *quality* of wood, i.e., its suitability for a particular use, is judged by one or more of the variable factors which affect its structure and in turn its physical properties. For example, seemingly minor changes in cell length, cell wall thickness, cell diameters, fibrillar angles in the walls, percentage of cell types, and cellulose to lignin ratios are important for assessment of pulp quality and are reflected in density changes, with all their related changes in physical properties. These measurable variables in wood arise from the fact that the physiological activities of the cambium are affected by several systems of influences:

1. Age or maturation changes in the cambium itself which are associated with variance within trees of a species.
2. Genetical factors that are one of the basic causes for between-tree variation.
3. Environmental factors such as rainfall, temperature, and silvicultural treatment which affect the net water and nutrient supply to the cambium. This type of influence affects both within- and between-tree variation.

I. VARIABILITY OF WOOD WITHIN A TREE

Patterns for the variation of structural elements and physical properties are fairly well established for normal trees grown under forest conditions. Unless otherwise noted, the statements which are made in this chapter apply to such trees. The discussion will be concerned mainly with the lengths of fibrous elements and the specific gravity of the wood. These characteristics are of primary interest as indices of mechanical properties and, along with the differences in cross section of fibers, are important indicators of the properties required for papermaking.

Variations of anatomical features or of physical properties in a tree stem, root, or branch can be described in terms of (1) those patterns of change which occur in the radial direction, i.e., across adjacent growth increments, and (2) changes that occur along the axis. The total variation of anatomical features and the corresponding modifications in physical properties in the tree trunk can be considered in terms of the above two axes, because of the radial symmetry of the stem.

A. Cell Dimensions

1. CELL LENGTHS WITHIN A CROSS SECTION

Because of the widespread interest in conifer fibers for papermaking, the lengths of tracheids have been widely investigated. Hardwood fiber lengths are not as well known. Sanio[51] established a generalized pattern for the variability of tracheid lengths within stems, branches, and roots of ***Pinus sylvestris*** L. He determined that for any given cross section of a trunk or branch, the tracheids are initially short near the pith, increase rapidly in length for the early years of growth, and then level off to a constant length as the tree matures (Fig. 7-2). In conifers the length of tracheids in the wood near the pith varies from 0.5 to 1.5 millimeters, while in hardwoods the fibers range from 0.1 to 1.0 millimeters in length. The period of rapid increase in length usually covers the first 10 to 20 years of growth. However, the limits of this period of development in length of initials are sometimes uncertain, since there may be a gradual transition to the zone in which maximum fiber length is attained. The age at which the fibrous cells reach maximum length is related to the expected life span of the species. For example, the maximum fiber length in quaking aspen (***Populus tremuloides*** Michx.), which has a life span of 60 to 70 years, appears to be reached shortly after the period of rapid initial increase. In most species the maximum length of fibrous elements is reached well before the tree is 100 years of age.[12] As an extreme, redwood trees [***Sequoia sempervirens*** (D. Don) Endl.], which are known to live more than 1000 years, do not attain maximum tracheid length until the tree is 200 to 300 years old.[3] As trees pass the normal span of life expectancy and become overmature the cells tend to become shorter.[3]

The condition of constant length in the mature periods of growth, as stated by Sanio, has been found in a number of different kinds of trees. For instance, Dinwoodie[12] cites ***Carya ovata, Eucalyptus regnans, Picea abies, Pinus densiflora,***

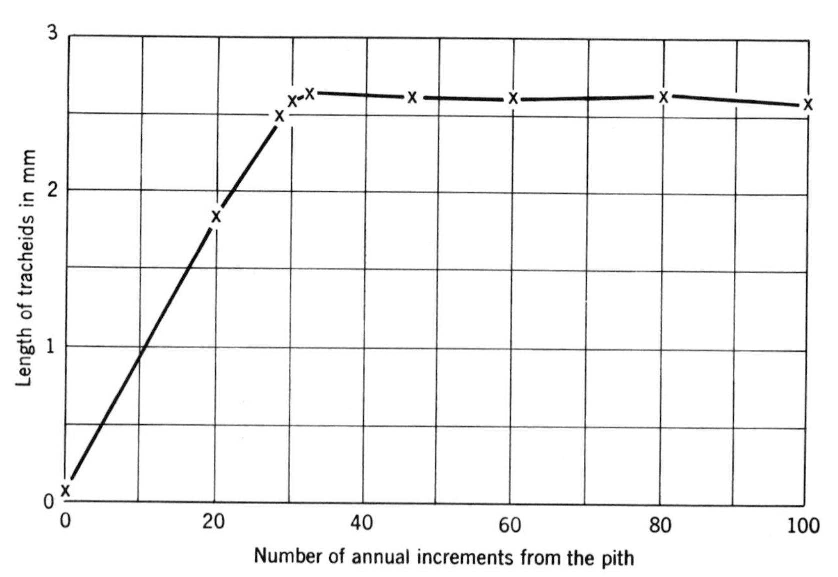

FIG. 7-2 Tracheid-length variation with number of increments from the pith in Scotch pine (***Pinus sylvestris*** L.). [*Drawn from data given by Karl Sanio in Jahrb. wiss. Bot.*, **8**:410 (1872).]

Pinus radiata, Pinus taeda, Pseudotsuga menziesii, and *Sequoia sempervirens.* The same trend has been shown in *Larix* spp.,[40] in *Pinus elliottii,*[56] and in *Pinus merkusii.*[54] However it is also apparent at present that some kinds of trees do not attain a constant cell length for the fibrous cells, or at least this point of development is greatly delayed (Fig. 7-7). Continued increase in fiber lengths in the mature periods of growth has been reported for *Liriodendron tulipifera,*[57] *Picea sitchensis,*[13] *Pinus ponderosa,*[60] *Pinus resinosa, Pinus strobus,*[19] *Pinus taeda,*[21] and *Thuja plicata.*[62]

Within a growth increment there are consistent differences between the lengths of fibrous elements in the early wood and late wood. In the conifers a sharp decrease in cell length takes place at the beginning of the growth increment and continues until a minimum length is reached at the boundary between early and late wood. The maximum length of fibers or tracheids in the growing ring is attained in the late wood, usually near the end of the seasonal growth. The increase in tracheid length from early to late wood ranges from about 12 to 25 percent. In the hardwoods percentage increase in length of the late-wood fibers over those in the early wood bears an inverse relationship to the length of the fiber. Fibers 1 millimeter and more in length show an increase of about 15 percent, while those less than 1 millimeter in length increase 75 to 80 percent[12] (Figs. 7-3 to 7-6).

The shapes of the curves representing variation in cell lengths across the increment are related to the character of the transition from early to late wood. In conifers an abrupt transition from early wood to the dense late wood results in sharp peaks, related to the long tracheids in the late-wood zone, as indicated in the curve for Douglas-fir [*Pseudotsuga menziesii* (Mirb.) Franco] (Fig. 7-3). On the other hand, plotted data for an increment having a gradual transition from early to late wood yield a diagram resembling the tooth outline of a circular saw. This pattern is illustrated in Fig. 7-4 for an Australian plantation-grown tree of *Pinus radiata* D. Don with a gradual transition from early to late wood, in spite of the fact that this species is a hard pine.

Ring-porous hardwoods exhibit a rapid increase in the lengths of fibers from the narrow zone of early-wood vessels to the outer portion of the growth increment (Fig.

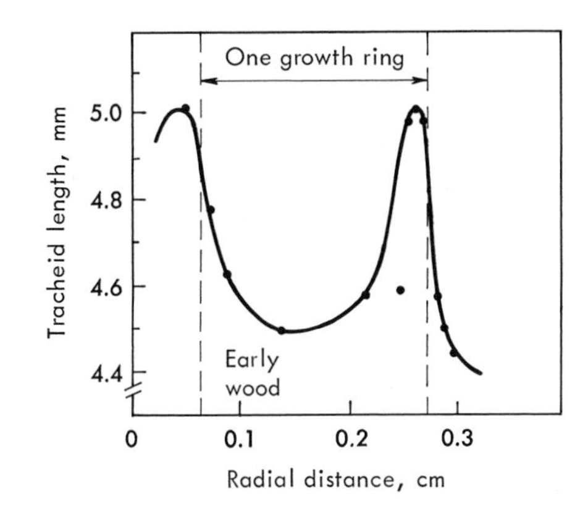

FIG. 7-3 Tracheid lengths across one growth increment of *Pseudotsuga menziesii* (Mirb.) Franco. [*From I. J. W. Bisset and H. E. Dadswell, Australian Forestry,* **14**(1):(1950), *by permission.*]

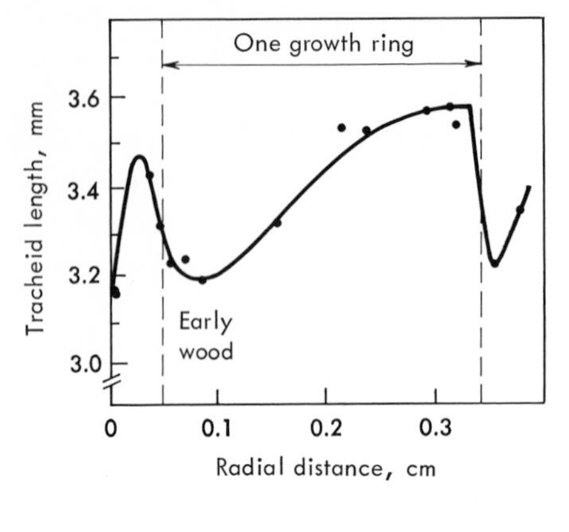

FIG. 7-4 Tracheid lengths across one growth increment of *Pinus radiata* D. Don. [*From I. J. W. Bisset and H. E. Dadswell, Australian Forestry*, **14**(1):(1950), *by permission.*]

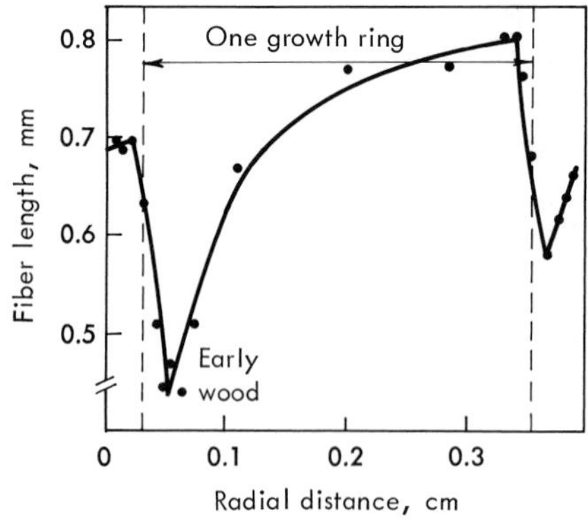

FIG. 7-5 Fiber lengths across one growth increment of *Catalpa bignonioides* Walt. [*From I. J. W. Bisset and H. E. Dadswell, Australian Forestry*, **14**(1):(1950), *by permission.*]

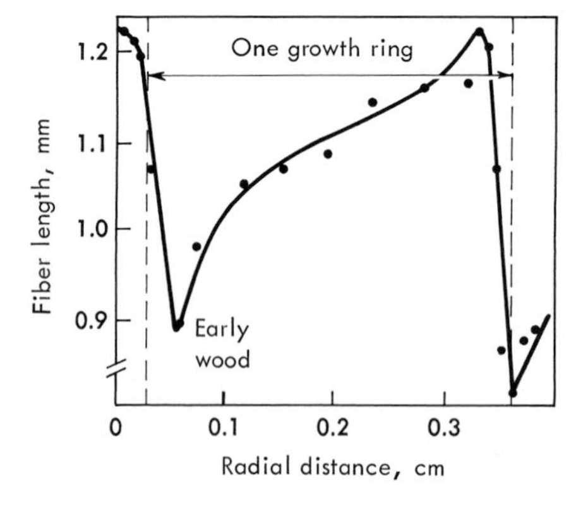

FIG. 7-6 Fiber lengths across one growth increment of *Populus tremuloides* Michx. [*From I. J. W. Bisset and H. E. Dadswell, Australian Forestry*, **14**(1):(1950), *by permission.*]

7-5). In semi-ring-porous wood increase in the cell length from the early to late wood is more gradual than that for the ring-porous woods (Fig. 7-6). Hardwoods of the diffuse-porous type, grown under conditions that produce little seasonal differentiation in visible structure, display almost no variation in fiber length within a ring.[7]

Vessel elements are significantly shorter in the early wood of ring- and diffuse-porous woods than in the late wood.[12] According to Bisset and Dadswell[7] the pattern of variation between early and late wood in the growth increment is similar to that for fiber lengths of ring-porous hardwoods. This general relationship was also confirmed by Bergman's study of fiber and vessel lengths in 49 hardwoods native to the United States, in which he found a high degree of correlation (+0.706) between the vessel and fiber lengths in the same samples.[6]

When the lengths of early- and late-wood tracheids in conifers are plotted separately for a series of increments along a given radius, the graph of the peak values for both groups of measurements corresponds to the curve of tracheid-length variation already described in a previous paragraph. However, the rate of increase in length is greater for the late-wood tracheids than for those in the early-wood zones, and the percentage difference increases gradually with age until a maximum is reached in the mature tree[7,61] (Fig. 7-7).

As a consequence of the general increase in the length of fibrous elements with age, there is greater variability between increments than there is within a growth increment. For both softwoods and hardwoods, the longest fibrous elements are generally at least twice as long as the shortest found near the pith; usually they are three to five times longer.[3,7,12]

The late-wood vessel elements in *Fraxinus excelsior* L. exhibit practically no variation in length with increasing distance from the pith.[8] In contrast, vessel elements in the early wood, while they are always shorter than those in the late

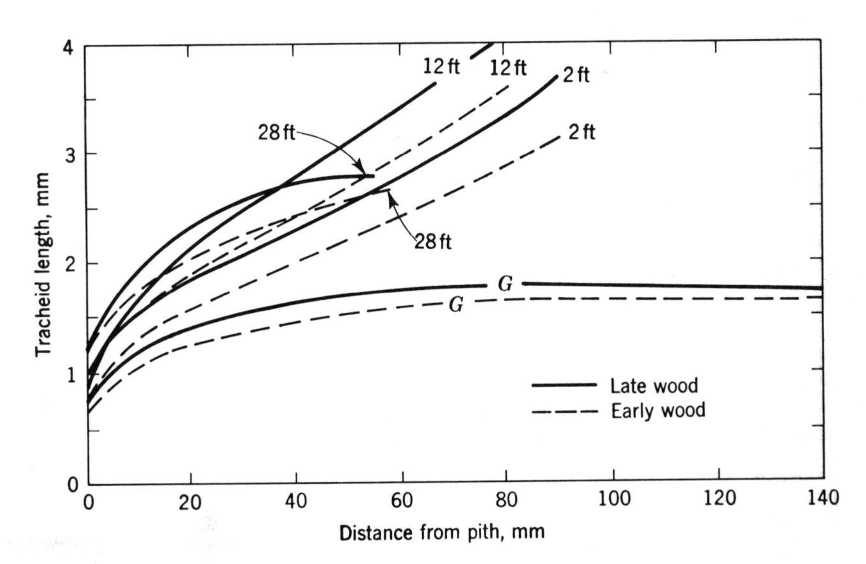

FIG. 7-7 Tracheid-length variation with height in the stem for a 35-year-old red pine (*Pinus resinosa* Ait.). The shortest tracheids are found at ground level (*G*). Tracheid lengths increase rapidly to stump height (2 feet). Increase in length continues to a maximum below the crown (12 feet), and then decreases in the upper crown. The early-wood curves are similar to those for the late wood but indicate shorter tracheids in all cases.

wood, vary with increasing distance from the pith in the same manner as tha described for the fibrous cells. The vessel elements in *Carya ovata* have been showr to be almost the same length across the radial section.[48] Lack of evidence from othe studies prevents the extension and clarification of these trends to include ring-porou woods in general.

Vessel elements in hardwoods with storied cambia apparently differ from the general pattern of length changes given for *Fraxinus*. This is indicated in a study or black locust (*Robinia pseudoacacia* L.), which reveals that there is no variation ir vessel-element length within an increment and only a very slight increase with increasing distance from the pith.[26]

A generalized explanation of the underlying causes for cell length changes across a transverse section of a stem can be based on the maturation or age changes associated with the cambial initials and the xylem mother cells. Pseudotransverse (anticlinal) divisions for the production of new initial cells in the cambial region theoretically should yield two initials of equal length after division. However, Bannan[5] has shown that the derived cells are not always equal in length and that the longest are favored for continuation as cambial initials because they exhibit the greatest number of ray contacts and therefore produce more efficient functioning tracheids. The mean length of the cambial initials at any one period is a function of the rate of pseudotransverse division and the percentage of survival. (For more detailed discussion of this subject see Chap. 2.)

In the first few years of growth after the initiation of a cambium, the rate of pseudo-transverse divisions is rapid and the survival rate is high,[37] with the result that the mean length of the initials and derived cells is short. This is known as the "juvenile pattern." The rate of anticlinal division and the rate of survival of derived cells both decrease with increasing age of the cambium and result in longer initials. In addition, the rate of lengthening of the initials is affected by the fact that the pseudotransverse divisions occur throughout the growing season in the first few years of cambial activity; thereafter, as the period of maturity is reached, these divisions occur mainly in the later part of the growing season during formation of the late wood. In trees which are past maturity, the pseudotransverse divisions are confined to the period when the last-formed late wood is produced.[37]

Growth rate of the tree also affects the increase in length of the cambial initial cells. Fast growth retards the rate of length increase in cambial initials during the early years of activity of the cambium and delays the time of production of maximum length of cells.

Within the growth increment the xylem mother cells, which are initially nearly identical in length to the cambial initials from which they were derived, divide with the "juvenile pattern" in the rapidly growing early wood and produce cells which are shorter than the cambial initials. The reduction in length is proportional to the rate of growth, i.e., slower growth will tend to result in longer cells. In the late-wood zones the time for maturation is longer and the derived cells can attain the maximum elongation from the fusiform initials in the cambium at that position in the cross section.

2. CELL LENGTHS IN THE AXIAL DIRECTION

Within a growth increment the lengths of fibrous cells increase directly with increasing height in the stem to a maximum part way up the trunk. Above this level the lengths of the fibers decrease with increasing height until the top of the cone formed by a given growth increment is reached. The fibrous cells are generally

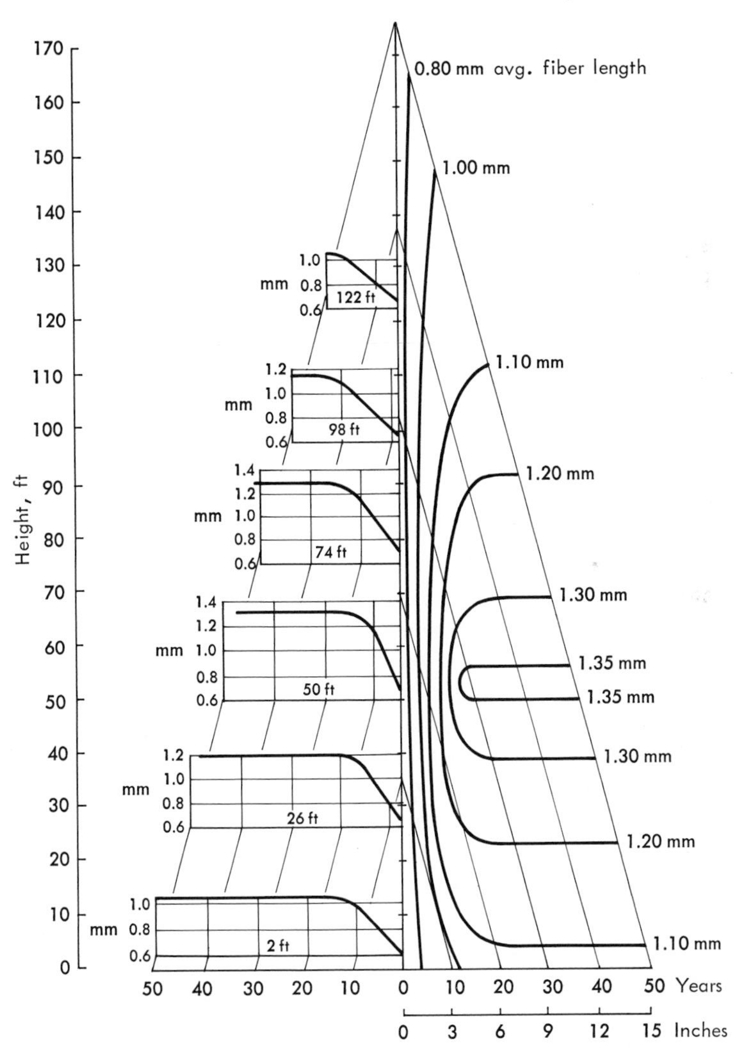

FIG. 7-8 Variation of fiber length in the radial and vertical directions in the trunk of a tree of *Eucalyptus regnans* F.v.M. The two halves of the diagram represent the same data, drawn as a topographic map on the right and as a series of curves for specific heights on the left. (*Adapted by permission from Bisset and Dadswell.*[7])

shorter at that point than at any other level in the tree height. For example, in the right-hand half of Fig. 7-8, along the line indicating the fiftieth increment, the fibers at the base of the tree are between 1.00 and 1.10 millimeters in length; at the 50-foot elevation in the stem the lengths of the fibers have increased to a maximum of 1.35 millimeters; above the 50-foot level, the fiber lengths decrease rapidly until they are less than 0.80 millimeter at the top of the tree.

The maximum length for the fibers in the diagram in Fig. 7-8 occurs at the 50-foot level in the trunk. However, in the increments formed before the tree reached 30 years of age the longest fibers in a given increment are shorter than the maximum of 1.35 millimeters for the entire tree. For example, the longest fibers developed in the tenth increment are barely 1.00 millimeter and occur at the 10-foot level; the longest fibers in the twentieth ring are 1.20 millimeters long at a height of 30 feet above ground; at 30 years of age the longest fibers in the increment are located 50 feet above ground and are of a maximum length for the tree.

The graphs in the left half of Fig. 7-8 represent the variation in fiber length in a horizontal plane at six different heights in the tree. It will be noted from these graphs that the rate of initial elongation of the fibers increases from the 2-foot level up to a maximum at the 50-foot level; this is indicated by the steeper slope in the graph. Above this height the elongation rate decreases and is less at the top of the tree than at the base of the trunk.

The general system of cell-length variation within the tree that has just been described applies to both conifer tracheids and hardwood fibers. This postulation of cell-length variation with height in the tree was made in 1872 by Sanio and has since been supported by a considerable volume of evidence for conifers, but less extensively for hardwoods.[12,13,21,54]

A few studies indicate that fibrous cells in some kinds of conifers and hardwoods do not conform to the above generalization for variability with height. Anderson[1] in a study of white fir *Abies concolor* (Gord. & Glend.) Lindl. showed that the minimum tracheid length occurs at the stump height; above the 22-foot elevation in the trunk there is statistically no significant change in tracheid length with height in the tree. Recent work in young-growth *Pinus ponderosa* has shown that there is no change in tracheid length above the 5-foot height in the stem.[60] Nicholls and Dadswell[43] report that in *Pinus radiata* D. Don there is a uniform reduction in tracheid length from the base to the top of the trunk in any given growth increment. This same pattern is shown in *Carya ovata*,[48] *Liriodendron tulipifera*,[57] and *Thuja plicata*.[62]

3. DISTRIBUTION OF CELL LENGTHS IN BRANCHES AND ROOTS

In branches of conifers the lengths of elements conform to patterns of variability which are approximately similar to those found in trunks (Fig. 7-9). The lengths of tracheids in branches are consistently shorter than those in the wood of the stem adjacent to the branch.[12,18,21,51] The data in Fig. 7-9 for *Pinus resinosa* show quite clearly that the lengths of all tracheids near the pith at the base of the branch are

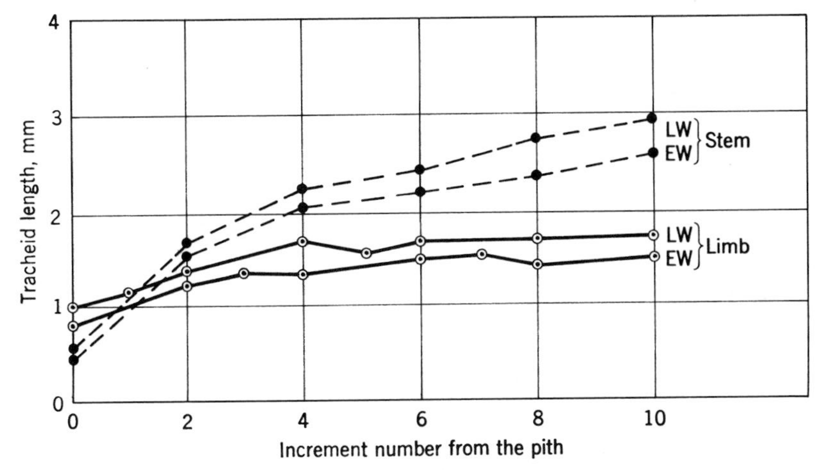

FIG. 7-9 Tracheid lengths in relation to number of increments from the pith in red pine (*Pinus resinosa* Ait.). The two lower curves (solid lines) are for the basal end of a limb from the lower crown of a 35-year-old tree. The curves shown with broken lines are for tracheids in the stem adjacent to the limb.

comparable to those at the pith in the adjacent stem. In the last-formed growth increments of the branch the tracheids are 40 to 50 percent shorter than those in the outer increment of the adjacent stem wood. Jackson's[29] data for *Pinus taeda* L. agree with these conclusions. Other work has shown that there is a highly significant relation between branch- and stem-tracheid lengths regardless of whether the stem tracheids are long or short in comparison to the species average.[21]

Within the outer growth increment of the branch, tracheid lengths increase toward the branch tip to a maximum of 10 percent at about one-fifth the branch length, then decrease to the tip. Tracheid lengths at the branch tip are about 40 percent shorter than at the branch base in the same increment.

For hardwoods much less is known about the relationship between branches and stem wood. Fegel[18] indicates that fibers in branches are about 25 percent shorter than those in the stem but gives no information on distribution or on the part of the branch which was measured. In contrast, data on *Castanea crenata* S. & Z. indicate that while branch fibers are consistently shorter at all periods of growth, those in early wood are about 5 percent shorter and in late wood 2 percent shorter than in the stem wood.[28]

In roots the available information indicates relationships that are variable among the conifers but relatively consistent in the hardwoods. *Pinus taeda* L.[18] is reported to have root tracheids about 10 percent shorter than those in the stem, i.e., intermediate between the short ones in the branches and those in the stem. *Pinus palustris* Mill. and *Pinus strobus* L. are stated to have root tracheids as long as or longer than those in the trunk.[20] These inconsistencies may be due to the variation pattern and the part of the root which was measured. In *Pinus resinosa* Ait. the length of the late-wood tracheids in the first ring next to the pith increases from 0.75 millimeter at the root collar to 3.0 millimeters at the small ends of the roots. The rate of increase in length is rapid at first; after the initial foot of length of the root, it changes more slowly. In the radial direction the increase in length from the pith is abrupt and of short duration, with more or less constant lengths thereafter. The increase in length from pith outward is about 30 percent. In the hardwoods root fibers are reported to be about 10 percent shorter than those in the stem.[18]

4. CELL DIAMETERS AND WALL THICKNESS

Some of the variations in wall thickness and radial diameters which are a feature of incremental growth are quite evident. Conifers in temperate climates generally show extremes of differences in both radial diameters and wall thickness between the early- and late-wood portions of the growth ring. The ratio of diameters may be as much as four or five to one from the wide early-wood cells to the tabular late-wood tracheids. The contrast in wall thickness within a ring is normally not as great as for cell diameters. An increase of one and one-half to two times from early- to late-wood thickness is fairly normal. In the hardwoods, diameter variation in cells is confined to the vessel elements, with a decrease in diameter, or volume, or both from early to late wood.

There is little information about trends for cell diameter and wall thickness changes in the radial direction from the pith. The tangential diameter of tracheids increases with distance from the pith to the bark for a number of conifers, according to Bailey;[2] this has been more recently verified in ***Pinus taeda*** L.[63] The wall thickness also increases in the radial direction toward the outside of the trunk. This has been demonstrated in ***Fraxinus pennsylvanica*** Marsh.,[52] ***Picea sitchensis*** (Bong.) Carr.,[46] ***Pinus resinosa*** Ait.,[38] and ***Pinus echinata*** Mill.[41] In general the early-wood wall thickness appears to increase about 15 percent from the pith to the bark in 30 years. The late wood shows about the same increase as the early wood in ***Pinus echinata***, but for ***Pinus resinosa*** the increase is more than double this amount in the 30-year period.[38] In contrast to the changes in the radial direction, Larson also has shown that in the vertical direction along the trunk there is little difference in either early- or late-wood wall thickness between the 6-foot and 18-foot levels in ***Pinus resinosa*** Ait.[38]

Conifers grown under tropical or semitropical conditions do not exhibit the marked differences between early and late wood that are found in the temperate zones. They are reported to exhibit tracheids that have only about 50 percent reduction in radial diameter from early to late wood and have walls which are thick and almost uniform in width across the increment. The thickness of these walls is about equal to that found in ***Larix*** and hard pine late wood.

As was indicated previously in Chap. 2, the changes in wall thickness and radial cell diameters which are associated with growth increments have been related by Larson to the physiology of growth.[37] Early-wood formation appears to be induced by growth hormones (auxins) that are produced by growth centers in the tree crown.

Large radial diameters of conifer early-wood tracheids are produced in the period of active internodal elongation when auxin production is at a maximum. As elongation slows, the radial diameters decrease and the transition to late wood coincides with cessation of internodal elongation. In the hardwoods, this same mechanism controls the production of the large early-wood vessels.

A second physiological system controls the wall thickness in tracheids. The amount of wall material produced in a developing cell depends upon the amount of photosynthesis products received. In the early part of the growing season the competition between internodal elongation, production of new needles, and development of secondary xylem and phloem is intense, a minimum of photosynthesis products is received in the cambial region, and the secondary walls are a minimum thickness. As the crown development terminates, the net amount of elaborated material from the leaves becomes greater in the cambial region and wall thickness increases, becoming a maximum in the late part of the growing season. This theory for varying cell wall thickness accounts for the almost unvarying heavy wall thickness previously reported for tropical pines grown under very favorable conditions.

5. FIBRILLAR ANGLE

The fibrillar angle in tracheids and other fibrous cells is inversely related to the cell length, as has been discussed elsewhere. Because of this relationship the fibrillar angle varies from pith to bark as a direct reflection of tracheid-length changes. The angle is large in the cells near the pith and decreases rapidly in the increments near the pith, to become almost parallel to the cell axis near the outside of the trunk. Investigations have shown this relationship to exist in *Pinus elliottii* Engelm., and *Pinus taeda* L., in *Pinus echinata* Mill.,[41] and in *Araucaria* sp. The same trends in fibril angles apparently exist in the hardwoods and have been demonstrated in *Fraxinus americana* L.

B. Specific-gravity Variation

Specific gravity of wood is principally a reflection of the proportionate volume of the wood which is occupied by wood substance. The presence of extractives can mask the true specific gravity and confuse the usual relationships with strength properties. Cell wall thickness and cell cross-section dimensions are directly related to specific gravity of the wood and, together with ring widths and early-wood–late-wood proportions, define specific-gravity variations. Much of the information available on specific-gravity differences within trees has been obtained by measuring cross-section disks, wedges from such disks, or blocks representing a group of growth increments. Detailed analysis of individual rings and early wood or late wood for separate rings has been lacking until recently, when several methods for continuous trace analysis have been developed. One method uses differential absorption of beta rays in thin strips of wood, another uses a microdensitometer to measure densities of x-ray negatives from radial wood samples, a third uses a microphotometer and light transmitted through a cross section. These semiautomatic methods have greatly increased the amount of data available.

1. WITHIN-INCREMENT VARIATION

In general, the patterns of intraincrement variation in specific gravity resemble those for tracheid- and fiber-length variation within increments. Figure 7-10 is an example of a density variation diagram across several adjacent increments in *Pinus palustris* Mill. It can be seen from the tracing that the density ranges from 0.3 grams per cubic centimeter in the early wood to a maximum of 0.9 in the late wood. It is apparent from these figures that the late wood is approximately three times denser than the early wood. Reference to Fig. 7-11 for *Pinus resinosa* Ait. shows that these two

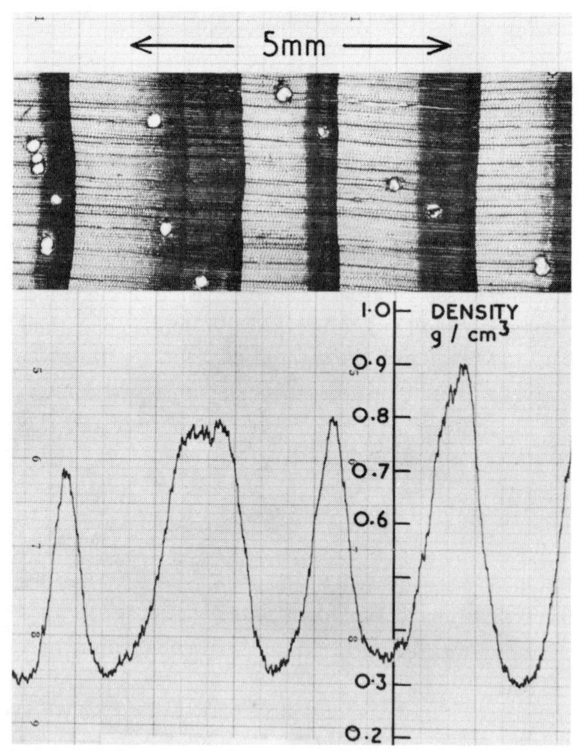

FIG. 7-10 Density variation within the growth increment of longleaf pine (*Pinus palustris* Mill.). Density was determined by beta-particle radiation using carbon-14. [*From E. W. J. Phillips et al., J. Inst. Wood Sci.,* **10:**11–28 (1962). *Original crown copyright and reproduction by permission of the Controller, H. M. Stationery Office, London, England.*]

woods having abrupt late-wood transitions have very similar patterns of density and tracheid-length variations.

A wide variation in density between early and late wood within the growth increment is not necessarily related to an abrupt transition such as that cited for *Pinus palustris*. For example, data for *Picea sitchensis* (Bong.) Carr., which has a fairly gradual transition from early to late wood, indicate a range of density within the increment from 0.2 to 0.5 gram per cubic centimeter and an increase of 2.5 for late wood over the early wood. Other conifers exhibit comparable or even greater increases within the growth increment. For example, Wilson[64] cites early- to late-wood specific-gravity increase in ratios as: 3.5 in *Pseudotsuga menziesii* (Mirb.) Franco, 5.0 in *Thuja plicata* Donn, 2.4 in *Abies amabilis* (Dougl.) Forbes, 2.5 in *Picea sitchensis* (Bong.) Carr., and 2.5 in *Tsuga heterophylla* (Raf.) Sarg.

2. VARIATION IN A RADIAL DIRECTION

Changes in mean specific gravity of wood within a cross section of a stem can be classified into a few patterns which indicate that in general terms the specific gravity either tends to increase from the pith outward or shows a decreasing trend toward the bark side. A general classification may be made as follows into four patterns or types of trends:

Type 1. Specific gravity of the wood increases from the pith to the bark.
Type 2. Specific gravity of the wood is high at the pith, decreases outward for the first few years and then increases to a maximum at the bark.
Type 3. Specific gravity increases in the increments near the pith, then remains more or less constant; or sometimes the specific gravity may even decrease in the last-formed increments next to the bark.
Type 4. Specific gravity of the wood exhibits a general decrease from pith to bark in the stem.

An examination of the tabulation of reported trends, as in Table 7-1, shows that the conifers are preponderantly in the group of increasing specific gravity from pith to bark. *Chamaecyparis* and *Thuja* are notable exceptions among the conifers listed in exhibiting minimum density at the outside of the stem. Other exceptions to the usual trends, as in *Pinus echinata*, may be due to unreported differences in silvicultural treatments or to methods of analysis of data. Among the hardwoods there is an almost even division between reported increases and decreases in specific gravity from pith to bark.

Trends of specific-gravity variation in the radial direction for early and late wood, considered separately, are not alike. A graphic display of the intra- and inter-increment density changes, as in Fig. 7-11, illustrates the usual relationships in conifers. Separate trends for the early- and late-wood portions of the radial variation may be revealed by drawing curves through the successive minimal or maximal points of the graph. The curves produced in this way show that the early wood in a

TABLE 7-1 Radial Specific Gravity Variation Patterns in Stems of Conifers and Hardwoods

Species	Pattern*	Reference
Conifers		
Abies grandis	2	Polge, H.: *Rev. Forest. Franc.,* **16**(6):480 (1964).
Abies nordmanniana	2	Polge, H.: *loc. cit.*
Abies pectinata	2	Polge, H.: *loc. cit.*
Chamaecyparis lawsoniana	4	Polge, H.: *loc. cit.*
Larix europaea	1	Polge, H.: *loc. cit.*
Larix leptolepis	1	Pechman, H. von.: *Schweiz. Z. Forstw.,* **109**(11):615–647 (1958).
Larix leptolepis	2	Polge, H.: *loc. cit.*
Libocedrus decurrens	1	Resch, H., and S. Huang: *Calif. Agr. Exp. Sta. Bull.* 833, 1967.
Picea abies	2	Hakkila, P.: *Commun. Inst. For. Fenn.,* **61**(5):1 (1966).
Picea abies	1	Pechman, H. von: *loc. cit.*
Picea abies	2	Polge, H.: *loc. cit.*
Picea abies	2	Tamminen, Z.: *Inst. for. Virkeslara Skogkolan Rapporter Nr.* R47, 1964.
Picea mariana	2	Risi, J., and E. Zeller: *Laval Univ., Contrib. No.* 6, 1960.
Picea sitchensis	2	Jeffers, J. N. R.: *J. Inst. Wood Sci.,* **4:**44 (1959).
Pinus contorta	1	Collett, R. M.: M.S. thesis, Colorado State University, 1963.
Pinus echinata	4	Gilmore, A. R.: *J. Forestry,* **61**(8):596 (1963).
Pinus ponderosa	2	Polge, H.: *loc. cit.*
Pinus radiata	2	Nichols, J. W. P., and H. E. Dadswell: CSIRO Australia, Div. Forest Prod. Tech. Paper 37, 1967.
Pinus resinosa	1	Cooper, G. A.: *Iowa State J. Sci.,* **34**(4):693 (1960).
Pinus resinosa	1	Jayne, B. A.: Tappi **41**(4):162 (1958).
Pinus resinosa	1	Jimenez, A.: M.S. thesis, SUNY College Forestry, Syracuse, N.Y., 1967.
Pinus resinosa	2	Peterson, T. A.: Ph.D. dissertation, University of Wisconsin, Madison, Wis., 1967.
Pinus strobus	2	Foulger, A. N.: *Forest Prod. J.,* **16**(12):45–47 (1966).
Pinus strobus	2	Polge, H.: *loc. cit.*
Pinus sylvestris	1	Hakkila, P.: *loc. cit.*
Pinus spp. (Mexican)	1	Zobel, B. J.: *Silvae Genet.,* **14**(1): 1965.
Pinus taeda	1	Zobel, B. J., and R. McElwee: *Tappi,* **41**(4):158 (1958).
Pseudotsuga menziesii	3	Phillips, E. W. J., *et al.*: *J. Inst. Wood Sci.,* **10:**11–28 (1962).
Pseudotsuga menziesii	2	Polge, H.: *loc. cit.*
Pseudotsuga menziesii	1	Sastry, C. B. R.: M.S. thesis, University of British Columbia, 1967.

TABLE 7-1 (Continued)

Species	Pattern*	Reference
Thuja plicata	2	Wellwood, R. W., and P. E. Jurazs: *Forest Prod. J.*, **18**(12):45 (1968).
Thuja plicata	4	Polge, H.: *loc. cit.*
Tsuga heterophylla	2	Krahmer, R. L.: *Tappi*, **49**(5):227 (1966).
Tsuga heterophylla	2	Polge, H.: *loc. cit.*
Tsuga heterophylla	2	Wellwood, R. W.: *Pulp Paper Mag. Can.*, **63**(2):T61–T67 (1962).
Hardwoods		
Betula pubescens	1	Hakkila, P.: *loc. cit.*
Betula verrucosa	1	Hakkila, P.: *loc. cit.*
Eucalyptus marginata	3	Rudman, P.: *Holzforschung*, **18**(6):172 (1964).
Eucalyptus obliqua	1	Report Forestry Comm. New South Wales, Australia, 1967.
Eucalyptus viminalis	1	Report Forestry Comm. New South Wales, Australia, 1967.
Fagus sylvatica	4	Grossler, W.: *Holz Roh-Werkstoff*, **6**(3):81 (1943).
Liriodendron tulipifera	3–4	Taylor, F. W.: *Forest Prod. J.;* **18**(3):75 (1968).
Liriodendron tulipifera	1	Thorbjornsen, E.: *Tappi*, **44**(3):192 (1961).
Nyssa aquatica	3	McElwee, R. L., and J. B. Faircloth: *Tappi*, **49**(21):538 (1966).
Populus deltoides	1–2	Farmer, R. E., and J. R. Wilcox: Tappi 3rd Forest Biol. Conf., Madison, Wis., 1965.
Populus nigra	3	Hirai, S., and E. Aizawa: *Tokyo Univ. Forest Bull.* 62, 1966.
Populus spp. (hybrids)	4–2	Gohre, K.: *Wiss. Abhandl. Deut. Akad. Landwiss.*, **44:**51–79 (1960).
Prunus serotina	4	Koch, C. B.: *J. Forestry*, **65**(3):200 (1967).
Quercus falcata	4	Hamilton, J. R.: *Forest Prod. J.*, **11**(6):267–271 (1961).
Shorea almon	1	Cruz, R. de la: Forest Prod. Res. Inst., Laguna, Philippines, 1967.
Terminalia superba	1	Mottet, A.: Proc. Internat. Union Forest Res. Organ, Sect. 41, Madison, Wis., 1968.

* *Type* 1. Specific gravity of the wood increases from the pith to the bark.

Type 2. Specific gravity high near the pith, decreases outward for the first few years, and then increases to a maximum at the bark.

Type 3. Specific gravity increases in the increments near the pith, then remains more or less constant, or sometimes the specific gravity decreases in the last-formed increments at the bark.

Type 4. Specific gravity of the wood exhibits a general decrease from pith to bark.

series of increments rapidly decreases in density in the first few years from the pith, and then maintains a minimum density with little fluctuation or perhaps increases slightly in the later years of growth. In contrast, the density of the late wood from the pith outward exhibits one of the radial trends of variation previously given for mean specific gravity. Thus, the late wood trend may be unlike that for the early wood

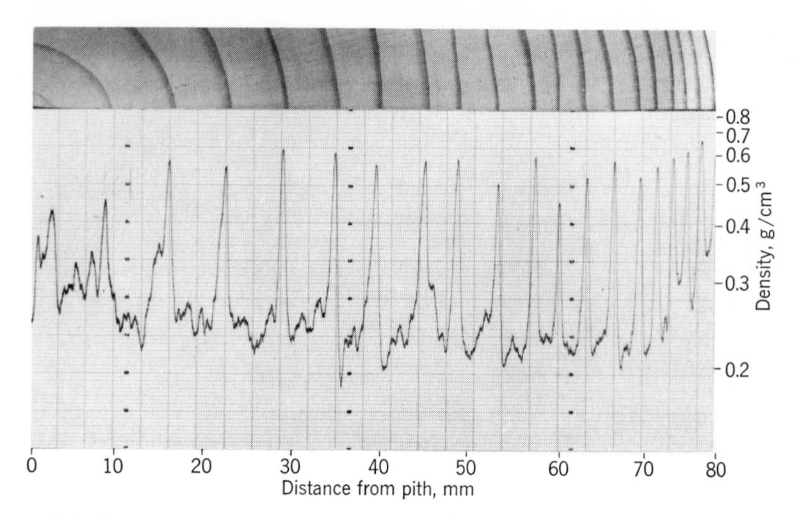

FIG. 7-11 Density variation in the radial direction in a cross section of red pine (*Pinus resinosa* Ait.) cut from the stem below the live crown. The density was measured by beta-ray absorption using carbon-14 as a source of the beta rays.

trend, or similar as in Fig. 7-11. It is evident, from these trends for early and late wood, plotted separately, that the early wood is not greatly affected by age or distance from the pith after the first few years. The late-wood increments, however, do reveal trends of change with age and distance from the pith that are to be expected from previous discussion of morphology.

3. VARIATIONS IN THE AXIAL DIRECTION

The trends in specific gravity variation along the axis of the trunk from the base of the tree to the top have been reported as (a) decreasing uniformly; (b) decreasing in the lower trunk and increasing in the upper trunk; (c) increasing in the stem from base to top in a nonuniform pattern.

Decrease in specific gravity from the base upward is the most commonly reported trend of variation in the conifers. It appears to be typical of the pines and is reported also in *Libocedrus decurrens* Torr., *Picea abies* (L.) Karst., *Pseudotsuga menziesii* (Mirb.) Franco, and *Tsuga heterophylla* (Raf.) Sarg. In the hardwoods this type of trend in specific gravity is not common, but it has been shown to exist in *Acer rubrum* L. and *Acer saccharinum* L.

The second type of trend, i.e., decrease in specific gravity for the lower part of the trunk and increasing trends in the top, has been reported among the conifers in *Pinus contorta* Dougl., and in *Pinus strobus* L. For hardwood trees this pattern of variation has been shown in *Liriodendron tulipifera* L. and *Tectona grandis* L. (teak). It would appear that this pattern of axial variation may be only a modification of the general decrease from base to top in the tree caused by the inclusion of greater proportions of high-density knots in the mean specific gravity of crown wood.[36] However, some evidence exists to indicate that wood in the crown region is basically denser, but no reasons for this have been shown.

In hardwoods, the third trend (type *c*) in specific gravity increase is reported more commonly than any other trend. For example this type of trend is found in *Fagus sylvatica* L., *Fraxinus pennsylvanica* Marsh., *Nyssa aquatica* L., *Liquidambar styraciflua* L., and *Quercus falcata* Michx.[25] In conifers the regular increase in specific gravity from base to top of the tree is apparently uncommon. It has been reported only in *Picea sitchensis* (Bong.) Carr.[31] and in *Thuja plicata* Donn.[61]

Other reports of variation in specific gravity along the axes of stems indicate a variety of patterns within a species. For example Baker[4] states that, whereas forest-grown red pine (*Pinus resinosa* Ait.) exhibits a regular decrease in specific gravity with increasing height in the trunk, plantation-grown red pine increased continuously from the base upward or decreased in the lower trunk and increased in the upper trunk. In *Tsuga heterophylla* (Raf.) Sarg., Krahmer[34] found that the general vertical decrease in specific gravity toward the top is associated with growth rates of 10 rings or more per inch, but faster growth rates (less than 10 rings per inch) reversed the trend so that specific gravity increased from base to top of the tree. It is apparent from these and many other reports that gross specific gravity within the stem is a resultant of a number of inherent factors modified by growth conditions. The interactions of these various factors are only partly understood.

All the previous statements concerning specific gravity variation along the trunk axis have been based on mean values for early and late wood combined. Very little is known about variation for the early and late wood as independent systems. Taras,[56] investigating specific-gravity variations with height in the tree for *Pinus elliottii* Engelm., found that while the early wood decreased from the base to breast height or somewhat higher and then increased to the top, the late wood increased to breast height and then decreased in the upper parts of the tree.

C. Variations in Chemical Composition

1. EARLY- TO LATE-WOOD VARIATION

In the older parts of stems it is apparent that the late wood has a higher percentage of cellulose and a lower percentage of lignin than the early wood in the same increment.[38,65] Lignin in the early wood of *Pinus resinosa* Ait. shows percentages that are consistently from 2 to 3 percent higher than in the late wood.[38] Wilson and Wellwood[65] found approximately the same relationships and indicated that the statistical differences for these zones within the increment are highly significant in *Abies amabilis* (Dougl.) Forbes, *Picea sitchensis* (Bong.) Carr., and *Pseudotsuga menziesii* (Mirb.) Franco, but are not statistically significant for *Tsuga heterophylla* (Raf.) Sarg.

Holocellulose (chlorite) in *Pseudotsuga menziesii* (Mirb.) Franco has been shown to be at a maximum (77 to 78 percent) at the point of late-wood initiation and at a minimum (72.5 to 73.5 percent) at the increment boundary.[65] Very similar values for holocellulose in *Pseudotsuga* have been reported by Hale and Clermont.[24] These authors also cite 73.2 percent for early-wood and 74.8 percent for late-wood holocellulose in *Pinus resinosa* Ait.

Fragmentary work in a few species of conifers indicates that the cellulose in the late wood not only is higher in terms of percentage but also has a greater degree of polymerization, higher packing density, and a higher degree of crystallinity.

If the polysaccharides in *Pinus resinosa* Ait. are considered in terms of hydrolyzed simple sugars, xylose in the early wood appears to be 1 or 2 percent higher than in the late-wood portion of the increment. The reverse situation is true for mannose. No difference is evident between early and late wood for galactose, arabinose, and glucose in young trees.[38]

2. VARIATIONS WITH POSITION IN THE TRUNK

Cellulose content increases from the pith to the bark and can be plotted with curves which approximate those for changes in tracheid length from pith to bark. Actual percentage values increase approximately 3 percent in *Pinus contorta* Dougl. and *Pinus taeda* L., 6 to 10 percent in *Pseudotsuga menziesii* (Mirb.) Franco, and 8 to 20 percent in *Pinus radiata* D. Don. Alpha cellulose has been shown to vary by 7.5 percent between pith and bark in *Pinus taeda* L., even with equal specific gravities in the two areas sampled. Most of this increase occurs in the first few rings outward from the pith during the period of maximum increase in late-wood and in

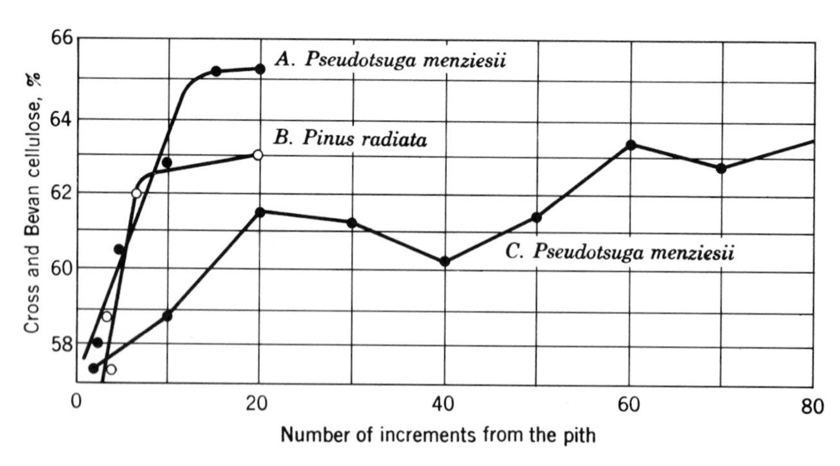

FIG. 7-12 Variation of Cross and Bevan cellulose content with distance from the pith in three trees. [*A and B adapted by permission from A. B. Wardrop, Australian J. Sci. Res.*, **B4**(4):391–414 (1951). *Curve C from R. W. Kennedy and J. M. Jaworsky, Tappi*, **43**(1):25–27 (1960).]

growth-increment specific gravity. Cellulose percentage fluctuates about a maximum during the period of mature growth of the tree but decreases again as the tree becomes overmature and specific gravity of the growth increments decreases. Cellulose decreases associated with overmaturity begin at different ages depending on the species of tree: Hale and Clermont[24] cite this as 150 years in *Picea glauca* (Moench) Voss, and 300 to 400 years in *Pseudotsuga menziesii* (Mirb.) Franco.

In pines, decrease in cellulose content along the tree trunk axis is reported to be small but significant in some species, but insignificant in others. In the hardwoods the only information is available for hybrid poplars, in which no apparent relationship has been found between height in the tree and cellulose distribution.

Analysis of the simple sugars produced by hydrolyzing the polysaccharides in *Pinus resinosa* Ait. wood[38] shows that glucose increases with age from the pith in a flat curve resembling the change in length of tracheids in the same radial direction, mannose increases linearly with rings from the pith, while galactose, xylose, and arabinose decrease slightly and uniformly with age from the pith. For the same group of hydrolyzed sugars, Larson[38] indicates no appreciable difference for any of the sugars with increase in height in the stem for heights from 6 to 18 feet.

Lignin distribution in the radial direction shows a general decrease of approximately 2 to 3 percent from pith to bark, with parallel trends for early and late wood.[38] There is no sharp decrease associated with the zone near the pith, but overmaturity in *Pseudotsuga* apparently causes an increase in lignin content.[24] Changes with height along the stem axis are not at all clear. A few pieces of evidence indicate that there is a slight increase upward in some of the pines and poplar hybrids.

Variations in extractives within the stem are principally associated with their presence in heartwood, except for resins. In the conifers, resin content is reported to be highest near the pith at the butt of the tree, decreasing outward and upward in the stem. A secondary increase in resins occurs in the outer heartwood and the top of the tree. Resin contents are usually much lower in the sapwood than in the heartwood. In general the resin contents of the late-wood zones in conifers are slightly higher than those for the early wood because of the usual tendency for concentration of resin ducts in the late wood.

Many of the heartwood extractives increase from the pith to the outer heartwood boundary, e.g., thujaplicin in *Thuja plicata* Donn and arabinogalactan in *Larix* spp. Vertically along the stem the distribution varies with the extractive in some cases and with the species of tree in other cases; e.g., arabinogalactan is reported to decrease upward from a maximum at the butt in *Larix occidentalis* Nutt. and to increase upwards in the stem in *Larix laricina* (DuRoi) K. Koch.[10]

Another type of variability in extractives of heartwood is related to differences in toxicity or repellant properties of compounds that are known to impart termite or fungal resistance to wood. It is known for instance that the inner heartwood of some *Eucalyptus* spp. and *Thuja plicata* Donn is less resistant to biological deterioration than the outer heartwood because of loss or alteration of the toxic compounds with time.

The ash content of loblolly pine in central Louisiana is reported* to increase with distance from the pith in the stem, with increased rate of stem diameter growth, and to increase from early to late wood.

D. Variations in Physical Properties

Earlier discussions in Chap. 6 have shown that physical behavior is very closely related to the anatomical characteristics of the wood, to the chemical composition of the cell walls, as well as to the nature and extent of the extractives which may be present. It is to be expected that variations in the factors affecting a given physical property will result in appropriate differences in that property.

1. STRENGTH AND ELASTICITY

The tensile properties of wood are directly related to fiber morphology, i.e., cell length,[53] diameter, wall thickness, and fibril angle,[15] as well as to the lignin-cellulose ratio[22] and the molecular organization of the cellulose. Tests of microsections cut serially within a single increment have shown that in *Pseudotsuga menziesii* (Mirb.) Franco[27] and in *Picea abies* (L.) Karst[32] the tensile strength and modulus of elasticity increase from the inception of early wood across the increment, then increase sharply in the transition zone, and reach a maximum in the late wood. This pattern resembles the curves for tracheid-length and specific-gravity variation within increments, except in the early-wood zone. The minimum values for both tracheid length and specific gravity occur about midway across the early-wood zone, but tensile strength shows no such minimum value at a corresponding position. This apparent anomaly is probably the result of the decreasing fibril angle outward in the increment, plus the rising proportion of cellulose, as well as the increasing degree of crystallinity of the cellulose from early to late wood. All three of these factors would tend to more than counteract the decrease of specific gravity in the central early wood and result in constant rate of increase in tensile strength. When these same tensile tests within increments are plotted as specific tensile strength, i.e., on unity specific-gravity basis, the variation is parabolic from early to late wood, with the maximum at the early-late wood boundary. The location of a maximum at this point in the increment is an indication that the optimum conditions for tensile strength exist at this point and that an increased proportion of this part of the ring would improve the overall strength properties of the wood.

Variations of strength and elastic properties within the cross section are also reflections of the changes in specific gravity and tracheid lengths, among other factors. This has been demonstrated for tensile strength in *Tsuga heterophylla* (Raf.) Sarg.[61] (Fig. 7-13). An examination of these curves shows that the tensile strength, tracheid length, and specific gravity increase from the pith outward and are all approximately

* C. W. McMillan, Ash Content of Loblolly Pine Wood as Related to Specific Gravity, Growth Rate, and Distance from Pith, *Wood Science*, **2**(1):26–30 (1969).

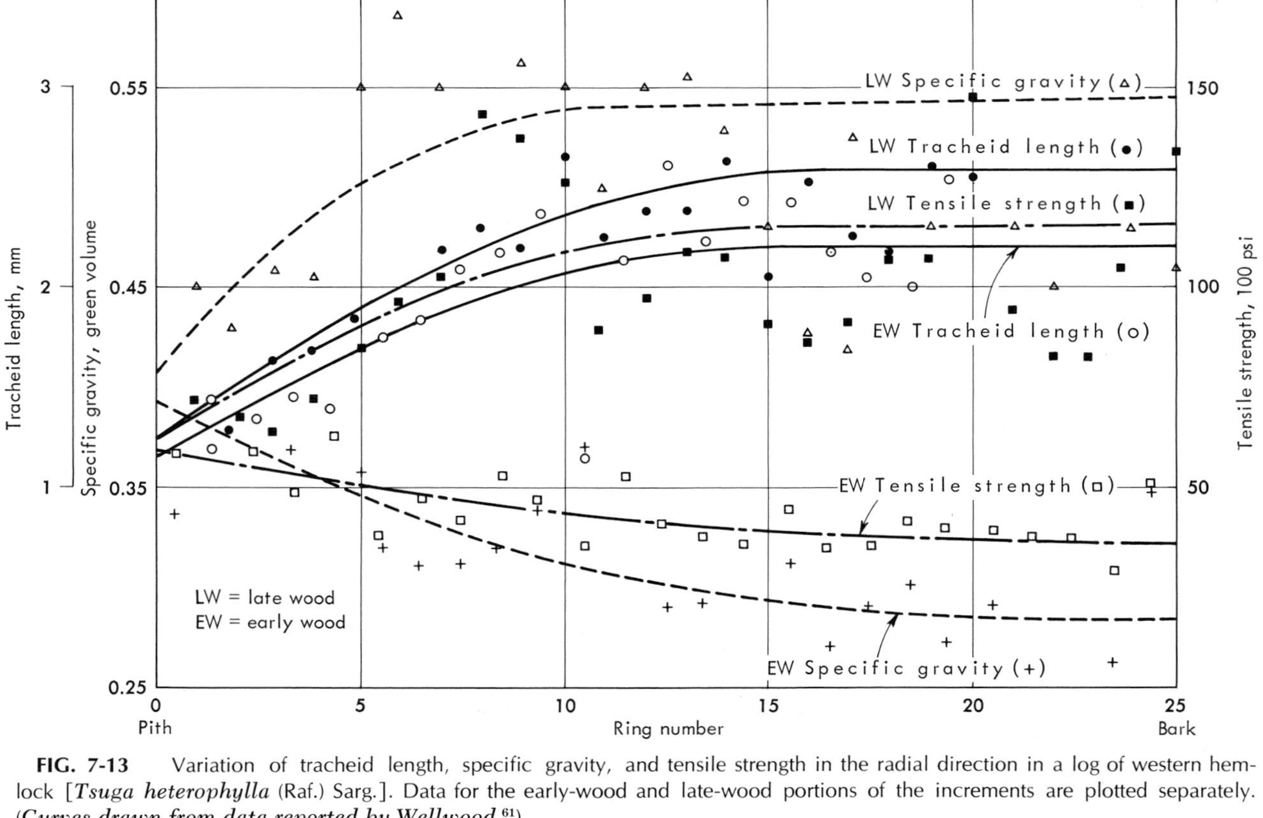

FIG. 7-13 Variation of tracheid length, specific gravity, and tensile strength in the radial direction in a log of western hemlock [*Tsuga heterophylla* (Raf.) Sarg.]. Data for the early-wood and late-wood portions of the increments are plotted separately. (*Curves drawn from data reported by Wellwood.*[61])

parallel for the late wood, while early-wood tensile strength decreases radially and is most closely related to its specific gravity. These relationships between early and late wood and variation radially have been confirmed in *Pseudotsuga menziesii* (Mirb.) Franco for static bending and toughness, and for tensile strength in *Pinus radiata* D. Don.

Variation of strength properties with respect to position in the tree, but without regard to differences within growth increments, has been shown for some other kinds of trees. Static bending properties increase from the pith outward and decrease upward in the stem for *Pinus resinosa* Ait.[47] and *Shorea almon* Foxw.[11] However, in *Liriodendron tulipifera* L. the maximum values are found near stump height, decrease upward to about 16 feet, and then increase toward the top but never become as great as at stump height.

Other data for strength, elasticity, and impact variations with position in the tree are rather sparse. However, all the evidence available confirms that these variations are also reflections of the structural changes, and amounts and distribution of the tissues within the tree.

2. NONMECHANICAL PHYSICAL PROPERTIES

Shrinkage variations with position in the tree have been studied by relatively few persons, but the patterns which have been demonstrated conform to expected relationships with cell morphology and chemical composition. Within an increment the early wood appears to have consistently greater longitudinal shrinkage than the late wood and a reduced tangential shrinkage.

The shape of the curve relating tangential shrinkage to position within the increment is also of the sawtooth type which has been shown for specific gravity and tracheid length.[17] Variation within the stem is different for longitudinal shrinkage and for dimensional changes in the lateral directions. Foulger[19] has demonstrated in *Pinus strobus* L. that longitudinal shrinkage is greatest near the pith, decreases rapidly in the first 10 rings outward from the pith, and then decreases irregularly toward the bark. This pattern can be explained on the basis that the inner core around the pith has short tracheids, large fibrillar angle with the cell axis, low cellulose content, and low cellulose crystallinity. All these factors contribute to the abnormally high longitudinal shrinkage near the pith of this and all other woods which have been investigated.

Radial, tangential, and volumetric shrinkages increase linearly as the specific gravity increases; however, within tree stems the presence of heartwood and its associated extractives complicates the basic relation to specific gravity. The presence of heartwood extractives reduces the shrinkage proportionately to the amount present, below the level expected from the extracted specific gravity of the wood. For this reason, shrinkage variations for unextracted wood are related more to patterns of extractives than to those for specific gravity of the wood. Very little is known concerning variations of shrinkage with height along the tree axis. Work with *Picea abies* (L.) Karst in Sweden[55] indicates that regardless of the specific gravity of the wood, volumetric shrinkage increases markedly from ground level to breast height in the tree, but does not vary much in the upper levels of the stem. Longitudinal shrinkage in *Pinus ponderosa* Laws,* is abnormally large near the pith up to stump height. Its magnitude decreases somewhat in the core of wood near the pith above the stump but still remains excessive.

Investigations in *Picea abies* (L.) Karst[55] have shown that the fiber saturation point in the outer sapwood increases slightly from stump height to halfway up the stem and then decreases toward the top of the tree (mean value is 34 to 36 percent M.C.). In contrast, the fiber saturation point in the heartwood increases sharply from stump height to about 10 percent of the tree height and then is more or less constant to the top of the tree (mean value is 27 to 30 percent M.C.). Differences in the fiber saturation point of sapwood and heartwood are explainable on the basis of infiltration products in the cell wall for the heartwood, but no reasons have been advanced for variations in this property in either of these regions with height along the stem axis.

Comstock,[9] working with the longitudinal permeability of *Tsuga canadensis* (L.) Carr., has found that for sapwood the permeability is about 100 times greater than that for the heartwood: furthermore it decreases with distance inward from the bark and increases with increasing height in the tree. For the heartwood the systematic variation with radial position or height in the stem was not obvious.

E. Growth-Increment Variation

The width of growth rings and the percentage of proportionate amounts of late wood are criteria which are often used to judge the quality of wood. Because these characteristics are used so frequently, it becomes important to know the nature of their variation within trees.

1. GROWTH-RING-WIDTH VARIATION

The radial width of an increment in a tree stem varies with the height above ground. Minimum width is found near the base of the tree; the width increases with increasing height in the stem to a maximum in the general vicinity of the base of the live crown and then decreases to the top of the tree. In young conifers the maximum width of ring occurs in the region of the fourth or fifth internode below the tip.[14]

For conifers, reports of ring-width relationships to the various wood properties are not in agreement. A considerable number of investigations have shown that there is a decrease in specific gravity with increasing ring width. Many of these conclusions are based on statistical analyses. However, many studies, which appear to be equally valid and as carefully analyzed, claim no relationship between specific gravity and ring width, or at most a very limited correlation. A possible reason for part of this

* R. A. Cockrell and R. A. Howard, Specific Gravity and Shrinkage of Open-Grown Ponderosa Pine, *Wood Sci. and Technol.*, **2**(4):292–298 (1969).

discrepancy may arise because some of the studies have been based on old-growth wood and others on young trees. Knigge[33] brings out this possibility by showing that in the first decades of growth in ***Pseudotsuga menziesii*** (Mirb.) Franco, specific gravity of the wood is significantly related to ring width, but in the mature growth periods age of the cambium becomes the significant related factor.

Tracheid dimensions have been shown to be related to ring widths in some conifers. For the most part maximum tracheid lengths are associated with optimal ring width, which is usually from 0.5 to 2 mm wide, depending on the species.[5,35] Ring widths either lesser or greater than this optimum are associated with shorter tracheid lengths.

In ring-porous hardwoods the ring width is directly related to the average specific gravity of the wood. In general, the early-wood volume is fairly constant and the percentage of late wood tends to increase as the width of growth increments increases (Fig. 7-14). Since the late wood contains the bulk of the fibrous tissue, proportionate increases in the amount of late wood raise the average specific-gravity values and thus favorably affect the mechanical properties of wood with wide growth increments. This statement must again be accepted with caution because wood formed near the pith and reaction wood (see Chap. 8) are notably different. Furthermore, hereditary differences between trees also may influence the average specific gravity of the wood; so that growth increments of comparable widths in different trees may result in differing specific gravities of the wood.

In diffuse-porous hardwoods there is little relationship between the specific gravity of the wood and the ring width for mature timber. Exceptions are found in trees with extremely slow growth for long periods and in wood with growth increments containing unusually large percentages of vessel volume. As a result, this type of growth has extremely low specific gravity. Overmature timber and wood from near the pith must also be excluded from consideration in all types of hardwoods and in the conifers.

2. EARLY- AND LATE-WOOD PERCENTAGE

The proportions of early and late wood within the increment are not necessarily the same for equal ring widths. The two zones in the growth increment vary independently. Maximum width in the early wood occurs in the upper part of the live crown and decreases down the trunk to the minimum width 5 to 10 feet above the ground. The late-wood zones are reported to be of a maximum width in the basal portion of the stem below the live crown.

In conifers there is a definite relationship between specific gravity of the wood and late-wood percentage. This interdependence is shown best in wood with reasonably uniform ring widths formed in the mature periods of growth for the tree. Within any one increment the late-wood percentage is greatest at the base of the tree and decreases upward to a minimum value near the pith at the top of the tree.[36] The general pattern for variation in the amount of late wood in conifer stems appears to bear a direct relationship to age,[35] and to be inversely related to height in the tree.

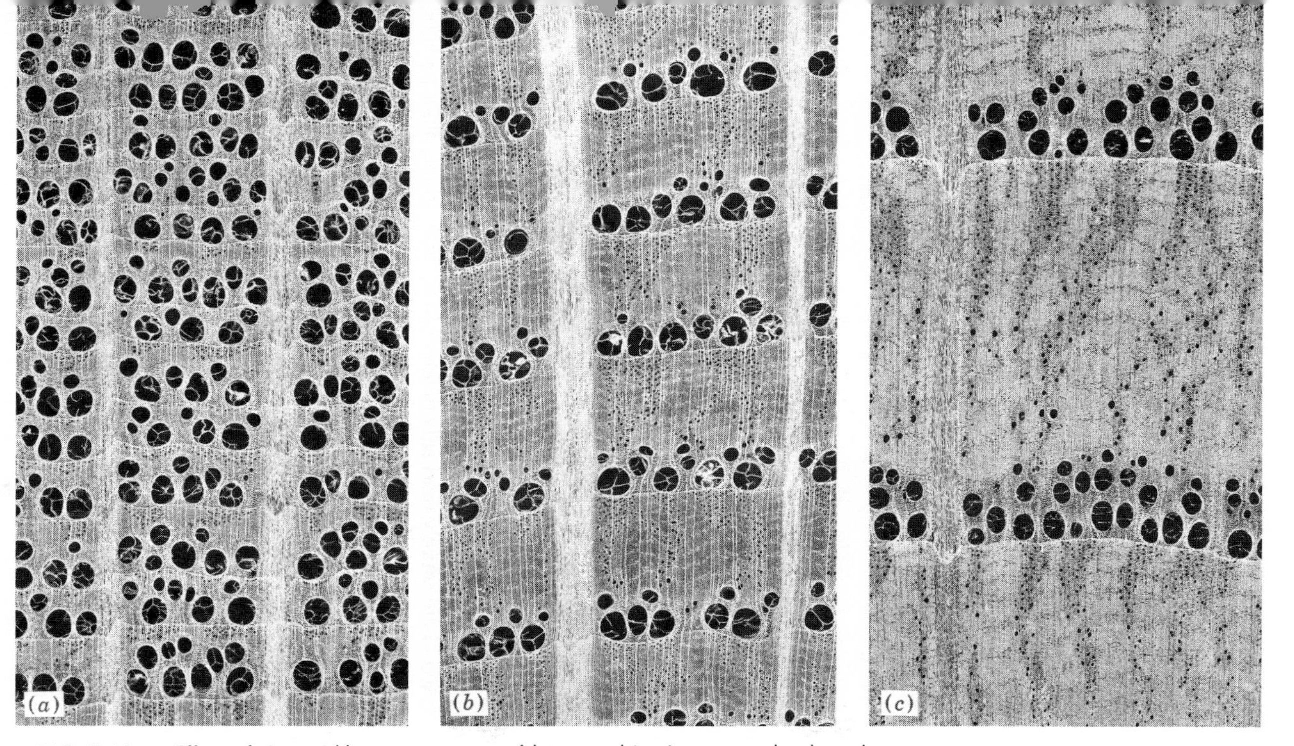

FIG. 7-14 Effect of ring width on percentage of late wood in ring-porous hardwoods.
(*a*) Narrow-ringed sample of post oak (*Quercus stellata* Wangenh.) with greatly reduced late wood. (10×.)
(*b*) An average width of ring in white oak (*Q. alba* L.) with normal amount of late wood. (10×.)
(*c*) An unusually wide-ringed sample of white oak (*Q. alba* L.) with high percentage of late wood. (10×.)
(*Photomicrographs by S. Williams.*)

For hardwoods of the ring-porous type, late-wood percentages appear to follow the same general pattern as that given for specific gravity. Hamilton[25] reports that the highest percentage of late wood is at the center of the base of the trunk and the lowest percentages occur in the outer periphery of the trunk, in the region below the crown. Very little information is available with regard to the total variation of late wood in the diffuse-porous hardwoods, precluding any general statements on this point.

It is important to remember that ring width alone has little meaning as an indicator of the properties of the wood; width of growth increment must always be considered in connection with the percentage of late wood. This interdependence has been recognized for a number of years in the grading rules for southern pine and Douglas-fir lumber, in which dense structural material is required to have not less than six rings per inch and one-third late wood. The latter must also be distinctly contrasted in color from the early-wood zone to eliminate compression wood.

F. Juvenile and Adult Wood

Reference to the trends of variation in characteristics of wood in the tree indicates that in all cases the portion of the stem that is close to the pith possesses distinctly different cellular structure and properties than the wood in the outer part of the trunk. For example, fibers in hardwoods in the first-formed wood near the pith are less than 1 millimeter long, regardless of the final fiber length (Fig. 7-8); the specific gravity, late-wood percentage, and tensile strength in conifers is also low (Fig. 7-13). The cellulose content of wood from near the pith of conifers is well below the level found in wood nearer to the bark (Fig. 7-12), and longitudinal shrinkage is excessive, causing unusual warping in lumber sawn from the region near the pith, even when reaction wood is not present. In addition, in hardwoods, the pattern and size of vessels in such wood are usually markedly different from those in the older parts of the stem and resemble those of wood found in branches.

Wood formed in the region near the pith is called *juvenile wood*. Rendle[49] defines juvenile wood as "Secondary xylem produced during the early life of the part of the tree under consideration and characterized anatomically by a progressive increase in the dimensions and corresponding changes in the form, structure and disposition of the cells in successive growth layers." Formation of juvenile wood is associated with the prolonged influence of the apical meristems in the regions of the active crown during the growing season.[36] As the tree crown moves upward in the older tree, the cambium at a given height becomes less subject to the direct influences of the elongating crown region and *adult wood*[49] is formed.

It should be noted that the primary basis for defining juvenile wood is the cell structure and properties of the wood. Juvenile wood in plantation-grown trees can be related to fast growth near the pith, but wide rings are not necessarily associated with juvenile wood in all trees. For example, young stems growing under heavy competition form narrow rings in the region near the pith, while wide increments may form any time that growth conditions are favorable. The duration of the juvenile period is quite variable between species, ranging from 5 to 20 years. Termination of the juvenile period of growth is abrupt in some kinds of wood and in others is marked by a transition period to the adult condition. As a general rule the low quality of juvenile wood is more marked in conifers than in hardwoods.

Alternate terms for these two regions in the stem are *core, crown-formed,* or *pith* wood for juvenile wood, and *outer, stem-formed,* or *mature wood* for adult wood. Since the terms juvenile and adult wood relate to the maturity of the cambial region, which gives rise to the successive increments in the two zones, they may be considered to be more appropriate than the alternate terms.

G. Summary of Variation within Trees

The stem of the tree may be divided into two regions on the basis of fundamentally different wood properties and structure: (1) the juvenile wood, which consists of a cylindrical column of wood, formed about the pith before the cambial initials reach their optimum size, and (2) the adult wood formed in the outer portion of the trunk by the physiologically mature cambium. The latter possesses characteristics which are considered to be normal for the species. The juvenile wood, on the other hand, contains cells with short lengths and has physical properties which are inferior to those of adult wood in the same tree.

Curves relating late-wood fiber length, specific gravity, mechanical properties, and cellulose content to position with respect to the pith, at any given elevation in the tree, all conform to a similar pattern, when the variability within the tree arising from environmental factors is eliminated. An abrupt initial rise of the curve marks the juvenile period, during which most of the cambial maturation occurs. As the tree becomes physiologically mature, the slope of the curves relating properties of the wood to age gradually changes and becomes horizontal when the maximum level is attained in the adult wood. Extreme old age in some cases is characterized by reduction in fiber length, specific gravity, and mechanical properties from the maximum values.

When data for fiber length in the early- and late-wood zones are plotted separately, they reveal a similar relationship. Both increase, but the increase for fiber length in the early wood, from the pith outward, is smaller and less variable. Specific gravity and tensile strength of the early wood decrease with age to a constant minimal level; in contrast, in the late wood these properties follow the trend indicated for fiber length and increase with age.

With respect to vertical position in the tree, fiber lengths in the trunk generally increase gradually within a growth increment to a maximum below the crown, then decrease toward the top of the tree to the minimum lengths typical of juvenile wood. Differences in average lengths of fibrous elements have been demonstrated between wood from the roots, trunk, and branches within the same tree. The fibers of hardwoods and tracheids of some conifers reach a maximum length in the trunk of a tree. Fibers in the roots of hardwoods are somewhat shorter than those in the stem, and those in the branch wood are the shortest at any location in the tree. Tracheids in the root wood in conifers follow the same pattern as fibers in hardwoods for the parts of the principal roots near the root collar; however, the tracheids tend to increase in length from the large end of the roots to the small ends of the same roots.

Specific gravity in the majority of conifers increases with age in the radial direction and decreases with increasing height in the tree. Notable exceptions are a few conifers in which these trends are reversed for both directions in the stem. The hardwoods exhibit both patterns of specific-gravity distribution as above, but a greater proportion form the densest wood near the pith at all heights and display specific gravity decreasing outward toward the bark and increasing with height in the tree.

Ring width of any given growth increment increases from the top of the tree downward to a maximum in the lower part of the crown region. Below the level of the crown the width of a given increment slowly decreases to the base of the tree. In the normal wood of conifers there is a limited relationship between the ring width, late-wood percentage, and the specific gravity of the wood. In the conifers, when mechanical properties are important, the width of the growth increment should always be considered in connection with the percentage of late wood in the ring, since the latter factor is directly related to the specific gravity and the mechanical properties. In the hardwoods an increase in the width of the growth ring may result in higher specific gravity of the wood, in the ring-porous woods, but has little effect on the diffuse-porous hardwoods.

II. VARIABILITY OF WOOD BETWEEN TREES OF THE SAME SPECIES

The variability patterns in wood that have been discussed in terms of single trees arise from cambial changes with increasing age of the tree, modified by changes in growing conditions both within the same year and for successive years. Local site and climate variability may also produce major modifications in the characteristics of wood formed, among, as well as within, individual trees of a species. In addition to the growth-related variability, individuals within a species will exhibit inherited differences in properties which distinguish individuals even under almost identical growth conditions. For example it is known that individual pine trees can have cambial initials which are markedly different in length, and as a result the lengths of mature tracheids for the entire life of these trees will be proportionately different.

A. Variability Resulting from Growth Conditions

In the last hundred years many studies have been carried out in an attempt to determine the effects of the various growth factors on specific gravity, tracheid length, mechanical properties, tree form, and other features of interest. It would seem that a consistent picture of these relationships should be available on the basis of the mass of collected evidence. However, this is not the case, since these data were not collected and analyzed as systems of interrelated factors. In most cases only one or two variables have been measured, while others have been ignored or assumed to be constant. As a result the literature is full of conflicting reports based on these incomplete studies. Only the bare outlines of the effects of tree growth on variation in wood characteristics are available. Since this subject is so extensive it will be covered only in outline.

1. SILVICULTURAL TREATMENT

In general the treatments which have been shown to have major effects on some wood properties are either those designed to manipulate the spacing of trees, or length of clear bole, or to stimulate the growth by supplying needed nutrients or water.

a. Crown size and tree spacing in conifers have long been recognized as important in controlling growth rate. Comparisons of dominant, co-dominant, and suppressed trees within stands have shown that specific gravity increases with suppression for the same age and position of the sample within the stem.[7] Dominant trees exhibit larger tracheid diameters, shorter lengths of tracheids, and lower specific gravity than those in comparable suppressed trees. However, cell wall thickness is a maximum in co-dominant trees when comparison is made with wood of similar ages and stem positions in dominant or suppressed trees. By controlling the numbers of trees in a stand, it is possible to manipulate the spacing of trees to produce the type of growth rate desired. Slowing growth in stands of young trees, for example, will reduce the core of juvenile wood to a minimum and decrease the amount of the low-density material that is related to abnormal shrinkage. Thinning the stand of young or mature trees to allow optimum crown spacing results in increased growth rates, increase in proportion of early wood, and a drop in the mean value of the increment specific gravity. Thinning of overmature conifer stands on the other hand, results in an increase in late-wood percentage and specific gravity in many species.

b. Application of fertilizers to growing trees can result in increased growth, if the nutrient elements of the site are less than optimum for the tree. In general, medium to fast growth rates, caused either by fertilization or thinning, give equal responses in properties of the wood produced. Spruce and pine stands in northern Europe will commonly show 30 to 50 percent increases in the volume increment rate after fertilization. At the same time mean specific gravity decreases about 5 percent, dry weight yields increase about 35 percent, and microtensile strengths decrease about 20 percent after fertilization. It can be seen that if weight yields are the most important consideration in managing a stand, fertilization is a useful treatment method.

c. Available water, either as rainfall or in terms of soil moisture, has been shown to influence the percentage of late wood. Studies in all but a few of the conifer species indicate that optimum available moisture throughout the growing season, with low moisture stress in the tree, promotes wide incremental growth in both early and late wood, maximum percentage of late wood, and increased mean specific gravity of wood for the same stem position and kind of tree. This holds true whether the moisture is derived from rainfall, irrigation, or ground water sources. The increase in the late-wood formation with these optimal conditions is due to increased production of the thick-walled cells with relatively large diameters in the first-formed late wood.

High moisture stress, i.e., too little moisture available in the early part of the growing season, curtails growth activity in the tree crown and reduces auxin production, with consequent early onset of late-wood formation and a narrowing of the increment. Continued dry conditions throughout the growing season also reduce the percentage of late wood, because only the flat type of late-wood cells form under such conditions. Not all kinds of conifers appear to react to moisture in this manner, however. Studies in *Libocedrus decurrens* Torr.[50] have shown that neither growth-ring width nor late-wood percentage is correlated with spring rainfall or total annual precipitation. In contrast *Pseudotsuga menziesii* (Mirb.) Franco, growing on the western slopes of the Cascade mountains in Oregon and Washington, exhibits trends of decreasing specific gravity with increase in summer rainfall and in elevation.[39] The trends are related to both the late-wood percentage and tracheid-wall thickness changes.

d. Geographic location, plus the climatic effects relating to temperature and rainfall, have been shown to cause variation between trees in a species. Wood-density surveys in the southern pines[42] have shown that there is a trend toward increasing specific gravity in several species from northwest to southeast, within the ranges of their normal occurrence in the coastal plains regions, in direct relation to warm-season rainfall. For example, average warm-season rainfall in the State of Mississippi increases from 22 inches in the northwest part of the state to 34 inches in the southeast. Average specific gravity in *Pinus echinata* Mill. over the same range increases from 0.49 in the northwest to 0.54 in the southeast part of the state. *Pinus taeda* L. varies similarly with mean specific gravity, increasing from 0.47 to 0.51 in the State of Mississippi, and also exhibiting increase in average specific gravity of 0.48 to 0.58 from north to south in the State of Florida.* Other studies have emphasized that the differences which exist over species ranges are due to other factors than rainfall alone. For example, *Pinus taeda* L., grown in Georgia from seed lots representing locations over its natural range, showed increases in average specific gravity for the various localities of seed source, from a minimum of 0.510 in the southeast part of the geographic range to 0.519 in the northern part of the range (North Carolina), and a maximum of 0.536 for the western range in Texas.[30] Studies in variation over the wide natural range in latitude and elevation of *Pseudotsuga menziesii* (Mirb.) Franco have shown that about 5 percent of the total variation in specific gravity can be related to latitude and about 1 percent to elevation of the site on which the trees were grown.

Other variations associated with geographic locations have shown that *Pseudo-*

* A. G. Hunter and J. F. Goggans, Variation of Fiber Length of Sweetgum in Alabama, *Tappi*, **52**(1):1952–1954 (1969), report increase in fiber length from north to south Alabama, correlated with length of growing season and summer rainfall.

tsuga menziesii (Mirb.) Franco has the greatest mean length of tracheids for trees from the coastal regions of western North America and least in the interior parts of its range. The longest tracheids for *Picea sitchensis* (Bong.) Carr. are found in trees grown in northern California, and the shortest mean lengths in trees growing in Alaska.

B. Variability Resulting from Genetic Factors

Evidence from multivariate analysis of a number of extensive studies has shown that the between-tree component of the variation within many species of conifers, as well as in *Populus* and *Liriodendron* among the hardwoods, is much larger than the within-tree component and is often larger than the variation associated with geographic races within the species. This situation has led to the conclusion in a number of instances that tree breeding or tree selection for improvement in given characteristics is possible. The desirability of such improvement can be appreciated if one considers that an average change of only 0.02 in specific gravity of the southern hard pines results in a change in modulus of rupture of about 1000 psi and a dry weight difference of 100 pounds or more per cord of pulpwood.[42]

The inheritance of several characteristics important in wood quality has been demonstrated in recent years. Length of tracheids has been shown to be an inherited feature, at least in young trees, for *Pinus radiata* D. Don, the southern hard pines, and *Picea sitchensis* (Bong.) Carr., but has not been statistically related in studies of *Picea abies* (L.) Karst clones or for certain *Populus* hybrids. On the whole, length of fibrous elements appears to be more heritable than diameter. Mean specific gravity of the increment has also been well established as an inherited feature, as shown by studies in *Liriodendron tulipifera* L., *Pinus radiata* D. Don, *Populus* hybrids, and in *Picea abies* (L.) Karst. For the latter, both mean specific gravity of the entire increment and that for the early- and late-wood zones separately have been shown to be inherited characters. Cellulose content of the wood has been shown to be an inherited character by studies of the southern hard pines and *Pseudotsuga menziesii* (Mirb.) Franco. Packing densities of the wood in *Pinus taeda* L. are known to be genetically controlled. Some other factors which are known to be genetically controlled are percentage of late wood in *Picea abies* (L.) Karst; branch diameter and knot volume in *Pinus taeda* L.; and duration of juvenile-wood formation in *Pinus elliottii* Engelm.

Efforts to improve the quality of wood began with selection of trees of apparent high quality. In species with wide ranges and diverse environment, the population tends to be genetically variable. As a result, in species such as *Pinus sylvestris* L., the inherited differences in adaptability to geographic and altitudinal differences evolve groups which are known as *races*. These races are not sharply distinct on the basis of external appearance, site requirements, or characteristics of the wood produced. Larson,[35] for instance, points out that of three studies of the races of Scotch pine (*Pinus sylvestris* L.), only one confirmed existence of significant race characteristics, and the other two revealed no differences of significance. Other studies dealing with European spruce [*Picea abies* (L.) Karst] and *Pinus resinosa* Ait. indicate the possibility of group differences in seed sources.

At the present time attention is being centered more on the selection of outstanding or "elite" trees which differ significantly from the norms for the species in some trait or combination of characters. For example, both the Western and Southern Wood Density Surveys of the United States Forest Service[58,59] revealed individual trees with outstanding specific gravities. These individuals in many cases have since been utilized as seed sources in tree-improvement programs and in tree breeding. A method for ascertaining the relative quality of a given tree with respect to the species mean has been developed in Australia.[44] Regression curves for tracheid length, density, or other properties and characteristics are drawn from equations developed for the species. Each set of regression values consists of a curve of means and curves indicating upper and lower limits of expected values (Fig. 7-15). Data for the individual tree are superimposed on the regression curves for the species. The relative position of the data for the single tree with respect to the species curves indicates quite clearly whether a given tree is above or below average. The summation of a number of such comparisons yields a good picture of the structural and physical properties of the tree with respect to others in the same species.

A related method of assessing wood quality in trees from a series of differing sites and age classes has been developed in Canada.[23] A series of age-class regression curves is drawn for a given species and site, using specific gravity and tree diameters as the axes of the chart (Fig. 7-16). This set of curves then expresses the normal variation of specific gravity by size of tree for different age classes. Comparison of data for single trees with these "norms" allows the segregation of significantly high or low individuals for use in genetic studies.

The most sophisticated means for determining outstanding individuals or groups of trees is the use of multiple-regression studies. By this mathematical method a large number of simultaneously varying factors can be related and components can be segregated in order of their magnitude. The mathematics are cumbersome and require the use of a computer, but the method of analysis is powerful and reveals more information than can be shown by other methods of analysis.

The breeding of trees for specific sets of characters is still in the early stages of development. Very few results of tree breeding under controlled conditions are available to show whether such crosses actually produce differences in the progeny. Jackson and Green,* in a study of controlled crosses between *Pinus elliottii* Engelm. and *Pinus taeda* L., found that five of seven possible combinations produced tracheid lengths intermediate between the two parents. Similarly, pine hybrids in Australia exhibit physical properties that are intermediate between the parents.[44] Controlled pollination between selected long-fibered trees of *Pinus elliottii* Engelm. produced a greatly improved average tracheid length in the progeny,[15] indicating that certain characteristics can be improved by careful breeding. This

* L. W. R. Jackson and J. T. Green, Tracheid Length Variation and Inheritance in Slash and Loblolly Pine, *Forest Sci.*, **4**(4):316–318 (1958).

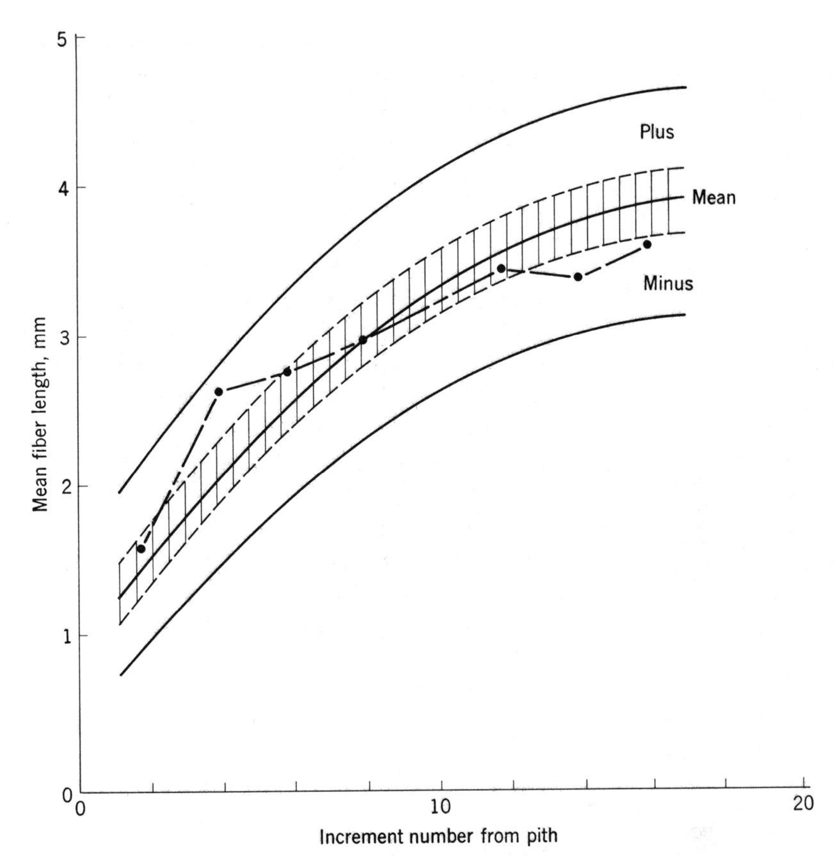

FIG. 7-15 Graphic system for evaluation of individual trees against the mean characteristics for the species. The species mean is shown as a heavy curve centrally located in a shaded region representing lengths of fibers associated with the normal size expected for the age of the tree. The broken line represents plotted values for an individual tree. It is evident that in the period up to 6 years of age the fiber lengths are longer than average, i.e., they are above the shaded area. For the older trees, however, the fibers are normal or shorter than average. (*Adapted from J. W. P. Nicholls and H. E. Dadswell, Div. Forest Prod. Tech. Paper No.* 37, *CSIRO, Melbourne, Australia,* 1965.)

evidence is strong indication that hybridization in trees follows the usual laws governing inheritance in plants. Assessment of some hybrid forms of trees, which have been long known and planted for a sufficient time to produce normal mature wood, indicates that the well-known hybrid vigor applies to trees as well as to cereal crops. This has been well documented by studies of hybrid poplars and crosses between *Larix decidua* and *Larix leptolepis*. These crosses show increased vigor in both height and in diameter growth, as well as increase in wood density and fiber length. A major difficulty in tree-breeding work is that the time between seed production and mature wood formation in the crosses is so long that assessment of results is

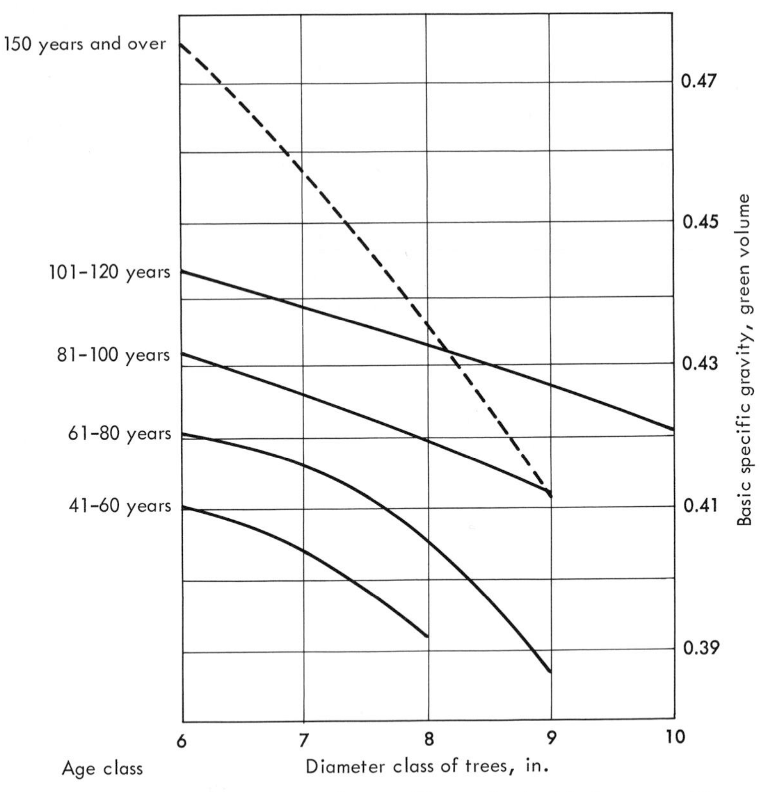

FIG. 7-16 Regression curves for specific gravity in black spruce [*Picea mariana* (Mill.) B. S. P.] by age classes. The data include trees from both swamp and slope sites. (*Adapted from Hale.*[23])

very slow. In order to reduce this time lag, attempts have been made to assess the desired qualities in the wood produced by seedlings and branches of young trees. These methods, however, are of only limited success because of the poor correlations between the young material of the progeny for a given character and the same character in a mature tree with identical genetic composition.

SELECTED REFERENCES

1. Anderson, E. A.: Tracheid Length Variation in Conifers as Related to Distance from the Pith, *J. Forestry*, **49**(1):38–42 (1951).
2. Bailey, I. W.: The Cambium and Its Derivative Tissues, II, Size Variations of Cambial Initials in Gymnosperms and Angiosperms, *Am. J. Botany*, 7(9):355–367 (1920).
3. ——— and A. F. Faull: The Cambium and Its Derivative Tissues, IX, Structural Variability in the Redwood, *Sequoia sempervirens*, and Its Significance in the Identification of Fossil Woods, *J. Arnold Arb.*, **15**:233–254 (1934).
4. Baker, G.: Estimating Specific Gravity of Plantation-grown Red Pine, *Forest Prod. J.*, **17**(8):21–24 (1967).

5. Bannan, M. W.: Anticlinal Divisions and Cell Length in Conifer Cambium, *Forest Prod. J.*, **17**(6):63–69 (1967).
6. Bergman, S. I.: Lengths of Hardwood Fibers and Vessel Segments, A Statistical Analysis of Forty-nine Hardwoods Indigenous to the United States, *Tappi*, **32**(11):494–498 (1949).
7. Bisset, I. J. W., and H. E. Dadswell: The Variation of Fibre Length within One Tree of *Eucalyptus regnans* F.v.M., *Australian Forestry*, **13**(2):86–96 (1949).
8. Bosshard, H. H.: Variabilität der Elemente des Eschenholzes in Funktion von der Kambiumtätigkeit, *Schweiz. Z. Forstw.*, **102**(12):648–655 (1951).
9. Comstock, G. L.: Longitudinal Permeability of Green Eastern Hemlock, *Forest Prod. J.*, **15**(10):441–449 (1965).
10. Côté, W. A., Jr., A. C. Day, B. W. Simson, and T. E. Timell: Studies of Larch Arabinogalactan, I, The Distribution of Arabinogalactan in Larch Wood, *Holzforschung*, **20**(6):178–192 (1966).
11. Cruz, R. de la: The Effect of Height and Pith Orientation on Wood Strength, I, Almon, unpublished report, Forest Products Research Institute, Laguna, Philippines, 1967.
12. Dinwoodie, J. M.: Tracheid and Fibre Length in Timber, a Review of Literature, *Forestry*, **34**(2):125–144 (1961).
13. ———: Variation in Tracheid Length in *Picea sitchensis* Carr., Forest Products Research, Special Report 16, Dept. of Scientific and Industrial Research, London, 1963.
14. Duff, G. H., and N. J. Nolan: Growth and Morphogenesis in the Canadian Forest Species, *Can. J. Botany*, **31:**471–513 (1953).
15. Echols, R. M.: Linear Relation of Fibrillar Angle to Tracheid Length and Genetic Control of Tracheid Length in Slash Pine, *Trop. Woods*, **102:**11–22 (1955).
16. Elliott, G. K.: The Distribution of Tracheid Length in a Single Stem of Sitka Spruce, *J. Inst. Wood Sci.*, **5:**38–47 (1960).
17. Erickson, H. D.: Tangential Shrinkage of Serial Sections within Annual Rings of Douglas-fir (*Pseudotsuga menziesii*) and Western Red Cedar (*Thuja plicata*), *Forest Prod. J.*, **5**(8):241–250 (1955).
18. Fegel, A. C.: Comparative Anatomy and Varying Physical Properties of Trunk, Branch and Root Wood in Certain Northeastern Trees, *N.Y. State College Forestry Tech. Bull.* 55, 1941.
19. Foulger, A. N.: Longitudinal Shrinkage Pattern in Eastern White Pine Stems, *Forest Prod. J.*, **16**(12):45–47 (1966).
20. Gerry, E.: Fiber Measurement Studies: Length Variations; Where They Occur and Their Relation to the Strength and Uses of Wood, *Science*, **61**(1048):179 (1915).
21. Greene, J. T.: Selection for Tracheid Length in Loblolly Pine (*Pinus Taeda* L.), Georgia Forest Res. Council, Report 18, 1966.
22. Grozdits, G. A., and G. Ifju: Development of Tensile Strength and Related Properties in Differentiating Coniferous Xylem, *Wood Sci.*, **1**(3):137–147 (1969).
23. Hale, J. D.: Minimum Requirements for Defining Species Norms for Quality of Variable Woods, *Tappi*, **45**(7):538–542 (1962).
24. ——— and L. P. Clermont: Influence of Prosenchyma Cell-wall Morphology on Basic Physical and Chemical Characteristics of Wood, *J. Polymer Sci.*, **C2:**253–261 (1963).
25. Hamilton, J. R.: Variation of Wood Properties in Southern Red Oak, *Forest Prod. J.*, **11**(6):267–271 (1961).
26. Hejnowicz, A., and Z. Hejnowicz: Variation in Length of Vessel Members and Fibers in the Trunk of *Robinia pseudoacacia*, *Acta Soc. Botan. Polon.*, **28**(3):453–460 (1959); also *Proc. IX Intern. Botan. Congr.*, 1959.
27. Ifju, G., R. W. Wellwood, and J. W. Wilson: Relationship between Certain Intra-increment Physical Measurements in Douglas-fir, *Pulp & Paper Mag. Can.*, **66**(9):T475–T483 (1965).
28. Ito, M.: The State of the Seasonal Variation of Xylem Elements in One Growth Ring of Chestnut Tree (*Castanea crenata* S. & Z.), *Sci. Rept. Fac. Liberal Arts Educ., Gifu Univ., N. S.* **2**(1):74–79 (1957).

29. Jackson, L. W. R.: Loblolly Pine Tracheid Length in Relation to Position in Tree, *J. Forestry*, **57**(5):366–367 (1959).
30. ——— and R. K. Strickland: Geographic Variation in Tracheid Length and Wood Density of Loblolly Pine, Georgia Forest Research Paper No. 8, Georgia Res. Council, 1962.
31. Jeffers, J. N. R.: Regression Models of Variation in Specific Gravity in Four Provenances of Sitka Spruce, *J. Inst. Wood Sci.*, **4:**44–59 (1959).
32. Kennedy, R. W.: Intra-increment Variation and Heritability of Specific Gravity, Parallel to Grain Tensile Strength, Stiffness and Tracheid Length in Clonal Norway Spruce, *Tappi*, Third Forest Biology Conference, Paper 1–5, 1965.
33. Knigge, W.: Untersuchungen über die Abhangigkeit der Mittleren Rohdichte Nordamerikanischer Douglasienstamme von unterschiedlichen Wuchsbedingungen, *Holz Roh-Werkstoff*, **20:**352–360 (1962); also as University of British Columbia translation 22, 1963.
34. Krahmer, R. L.: Variation of Specific Gravity in Western Hemlock Trees, *Tappi*, **49**(5):227–229 (1966).
35. Larson, P. R.: Effect of Environment on the Percentage of Summerwood and Specific Gravity of Slash Pine, *Yale School Forestry Bull.* 63, 1957.
36. ———: A Biological Approach to Wood Quality, *Tappi*, **45**(6):443–448 (1962).
37. ———: Microscopic Wood Characteristics and Their Variations with Tree Growth, Proc. Int. Union Forest Res. Organ., sec. 41, Madison, Wis., 1963.
38. ———: Changes in Chemical Composition of Wood Cell Walls Associated with Age in *Pinus resinosa*, *Forest Prod. J.*, **16**(4):37–45 (1966).
39. Lassen, L. E., and E. A. Okkonen: Effect of Rainfall and Elevation on Specific Gravity of Coast Douglas-fir, *Wood and Fiber*, **1**(3): 227–235 (1969).
40. Liang, S.: Variation in Tracheid Length from the Pith Outwards in the Wood of the Genus *Larix*, *Forestry*, **22:**222–237 (1948).
41. McGinnes, E. A., Jr.: Growth-quality Evaluation of Missouri-grown Shortleaf Pine (*Pinus echinata* Mill.), *Univ. Missouri, Agr. Expt. Sta., Res. Bull.* 841, 1963.
42. Mitchell, H. L.: Specific Gravity Variation in North American Conifers, Proc. Intern. Union Forest Res. Organ., sec. 41, Madison, Wis., 1963.
43. Nicholls, J. W. P., and H. E. Dadswell: Tracheid Length, in *Pinus radiata* D. Don, *Div. Forest Prod. Tech. Paper* 24, Commonwealth Scientific and Industrial Research Organization, Australia, 1962.
44. ——— and ———: Assessment of Wood Qualities for Tree Breeding, III, in *Pinus radiata* D. Don, *Div. Forest Prod. Tech. Paper* 37, Commonwealth Scientific and Industrial Research Organization, Australia, 1962.
45. Olinmaa, P. J.: Vertailevia Tutkimuksia Puusyiden Pituudesta Hieskoivun Kevat ja Kesapuussa (Comparative Studies on the Length of Wood Fibers in the Early and Late Wood of Birch), *Pap. ja Puu*, **40**(11):599–601 (1958).
46. Packman, D. F., and R. A. Laidlaw: Pulping of British-grown Softwoods. Part IV: A Study of Juvenile, Mature, and Top Wood in a Large Sitka Spruce Tree, *Holzforschung*, **21**(2): 38–45 (1967).
47. Perem, E.: The Effect of Compression Wood on the Mechanical Properties of White Spruce and Red Pine, *Forest Prod. Lab. Can., Tech. Note* 13, Ottawa, Canada, 1960.
48. Prichard, R. P., and I. W. Bailey: The Significance of Certain Variations in the Anatomical Structure of Wood, *Forestry Quart.*, **14**(4):662–670 (1916).
49. Rendle, B. J.: Juvenile and Adult Wood, *J. Inst. Wood Sci.*, **5:**58–61 (1960).
50. Resch, H., and S. M. Huang: Variation in Wood Quality of Incense Cedar Trees, *Bull. Calif. Agr. Expt. Sta.* 833, Berkeley, Calif., 1967.
51. Sanio, K.: Über die Grösse Holzzellen bei der gemeinen Kiefer (*Pinus silvestris*), *Jahrb. wiss. Botan.*, **8:**401–420 (1872).
52. Saucier, J. R., and J. R. Hamilton: Within Tree Variation of Fiber Dimensions of Green Ash (*Fraxinus pennsylvanica*), Res. Paper Georgia Forest Res. Council, No. 45, 1967.

53. Spurr, S. H., and M. J. Hyvarinen: Wood Fiber Length as Related to Position in Tree and Growth, *Botan. Rev.*, **20**(9):561–575 (1954).
54. Sukotjo, Wiratmoko: Specific Gravity, Fiber Length and Holocellulose Variations in *Pinus merkusii*, M.S. thesis, Iowa State University, Ames, Iowa, 1964.
55. Tamminen, Z.: Fuktighet, Volymvikt m.m. hos Ved och Bark, II, Gran (Moisture Content, Density and Other Properties of Wood and Bark, II, Norway Spruce), Institutionen för Virkeslära, Skogshogskoloan, Rapporter Nr.R. 47, Stockholm, 1964.
56. Taras, M. A.: Some Wood Properties of Slash Pine (*Pinus elliottii* Engelm) and Their Relationship to Age and Height within the Stem, Ph.D. dissertation, North Carolina State University, Raleigh, 1965.
57. Taylor, F. W.: Variation of Wood Elements in Yellow-poplar, *Wood Sci. & Technl.*, **2**(3):153–165 (1968).
58. United States Department of Agriculture, Southern Wood Density Survey—1965 Status Report, *U.S. Forest Service Res. Paper* FPL-26, 1965.
59. United States Department of Agriculture, Western Wood Density Survey—Report No. 1, *U.S. Forest Service Res. Paper* FPL-27, 1965.
60. Voorhies, G., and D. A. Jameson: Fiber Length in Southwestern Young-growth Ponderosa Pine, *Forest Prod. J.*, **19**(5): 52–55 (1969).
61. Wellwood, R. W.: Tensile Testing of Small Wood Samples, *Pulp & Paper Mag. Can.*, **63**(2):T61–T67 (1962).
62. ——— and P. E. Jurazs: Variation in Sapwood Thickness, Specific Gravity and Tracheid Length in Western Red Cedar, *Forest Prod. J.*, **18**(12):37–46 (1968).
63. Wheeler, E. Y., B. J. Zobel, and D. L. Weeks: Tracheid Length and Diameter Variation in the Bole of Loblolly Pine, *Tappi*, **49**(11):484–490 (1966).
64. Wilson, J. W.: Wood Characteristics, III, Intra-increment Physical and Chemical Properties (Summary of Studies in Progress at University of British Columbia), *Res. Note* 45, Pulp and Paper Research Institute of Canada, Montreal, 1964.
65. ——— and R. W. Wellwood: Intra-increment Chemical Properties of Certain Western Canadian Coniferous Species, in W. A. Côté, Jr. (ed.), "Cellular Ultrastructure of Woody Plants," Syracuse University Press, Syracuse, N.Y., 1965.

8

Growth-related Defects in Wood

The term *defect,* as applied to wood, refers to any irregularity or deviation from the qualities that make wood suitable for a particular purpose. Growth-related defects are imperfections in the wood of living trees which arise from tree growth or irregularities of growth. The defects which develop in wood after it has been cut, as the result of treatment or foreign organisms, are discussed in chapters that follow.

I. CROSS GRAIN IN WOOD

When the fiber alignment in a piece of wood does not coincide with the longitudinal axis of the piece, the wood is said to be *cross-grained*. The irregular forms of cross grain such as curly, wavy, interlocked, and combinations of these, when used for decorative purposes, are considered to be an advantage because of the figure that is produced. However, any form of cross grain which occurs in structural lumber is a defect because of the reduction in the strength of the member in which it occurs.

A. Types of Cross Grain

1. SPIRAL GRAIN

This term is applied to the helical orientation of the fibers in a tree stem, which gives a twisted appearance to the trunk after the bark has been removed. The twisted appearance is accentuated by surface checks, parallel to the fiber direction, that make spiral grain very evident in some standing dead trees and in such items as telephone poles and posts. The spiraling may be either right- or left-handed; the slope may be constant in a given tree or may change with age of the tree. In addition, some trees show a regular reversal of spiral grain in successive zones of growth increments to form interlocked grain.

Because of the widespread presence of spiral grain, both in conifers and hardwoods, this type of grain may be considered as a normal growth pattern, rather than straight-grained wood. Though no conclusive evidence exists, it is believed that the basic patterns in the spiral-grain arrangements are hereditary, but that the extent and distribution of spirality in an individual tree are affected by environmental factors.

In conifers the prevailing orientation is in the left direction near the pith, with the angle increasing sharply in the first-formed rings in the juvenile wood, and then gradually decreasing to the straight-grained condition. This is followed by a gradual change to a right-angle spiral, which tends to increase in magnitude with age of the tree.[71] Another pattern of spirality in conifers shows continuing increase of left spiral from the pith to the cambium, as a tree becomes older. Lowery and Erickson[41] report that coniferous poles with left spiral grain twist more with changes in moisture content than those containing both left- and right-spiraled grain, because in the latter the tendency to twist to the left tends to be counterbalanced by the tendency to twist to the right.

Not enough information is at present available to make a generalized statement about the spiral-grain arrangement in hardwoods. Indications are that the grain pattern in hardwoods tends to be the reverse of that in the coniferous species, i.e., it is in the right direction near the pith. It may increase gradually in magnitude in the same direction from the pith to the cambium, or it may gradually decrease in the juvenile wood to a zero angle with the axis of the stem, and then reverse to the left as the tree increases in diameter. In hardwoods, not uncommonly an interlocked grain pattern is superimposed on these patterns of spiral-grain arrangement.

The general grain pattern and the magnitude of spiraling vary widely in individual trees, especially with regard to the age at which the changeover in direction of the spiral takes place. Likewise, the spiral grain pattern from the pith to the cambium is not the same along different radii in the stem, or at different heights in the trunk of the tree.

The origin of spiral grain in tree stems appears to be related to the pseudotransverse divisions of the fusiform cambial initials. The anticlinal division of the fusiform initials tends to be unidirectional and is amplified by intrusive elongation of the fusiform initials, as well as by a loss of some of the cells.[1,27] From time to time changes occur in the direction of inclination in the pseudotransverse division of the fusiform initials. These are evident either as a change in the angle or as a reverse in the direction of the spiral around the stem.[1] In conifers the spacing of these reversals, in terms of thickness of xylem formed, is related to the frequency of the anticlinal divisions; as the rate of divisions rises, the intervals between changes shorten. The severity of spirality is controlled by the degree of inclination of the fusiform initials and by the balancing effect of spiral reversal. Vité[57] claims that dynamic distribution of water transportation throughout the crown is responsible for these changes.

The effect of the spiral grain on the utility of wood depends on the severity of the spiral grain and the intended use of the wood. Because of the anisotropic nature of wood, spiral grain may seriously reduce its strength and stiffness. It may also be a major contributing cause to twisting of poles and other round products, lumber, and plywood. Spiral grain is also a cause of surface roughness in planed lumber.

2. DIAGONAL GRAIN

When stock is sawn so that the grain of the wood intersects the surface at an angle, the piece has diagonal grain. The principal cause of diagonal grain is the practice of sawing lumber parallel to the pith. This defect can easily be eliminated in straight logs by sawing parallel to the bark surface. Diagonal grain can also originate from crooked logs, by sawing irregular pieces from straight-grained stock, or resawing and ripping straight-grained material at an angle to the long axis.

3. CROSS GRAIN

Grain deviations in the flat-sawn faces of boards or timbers may be the result of spiral grain in the tree, or they may be due to crook and sweep in the log and to localized grain disturbance around large knots. As stated in the preceding paragraph, grain deviation may be due to the manner of sawing lumber (*diagonal grain*). In sawn material, all these types of grain deviation appear the same and are simply designated as *cross grain,* regardless of the true origin of such grain. The term *spiral grain* should be limited in its application to grain direction in tree stems, logs, and poles.

4. INTERLOCKED GRAIN

A regular reversal of right and left spirality in a tree stem produces the condition known as *interlocked grain* in wood. Cut radial surfaces in this type of wood produce the ribbon-stripe figure (see p. 411). Interlocked grain also markedly increases the resistance to splitting in the radial plane, and affects both bending strength and elasticity. Weddell[64] indicates that both static bending strength and stiffness in bending are greatly decreased by the presence of well-defined interlocked grain in beams.

5. DETERMINATION OF GRAIN DIRECTION

The best way to determine the direction of the grain in wood is to split the piece. In most cases, however, splitting is not a practical procedure. The direction of the grain may also be determined by observing the orientation of resin canals or vessels. Seasoning checks on the tangential face, or growth-increment boundaries on the radial faces, should be used as indicators of grain direction only when the piece shows true flat-sawn or radial surfaces. One of the most commonly employed tools for determining grain direction is a needle-pointed scribe on a swivel handle. However, scribe lines can be used only on surfaces where damage to the finish is unimportant. Another method for determining grain direction is to apply free-flowing ink from a pen or eye dropper and observe the direction of the fine lines on either side where the ink runs with the grain. For methods of measurement of grain angle and effects of grain slope on strength properties see Chap. 6, pp. 225 to 226.

II. KNOTS

A knot is a branch base that is embedded in the wood of a tree trunk or of a larger limb or branch (Fig. 8-1); such branch bases are gradually included in the wood of

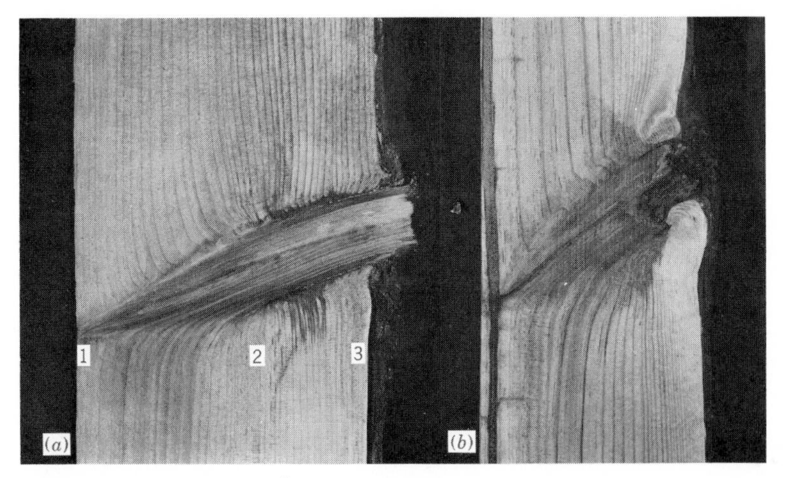

FIG. 8-1 Knots in wood.

(*a*) Spike knot with projecting spur in eastern white pine (*Pinus strobus* L.). The branch was living during the period when the series of annual increments between 1 and 2 were added to the main stem; dead between 2 and 3. These zones indicate where the knot is "tight" and "loose," respectively.

(*b*) Spike knot with partially decayed spur in ponderosa pine (*Pinus ponderosa* Laws.). The knot is in process of being covered by callus.

(*Photographs by permission of U.S. Forest Products Laboratory.*)

the larger member through addition, year by year, of successive increments of woody tissue.

Each branch is in reality a secondary axis with a growing point, or bud, at its end, and a lateral meristem. The organization of the branch is similar to that in the main stem, and the tissue systems of the two structures are interconnected. A continuous growth increment is laid down simultaneously over the stem and the branch by their cambia. The continuity of these increments is most easily traced on the underside of the branches, as in Fig. 8-1*a*, 1 to 2. On the upper side of the branches, the angle between the branch and the trunk is acute; as a result, the cell arrangement in this transition zone is irregular. A cross section through the knot in the tangential plane of the stem shows that the grain orientation in the trunk wood is laterally distorted around the knot and passes in a wide sweep on either side. This local distortion causes cross grain in the region of the knot; the extent of this cross grain is in direct relation to the size of the knot (Fig. 8-2). The continuity of growth between the stem and the embedded branch base that is formed while the branch is living causes ***intergrown, tight,*** or ***red knots*** in lumber (Fig. 8-2).

When a branch dies, the cambium in the branch ceases to function. No new increments are formed in the branch, although they continue to develop in the living member around the branch. From this time on there is a break in the continuity of the woody tissue between the branch and the stem (Fig. 8-1*a*, 2 to 3). The projecting end of the dead branch drops off sooner or later, leaving a short or long stub depending on the kind of tree. The dead stub becomes gradually embedded in the new wood as successive growth layers are added. Eventually the end of the stub is overgrown by wound tissue or callus (Fig. 8-1*b*), and successive growth increments in the stem laid over it show no evidence of the knot which lies deeper in the wood. The portion of the branch base that is embedded in the tree trunk after the branch dies causes a ***loose, encased,*** or ***black knot*** in lumber (Fig. 9-3*a*). Since the wood of the tree trunk is not continuous with that of the encased knot, there is less distortion of the grain around it than is the case with intergrown knots.

The embedded branch base, formed while the branch was living, has a generally

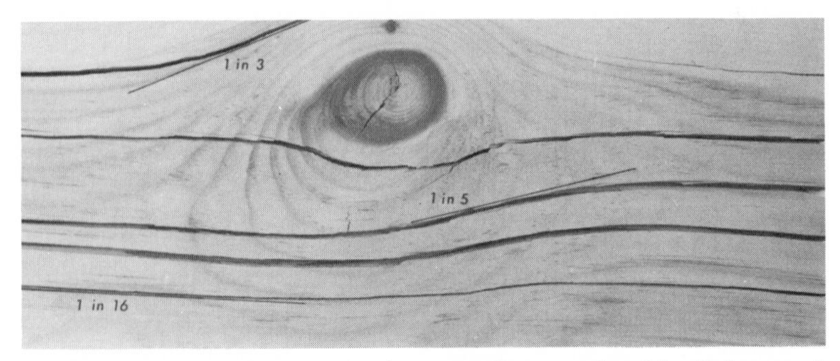

FIG. 8-2 A board of eastern white pine (*Pinus strobus* L.) split in several places to show the grain deviation around a knot. Slope of grain is indicated at three points.

conical shape with the small end connected to the pith. The encased portion of the knot toward the outside is approximately cylindrical. When a sawn face cuts lengthwise through the branch base, the surface will show a ***spike knot***. If the direction of the cut is more or less at right angles to the long axis of the branch base, the resulting knot is a ***round knot***. Round knots may actually range from circular to oval in shape, depending on the plane of the cut with respect to the axis of the branch base, or the shape of the cross section of the branch base itself. Loose, round knots frequently fall out of the lumber on drying and leave a ***knothole***. Loose, oval knots are more likely to remain in place in the board. In grading rules knots are classified on the basis of quality, as ***sound, unsound, decayed, intergrown,*** or ***encased,*** and on the basis of size as ***pin, small, medium*** (***standard***), and ***large***.

The number, size, and type of knots formed in the wood of a tree depend on the number and size of the limbs from which they originate, the age at which the limbs die, and the length of time the dead limb stubs remain on the tree. All these factors vary according to differences between species, site conditions, climate, stand composition, and density. For instance, an inherent difference between species is indicated by the fact that eastern white pine (***Pinus strobus*** L.) retains dead branches much longer than red pine (***Pinus resinosa*** Ait.) growing in the same stand.[37] However, both kinds of trees are so slow in shedding their branches that the stands must be pruned at an early age in order to produce any appreciable volume of clear lumber. The southern pines lose their lateral branches rapidly, and for this reason there is much less need for pruning in these stands.[37] The differences in branch losses between the northern and the southern pines may be due to the inherent differences between the groups of trees, or they may simply reflect the more rapid decay of dead wood in the south. The spruces are typical of trees which retain large numbers of small dead branches on the lower stem. In such trees the resulting encased branch stubs cause a core of small, loose knots to form toward the center of the log.

Actual volumes of knot wood in trees have been investigated to a limited extent. A recent study of six 11-year-old loblolly pines (***Pinus taeda*** L.)[65] showed that the average knot volume was 1.0 percent, with a range from 0.7 to 1.4 percent. The distribution of these knots, on the basis of percentage of knot volume to volume of 5-foot bolts cut at different heights in the tree, showed that the volume varied from 0.3 percent at the base of the tree to more than 2.0 percent at heights above 35 feet from the ground.

It is evident from the previous discussion that the size and character of the knots in a log, as well as their number, will vary with the distance from the pith. Because knots are one of the primary features used in classifying lumber into grades, the grades of lumber produced from a log will vary with the position in the log. In sawing a typical surface-clear log into lumber, the first boards inside the bark will show at least one clear face and will fall into the best grades. The first boards produced from the knotty interior will contain large, encased, or loose knots resulting from the inclusion of the dead branch stubs. The size, looseness, and the numbers of knots from this zone in the log generally cause the lumber to be placed in the poorer grades. Boards cut from the region of intergrown knots, still deeper in the log, will fall into the intermediate grades.

Normally the presence of knots in wood is a detriment and causes a loss in value. The only exception is in the *knotty paneling* grades of western pine (*Pinus monticola* Dougl.), western redcedar (*Thuja plicata* Donn), and eastern white pine (*Pinus strobus* L.). Boards for these special grades must show sound, tight, well-shaped knots that will not interfere with milling operations. This material is sorted out from ordinary yard grades and sold at a higher price than would be received for ordinary lumber containing the same size of knots and other defects.

The presence of knots in lumber or timbers has a direct relationship to the mechanical behavior of the member. The knot itself is harder, more dense, often more resinous, and shrinks in a different manner than the surrounding tissue. Because of these characteristics, the presence of sound, tight knots increases the compression strength, hardness, and shear characteristics of the wood. However, these same knots cause uneven wear on the surfaces on which they occur, give trouble because of the checking with moisture changes, make difficulty in painting, because of high resin concentrations or excessive absorption of oil by the end grain, and increase the forces required in cutting by as much as 100 percent.[56] The major loss in mechanical properties is associated with the grain distortion in the wood around the knot (Fig. 8-2). This localized cross grain may be very minor if the knot is small, but knots an inch or more in diameter can cause very steep slopes of grain in the surrounding wood. The changes in strength properties associated with cross grain have already been discussed. The greatest loss of strength from the presence of knots occurs in members subjected to bending stresses. This is true because tensile strength is one factor controlling the bending behavior of wood and it is the most sensitive of the mechanical properties to adverse slope of grain. The presence of knots at, or near, any edge of a bending member will produce local cross grain at that edge and reduce the mechanical properties of the wood. For this reason, in planks and timbers which are *stress-graded* for structural purposes, the knots are classified according to soundness, size, and position on the center or edge of a face.

III. GROWTH STRESSES IN TREES

The woody tissue in growing tree stems is subjected to stresses which arise from a slight shrinkage in length of the individual cells in the last phase of maturation.[9] These stresses are distinct from those occurring in wood as a result of removal of water in seasoning. These growth stresses are a primary cause for shakes in standing trees and felled logs; the "spring" of sawn timber, i.e., bowing and crooking of pieces as they are sawn from the log; and brittle heart and compression failures in standing timber.

Measurements in growing trees have shown that at the end of the annual growth in an increment there are stresses in the increment in the three major axes of stem symmetry: (1) longitudinal tensile stresses parallel to the grain, (2) compressive stresses in the tangential direction perpendicular to the grain (*ring stresses*), and (3) tensile stresses in the radial direction perpendicular to the grain. Theses stresses are cumulative as the trunk of the tree increases in diameter and result in characteristic stress gradients varying from the bark to the pith.

The longitudinal tensile stress in the outer surface of the trunk compresses the older tissues toward the interior of the stem, producing compressive stresses of increasing magnitude toward the pith as the stem increases in diameter. It has been demonstrated that boards sawn from the outer shell of large logs shrink along the grain after they are cut from the log, and that boards cut from the central part of the same log expand in length after cutting.[5,31,32] Boyd has shown by measurements and theoretical calculations that the longitudinal stresses are a maximum in tension at the surface of the log, with measured stresses in stems at least 24 inches in diameter varying from 1000 psi to a maximum of almost 3000 psi in some *Eucalyptus* spp.,[5] and that they decrease inward to zero at about one-quarter of the radial distance to the pith. From this point inward to the pith the stresses are compressive in nature and increase in magnitude to a maximum at the pith. Actual compressive stresses toward the pith are those resulting from long-time stress relaxation and are estimated to be approximately 3000 psi in *Eucalyptus regnans* F.v.M. 24 inches and over in diameter.[5] Measurements have also shown that the longitudinal growth stresses in the gymnosperms are never as great as those found in the hardwoods and do not exhibit consistent tensile stresses at the bark until the stem is over 3 inches in diameter.[5,23]

In the transverse direction, measurements on log cross sections show that the tangential stresses within the ring and perpendicular to the grain are a maximum in compression at the surface of the stem, reduce inward to zero at a point very close to the pith, and develop tension stresses in a central core whose diameter is about one-sixth of the stem diameter.[5,31,38] Boyd[5] shows compression and tensile maxima in the tangential direction of approximately 350 psi in *Eucalyptus regnans* logs. The radial-transverse stresses are tensile in nature across the entire radius of the stem; they increase in magnitude inward but are never very large.

A. Effects of Growth Stresses in Standing Trees

The system of growth stresses which has been described is in equilibrium in stems of standing trees, and is usually too low in magnitude to cause failures in the wood. However, certain conditions can produce characteristic defects that are the result of these growth stresses.

1. HEART CRACKS AND SHAKES

These are longitudinal separations of the wood which appear in the standing tree. If the cracks are radial and pass through the pith, they are called *heart shakes* or *rift cracks*. If such cracks are single, the defect is a *simple heart shake;* if several cracks radiate from the pith, the defect is known as *star shake*. Heart shakes originate in woods in which the tensile strength perpendicular to the rays is lower than the tensile-growth stresses in the stem in the tangential-transverse direction. Large rays (as in the oaks) and maximum growth stresses (as in butt logs of large, overmature trees) tend to favor formation of this type of defect.

Separations of the wood in standing trees in the plane of the growth increment are called *ring shakes, cup shakes,* or *round shakes.* They may occur in all kinds of hardwoods and conifers but are most common in hemlocks, true firs, western larch, and bald-cypress among the conifers, and in the sycamores and oaks among the hardwoods. Ring shakes have been studied in detail in western hemlock, western true firs, and Douglas-fir by Meyer and Leney,[44] and in walnut by McGinnes.[42] These investigators have found that ring shakes formed in standing trees occur as separations of late-wood cells along the compound middle lamellae (Fig. 8-3). The line of failure occurs within the outer portion of the late-wood zone in conifers and may occasionally follow a radial path to an adjacent ring and continue tangentially within the late wood of that ring. The predominance of failures occurring by separation of cells along the middle lamella has also been confirmed by the use of electron microscopy technique in a scarlet oak tree.*

Meyer and Leney[44] report that the interior of ring shakes which they examined exhibited smooth faces with varying amounts of loose fibers present. The extremes of this condition are shown in Fig. 8-3. These loose fibers may be present as mats of single fibers either completely free or attached at one end, or they may be present in aggregates. In addition the cavity may contain masses of resin, a white compound identified as matairesinol, or dark amorphous compounds

Little is known of the causes for formation of ring shakes. It has been suggested that they arise because of zones of tissue damage which result in traumatic resin ducts, or because of frost-damage zones, or they may be due to wind action. If the radial-transverse stresses in tree stems, which have been shown by Boyd,[5] are combined with the longitudinal shear stresses developed in tree trunks bent under wind or snow action, shear stresses could be produced at the growth-increment boundary of sufficient magnitude to cause failure in the wood. The location of these failures in the late-wood zone is to be expected, since crack propagation in such failures tends to progress in the densest material.

2. COMPRESSION FAILURES

The long-time longitudinal compression stresses which are present in large stems produce a widespread formation of microscopic shear failures transverse to the cell axes. These minute slip planes lie at angles from 30 to 45° to the fiber axis, extend through the cell walls of adjacent cells, and form a line of distinct tensile weakness. These failures are too small to be observed with an ordinary microscope but are readily visible in polarized light. In certain hardwoods the slip lines are so numerous and well developed that the central core of the stem exhibits a defect called *brittle heart.* This condition is distinguished by a characteristic carroty surface in the

* S. A. Kendel and E. A. McGinnes, Jr., Ultrastructure of Ring Shake in Scarlet Oak (*Quercus coccinea,* Muenchh), *Wood Science,* **2**(3): 171–178 (1970).

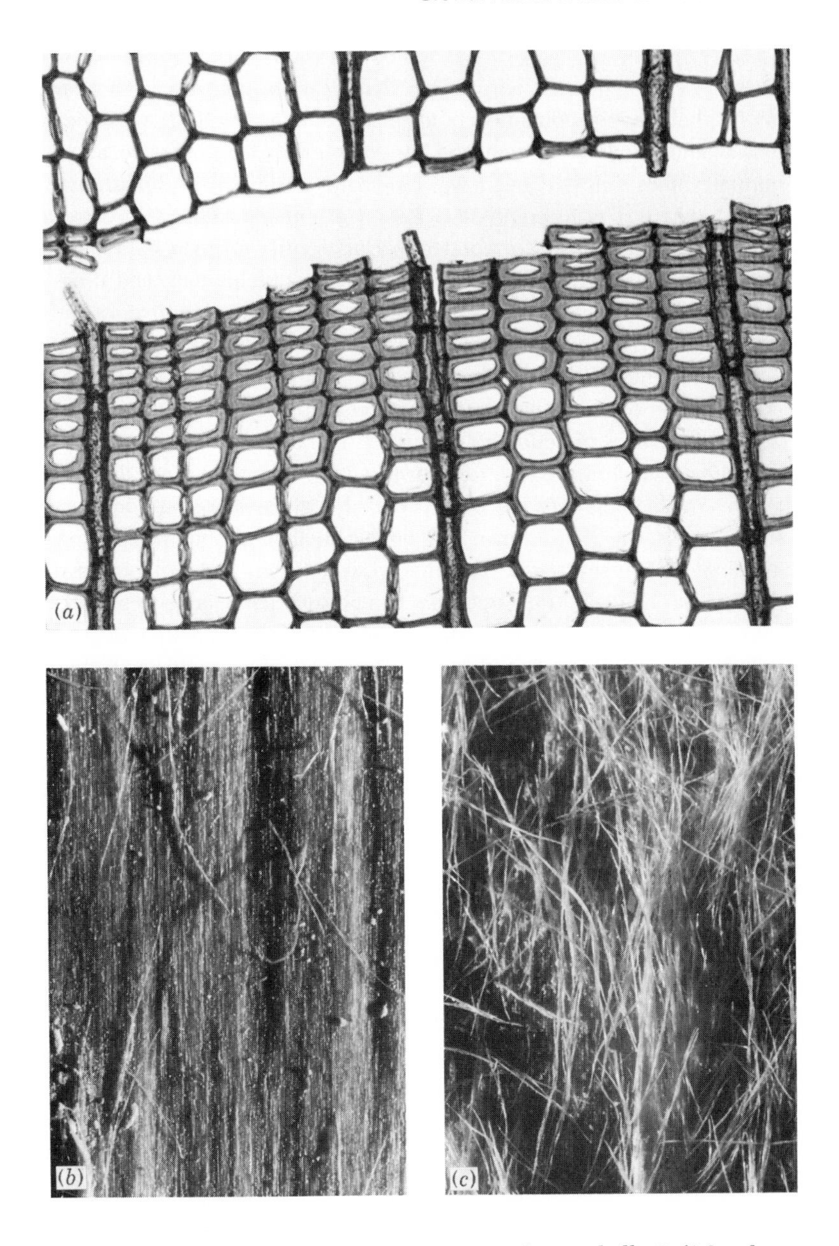

FIG. 8-3 Shakes in western hemlock [*Tsuga heterophylla* (Raf.) Sarg.].

(*a*) Cross section showing a shake formed in the standing tree by separation of cells at the middle lamella. (170×.)

(*b*) Longitudinal surface of a shake cavity on the side toward the pith showing limited development of loose fibers. (25×.)

(*c*) Longitudinal surface of a shake cavity on the side toward the bark with masses of loose fibers projecting into the cavity. (25×.)

[*Courtesy of R. W. Meyer and L. Leney, Forest Prod. J.*, **18**(2):51–56 (1968).]

broken cross section and by large numbers of broken fibers in macerated wood, but is not necessarily associated with low-density wood. Because of these microscopic failure planes, wood containing brittle heart is characterized by a marked decrease in tensile and bending strength and by reduction in toughness to about one-third normal values.[53] Brittle heart was originally described in *Eucalyptus regnans* F.v.M., where it is of common occurrence and often found in association with tension wood.[20,61] It is also reported in *Eucalyptus robusta* from Hawaii[53] and in large trees of the red lauans (*Shorea* spp.) of the Philippines, and it has also been reported in European beech.[10] This defect is undoubtedly much more prevalent than has been realized in the past.

Long-continued loading of zones containing microscopic failures in cells results in the extension of these compression failures across a number of adjacent cells until well-defined lines of double folding are visible at right angles to the grain.[36] These lines are called *compression failures* (Fig. 8-4). The formation of these compression failures in standing trees has been ascribed to stresses arising from heavy snow or strong winds. Boyd[5] has shown mathematically that the superpositioning of the longitudinal growth stresses in compression in the stem and the compressive stresses arising from bending of the tree stem can be sufficiently large to produce compression failures in the tree stem. Furthermore, the theoretical pattern of greatest magnitude of stress near the pith and decrease toward the bark predicted by his theoretical study agrees with the actual distributions of stresses found in the tree. These compression failures may be difficult to detect; nevertheless they cause a severe loss in bending and tensile strength in the wood in which they appear and are especially important in reduction of resistance to applied impact loads. For these reasons compression failures constitute one of the principal defects associated with failure in rails and rungs of wooden ladders.

B. Effects of Growth Stresses on Logs and Sawn Timbers

When trees are felled, the ends of the logs often develop radial cracks. Boyd[5] explains this as the result of transverse tensile stresses arising from strain energy developed as the longitudinal stresses at the cut ends are released and the cells return to their unstressed shapes. The tensile stresses in the radial plane arising from this cause, plus the tangential ring stresses already existing in the log, are sufficiently large to cause failure in a radial direction parallel to the rays. This type of defect, called a *heart check,* is readily extended by the drying which takes place at the ends of the logs.

Planks or cants cut from logs will no longer have a balance of longitudinal growth stresses and will bow, warp, or twist as they are cut from the log, depending on the relative changes in the lengths of the surfaces of the pieces as the growth stresses are relieved. Flat-sawn boards cut from the log will bow or "spring" outward from the log face as long as the bark side is to the outside. If sawn with the pith side away from the knees of the saw carriage, the bow will be toward the log and will pinch the saw.

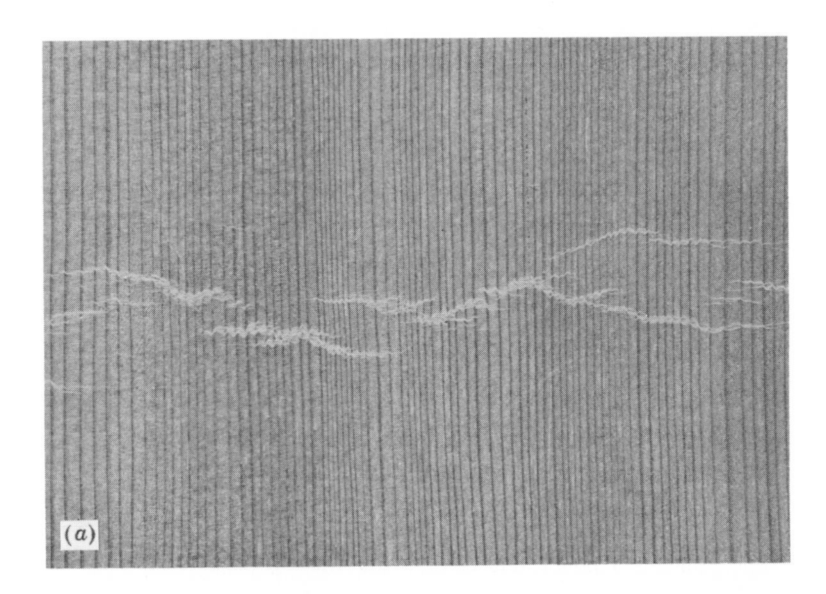

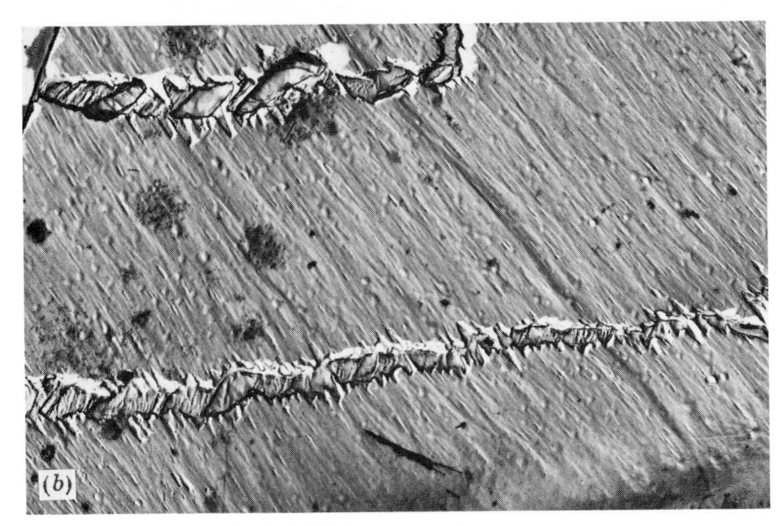

FIG. 8-4 Compression failures in wood.

(*a*) Surface appearance of compression failures in an edge-grain board of Sitka spruce [*Picea sitchensis* (Bong.) Carr.]. Natural size. (*By permission from Eric A. Anderson, Detection of Compression Failures in Wood, Forest Prod. Lab. Mimeo Rept.* 1588, 1944.)

(*b*) Electron micrograph of a replica of slip lines produced in the cell wall of ponderosa pine (*Pinus ponderosa* Laws.) by compression stress. The lateral displacement of the microfibrils in the S3 layer of the cell wall can be observed crossing the lines of fracture of the inelastic warty layer on the surface. (4700×.)

(*Photograph courtesy of W. A. Côté, Jr.*)

Planks that are cut diametrically in the log will split through the pith at both ends, with the two halves curving outward from each other. This type of defect arises because the longitudinal growth stresses are no longer balanced by the stresses in the side slabs that have been removed, and the opposing stresses in the two halves of the planks produce a large component of tensile stress perpendicular to the grain at the pith on the ends of the plank.

IV. REACTION WOOD

In logs that have an eccentric pith, the wood on the side showing greater growth usually does not conform to the pattern of structure and physical behavior that has been described in previous sections of this text. The general term ***reaction wood*** is applied to the specialized type of wood that is produced on the wide side of such eccentric cross sections. The reaction wood of gymnosperms is called ***compression wood,*** because it is formed on the wide side of eccentric cross sections corresponding to the lower or compression stress side of leaning trunks or branches. The reaction wood in angiosperms is known as ***tension wood***. The usual location for tension wood in a stem cross section is on the upper side of leaning trunks or branches. However, it may occur in a diffuse arrangement in the stem cross section or be present with little evidence of stem eccentricity.

Reaction-wood formation is thought to be mainly a mechanism for restoring leaning tree stems to their normal vertical orientation, or for maintenance of a preferred angular orientation of branches. Originally it was believed that the unbalance in stresses between the upper and lower sides of leaning trunks or inclined branches caused a stimulation of localized growth in the cambial region that either pushed the stem back to the vertical or pulled it up, depending on whether the tree was a conifer or a hardwood. However, careful experiments have shown that simple stress is not a causative factor in the development of reaction wood.[47,54,67] There does appear to be a relation between gravity and the location of reaction-wood formation in the trunk, although the relationship has not been satisfactorily explained. There is undoubtedly an asymmetric response to growth hormones in the regions of the stem which are bending to restore the crown uprightness. Probably more than one hormone is involved in the growth changes, since stimulation of one side of the stem is almost invariably associated with reduced growth rate on the opposite side. Evidence exists that growth-promoting auxins, such as indoleacetic acid, will cause stem curvature in conifers and that this same auxin acts as an anti-auxin, or growth suppressant, in the hardwoods.[67] Uneven distribution of growth hormones in the stem has been offered as an explanation of the asymmetric growth of reaction wood, but this is apparently not valid according to recent investigations. Westing[67] states: "Perhaps the best explanation is that one side of the inclined stem develops a localized sensitization to auxin possibly from production of some immobile compound in an appropriate localized position." All the theories advanced to date fail in some way to explain known behavior. It would appear that a single unified theory may eventually be developed to explain all reaction-wood formation. Evidence to support a unified theory for reaction-wood development can be found in the fact that certain of the properties of the wood on the suppressed-growth side of eccentric conifer stems resemble those of tension wood, and the same sort of correspondence can be observed between the suppressed-growth side of tension wood and compression wood. Furthermore, some primitive angiosperms form xylem similar to compression wood tissue in leaning stems, and some conifers develop tension-wood fibers in the bark of leaning stems.[29]

Westing[68] has outlined a mechanism for straightening of leaning stems which occurs through a combination of simultaneous activities: (1) increase of radial growth on the lower side in conifers and the upper side in angiosperms; (2) an expansion in length for conifer tracheids in the last stages of development and a shrinking in length for hardwood fibers in the zones of reaction-wood formation; (3) suppression of radial growth on the side opposite the reaction wood; (4) the usual shrinking in length of the normal tissue forming on the side opposite the reaction wood in the stem, of the type that was discussed earlier in this chapter under Growth Stresses in Trees; and (5) perhaps an increase in osmotic pressure for the cambial zone on the lower side of the stem for conifers. The first three of these factors that influence stem straightening are known to be controlled by hormones, so that chemical control of reaction-wood formation seems fairly evident.

Reaction wood is a modification of the normal cellular structure in woody plants that can vary from the most obvious and extensive condition to a scarcely noticeable change in cell character. The nature and appearance of compression wood are quite constant in all the species of Coniferales, Ginkgoales, and Taxales.[47,68] Furthermore, examples of fossilized compression wood estimated to be 180 million years old appear the same as that formed today. On the other hand, tension wood is much less uniform in its appearance and varies markedly in its nature among species and genera, with a few kinds of hardwoods such as *Drimys, Buxus,* and *Cotinus* producing reaction xylem that resembles compression wood in location and appearance. However, regardless of any of these possible variations, experience has shown that the presence of small amounts of reaction wood, or very slight modification of normal tissues, will produce obvious changes in the qualities of wood. For this reason and because of its widespread appearance in trees, reaction wood must be considered as one of the serious defects of wood.

A. Compression Wood

1. GENERAL APPEARANCE

Compression wood in logs is usually indicated by eccentric growth rings which appear to contain an abnormally large proportion of late wood in the region of fastest growth (Fig. 8-5). When compression wood is obvious, the portions of the cross section showing the fastest growth rate are much more red than normal, especially in

FIG. 8-5 Compression wood in an eccentric cross section of a southern pine log (*Pinus* sp.). (*By permission from U.S. Forest Products Laboratory.*)

pines. This characteristic color explains the origin of the name *rotholz,* as used in the German literature.

The presence of compression wood in conifers causes the transition from early to late wood to become quite gradual, both on the cross section and along the grain (Fig. 8-6). The distinctness of the late-wood zone also changes when compression wood is present; if normal wood shows a distinct and sharp late-wood band, this becomes indistinct in compression wood. On the other hand, if late wood is usually indistinct, as in eastern white pine (*Pinus strobus* L.), the late-wood band in

---

FIG. 8-6 Variation in gross appearance of compression wood.

(*a*) Douglas-fir [*Pseudotsuga menziesii* var. *glauca* (Beissn.) Franco] normal wood in cross section. (10×.)

(*b*) Douglas-fir [*Pseudotsuga menziesii* var. *glauca* (Beissn.) Franco] compression wood in cross section (10×), with a row of traumatic canals in the early wood of the upper growth increment.

(*c*) Eastern white pine (*Pinus strobus* L.) normal wood in cross section. (10×.)

(*d*) Eastern white pine (*Pinus strobus* L.) compression wood in cross section. (10×.)

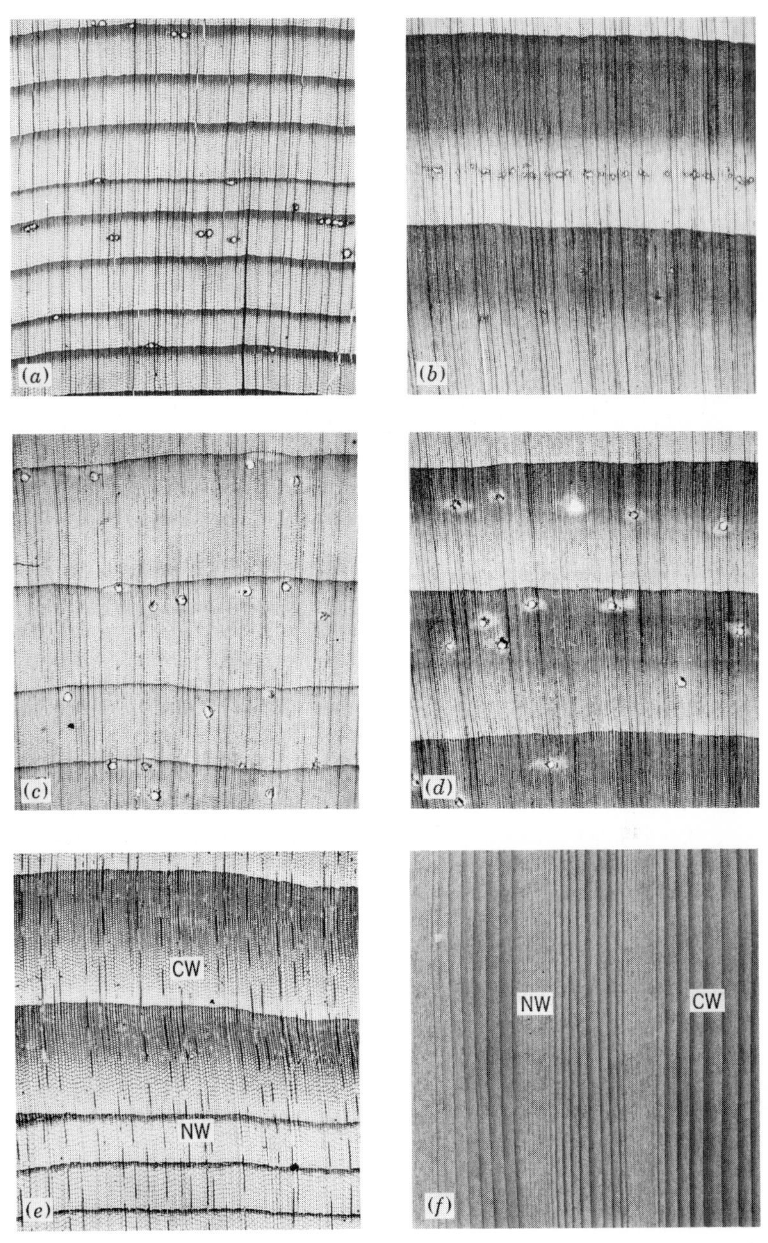

(*e*) Redwood [*Sequoia sempervirens* (D. Don) Endl.] cross section (10×) showing compression wood (CW) and normal wood (NW) in one sample.

(*f*) Redwood [*Sequoia sempervirens* (D. Don) Endl.] edge grain with alternating bands of normal (NW) and compression wood (CW). Note the increased width of growth increments, wider late wood, and more gradual transition in the compression wood bands. Normal size.

(*All photographs courtesy of H. A. Core and W. A. Côté, Jr.*)

compression wood is more evident[12] (Fig. 8-6). As a general rule, the luster of compression wood on longitudinal surfaces is less evident than that shown by normal wood. Density is almost invariably high in compression wood and is associated with lower strength properties than would be predicted on the basis of its weight. Well-developed compression wood usually is harder than would be expected for a given species and is more difficult to work with tools than normal wood.

Compression wood of conifers, with the exception of that found in the genera *Agathis* and *Araucaria* of the Araucariaceae, is quite opaque to transmitted light along the grain of the wood.[60] Cross sections of wood cut about ⅛ inch thick, and placed over a light source, will reveal the compression wood as areas that are darker than the surrounding normal wood. This method is very useful for the determination of intermediate stages that are difficult to detect otherwise, or for marking the full extent of compression wood in a given piece.

2. MICROSCOPIC STRUCTURE

Accurate determination of compression wood depends upon a microscopic examination; this is especially true when the wood has been only mildly affected. The principal features which are typical of compression wood may be stated as follows:

1. Growth rings consist chiefly of tracheids with rounded cross sections, in contrast with the angular outline of normal tracheids (Fig. 8-7). This feature is most evident in the darker-colored portion of the growth increment. It is apparent in the early stages of postcambial development.
2. Intercellular spaces are apparent on the cross sections where three or four tracheids meet (Fig. 8-7). These spaces are most prominent in portions of the growth increments where the thickest cell walls occur. In the first 5 to 10+ rows of cells in the early wood the cells are square in outline and do not show intercellular spaces. A few layers of flattened late-wood cells at the end of the ring also show no intercellular spaces.[14] In some kinds of conifer woods the intercellular spaces may be partly or wholly filled with lignin or pectic material.
3. Wall thickness of compression-wood tracheids in the dark-colored part of the increment is approximately twice as great as in comparable normal tracheids (Fig. 8-7). Tracheids in the first-formed early-wood zones of the increments have approximately the same wall thickness as normal cells or are perhaps even a little thinner.[14]
4. The radial cell diameters show very little difference across the increments except at their boundaries.[14] Both the average radial and tangential diameters are less than would be evident in a comparable normal-wood increment.
5. In compression-wood tracheids the secondary cell wall layers are reduced to the $S1$ and a modified $S2$ layer.[46,60]
6. The modified $S2$ layer in compression wood has a microfibrillar angle of about 45° to the cell axis. This angle is much larger than is found in the $S2$ layers of normal wood. The spiral angle is quite evident in the cell walls of many kinds of conifers

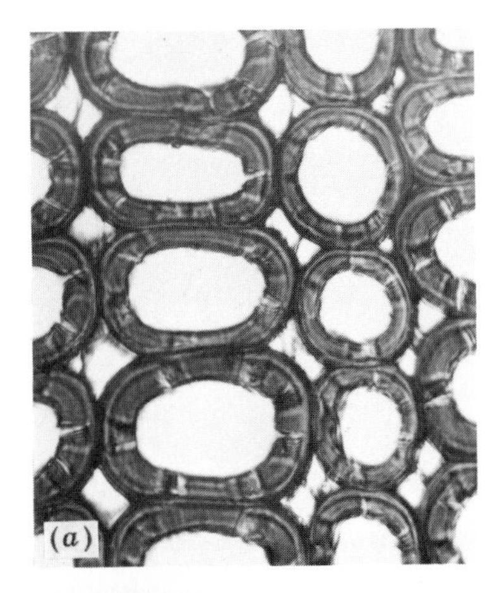

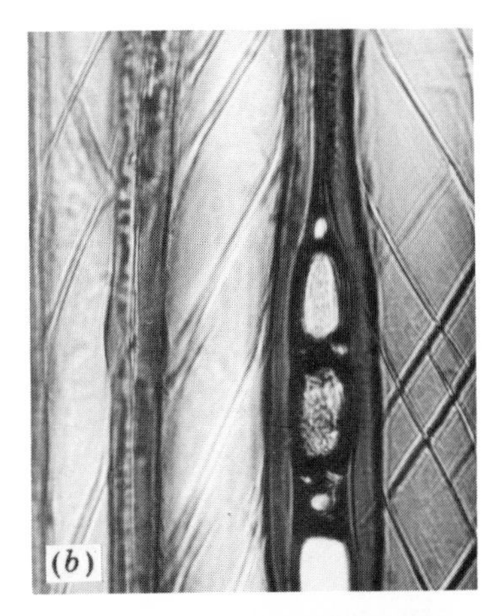

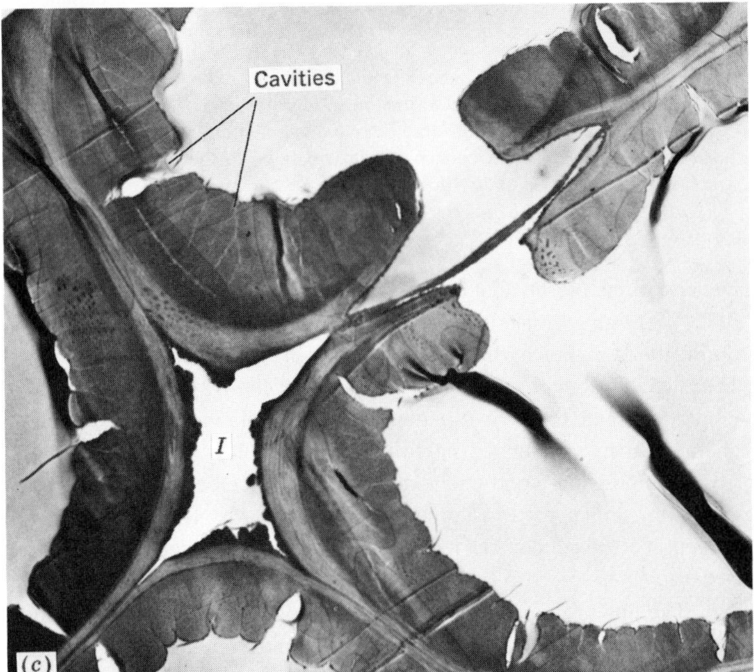

FIG. 8-7 Spiral cavities in compression-wood tracheids of redwood [*Sequoia sempervirens* (D. Don) Endl.].

(*a*) Cross section exhibiting pronounced radial-longitudinal cavities in the rounded tracheids. The characteristic intercellular spaces are evident and show the remains of the parent cell wall at several places. (400×.)

(*b*) Spiral cavities in the tangential-longitudinal walls of compression-wood tracheids. (400×.)

(*c*) Electron micrograph of an ultra-thin cross section. Radial-longitudinal cavities of several sizes are shown; an intercellular space (*I*); and a bordered pit with an aspirated torus.

(*Photographs courtesy of H. A. Core and W. A. Côté, Jr.*)

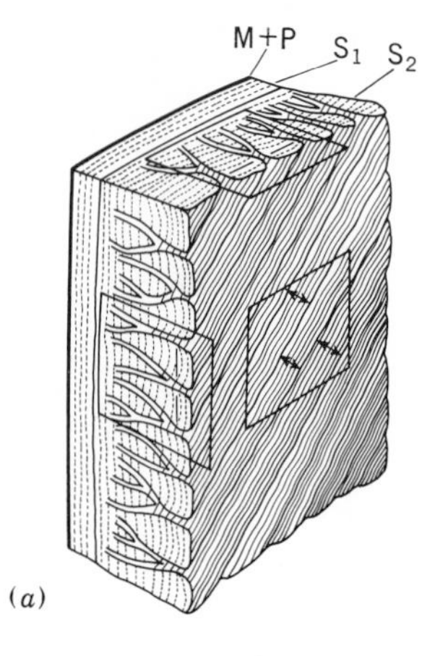

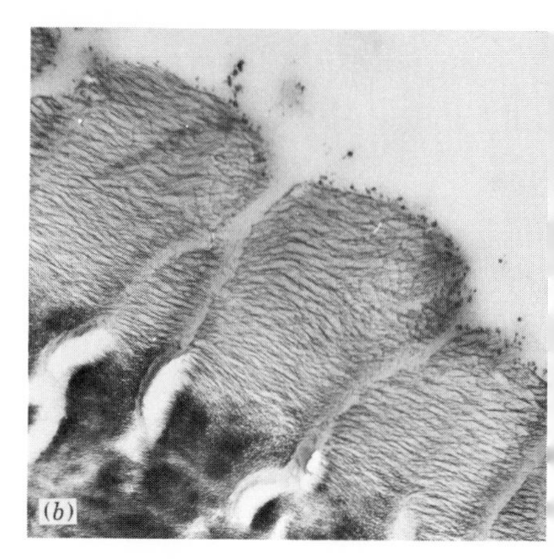

FIG. 8-8 Helical ridges and cavities in compression wood of balsam fir [*Abies balsamea* (L.) Mill.].

(*a*) Diagrammatic representation of cavities and ridges in a portion of the cell wall of a compression-wood tracheid. Arrows indicate the direction of the probable contraction of cell wall material during cavity formation (*Diagram after Casperson and Zinsser,* 1965.)

(*b*) Mature ridges and cavities in the compression-wood tracheid wall. Ridges are rounded adjacent to the lumen and contain lamellae parallel to the lumen surface. Electron micrograph of a cross section at 12,400×.

[*Courtesy of W. A. Côté, Jr., et al., Holzforschung,* **22**(5):138–144 (1968).]

because of the presence of helical checks which follow the microfibrillar angle (Fig. 8-7). These checks are in reality a system of radial cavities originating in the *S*2 during the formation of this wall layer,[59] and are not caused by drying of the cell wall. These cavities begin as a lateral shrinkage of the *S*2 wall lamellae; thereafter the ridges are built up with rounded layers of cell wall material separated by the characteristically branched system of helical cavities[8,16,63,66] (Fig. 8-8).

7. Compression-wood tracheids are 10 to 40 percent shorter than normal-wood cells from similar parts of the growth increment.[47,60] This shortening is primarily a reflection of the general cell length reduction that accompanies accelerated growth[60] (see also Chap. 7).

8. The tips of compression-wood tracheids are frequently very much distorted, while normal tracheids are usually quite simple in outline.

3. CHEMICAL COMPOSITION OF COMPRESSION WOOD

There is a marked change in the chemical composition of compression wood from that described previously for normal wood. The average lignin content, on a dry weight basis, is increased by about 9 percent; at the same time the percentage of cellulose in compression wood is reduced in the order of 10 percent from normal levels, and the galactose content of the compression wood is increased 7.8 percent over that in normal wood[17] (Table 8-1). These major changes in the chemical

TABLE 8-1 Chemical Composition of Compression Wood and Comparable Normal Wood

Kind of wood	Lignin, % CW*	NW*	Cellulose, % CW	NW	Galactose, % CW	NW	Reference
Abies balsamea Mill.	33.6	24.6	39.4	51.6	. . .	. . .	*a*
	39.8	31.7	32.4	41.2	7.1	1.0	*b*
Larix decidua Mill.	38.3	26.2	32.1	42.6	9.4	2.0	*b*
Larix laricina (DuRoi) K. Koch	38.1	27.0	31.5	42.2	10.0	2.4	*b*
Picea abies (L.) Karst.	40.8	29.1	30.9	40.1	7.7	2.3	*b*
	38.0	28.0	27.3	41.5	. . .	. . .	*c*
	39.0	28.0	30.0	44.0	. . .	. . .	*d*
Picea glauca (Moench.) Voss	39.4	28.6	31.9	40.4	10.9	1.9	*b*
Picea mariana (Mill.) BSP	39.5	30.2	31.2	41.1	8.7	2.0	*b*
Picea rubens Sarg.	37.0	28.1	32.8	40.4	11.5	2.2	*b*
Picea sitchensis (Bong.) Carr.	30.9	25.8	53.7	60.6	. . .	. . .	*e*
Pinus densiflora S. et Z.	36.3	26.6	46.1	56.2	. . .	. . .	*f*
Pinus radiata D. Don	34.4	24.2	. . .	. . .	. . .	. . .	*g*
Pinus resinosa Ait.	39.7	29.1	. . .	. . .	11.6	1.8	*b*
Pinus rigida Mill.	37.6	28.1	32.7	43.9	11.5	1.4	*b*
Pinus strobus L.	39.4	29.0	30.9	40.9	11.0	3.8	*b*
Pinus sylvestris L.	37.9	27.4	31.8	40.3	11.6	3.1	*b*
	36.0	31.0	41.0	46.0	. . .	. . .	*h*
Pinus taeda L.	35.2	28.3	46.1	58.7	. . .	. . .	*i*
Sequoia sempervirens (D. Don) Endl.	37.7	32.2	38.0	44.3	. . .	. . .	*e*
Tsuga canadensis (L.) Carr.	40.9	33.7	29.0	39.6	12.9	1.8	*b*
Juniperus communis L.	37.2	31.4	32.2	37.7	8.2	3.0	*b*
Thuja occidentalis L.	39.4	33.3	31.8	42.7	7.2	1.5	*b*

* *CW* = compression wood; *NW* = normal wood.

a. B. Johnson and R. W. Hovey, *J. Soc. Chem. Ind. Trans.*, **37**:132T (1918).

b. W. A. Côté, Jr., B. W. Simson, and T. E. Timell, *Svensk Papperstidning,* **69**(17): 553–555 (1966).

c. E. Hägglund and S. Ljungren, *Papier-Fabrikant,* **31**(27A):35–38 (1933).

d. H. Meier, in M. H. Zimmermann (ed.), "Formation of Wood in Forest Trees," Academic Press, Inc., New York, 1964.

e. H. E. Dadswell and E. F. Hawley, *J. Ind. Eng. Chem.*, **21**(10):973–975 (1929).

f. K. Hata, *Japanese Forestry Soc. J.*, **33**(4):136–140 (1951).

g. H. E. Dadswell et al., in Bolam (ed.), "Fundamentals of Papermaking Fibers," British Paper and Board Makers Association, London, 1961.

h. L. P. Zherebov, *Bumazhn.-Prom.*, **21**:14–26 (1946).

i. M. Y. Pillow and M. W. Bray: *Paper Trade J.* (*Tech. Sect.*), **101**:361–364 (1935).

composition of compression wood reduce its suitability for pulp and paper manufacture and alter its shrinkage characteristics, as well as its mechanical properties.

Studies of the structure of compression-wood tracheids show that the compound middle lamella resembles that of normal wood; the $S1$ layer is thicker than usual and has approximately the same microfibrillar organization, but its lignin content is lower than in normal wood. The modified $S2$ layer is thickened; not only does it have less cellulose than usual, but the cellulose is of a lower order of crystallinity than in normal wood.[39] The lignin distribution has been shown to be largely concentrated in the $S2$ layer in compression wood.[15] An outer zone in the $S2$ layer has been shown to contain 40 percent of the total lignin, with an additional 40 percent uniformly distributed over the remaining portion of the $S2$. In those species showing helical cavities in the compression-wood tracheids, the lignin-rich zone of the $S2$ marks the outer termination of these cavities.[18]

Examination of the cellulose in compression wood has shown that it has a lower degree of crystallinity and that some of the hemicellulose components are strongly retained in the purified cellulose. For example, Schwerin[51] reports that about half the galactose is retained in the purified alpha cellulose. The lignin in compression wood is relatively high in volume, as is shown in Table 8-1. In addition Bland[4] has determined that the lignin is abnormal owing to failure of the methylation mechanism during its deposition.

4. PHYSICAL CHARACTERISTICS

The most evident physical characteristic of compression wood is the increase in specific gravity over that shown by comparable normal wood. The increase is the result of the generally thicker walls in compression wood and results in about one-third more cell wall material than in normal wood. In terms of a ratio, the difference between compression- and normal-wood specific gravities is approximately 4 to 3.

Shrinkage and swelling for well-developed compression wood in the longitudinal direction are much larger than normal, as can be seen in Table 8-2. An increase of ten times in longitudinal shrinkage is not at all unusual, and an instance of 8.6 percent shrinkage along the grain in compression wood has been reported.[73] These increases are accompanied by reduced lateral shrinkage, with radial and tangential changes in compression wood about half as great as normal (Table 8-2).[24] The unusually large shrinkage along the grain causes boards containing compression wood to warp and twist badly, with frequent splits at the junctions between compression and normal wood; it also causes transverse tension breaks to form in the compression-wood bands that may be present (Fig. 8-9). These altered characteristics can be explained on the basis of the increase in the microfibrillar angle in the $S2$ layer of compression-wood cell walls, which promotes a greater longitudinal component of change in dimension. Other factors may be just as important; e.g., Wooten et al.[72] show that longitudinal shrinkage in compression wood of *Pinus taeda* L. seedlings is dependent on the thickness of the $S1$ layer, not on the microfibril angle. However, this conclusion was drawn on the basis of juvenile tissue, which already has abnormally high microfibril angles, and may not apply to mature wood.

The equilibrium moisture-content level for air-dry compression wood is slightly higher than in comparable normal wood under the same temperature and relative humidity conditions. However, the fiber saturation point for compression wood is lower than it is in normal wood, because of a low absorption per unit volume induced by the excessively high lignin content. Longitudinal permeability is also much altered, with about half as much water movement in a given time, probably because of reduced lumen size and smaller pits in the compression-wood tracheids.

Compression wood, in the green condition, is stronger than normal wood in bending and compression; it is also tougher. However, this same compression wood shows a lower rate of increase in strength when it is dried than is shown by comparable normal wood. The tensile properties and elastic moduli of compression wood are lower than would be predicted from its specific gravity. The low tension strength along the cell axis is explainable on the basis of the presence of spiral cavities in the cell walls of the compression-wood tracheids.

When compression wood is compared with normal wood on the basis of equal amounts of cell wall material (specific strength, see Chap. 6, page 220), the compression wood is shown to be definitely weaker than normal wood and lower than normal in elastic properties. This conclusion is supported by the evidence from chemical analysis of compression and normal wood. The approximately 30 percent reduction in cellulose for compression wood, coupled with the high microfibrillar angle and spiral checking, offers a good explanation of the loss in all mechanical properties, except compression parallel to the grain. The fact that compression wood is high in strength as a column can be partly explained on the basis of the large volume of cell wall material which it exhibits. However, the major increase in lignin also plays a part in the greater compression strength of reaction wood in conifers. In compression wood the spaces between microfibrils are more completely filled with lignin than normal; this reduces the buckling tendency of the cell wall, and thus allows a greater compression load than normal for this type of wall organization before failure occurs.

Compression-wood formation is much more common in trees of vigorous growth than in those with slow growth, and suppressed trees will often show very little

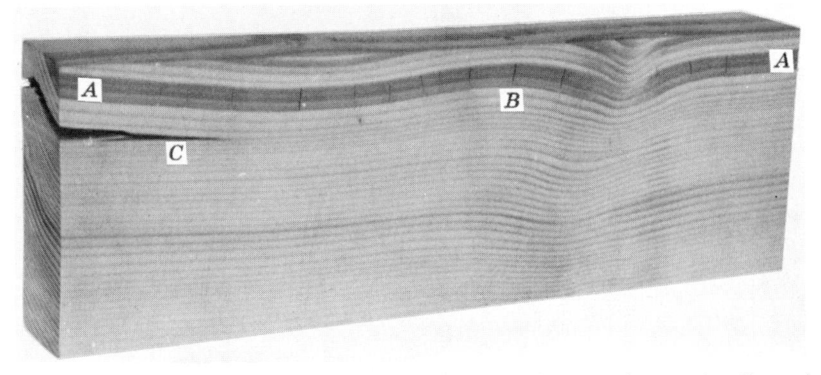

FIG. 8-9 Compression wood in a 2 × 4 of true fir (*Abies* sp.) showing the effects of differential dimensional changes between the compression-wood zones and the normal wood. Greater longitudinal shrinkage in the compression wood (*A-A*) than in the normal wood has caused the formation of cross breaks at *B* and a split at *C*.

TABLE 8-2 Physical Properties of Reaction Wood and Comparable Normal Wood

Kind of wood	Type of wood[a]	Moisture content %	sp gr (G)	Shrinkage to ovendry, %			Modulus of rupture R, psi		Modulus of elasticity bending E, psi × 1000		Compression parallel to grain C, psi		Tension parallel to grain T, psi		Tough-ness, in.-lb
				Long.	Rad.	Tang.	R	$R/G^{1.5}$	E	E/G^2	C	C/G	T	T/G	
Compression wood (CW)	CW	109	0.470	0.54	...	...	7,570	23,500	984	4452	3580	7,620	....	....	141
Abies concolor Mill.[b]		**11.9**	**0.509**	...	...	...	**12,700**	**34,990**	**1108**	**4280**	**5900**	**11,590**	....	....	**114**
	NW	187	0.346	0.12	...	...	6,040	29,600	1180	9830	2780	8,030	....	....	130
		11.7	**0.375**	...	...	...	**10,460**	**45,480**	**1327**	**9410**	**5220**	**13,920**	....	....	**116**
	CW	Green	0.387	...	...	...	5,290	21,950	...	....	2395	6,190	....	....	271
Picea glauca (Moench.)		**10**	**0.392**	...	...	...	**9,620**	**39,110**	...	....	**5270**	**13,440**	....	....	**121**
Voss[c]	NW	Green	0.316	...	...	...	4,610	25,900	...	....	1980	6,270	....	....	208
		10	**0.332**	...	...	...	**9,270**	**48,530**	...	....	**4990**	**15,030**	....	....	**186**
	CW	87	0.467	0.80	2.2	5.1	6,120	19,180	842	3860	3300	7,060	9,690	20,750	173
Pinus ponderosa Laws.[b]		**12.6**	**0.499**	...	...	...	**11,710**	**33,170**	**1019**	**4090**	**5970**	**11,960**	....	....	**100**
	NW	133	0.354	0.21	3.9	6.4	4,640	21,990	1074	8590	2340	6,610	11,780	33,280	101
		12.0	**0.372**	...	...	...	**9,840**	**43,350**	**1345**	**9750**	**5210**	**14,000**	....	....	**79**
	CW	Green	0.415	...	...	...	5,355	20,060	...	....	2390	5,760	....	....	385
Pinus resinosa Ait.[c]		**10**	**0.448**	...	...	...	**10,280**	**34,270**	...	....	**5590**	**12,480**	....	....	**161**
	NW	Green	0.376	...	...	...	4,830	20,900	...	....	2130	5,660	....	....	351
		10	**0.400**	...	...	...	**10,530**	**41,620**	...	....	**5400**	**13,500**	....	....	**186**
	CW	43.3	0.513	0.67	2.5	4.2	8,010	21,800	1016	3860	4150	8,090	10,880	21,200	182
Pseudotsuga menziesii		**12.1**	**0.527**	...	...	...	**12,500**	**32,640**	**1188**	**4270**	**7140**	**13,550**	**12,800**	**24,290**	**89**
(Mirb.) Franco[b]	NW	58.3	0.428	0.17	3.4	5.9	6,780	24,200	1369	7480	3280	7,660	13,850	32,360	185
		11.5	**0.459**	...	...	...	**12,950**	**41,640**	**1666**	**7900**	**7230**	**15,750**	**13,200**	**28,760**	**204**
	CW	102	0.506	1.19	1.4	2.4	7,470	20,750	685	2680	4640	9,170	5,910	11,680	70
Sequoia sempervirens		**10.5**	**0.510**	...	...	...	**8,890**	**24,420**	**788**	**3030**	**7250**	**14,220**	**7,560**	**14,820**	**64**
(D. Don) Endl.[b]	NW	113.7	0.380	0.14	1.5	3.5	7,310	31,240	1110	7700	3950	10,400	10,140	26,680	83
		9.9	**0.380**	...	...	...	**10,210**	**43,630**	**1253**	**8700**	**7160**	**18,840**	**8,850**	**23,290**	**65**

Tension wood (TW)	TW	Green	0.59	0.545	6.80	7.65	8,190	18.080	1088	3130					
Acer saccharum Marsh.[d]		**10**	. . .	. . .	. . .	. . .	**14,930**	. . . .	**1576**	. . .	**7079**				
	NW	Green	0.60	0.213	6.19	10.65	10,900	23,440	1536	4270					
		10	. . .	. . .	. . .	. . .	**17,755**	. . . .	**1862**	. . .	**7650**				
Betula pubescens and	TW	Green	0.69	0.64	4.45	6.82	. . . .	. . . .	. . .	. . .	2873	4,160			
B. verrucosa[e]	NW	Green	0.64	0.31	4.30	6.30	. . . .	. . . .	. . .	. . .	3286	5,130			
Populus deltoides Bartr.[f]	TW	**12**	**0.44**	. . .	. . .	. . .	. . . .	. . . .	. . .	. . .	**4704**	**10,690**	. . . .	. . . .	**375**
	NW	**12**	**0.43**	. . .	. . .	. . .	. . . .	. . . .	. . .	. . .	**5063**	**11,770**	. . . .	. . . .	**167**
	TW	Green	. . .	. . .	. . .	. . .	. . . .	. . . .	. . .	. . .	. . .	. . . .	7,600		
Populus regenerata[g]		Air-dry	**0.44**	. . .	. . .	. . .	. . . .	. . . .	. . .	. . .	. . .	. . . .	**15,000**	**34,090**	
	NW	Green	. . .	. . .	. . .	. . .	. . . .	. . . .	. . .	. . .	. . .	. . . .	9,300		
		Air-dry	**0.39**	. . .	. . .	. . .	. . . .	. . . .	. . .	. . .	. . .	. . . .	**9,500**	**24,360**	

[a] CW = compression wood; TW = tension wood; NW = normal wood.

[b] M. Y. Pillow and R. F. Luxford, *U.S. Dept. Agr. Tech. Bull.* 546, 1937.

[c] E. Perem, *Forest Prod. J.*, **8**(8):235–240 (1958). (Specific gravity is weighted average of three groups.)

[d] A. Marra, Characteristics of Tension Wood in Hard Maple, thesis, N.Y. State College of Forestry, 1942.

[e] P. J. Olinmaa, *Acta For. Fenn.*, **64**(3):1–170 (1956). (Data are average of two species.)

[f] L. E. Lassen, *Forest Prod. J.*, **9**(3):116–120 (1959). (Values are grouped from the data given.)

[g] H. von Pechmann, *Holz Roh- Werkstoff*, **11**(9):364 (1953).

reaction wood. In terms of percentage of total merchantable volume in *Pinus taed*
L., Haught[25] found 6 percent of compression wood in reasonably straight trees, 9.
percent in rather more crooked trees, and up to 67.1 percent in a very crooked tree
Compression wood is associated with knots in trunks of trees whether upright o
leaning. In *Pinus taeda* L. this knot-related compression wood amounts to 1
percent and is about equal in volume to the knot volume.[25] Compression-wood for
mation has been reported in the roots of some of the conifers. Certain specie
develop compression wood in subterranean roots, others only in surface root
exposed to light.[67,68]

An interesting type of abnormal wood which resembles compression wood ver
closely has been discovered in recent years. The balsam woolly aphid (*Adelge*
piceae Ratz.) in its attack on stems of *Abies balsamea, Abies grandis,* and *Abie*
amabilis induces complete increments whose appearance, structure, and propertie
are almost identical with compression wood formed in leaning stems of the same
species.[21,22,49] The mechanism for this abnormality is probably some hormona
secretion which is injected into the stem by the insects.

B. Tension Wood

1. GENERAL CHARACTERISTICS

Tension wood is usually formed on the upper side of leaning stems or branches o
hardwoods, but a few kinds of trees form tension wood on the lower side.[47] The cros
sections of these leaning trunks or branches are quite often eccentric in outline, with
the major radius on the upper side of the lean and with suppressed growth on the
lower side (Fig. 8-10). The tension wood usually occurs in the region of fastest
growth in such eccentric stems or branches. However, the location of tension wood
is not always in the region of accelerated growth. It may occur on the
suppressed-growth side, as was reported for a branch of *Sassafras albidum* by
White[69] and in stem wood of *Tilia americana*. In many kinds of trees, tension wood
may be formed with little evidence of eccentricity in the mature trunk; therefore
irregularity of cross section is not a constant indication of the presence of this type of
reaction wood. This is particularly evident in a few genera such as *Pawlonia* and
Catalpa and in roots which exhibit tension wood without any evident orientation
with respect to gravitational axis.[29,30,33,48]

Tension wood, as has been pointed out earlier, is formed as a mechanism to
correct lean in stems and probably also has a function in maintaining branch angle.
However, in tropical woods there is evidence that tension wood may form in
nonleaning trees as a means for crown movement in an attempt to obtain sufficient
light in dense forest. Taylor's data[55] on nonleaning *Liriodendron tulipifera* L. can
perhaps be interpreted in the same way. Regardless of the ultimate purpose of the
reorientation process, the mechanism is the same and follows the general scheme
outlined earlier. There is definite evidence that tension wood shrinks in the final
developmental stages. Aerial roots of *Ficus benjamina* L. containing tension wood

FIG. 8-10 Cross section of a log of sugar maple (*Acer saccharum* Marsh.) showing tension wood on the upper side. (*Photograph courtesy of A. A. Marra.*)

ave been shown to lift pots, in which they were rooted, completely free of the round.[73] Jacobs[31] in his work on growth stresses has also shown shrinkage to be resent in developing tension wood. The mechanism of this shrinkage has not been larified, but it has been suggested that it is related to hydration of the specialized bers formed in tension wood.[26]

Tension wood occurs far more commonly than has been realized in the past be-ause of the general difficulty in recognizing this type of tissue and the variability n its location in the stem. Relatively few families that have been investigated do ot show tension-wood formation, but the most common occurrence is in families vith unspecialized anatomical structures.

There are several features in the surface appearance of wood that indicate the resence of tension wood. The most characteristic of these is the appearance of a voolly surface in boards that are sawn green.[19,43,62] In such boards the tension-wood ssue tears loose in bundles of fibers that can choke and overheat a saw in extreme onditions (Fig. 8-11). When cutting veneers from tension wood the surfaces are ough and the veneers buckle. Some kinds of trees give evidence of the same voolliness on the cross section. In *Fagus* sp., for example, the ends of the fibers in

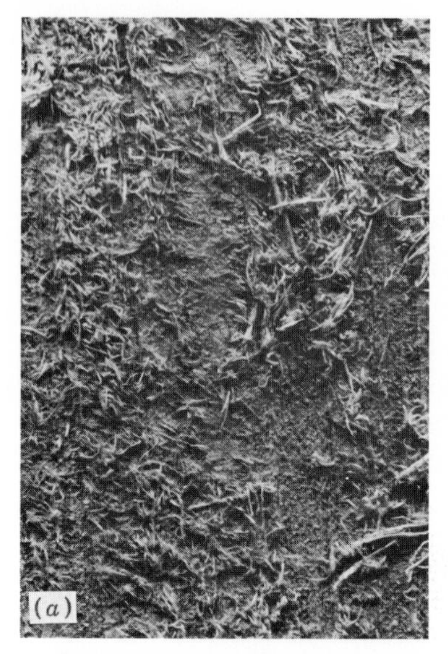

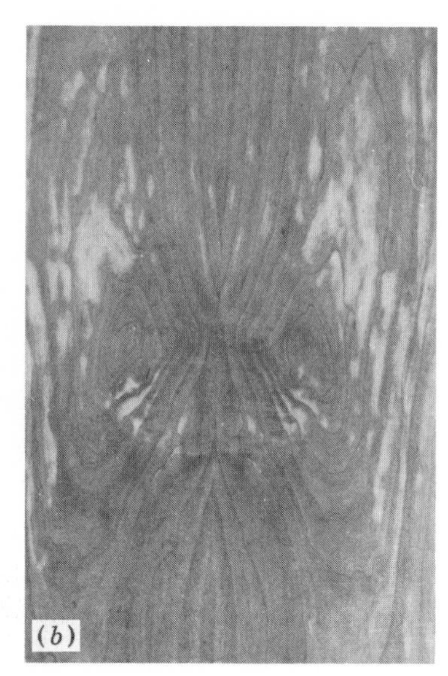

FIG. 8-11 Surface appearance of tension wood.
(*a*) Woolly appearance of a longitudinal sawn surface in sugar maple (*Acer saccharum* Marsh.) traceable to areas of gelatinous fibers in tension wood. Natural size. (*Courtesy of A. A. Marra.*)
(*b*) Silvery appearance of tension-wood areas on a finished surface of sugar maple (*Acer saccharum* Marsh.). One-sixth natural size.

the tension-wood zones do not cut cleanly and show on the surface with a silver sheen that contrasts with the surrounding tissue. Other kinds of wood give indicatio of tension-wood zones on the smooth cross section by darker color in these region than is normal.[62] Dry lumber containing tension wood does not tear so much i sawing and machining as the same lumber in a green condition, but the fibers d tend to tear out to some extent. The slight woolliness of the cut surface tends t remove the set of the saw blade rapidly and produces a "dulling" of the saw.[62]

Other features which are not quite so obvious in indicating the presence of tensio wood are unusual bowing and irrecoverable collapse; both are the result of exces sive longitudinal shrinkage.

Positive identification of tension wood depends upon microscopic examination o thin sections of wood. Differential coloration, by the use of one of several stai combinations, is the best general indicator for all types of tension wood and all level of severity. The use of safranin followed by light green on thin sections of woo shows the normal lignified tissue as dark red in color, while the secondary walls o tension-wood fibers stand out as a bright green. This dye combination has the advan tage of indicating single tension-wood fibers, as well as larger groups, and in additio identifies some types of unusual tension-wood tissue that occur in certain taxonomi groups.

Another stain which is widely used is chlor-zinc-iodide reagent that can be eithe applied to thin sections or brushed on the smooth surfaces of larger pieces. Tension wood is stained a blue-gray or bluish violet with chlor-zinc-iodide and the normal tissue a yellowish brown. Where tension wood is severe in some kinds of lumber, this type of stain can be effective, but in thin sections the differentiation is not so good as the safranin-light green combination. Phloroglucinol-HCl lignin indicator solution has also been used in the same way as chlor-zinc-iodide reagent.

2. ANATOMICAL FEATURES OF TENSION WOOD

The principal anatomical differences associated with tension wood are related to the fibers. The vessels in definite zones of tension wood are unchanged in character from those found in normal wood, but they are smaller in diameter and less numerous.[2,10,35] The rays in tension wood are decreased in size, at least in *Populus* and *Alnus*,[35] and axial parenchyma is decreased in both size and amount.[28]

The modifications in fibers that accompany the development of tension wood reflect the principal changes in woody tissue associated with reaction-wood formation.

1. The zones of modified fibers in tension wood of deciduous trees are usually confined to the early wood, but in evergreen hardwoods, such as *Eucalyptus*, they may occur throughout the increment. As a general rule the percentage of fibers is distinctly greater than in normal wood; also these fibers have somewhat smaller diameters, they usually have greater lengths, and they have fewer pits than those in normal wood;[35,47] furthermore they exhibit thicker walls with somewhat more rounded outlines. In contrast, the suppressed-growth zones opposite to tension-wood development have been reported to have fewer and shorter fibers, with more highly lignified walls, than in normal wood.

2. The most characteristic modification of fibers in the tension wood is the development of a *gelatinous layer* (G layer); in fact, most definitions of tension wood are based on the presence of this type of wall in the fibers. The gelatinous layer is a sheath of cellulose microfibrils oriented about 5° from the long axis of the cell.[13,43,46,47,61] This special layer is usually equal to or greater in thickness than the S2 layer of normal cell walls; it always appears on the lumen side of the cell wall and frequently has a buckled and swollen appearance, with partial separation from the layers beneath. The gelatinous layer is more refractory to light than normal wall layers, and can give the appearance of a clear gel that nearly fills the cavity (Figs. 8-12 and 5-8).

There are three types of wall organization for gelatinous fibers.[47,61] (*a*) A gelatinous layer may be present in tension-wood fibers in addition to the three layers of the normal lignified secondary cell wall. (*b*) The gelatinous layer may replace the S3 layer in tension-wood fibers, with the S1 and S2 regions of the normal secondary wall remaining unchanged. (*c*) The wall layers may be reduced to a thick gelatinous layer, and a zone corresponding to the S1 layer of normal fibers. The presence of a gelatinous layer, in any of these forms, results in a cell wall that is almost always

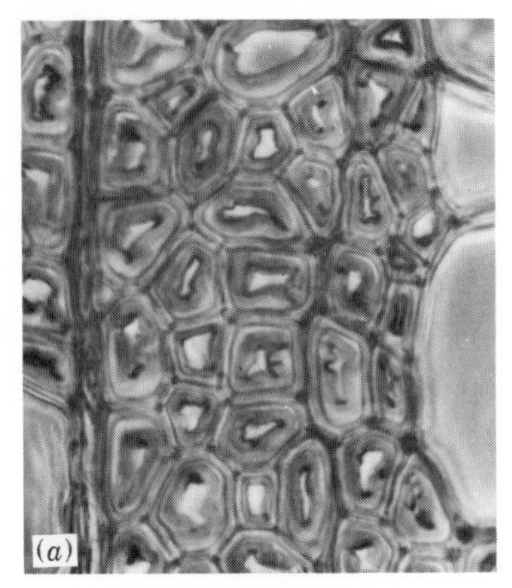

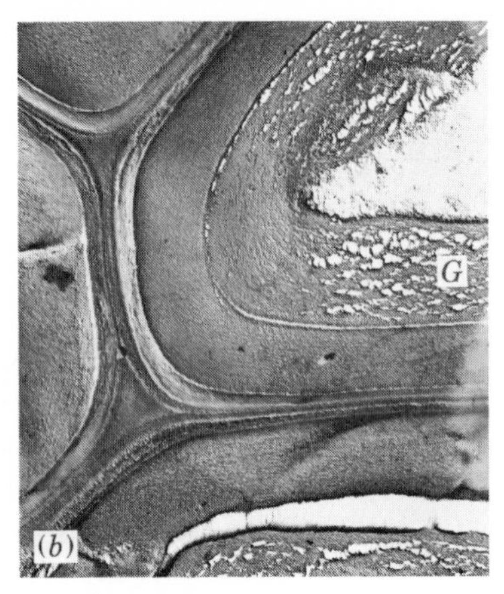

FIG. 8-12 Gelatinous fibers in tension wood.

(*a*) Gelatinous fibers in a cross section of a tension-wood area in *Populus* sp. Stained with safranin and fast green and photographed by phase contrast. (550×.)

(*b*) Electron micrograph of an ultra-thin section of gelatinous fibers in hackberry (*Celtis occidentalis* L.). The gelatinous layer (*G*) is attached to the secondary wall in the upper fiber and has pulled away in the fiber at the bottom. (6300×.)

(*Photographs courtesy of W. A. Côté, Jr., and A. C. Day.*)

thicker than normal, often appears loose and buckled under the microscope, and shows no indication of lignin by chemical stains or by phase contrast microscopy.

The initiation of tension wood occurs soon after the stem is tilted. For example, gelatinous fibers appear in stems of *Populus* seedlings which have been stimulated by tilting the plant for as short a period as 24 hours.[45] The initial appearance of the *G* layer is convoluted in cross section; later it flattens out to the conformation of the other secondary-wall layers in many kinds of wood, but in others it remains as a crumpled layer at maturity. In the green wood this *G* layer is rather poorly attached to the other cell wall layers. In well-developed tension wood the fibers are almost pure cellulose, which analyzes to 98.5 percent glucose and 1.5 percent xylose in *Populus tremula* L.,[45] for example. The *G* layer apparently has an extensive system of micropores,[52] but the microfibrils are not widely dispersed in a honeycomb system, as has been stated by some investigators. This dispersion is a swelling artifact produced by the embedding medium used in making electron microscope sections.

The degree of lean in the stem has been shown to be the most important factor in the development of numbers of gelatinous fibers in *Populus* and *Salix*.[3] Both lean and rate of radial growth on the tension-wood side are positively correlated with numbers of gelatinous fibers. This indicates that under most circumstances tension wood effects will increase with increased lean of the trunk and eccentricity of the cross section.

3. A number of modifications of the typical gelatinous fiber are known to exist in hardwood trees which should form tension wood by the nature of the cross section but do not show gelatinous fibers by the usual methods of examination. Casperson[7] reports that *Quercus robur* L. has gelatinous fibers whose *G* layers contain bands

of lignin in sufficient quantity so that normal stain methods do not differentiate these fibers. In other genera, such as *Fraxinus* and *Tilia* and some specimens of *Liriodendron*, there is no evident *G* layer but there is a general lack of lignification.[2,11,70] The most suitable method for demonstrating this condition is by the use of phase contrast microscopy.[34]

3. CHEMICAL COMPOSITION OF TENSION WOOD

Chemical analyses of tension wood and comparable normal wood show that tension wood is higher than normal in cellulose and ash content, but lower in lignin and hemicellulose fractions (Table 8-3). The low hemicellulose content, combined with a lesser degree of hydrolysis in tension wood, indicates that the cellulose in this type of

TABLE 8-3 Chemical Composition of Tension Wood and Comparable Normal Wood

Kind of wood	*Tension T or normal wood N*	*Lignin, %*	*Cellulose, %*	*Pentosans P or hemicellulose H*	*Ash, %*
Acer saccharum Marsh.[a]	T	21.1	72.9	. . . .	0.39
	N	21.9	70.3	. . . .	0.40
Betula spp.[c]	T	17.6	57.7	17.6 (P)	0.56
	N	21.3	47.0	27.1 (P)	0.35
Eucalyptus goniocalyx F.v.M.[h]	T	10.3	67.2	11.2 (P)	
	N	25.2	50.0	19.7 (P)	
Eucalyptus regnans F.v.M.[d]	T	16.0	63.5	4.3 Xylan	
	N	22.2	55.8	11.5 Xylan	
Eucalyptus regnans F.v.M.[h]	T	12.4	68.1	9.7 (P)	
	N	21.4	55.1	17.3 (P)	
Fagus sylvatica L.[e]	T	15.3	62.1	19.0 (P)	0.43
	N	19.6	57.2	26.3 (P)	0.29
Fagus sylvatica L.[b]	T	13.0	57.0	30.0 (H)	
	N	20.0	38.0	42.0 (H)	
Populus regenerata[f]	T	16.9	59.7	23.4 (H)	
	N	25.7	41.1	32.2 (H)	
Populus tremuloides Michx.[g]	T	16.5	53.6	21.8 (H)	
	N	17.6	51.0	24.1 (H)	
Ulmus americana L.[g]	T	27.3	45.4	22.3 (H)	
	N	29.4	42.0	23.8 (H)	

[a] A. Marra, see ref. 43.

[b] H. Meier, in Zimmermann (ed.), "Formation of Wood in Forest Trees," Academic Press, Inc., New York, 1964.

[c] P. J. Olinmaa, see ref. 46 (combines data on *Betula pubescens* and *B. verrucosa*).

[d] A. B. Wardrop and H. E. Dadswell, *Australian J. Sci. Res. Ser.* B, **1**(1):3–16 (1948).

[e] K. Y. Chow, see ref. 10.

[f] W. Klauditz and I. Stolley, *Holzforschung*, **9**(1):5–10 (1955).

[g] L. P. Clermont and F. Bender, *Pulp Paper Mag. Can.*, **59**(7):139–143 (1958).

[h] H. E. Dadswell et al., in Bolam (ed.), "Fundamentals of Papermaking Fibers," British Paper and Board Makers Association, London, 1961.

reaction wood is more crystalline than in normal wood.[62] X-ray studies confirm this conclusion.

The distribution of lignin within the cell wall of tension-wood fibers has been studied by means of indicating stains and phase contrast microscopy. Using these techniques, it has been concluded that lignin is absent in the gelatinous layer, and may be reduced or absent in one or more of the other secondary-wall layers in tension-wood fibers. This general lack of lignin in the secondary walls of fibers is the primary characteristic of tension-wood development in all kinds of hardwoods.

The stored starch and sugars in tension-wood tissue are present in lesser amounts than in the opposite side of the same cross section in the suppressed growth.[28] This could affect the extent of biological attack by organisms that depend on these substances for their development or survival.

Pulping tests have shown that yields of chemical pulp are greater from tension wood than from normal wood, and that tension wood can be defibrated mechanically more easily than normal wood. However, paper produced from tension-wood fibers is lower in strength than that made with normal-wood fibers. Tension wood is very well suited for dissolving pulps, since yields are high and strength of the original fiber is unimportant in these types of pulps.

4. PHYSICAL PROPERTIES

Tension wood differs from the adjacent normal wood in many of its physical characteristics. However, the differences are much less marked than the deviations shown between compression wood and normal wood in conifers.

1. The difference between the specific gravity of tension wood and adjacent normal wood is dependent upon the type of wall organization in the tension-wood cells. Thick-walled gelatinous fibers can increase the specific gravity of the tension-wood cells as much as 30 percent over normal wood. The increase in specific gravity for tension-wood formation is normally much smaller and falls in the range of 5 to 10 percent for woods with relatively thin gelatinous walls (Table 8-2). The extreme case has been found in *Tilia,* which produces tension wood that has lower density than normal wood.

2. The shrinkage and swelling of tension wood differ from those of comparable normal wood, but the differences are not so great as those shown in the reaction wood of conifers. The longitudinal shrinkage of tension wood seldom exceeds 1 percent and is usually much less (Table 8-2). Increase in longitudinal shrinkage from normal to tension wood may be as low as 40 percent in aspen (*Populus tremuloides* Michx.), but 100 to 200 percent in other hardwoods, as shown in the values given in Table 8-2; it can be as large as 300 percent in white oak and 500 percent in cottonwood (*Populus deltoides* Bartr.). The increase in longitudinal shrinkage for tension wood is directly related to the numbers of gelatinous fibers present.[46,58]

It has been proposed that the low amounts of lignin in tension wood account for the observed increases in longitudinal shrinkage by allowing the microfibrils to draw close together when water is removed from the cell wall. However, this supposes a large microfibrillar angle with respect to the cell axis. The most evident layer in the cell wall is the gelatinous layer, which has microfibrils oriented at a very small angle to the cell axis and which therefore cannot contribute much to longitudinal changes. Observations of isolated *G*-layer fragments from ***Populus tremula*** L. have shown that they have practically no longitudinal shrinkage but do have a large transverse shrinkage.[45] As was noted earlier, the *G* layer is quite loosely attached to the other wall layers in the green condition and for this reason probably plays little part in the longitudinal shrinkage. The abnormally high shrinkage along the grain can perhaps be better explained by considering that the *S*1 layer in tension wood is often thicker than in normal wood and is not counteracted by the *S*2 layer, as is usually the case in the normal cell wall. This condition would produce a large component of longitudinal shrinkage because of the low fibril angle in the *S*1.

Information on the transverse shrinkage of tension wood is sparse and contradictory. Table 8-2 summarizes most of the comparative evidence on the transverse shrinkage in tension and normal wood. It is apparent from these figures for tension wood that the gross radial shrinkage in maple and both the radial and the tangential shrinkage in birch increase. Additional data will probably show that tangential shrinkage is generally larger in tension wood than in normal wood.

The excessive shrinkage of tension wood is associated with collapse in many kinds of wood. In eucalyptus, the collapse associated with tension wood is not recoverable. The inability to obtain recovery in this kind of collapsed eucalyptus lumber is explained on the basis that hydrogen bonding can take place between the lamellae of the wall when shrinkage brings the microfibrils into close proximity, since there is little or no lignin present in the cell wall to block the bond formation.[62] Once the hydrogen bonds have been formed, the wall is stabilized and loses most of its ability to swell with the addition of moisture.

3. There is evidence that the equilibrium-moisture-content behavior of tension wood in ***Acer saccharum*** Marsh. follows the same pattern as that given previously for compression wood, and has a slightly higher value than found in normal wood at the same temperature and relative humidity conditions.[43]

4. Information on mechanical properties of tension wood is quite incomplete, as can be seen by reference to Table 8-2, which summarizes most of the available numerical data. It appears that compression parallel to the grain is lower in tension wood at all moisture levels than in comparable normal wood, whether they are compared on an actual-strength or an equal-density basis. The same statement can be made for compression perpendicular to the grain, modulus of rupture in static bending, longitudinal shear, and modulus of elasticity in bending, on the basis of the small amount of evidence that is available.

Tension wood of sugar maple is tougher in impact bending, in the air-dry condition, than comparable normal wood. The relationship can probably be extended to other woods, both green and dry, although no supporting data are available.

The available evidence on tension parallel to the grain for tension wood indicates that it is weaker than normal wood in the green condition, because at this moisture condition the gelatinous layer that is present in these woods is not well bonded to the rest of the secondary wall and contributes little to the tensile strength. As the wood dries, the gelatinous layer bonds to the remainder of the secondary wall and can contribute to the tensile strength of the tension wood. This situation explains the strength increase shown for dry ***Populus*** tension wood, which is stronger than dry normal wood (Table 8-2). The contradictory evidence on low tensile strength of dry tension wood in eucalyptus[19] is undoubtedly due to the fact that these woods contain a defect called ***brittle heart*** (see page 284) that lowers the tensile strength.

The general strength relations shown by tensile wood in comparison with normal wood can be qualitatively explained on the basis of the cellulose-lignin ratio changes. The decrease in lignin content in tension wood indicates that there is a decreased lateral support between the cellulose microfibrils in the cell wall. As a result, the microfibrils and lamellae in tension wood act as long unsupported columns under compression parallel to the grain, and buckle at lower loads than is the case with normal wood containing more lignin. This same lack of stiffness in tension wood shows up in its pliability and increased toughness. The increase in tensile strength in dry tension wood is also related to the increase in cellulose content over comparable normal wood.

V. BRASHNESS

Brashness is an abnormal condition that causes the wood to break suddenly and completely across the grain at stress levels lower than expected. The surfaces of a brash break are relatively smooth and show the structure of the wood in cross section quite cleanly. In contrast, the normal type of failure results in a more jagged surface (Fig. 8-13). The most objectionable feature of brash wood is the sudden failure without previous warning, especially when shock-loaded.

Brash wood usually can be distinguished from normal wood of the same species because of its unusually light weight. This light weight in brash wood is a reflection of a decrease in the amount of cell wall material and is the result of decreased fiber volume, increased proportion of thin-walled elements such as parenchyma or vessels, or an overall decrease in wall thickness. Unusually narrow-ringed growth in both conifers and hardwoods commonly results in brash wood. In the conifers exceptionally wide growth rings are indicative of brashness in most cases because of the low late-wood percentage.

Brashness in wood can arise from other causes than a decrease in specific gravity, and may be present in wood of normal density. A decrease in the cellulose content of compression wood causes a decrease in impact strength and a brash condition. However, in compression wood no single factor is the sole cause for the reduced strength. For example, compression wood not only has low cellulose content, but it has an abnormally low spiral angle in the secondary wall ($S2$), which also reduces

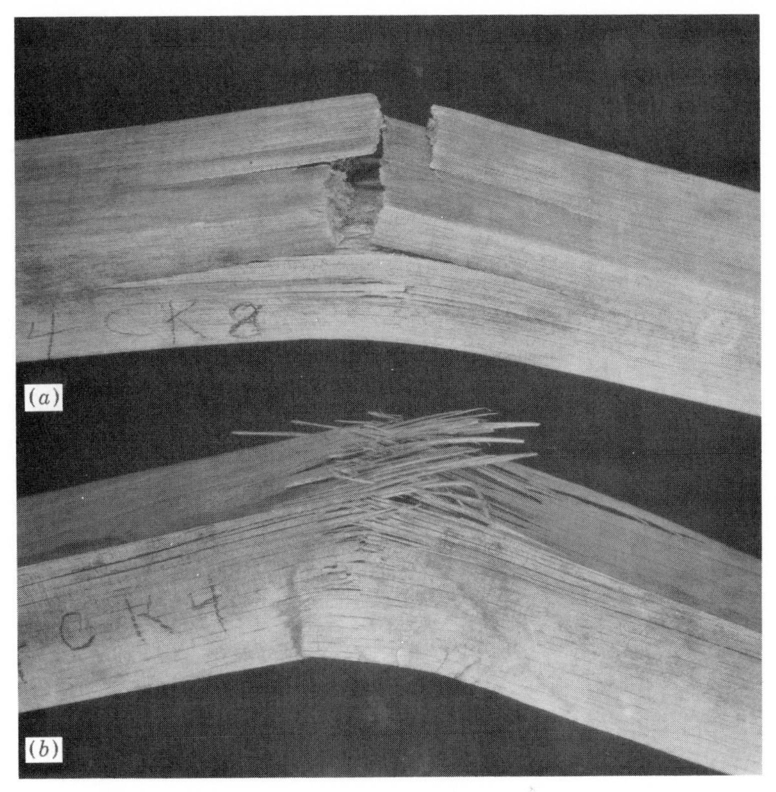

FIG. 8-13 Impact bending fractures in hickory (*Carya* sp.).
(*a*) Typical brash fracture of a piece of hickory which failed under an 18-inch drop of the testing machine hammer.
(*b*) Typical fracture of a normal piece of hickory which failed under a 50-inch drop of the same testing machine hammer used in *a*.
(*By permission from Arthur Koehler, Causes of Brashness in Wood, U.S. Dept. Agr. Tech. Bull.* 342, 1933.)

the tensile strength. Degradation of the cellulose through decay or as the result of abnormal heating of the wood can also cause brashness. Lastly, the presence of "brittle heart" and gross compression failures in wood are important causes of brashness.

VI. FROST INJURIES

Two types of defects develop in the wood of living trees, supposedly as the result of freezing temperatures. These are known as *frost rings* and *frost cracks*.

Frost rings appear to the naked eye as brownish lines within and parallel to the boundaries of growth rings; their general appearance simulates false rings (Fig. 8-14). These zones of discoloration result from frost injury to the cambium and the imma-

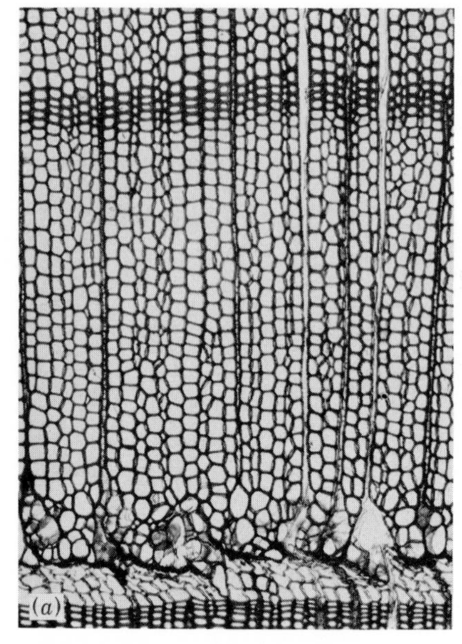

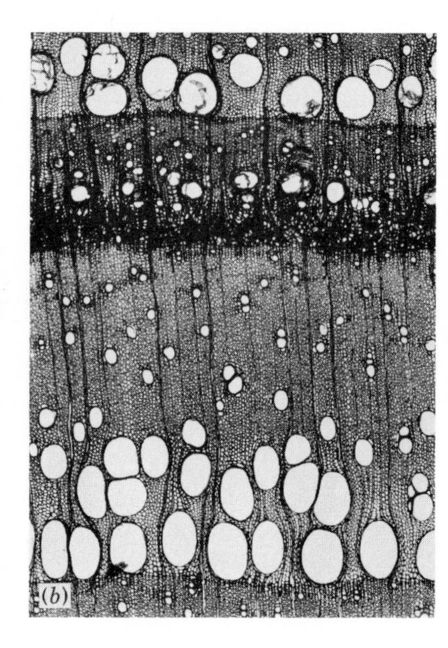

FIG. 8-14 Cambial damage.

(*a*) Cross section of jack pine (*Pinus banksiana* Lamb.) at 80×, showing the production of wound tissue in the early wood, probably due to frost injury. (*Photograph from U.S. Forest Products Laboratory.*)

(*b*) Cross section of white ash (*Fraxinus americana* L.) showing a zone of abnormal tissue (above the center) which has replaced almost all the normal early wood in the increment. (22×.)

ture xylem cells, after the cambium has become active in the spring, and before it becomes dormant in the autumn.[50] Examination of frost rings with a microscope reveals that there is collapse and distortion of cells that were immature at the time the freezing occurred, parenchyma appears in abnormal amounts, and the rays are wider at that point than normal.

Frost cracks develop as radial splits in the wood and bark near the base of the tree. They are found in all species growing in cold climates but are most frequent in hardwoods. Frost cracks are most common in old trees with stout primary roots and broad crowns and are absent in very young trees.[6] This type of crack always develops between two primary roots or the collars of two such roots. Frost cracks are usually bridged the following growing season by callus tissue which is formed by the cambium. However, this bridging is weak and usually breaks the next winter, when very low temperatures occur. Repeated healing and opening of frost cracks results in the formation of protruding lips of callus along the edges of the cracks. The prominent ridges formed by this type of growth are called *frost ribs*.

The development of frost cracks has been explained in several ways. One theory claims that when a tree is exposed to very low temperatures, the low heat conductivity of wood causes the outer region of the trunk to contract before the center of the trunk is affected by the temperature change. This differential shrinkage sets up a tensile stress on the outside of the trunk and results in radial cracks. Another theory considers that the mechanical action of wind on the frozen wood is the most important factor in frost-crack formation.[6] Wetwood has also been advanced as the principal cause of frost-crack formation.

VII. PITCH DEFECTS

A number of defects known as *pitch, pitch streaks,* and *pitch pockets* are found in softwoods in which resin canals are normal in the wood. These defects are common in spruce, Douglas-fir, pines, and larches. The other domestic conifers are remarkably free of pitch blemishes, although they may develop in any coniferous wood as a result of injury to the cambium.

Pitch and *pitch streaks* are defects that develop through the accumulation of resin in excessive amounts in localized regions of the wood. In these local areas the abundant resin fills the cell lumina and permeates the cell walls so that the wood takes on a resin-soaked appearance and changes color. If the patches are irregular in shape or poorly defined, they are termed *pitch;* if sharply outlined, they are called *pitch streaks*. This type of defect probably results from one of several causes, e.g., injuries due to turpentining in the southern pines, or insect attack.

Pitch pockets are planoconvex cavities in the wood with their flat faces parallel to the plane of the growth increment and facing the pith of the tree. Pitch pockets are usually confined within the boundaries of a single growth increment. On a cross section, pitch pockets have a planoconvex shape; this is true also for their appearance on the radial surfaces. However, on the flat-sawn face of the wood, pitch pockets appear as oval or elliptical patches with their long axes parallel to the grain (Fig. 8-15).

Pitch pockets usually contain resin in liquid or solid form, although this is not always the case. Some pockets are empty or may contain some bark inclusions as well as pitch. The usual size for pitch pockets is less than ½ inch in the radial opening, several inches in the horizontal distance parallel to the growth increments, and proportionally larger in the vertical axis. The range in size for pitch pockets is quite great and may reach very large dimensions in Douglas-fir, where single pockets are reported to contain several gallons of resin.

Microscopic examination of pitch pockets shows that the surfaces of the cavity are covered with epithelial tissue similar to that found in resin canals. Whether this epithelial tissue arises by spontaneous action of the cambium which develops epithelium rather than tracheids for a short time, or whether wind action causes a break in the cambium and the epithelium subsequently forms on the surface of the break is not known. It would appear from the shape of the cavity that the pocket is postcambial in origin, because the flat face is oriented toward the pith and the convex face is on the cambium side, where adjustment can be made for the growth in cavity size of the pocket as resin is secreted from the epithelial cells.

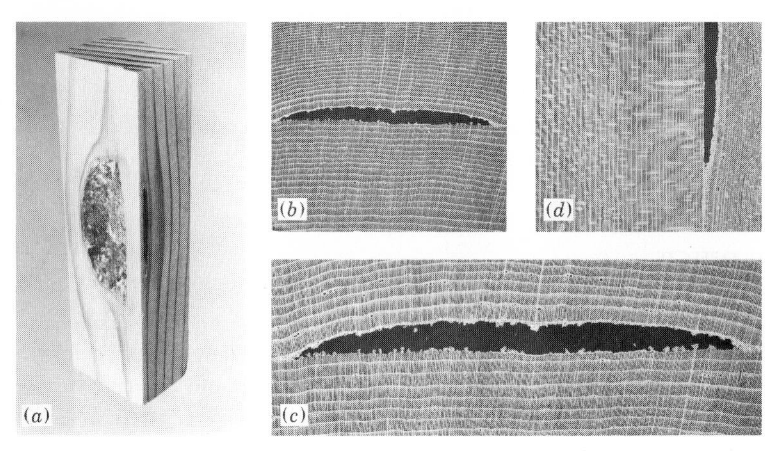

FIG. 8-15 Pitch pockets in conifers.

(*a*) A pitch pocket in red spruce (*Picea rubens Sarg.*) in tangential (surface) and in radial (edge-grain) views. (2/5×.)

(*b*) End-grain view of a pitch pocket in black spruce [*Picea mariana* (Mill.) B.S.P.]. (2½×.)

(*c*) The same as *b*. (5×.)

(*d*) Radial view of a pitch pocket in black spruce [*Picea mariana* (Mill.) B.S.P.]. (2½×.)

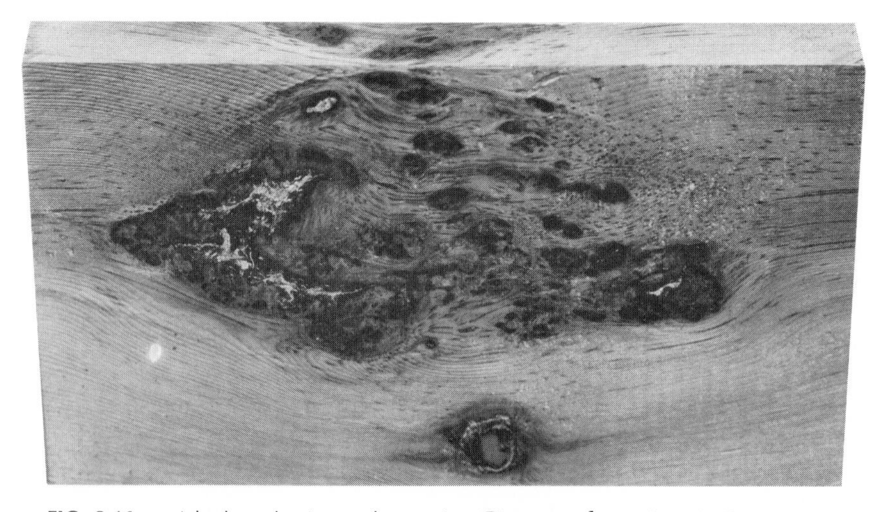

FIG. 8-16 A bark pocket in ponderosa pine (*Pinus ponderosa* Laws.). (*By permission from H. W. Eades, British Columbia Softwoods, Their Decay and Natural Defects, Can. Forest Serv. Bull.* 80, 1932.)

VIII. BARK POCKETS

These are small patches of bark that are embedded in wood (Fig. 8-16). They apparently develop from some injury to the tree, resulting in the death of a small area of the cambium; meanwhile, the surrounding tissue continues to function, and a new cambium forms over the gap in the inner bark, thus embedding a portion of the bark in the wood. Not infrequently, as stated before, portions of bark are found in pitch pockets.

Bark pockets frequently arise as a result of bird pecks (work of sapsuckers, etc.) and from injuries inflicted on the tree by insects (bark beetles, etc.). The shallow tunnels of certain insects may become partly overgrown with bark and subsequently embedded in the wood by the continued activity of the cambium. Bark pockets of this sort, generally containing an appreciable amount of resin, are very common in western hemlock; in this species, this defect is known as *black check* or *black streaks.*

SELECTED REFERENCES

1. Bannan, M. W.: Spiral Grain and Anticlinal Divisions in the Cambium of Conifers, *Can. J. Botany,* **44:**1515–1538 (1966).
2. Barefoot, A. C., Jr.: Influence of Cellulose, Lignin, and Density on Toughness of Yellow Poplar, *Forest Prod. J.,* **15**(1):46–49 (1965).
3. Berlyn, G. P.: Factors Affecting the Incidence of Reaction Tissue in *Populus deltoides* Bartr., *Iowa State J. Sci.,* **35**(3):367–424 (1961).
4. Bland, D. E.: The Chemistry of Reaction Wood. Part III: The Milled Wood Lignins of *Eucalyptus goniocalyx* and *Pinus radiata, Holzforschung,* **15**(4):102–106 (1961).
5. Boyd, J. D.: Tree Growth Stresses, II, The Development of Shakes and Other Visual Failures in Timber, *Australian J. Appl. Sci.,* **1:**296–312 (1950).
6. Busse, W.: Frost-, Ring-, und Kernrisse, *Forstwiss. Zentr.,* **32:**74–84 (1910).
7. Casperson, G.: Über die Bildung von Zellwanden bei Laubholzern, 4, Untersuchungen an Eiche (*Quercus robur* L.), *Holzforschung,* **21**(1):1–6 (1967).
8. ——— and A. Zinsser: On the Cell Wall Formation in Reaction Wood. Part III: Formation of Cavities in Compression Wood of *Pinus sylvestris* L., *Holz Roh-Werkstoff,* **23**(2): 49–55 (1965).
9. Chalk, L.: Wood Anatomy, *Advances Sci.,* **18**(75):460–463 (1962).
10. Chow, K. Y.: A Comparative Study of the Structure and Chemical Composition of Tension Wood and Normal Wood in Beech (*Fagus sylvatica* L.), *Forestry,* **20:**62–77 (1946).
11. Clarke, S. H.: "The Growth, Structure and Properties of Wood," Forest Prod. Res., Spec. Rept. 5, London, 1939.
12. Core, H. A., W. A. Côté, Jr., and A. C. Day: Characteristics of Compression Wood in Some Native Conifers, *Forest Prod. J.,* **11**(8):356–362 (1961).
13. Côté, W. A., Jr., and A. C. Day: The G Layer in Gelatinous Fibers–Electron Microscope Studies, *Forest Prod. J.,* **12**(7):333–335 (1962).
14. ———, ———, and T. E. Timell: Studies on Compression Wood, V, Nature of the Compression Wood Formed in the Early Springwood of Conifers, *Holzforschung,* **21**(6): 180–185 (1967).
15. ———, ———, and ———: Studies on Compression Wood. Part VII: Distribution of Lignin in Normal and Compression Wood of Tamarack [*Larix laricina* (DuRoi) K. Koch], *Wood Sci. & Technol.,* **2**(1):13–37 (1968).

16. ———, N. P. Kutscha, and T. E. Timell: Studies on Compression Wood, VIII, Formation of Cavities in Compression Wood Tracheids of *Abies balsamea* (L.) Mill., *Holzforschung,* **22**(4):138–144 (1968).
17. ———, B. W. Simson, and T. E. Timell: Studies on Compression Wood, II, The Chemical Composition of Wood and Bark from Normal and Compression Regions of Fifteen Species of Gymnosperms, *Svensk Papperstidning,* **69**(17):547–558 (1966).
18. ———, T. E. Timell, and R. A. Zabel: Studies on Compression Wood, I, Distribution of Lignin in Compression Wood of Red Spruce (*Picea rubens* Sarg.), *Holz Roh-Werkstoff,* **24**(10):432–438 (1966).
19. Dadswell, H. E., and A. B. Wardrop: What Is Reaction Wood? *Australian Forestry,* **13**(1):22–33 (1949).
20. Dinwoodie, J. M.: Failure in Timber. Part I: Microscopic Changes in Cell-wall Structure Associated with Compression Failures, *J. Inst. Wood Sci.,* **21**:37–53 (1968).
21. Doerksen, A. H., and R. G. Mitchell: Effects of the Balsam Woolly Aphid upon Wood Anatomy of Some Western True Firs, *Forest Sci.,* **11**(2):181–188 (1965).
22. Foulger, A. N.: Effect of Aphid Infestation on Properties of Grand Fir, *Forest Prod. J.,* **18**(1):43–47 (1968).
23. Hallock, H.: Growth Stresses and Lumber Warp in Loblolly Pine, *Forest Prod. J.,* **16**(2): 48–52 (1966).
24. Harris, J. M., and B. A. Meylan: The Influence of Microfibril Angle on Longitudinal and Tangential Shrinkage in *Pinus radiata, Holzforschung,* **19**(5):144–153 (1965).
25. Haught, E.: "Further Study on Compression Wood in Loblolly Pine," 2d Report Forest Tree Improvement Program, School of Forestry, North Carolina State College, Raleigh, 1958.
26. Hejnowicz, Z.: Some Observations on the Mechanism of Orientation Movement of Woody Stems, *Am. J. Bot.,* **54**(6):684–689 (1967).
27. ———: "Changes in Anatomy and Physiology of the Cambium as Related to Spiral Grain Development," Proc. 14th Congr. Int. Union Forest Res. Organ., IX (22/41):352–362, Munich (1967).
28. Hillis, W. E., F. R. Humphreys, R. K. Bamber, and A. Carl: Factors Influencing the Formation of Phloem and Heartwood Polyphenols. Part II: The Availability of Stored and Translocated Carbohydrates, *Holzforschung,* **16**(4):114–121 (1962).
29. Hoster, H., and W. Liese: Über das Vorkommen von Reaktionsgewebe in Wurzel und Asten der Dikotyledonen, *Holzforschung,* **20**(3):80–90 (1966).
30. Hughes, F. E.: Tension Wood, a Review of Literature. Part I: *Forestry Abstr.,* **26**(1):1–9 (1965); Part II: *Forestry Abstr.,* **26**(2):179–186 (1965).
31. Jacobs, M. R.: "The Growth Stresses of Woody Stems," Commonwealth Forestry Bureau, Bull. 28, 1945.
32. ———: "Stresses and Strains in Tree Trunks as They Grow in Length and Width," Commonwealth of Australia, Forestry and Timber Bureau, Leaflet No. 96, Canberra, 1965.
33. Jagels, R.: Gelatinous Fibers in the Roots of Quaking Aspen, *Forest Sci.,* **9**(4):440–443 (1963).
34. Jutte, S. M., and J. Isings: The Determination of Tension Wood in Ash with the Aid of the Phase Contrast Microscope, *Experientia,* **11**:386 (1955).
35. Kaeiser, M., and S. G. Boyce: The Relationship of Gelatinous Fibers to Wood Structure in Eastern Cottonwood (*Populus deltoides*), *Amer. J. Bot.,* **52**(7):711–715 (1965).
36. Keith, C. T., and W. A. Côté, Jr.: Microscopic Characteristics of Slip Planes and Compression Failures in Wood Cell Wall, *Forest Prod. J.,* **18**(3):67–74 (1968).
37. Koehler, A.: A Method of Studying Knot Formation, *J. Forestry,* **34**(12):1062–1063 (1936).
38. Kubler, H.: Studien über Wachstumsspannungen des Holzes. Erste Mitteilung: Die Ursache der Wachstumspannungen und die Spannungen quer zur Faserrichtung, *Holz Roh-Werkstoff,* **17**(1):1–9 (1959).
39. Lee, C. L.: Crystallinity of Wood Cellulose Fibers Studied by X-ray Methods, *Forest Prod. J.,* **11**(2):108–112 (1961).

40. Low, A. J.: Compression Wood in Conifers—a Review of Literature, *Forestry Abstr.*, part 1, **25**(3):1–13; part 2, **25**(4):1–7 (1964).
41. Lowery, D. P., and E. C. O. Erickson: The Effect of Spiral Grain on Pole Twist and Bending Strength, Intermountain Forest and Range Expt. Sta., Res. Paper INT-35, U.S. Forest Service, 1967.
42. McGinnes, E. A., Jr.: Extent of Shake in Black Walnut, *Forest Prod. J.*, **18**(5):80–82 (1968).
43. Marra, A.: Characteristics of Tension Wood in Hard Maple, *Acer saccharum* Marsh., M.S. thesis, Department of Wood Technology, New York State College of Forestry, Syracuse, N.Y., 1942.
44. Meyer, R. W., and L. Leney: Shake in Coniferous Wood—an Anatomical Study, *Forest Prod. J.*, **18**(2):51–56 (1968).
45. Norberg, H., and H. Meier: Physical and Chemical Properties of the Gelatinous Layer in Tension Wood Fibers of Aspen (*Populus tremula* L.), *Holzforschung*, **20**(6):174–178 (1966).
46. Olinmaa, P. J.: Reaktiopuutukimuksia (Study of Reaction Wood), *Acta Forest. Fenn.*, **72**(1):1–54 (1961).
47. Onaka, F.: Studies on Compression and Tension-Wood, *Wood Res. Bull.* 1, Wood Research Institute, Kyoto University, Kyoto, Japan, 1949. (Also as Translation 93 of Department of the Secretary of State of Canada, 1956.)
48. Patel, R. N.: On the Occurrence of Gelatinous Fibers with Special Reference to Root Wood, *J. Inst. Wood Sci.*, **12**:67–80 (1964).
49. Perem, E.: The Structure and Properties of Reaction Wood Formed in Trees Infested by Balsam Woolly Aphid, Forest Products, Res. Lab., Information Report OP-X-3, Ottawa, Canada, 1965.
50. Rhoads, A. S.: The Formation and Pathological Anatomy of Frost Rings in Conifers Injured by Late Frosts, *U.S. Dept. Agr. Bull.* 1131, 1923.
51. Schwerin, G.: The Chemistry of Reaction Wood. Part II: The Polysaccharides of *Eucalyptus goniocalyx* and *Pinus radiata*, *Holzforschung*, **12**(2):43–48 (1958).
52. Scurfield, G.: The Ultrastructure of Reaction Wood Differentiation, *Holzforschung*, **21**(1):6–13 (1967).
53. Skolmen, R. G., and C. C. Gerhards: Brittleheart in *Eucalyptus robusta* grown in Hawaii, *Forest Prod. J.*, **14**(12):549–554 (1964).
54. Spurr, S. H., and M. J. Hyvarinen: Compression Wood in Conifers as a Morphogenetic Phenomenon, *Botan. Rev.*, **20**(9):551–560 (1954).
55. Taylor, F. W.: Specific Gravity Differences within and among Yellow-Poplar Trees, *Forest Prod. J.*, **18**(3):75–81 (1968).
56. Thunell, B., and G. Wallin: The Effect of Knots on the Cutting Forces in Frame Sawing, *Pap. ja Puu*, **49**(2):71–73 (1967).
57. Vité, J. P.: Water Conduction and Spiral Grain: Causes and Effect, Proc. 14th Congr. Intern. Union Forest Res. Organ., sect. 22/41:338–351, Munich, 1967.
58. Wahlgren, H. E.: Effect of Tension Wood in a Leaning Eastern Cottonwood, *Forest Prod. J.*, **7**(6):214–219 (1957).
59. Wardrop, A. B.: The Reaction Anatomy of Arborescent Angiosperms, in M. H. Zimmermann (ed.), "The Formation of Wood in Forest Trees," Academic Press, Inc., New York, 1964.
60. ——— and H. E. Dadswell: The Nature of Reaction Wood, II, The Cell Wall Organization of Compression Wood Tracheids, *Australian J. Sci. Res. Ser.* B, *Biol. Sci.*, **3**(1):1–13 (1950).
61. ———, and ———: The Nature of Reaction Wood, IV, Variations in Cell Wall Organization of Tension Wood Fibers, *Australian J. Botany*, **3**(2):177–189 (1955).
62. ——— and ———: The Structure and Properties of Tension Wood, *Holzforschung*, **9**(4):97–103 (1955).
63. ——— and G. W. Davies: The Nature of Reaction Wood, VIII, The Structure and Differentiation of Compression Wood, *Australian J. Botany*, **12**(1):24–38 (1964).

64. Weddell, E.: Influence of Interlocked Grain on the Bending Strength of Timber with Particular Reference to Utile and Greenheart, *J. Inst. Wood Sci.,* **7**:56–72 (1961).
65. Wedel, K. W. von, B. J. Zobel, and C. J. A. Shelbourne: Prevalence and Effects of Knots in Young Loblolly Pine, *Forest Prod. J.,* **18**(9):97–103 (1968).
66. Wergin, W., and G. Casperson: Über Entstehung und Aufbau von Reaktionsholzzellen. 2. Mitt. Morphologie der Druckholzzellen von *Taxus baccata* L., *Holzforschung,* **15**(2): 44–49 (1961).
67. Westing, A. H.: Formation and Function of Compression Wood in Gymnosperms, I, *Botan. Rev.,* **31**(3):381–480 (1965).
68. ———: Formation and Function of Compression Wood in Gymnosperms, II, *Botan. Rev.,* **34**(1):51–78 (1968).
69. White, D. J. B.: Tension Wood in Sassafras, *J. Inst. Wood Sci.,* **10**:74–80 (1962).
70. ——— and A. W. Robards: Gelatinous Fibers in Ash (*Fraxinus excelsior* L.), *Nature,* **205**(4973):818 (1965).
71. Woodfin, R. O., Jr.: Spiral Grain Patterns in Coast Douglas-fir, *Forest Prod. J.,* **19**(1): 53–60 (1969).
72. Wooten, T. E., A. C. Barefoot, and D. D. Nicholas: The Longitudinal Shrinkage of Compression Wood, *Holzforschung,* **21**(6):168–171 (1967).
73. Zimmermann, M. H., A. B. Wardrop, and P. B. Tomlinson: Tension Wood in Aerial Roots of *Fiscus benjamina* L., *Wood Sci. & Technol.,* **2**(2):95–104 (1968).

9

Defects Due to Seasoning and Machining

In this chapter are described various defects which, though not present in the standing trees, may occur in the wood after a tree is cut. Some of them develop because of the anisotropic nature of wood and the hygroscopicity of the wood substance. Since these failures are associated mainly with changes in the moisture content in the wood they are called *seasoning* or *drying defects;* they include checks, warping, collapse, casehardening, and honeycombing. Other wood defects result from damage during machining; they may or may not be associated with changes in moisture content. Machining damage, other than that due to sheer carelessness, includes planer splitting; breakage of knots; fallen knots resulting in knotholes; chipped, torn, and raised grain; and machine burns.

I. SEASONING DEFECTS

As wood comes from the tree it contains appreciable amounts of water, frequently as much as 100 to 200 percent in terms of its ovendry weight. Wood begins to lose moisture as soon as it is exposed to atmospheric conditions.

When green wood dries, the first water to leave it is the free moisture in the cavities (lumina) of the cells. No normal dimensional changes accompany this stage in the drying process. When, however, the drying is continued below the fiber saturation point, shrinkage takes place.

Unfortunately, as wood continues to dry below the fiber saturation point, the shrinkage is not equal in all directions; this is because of the anisotropic nature of wood. The shrinkage along the grain, i.e., longitudinally, is negligible, except in reaction wood (see page 298) and in the material with diagonal and spiral grain (see page 277), in which it can be appreciable. Across the grain wood shrinks considerably and about twice as much in the tangential as in the radial direction. This inequality in shrinkage in three directions at right angles to one another sets up strains which are unavoidable and which, if they become too great, cause actual fractures in the wood tissue.

Most seasoning defects can be minimized by careful air seasoning, and they can be largely eliminated by proper control of drying conditions in the kiln. It is even possible, within limits, to correct seasoning defects by further treatment, once they have occurred. In the main, the proper techniques of drying attempt to minimize the damage arising from the stresses that necessarily arise in drying wood, by keeping them below the point at which failures would develop.

A. How Wood Dries

Wood is a hygroscopic material. Because of this it loses liquid water and water vapor when it is exposed to a surrounding medium in which the relative vapor pressure is less than that within the wood itself. Water absorption takes place when these conditions are reversed. This exchange of moisture between the wood and the surrounding medium continues until a state of balance, called the ***equilibrium moisture content,*** is reached.

The rate at which moisture moves in the wood depends on (1) the relative humidity of the surrounding air, (2) the steepness of the moisture gradient, and (3) the temperature of the wood. Of these, relative humidity, i.e., the difference between the relative vapor pressure of the air and that in the wood, is of utmost importance. Low relative humidity increases the capillary flow of moisture from the wood and stimulates diffusion of water by lowering the moisture content at the surface. This results in steepening of the moisture gradient from the surface to the interior of the piece, accompanied by development of stresses traceable to the strains arising from dimensional changes.

Schniewind[19] suggests that the drying stresses in wood could be classified into three orders on the basis of the scale on which these stresses operate. The first-order stresses occur in the individual cells because of the unequal shrinkage potential of the cell wall layers and hydrostatic tension in the cell cavities. The second-order stresses occur from unequal shrinkage potential of the various tissues in the wood. The third-order stresses occur in larger pieces of wood as a result of the moisture gradient that develops in normal drying and from the gross anisotropic nature of wood. When these combined drying stresses exceed the strength properties of wood, various drying defects occur. If the relative humidity is too low during the early stages of drying of wood with high moisture content, excessive end and surface checks may develop.

A rise in the temperature of wood also accelerates the rate at which moisture moves from the regions of higher moisture content to those with lower. When temperature is too high, collapse, honeycombing, and reduction in strength properties may result.

B. Checks

Checks are ruptures in wood along the grain which, with the exception of heart checks (see page 283), develop during seasoning (Fig. 9-1*a*). Defects of this type arise for two reasons: (1) because of a difference in radial and tangential shrinkage, resulting in stresses of sufficient magnitude to cause the failure of the wood along the planes of greatest weakness, usually at the juncture of longitudinal tissue with the rays; (2) because of differences in shrinkage of the tissue and development of stresses of different magnitude in adjacent portions of the wood, occasioned by varying moisture content.

Seasoning checks are of two types, *end checks* and *surface checks*. As the terms imply, the former are confined to the ends of the piece. Such checks normally follow the rays, although occasionally they may extend along the growth rings; unless their development is arrested, end checks may develop into extensive *end splits* (Fig. 9-2). Surface checks result from the separation of the thinner-walled early-wood cells; they also follow the rays and therefore are confined largely to the tangential surfaces. Surface checks extend into the wood for varying distances. Schniewind suggests[18] that at room temperature surface checks develop within the rays when, as a result of dimensional changes caused by drying, the tensile stresses within the rays exceed the compressive stresses in the adjacent longitudinal tissues. At elevated temperatures additional stresses, arising from thermal expansion, also occur. When these exceed those caused by loss of moisture, the checks form just outside the ray.

As drying progresses deeper into the wood, many of the checks (especially those of the surface type) close; but of course such ruptures are never bridged by wound tissue, and the checks may reopen if the wood surface is allowed to undergo further moisture-content changes. In woods with large rays (oaks, beech) the closing of the surface checks may lead to honeycombing (see page 329). The formation of surface checks in certain stages of drying may also indicate the presence of drying conditions favorable to the development of casehardening (see page 325).

Superficial surface checks may be removed by planing off the affected surface. Deeper checks may affect the finishing characteristics of the wood and considerably decrease the strength of wood, particularly in shear. The latter effect is due to reduction in the area of the surface that can offer resistance to shear.

In air seasoning both types of checking can be minimized by following procedures that tend to promote the less-rapid but uniform evaporation of moisture. These include reduction of circulation in the lumber pile in the case of surface checking, and shading the ends of boards and timbers or painting them with moisture-resistant

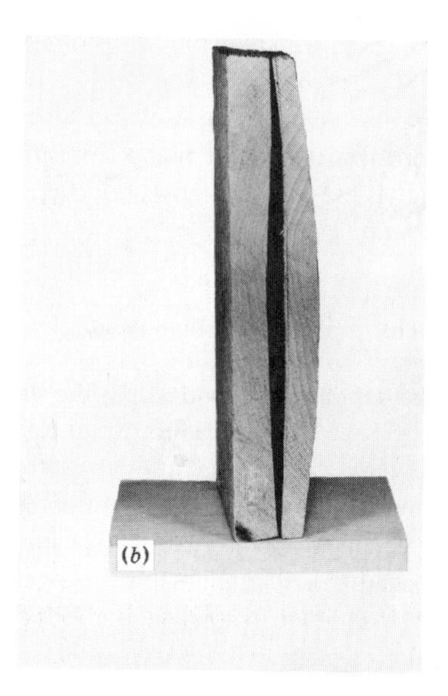

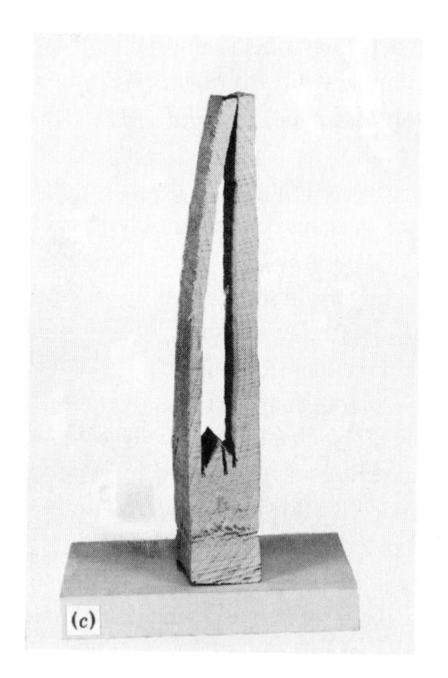

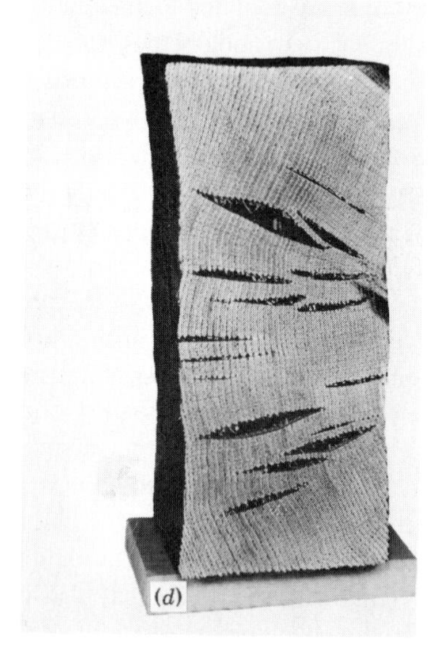

FIG. 9-1 Defects due to seasoning.
(*a*) Surface checking on the surface of a board of oak (*Quercus* sp.).
(*b*) Section of resawn casehardened hardwood board, showing cupping.
(*c*) Test sample of a hardwood board showing late stages of casehardening; at this stage the core of the board is under tension, and the shell under compression.
(*d*) Section of a board of oak (*Quercus* sp.), showing honeycombing.
(*Photographs by permission, from H. L. Henderson, "The Air Seasoning and Kiln Drying of Wood," J. B. Lyons Co.*, 1936.)

FIG. 9-2 End checks and splits in oak planks. (*U.S. Forest Service, Forest Products Laboratory photograph.*)

coatings in the case of end checking. In kiln drying, checking is indicative of uneven or too-rapid drying and can be avoided by maintaining adequate circulation and proper temperature and humidity conditions.

Another type of checking frequently develops in wood but cannot be detected except by resorting to a compound microscope; these are known as *microscopic checks* (Fig. 8-7). Checking of this sort occurs between the fibrils in the secondary wall. The slope of such microscopic checks is therefore indicative of the alignment of the fibril helixes. Spiral checking frequently simulates spiral thickening of the type that characterizes the longitudinal tracheids of Douglas-fir; the slope of such checks is, however, steeper than the helixes formed as a result of spiral thickening.

Spiral checks frequently occur in the dense late wood of such conifers as southern pine and also in the tracheids of compression wood and the fibers of tension wood. In these two last instances they are not formed as the result of drying following the conversion of the wood, but develop in the living tree.

C. Checked and Loose Knots

Knots, especially the intergrown kind, frequently develop end checks and splits during the initial stages of drying. This is because of the differences in shrinkage parallel and across the growth rings within the knot, and more rapid drying of the end grain in the knot wood, as compared with the longitudinal grain of the board in which it is embedded (Fig. 9-3*b*).

Encased knots generally become loose during drying, because they are not

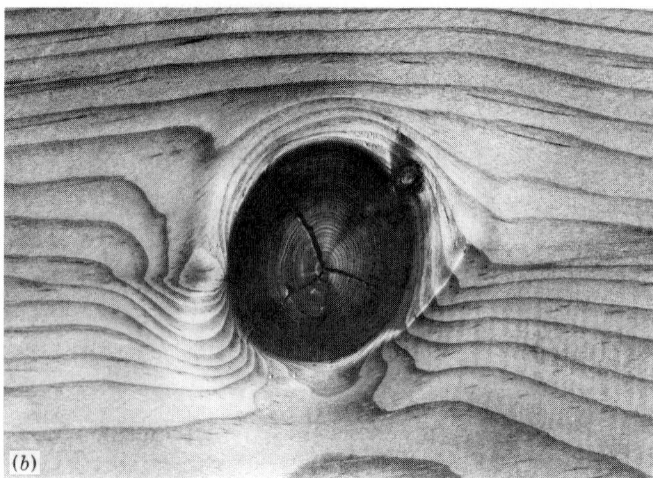

FIG. 9-3 (*a*) Loose knot in southern yellow pine. Note chipped grain in the board.
(*b*) Checked knot in sugar pine.
(*U.S. Forest Service, Forest Products Laboratory photograph.*)

hysically attached to the surrounding wood, and their wood usually is denser and ence shrinks more than the surrounding tissue. Furthermore, knots shrink onsiderably in the cross section, while the board in which they are located shrinks ppreciably only in width, but little in length. As a consequence, knots become maller than the knothole and easily fall out of the lumber in machining and in andling (Fig. 9-3*a*).

D. Warping (Fig. 9-4)

n a broad sense, the term *warping* is used to describe any distortion from the true lane that may occur in a piece of wood during seasoning. Several different types of

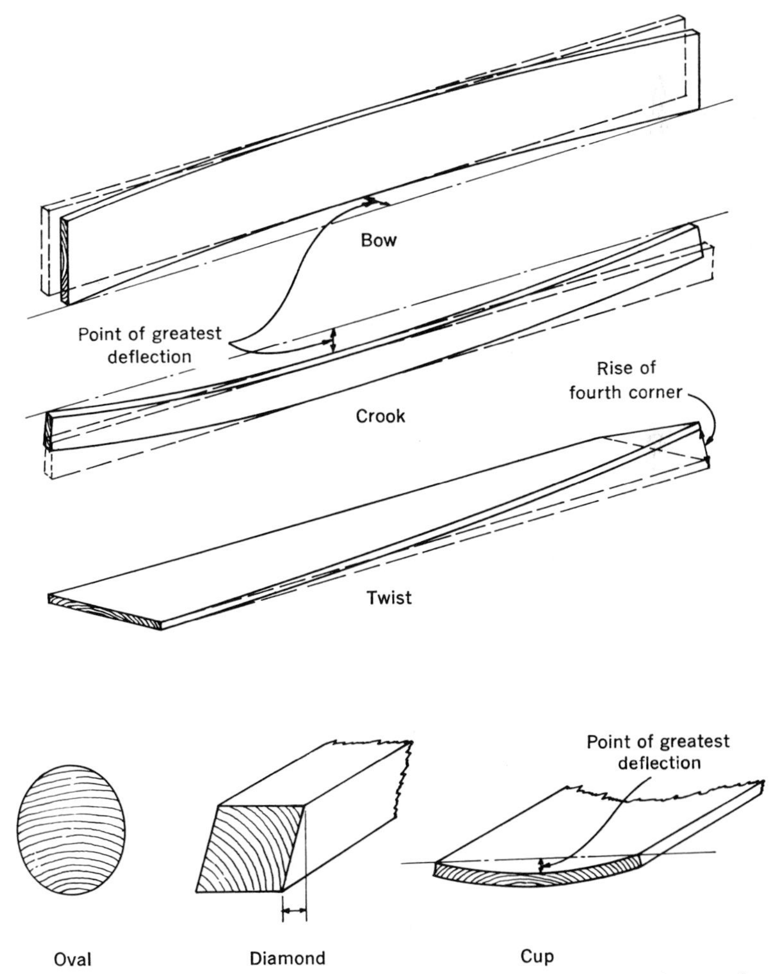

FIG. 9-4 Various types of warp. (*U.S. Forest Service, Forest Products Laboratory drawing.*)

warping are recognized and are easily detected by the appearance of the deformed piece, each arising from a different cause. The principal types of warping are bowing, crooking, cupping, twisting, and diamonding.

1. *Bowing* is defined as longitudinal curvature, flatwise, from a straight line drawn from end to end of the piece.

2. *Crooking,* by comparison, is longitudinal curvature, edgewise, from a straight line, drawn from end to end of the piece.

Both the above-mentioned types of warping arise as a consequence of differences in longitudinal shrinkage, the first from a discrepancy in the shortening on the faces of a board, the second from a difference in shrinkages on the edges of a board.

Bowing and crooking, for obvious reasons, are very common defects in lumber in which the grain is irregular, e.g., in curly-grained stock or where boards are so sawed that the grain on one side or edge is canted more than on the other. Straight-grained pieces of wood are less subject to these defects, although bowing occasionally occurs in what appears to be perfectly straight-grained stock and then invariably toward the bark side. The following reasons have been offered to explain this: (*a*) tension along the grain which may have existed in the outer portion of the tree producing the timber, (*b*) in conifers greater longitudinal shrinkage of the wood toward the periphery of the trunk than of the core, explainable by the fact that the former was less dense, and (*c*) the presence of reaction wood (compression or tension wood). In pieces containing both reaction and normal wood, the excessive longitudinal shrinkage of the former is retarded by the lower longitudinal shrinkage of the latter. This sets up stresses, which frequently are of sufficient magnitude to cause the piece to bow and twist.

3. *Cupping* signifies the curving of the face of a plank or a board so that it assumes a troughlike shape, the edges meanwhile remaining approximately parallel to each other. There are two primary causes for cupping: (*a*) the more rapid drying of one face of a board than of the other, and (*b*) the discrepancy between radial and tangential shrinkages which causes one side of the piece to shrink more than the other.

In the first instance this defect may arise when one surface of a board is in contact with, or near the surface of, the ground or is in touch with another object and hence cannot dry at the normal rate; if the other surface is freely exposed to the sun or is merely better aerated, cupping may result on the side that is exposed. Curvatures on lumber of this kind are usually only temporary and tend to correct themselves as the wood becomes uniformly dry.

Cupping arising as a result of the discrepancy existing between tangential and radial shrinkages across the grain frequently develops in plain-sawn lumber in which one face of the wood approaches the quarter-sawn condition and the other is flat-grained. The board, of course, shrinks less on the quarter face, causing cupping on the opposite side, i.e., on the face closest to the bark. The greater the difference between tangential and radial shrinkage, the greater the degree of cup. Because of this a flat-sawn board cut near the bark cups less than a similar board cut near the pith.

Cupping may also result when casehardened lumber is resawn or is dressed more n one side than on the other. It may likewise occur in resawing lumber that is much etter in the interior than on the outside; such stock will usually straighten out or ven curve in the opposite direction as the exposed interior dries. Finally, any piece wood which has been painted or varnished on one side and not on the other or hich has received a heavier coat of finish on one side, if exposed to sudden nanges in atmospheric conditions, may cup. In such instances, the better-protected urface loses and gains moisture less quickly than the other face, and cupping may evelop on the less protected side, if moisture is lost, and on the opposite, more otected surface, if moisture is gained.

4. *Twisting* indicates a condition in which one corner of a piece of wood twists ut of the plane of the other three; i.e., it denotes a situation in which the four corners f a surface are no longer in the same plane. Twisting is usually a concomitant of oss or irregular grain; but, as in bowing and crooking, it may also develop in a raight-grained piece of wood as a result of uneven drying or because of inherent resses in the standing tree.

5. *Diamonding* describes uneven shrinkage, usually developing in squares in hich the growth increments extend diagonally so that the faces of the piece are either flat- nor edge-grained. Such pieces, although square when green, become nomboidal or diamond-shaped in cross section on drying. The reason for diamond-ng is, of course, the discrepancy existing between radial and tangential shrinkage.

E. Casehardening[12,13]

asehardening is a term applied to dry lumber with uniform moisture content but haracterized by the presence of residual stresses, tension in the interior of the piece core), and compression in the outer layers of cells (shell). These stresses are due to hrinkage strains, which are inevitable during drying of wood but whose magnitude epends to a large extent on the severity of the prevailing drying conditions.

Contrary to the implied meaning of this term, the surface of a casehardened piece f wood is softer than the core, not harder. This is because the cells in the surface ayers of the casehardened wood dry in a partially stretched condition. As a result the vood in the shell is somewhat less dense than it would have been if a full amount of hrinkage were allowed to occur.

The sequence in the development of casehardening stresses may be stated briefly s follows. When wood is seasoned, the outer portion of the piece dries below the ber saturation point first but is restrained from shrinking fully, because the adjacent nner layers, being wetter, lag in shrinking. In consequence, the wood in the surface ayers is stretched, i.e., put under tension, while that of the core is compressed. These onditions are normal during the initial stages of drying. When the drying conditions re not too severe, the tensile stresses are relieved as soon as the adjacent inner ayers dry and shrink, thus allowing the outer layers to shrink further. On the other and, if drying proceeds at too rapid a rate, the differences in moisture content, and ence in shrinkage, between the shell and the core may become sufficient to cause a "permanent set" of the outer layers, i.e., to allow them to dry in a partly stretched condition, without attaining full shrinkage. If stresses that develop exceed th maximum strength of the wood perpendicular to the grain, surface checks will forr at this stage of drying.

As the drying of such a piece progresses, more moisture is lost from the core, an the tissue composing this tends to shrink further. Since, however, the shell has "set but is organically connected with the core, it now prevents the normal shrinkage c the core. Hence, the stresses are reversed; i.e., the shell is now in compression an the core is in tension. Honeycombing (see page 329), a type of defect in whic numerous pockets form in the wood, often develops as a result of the stresses set u during this second and final stage of casehardening.

Drying stresses may be relieved in a kiln by subjecting the lumber to a condition ing treatment at an elevated temperature above that used in the final stages of dryin and at a relative humidity chosen so as to create an equilibrium moisture condition to 4 percent higher than the average moisture content of the stock.* These condition should be maintained until stresses are relieved, but care must be taken not to con tinue this treatment too long, since otherwise reverse casehardening (see F section may result. Casehardening also can be removed by the application of elevate temperatures alone, toward the end of the drying.[1] In softwoods, but not in hard woods, the stresses may be relieved by prolonged storage at air temperatures.

When the residual drying stresses are not relieved at the completion of drying or b storage, the lumber *is said to be casehardened.* Whether casehardening i considered a defect depends on the final use of the seasoned material. When lumbe is used without further machining, casehardening stresses do no harm, provided tha no honeycombing or severe surface checking has taken place. Since casehardene lumber dries without attaining the full amount of shrinkage, the resulting additiona volume of wood may be considered an advantage. However, if severel casehardened lumber is subjected to further fabrication, a number of difficulties ma be experienced. The most common of these are end checking, planer splitting, an warping.

End checks may develop when a freshly crosscut casehardened board is expose to low relative humidity. When this happens the stresses brought about by en drying, coupled with the stresses already present in the core, may exceed the strengt of the wood, causing checking. Splits may occur on the surface of relatively dr casehardened boards during planing. This is because of the internal stresses in th board, combined with the forces applied by the knives in the cutterhead.

The most objectionable feature of casehardened lumber is its tendency to cup an warp (Fig. 9-1*b*) when stresses are unbalanced in a board by resawing, planing mor heavily on one side, sanding, or by any other machining operation that remove more wood from one side of the casehardened board than from the other. An subsequent distortion of the stock is a source of trouble in further fabrication and i gluing.

* For detailed information on relieving casehardening, see Ref. 17.

F. Reverse Casehardening

Reverse casehardening is a condition that develops in lumber as a result of oversteaming, generally in an attempt to relieve casehardening. If, during conditioning at high humidity, the surface layers are permitted to absorb moisture considerably in excess of that necessary to even up the differences in the moisture content of the shell and of the core, the surface layers tend to swell. This tendency, however, is resisted by the drier interior and results in permanent compression of the outside layers. When, subsequently, the surface layers dry, they shrink more than normal wood, thus producing tension at the surface and compression in the interior of the piece. Reverse casehardening causes outward cupping on resawing; in extreme cases, the initial swelling of the surface layers may also cause honeycombing by developing internal tensile stresses that exceed the maximum tensile strength of the wood perpendicular to the grain.

G. Collapse

Collapse is a defect that sometimes develops when very wet heartwood of certain species is dried. It is usually evidenced by abnormal and irregular shrinkage. Collapse generally appears merely as excessive shrinkage (and is frequently so interpreted). In more severe cases, the sides of the lumber may cave in (Fig. 9-5), resulting in irregular depressions and elevations on the surfaces of a board and in internal honeycombing. Among native species, collapse is most frequent in redwood, western redcedar, baldcypress, redgum, bottomland oaks, and cottonwood; it also occasionally develops in other domestic woods, such as hickory and black walnut. Certain tropical woods, especially those produced by the Australian genus, *Eucalyptus*, are especially susceptible to this defect.

Unlike other seasoning defects, collapse begins before all the free water has disappeared from cell cavities, i.e., before normal shrinkage starts. It differs from normal shrinkage in that the latter is caused by the drawing together of the cell wall particles as the moisture dries out of the walls, while in collapse the walls of the cells are actually pulled together, causing the cells to buckle (collapse) in more or less extensive tracts throughout the wood (Fig. 9-5*b*).

Two theories attempt to explain the cause of collapse. According to one, it is produced by the hygroscopic tension exerted by water, as it is withdrawn from cell cavities, on wet and plastic cell walls.[21] The tension results from the evaporation of water from completely enclosed fiber cavities (not vessel elements that are perforate), through minute openings in the pit membranes. Such tension can be set up only when the fiber cavities are completely filled with water. An air bubble with a diameter exceeding that of the largest opening through which evaporation takes place would tend to relieve the stress on a cell by expanding under tension. The other theory[20] postulates that collapse occurs when the compressive stresses,

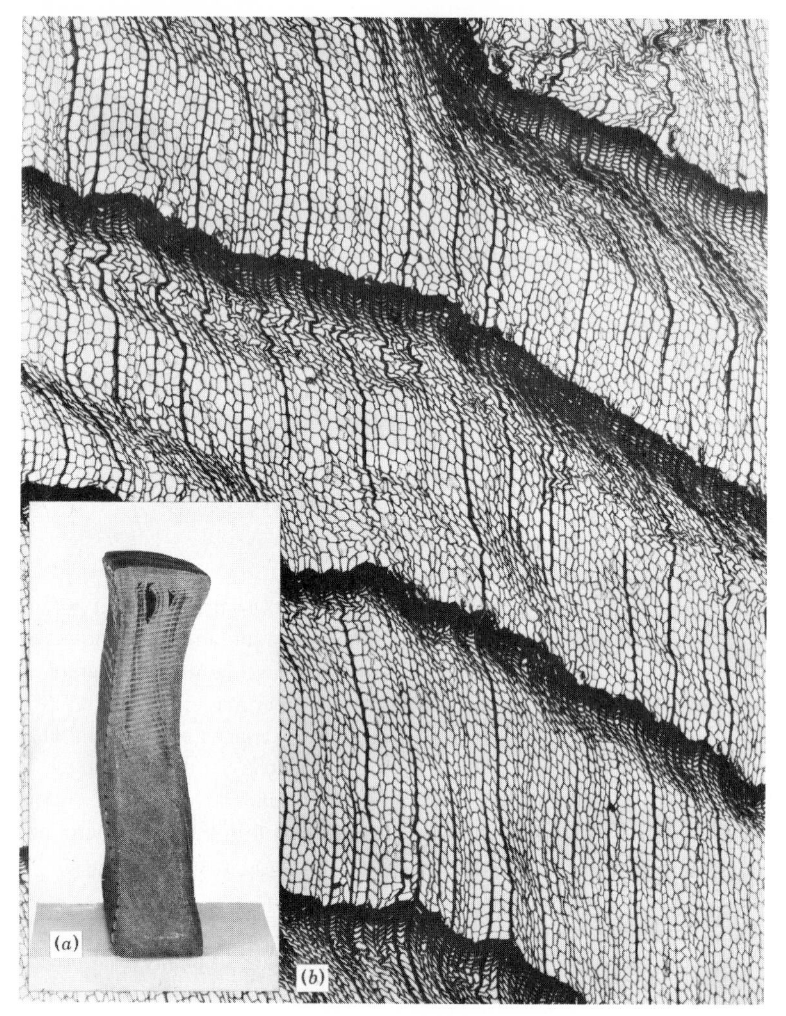

FIG. 9-5 Defects due to seasoning, collapse.
(*a*) End-grain view of a board of redwood [*Sequoia sempervirens* (D. Don) Endl.], showing collapse. (*By permission, from H. L. Henderson, "The Air Seasoning and Kiln Drying of Wood," J. B. Lyons Co.*, 1936.)
(*b*) Transverse section of the collapsed wood of western redcedar (*Thuja plicata* Donn). (20×.) (*U.S. Forest Service, Forest Products Laboratory photograph.*)

developed during drying, exceed the compressive strength of the wood; as a result, the fiber walls collapse into the fiber cavities.

Both these conditions, i.e., liquid tension in the cell cavities and compression drying stresses of considerable magnitude, occur early in the drying process. On the basis of experimental work with *Eucalyptus regnans* F.v.M., Kauman[8] concluded that liquid (hydrostatic) tension is the principal cause of collapse, but that the intensity of collapse is markedly affected by drying stresses. This led Ellwood and his associates[3] to propose that the substitution of water in wood with liquids having lower surface tension than water should reduce or eliminate collapse. Their experiments with the collapse-susceptible hardwoods proved this to be the case. The most satisfactory results were obtained with ethylene glycol monoethyl ether, methanol, and ethanol.

Extensive investigations in Australia[5] have led to the development of a method for reconditioning collapsed eucalyptus timber. It was found that the cells can be restored to their original shape, if the walls have not broken down completely, by remoistening the collapsed wood at a high temperature, after which the stock is redried. In this procedure, the lumber is first dried to about 10 to 15 percent of moisture; it is then subjected to a high-humidity treatment at about 180 to 200°F for several hours, to permit it to reabsorb 2 to 6 percent of moisture and to return to its original shape. The stock is then redried to its original moisture content, this time without the recurrence of collapse since (1) the redrying process was started below the fiber saturation point, i.e., without water in a free state being present in the wood or at least in a quantity sufficient to fill the cell cavities completely, and (2) the cell walls may have been rendered more pervious.

During the later stages of drying, collapse is frequently accompanied by honeycombing (described below). This occurs when the outer layers (shell) of a stick of wood dry below the fiber saturation point and set. Under these conditions, if the wetter core begins to collapse, internal checks may occur of a type to produce severe honeycombing.

H. Honeycombing

This defect, also called *hollow horning,* is traceable to internal checking and splitting, generally along the rays, occurring in some lumber as it dries (Fig. 9-1*d*). The splits generally do not extend all the way to the surface, although one form of honeycombing develops through the closing of surface checks, the cracks meanwhile deepening and broadening in the interior of the wood. Honeycombing invisible on the surface may not be apparent until the wood is machined.

The most common causes of honeycombing are the internal stresses that develop in casehardening and collapse. Honeycombing therefore can be minimized by avoiding excessively high dry-bulb temperatures until all the free moisture has been evaporated from the entire piece.

I. Ring Failures

These failures occur parallel to the growth ring and frequently show in several adjacent rings. Ring failures appear to be similar to shakes, but whereas the latter develop in the standing trees, ring failures occur in perfectly sound stock during drying. They may occur in the end grain in the initial stages of drying and then extend in depth and in length as drying progresses, or they may develop internally, especially in severely casehardened stock. Ring failures can be controlled by end coating the stock and by using initially lower temperatures and higher relative humidity.

J. Box-heart Splits

These splits originate in the wood surrounding the pith during initial stages of drying and extend for a considerable distance into the piece as drying progresses. Box-heart splits are due to the stresses set up by the difference in tangential and radial shrinkage of the wood around the pith, causing the wood to split; such splits are difficult to prevent.

II. MACHINING DEFECTS

A number of defects which may or may not be related to improper seasoning or storage, but which become evident during fabrication of wood, are described, for convenience of discussion, under the single title of *machining defects*.

A. Raised Grain

According to the American Lumber Standards, *raised grain* denotes the "roughened condition of the surface of dressed lumber in which the hard summerwood is raised above the softer springwood, but not torn from it." The surface of a board with this defect has a corrugated feel and appearance. In practice, however, the term raised grain is also extended to include the *surface fuzzing* that is traceable to the breaking or loosening of individual fibers and groups of fibers in sanding operations. These two types of defects are discussed separately.

1. RAISED GRAIN[11,16]

Raised grain (Fig. 9-6) occurs mainly in lumber machined at a moisture content of more than 12 percent. In plain-sawn lumber the corrugated appearance of the wood surface is due largely to the crushing of the hard late wood into the softer early wood beneath by the action of planer knives (Fig. 9-7). Mechanically, compressed wood tends to swell more than normal wood under the same conditions of moisture. When conditions favor the absorption of moisture, the late wood in a planed piece rises as the cells in the early wood beneath them gradually resume their original size.

Although raised grain of the type described above occurs in the lumber of all species, it is most common in softwoods such as southern pine and Douglas-fir in which the nature of the early wood and the late wood is quite different, and least pronounced in the species with little contrast between these tissues.

The corrugation of plain-sawn lumber is more common and more pronounced on

FIG. 9-6 The pattern of the raised grain showing through a coating of paint on a southern yellow pine board. (*U.S. Forest Service, Forest Products Laboratory photograph.*)

the pith side of a board. This is because of greater resistance to cutting offered by the late wood on the pith side than on the bark side caused by the curvature of the growth rings. As a result there is more crushing of late wood into early wood on the pith side.

In quarter-sawn lumber of species with pronounced late wood, the corrugation of the surface is due to the greater transverse (across the grain) dimensional changes in the late wood in comparison with those in the early wood. For the reason stated, the extent of the corrugation is in direct proportion to the changes in moisture content that occur after the lumber is dressed. If the lumber is planed while still at a relatively high moisture content and then allowed to dry, the late-wood bands shrink below the level of the early wood; if the opposite is true, i.e., if the quarter-sawn lumber is planed while dry and then permitted to absorb moisture, the late wood rises above the early wood. In either case the result is a corrugated surface.

The corrugation of the surface is increased by dull planer knives, and therefore its extent can be reduced by keeping planer knives sharp. Raised grain is also affected by the magnitude of the moisture-content changes that occur after the lumber is planed. The most distortion takes place in planed stock that picks up some moisture

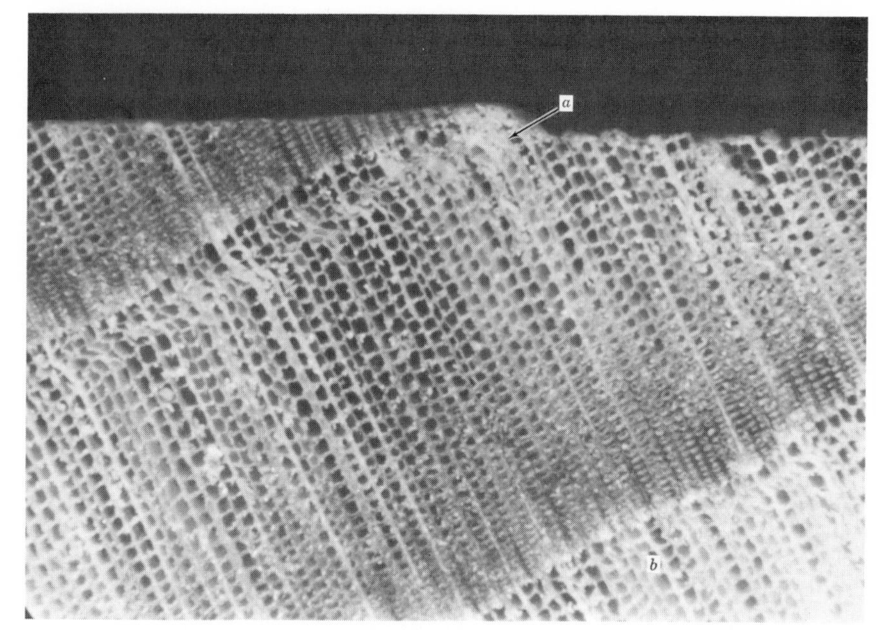

FIG. 9-7 Corrugation of the surface in a white fir board, due to the planer knives crushing the hard late wood into the softer early wood at (*a*). The early-wood cells at (*b*) in the growth ring below are not crushed. (*U.S. Forest Service, Forest Products Laboratory photograph.*)

and swells, because compressed wood swells more than the normal type; the least distortion occurs in lumber in which changes in the moisture content, subsequent to planing, are kept at a minimum.

2. LOOSENED GRAIN[16]

The term *loosened grain* refers to separation and curling of the tips of growth rings on the surface of flat-grained lumber (Fig. 9-8). This defect is due primarily to the pounding action of the planer knives, or to pressure in sanding sufficient to crush the early wood in one or more growth rings under a layer of late wood exposed on the face of a board. The crushing of early wood is the initial cause of this defect, but the actual separation of the growth increments near the surface is due to the shrinkage stresses that develop where the late wood of one increment adjoins the early wood of another. Such stresses arise because late wood tends to shrink more transversely and less longitudinally than early wood. The curling of the loosened splinters can be attributed to the fact that individual growth increments, were they entirely free, would tend to curve endwise in drying, with the concave side toward the pith.

The grain loosening may also develop through hard usage, as, for instance, in flat-sawn flooring, especially in that made from softwoods. The Indians and the early

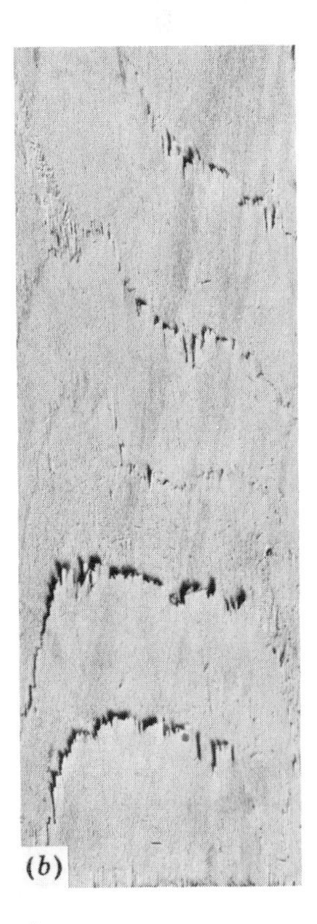

FIG. 9-8 (*a*) Loosened grain in southern pine flooring.
(*b*) Loosened grain in soft maple flooring.
[*By permission, from A. Koehler, More about Loosened Grain, Southern Lumberman,* **161:**171–173 (1940).]

settlers were well aware that pounding separates growth rings, and used this method to obtain black ash splints for basket weaving.

Loosened grain is objectionable because of the difficulties encountered in machining and painting surfaces with this defect. When the wood is in use, its objectionable features likewise are quite evident. For example, softwood floors with slivery surfaces cannot be finished satisfactorily and are difficult to mop.

3. FUZZY GRAIN

Surface fuzzing develops when individual fibers or small groups of fibers become loosened on the surface of a board in large numbers (Fig. 9-9). The injury to the cells is brought about in sanding or occasionally in planing, and fuzzing occurs when such cells swell under the action of swelling agents, such as stains, or as a result of the atmospheric humidity. Fuzzy grain is objectionable because of the difficulties experienced in obtaining a smooth surface when this condition prevails.

The hardwoods, as a group, fuzz more than the softwoods. Some species fuzz more than others, and considerable variation in the amount of fuzzing may be encountered in the same wood, depending on its characteristics and on working conditions.

FIG. 9-9 Fuzzy grain caused by projecting groups of tension-wood fibers on the sawn surface of a mahogany board. (*U.S. Forest Service, Forest Products Laboratory photograph.*)

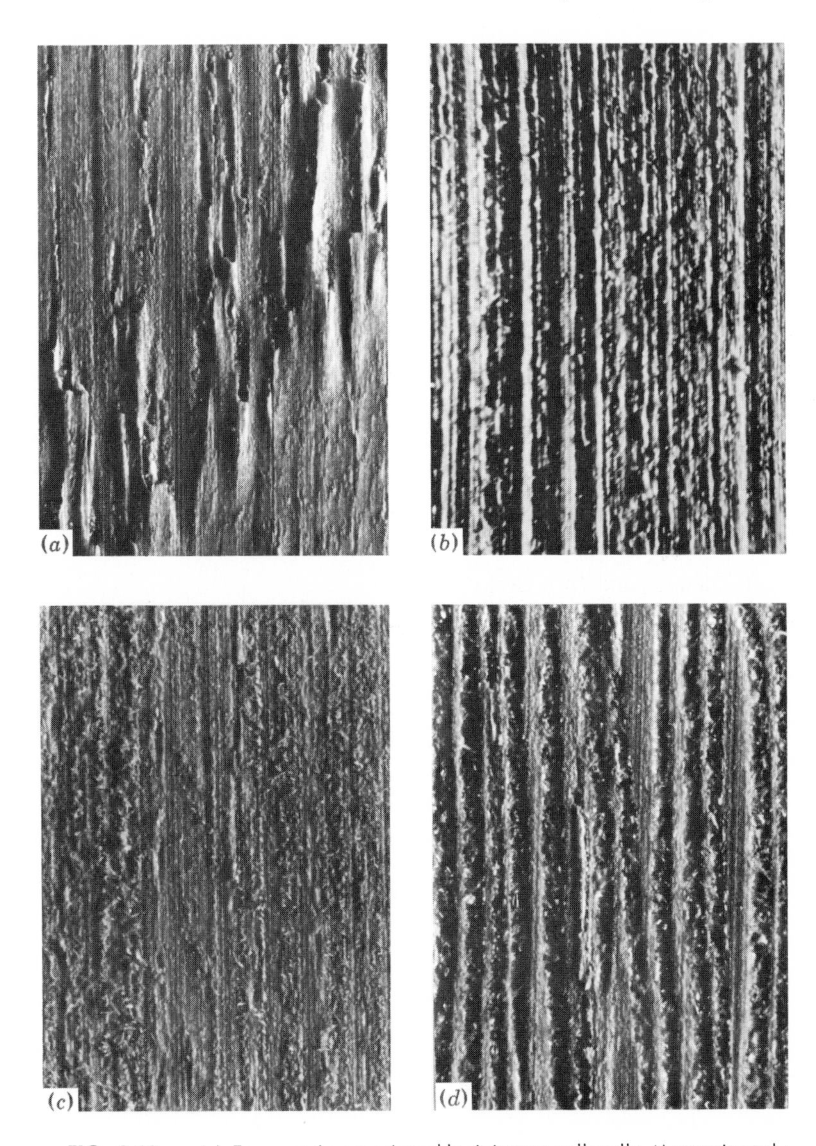

FIG. 9-10 (*a*) Fuzzy grain occasioned by injury to cell walls. (Approximately 4½×.)

(*b*) Fuzzy grain associated with scratches produced by abrasives. (Approximately 11×.)

(*c*) Fuzzy grain due to fiber separation in the sanding operation.

(*d*) Flat-sawn straight-grained oak block showing abundant tyloses and lack of collapsed walls. (Approximately 9×.)

(*From G. G. Marra.*[14])

In hardwoods fuzzing is commonly associated with tension wood (see page 301), and insofar as this type of wood may occur in any species, fuzzy grain may develop in wood which is normally free of this defect. Fuzzy grain in tension wood is due to the peculiar structure of the cell wall in this type of wood, characterized by the presence of a gelatinous layer, which becomes easily separated from the rest of the wall (see also page 304).

Marra[14] reports that in oaks, flat-sawn lumber is more subject to fuzzy grain than is quarter-sawn stock. This is explained by the fact that a large area of early wood, consisting of large thin-walled early-wood vessels, is exposed in flat-sawn material. Marra's experiments also brought out that the angle of the grain, with respect to the surface being sanded, has an important effect on the development of raised grain. Peak development occurs at angles between 3 and 15°, especially if the sand belt travels against the grain. Tyloses, when sufficiently abundant, are effective in restricting the amount of raised grain when this is occasioned by injury of the vessel walls (Fig. 9-10*d*).

B. Chipped and Torn Grain[16]

Wood surfaces from which pieces of wood have been scooped out or chipped by the action of cutting tools are said to contain *chipped* or *torn grain* (Fig. 9-3*a*). The occurrence of these defects may be due to the dullness and setting of the knife, the rate of feed of the stock into the machine, or the slope and variation in the grain of the wood. Lumber with a surface moisture content of 5 percent or less is more subject to tearing and chipping of the grain than that machined at a higher moisture content. Chipping may be minimized by reversing the direction of travel of the board through the planer.

SELECTED REFERENCES

1. Churchill, J. W.: The Effect of Time, Temperature and Relative Humidity on the Relief of Casehardened Stresses, *J. Forest Prod. Res. Soc.*, **4**(5):264–270 (1954).
2. Ellwood, E. L.: Properties of American Beech in Tension and Compression Perpendicular to the Grain and Their Relation to Drying, *Yale Univ. School Forestry Bull.* 61, 1954.
3. ———, B. A. Eclund, and E. Zavarin: Collapse in Wood and Exploratory Experiments to Prevent Its Occurrence, *Univ. Calif. Forest Prod. Lab. Rept.*, 1959.
4. Forest Products Laboratory: "Wood Handbook," U.S. Department of Agriculture Handbook 72, Washington, 1955.
5. Greenhill, W. L.: Collapse and Its Removal, *CSIRO, Div. Forest Prod. Tech. Paper* 24, Australia, 1938.
6. ———: The Shrinkage of Australian Timbers, I, *CSIRO, Div. Forest Prod. Tech. Paper* 21, Australia, 1936.
7. ———: The Shrinkage of Australian Timbers, II, *CSIRO, Div. Forest Prod. Tech. Paper* 35, Australia, 1940.
8. Kauman, W. G.: The Influence of Drying Stresses and Anisotropy on Collapse in *Eucalyptus regnans, CSIRO, Div. Forest Prod. Tech. Paper* 3, Australia, 1958.

9. Koehler, A.: Raised Grain—Its Causes and Prevention, *Southern Lumberman,* **137:**210M (1929).
10. ———: Some Observations on Raised Grain, *Trans. Am. Soc. Mech. Engrs.,* **54:**54–59 (1932).
11. ———: More about Raised Grain, *Southern Lumberman,* **161:**171–173 (1940).
12. McMillen, J. M.: What Is Casehardening in Lumber? Report presented at Technical Forum, Forest Products Laboratory, Feb. 1, 1956.
13. ———: Stresses in Wood Drying, *U.S. Dept. Agr. Forest Serv., Forest Prod. Lab. Bull.* 1652, 1963.
14. Marra, G. G.: An Analysis of the Factors Responsible for Raised Grain in the Wood of Oak, Following Sanding and Staining, *Trans. Am. Soc. Mech. Eng.,* **65:**177–185 (1943).
15. Pankevicius, E. R.: Collapse in Timber, Influence of Position in Tree on Collapse Intensity, *CSIRO, Div. Forest Prod. Progr. Rept.* 24, Australia, 1960.
16. Paul, B. H.: Raised, Loosened, Torn, Chipped and Fuzzy Grain in Lumber. *U.S. Dept. Agr. Forest Serv., Forest Prod. Lab. Rept.* 2044, 1955.
17. Rasmussen, E. F.: "Dry Kiln Operator's Manual," U.S. Department of Agriculture Handbook 188, Washington, 1961.
18. Schniewind, A. P.: Mechanics of Check Formation, *Forest Prod. J.,* **13**(11):475–480 (1963).
19. ———: On the Nature of Drying Stresses, *Holzforschung,* **14**(6):161–167 (1966).
20. Stamm, A. J., and W. K. Loughborough: Variation in Shrinkage and Swelling of Wood, *Trans. Am. Soc. Mech. Eng.,* **64:**379–386 (1942).
21. Tiemann, H. D.: Collapse in Wood as Shown by the Microscope, *J. Forestry,* **39:**271–282 (1941).

10

Wood Deterioration and Stains; Natural Durability of Wood

Deterioration of wood can be brought about (1) through mechanical wear, (2) by decomposition caused by physical agencies, such as prolonged heating or exposure to weather, (3) by chemical decomposition, and (4) by action of foreign biological agencies, such as fungi, bacteria, insects, and marine borers. The thermal decomposition of wood and deterioration by chemical agents are discussed briefly in Chap. 9 of this text. This chapter is concerned primarily with wood deterioration caused by the biological agents and with the various types of stains; natural durability of wood is also discussed.

I. DETERIORATION CAUSED BY FUNGI

Wood under certain conditions may sustain degradation through the action of numerous low forms of plant life known as *fungi.* Unlike green plants containing chlorophyll, wood-inhabiting fungi do not manufacture their own food but derive it from the digestion of organic matter (*substrate*), in this instance wood or the extraneous substances in wood, produced by the metabolism of another organism (the tree).

For convenience of this discussion wood-inhabiting fungi may be separated into two groups: (1) *wood-destroying fungi* and (2) *wood-staining fungi* and *true molds*. Members of the first group obtain the nourishment necessary for their growth and fruiting by disintegrating cell walls, thus causing breakdown (decay) in wood. Wood-staining fungi and molds, in contrast, derive most of their food from materials stored in cell cavities; consequently they have relatively little disintegrating effect on wood substance itself.

The life cycle of a fungus consists of a vegetative phase and a fruiting phase. Wood becomes infected either (1) by means of spores produced during the fruiting stage, which under favorable conditions germinate on the surface or in the cracks in the piece and produce filaments, called *hyphae* (singular *hypha*) that invade the wood, or (2) by spread of hyphae, collectively known as *mycelium*, from a source of previous infection. In either case the damage to wood by fungi occurs during the vegetative stage of the cycle of growth, from the initial point of attack through the pits and through the cell walls (Fig. 10-1).

The growth of a wood-inhabiting fungus depends on (1) favorable temperature, (2) a supply of oxygen, (3) an adequate amount of moisture, and (4) the presence of a suitable food supply. Since absence of any one of these requirements will prevent or greatly inhibit the growth, methods for controlling decay are based on elimination or modification of one or more of these conditions.

In most cases it is impractical to inhibit growth of fungi by controlling temperatures. Fungal growth is arrested by temperature approaching the freezing point, but no naturally existing low temperature will eliminate decay permanently.

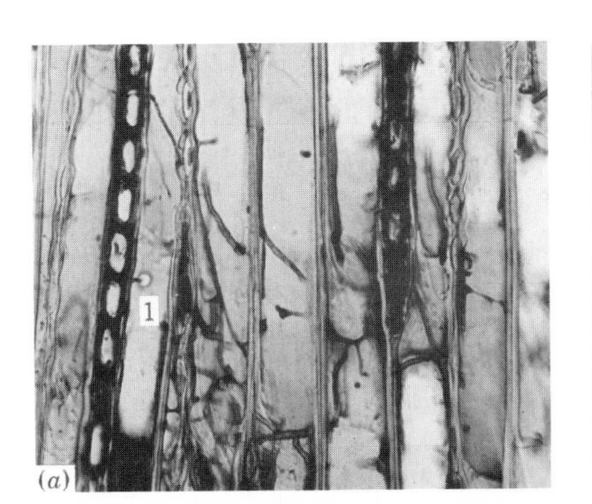

FIG. 10-1 (*a*) Highly magnified tangential section of eastern hemlock [*Tsuga canadensis* (L.) Carr.] containing hyphae of *Fomes pinicola* (Swartz) Cook. (260×.) At 1, a hypha passed through the back wall of a longitudinal tracheid; the opening was subsequently enlarged through further enzyme action.

(*b*) Transverse penetration of wood fibers by a colored zone-line hypha of *Fomes igniarius* (L.) Gill. (750×.)

[*By permission, from Henry Hopp, The Formation of Colored Zones by Wood-destroying Fungi in Culture, Phytopathology,* **28**(9):604 (1938).]

For instance, wood-destroying fungi occur in arctic regions in spite of prolonged low winter temperatures.

The optimum temperature for growth of 147 species and strains of fungi tested at the University of Michigan[5] was found to be between 76 and 86°F. Temperatures higher or lower than the optimum will retard or inhibit the growth of fungi, with the higher temperatures having a greater effect on fungus growth than the lower. In this study the minimum temperature for active growth was determined to be about 40°F. The maximum temperature for active growth of wood-decaying fungi is around 100°F. Indications are that between these two limits the activity of fungi doubles for every 20°F rise in temperature. Therefore wood exposed to outdoor conditions in the temperate climates will decay more rapidly in the summer than in winter, likewise decay will occur more rapidly in the tropical and subtropical regions than in the temperate-zone countries.[20]

Any wood-inhabiting fungus can be killed by resorting to elevated temperatures. Moist heat is more effective than dry heat. Snell,[50] working with five species, reports that none withstood a 12-hour exposure to a moist heat at 131°F; for the same period a dry heat of 221°F was required. Other studies[7] revealed that in commercial preservative treatments a temperature of 150°F maintained for 75 minutes was lethal provided the atmosphere around the wood was more or less saturated. The time necessary to kill a fungus decreased with increase in temperature, until only 5 minutes were needed at 212°F.

Temperatures lower than those cited can be used to kill wood-destroying fungi if the moisture content of the wood is not permitted to fall below the fiber saturation point. Higher temperatures or longer exposures are necessary if the conditions of sterilization are such that free moisture can escape. It should be kept in mind that sterilization of wood by elevated temperature does not guarantee against reinfestation, should moisture content and temperatures again become favorable to fungal growth.

Fungi require air because oxygen is consumed and carbon dioxide given off in their development. In general fungi tolerate quite a low pressure of oxygen, though considerable variation in tolerance to oxygen deficiency exists among different species. An excess of carbon dioxide around the hyphae decreases the rate of growth, and more thorough aeration accelerates it. The availability of oxygen and the accelerated rate of carbon dioxide diffusion out of the wood have been advanced as a probable reason why small pieces of wood decay more rapidly than similar larger pieces.

Fungi causing wood decay cannot grow in truly anaerobic conditions, e.g., those found in mud. Most wood-rotting or wood-staining fungi cannot grow in waterlogged pieces. Because of this the storage of wood completely submerged under water, or kept under a continuous spray of water, minimizes or precludes fungal infestation and spread of decay that may already be present. However, this method cannot be relied on to kill fungus already established in the wood, and its effectiveness is therefore limited to decay-free wood. In the instances where wood-destroying fungi were found to attack the surface of water-saturated wood, it was found that they generally belonged to the soft-rot fungi group (see page 347) and that their growth was affected by the degree of water aeration. Such attacks were reported both in fresh and saltwater but were found to be at a minimum in stagnant water.

Moisture in wood is essential for the germination of spores and for the growth of hyphae. The optimum moisture condition for the growth of fungi in wood appears to be somewhat above the fiber saturation point, i.e., when a film of free water covers the inner cell walls, facilitating enzymatic reactions, leaving sufficient air space for diffusion of gases.

On the basis of numerous tests conducted at the U.S. Forest Products Laboratory, it has been determined that the growth of the wood-rotting fungi is retarded at the ***average moisture content*** of 25 to 30 percent and is stopped at 20 percent, based on the ovendry weight of wood. This statement is based on the assumption that the fiber saturation point of the wood in question is about 30 percent. It is known, however, that in a number of woods, especially of tropical origin, the fiber saturation point is considerably lower than that figure. It is logical to assume that when this is the case the critical average moisture condition should be less than 20 percent.

A further assumption requiring careful verification is that the average moisture content of 20 percent or less signifies a uniform moisture gradient throughout the piece of wood. In practice there may exist areas within a given item in which moisture content could be considerably in excess of the critical moisture content. This may be especially true in well-insulated air-conditioned houses, in which moisture condensation occurs on the surface of wood members in the exterior walls. When this happens a sufficient amount of moisture may be present to allow spore germination or contact spread of hyphae.

The survival and spread of fungal infection depend to a large extent on the ability of the fungus to tolerate wide variations in moisture content and prolonged periods of desiccation. For instance some species of ***Lenzites*** and ***Poria*** can remain viable in a completely dry condition for a period up to 10 years or longer, and resume their activities when moisture conditions in wood become favorable again. On the other hand, dry rot (***Merulius lacrymans***) requires fairly constant moisture conditions.* Because of this ability of some fungi to remain dormant for long periods of desiccation, it is never safe to assume that a fungus has been eliminated by prolonged storage under dry conditions. For this reason wood used for construction and boatbuilding should be completely rot-free unless thoroughly pressure-treated with a preservative.

Once established, a fungus, particularly of a "dry-rot" variety (see page 344), may maintain itself and spread to the surrounding dry wood, by synthesizing water during respiration. This is accomplished by the breaking down of sugars derived from cellulose to carbon dioxide and water in quantities adequate to increase the humidity of the surrounding air sufficiently to allow decay to continue, even though the moisture content of the wood itself is below the critical point.

* W. P. K. Findlay, in "Holz und Organismen," pp. 200–211 in G. Becker and W. Liese (eds.), Duncker and Humblot, Berlin, 1966.

The maximum moisture content that can be tolerated by a fungus depends to a large extent on the ratio between the cell walls and air spaces. This ratio is higher for light woods, such as Balsa, in which decay may proceed at a moisture content above 200 percent, based on the oven dry weight; it is correspondingly lower for dense woods. Exact determination of the optimum moisture content for growth of a fungus is difficult because of the ability of the wood-rotting fungi to produce considerable quantities of water by their respiration.*

The principal source of nutrients for wood-inhabiting fungi are cell wall components and the various carbohydrates contained in the cell cavities. Wood-destroying fungi derive the bulk of their energy requirements for their growth from cellulose and other cell wall carbohydrates, though some, in addition, can also decompose lignin. Wood-staining fungi obtain most of their food from materials stored in the cell cavities.

Wood-destroying fungi also require nitrogen for their growth. Although the amount of nitrogen in wood is small, it is essential to the metabolism of large quantities of carbon-rich cellulose, hemicelluloses, and lignin. Merrill and Cowling† report that the nitrogen content of wood decreases from the first ring to the boundary of heartwood, where it remains stable except for some increase in the rings near the pith and in the pith itself. They found a direct correlation between the nitrogen content of individual rings and their rate of decay by *Lenzites trabea* and *Polyporus versicolor*.

Poisoning wood with preservatives is the most effective and reliable means of protecting wood against attack by fungi, provided the treatment is thorough and no untreated wood is exposed at any time during the service life of the treated material.

A. Wood-destroying Fungi[20]

On the basis of physical and chemical changes produced in wood and the resulting alterations in color of decaying wood, wood-destroying fungi are classified as *brown rots*, *white rots*, and *soft rots*. The first two kinds of decay are produced principally by the Basidiomycete fungi; soft rot is caused by Ascomycetes and Fungi Imperfecti.

1. BROWN AND WHITE ROTS

Brown rots are more commonly associated with softwoods and white rots with hardwoods. Some Basidiomycete fungi will attack a variety of conifers and hardwoods. Others will attack certain species only, probably because of their tolerance to the specific toxic action of the extraneous materials contained in these woods.

Brown rots attack primarily the cell wall carbohydrates, leaving behind a network consisting of modified lignin, with small amounts of more resistant crystalline cellulose. The affected wood develops a brown color. On drying, the decayed wood

* *Ibid.*

† In G. Becker and W. Liese (eds.), *op. cit.*, pp. 263–268; see footnote p. 341.

surface-checks in a characteristic cubelike pattern. In the final stages of brown rot, the decayed wood is converted into a powdery mass of varying shades of brown.

White-rot fungi attack both cellulose and lignin, leaving behind a spongy or stringy mass, with the wood surface on drying remaining unchecked. The affected wood is usually white or grayish white in color, but it may assume various shades of yellow, tan, and light brown.

In both types of decay, hyphae penetrate the cell walls transversely through pits or by formation of boreholes and then ramify through the cell cavities.* However, Wilcox[55] reports that while the hyphae of the white-rot fungus (*Polyporus versicolor* L.) penetrated the wood of sweetgum and southern pine through the pits as well as by means of numerous boreholes through the cell wall, the hyphae of the brown-rot fungus (*Poria monticola* Murr.) penetrated almost exclusively through the pits. In the later stages of decay, both fungi enlarged pit canals to such an extent that they could no longer be distinguished from true boreholes.

The penetration of the hyphae of wood-destroying fungi is mostly a chemical rather than a mechanical action. It is accomplished by the secretion of enzymes† which are capable of converting insoluble wood substance into soluble forms,‡ in advance of the actual penetration of hyphae.[13] It was shown by Proctor[38] that enzymes are secreted by the tip of the advancing hyphae. This enzymatic action dissolves minute openings (*boreholes*) in cell walls, through which the threads of the fungus pass. The hyphae are greatly constricted where they pass through the wall (Fig. 10-1) but resume normal size when they reach the opposite lumen. Later on, the size of the boreholes may increase to many times the size of the hyphae, indicating that enzyme secretion is not necessarily confined to the tips of the hyphae.

Cowling,[14] Wilcox,[54] and others have shown that considerable differences exist in the manner of degradation of the cell walls by the brown- and white-rot fungi. The hyphae of the brown-rot fungi liberate enzymes which diffuse from the cell lumina, where the hyphae are located, through the secondary cell wall. In the hardwoods they appear to attack carbohydrates in the *S*2 layer of the secondary wall first, followed by the *S*1 and finally the *S*3 layers, leaving the compound middle lamella more or less untouched. This preferential decomposition points to some correlation between the amount and possibly the characteristics of the available carbohydrates, the extent of lignification of the various cell wall layers, and the decay resistance of the cell walls.

* The cultures of white and brown fungi, grown in the laboratory on malt agar, can be readily separated by dropping an alcoholic solution of gum guaiac on the surface of the cultures; within 2 or 3 minutes those of white-rot fungi are stained blue, while no color change occurs in the brown-rot cultures.[7]

† Enzymes are protein molecules produced by living cells; they act as organic catalysts, making it possible for the biochemical reactions necessary for physiological processes to take place.

‡ E. T. Reese ("Advances in Enzymic Hydrolysis of Cellulose and Related Materials," Pergamon Press, 1963) states that the enzymes of all the fungi studied to date cleave polysaccharide molecules by hydrolysis of the glucosidic links between monosaccharide units. According to Cowling (*ibid.*, pp. 91–102, see footnote, p. 341), "The mechanism by which lignin molecules are depolymerized by the enzymes of wood-destroying fungi is entirely unknown."

In softwoods there is considerable variation in the order in which the three layers of the secondary wall are attacked. However, some evidence exists that, as in the case of hardwoods, decay starts in the $S2$ layer and then spreads to the $S1$ and $S3$ layers.

In contrast, enzymes secreted by the white-rot fungi, both in conifers and hardwoods, produce a gradual erosion of all the cell wall constituents from the lumen outward. The $S3$ layer is attacked first, followed progressively by the other layers as each becomes eroded. In some instances the lignin is attacked first, in advance of the cellulose; in others the lignin and cellulose are removed simultaneously. In hardwoods the compound middle lamella and the secondary wall of ray parenchyma and vessels appear to be more resistant to attack by both white- and brown-rot fungi than the fiber walls.

In the early stages of infection the wood appears to be quite normal; the only evidence of attack, if any, is a discoloration in the wood, which is frequently mistaken for chemical stain (see page 364). This early stage of decay is known as ***incipient decay.*** In this stage the spread of decay may go on unnoticed for a long time, since it can be identified positively only by observation under a microscope or by cultural diagnosis. Meanwhile, hyphae of the fungus may extend longitudinally for a considerable distance in advance of any recognizable evidence of rot.

Incipient decay is dangerous because unless arrested or eliminated it will eventually pass into the ***advanced stage.*** Besides, even in the incipient stage of decay the strength of the affected wood may be seriously impaired by some fungi, such as brown rot.

As decay progresses into the advanced stage, the cell walls begin to exhibit increasing evidence of disintegration (Fig. 10-2) and the wood undergoes definite changes in appearance, structure, and physical and chemical characteristics. Depending on the fungus, and the manner in which it works, the decayed wood becomes friable, spongy, stringy, pitted, or ring-shaked. The destruction may progress to a point where the affected wood breaks down completely, leaving hollows in trees or in affected pieces of the wood.

2. DRY ROT

A special type of brown rot, known as ***dry rot,*** causes damage in buildings in Europe and in the United States. In this country the causal organism is ***Poria incrassata*** (B. & C.) Curt.; in Europe the damage is done principally by ***Merulius lacrymans*** (Wulf.) Fr. The term ***dry rot*** is misleading in that, although the decay may occur in wood of low moisture content, dry-rot fungi are able to transport the necessary moisture considerable distances by means of ***rhizomorphs,*** tubelike conducting veins embedded in sheets of mycelium. These tubes enable the fungus, once established in a damp part of the house, to spread to adjacent woodwork, which may be comparatively dry. It has also been shown that, at least in the case of ***M. lacrymans,*** an adequate amount of moisture for the fungous growth may be produced by the decomposition of the wood itself, without any external source of water being

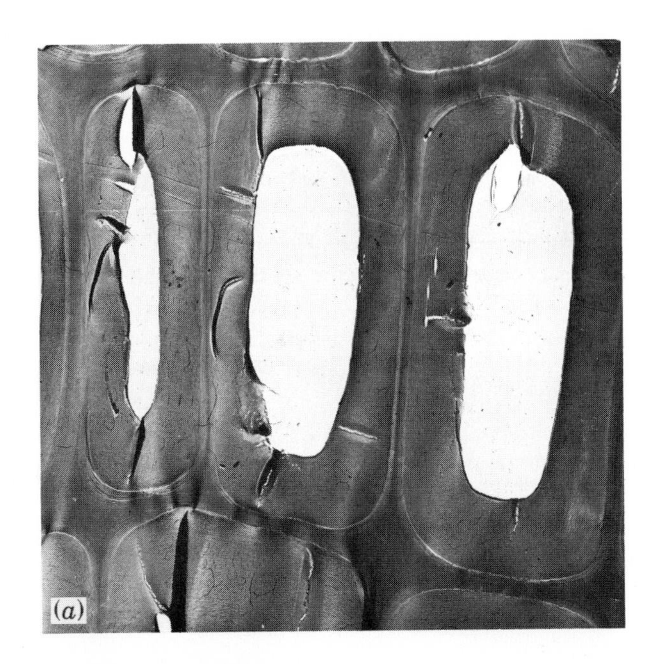

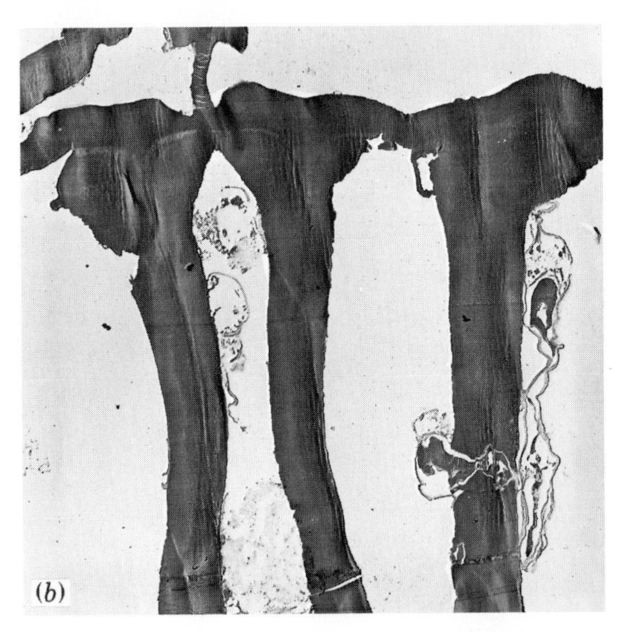

FIG. 10-2 Brown rot in Sitka spruce [*Picea sitchensis* (Bong.) Carr.]; wood blocks matched for specific gravity. The control (*a*) is shown on top. (4200×.) Shown beneath (*b*) is erosion amounting to 65.7 percent weight loss, caused by brown rot. (5000×.) (*Thin-section electron micrographs courtesy E. B. Cowling and W. A. Côté, Jr.*)

present. The rhizomorphs of *M. lacrymans* can penetrate through plaster masonry, brickwork, and other nonwoody materials, not containing food materials, which makes eradication of this fungus difficult once it has established a foothold in a building.

Dry-rot fungi attack both softwoods and hardwoods. The attack is confined to wood in buildings and to stored lumber; wood exposed to external atmospheric conditions is rarely affected. The European dry rot seldom occurs in the United States because of its extreme sensitivity to the higher temperatures prevailing in

FIG. 10-3 The water-conducting fungi (dry rot).
(*a*) In the untreated oak piers of a 5-year-old war emergency building.
(*b*) In the foundations of a Chicago building (*Merulius lacrymans*).
(*Photographs courtesy U.S. Forest Service, Forest Products Laboratory.*)

centrally heated American houses. It is also never found in the tropics or in mines where the temperature exceeds 25°C.

The first indications of dry rot are paper-like fan-shaped sheets of light-colored mycelium on the surface of the wood. In the advanced stages the decay is a typical brown rot, cubelike in appearance on the surface and crumbly upon drying (Fig. 10-3).

3. PECKY DRY ROT

This defect is common in baldcypress and incense-cedar; it is also called *peckiness* and *pocket dry rot*. In incense-cedar, peckiness is caused by *Polyporus amarus* Hedg.; the causal organism in baldcypress, formerly believed to be *Fomes geotropus* Cke., has more recently been identified as a species of *Stereum*. Both fungi attack the heartwood of living trees, although *P. amarus* sometimes continues to grow in dead trees that were infected while living. Further development of this fungus is arrested when the trees are felled and converted into lumber, and therefore such lumber is perfectly safe to use.

Peckiness is characterized by finger-sized pockets running with the grain, sometimes for a distance of 6 inches to 1 foot or more (Fig. 10-4). In the early stages of decay the affected areas in the xylem are firm and only faintly discolored. As the decay progresses, pockets of a dark-brown friable mass of decayed wood develop. The wood between them apparently remains unaffected in both general appearance and strength. Since all further growth of fungi causing peckiness ceases after the trees are felled, pecky wood is usable wherever the reduction in strength caused by this defect and its unsightliness are not objectionable.

4. SOFT ROT

More recently it has been recognized that fungi other than Basidiomycetes are capable of causing decay.[12,15,30,31] Since the surface of the affected wood is typically

FIG. 10-4 Pecky dry rot in California incense-cedar (*Libocedrus decurrens* Torr.) caused by the fungus *Polyporus amarus* Hedg. (Two-fifths natural size.) Finger-shaped pockets are formed which are directed along the grain and are filled with a brown friable mass of decayed wood.

softened, the term *soft rot* has been applied to this type of decay caused by the Ascomycetes and Fungi Imperfecti groups of fungi. Both softwoods and hardwoods are susceptible, but the latter disintegrate more rapidly and extensively when attacked by soft rot. This may be because of the higher concentration of lignin in the secondary walls of coniferous woods, which is more resistant to the enzymatic action of the soft-rot fungi than is cellulose.

Soft rot is characterized by hyphae which, unlike those of the Basidiomycete fungi, produce tunnels in the cell walls that run along the grain and are generally confined to the less-lignified *S*2 layers of the secondary walls (Fig. 10-5). As seen longitudinally, the hyphae appear as spirals contained within the eroded areas in the secondary walls. The eroded areas are in the form of cavities with pointed ends, aligned in

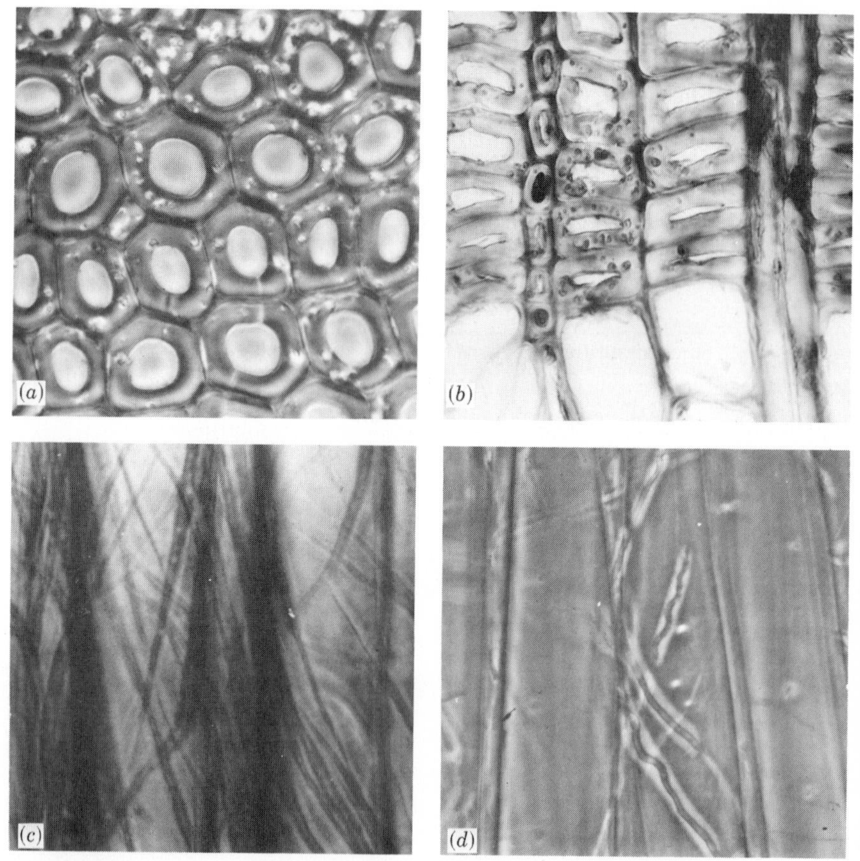

FIG. 10-5 Soft rot; the hyphae within the secondary wall.
(*a*) Transverse section, ***Mimusops*** sp.
(*b*) Transverse section, redwood [***Sequoia sempervirens*** (D. Don) Endl.].
(*c*) Longitudinal section of pine, showing the typical spiral course of the hyphae.
(*d*) Longitudinal section, ***Mimusops*** sp., showing the hyphae lying within larger spindle-shaped cavities.
(***Photographs courtesy of U.S. Forest Service, Forest Products Laboratory.***)

the same direction as the microfibrils. In cross section the cavities produced by hyphae appear as holes equal to or exceeding the diameter of the hyphae.[15] In some woods, especially in hardwoods, soft-rot may also break down the $S3$ and $S2$ cell wall layers from the cell lumen side.

This type of deterioration is generally found on surfaces exposed to persistently damp conditions, resulting in a moisture content considerably above that tolerated by the Basidiomycetes. In hardwoods soft-rot fungi attack fibers in preference to other types of cells and the early-wood fibers in preference to those of the late wood, with the result that in severely decayed specimens the fibers of the early wood may be completely destroyed, while the late-wood cells may remain relatively intact.[54]

Soft rot is common in wood placed in contact with the soil, e.g., in the below-ground portion of fence posts and poles, and in warm humid places such as greenhouses and humidity rooms. It assumes economic importance mainly in situations where for some reason the attack of the Basidiomycete fungi is inhibited, e.g., in water-cooling towers or in logs submerged in water. In such cases decay remains superficial unless the wood is allowed to dry periodically; the depth of soft-rot penetration may then become considerable. Soft rot may also occur in situations where wood is only occasionally wet, e.g., on the above-ground surfaces and tops of fence posts.

In its earlier stages soft rot is difficult to detect without the aid of a microscope. In the later stages the surface of the wood affected with soft rot generally becomes discolored. When wet, the infected surface is soft and can be scraped off, revealing relatively sound wood underneath. When dry, the affected surface feels spongy and eventually develops cracks, assuming an appearance of slightly charred wood (Fig. 10-6). Wood containing soft rot tends to break with a brash fracture. Wood items with high surface-to-volume ratio, such as plywood, are affected by soft rot more

FIG. 10-6 Soft rot in above-ground (left) and below-ground (right) segments of a preservative-treated pole. (*Photographs courtesy of U.S. Forest Service, Forest Products Laboratory.*)

readily than thicker pieces, apparently because such surfaces respond more quickly to changes in external atmospheric conditions.

5. MARINE FUNGI

As was pointed out before, a number of fungi, especially those belonging to the soft-rot group, are capable of attacking wood in fresh and salt water, as long as the water is sufficiently aerated. Kohlmeyer* reports that microorganisms were found on a variety of wood species at depths of up to 3700 meters in various locations in the Caribbean Sea and Pacific Ocean. The fungi attacking wood in deep seawater are morphologically different from the terrestrial forms. When the dissolved oxygen content of water becomes low, fungal growth is inhibited but bacterial activity is promoted.

6. CHARACTERISTICS OF DECAYED WOOD

Advanced stages of decay are always accompanied by readily recognizable changes in color, and sometimes by characteristic odor. Other important changes in physical properties of the affected wood are more difficult to recognize.

a. Color. The incipient stages of decay may be accompanied by various changes in the natural color of the wood, or color changes may be absent. When color changes occur during the incipient stages of rot, they are more pronounced in the case of white rot; such discolorations may vary from white streaks and pockets to shades of gray, olive-brown, and reddish. Some of these discolorations may be readily confused with nonbiological stains, such as mineral streaks and oxidative stains (see page 362). With a few exceptions the incipient stages of brown rot are seldom accompanied by extensive discoloration.

A conspicuous feature of many white- and brown-rot-causing fungi are brown or black lines or zones caused by a concentration of gnarled masses of dark-colored hyphae in the cell lumina, associated with dark-colored pigments produced by them. Black zones are more common in white than in brown rots and in hardwoods than in softwoods.

Color changes in wood resulting from the advanced stages of decay are quite conspicuous. It is not always possible, however, to identify the fungus involved by the appearance of advanced decay. These color changes are due to the color and concentration of the invading hyphae, to destruction, or to a chemical alteration of one or more of the principal cell wall components and the pigmented materials in the wood, or in some cases to the formation of distinctive coloring substances by fungi.

b. Odor. The advanced stages of decay may be accompanied by characteristic odors. These odors are variable in intensity and character, with those of anise and wintergreen being quite common. In some cases these odors may be of diagnostic

* Jan Kohlmeyer, "Materials Performance in the Deep Sea," ASTM, STP445, 1969.

value. In most instances they are only a nuisance, but in some forms of wood utilization, for instance, in containers for foodstuffs, they may be quite a serious problem.

c. Water Conduction and Moisture-holding Capacity. Decayed wood absorbs moisture more rapidly than sound wood, presumably because the boreholes in the cell walls made by the hyphae allow for more rapid entry of water. Decayed wood is also capable of maintaining a substantially higher moisture content when exposed intermittently to liquid water. As a result the moisture conditions in a piece of wood affected by decay will remain favorable for fungal growth longer than in a piece of sound wood exposed to similar conditions.

Greater absorptive characteristics of decaying wood facilitate penetration of the water-borne preservatives as well as of creosote. On the other hand the unequal distribution of water, which results from the difference in the rate of moisture loss in sound and decayed wood, presents special problems in seasoning of wood.

d. Dimensional Changes. Wood containing decay in the advanced stages shrinks more on drying than sound wood. The differences in the rate of shrinkage are more pronounced in brown rots than in white rots at a comparable stage of decay. It has been found that collapse (see page 327) may be attributable to incipient decay, since in some instances it occurs only in the discolored areas resulting from a fungous attack.

e. Density. Because of destruction of wood substance, wood becomes less dense as decay progresses. The loss of dry weight, expressed as a percentage of the original dry weight, may be used as a quantitative measure of the amount of destruction caused by decay. Low density alone is not, however, a sufficient indication of the presence of rot, since low density may be a result of adverse growth conditions.

f. Mechanical Properties. The mechanical properties of wood may or may not be significantly impaired. Wood with the incipient stages of brown rot suffers a considerable reduction in toughness, and becomes brash. Wood affected with white rot shows little weakening until the decay reaches more advanced stages. It has been suggested[15] that this difference may be due to a more rapid rate and a greater extent of cellulose degradation by the brown than by white rots. According to Cartwright and Findlay* toughness or resistance to impact is the strength property that is affected first by fungal infection; it is followed, in the approximate order of susceptibility, by reduction in bending strength, compression strength, hardness, and bending elasticity.

g. Others. Dry decayed wood ignites more readily than sound wood. However, the calorific value of decayed wood is reduced by the amount of wood substance destroyed by the fungus.

Wood containing decay is more readily attacked by some insects. In part this is because of the softened condition of the affected wood, which makes access into the wood by insect larvae easier, and in part because of the conversion of wood substance by fungi into products that can be more readily assimilated by insects.

* K. St. G. Cartwright and W. P. K. Findlay, "Decay of Timber and Its Prevention," 2d ed., Dept. Sci. & Ind. Res., *Forest Prod. Lab.*, London, 1958.

Removal of the cell wall components by fungi may also enhance the nutritional value of wood for insects by increasing the nitrogen content of the decaying wood; the reverse may also occur, i.e., the initial insect attack on wood may be followed by the entrance of staining or wood-destroying fungi.

B. Sap Stains[19,48]

Abnormal discolorations in wood, other than those caused by decay-inducing fungi, may result from growth of fungi that discolor the wood without appreciably decomposing it, and sometimes through bacterial action. Other types of stains may also arise from natural causes or be brought about by chemical action between wood and other materials coming in contact with it. The last two classes of stains are discussed in Sec. III (pages 361 to 365) of this chapter.

Discolorations in wood caused by fungi are by far the most important type of stains from the economic point of view. Since the activity of this class of fungi is almost exclusively confined to the sapwood, the resulting discolorations are called *sap stains* and the fungi causing them *sap-stain fungi*.

Sap stain may occasionally occur in the sapwood of living trees. Generally, however, it develops in the sapwood of dead trees, in logs and in lumber, and under favorable conditions in wood which has been seasoned and then allowed to become wet.

Sap-stain fungi fall into two classes; *molds*, which grow on the surface of wood and cause only superficial discolorations that in most cases can be brushed off, and *true sap-stain fungi*, which penetrate into the sapwood and cause staining too deep to be easily removed.

1. MOLDS

These fungi are characterized by cottony or downy growth, varying in color from white through shades of yellow, brown-red, purple-blue, and green to black. The stain produced by molds comes mainly from the colored spores of the fungi. Mild temperatures, an abundant supply of moisture, and still air, such as results from poor ventilation, favor their development. Such conditions frequently prevail when lumber is improperly piled for air seasoning, when green lumber is shipped, in dry kilns when stock is subjected to prolonged treatment at high humidities and temperatures low enough to favor fungal growth, and in highly humid buildings.

Cottony mycelium may become so luxuriant during prolonged drying in the kilns at low temperatures and high humidity that the spaces between the courses of boards may become filled, impeding air circulation. As the mycelial mat dries up, the wood surfaces covered by it are likely to develop surface checks, because of higher moisture content retained in these areas, compared with that of the surrounding wood. Molds apparently do not affect the strength properties of wood; they may constitute a serious defect when wood is utilized for food containers, because of the possibility of contamination.

Molds are generally confined to sapwood surfaces only. However, when logs or wood items containing both sapwood and heartwood are steamed or soaked in hot water, molds may develop also on the heartwood, which otherwise is immune. This is because of leaching of soluble materials from the sapwood which subsequently impregnate the heartwood, thus rendering it susceptible to molding.

a. Effect on Permeability of Wood. Lindgren[32] reports that studies conducted on southern pine posts indicate that the permeability of posts was increased by molding. The greatest effect was produced by the common green mold (*Trichoderma viride* Fr.). Growth of molds was greatly stimulated by spraying with fluorine-containing solutions, such as sodium fluoride, which in themselves are toxic to the common stain and decay-producing fungi. With a 5-minute soaking of end-coated posts in 5 percent pentachlorophenol solution, absorption of 5 to 9 pounds per cubic foot of the wood was obtained for moderately and heavily molded posts, respectively. Penetration was complete even though only side penetration was involved. In contrast, an average absorption of only 1 pound per cubic foot, and penetration of only 0.1 inch, was obtained in the mold-free wood.

Schultz[49] reports similar results for aspen and spruce. Heavy mold infection increased the permeability of aspen three to five times and that of spruce two to three times, compared with the mold-free specimens. Increased permeability occasioned by mold infestation is apparently due to the partial or complete breakdown of the ray-parenchyma cell.

These findings suggest that erratic results obtained in impregnation of wood with preservatives may sometimes be due to molds, leading to either undertreatment or overtreatment. For example, in heavily molded wood, dried under cover, overtreatment may result, while wood exposed to rain may be undertreated because of heavy absorption of rain water. Furthermore, uneven growth of mold will also result in different degrees of permeability in adjacent areas of the surface of the piece. Finally, there is a possibility that the increased porosity of the infected wood might also bring about more rapid leaching out of preservatives.

b. Control Measures. In air seasoning, molding can be controlled by adequate circulation of air. Molds on wood can be destroyed by steaming it at 170°F and 10 percent humidity for about 1 hour. Longer periods of time are necessary if lower temperature and humidity conditions are used. Dipping and spraying freshly cut stock in a chemical solution such as sodium pentachlorophenate or ethyl mercury phosphate are also effective in preventing mold infection.

2. TRUE SAP-STAIN FUNGI

The main difference between true sap-stain fungi and wood-destroying forms is that the former do not decompose the wood substance proper, or they deteriorate it only slightly; they derive nourishment from food materials stored in cells of the sapwood, chiefly in the parenchyma. Their hyphae are usually larger than those of a

Table 10-1 Common Sapwood Discolorations and Their Causes*

Color of stain	Designation and miscellaneous characteristics	Sapwood in which it commonly occurs	Common cause
Bluish black to steel gray principally; brown shades common	Blue stain—occurs in spots, streaks, or patches which cover all or part of the sapwood. Moldlike growths of causal fungi often present on surface of stained areas. May penetrate deeply	Lumber and logs of practically all commercial wood species	Dark hyphae of species of *Ceratostomella, Endoconidiophora, Diplodia, Cladosporium, Hormodendron,* and *Graphium,* principally
Inky blue	Iron tannate stain—inklike streaks or blotches where nails or iron equipment have contacted freshly cut stock. Without precautions may develop as general discoloration when dipping oak lumber with iron equipment. Usually shallow penetration	Various products of oak, chestnut, redgum, and other species with necessary tannin content	Chemical reaction between iron and tannins in the wood
Pale blue and brown	Chemical stain of hardwoods—most common as a general interior stain and may not be observed until lumber is surfaced. Frequently resembles light blue stain. Sometimes appears only under seasoning stickers. May penetrate deeply	Oak, birch, maple, basswood, tupelo gum, magnolia, and other hardwood lumber	Oxidation of certain wood substances during air seasoning or kiln drying
Greenish brown to greenish black	Mineral stain—occurs in lenticular streaks of all sizes or as a general discoloration	Living hardwood trees—hard maples principally†	Unknown. Possibly initiated by injuries
	Chemical stain of persimmon—occurs as a general discoloration. May penetrate deeply	Persimmon	Oxidation of certain wood substances upon contact with air during seasoning
Dull brown to gray	Weather stain—occurs as general surface discoloration on exposed portions of wood. Usually shallow penetration	Lumber of all commercial wood species†	Action of air, dust, rain, sunlight ("weathering")
	Mold stain—discoloration persisting in surface layers of molded wood after surfacing or brushing. Usually shallow penetration	Lumber of all commercial wood species†	Sporulation of mold fungi in vessels and resin ducts near the surface of the wood
Various, green predominating, black common	Molds—colored fungus growths present on surface of wood. Generally surfaces off	Various products of all commercial wood species	Presence of *Trichoderma, Penicillium, Gliocladium, Aspergillus, Monilia,*

shades principally	simply as a deepening of color. Essentially a chemical stain but generally not objectionable. May penetrate deeply	cipally	during seasoning
Yellow-brown to dark brown	Brown seasoning stain of western and eastern pines—narrow margins of bright wood common at board surfaces and at juncture of heartwood and sapwood. May penetrate deeply	White pine, ponderosa pine, and sugar pine lumber	Oxidation of certain wood substances during air seasoning or kiln drying
	Kiln burn and machine burn—surface of wood has a scorched appearance. Usually shallow penetration	All lumber†	Light burning of wood due to excessive kiln temperature or to heat developed by planer knives
Red	Bright-colored chemical stain of hardwoods—appears more or less as a general discoloration. May penetrate deeply	Western alder lumber and other hardwoods	Oxidation of certain wood substances upon exposure to air
Reddish yellow to rusty		Birch logs	Oxidation of certain wood substances upon exposure to air. Promoted by warm weather
Purple to pink	Bright-colored fungus stain—occurs usually as blotches or small streaks. May penetrate deeply	Southern pine and redgum lumber and logs	Soluble pigment and colored hyphae of *Fusarium moniliforme, F. solani, F. viride,* and *F. roseum*
Crimson to orange	Bright-colored fungus stain—occurs usually as blotches or small streaks. May penetrate deeply	Southern pine, gum, oak, and other hardwood lumber and logs	Soluble pigment of *Penicillium roseum* and *P. aureum*
		Southern pine, southern cypress, and oak lumber†	Soluble pigment of *Geotrichum* sp.
Pale yellow	Bright-colored fungus stain—occurs usually as blotches or small streaks. May penetrate deeply	Lumber and logs of oak, birch, hickory, and maple†	Soluble pigment of *Penicillium divaricatum*
Deep yellow		Southern pine and redgum lumber and logs	Colored hyphae and to some extent soluble pigment of a *Gymnoascus*-like fungus

* By permission from T. C. Scheffer and R. M. Lindgren, Stains of Sapwood and Sapwood Products and Their control, *U.S. Dept. Agr. Tech. Bull.* 714, 1940.

† Also occurs in heartwood.

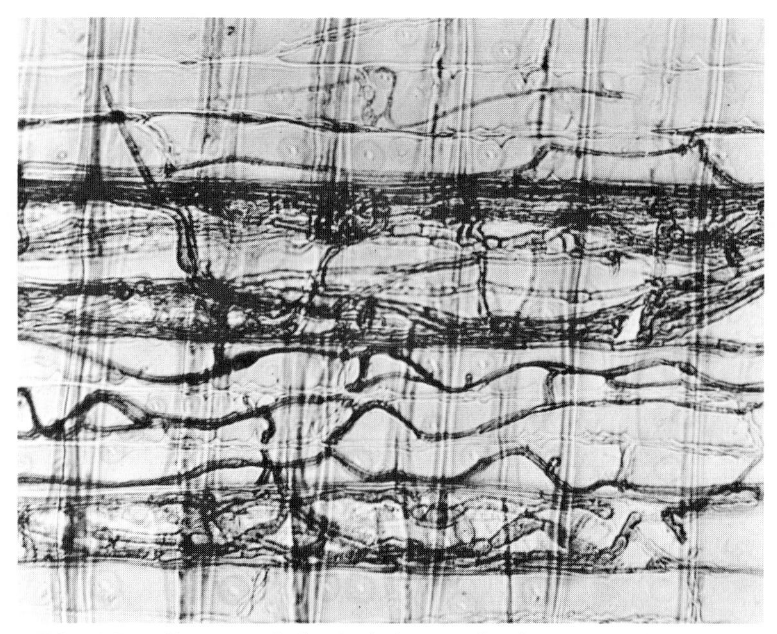

FIG. 10-7 True sap-stain fungi which cause discoloration. Highly magnified section of loblolly pine (*Pinus taeda* L.) containing hyphae of blue stain [*Ceratostomella* (*Ceratocytis*) *pilifera* (Fr.) Wint.] (Radial section with concentration of the threadlike hyphae in one of the wood rays). (370×.) (*By permission, from Theodore C. Scheffer and Ralph M. Lindgren, Stains of Sapwood and Sapwood Products and Their Control, Plate* 1, *U.S. Dept. Agr. Tech. Bull.* 714, 1940.)

wood-destroying fungus (Fig. 10-7), and they penetrate from cell to cell mostly by passing through the pits in the walls, in contrast to those of a wood-decaying fungus. In cases when hyphae of a stain fungus penetrate directly through the cell wall, the boreholes are extremely minute, several times narrower than the normal width of the hyphae. Another significant point of difference between these two classes of fungi is that perforation by stain fungi takes place mostly mechanically, without prior enzymatic action as is the case with the wood-destroying kinds.

Krapivina,* reporting on research conducted on 16 species of blue-stain fungi, states that some of these fungi are responsible for a greater amount of destruction of the cell wall substance than has been previously suspected. When cell walls are invaded by the blue-stain-causing fungi, their hyphae branch out through the secondary wall and follow the general direction of the microfibrils, in much the same manner as those of the soft-rot fungi. There is also some indication that this penetration may be of a chemical rather than of mechanical nature.

In temperate zones sap stains are more common in softwoods than in hardwoods,

* I. G. Krapivina, Investigations of the Destructive Action of Blue-stain Fungi on the Anatomical Elements of Wood (in Russian), *Questions of Wood Protection*, Central Scientific Research Institute of Mechanical Wood Conversion, U.S.S.R., 1961.

with some notable exceptions that include sweetgum, poplar, and buckeye, in which sap stains are quite troublesome.[9] Staining is a serious problem in many light-colored tropical hardwoods, seriously affecting their exploitation.

The discolorations in the sapwood caused by the sap-staining fungi are of various colors, depending on the host and the causal organism. However, from the standpoint of economic loss blue stains are by far the most important.

Examples of stains other than blue are red stains in boxelder, green stains in oaks,* yellow stains in hardwoods, caused by the *Penicillium* group, brown stains in pines caused by *Cytospora* sp., and grayish olive stains, induced by *Lasiosphaeria pezizula* (B. & C.) Sacc. in sapwood and heartwood of stored logs of beech, sweetgum, tupelo, and persimmon.

a. Blue Stains. The discolorations caused by the blue-stain fungi range from grayish through dark blue to blackish; the stained areas may appear as spots, streaks, and irregularly shaped areas on the surface or ends of boards and logs (Fig. 10-8). Under conditions favorable to fungal growth the entire sapwood may become discolored. The characteristic color of blue stain apparently is caused by a concen-

* The stain causing "green oak," *Chlorosplenium* (*Chlorociboria*) *aeruginosum* (Ded.) de Not., is the only staining fungus subject to a patent, taken out by Prof. F. T. Brooks in 1911.

FIG. 10-8 Sap stain in southern yellow pine. (*Photograph courtesy of U.S. Forest Service, Forest Products Laboratory.*)

tration of hyphae. But since examination under a microscope reveals no blue tint in the hyphae, it is a matter of speculation how brown hyphae cause a blue color in wood.

Blue stains are caused by a number of different fungi, the most important of which are species of *Ceratocystis* (formerly known as *Ceratostomella*). Blue stains may appear in dying trees, but usually they develop in dead trees, in logs, and in green lumber, as well as in manufactured wood products, including mechanical pulp, as long as these contain a sufficient amount of moisture. Wood dried without discoloration may develop blue stain once it becomes moist. Occasionally blue stain may occur in heartwood of such species as Sitka spruce, ponderosa pine, and southern pines.[7]

Once the wood is stained, its normal color cannot be restored except by resorting to strong bleach. The treatment, however, may have to be so severe that much of the natural color of the wood is also lost. Likewise pulp obtained from the blue-stained wood is dark-colored and hence less desirable than that obtained from bright wood, because it requires an additional amount of bleach to achieve the desired whiteness.

Strength tests on blue-stained softwoods have established that bending and compression strength of stained wood are not significantly reduced, while toughness may be up to 25 percent lower in the heavily stained pieces. Similarly to molds, staining fungi increase the permeability of wood to liquids, after the stained wood has been seasoned. Contrary to widespread notion, there is no conclusive evidence that blue-stained wood is more susceptible to decay than unstained wood.

b. Interior Stains. A form of stain deserving special attention is known as *interior stain.*[48] It differs from ordinary sap stain only in that it is confined to the interior of an infected piece and is not evident on the surface; it is caused by the same fungi.

Interior stains develop when, shortly after infection, the surface of the affected piece will no longer support growth of the fungus because of the rapid development of unfavorable surface conditions. Such conditions may result either from rapid surface drying or from superficial treatment with a chemical intended to prevent stain. As a result the growth of the fungus may be inhibited in the outer zone of the infected piece, before the hyphae acquire the normal dark color, while at some distance from the surface the fungus may continue its development as long as adequate moisture is present.

c. Control Measures.[43,48,53] The sap-stain fungi are spread principally by means of spores which may be wind- or rain-borne, though bark beetles also play an important role in disseminating spores, especially in standing trees and in logs. It is believed that fragments of mycelium dislodged in sawing stained logs may likewise be a factor in spreading the infection.

Fungi causing stain develop most rapidly in freshly cut wood, during humid weather, when the temperature ranges from 75 to 85°F; they cease growing at temperatures below 40°F and above 95°F. True sap-stain fungi survive long periods of exposure to extremely low temperatures and some can grow and spread below the freezing point; they are also rather resistant to short exposures to high temperatures in the range of 150 to 160°F but can be killed by exposure of several days to temperatures as low as 120 to 130°F.

The minimum moisture content for the growth of most sap-staining fungi is in the vicinity of the fiber saturation point, which for most native woods is approximately 26 to 30 percent. The maximum moisture content tolerated by the staining fungi is limited by the supply of oxygen; this means the lower the density of the wood, the higher the critical moisture content.

Controlled experiments do not support the view that timber cut in the autumn or in winter is less susceptible to stain than wood cut in the spring or summer. A lesser amount of staining found in fall- and winter-felled timber may be attributed principally to partial seasoning of such material during the cold months, when temperatures are not favorable to fungal growth, thereby making it less susceptible to staining when warm weather arrives.

In newly cut logs the bark affords some protection against the entry of stain-causing fungi. The ends of logs, however, are exposed to fungal infection. In order to minimize the damage, the logs should be converted into lumber as soon as possible, especially during warm weather. If this is not possible, storing logs under water or treating them with a combination of fungicide and insecticide is recommended. A water-impermeable end coating should be used, after the toxic spray has been applied, if land storage for long periods is expected. This prevents seasoning checks, which, if they occur, may allow spores to penetrate deep into the logs through the treated ends. However, in the Southeast, even with these precautions, it is difficult to keep logs stain-free.

The most effective way to prevent staining of sawn timber is rapid drying to reduce the moisture content of the surface below the fiber saturation point, before the spores can germinate. In some regions where high humidity and temperatures prevail over long periods of time prompt treatment of green lumber by dipping or spraying with a fungicide-insecticide solution is strongly recommended. Application of chemicals to the surface of green stock does not provide permanent protection. The chief value of such treatment is in the temporary protection it affords to the surface of the wood for a long-enough time to allow it to dry below the fiber saturation point. Seasoned wood, whether dipped in a toxic solution or not, becomes susceptible to fungous attack, if it is allowed to become wet.

Kiln drying of lumber green from the saw is also an effective preventive measure. The temperature used must be above 120°F, since lower temperature, when coupled with high humidity, not only is not effective in killing the stain-causing fungi but may actually foster rapid spread of the stain.

II. DETERIORATION CAUSED BY BACTERIA

Little information is available on the role bacteria may play in the breakdown of the cell wall in its natural lignified state. It is known, however, that when isolated, each of the major components of wood, including lignin, can be readily decomposed by bacteria.

1. WETWOOD

Bacteria cause a number of diseases in plants, including trees. Of these, wetwood deserves special attention. This term is applied to the heartwood, usually to the outermost layers of it, or to the intermediate zone between the inner sapwood and heartwood, and to localized areas in the sapwood, especially in oaks, with obviously higher moisture content than the adjacent sapwood. This zone of wetter wood is generally somewhat darker in color, less acid than the sapwood, and in a number of species distinctly alkaline. The excessive wetness is often associated with unpleasant fermentation odors and a positive gas pressure, the gas being about half methane.

Wetwood is found in many hardwoods and less frequently in softwoods, e.g., in true firs, eastern white pine, and hemlocks. Since bacteria have been found in the cultures made from wetwood in all the species studied, it has been concluded that "most of the properties peculiar to wetwood are due to saprophytic or weakly parasitic bacteria in the standing trees."[24] These conclusions have not been universally accepted in the case of coniferous woods. Some workers claim that in the coniferous species wetwood indicates only a condition of excessive accumulation of water and is not a disease symptom.

It has been suggested that since wetwood is frequently associated with shakes and frost cracks (see page 283), it may be a factor in the initiation and extension of these conditions. Wetwood is subject to excessive checking and collapse in woods such as aspen and western firs. Veneers containing wetwood zones dry unevenly, causing gluing difficulties. Some reduction in toughness of wetwood in aspen was noted by Clausen and Kaufert.[11] Haygreen and Wang[25] report average reduction in strength properties of wetwood in aspen of 18 to 28 percent, and 52 percent in the modulus of elasticity. On the other hand Wilcox[54] found that in white fir, wetwood had little effect on wood properties, other than that derived from the higher moisture content.

2. BROWN STAIN IN HEMLOCK

The stain in western hemlock, ranging from orange to almost black, is referred to in the trade as ***hemlock brown stain.*** It occurs both in air-dried and kiln-dried hemlock lumber, but it can be removed by planing, provided the lumber remains dry. The stain is confined to the sapwood and is most prevalent in the outer sapwood adjacent to the cambium and at the heartwood-sapwood boundary.[4]

Hemlock brown stain is believed to be caused by condensation of phenolic compounds in the sapwood, principally catechin and to a lesser extent leucocyanidin, in the presence of bacterial enzymes, forming soluble compounds. These polymers migrate to the surface of the wood during drying and are oxidized to dark-brown substances when in contact with the air. The stain precursors and the bacterial infection are believed to develop within the logs during storage in the woods.[17] Application of antioxidants, such as thiourea, has been suggested as a stain control. To date, however, none has proved to be effective in field tests because of the continued movement of the stain-forming compounds to the surface and formation of fresh enzymes during drying.[4]

3. BACTERIAL DETERIORATION IN LOGS

Paint blistering on ponderosa and sugar pine millwork was traced by Ellwood and Ecklund[16] to excessive absorption of preservatives by the sapwood of pond-stored logs. The increased porosity of the sapwood, responsible for overabsorption of water repellents, was linked to the presence of a large number of bacteria, identified as *Bacillus polymyxa*, and perhaps to other related species of bacteria.

Extreme porosity in the sapwood of ponderosa and sugar pine logs developed only on the immersed portions of the logs, stored in virtually stagnant water. The development of porosity was most prominent in the logs stored under summer conditions, when temperature of the water reached 90°F. Similar logs decked under water spray for the same length of time showed no evidence of excessive porosity. From this it was concluded that, though the causative bacteria may be present in logs before ponding, the bulk of attack is caused by bacteria in the water.

Further tests conducted at the University of Wisconsin[28] confirmed that the increase in porosity of the sapwood was due to hydrolysis of pectic compounds, starch, and possibly also of hemicelluloses and cellulose in the thin-walled parenchyma cells, resulting in their disintegration.

Similar bacterial deterioration of sapwood, resulting in destruction of parenchyma cells and in increase in permeability, was reported in southern pine stored in warm water and under spray in warm weather. Little adverse effect on mechanical properties was noted.[33]

III. NATURAL AND CHEMICAL STAINS

Numerous kinds of stains in wood, both in the sapwood and in the heartwood, cannot be definitely attributed to fungi or bacteria. Since in appearance these stains frequently resemble one another, it is essential to be able to distinguish them, in order to provide suitable control measures.

1. MINERAL STAINS

This term has been loosely used in the trade to denote stains of various kinds in lumber without regard to their origin. In a more restricted sense, the expression *mineral streaks* or *mineral stain* should refer only to the olive and greenish-black, usually lenticular, areas common in otherwise normal wood of hard maple and occasionally also in other hardwoods. The discolorations are traceable to crystalline inclusions in the lumen of the xylem cells, composed of various minerals in association with carbonates, dispersed among noncrystalline phenolic substances. No apparent damage to the cell walls was found in the cells containing such inclusions.* These extraneous materials are apparently responsible for the higher mineral content of such areas compared with that of the surrounding wood, averaging 5.2 percent ash content in hard maple, as contrasted with only 1.2 percent in the normal bright wood.[44] When wood containing mineral streaks is seasoned, cracks frequently form where the discoloration is deepest. Millmen contend that wood with mineral streaks is harder than normal stock and that it has a pronounced dulling effect on cutting tools. The discolorations caused by mineral stain are also considered objectionable because of the problems such local stains present in wood finishing.

No acceptable explanation has yet been found to account for the development of mineral streaks. There is some evidence that the mineral discolorations may be initiated by obscure injuries, which in some manner interfere with the normal physiological functioning of the cells proximate to such areas. Likewise, a possibility of bacterial infection as a contributing cause to formation of mineral stains cannot be dismissed (see page 360).

2. OXIDATIVE STAINS

Sapwood of all trees contains appreciable amounts of organic compounds, as well as of oxidizing enzymes. When temperature and moisture conditions of the air are favorable, the action of enzymes on the organic materials may give rise to colored substances causing discolorations that may be superficial or may penetrate deeply into the wood. Such stains are known as ***oxidative stains***; they vary in color, ranging from shades of yellow, gray, greenish, and orange to brown. These stains develop both in softwoods and in hardwoods, sometimes in logs during storage, but more frequently during the seasoning of lumber. Stains of this nature do not affect the strength of wood.

As a group, hardwood species are more susceptible to oxidative discolorations. Many of these are of little economic importance; others may cause considerable degrade. For instance, a deep greenish-brown stain in persimmon, which commonly develops during air seasoning, may penetrate through the thickness of the sapwood and hence render the stock unfit for certain special uses requiring bright wood. This type of stain can be prevented by mild steaming, which destroys the oxidizing enzymes.

Another example is a bluish discoloration, resembling blue stain (see page 357), frequently occurring in tupelo and magnolia. Examination of discolored wood under a microscope indicates that starch in the parenchyma cells is converted to gumlike

* A. J. Mia, Light and Electron Microscope Studies of Crystalline Substances in *Acer Saccharum* Mineral Stain, *Wood Science*, **2**(2):120–124 (1969).

deposits, oxidized to a brownish color. No satisfactory method of control has been found for the thicker stock affected with this type of stain, but more rapid drying of thinner lumber is effective in reducing the damage.

Though less widely distributed, chemical stain in the softwoods causes considerable financial losses to producers of lumber. The most important are brown stains in western and eastern pines, principally in sugar, ponderosa, and white pines. Brown stains develop in lumber during drying. They are more pronounced in kiln drying than in air seasoning; the former is called ***kiln brown stain*** and the latter ***yard brown stain***.

Brown stains may be confined to the surface or develop throughout the thickness of sapwood. Though brown stains affect appearance, they do not impair strength. They are believed to be caused by chemical reaction that takes place as the water-soluble cell contents are transferred to the surface by water movement, and are deposited as solids when the water evaporates. These solids may be colored in themselves or become colored through oxidation or polymerization, enhanced by elevated temperatures.[36] The longer the time between felling the tree and its conversion into lumber, and the time between sawing and piling lumber and air seasoning or kiln drying, the more pronounced the stain. In kiln drying staining may be reduced by employing an initial dry-bulb temperature of less than 130°F, with relative humidity as low as the stock will tolerate without excessive checking; a wet-bulb temperature not to exceed 120°F must be maintained throughout the drying. According to Ceth,[10] brown stain in white pine lumber can be eliminated almost entirely by treating it with sodium fluoride solution immediately after sawing.

A special case of oxidative stain is that of ***sticker markings***, which occur during air seasoning and kiln drying. Sticker markings, variable in color, develop on and beneath the surface of the board, where stickers come in contact with it (Fig. 10-9). It is believed that these stains develop as a result of chemical changes in wood extractives during drying. No means of preventing such discolorations are known, but their extent can be reduced by the use of dry narrow stickers and by starting drying of green lumber as rapidly as possible.

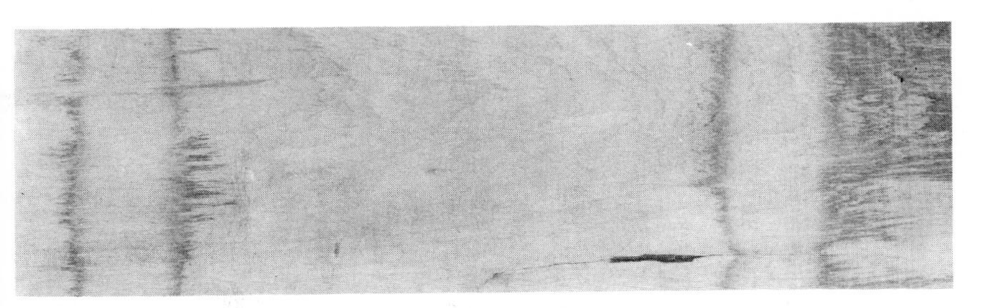

FIG. 10-9 Sticker marks in hard maple. (*Photograph courtesy of U.S. Forest Service, Forest Products Laboratory.*)

3. CHEMICAL STAINS

The most common of the chemical stains, though usually of minor economic importance, is that produced by reaction between wood containing tannins and iron. The resulting iron tannate is a blue-black compound resembling ink stain. This stain is superficial and generally can be removed by surfacing the lumber. Iron stains may be objectionable in wood products constructed with iron fasteners. Since acid conditions are required for formation of iron tannate stains, discolorations can be prevented by applying an alkaline solution to the surface of the wood before it comes in contact with the iron. It has also been suggested that formation of iron tannates may reduce decay resistance of wood by removing some fungitoxic chemical protection.

4. WHITE SPOTS OR STREAKS ("FLOCCOSOIDS") IN WESTERN HEMLOCK

Floccosoids is a term coined by Grondal and Mottet[23] for the whitish spots that quite frequently occur in western hemlock [*Tsuga heterophylla* (Raf.) Sarg.]. They are due to whitish deposits in the wood, mainly crystalline α-conidendrin-4-glucoside, a normal minor organic constituent of hemlock wood.*

Floccosoids sometimes resemble the white decay pockets of certain fungi, e.g.,

* G. M. Barton and C. R. Daniels, An Explanation of Floccosoid Formation in the Wood of Western Hemlock, *Wood Sci.*, **1**(4):238-240 (1969).

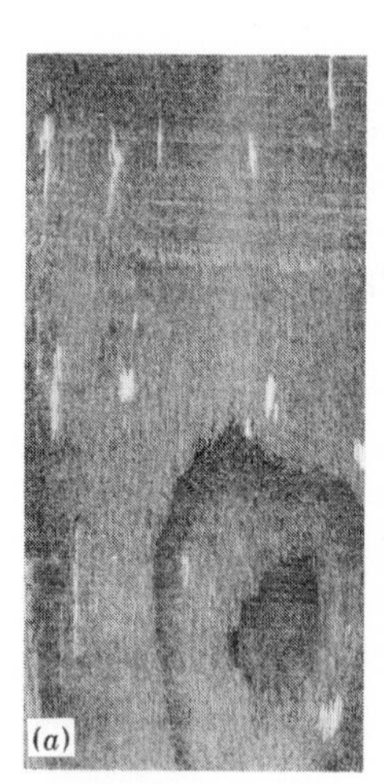

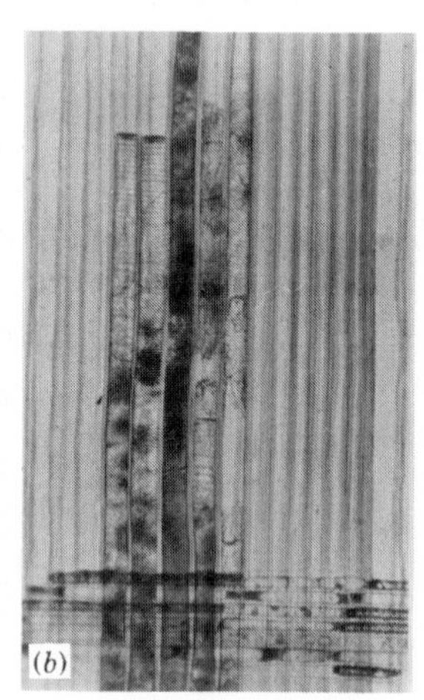

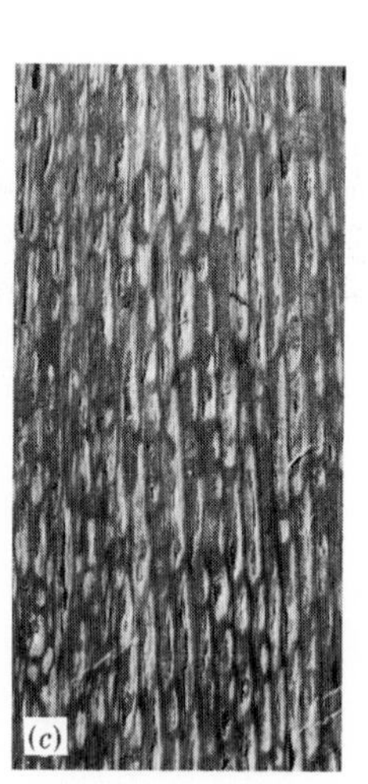

FIG. 10-10 Floccosoids (white specks) in western hemlock [*Tsuga heterophylla* (Raf.) Sarg.].

(*a*) Floccosoids in rotary-cut veneer.

(*b*) Floccosoids as seen along the grain through the microscope. (Approximately 120×.)

(*c*) Decay pockets in all stages of development in sliced, edge-grained veneer. Such pockets are sometimes mistaken for floccosoids.

[*By permission from Eloise Gerry, Western Hemlock "Floccosoids" (White Spots or Streaks), U.S. Forest Prod. Lab. Rept.* 1392, 1943.]

those of *Fomes pini* (Thore) Lloyd (Fig. 10-10). The wood in a floccosoid, however, is as sound as normal wood; the light color of the tissue is traceable to infiltrations in excess of normal. In decay pockets of the type portrayed, in contrast, the bleaching of the wood is due to a white rot, and the partially decomposed tissue is softer than normal wood. A positive test for floccosoids can be made by treating a small, thin, longitudinal section through an involved area with a few drops of caustic alkali (potassium or sodium hydroxide) on a glass slide. A cover glass is then added and the mount is examined under a microscope. If the spot is a floccosoid, it disappears gradually but completely in a relatively short time; a decay pocket remains unchanged.

5. MACHINE BURNS

Brown stains caused by excessive friction developed when progress of the piece is impeded in planing, or some other woodcutting operation, are called *machine burns*. High temperatures developed because of friction result in scorching of the surface. Generally such stains are superficial and can be easily removed by further light planing or sanding.

IV. WEATHERING OF WOOD

When wood, unprotected by paint or any other means, is exposed to the weather, its surface undergoes changes, in part physical and in part chemical, the cumulative effects of which are termed *weathering*.

The first indication of weathering is change in color. Initially, the dark-colored woods tend to fade and the light-colored varieties to darken somewhat. However, as the weathering continues, all woods assume a silvery-gray color, with the gray layer extending 0.003 to 0.01 inch in depth. In humid climates the final appearance of weathered surfaces may be further modified by the surface growth of spores and mycelia of fungi, resulting in unsightly, blotchy discolorations, usually dark gray in color. At this stage the process of weathering is mainly photooxidative in nature. The surface appearance is greatly affected by the wavelength of light, with the ultraviolet light being most destructive, resulting in degradation of lignin and extractives, followed by removal of the extractive materials by the action of atmospheric moisture. As a result, the partially loosened wood fibers in the gray layer consist mostly of the more leach-resistant fractions of cellulose.

Since wood is hygroscopic in nature, its unprotected surface tends to absorb moisture and to swell during humid and rainy weather, and to lose moisture and shrink in periods of dry weather. This, coupled with the differential shrinkage of early and late wood, leads to formation of raised grain. Owing to the slow rate of moisture transfusion through wood, these changes in moisture content, and hence in wood dimensions, are for the most part confined to the surface layers. The alternate compression and tension stresses set up in the shell of a piece of wood eventually result in formation of microscopic checks on the exposed surface.

If weathering is permitted to proceed unimpeded, most wood will develop larger and deeper checks, which become visible. This, combined with the abrasive effect of rain, hail, wind-borne particles, and freezing and thawing, causes the wearing away of surface layers and a roughened appearance of the surface. The differential dimensional changes between the surface and the interior of a board, combined with the surface abrasion, frequently lead to warping of weathered boards and the loosening of fasteners. Warping is more pronounced in the denser woods and in wide boards the width of which exceeds their thickness by more than eight times.

The effect of normal weathering may be accelerated by polluted atmosphere containing sulfur dioxide, which may be responsible for partial hydrolysis of cellulose and disintegration of lignin. Surface molds may also contribute to weathering. Finally, it should be noted that the surface conditions commonly attributed to nonbiological weathering may in fact be due to the attack of soft-rot fungi (see page 347).

In the case of the more popular species used for siding, e.g., redwood, western redcedar, or baldcypress, more uniformly attractive, weathered surfaces can be achieved by treating the exposed surfaces with a water-repellent preservative, which tends to retard leaching of wood extractives and formation of mildew on the surface

V. DETERIORATION CAUSED BY INSECTS

From the standpoint of wood utilization the insects that damage wood can be segregated roughly into those whose attacks are confined to wood before it is utilized and those whose damage is mainly restricted to wood in service. Pith flecks, pinholes, and grub holes result from the activities of insects belonging to the first category; powder-post beetles and termites are the most important examples of insects that attack converted wood. It should also be recognized that in some cases insect attack, of minor significance in itself, is a major factor in spreading fungous diseases, e.g., sap stain.

A. Insect Damage in Wood before It Is Utilized

1. PITH FLECKS.[8] (Fig. 10-11)

Pith flecks, or medullary spots, are confined to hardwoods. On transverse surfaces they appear as small areas of wound tissue which are usually darker than the surrounding tissue and are wholly within the limits of a growth ring.

Pith flecks are usually either semicircular or lunate (*g*) and so oriented that the long diameter is directed tangentially, ranging for the most part between $^1/_{16}$ and $^1/_8$ inch. Along the grain, pith flecks appear as dark streaks of varying length (*a*). The degree of prominence of such dark lines varies considerably.

This defect results from injury to the cambium by the larvae of flies (*f*) belonging the genus *Agromyza*. The female adult insect perforates the periderm of a young ranch with her ovipositor and deposits an egg in the living tissue beneath. This atches into a filiform larva (*d*) which invades the cambium during the early part of e growing season and mines downward, leaving a burrow (*c*) about the thickness f a darning needle. Eventually the larva emerges underground and pupates (*e*).

Some of the cambial cells, and neighboring phloem and immature xylary cells, are estroyed as the larva travels downward; at the same time, the continuity of the rays rough the cambial zone is broken. Healing processes start shortly after the passage f the larva. The living phloem ray cells on the outside of the tunnel begin to roliferate and soon occlude the tunnel with a mass of parenchymatous cells con- aining dark contents (*g*, *h*, and *i*). A short time later the cambium closes in over the ass of wound tissue, and normal wood is produced thereafter. Since the wound ssue occluding the mine resembles that of pith, or medulla, the spots formed as escribed above have been called *pith flecks* or *medullary spots*.

Pith flecks are fairly common in some hardwoods and unusually abundant in thers. Birch, maple, basswood, willow, and cherry frequently exhibit these defects; pots of this sort are especially abundant in gray and in river birch. Pith flecks do not aterially affect the strength of wood although they do detract from its appearance; e last has a bearing, of course, where the wood is to be given a natural finish.

. PINHOLES

inholes are small, round, and usually unoccluded holes $^1/_{100}$ to $^1/_4$ inch in diameter, esulting from the mining of ambrosia beetles (Fig. 10-12). These insects belong to he families Scolytidae and Platypodidae, with more than 1000 species of insects nvolved.

Ambrosia beetles mainly attack recently killed trees, logs, bolts, and green lumber. o species of timber is apparently immune. These insects attack chiefly the sapwood f both softwoods and hardwoods, though sometimes they also damage the heart- vood. In domestic species pinholes are especially common in oak and chestnut, in vhich a special "sound wormy" grade of lumber is recognized as suitable for uses vhere permeability and impairment of finishing characteristics of the wood are ot important. Unless pinholes are extremely numerous, there is no reduction in trength. Since the attack by the ambrosia beetles is confined to the green lumber, hese insects cause no further damage in seasoned wood if it is kept dry.

Ambrosia beetles do not consume wood; they derive nourishment from a nold-type fungus, introduced by the beetles into their tunnels, on the walls of which t grows. It is to this mold that the name *ambrosia* was applied, and later extended to he beetle.

The damage caused by ambrosia beetle tunnels may be further intensified by levelopment of greenish-gray or bluish-black stained areas in the vicinity of insect unnels. These areas resemble mineral stains in appearance. They can be caused by taining fungi, which are also carried into the wood by beetles, or stains may develop

FIG. 10-11 Pith flecks in wood.

(*a*) Pith-fleck streaks on the face of a flat-sawn board of silver maple (*Acer saccharinum* L.) (one-half natural size). (*Photograph by the U.S. Forest Service.*)

(*b*) Pith flecks on the transverse surface of a limb of river birch (*Betula nigra* L.) (one-half natural size). (*Photograph by the U.S. Forest Service.*)

(*c*) Mines caused by the larvae of *Agromyza pruinosa* Coq., under the bark of *Betula nigra* L.; they are occluded with dark wound tissue and are conspicuous on the surface of the wood when the bark is removed. (*Photograph by the U.S. Forest Service.*)

(*d*) Cylindrical larva of *Agromyza pruinosa* Coq., showing details of structure (the larva is white except for the shiny black mouth parts), shiny mouth parts (1, enlarged) (2, in position); the anterior spiracles (3); the posterior spiracles (4). (3½×.) (*U.S. Bureau of Entomology.*)

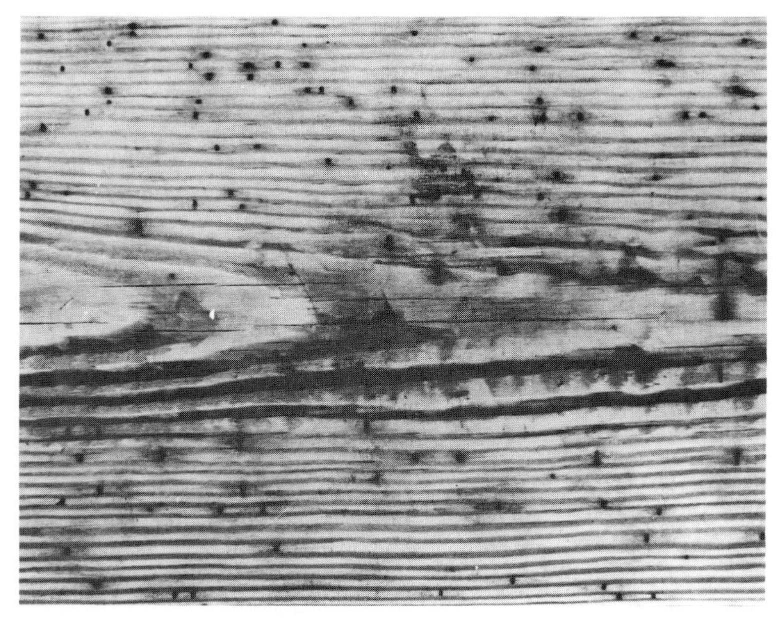

FIG. 10-12 Ambrosia beetle damage in sapwood and heartwood of southern yellow pine. (*Photograph U.S. Forest Service, Forest Insects Laboratory.*)

n the absence of sapwood-staining microorganisms through mechanical damage by the adult beetles, as in the case of the Columbian timber beetle (*Corthylis columbianus* Hopkins).* This species of the Ambrosia beetle is noted for attacking only living trees, without killing the host. Repeated invasions of this insect result in a large volume of wood degradation in the infested trees, by tunneling and extensive sapwood staining, greater than that caused by other North American species of ambrosia beetles. It is reported to have caused losses in the past 10 years amounting to 40 percent in the timber value of yellow poplar and white oak veneer and cooperage in the Southeastern states, and 30 to 35 percent reduction in the value of soft maple lumber in the Central states.

Ambrosia beetle damage can be prevented or at least minimized by prompt conversion of the extracted timber, followed by drying or by chemical control, involving

* K. L. Giese, *op. cit.*, pp. 361–370; see footnote p. 341.

(*e*) Pupa of *Agromyza pruinosa* Coq. (3½×.) (*Courtesy of the U.S. Bureau of Entomology.*)

(*f*) Adult male of *Agromyza pruinosa* Coq. (5½×.) (*Courtesy of the U.S. Bureau of Entomology.*)

(*g*) Pith flecks on the transverse surface of gray birch. (*Betula populifolia* Marsh.) (3×.)

(*h*) A pith fleck in red maple (*Acer rubrum* L.), transverse view. (10×.)

(*i*) Occluded tunnel in red maple (*A. rubrum* L.), made by the larva of a pith-fleck insect. Radial view. (10×.)

application of the insecticide-containing products by spraying, dusting, or dipping. Storage of logs under water is also recognized as an effective method of preventing infestation, as long as the entire log is submerged; exposed surfaces may still be freely attacked.

3. GRUB HOLES

Insect mines, exceeding 1/4 inch in diameter, are generally called *grub holes*. Defects of this type are oval or irregular openings 3/8 to 1 inch in diameter (Fig. 10-13). The holes are sections of the tunnels of some adult insects and of the larvae of others that deposit eggs in the living trees, felled trees, logs, and in unseasoned lumber. Grub holes may occur in the wood of any species, in either the sapwood or the heartwood.

Many different types of insects are involved, including the horntails and the flat- and round-headed borers. One of the more common is the horntails, the larvae of which, unlike ambrosia beetles, utilize wood as food. The adults of horntails are wasplike insects which fly in the spring and summer months. Horntails appear to be

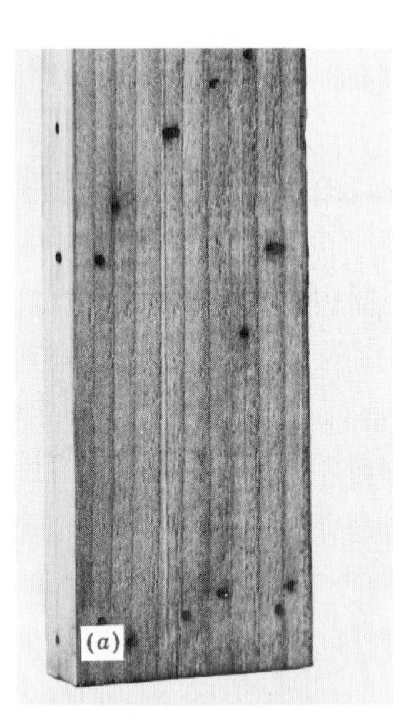

FIG. 10-13 Insect damage appearing in wood before it is utilized.
(*a*) Pinholes in western hemlock [*Tsuga heterophylla* (Raf.) Sarg.] caused by ambrosia beetles (adult stage). The dark linings of the tunnel walls are characteristic of the borings of these insects.
(*b*) A grub hole in western fir (*Abies* sp.) caused by the pine sawyer beetle (*Monochamus* sp.).
(*Photographs by permission, from H. W. Eades, British Columbia Softwoods, Their Decay and Natural Defects, Can. Forest Serv. Bull.* 80, 1932.)

especially attracted by recently fire-killed trees but will attack any recently killed or freshly cut logs. The larvae may continue to work in lumber sawn from infested logs, but this insect will not reinfest wood once it is dried. Kiln drying will kill larvae in sawn material.

The future usefulness of the wood with grub holes depends upon the extent of the damage and the manner in which the stock is to be used. Not infrequently material containing grub holes is rejected, when strength, impermeability, and appearance are primary considerations; at other times, when less-exacting service is required, it is only downgraded.

4. BLACK STREAK IN WESTERN HEMLOCK

The black streaks in western hemlock [*Tsuga heterophylla* (Raf.) Sarg.] (Fig. 10-14), also known as *black checks*, are caused by the maggots of a small black fly (*Chilosia alaskensis* Hunter) which live under the bark and feed at the surface of the newly formed wood; the larval stage lasts from 1 to 5 years. As a result of the injury and destruction of portions of the cambium by the growing maggot, a chamber is formed, above and below which a black streak, resembling a pencil line as viewed in edge-grain stock, extends from several inches to as much as 3 feet along the grain. In flat-sawn lumber the chamber is irregular in shape.

The chamber formed as described above may remain open or become considerably narrowed by the subsequent formation of wound tissue. In the latter instance, the traumatic zone is black and not only encircles the chamber but extends along the grain above and below it; it consists of numerous short parenchymatous cells with moderately thick walls and usually with dark resinous contents (Fig. 10-14*b*). Included within the black streak are numerous small, traumatic resin canals or short cystlike openings, arranged in a tangential row (Fig. 10-14*c*).

Mechanical tests of western hemlock wood containing black streak indicate that stock with this defect, except when maggot chambers are present, is suitable for all purposes to which this wood is put.

B. Insect Damage to Wood in Service

1. POWDER-POST BEETLE DAMAGE[41]

The term *powder-post beetles* is applied to the beetles of two related families, the Bostrychidae and Lyctidae. In the United States most damage is done by the *Lyctus* species (Lyctidae).

The attack of the American powder-post beetles is confined mainly to the hardwoods, though the beetles, especially those belonging to the Bostrychidae family, are capable of attacking softwoods as well. Powder-post beetles attack sapwood only, although the adult beetles may emerge through heartwood. The larvae of the beetle, or sometimes the beetles themselves, derive food from the reserve foodstuffs in the parenchyma, largely starch; the latter is absent, in the main, in the parenchyma cells

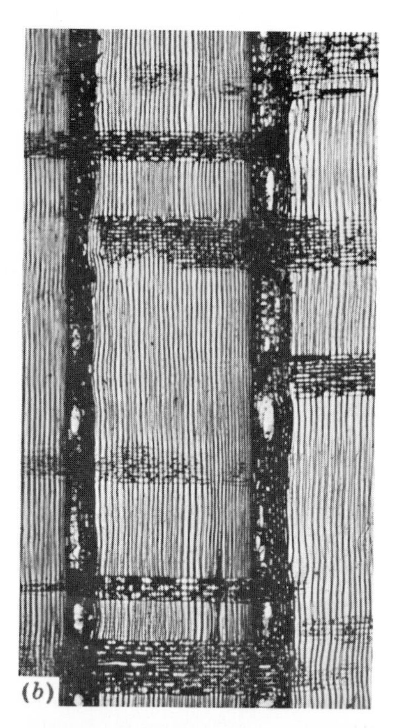

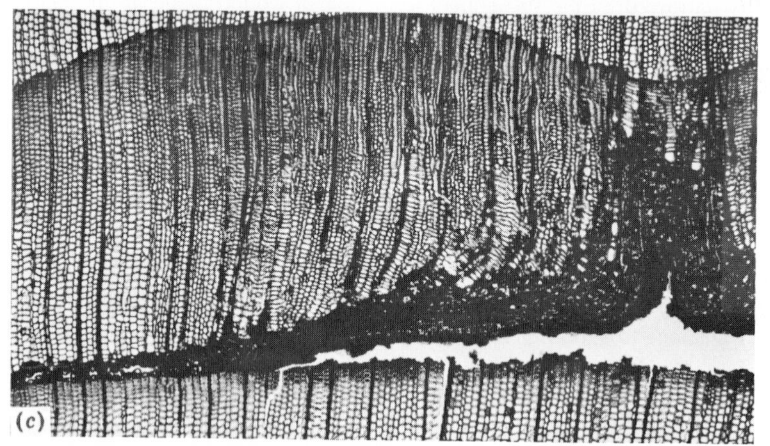

FIG. 10-14 Insect damage occurring in wood before it is utilized. Black streak in western hemlock [*Tsuga heterophylla* (Raf.) Sarg.].

(*a*) Two black streaks in an edge-grained board, with maggot chambers (wider portions of streaks).

(*b*) Two black streaks on edge-grained surface, with resin cysts. (Approximately 30×.)

(*c*) Maggot chamber and black streak with resin cysts, on end-grain. (Approximately 30×.)

(*By permission from R. F. Luxford, L. W. Wood, and Eloise Gerry, Black Streak in Western Hemlock; Its Characteristics and Influence on Strength, U.S. Forest Products Laboratory Rept.* 1500, *December*, 1943.)

of heartwood. Powder-post injury is practically limited to air-seasoned and kiln-dried lumber, though cases are on record where these beetles have invaded wood containing as much as 40 percent moisture.

The term powder-post beetles was applied to these insects because the larvae bore through the wood and leave residual wood substance in a finely pulverized condition; the flourlike residue sifts out from the tunnels when the adults emerge during late spring and summer, leaving holes 1/16 to 1/12 inch in diameter (Fig. 10-15). The interior of the wood may be riddled with tunnels, with little visual evidence on the outside of the piece to indicate the extent of destruction.

The adults deposit their eggs in the vessels (pores) of hardwoods. Because of this the attack is largely confined to woods the pores of which are sufficiently large to receive the eggs. In the United States powder-post damage is most frequent in ash, hickory, oak, and California-laurel, although cherry, elm, maple, poplar, and black walnut are also affected. However, the powder-post beetles may also start infestation by depositing eggs in the cracks and crevices of hardwoods with small pores and in the softwoods. The same piece may be reinfested by several generations of larvae, until it has been rendered useless.

Powder-post beetles cause serious damage in dry lumber, in buildings, and in a variety of hardwood products in the various stages of manufacture.

Prevention is the key to powder-post damage; once infestation has started, elimination of the beetles is expensive and frequently uncertain.

Since the larvae of Lyctus beetles derive their food from starch, any treatment that will reduce the starch content in the parenchyma cells of the sapwood will minimize the danger of powder-post beetle infestation. For instance, storing logs for several months after cutting will tend to reduce the reserves of starch because of the continuing respiration of the living parenchyma cells in the sapwood. Such treatment,

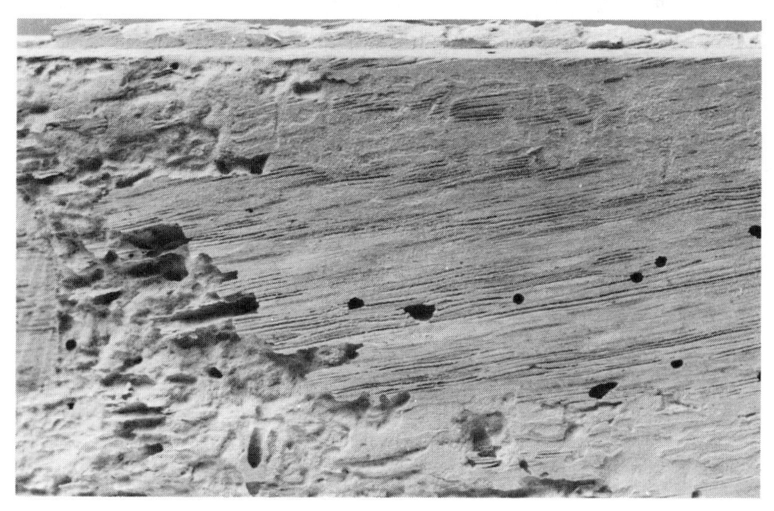

FIG. 10-15 Damage in oak flooring by *Lyctus planicallis* Lec. (*Photograph courtesy of U.S. Forest Service, Beltsville Insect Laboratory.*)

however, introduces the danger of infestation by other insects and fungi, and therefore is not practical in most cases.

Effective prevention methods should include periodic inspection of suspected stock and proper sanitation, requiring prompt burning of infested wood. Heat sterilization by exposing wood to live steam at not less than 135°F, or by drying at 180°F, offers an effective method for the destruction of beetles and their larvae in the already infested material.

A thorough surface treatment, or dipping when practical, with insecticides, such as dieldrin or DDT, will provide effective protection against attack. Such treatment is of little value for destroying larvae already in the wood, unless large amounts of insecticide are used and the chemical is allowed to penetrate deeply into the wood. Even then, several applications may be required and it may take several months before all the larvae are destroyed. It should also be kept in mind that the effectiveness of chemicals may be nullified if the treated material is surfaced or is resawn. Complete protection with insecticides is, of course, possible if chemicals are injected into the wood by a pressure method.

A pest-control operator may be able to obtain comparable results by injecting the chemical under pressure through holes bored at short intervals in the infested wood that is an integral part of a building. But by far the most practical method of eliminating powder-post beetles in buildings and in furniture is by fumigation with hydrocyanic acid or methyl bromide gases. These gases are extremely poisonous, and therefore *fumigation should be done only by a licensed operator.*

Finished products made of wood can be protected against *Lyctus* damage by coating thoroughly *all surfaces* with a substance that will seal the pores, thus preventing the adult female insects from depositing eggs in the wood. This is not practical in most cases. Furthermore, experimental work at the Forest Products Research Laboratories in England and in Australia has shown that some powder-post beetles are capable of cutting through the finish, exposing the vessels underneath, by a series of "tasting marks."[6]

2. COMMON FURNITURE BEETLES[3,20]

In spite of the name, the attack of this beetle (*Anobium* sp.) is not confined to furniture but extends to structural timber, plywood, and other types of seasoned wood products. Because of the size of the escape holes, the damage is frequently mistaken for that caused by powder-post beetles.

These beetles infest chiefly the sapwood of both the softwoods and the hardwoods; they will also attack the heartwood of some species if nitrogen is present in the quantities necessary to meet the nutritional requirements of the insects. Larvae of this beetle grow more rapidly in wood affected with soft and white rot, and in situations where relatively high humidity and low maximum temperatures prevail. Analysis of the digested wood indicates that in addition to nitrogenous materials these beetles remove carbohydrates, leaving the lignin unaffected.

The same remedial treatment used in preventing powder-post-beetle damage is applicable for control of common-furniture-beetle infestations.

3. TERMITES

Termites have a worldwide distribution through the tropical and temperate zones. In the United States they are found in every state of the Union, with the possible exception of Alaska. They are most destructive in the southeastern states and in California.

Originally a forest-inhabiting insect, confined mostly to the South, termites have invaded man-made structures and have spread northward with the general adoption of central heating. Heated basements and the extensive use of lumber cut from young trees with wide sapwood account for the rapid spread of this insect. The diversion of forested areas for suburban developments has also contributed to the seriousness of the termite problem.

Termites attack all species of wood, either sapwood or heartwood, and will infest sound or decayed wood. This latter fact led to an assumption that all termites are dependent on fungi or other microorganisms for their survival. There is no clear-cut proof that this is the case. All termites use cellulose and other carbohydrates in the wood for food. Some species of termites found in the United States convert cellulose into a digestible form by the action of protozoa inhabiting their intestinal tracts. Others apparently are capable of initiating the attack on wood and of maintaining themselves in it without the presence of any microorganisms. In fact some wood-destroying fungi and bacteria are repellent or even toxic to termites.

There are three groups of termites in the United States: (1) the subterranean, (2) the dry-wood, and (3) the damp-wood types. Of these the subterranean termites are by far the most common and the most destructive. The dry-wood termites are restricted mainly to the Deep South, near the coast from Cape Henry to the Florida Keys and westward along the coast of the Gulf of Mexico to the Pacific Coast, as far as northern California. A local infestation has also been reported in Tacoma, Wash. The damp-wood termites are of relatively little economic importance, since their attack is confined mainly to wood buried in the ground.

All termites, like ants, are social insects, living in colonies made up of reproductive members, workers, and soldiers. Subterranean termites establish their colonies in the ground, from which the workers move out in search of food and to which they return with it. Usually built near the source of food, termite nests may sometimes be yards distant from the affected structures.

The workers penetrate the wood, destroying the interior of infested pieces but leaving a thin unbroken shell on the outside for protection (Fig. 10-16). This method of attack frequently renders detection difficult until the damage has reached a point where the affected part collapses. To pass over an obstruction or to reach wood not in contact with the soil, termites construct covered runways, or shelter tubes, $1/4$ to $1/2$ or more inches wide (Fig. 10-17). These tubes consist of partially digested particles of wood cemented by excretion. In addition to being passageways connecting the

FIG. 10-16 Termite damage follows the grain. In the middle photograph a portion of the wood was removed to expose tunnels cluttered with gray-brown excretions. (*a*) An adult "worker" termite (*x*). (*Photographs courtesy of U.S. Forest Service, Beltsville Insect Laboratory.*)

wood and the soil, the tubes serve to provide the essential moisture and to protect termites against the drying effect of direct exposure to air.

Dry- and damp-wood termites require no contact with the soil; they enter the wood directly from the air at the time of swarming. Dry-wood termites use wood for shelter as well as for food. Damage caused by dry-wood termites can be recognized by cavities cut across the grain as well as with it. These cavities contain pellets of partially decomposed wood. Further indication of infestation is small piles of fecal matter that fall through tiny openings in the affected wood and accumulate on the surfaces below. Dry-wood termites are able to work in wood with a moisture content of only 10 percent and require no outside moisture as do subterranean types.

The best time to provide the most effective and economic protection against subterranean termites is in the course of planning and construction of a building. These measures can be briefly summarized as follows: (1) sanitation, i.e., complete removal of all tree roots, stumps, and wood debris from the building site; (2) provision for adequate soil drainage beneath and around the building; (3) chemical treatment of the soil under the building and around the foundation, especially when slab construction is used; (4) adequate ventilation under the building; (5) keeping all wood in the building at least 8 inches from the ground; (6) installation of metal shields; (7) making all foundations as impervious to termites as possible; and (8) use of pressure-treated lumber, especially in the foundations.[51] The slow-growing heartwood of redwood, tidewater baldcypress, eastern redcedar, and very pitchy southern pine is considered quite resistant to termites, but this material is not immune to their attack and is not so resistant as wood well treated with preservatives applied under pressure.

FIG. 10-17 Termite shelter tube built to a height of 15½ inches from the ground to reach wood joists in the house. (*Photograph courtesy of Shell Chemical Company.*)

The elimination of termites in an infested building is difficult and expensive but can be done by the application of chemicals, destruction of the nest, and replacement of the affected parts. All these measures are futile, however, unless steps are taken to eliminate conditions which favored the attack in the first place. These measures are identical with those recommended for new construction.

4. CARPENTER ANTS*

These are a group of large, jet-black ants, belonging mostly to the species of genus *Camponetus*. Originally forest-inhabiting insects, they may be considered beneficial in their native habitat because of the role they play in reduction of wood debris. However they also invade living trees through wounds and stubs of dead branches and establish themselves in the sound heartwood, thereby weakening the

* J. B. Simeone, Carpenter Ants and Their Control, *State Univ. New York, College of Forestry at Syracuse*, Bull. 34, 1954.

tree and reducing the quality of its wood. But it is because of damage the carpenter ants cause to poles, posts, and structures that they are regarded as a serious pest in some parts of this country and Canada.

Carpenter ants live in colonies consisting of a variety of individuals. The damage to wood is done by the workers, which extend galleries in the infested wood to provide more room for the colony. These galleries usually follow the softer portions of the early wood, with the inner surfaces of excavations having a characteristic sand-papered appearance (Fig. 10-18).

Although nesting in wood, carpenter ants do not utilize it for food. Their food is derived from secretions of other insects such as aphids, mealy bugs, and tree hoppers. The ant nest may generally be detected by the activity of the workers and by the accumulation of piles of discarded, sawdustlike wood fragments at the base of the nest.

The new colonies of ants are established in the summer months by the winged males and females which emerge from the old colony. After mating the male dies and the female excavates a brood chamber in suitable wood, in which she lays eggs. The population of the new colony increases rapidly after the first year and may reach two to three thousand members under favorable conditions. After 4 or 5 years new swarms of winged ants are produced every year in the summer, thereby spreading the infestation.

Damage to structures by carpenter ants may be prevented by the use of wood with a moisture content below 15 percent and by maintaining conditions which will not result in increase in the moisture content. Use of wood treated with preservatives is highly recommended, especially where low moisture content of the wood cannot be guaranteed. Existing colonies can be exterminated by a thorough application of insecticide such as chlordane or DDT. When infestation is extensive, the services of pest-control operators may be advisable. Carpenter ants may be attracted by food in the kitchen areas; in such cases they can be controlled with the same chemical recommended for their extermination in structures.

VI. DEFECTS RESULTING FROM THE ACTIVITIES OF MARINE BORERS

Marine borers are animals belonging to the Mollusca (molluscs) and the Crustacea (crustaceans). These organisms inhabit salt and brackish water and inflict extensive damage on submerged wooden structures or parts exposed to low tide. Untreated piling and other wood objects, especially those in protected situations such as harbors, are subjected to these attacks. In the United States marine borers are destructive all along the Pacific, South Atlantic, and Gulf Coast, especially in warm water. Ruth D. Turner,[39] after observing that "Man's fight against the shipworms probably goes back to the time primitive man first learned to use rafts and small boats in sea water," goes on to say, "It appears certain that many of the early explorers who set out on voyages of discovery never to be heard from again probably were

FIG. 10-18 Carpenter ant damage. Inner surface of excavations has a characteristic sandpapered appearance. Normal size.

wrecked not because of poor navigation, but because their vessels were so weakened by shipworms that they could not withstand the buffeting of heavy storms."

Three genera of molluscs—*Teredo*, *Bankia*, and *Martesia*—and three of the crustacean genera—*Limnoria*, *Sphaeroma*, and *Chelura*—are represented in American waters. Members of the first two genera of molluscs are known as *shipworms*, largely because of the vermiform shape assumed by the adults. Shipworms may attain a length of 1 to 4 feet and a diameter up to 1 inch, although they are generally much smaller. In early life, they are minute and free-swimming but soon become attached at the caudal end to some wooden object. They proceed then to burrow in the wood, meanwhile increasing in length and diameter. The burrow is made by the grinding action of a pair of toothed shell valves which develop on the head. Two siphons, situated at the caudal end, can be protruded at will beyond the surface of the wood; one of these is used to suck in water and microscopic organisms which are used for food, the other to eject water and excrements. *Martesia*, the third genus of molluscan borers, consists of forms that are clamlike in appearance; they are generally not more than 2 to 2½ inches in length and about 1 inch in diameter. It is believed that wood-boring molluscs use wood not only for shelter but also for food, at least to some extent.

Crustacean wood borers have segmented bodies ranging from ⅛ to ¼ inch in length in *Limnoria*, to about twice this size in *Sphaeroma*. These organisms are equipped with several pairs of clawed legs by means of which they attach themselves to wood. The boring is done with a pair of toothed mandibles. *Limnoria* spp. are said to utilize wood not only for shelter but also for food (Fig. 10-20); the *Sphaeroma* spp., on the other hand, apparently make their homes in the wood but obtain their nourishment from other sources. Little is known about the habits of the sand fleas which belong to the genus *Chelura;* they jump like fleas, hence the name. Until recently sand fleas were considered to be largely tropical and subtropical in their distribution. It appears now that they cause damage along the Atlantic Coast.

There are striking differences in the life habits of molluscan borers of the shipworm type and the crustacean forms. The former, upon entering the wood, become attached to and imprisoned in it, extending their tunnels deeper into the core as their bodies elongate. As a result, the interior of infested piling is often thoroughly honeycombed though the wood appears sound on the outside (Fig. 10-19). The crustaceans, on the other hand, are motile throughout their whole life cycle; they can pass in and out of the wood at will, and their burrows seldom extend for any distance into it. But, since they attack wood in great numbers, the surface layers are riddled with their mines and are easily dislodged by wave action. The attack of crustacean borers is most severe between the half-tide and ebb-tide levels. In consequence the cumulative effect of repeated attacks, each resulting in the destruction of the surface layer exposed at a given time, wears away the infested portion of a pile. Eventually this may become hourglass in shape (Fig. 10-20). The diameter of a pile may be reduced as much as 2 inches in a single year.

No species of wood in its natural state is immune from the attack of marine borers. The heartwood of several tropical woods, however, is reputed to be unusually resis-

FIG. 10-19 Defects of wood resulting from the activities of marine borers.
(*a*) Transverse section of a sound pile of Douglas-fir [*Pseudotsuga menziesii* (Mirb.) Franco].
(*b*) Transverse section of a similar pile after 9 months' exposure to *Teredo* attack.
(*Photographs by permission, from T. A. McElhanney and associates, "Canadian Woods, Their Properties and Uses," Forest Products Laboratories of Canada,* 1935.)

FIG. 10-20 Untreated piling of Douglas-fir [*Pseudotsuga menziesii* (Mirb.) Franco] destroyed by *Limnoria*. The good piles which show in the photograph are replacements. (*By permission, from T. A. McElhanney and associates, "Canadian Woods, Their Properties and Uses," Forest Products Laboratories of Canada,* 1935.)

tant to their attack, among them greenheart [*Ocotea rodiaei* (Schomb.) Mez.] of South America. It has been suggested that some tropical woods exhibiting resistance to marine borers contain considerable quantities of silica inclusions in the ray cells. An interesting example of such a timber is Australian turpentine (*Syncarpia laurifolia* Ten.).

Considerable difference of opinion exists concerning the role wood plays as food for the marine borers. Some authorities maintain that all their food consists of plankton and that wood passes unchanged through their digestive tract. Chemical analysis of wood that passed through the gut of *Limnoria* indicates that these borers, at least, are capable of removing nearly all the hemicelluloses and about half the cellulose.[39] Similar results were also obtained for *Chelura*, *Teredo*, and *Bankia*.

Control of marine borers is difficult. Many methods of protecting wood have been tried with varying degrees of success. The best results, so far, have been attained by thorough pressure impregnation of the wood with coal-tar creosote and other suitable nonleaching preservatives. It should be noted, however, that different species of the marine borers exhibit a varying degree of tolerance to a preservative substance, and therefore one poison may not be effective in all cases. For instance, some species of *Limnoria* appear to be tolerant to creosote, which is toxic for other species of marine borers.[20]

VII. NATURAL DURABILITY OF WOOD

The natural durability of wood, as interpreted in this text, signifies its ability to resist the attacks of foreign organisms, i.e., fungi, insects, and marine borers. Although no native wood is entirely immune to the attack of such organisms, a number possess superior resistance. It must be kept in mind, however, that timber resistant to fungal attack may or may not be durable when subjected to attack by insects or marine borers. Furthermore, the durability of a given kind of wood may fluctuate between wide extremes.

Wood is more resistant to deterioration by microorganisms than most other plant tissues. Some of the native resistance of woody tissues to the biological agencies is due to the characteristics of its principal cell wall constituents. However, the main reason for the natural durability of some species of wood is the presence of toxic substances in the heartwood.

The natural resistance of wood to deterioration, that can be ascribed to reasons other than the toxicity of its extraneous material, may be summarized as follows:[47]

1. The woody cell walls consist of highly complex, insoluble polymers of high molecular weight; these substances must be altered (depolymerized), by enzymes produced by the attacking organisms, into simpler products that can be assimilated by them.

2. Wood lignification creates a physical barrier to enzymatic attack on the polysaccharides. Therefore only those organisms that possess enzymes capable of

destroying the lignin, or at least of altering its protective association with the polysaccharides, are capable of decaying wood.

3. The action of depolymerizing enzymes is restricted mainly to the noncrystalline regions of the cellulose. Since the cellulose in wood is more crystalline than that of most other plant tissues, it provides greater resistance to fungal and bacterial degradation.

4. Wood has a low nitrogen content, ranging from 0.03 to 0.1 percent by weight, in contrast to the herbaceous tissues that normally contain 1 to 5 percent. The low nitrogen content of wood reduces its susceptibility to decay.

5. For reasons not yet determined, much more moisture is required for initiating deterioration in wood than in other plant tissues. For instance, in cotton fibers 10 percent of moisture is adequate to initiate degradation by microorganisms, while no decay can be initiated in wood below the fiber saturation point, i.e., 26 to 30 percent of moisture.

A. Durability of Sapwood and Heartwood

The sapwood of all native species, even those in which the heartwood is highly durable, is susceptible to deterioration by biological agents, because it lacks extractives in sufficient quantity or of a toxicity to inhibit the growth of microorganisms. In fact, the presence of reserve foods in the parenchyma cells of sapwood may increase its susceptibility to decay and particularly to bacteria and fungal staining.

The greater durability of heartwood, in comparison with sapwood of the same species, is attributable largely to the presence in the former of toxic extraneous materials, such as essential oils, tannins, and phenolic substances. When these are present in sufficient amounts, they may prevent or at least considerably minimize the severity of the attack by destructive organisms. The validity of this statement has been tested repeatedly by leaching, with proper solvents, wood blocks taken from the heartwood of such durable species as black locust, redwood, and western redcedar; invariably the blocks so extracted decayed more rapidly than comparable unleached blocks. On the other hand, evidence also exists that some woods give off volatile substances that stimulate the growth of certain fungi.[5]

Other factors may be enumerated that also may explain the greater durability of heartwood. Among these are its lower moisture content, its lower rate of diffusion, and the blocking of cell cavities by gums, resins, tyloses in the vessels, and tylosoids in the resin canals. Any of these might conceivably affect adversely the balance between air and water necessary for the growth of fungi. Insects not infrequently attack sapwood and avoid heartwood, not because of the greater toxicity of the latter but because of the presence of reserve food in the parenchyma cells of the former.

From what has been said, it is obvious that, even in a tree species with relatively durable heartwood, the serviceability of an untreated piece of the wood is determined in no small measure by the amount of sapwood present. The disintegration of the less-durable sapwood, especially if it completely surrounds the heartwood, as in

TABLE 10-2 Comparative Resistance of Heartwood to Decay*

Resistant or very resistant	*Moderately resistant*	*Slightly or nonresistant*
Baldcypress (old growth)	Baldcypress (young growth)	Alder
Catalpa	Douglas-fir	Ashes
Cedars	Honeylocust‡	Aspens
Cherry, black	Larch, western	Basswood
Chestnut	Oak, swamp chestnut	Beech
Cypress, Arizona	Pine, eastern white	Birches
Junipers	Pine, longleaf	Buckeye‡
Locust, black†	Pine, slash	Butternut
Mesquite	Tamarack	Cottonwood
Mulberry, red†		Elms
Oak, bur		Hackberry
Oak, chestnut		Hemlocks
Oak, Gambel		Hickories
Oak, Oregon white		Magnolia
Oak, post		Maples
Oak, white		Oak (red and black species)‡
Osage-orange†		Pines (most other species)‡
Redwood		Poplar
Sassafras		Spruces
Walnut, black		Sweetgum‡
Yew, Pacific		Sycamore
		Willows
		Yellow-poplar

* From Comparative Resistance of Heartwood of Different Native Species When Used under Conditions That Favor Decay, *U.S. Dept. Agr. Forest. Serv., Forest Prod. Lab. Tech. Note* 229, rev. May, 1961.

† These woods have exceptionally high decay resistance.

‡ These species, or certain species within the groups shown, are indicated to have higher decay resistance than most of the other woods in their respective categories.

piles, posts, and mine props, may result in failure, even though the heartwood remains unaffected.

On the basis of many tests the U.S. Forest Products Laboratory has divided most of the more important timber species of the United States into three classes, indicative of the resistance of their heartwood to decay (see Table 10-2). It should be recognized, of course, that this classification is only a means for general comparison of the decay resistance of these woods under average conditions and should not be regarded as infallible.

1. VARIATIONS IN DURABILITY AMONG INDIVIDUAL TREES

The natural durability of the wood of individual trees of the same species may vary within wide limits. Such variability is thought to be largely genetically controlled, although tree vigor, size, and age may also have some effect, as indicated in the following section.

2. VARIATION IN DURABILITY OF HEARTWOOD ACCORDING TO THE POSITION IN THE SAME TREE

Evidence that any appreciable difference in durability of heartwood exists in the same growth increment of the tree, as determined at different heights of the stem, is fragmentary and inconclusive. For instance, MacLean and Gardner[34] found decrease in durability with increased distance from the base of the tree for western redcedar, while Boyce[7] reports that in white oak the heartwood in the upper part of the trunk is more resistant than that in the lower part.

On the other hand, a general tendency for the outer heartwood to be more resistant to decay than the inner zone, i.e., the heartwood nearer the pith, has been observed by many investigators, for the native as well as for the exotic species. Scheffer et al.[45] found that in the large logs of three species of white oak the difference may be "large enough to account for very considerable difference in service life of wood from different parts of large logs," while Zabel[56] recommends that white oak timber containing boxed-in heartwood should not be used under conditions favorable to decay. A similar suggestion is made by Findlay[20] for ship planking cut from the inner zones of heartwood of African mahogany. Anderson[1] notes: "The phenomenon of decreased decay resistance with aging of the wood in the tree occurs in many species, so necessarily there would be a variety of chemical changes involved." These changes may be brought about by enzymatic oxidation or by chemical transformation of the toxic compounds in the older heartwood, thereby reducing their fungicidal potency.

Scheffer and Cowling[47] offer the following generalization in reference to natural decay resistance:

> (*a*) Decay resistance decreases progressively from the outer heartwood to the pith; (*b*) decay resistance of the outer heartwood decreases progressively from the base of the tree upward, whereas the opposite is true for the inner heartwood; (*c*) outer heartwood at the base of the tree is more resistant and inner heartwood at the base is less resistant, heartwood in the upper bole being intermediate between these extremes; and (*d*) the larger the tree, the more resistant is the outer heartwood and the less resistant is the inner heartwood.

Scheffer and Cowling also report, "The bulk of the heartwood in older trees in general appears to be more resistant than that in younger trees of comparable size." The progressive reduction in decay resistance of heartwood from the outside to the pith is due to degradation of the toxic substances to less-active compounds by enzymatic or microbial action or through leaching.

B. Effect of Density and Rate of Growth on Durability

1. EFFECT OF DENSITY

Many heavy woods are highly durable. This may suggest that the density of wood, i.e., its weight per unit volume, may be a criterion of decay resistance. That this is not necessarily so is indicated by the fact that a number of light woods, such as redwood, cedars, or catalpa (see Table 10-2), are among the most durable. On the other hand, the heartwood of relatively heavy woods, such as beech, red and black oak, and maple, is among the least decay-resistant woods. This is because there is no significant difference in the decay resistance of the extractive-free wood substance, irrespective of the species; the superior durability of some woods, including those of light density, is traceable directly to the presence of toxic ingredients in sufficient quantities to inhibit deterioration.

Nor is there any conclusive evidence that variations in density within the species or a given tree have much effect on wood durability, unless the higher durability is correlated with the greater accumulation of toxic substances.

2. EFFECT OF GROWTH RATE

All attempts to correlate growth rate with durability have failed, since changes in the rate of growth entail variation in the amount of early wood and late wood in a growth increment. These tissues are decidedly different in density; hence variations in their amounts result in variation in density of the entire piece. But it has already been shown that density in itself is no measure of the durability of wood.

C. Silvicultural and Genetic Effects on Durability

There is no evidence that the geographic range of a tree species can be correlated with durability of that species. Scheffer,* for instance, in a study of decay resistance of oak wood, reported that he could detect no significant differences in the durability of white oak (*Quercus alba* L.) from such widely separated areas as the Upper Mississippi River Valley, the Central states, and the Northeastern states. On the other hand he did uncover strong evidence to the effect that, within a given region, the wood varied considerably in its decay resistance. He was unable to advance a logical reason for these differences, but it may be assumed that they are due to genetic variations. This view finds support in a suggestion made by Zabel[56] that variation in decay resistance among trees of the same species may, at least in part, be due to differences in composition of the extractives. It may be permissible to speculate that these, in turn, are of genetic origin.

Kennedy† suggests the possibility of propagating genetic types of Douglas-fir with more durable wood, containing larger amounts of taxifolin, by proper selection of parent trees displaying the highest degree of durability.

D. Effect of Climate on Durability

The effect climate may have on durability is determined largely by the amount of precipitation and by the temperatures and the relative humidities that prevail in a

* T. C. Scheffer, The Decay Resistance of Oak Wood, *U.S. Dept. Agr., Forest Path. Special Release* 13, 1943.

† R. W. Kennedy, Fungicidal Toxicity of Certain Extraneous Components of Douglas-fir Heartwood, *Forest Prod. J.*, **6**(2):80–84 (1956).

region. In a locality where the moisture is too scanty to promote the growth of fungi and/or the temperatures are unfavorable for their growth, a wood may be durable; elsewhere, where the conditions are different, the reverse may hold. To cite cases in point, the entire Eastern United States has a climate favorable for the development of wood-destroying fungi. But in the southern portion of this region, the South, the climate conditions that favor decay persist through a much longer period each year; for the reason stated, under comparable conditions of service, a given kind of wood tends to decay more rapidly in the South. In a northern climate, even when ample moisture is present, decay may proceed very slowly because of adverse temperatures; in an arid region, for example, in the Southwestern states, temperature is not the controlling factor, but rather the unusually low moisture content of the wood in service.

E. Effect of Season of Cutting on Durability

The statement is sometimes made that wood from trees felled during the winter months is more durable than that from trees cut at other seasons of the year. So far as is known, however, the season of cutting has no direct effect on the properties of wood, including its durability. Winter felling does offer certain advantages; e.g., the climatic conditions prevailing during the colder months are unfavorable for the activities of fungi and insects, thus making less critical the problem of caring for the lumber before it is used. However, winter-cut timber can become as badly decayed or insect-ridden during subsequent warmer periods as that from trees felled during the open months, if it is exposed to conditions favoring such attacks.

F. Relative Durability of Wood from Living and from Dead Trees

Contrary to a rather widely held opinion, wood from a dead tree, if sound, is fully as durable as that from a living tree. The reason for that is that only the parenchyma cells remain alive in sapwood and that heartwood consists entirely of dead cells. It therefore matters little in the case of sapwood, and none at all in the case of heartwood, whether the tree is living or dead at the time of felling; the wood, if sound, can be considered as equally durable for all practical purposes. This observation applies also to fire-killed trees.

In general, timber from dead trees requires more careful inspection than that from living trees because the likelihood of the presence of decay and insects is greater in the first instance. This holds especially when wood-destroying fungi were a contributing factor to the death of a tree or where dead trees are left standing long enough to permit of the ingress of wood-destroying agents.

Kimmey[27] carried out an extensive study in California on fire-killed ponderosa, Jeffrey and sugar pines, Douglas-fir, and white fir. He found that fungi and insects are so intimately associated in attacking fire-killed trees that their effects are best considered in combination rather than separately. This study revealed that the salvage of fire-killed trees, if economically feasible, must be carried out not later than the second year after the fire.

G. Effect of Various Treatments on Durability*

Considerable evidence exists that the natural durability of a species of wood may be seriously affected by the treatment to which this wood is subjected. An illustration of this is the reduction of durability of redwood brought about by drying. It was found that the decay resistance of redwood is decreased slightly by air drying, or by air drying followed by kiln drying. However, drying which involves presteaming, as in solvent seasoning, may lower the decay resistance of redwood sufficiently to force its reclassification from the highly resistant to the resistant class. A similar effect of steaming and solvent seasoning on durability has been observed for other species.

Prolonged soaking in water may result in leaching of the water-soluble toxic substances and hence in reduction in decay resistance. On the contrary, Suolahti[52] reports that the decay resistance of pine wood soaked for 6 months was increased in comparison with the unsoaked control samples. He attributes this increase in durability to the possibility of absorption by the wood of antibiotic substances produced by microorganisms in the water.

Reduction in decay resistance brought about by the application of heat during wood processing varies from insignificant to considerable, depending on the species of wood, the intensity and the duration of heat application, and the relative humidity of the air. Scheffer and Eslyn,[46] in a study conducted on the heartwood of three hardwoods and six coniferous species, have concluded that when the wood was heated in a dry atmosphere at temperatures ranging from 180 to 350°F, the decay resistance of all the woods tested was reduced only comparatively small amounts. However, when the same woods were subjected to the same high temperatures in an essentially saturated atmosphere, various but significant reductions in decay resistance, depending on the temperatures, duration of exposure, and wood species, were noted. These ranged from less than 10 percent when wet heat was maintained for 1 hour at 212°F to 17 to 32 percent at 300°F, maintained for the same length of time.

Wood treated with urea to accelerate seasoning was found to be more susceptible to decay than unsalted wood.[26] Likewise, it has been observed that more rapid decay of timber occurs in soils rich in nitrogen, mainly because of greater concentration of fungous population and also possibly through the infiltration of nitrogenous materials into the wood, thus increasing its susceptibility to decay.

As was previously reported, Krause[29] suggests that the reaction of iron with tannins, resulting in formation of iron tannate in white oak and mahogany, may reduce decay resistance of these woods by removing some fungitoxic chemical protection naturally present in the heartwood of these species. It is only logical to

* See also pp. 231 to 234.

ssume that similar reactions may occur between some other chemicals that come in ontact with wood and the extractives in the wood.

ELECTED REFERENCES

1. Anderson, A. B.: Biochemistry of Wood Extractives, *Proc. 5th World Forestry Congr.*, **3:**1390–1395 (1960).
2. Anon.: *Lyctus* Powder-post Beetles, *Dept. Sci. Ind. Res., Forest Prod. Res. Lab. Leaflet* 3 (England), revised 1963.
3. Anon.: The Common Furniture Beetle, *Dept. Sci. Ind. Res., Forest Prod. Res. Lab. Leaflet* 8 (England), 1959.
4. Barton, G. M.: Significance of Western Hemlock Phenolic Extractives in Pulping and Lumber, *Forest Prod. J.*, **18** (5):76–80 (1968).
5. Baxter, D. V.: "Pathology in Forest Practice," 2d ed., John Wiley & Sons, Inc., New York, 1952.
6. Bletchly, J. D.: Studying the Eggs of *Lyctus brunneus*, *Timber Technol.*, **68**(2247):29–31 (1960).
7. Boyce, J. S.: "Forest Pathology," 3d ed., McGraw-Hill Book Company, New York, 1961.
8. Brown, H. P.: Pith-ray Flecks in Wood, *U.S. Dept. Agr. Circ.* 215, 1913.
9. Campbell, R. H.: Fungus Sapstains of Hardwoods, *Southern Lumberman,* December, pp. 115–120, 1959.
0. Ceth, M. Y.: New Treatment to Prevent Brown Stain in White Pine, *Forest Prod. J.*, **16**(11):23–27 (1966).
1. Clausen, V. H., and F. M. Kaufert: Occurrence and Probable Cause of Heartwood Degradation in Commercial Species of Populus, *J. Forest Prod. Res. Soc.*, **2**(4):62–67 (1952).
2. Corbett, N. H.: Micro-morphological Studies on the Degradation of Lignified Walls by Ascomycetes and Fungi Imperfecti, *J. Inst. Wood Sci.*, **14:**18–29 (1965).
3. Cowling, E. B.: A Review of Literature of the Enzymatic Degradation of Cellulose and Wood, *U.S. Dept. Agr. Forest Serv., Forest Prod. Lab. Rept.* 2116, 1958.
4. ———: Comparative Biochemistry of the Decay of Sweetgum Sapwood by White-rot and Brown-rot Fungi, *U.S. Dept. Agr. Tech. Bull.* 1258, 1961.
5. Duncan, C. G.: Wood-attacking Capacities and Physiology of Soft-rot Fungi, *U.S. Dept. Agr. Forest Serv., Forest Prod. Lab. Rept.* 2173, 1960.
6. Ellwood, E. L., and B. A. Ecklund: Bacterial Attack on Pine Logs in Pond Storage, *Forest Prod. J.*, **9**(9):283–291 (1959).
7. Evans, R. S., and H. M. Halvorson: Cause and Control of Brown Stain in Western Hemlock, *J. Forest Prod.*, **13**(8):367–373 (1962).
8. Findlay, W. P. K.: Sap-stain of Timber, *Forestry Abstr.*, **20** (2):167–172 (1959).
9. ———: Timber Decay—A Survey of Recent Work, *Forestry Abstr.*, **17**(3):317–327; **17**(4):477–486 (1956).
20. ———: "Timber Pests and Diseases," Pergamon Press, New York, 1967.
21. Fisher, R. C., G. H. Thompson, and W. E. Webb: Ambrosia Beetles in Forest and Sawmill, Their Biological, Economic Importance and Control, *Forestry Abstr.*, **14**(4):147–155; **15**(1):3–15(1954).
22. Fracheboud, M., et al.: New Sesquiterpenes from Yellow Wood in Slippery Elm, *Forest Prod. J.*, **18**(2):37–40 (1968).
23. Grondal, B. L., and A. L. Mottet: Characteristics and Significance of White Floccose Aggregates in the Wood of Western Hemlock, University of Washington, *Forestry Club Quart.*, **16:**13–18 (1942).

24. Hartley, C., R. W. Davidson, and B. S. Crandall: Wetwood, Bacteria and Increased pH in Trees, *U.S. Dept. Agr. Forest Serv., Forest Prod. Lab. Rept.* 2215, 1961.
25. Haygreen, J. G., and S. S. Wang: Some Mechanical Properties of Aspen Wetwood, *Forest Prod. J.,* **16** (9):118–119 (1966).
26. Kaufert, F. H., and E. A. Behr: Susceptibility of Wood to Decay: Effect of Urea and Other Nitrogenous Compounds, *Ind. Eng. Chem., Ind. Ed.,* **34:**1510–1515 (1942).
27. Kimmey, J. W.: Rate of Deterioration of Fire-killed Timber in California, *U.S. Dept. Agr. Forest Serv., Circ.* 962, 1955.
28. Knuth, D. T., and E. McCoy: Bacterial Deterioration of Pine Logs in Pond Storage, *Forest Prod. J.,* **12**(9):437–442 (1962).
29. Krause, R. L.: Iron Stain from Metal Fastenings May Accelerate Decay in Some Woods, *J. Forest Prod. Res. Soc.,* **4**(2):103–109 (1954).
30. Levy, J. F.: The Soft Rot Fungi, Their Mode of Action and Significance in Degradation of Wood, *Advance Botan. Res.,* **2:**323–357 (1965).
31. ——— and M. G. Stevens: The Initiation of Attack of Soft Rot Fungi in Wood, *J. Inst. Wood Sci.,* **16:**49–55 (1966).
32. Lindgren, R. M.: Permeability of Southern Pine as Affected by Mold and Other Fungus Infection, *Proc. Am. Wood-Preservers' Assoc.,* **48:**158–174 (1952).
33. Lutz, J. F., C. G. Duncan, and T. C. Scheffer: Some Aspects of Bacterial Action on Rotary-cut Southern Pine Veneer, *Forest Prod. J.,* **16** (8):23–28 (1966).
34. MacLean, H., and J. A. F. Gardner: Distribution of Fungicidal Extractives (Thujaplicin and Water-soluble Phenols) in Western Red Cedar Heartwood, *Forest Prod. J.,* **6**(12):510–516 (1956).
35. McMillen, J. M.: Prevention of Pinkish-brown Discoloration in Drying Maple Sapwood, *U.S. Forest Serv. Res. Note* FPL-0193, 1968.
36. Millet, M. A.: Chemical Brown Stain in Sugar Pine, *J. Forest Prod. Res. Soc.,* **11**(3):232–236 (1952).
37. Oliver, A. C.: An Account of the Biology of *Limnoria*, *J. Inst. Wood Sci.,* **9:**32–91 (1962).
38. Proctor, P. B.: Penetration of the Walls of Wood Cells by the Hyphae of Wood-destroying Fungi, *Yale School Forestry Bull.* 47, 1941.
39. Ray, D. L. (ed.): "Marine Boring and Fouling Organisms," University of Washington Press, Seattle, Wash., 1959.
40. St. George, R. A., H. R. Johnston, and R. J. Kowal: Subterranean Termites, Their Prevention and Control in Buildings, *U.S. Dept. Agr., Home Garden Bull.* 64, 1960.
41. ——— and T. McIntyre: Powder-post Beetles in Buildings—What to Do about Them, *U.S. Dept. Agr. Leaflet* 358, rev. 1959.
42. Savory, J. G.: The Role of Microfungi in the Decomposition of Wood, *Record 1955 Annual Conv. Brit. Wood Preservers' Assoc.,* London, 1955.
43. Scheffer, T. C.: Control of Decay and Sap Stain in Logs and Green Lumber, *U.S. Dept. Agr. Forest Serv., Forest Prod. Lab. Rept.* 2017, 1958.
44. ———: Mineral Stain in Hard Maple and Other Hardwoods, *U.S. Dept. Agr. Forest Serv., Forest Prod. Lab. Rept.* 1981, 1954.
45. ———, G. H. Englerth, and C. G. Duncan: Decay Resistance of Seven Native Oaks, *J. Agr.,* **78:**129–152 (1949).
46. ——— and W. E. Eslyn: Effect of Heat on the Decay Resistance of Wood, *Forest Prod. J.,* **11**(10):485–490 (1961).
47. ——— and E. B. Cowling: Natural Resistance of Wood to Microbial Deterioration, *Ann. Rev. Phytopathol.,* **4:**147–170 (1966).
48. ——— and R. M. Lindgren: Stains of Sapwood and Sapwood Products and Their Control, *U.S. Dept. Agr. Tech. Bull.* 714, 1940.
49. Schultz, G.: Exploratory Tests to Increase Preservative Penetration in Spruce and Aspen by Mold Infection, *Forest Prod. J.,* **6**(2):77–80 (1956).

0. Snell, W. H.: The Effect of Heat upon Wood-destroying Fungi in Mills, *Proc. Am. Wood-Preservers' Assoc.*, **18:**25–29 (1922).
1. Snyder, T. E.: "Order Isoptera, The Termites of the United States and Canada," National Pest Control Association, New York, 1954.
2. Suolahti, O.: Dependence of the Resistance to Decay on the Quality of Scots Pine (Swedish and English summary), *Papper Tra.*, **23:**421–425 (1948).
3. Verall, A. F., and T. C. Scheffer: Control of Stain, Mold and Decay in Green Lumber and Other Wood Products, *Proc. Forest Prod. Res. Soc.*, **3:**480–489 (1949).
4. Wilcox, W. W.: Fundamental Characteristics of Wood Decay Indicated by Sequential Microscopic Analysis, *Forest Prod. J.*, **15**(7):255–259 (1965).
5. ———: Changes in Wood Microstructure through Progressive Stages of Decay, *U.S. Forest Serv. Res. Paper*, FPL 70, 1968.
6. Zabel, R. A.: Variations in the Decay Resistance of White Oak, *N.Y. State Coll. Forestry, Tech. Publ.* 68, 1948.

PART 2

Wood Identification and Descriptions of Woods by Species

11

Features of Wood of Value in Identification

Part 2 of this text deals with identification of the more important species of wood native to the temperate zone of North America. A few, presently less-valuable, species are also included because of their potential importance or because of special interest they present in terms of their anatomy.

Before attempting to identify a given piece of wood the reader must familiarize himself with the general (gross) and minute features that are common to all kinds of woods regardless of their botanical origin. Only then will it be possible to recognize those characteristics that are specific to a given kind of wood and which, therefore, will identify its botanical source.

In approaching wood identification the logical procedure is from the gross features, i.e., the characteristics evident with little or no magnification, to the minute, i.e., those that are evident only at magnifications that can be obtained with a compound microscope.

The gross features of wood are in general of two kinds: those traceable to its cellular structure, and those which fall in the category of physical or chemical properties, e.g., color, odor, weight, and hardness. In some cases the gross physical characteristics of the wood, such as its color or grain appearance, will provide all the information needed for its identification. In other instances these observations must be supplemented with information on wood structure that can be obtained with the

10× hand lens. The gross physical properties of wood are often less reliable for identification purposes than are the structural, i.e., anatomical features. This is because the former exhibit greater fluctuation than the latter. However, accurate identification can seldom be made on the basis of one feature alone; rather a number of wood characteristics must be considered together.

When positive identification is not possible on the strength of the gross anatomical and physical characteristics of wood, resort must be made to the minute anatomical features, i.e., those features which are discernible only with a compound microscope. When this becomes necessary, knowledge of wood anatomy should suggest which structural features are most important and reliable and on what surface, i.e., transverse, radial, or tangential, they can be seen to the best advantage.

To assist identification of woods grown in the temperate zone of North America a number of keys are presented in the following chapter. Keys I and IV are based on features visible with a 10× hand lens or discernible without magnification. Keys II, III, V, and VI require use of a microscope.

Before attempting to use Keys I and IV it should be helpful to review information presented in the sections of this chapter that follow. Use of Keys II, III, V, and VI, based on minute anatomical features of wood, presupposes intimate knowledge of wood anatomy; therefore the anatomical information presented in Chaps. 4 and 5 must be studied first.

I. GROSS STRUCTURAL FEATURES

A. The Planes in Which the Structural Features of Wood Are Studied

Because of the manner of the tree growth and the arrangement of wood cells within the stem, three principal planes are recognized in which wood is customarily examined. These planes, or surfaces, are ***transverse*** (also called ***cross section***), ***radial,*** and ***tangential*** (Fig. 11-1). The transverse surface is that which is exposed when wood is cut or sawed at right angles to the long axis of the tree stem; it is the surface that is presented at the end of a log. The radial and the tangential surfaces of wood are at right angles to the transverse section. The radial surface is exposed when the cut follows a radius of a cross section of the log, along the axis of the log, i.e., when it is made across the growth rings; if truly accurate to the direction, it would pass in a cross section of a log from the bark to the pith of the tree. Lumber cut so that the wide face of the board is principally radial is called quarter-sawn or edge-grained. The tangential surface of wood is exposed when the bark is peeled from a tree. In a log this surface is curved. In sawed or sliced material, when the cut is made tangentially, the exposed surface is flat, representing a chord of the curved surface under the bark; it is then so aligned that at least the central part of the wide surface of the board is approximately at a right angle to the wood rays (see Fig. 11-1). Lumber cut so that the wide face of the board is tangential is said to be plain or flat-sawn.

FIG. 11-1 A wedge-shaped block of red oak.
(*A*) The outer dead bark.
(*B*) The inner living bark.
(*C*) Sapwood.
(*D*) Heartwood.
(Approximately 70 percent original size.)

Rotary-cut veneers, i.e., sheets of veneer peeled from the surface of the log, are the only wood materials which are truly tangential at most points on their faces.

B. Growth Increments

In trees in the temperate zones, a new layer of wood is formed each year over the entire tree stem by the activity of the growing layer, called *cambium* (see Chap. 2).

Each new increment of growth of wood is in the shape of a hollow cone, with the entire tree stem therefore consisting of a series of cone-shaped layers of wood inseparably joined to each other (see Fig. 2-14). When viewed on the end of a log, these growth increments appear as concentric layers or rings; these are called *growth* or *annual rings*. On a transverse section of a piece of wood, or on the end of a board, these growth rings appear as more or less straight (or curved) parallel layers, since they represent but a part of the complete concentric layers of wood visible on the end of a log.

On the radial surface, growth increments appear as parallel striae, and on the tangential section they assume variable concentric and nested U- and V-shaped patterns. (For further discussion of growth increments, see pages 46 to 53.)

C. Early and Late Wood

Growth increments vary in width and distinctiveness in different species, in individual trees of the same species, and at different heights in a given tree. When individual growth rings are examined, it is generally found that each ring consists of a more porous and hence frequently lighter-colored portion, called *early* (or *spring-*) *wood,* formed early in the growing season, and the denser and often darker-colored portion, called *late* (or *summer-*) *wood,* because it is produced later in the season. In some woods the delineation between the early and the late wood is quite abrupt and distinct; in the others it is gradual and rather indistinct. (For further discussion of growth increments see pages 46 to 53.)

D. Sapwood and Heartwood

For a number of years the newly formed wood (*xylem*) not only provides mechanical support to a tree but also is physiologically functional. This is because some of the cells (mostly storage cells called *parenchyma*) in the outer layer of the trunk remain alive for some years after their formation. After an indefinite number of years, varying greatly in different kinds of trees and depending on the conditions of their growth, the protoplasm of the living cells dies, leading to formation of a core of physiologically dead wood. This change is generally accompanied by darkening of the wood through deposition of various kinds of infiltration products. The outer, usually lighter-colored portion of the stem, containing living parenchyma cells, is called *sapwood;* the inner and often darker part, consisting entirely of dead cells, is *heartwood* (Figs. 11-1, 11-2). (For more detailed information on sapwood and heartwood, see pages 53 to 59.)

II. THE ANATOMICAL FEATURES USEFUL IN IDENTIFICATION AT LOW MAGNIFICATION

A. Softwoods and Hardwoods

All woods described in this text fall into two groups, popularly known as *softwoods* and *hardwoods.* These terms frequently are somewhat confusing in that some softwoods, e.g., hard pines (*Pinus* spp.), are considerably harder than some hardwoods,

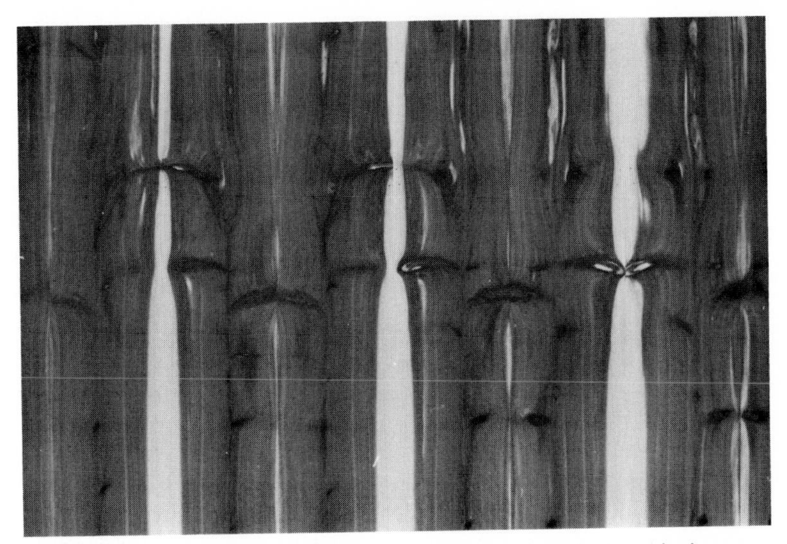

FIG. 11-2 Heartwood and sapwood figure in a panel of matched veneers of eastern redcedar (*Juniperus virginiana* L.).

such as basswood (*Tilia* spp.). These terms are therefore used to indicate a type of wood, not necessarily the physical characteristics of a given species or a piece of wood.

The softwoods are otherwise known as *coniferous woods,* because they are produced by coniferous trees such as pines, spruces, hemlocks, and cedars. The *hardwoods* are obtained from broad-leaved trees, such as ash, oak, maple, and poplar.

Softwoods are sometimes designated as *nonporous woods* and hardwoods as *porous woods.* The basis for this classification is that porous woods possess vessels—composite, tubelike structures which extend along the grain, i.e., with the long axis of the stem. These structures appear as minute openings, or *pores,* on the transverse surface and are often visible with the naked eye (see Figs. 12-75 to 12-140).

In some hardwoods the vessels in the early wood are much larger than those in the late wood of the same growth ring. The transition from one size of pores to another may be quite abrupt; such woods then are described as *ring-porous* [ex., oak (*Quercus* spp.), Figs. 12-126 to 131]. When transition in size of pores from the early to late wood is gradual, or when the pores do not appreciably vary in size throughout the ring, the woods are said to be *diffuse-porous* [ex., birch (*Betula* spp.) and maple (*Acer* spp.), Figs. 12-77, 12-86].

Vessels do not occur in the coniferous woods; hence the designation nonporous woods (see Figs. 12-1 to 12-30). Some softwoods may possess *resin canals,* which are tubular *intercellular spaces,* surrounded by secreting cells, called *epithelium.* Resin canals differ from vessels in that the vessels are formed through a fusion of cells in a longitudinal row. The resin canals, on the other hand, are tubular spaces formed by separation of the secreting cells.

Resin canals are a normal feature of the wood of pines, spruces, larch and tamarack, and Douglas-fir. They extend both longitudinally (along the grain) and transversely (across the grain). The former are more conspicuous and appear as small openings or white dots in the transverse section, for the most part in the central and the outer (late-wood) portions of the growth rings. They are most conspicuous in pines (Fig. 12-16), in which they may be numerous and visible with the naked eye. In other woods they are sparse and generally require careful observation with a hand lens for their detection (Figs. 12-10, 12-20). Although not always easy to see, resin canals provide an important diagnostic feature for the identification of coniferous woods.

B. Wood Rays

All species of wood contain structures called *wood rays*. These are ribbon-like aggregations of cells extending in the radial direction, i.e., from the bark toward the center of the tree. Rays take the form of lines of varying width, running at more or less right angles to the growth increments, when viewed on the transverse section (Figs. 12-23, 12-99, 12-109). On the radial surface rays appear as flecks (Fig. 11-1) and in the tangential plane, if visible, they show up as short (except in oaks, where they may reach a length of 1 inch or more, Fig. 12-126) and usually staggered lines, extending with the grain (Fig. 12-100).

Rays, their relative conspicuousness and spacing, as seen in the transverse section, constitute an important feature in the identification of some hardwoods. Rays are fine and inconspicuous in softwoods and hence are of little value in their identification.

C. Wood Parenchyma

Wood parenchyma consists of cells which serve for storage and conduction of food materials; these cells contain living protoplasm while they are in the sapwood, but lose their protoplasm when the sapwood is transformed into heartwood.

There are three kinds of parenchyma in wood: *ray parenchyma,* which constitutes the bulk of ray tissue; *epithelial parenchyma,* which surrounds the resin canals and therefore in the native woods is found only in certain coniferous woods; and *axial* or *longitudinal parenchyma*. The last type of parenchyma, as the term implies, extends along the grain in the form of strands, composed of individual cells of the axial parenchyma. It is inconspicuous on the longitudinal, i.e., radial and tangential, sections. On the transverse section, when abundant, it may be massed in lines or in bands running along the growth ring, e.g., in hickory [(*Carya* spp.), Fig. 12-89], or it may encircle the pores, making them appear thick-walled, as in California-laurel [*Umbellularia californica* (Hook & Arn.) Nutt.], Fig. 12-140. Well-defined axial parenchyma may be an important feature in the identification of woods with a hand lens.

III. THE PHYSICAL PROPERTIES OF WOOD OF VALUE IN IDENTIFICATION AT LOW MAGNIFICATION

In Keys, I, III, and IV reference is frequently made to a number of physical characteristics of wood. Among these are color, luster, odor, and occasionally taste, grain, texture, weight, and hardness.

A. Color

Color in wood is due mainly to extractives (see Extraneous Materials, pages 72 to 73). However, the major cell wall components, with the possible exception of cellulose, may also contribute to the color of exposed wood surface through oxidation. Wood color varies not only among different kinds of wood but also within a species and often even in the same piece of wood. This, in combination with the grain and figure in wood, makes each piece of wood unique, and may add materially to its decorative qualities.

It should be kept in mind that when color of wood is used for the purpose of identification it refers to that of the heartwood, unless otherwise indicated. The color of sapwood is seldom distinctive and hence is of little or no value in identifying a given kind of wood. Where some recognizable coloration is present in sapwood it may be gray or pale shades of yellow-white, pink, or red. When exposed to air, sapwood may turn shades of brown, because of enzymatic oxidations.

Color of heartwood, if striking or unusual enough, can be important in identification of wood; it may also be responsible, in part at least, for preferential uses. Examples of such woods are the rich chocolate- or purplish-brown heartwood of black walnut (***Juglans nigra*** L.) and the red-brown heartwood of cherry (***Prunus serotina*** Ehrh.), both long-time favorites for high-quality furniture. The jet black of the heartwood of genuine ebony (***Diospyros ebenum*** Koenig), from southern India and Ceylon, is another example of distinctive coloration. Shades of yellow are common in the wood of Rutaceae, e.g., in East Indian satinwood (***Chloroxylon swietenia*** DC.). There are no important trees of this family in the United States, and yellow colors are uncommon in our native woods; only yellowwood [***Cladrastis lutea*** (Michx. f.) K. Koch] produces heartwood with pronounced yellow coloration. Various shades of green and greenish brown provide a ready means of identification for yellow-poplar (***Liriodendron tulipifera*** L.). A special case of the use of color for identification is the yellow water-soluble pigment of Osage-orange [***Maclura pomifera*** (Raf.) Schneid.], which readily separates this wood from black locust (***Robinia pseudoacacia*** L.).

However, since in the majority of woods the color of heartwood is of a wide range of shades of brown, or grayish or reddish brown, precise description of the wood color is seldom possible. Such vague descriptions as "dark reddish brown" or "dull grayish brown" are not definitive enough to provide a positive means of identification, when used alone; in identification of woods by means of keys or similar devices, color descriptions must be supplemented with other, usually structural (anatomical) criteria.

More precise measurements of color can be obtained with the spectrophotometer, which records the amount of light reflected from a surface in the various parts of the spectrum.[3,7,8] Spectrophotometric measurements also may be used to indicate changes in color of wood that may have been caused by light, heat, and other environmental factors or brought about by the application of bleaches and finishes. Information obtained with a spectrophotometer provides a more accurate measurement of the color than visually based color descriptions, but with the present-day status of technology even spectrophotometric color description is of limited practical significance.

The heartwood of some trees darkens appreciably upon exposure, that of others tends to bleach, as a result of the action of the ultraviolet rays of light and because of leaching with water. Shades of dark red usually age to reddish brown. The heartwood of red mulberry (***Morus rubra*** L.) is orange-yellow to golden brown when first exposed to light but soon turns to a dull russet. The heartwood color of true mahogany (***Swietenia*** spp.) also changes with age and exposure to light, from a pinkish color in freshly cut heartwood to a rich reddish brown; in fact this characteristic is one of the reasons why mahogany is so much used in cabinetmaking. The light-yellow woods frequently fade on exposure to light. Color changes due to light could be minimized in some cases by incorporation in the surface finishes of ultraviolet-absorbent dye stuffs. Water-resistant surface films should restrict color changes due to leaching. However, the color of such film-forming materials, even when they are transparent in themselves, is affected by exposure to light, heat, and atmospheric moisture. Furthermore, when air in the exposed cell cavities is replaced by a finish having a higher index of refraction the wood surface tends to darken. A closer approximation of the natural color can be achieved by application of surface coatings which do not penetrate into the surface of the wood.

Some woods fluoresce when placed under ultraviolet light; for most woods the best wavelength is approximately 3650 angstroms. The most common color of fluorescence in heartwood is a bright yellow; most sapwood, however, turns some shade of blue. The effect of fluorescence is reduced by rough, dirty, or weathered condition of surfaces and by continuous exposure of wood to the light. Therefore, for the best effect test surfaces should be freshly planed or scraped; sanding, especially when heat is developed, may destroy most of the fluorescent effect. Among the native woods most vivid fluorescence is obtained in black locust (***Robinia pseudoacacia*** L.), honeylocust (***Gleditsia triacanthos*** L.), and the species of sumac (***Rhus*** spp.).

A photographic image of a varying degree of intensity for heartwood and sapwood is obtained when a photographic film is placed in contact with a wood specimen and then developed in the usual manner. This phenomenon is known as the ***Russel effect.*** A less-sharp image is obtained even when the film is placed above the wood specimen without coming in contact with it. This has led to the supposition that in the case of wood the Russel effect is due to the presence of reducing volatile substances of some yet-undetermined nature, one of which may be formaldehyde.[5]

B. Luster

Luster is the property of wood that enables it to reflect light, in other words, the property of exhibiting sheen. Whether woods are lustrous or dull depends on the extent to which they possess this characteristic. Luster is distinct from color and also from the ability of the wood to take a good polish. Many ornamental woods are comparatively lusterless in the unfinished state. Loss of luster, resulting in a "dead" appearance, is one of the first evidences of incipient decay in wood.

The luster of wood depends partly on the angle at which the light strikes the surface and the type of cells exposed on that surface.For example, quarter-sawn lumber generally reflects light more strongly than the tangential face, largely because of the presence of the numerous light-reflecting ray flecks. However, a more important cause of the presence or absence of luster is the nature of the heartwood infiltrations. Woods with comparatively small amounts of extractives may be lustrous; this often explains the difference in luster between sapwood and heartwood. Oily or waxy substances in the heartwood generally reduce the luster, e.g., in baldcypress [*Taxodium distichum* (L.) Rich.] and eastern redcedar (*Juniperus virginiana* L.).

The luster of wood is useful in the identification of wood only as a secondary feature. In some cases this characteristic can help to separate two woods that are otherwise similar in gross appearance. For instance, eastern spruce (*Picea* spp.) may be confused with eastern white pine (*Pinus strobus* L.) by the beginner in wood identification, but the spruce is readily distinguished by its greater luster. Among the porous timbers the luster helps to separate northern catalpa (*Catalpa speciosa* Warder), which is quite lustrous, from the dull wood of sassafras [*Sassafras albidum* (Nutt.) Nees]; white ash (*Fraxinus* spp.) can be distinguished from black ash (*Fraxinus nigra* Marsh.) by its greater luster. The majority of our native woods are intermediate in luster.

C. Odor and Taste

A number of woods have a distinctive odor, which may be caused by infiltration products in the heartwood or by the action of fungi, bacteria, or molds. In the latter case both the sapwood and heartwood may have odor, and because of the starch deposits the odor may be stronger in the sapwood.

Only the odors due to the presence of scented infiltration materials in the wood are of assistance in wood identification, and since these are found primarily in the heartwood, only the heartwood-containing portion of the piece should be used to determine whether a distinctive odor is present. Odors produced by the action of fungi and microorganisms are of no use in wood identification.

Since odor is due to emission of free molecules into the air, it soon disappears from an exposed surface. Therefore to ascertain whether or not odor is present it is essential to expose a fresh heartwood surface by cutting, or shaving, or scraping. Moistening and warming (by breath) the newly exposed surface will make the odor, if present, more obvious.

Odor in wood that is caused by infiltration products in the heartwood may be either pleasant or disagreeable. Aromatic compounds with persistent and agreeable odors are common in certain families producing hardwoods, such as the Lauraceae, Myristicaceae, and Santalaceae, and in cedars among the coniferous trees. The odors produced by aromatic extractives may be distinctive enough to be useful in identification of the wood. In our native trees examples of this sort of odor are found in sassafras [*Sassafras albidum* (Nutt.) Nees], California-laurel [*Umbellularia californica* (Hook. & Arn.) Nutt.], and Port-Orford-cedar [*Chamaecyparis lawsoniana* (A. Murr.) Parl.]. Many of the scents associated with these woods are strong and persistent, and pleasant to most people; they serve as a good means for identification but must be learned from known samples, since they cannot be described adequately in words. Some strongly scented woods may have disagreeable odors; e.g., dark, oily samples of baldcypress [*Taxódium distichum* (L.) Rich.] often smell rancid, and catalpa (*Catalpa speciosa* Warder) possesses a distinctive odor in the heartwood that is reminiscent of kerosene.

Since such odors are due to the presence of volatile materials, the scent is most pronounced in wood which has not been seasoned, or if seasoned has not been heated excessively in the process. The odor is generally confined to heartwood and is most evident on freshly cut surfaces. Prolonged exposure to air or water causes a loss of volatile material so that the odor is scarcely evident. For this reason, when using scent as a means for identification of the wood, a fresh cut should be made and the surface moistened and warmed slightly in order to make the odor most obvious.

The scent of wood may be an asset or a liability in utilization of a wood for a given purpose. For example, the strongly camphor-scented wood of camphor-tree [*Cinnamomum camphora* (L.) Nees & Eberm.] is said to be repellent to insects, and the wood is used in the Orient for chests to store woolens and furs. Heartwood of redcedar (*Juniperus virginiana* L.) has been reputed to possess the same property. However, tests conducted at the Savannah, Ga., laboratory of the U.S. Department of Agriculture indicated that, though redcedar has some inhibiting effect on the hatching of moth eggs laid in the chest, this cannot be relied upon to kill the moths and their larvae already present in the stored clothing.* Cigars are packed in veneer wrappers or boxes of Spanish cedar, toon, or calantas (*Cedrela* spp.) in the mistaken belief that the scent of the wood enhances the aroma of the tobacco. The strong, pleasant odor of burning sandalwood (*Santalum album* L.) has been used for centuries as incense in the Orient. On the other hand, certain food products, such as butter, are especially susceptible to taint, and the choice of woods for containers to be used in processing or shipping these products is critical. Wood of the true firs

* H. W. Nuttall, How Moth-proof Are Cedar Chests? *Wood Construction Building Materialist*, **43**(3):18–19 (1957).

(*Abies* spp.) and spruce (*Picea* spp.), when dry, is preferred for butter tubs, since the heartwood of these species contains very small quantities of infiltrations and hence does not taint the butter. Where wooden tanks are used in breweries at the present time, the interior is coated to prevent the beer from coming into contact with the wood.

Though many woods have taste, generally bitterish, it is seldom distinctive enough to be of use in wood identification. Among domestic woods only incense-cedar (*Libocedrus decurrens* Torr.) possesses a taste characteristic enough to be of assistance in separating this wood from western redcedar (*Thuja plicata* Donn).

D. Grain and Texture

Considerable confusion exists in the use of the terms *grain* and *texture* in the literature dealing with wood and its properties. When properly applied the term *texture* refers to the size and the proportional amounts of wood cells in a given piece of wood. To have specific meaning this term must be preceded by a qualifying adjective, such as fine or coarse.

The term *grain* should apply to the arrangement and direction of alignment of wood cells, when considered *en masse*. Like texture the term grain must be preceded by an adjective to connote special meaning. Such expressions as straight, spiral, or interlocked grain are commonly used. They denote the alignment of cells in relation to the long axis of the tree or a log. Woods such as maple, birch, or white pine, in which the transition between the early and the late wood is scarcely evident, may be called even-grained, while those with considerable variation between early and late wood, e.g., hard pines or oak, are said to be uneven-grained.

E. Weight

The weight of wood may help to identify it. However, since the weight of wood varies according to (1) the amount of wood substance present per unit volume, (2) the amount of infiltration, if a piece contains heartwood, and (3) the quantity of moisture present, certain conditions should be met before differences in weight could be considered in wood identification. For practical purposes only the (1) and (3) components of the wood weight need to be considered. This means that to use weight as a criterion in wood identification the pieces studied must be of approximately the same size (volume) and of comparable moisture content. This is because this test, as used in wood identification, is comparative and the decision as to the weight of a piece is made not by actually weighing it but by "hefting" it.

F. Hardness

The relative hardness or softness of wood is a useful indicator of its physical characteristics, since it depends mainly on the amount of the cell wall present. There is no way of ascertaining accurately the hardness of a piece of wood except by a test which requires special equipment. This information is presented in Tables 12-1 and 12-2. A useful idea of the approximate hardness of a wood can be gained, however, by testing it with a knife or the thumbnail.

IV. FIGURE IN WOOD

Wood exhibits a rich variation in decorative characteristics that originate from grain patterns and color markings on the longitudinal surfaces. When combined with its physical properties this makes wood unique among building materials. In a general sense, as used in this text, any distinctive markings of wood may be described as *figure*. Commercially, the term figure is restricted to the highly decorative patterns, resulting mainly from the appearance of the growth increments, irregularities in the orientation of cells, and uneven color distribution.

A. Growth Increments and Figure in Wood

Most kinds of trees exhibit some differences in the sizes and kinds of cells between early and late wood within the growth increment. The relative differences in appearance of the growth increments, resulting from these variations in size and uniformity of cell dimensions, constitute the *texture* of wood. Coarse texture can result from wide bands of large, open elements, as in the ring-porous hardwoods, or from uniformly distributed cells with large lumina, as in balsa (*Ochroma lagopus* Sw.). Even texture is the result of uniformity in cell dimensions and lumina, accompanied by inconspicuous rays. However, there is also a difference in even-textured woods; e.g., aspen and eastern redcedar are even- and fine-textured as well, while cottonwood and birch are even- and medium-textured woods.

The general direction of the fibers and other cells in wood, as well as the planes of cleavage produced by their orientation, is known as the *grain* of the wood. In common practice the terms grain and texture are often interchanged, and because the commercial usage of these terms cannot be altogether ignored, some of this confusion is carried over into the text. It should be reemphasized, however, that when properly employed, the term texture refers to relative sizes of the cellular components in wood and their arrangement within a growth increment, while grain is related to the direction of the cellular orientation.

The surfaces of wood may show grain and textural differences in two general patterns related to the longitudinal planes of symmetry in the log from which they were cut.

1. FIGURE RELATED TO THE TANGENTIAL PLANE OF CUT

Lumber produced by cutting the wide faces roughly parallel to the tangential-longitudinal plane of the log is called *plain* or *flat-sawn*. This same direction of cut in veneer is known as *plain or flat-cut* veneer.

The distinctive figure of flat-sawn boards consists of nested, angular, parabolic, or irregularly concentric patterns, formed by textural differences in successive growth increments (Figs. 11-3*a*, 11-4). Large logs with smooth surfaces and wide growth rings yield relatively unfigured, flat-sawn lumber, because few growth increments intersect the surface of the board. Irregularly shaped logs yield boards with random concentric patterns. Stock cut from logs of small diameter tends to show nested, parabolic figures whose tips all point toward the upper end of the log.

The general type of flat-sawn figure is modified when the boards are cut about halfway between bark and pith, as in Fig. 11-3*b*. It is obvious that only the central part of the board will show a flat-sawn figure and the sides of the piece will exhibit parallel lines marking the edges of the growth increments. Because of this normal

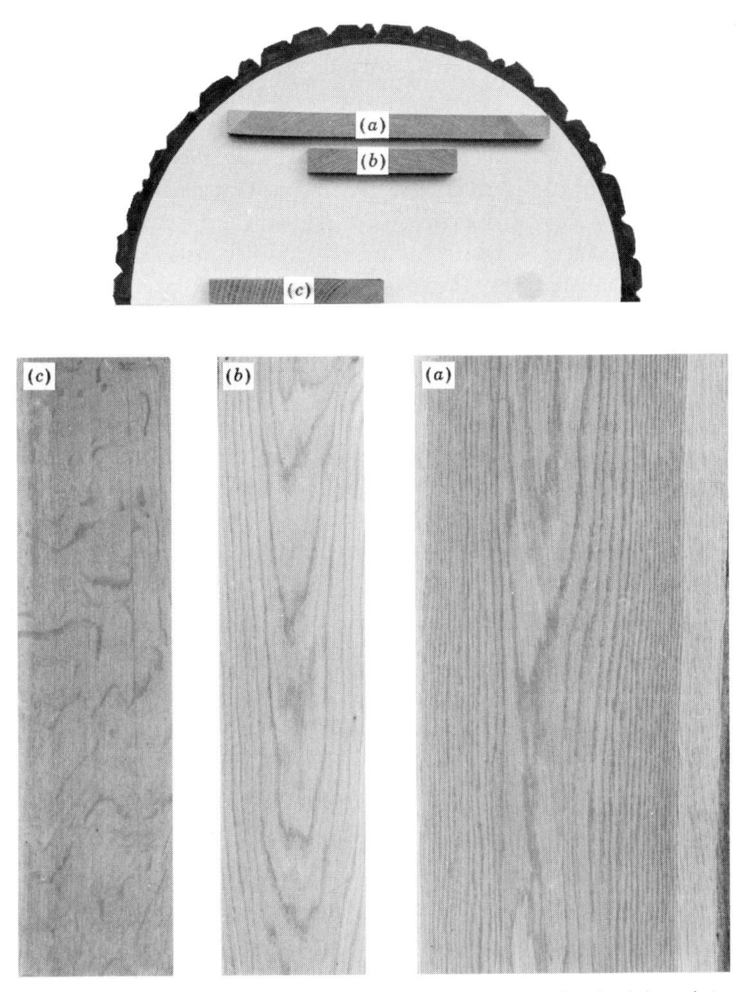

FIG. 11-3 Flat-sawn [(*a*) and (*b*)] and quarter-sawn (*c*) red oak boards in relation to position in the log from which they were sawn.

deviation of grain along the edges of flat-sawn boards, and also because many logs do not have circular outlines, commercial practice groups all boards in which growth rings form an angle of less than 45° with the wide surface of the piece as flat-sawn lumber.

In ring-porous woods the distinctive flat-sawn pattern stands out because of the band of large, early-wood vessels (Fig. 11-3). In conifers, with dark bands of late-wood tissue, flat-sawn figure is caused by a color contrast between early and late wood, such as is commonly found in the southern pines. Woods with slight differentiation in texture and color across the growth increments, such as *Pinus strobus* L. and *Tilia americana* L., show only an inconspicuous figure in flat-sawn lumber. An unusual structural feature, which causes patterns on the flat-sawn face, is the light-colored, irregular banding of included phloem in black-mangrove (*Avicennia nitida* Jacq.).

2. FIGURE RELATED TO THE RADIAL PLANE OF CUT

Lumber produced by cutting the wide faces approximately in the radial-longitudinal plane of the log is called *quarter-sawn*. Veneer produced by radial slicing or sawing is called *quartered veneer*. The characteristic appearance of this type of stock, when the rays are normal in size, is due primarily to the parallel stripes caused by the differences in texture between the early and late wood of the successive growth increments in edge view. The edgewise orientation of the growth increments in relation to the surface of the board is the reason that quarter-sawn lumber is also called *edge-grained* or *vertical-grained*. Edge-grained coniferous lumber is often used for flooring because of the more even wearing quality and the reduced transverse shrinkage.

Quarter sawing usually exposes a considerable amount of the ray tissue in wood. If the rays are conspicuous by their size, as in the oaks and sycamore, or by color differences, as in red mulberry, then a distinctive figure called a *ray fleck* is formed. The ray pattern may vary from the relatively inconspicuous figure in quarter-sawn maple to the large and strikingly lustrous *silver grain* in oak (Fig. 11-3*c*). Because of the irregularities in the direction of the rays in most kinds of wood, the board faces will seldom coincide exactly with the plane of the rays; consequently the ray figure will be irregular. Generally, the larger the rays, the more irregular the ray figure will be in quarter-sawn boards or in quartered veneer.

Lumber in which growth rings form an angle of 45° or more with the wide surface of the board is classified commercially as quarter-sawn. Oak lumber or veneer may be deliberately cut off the quarter to reduce the size of the ray figure. Oak boards produced by this type of sawing are called *rift-sawn*.

3. FIGURE RESULTING FROM METHOD OF CUTTING VENEER

It is possible to obtain some kinds of figure by veneer cutting which cannot be duplicated by sawing wood. This is true because veneers can be sliced with a knife to produce faces similar to those obtained by sawing, or also can be cut by rotating a log or a flitch against the knife in various ways that cannot be duplicated by sawing. The greatest volume of veneer is made by a process known as ***rotary cutting***. In this method the log is mounted on its centers in a lathe and rotated against a heavy, horizontal knife which peels off a continuous spiral sheet of veneer from the surface of the log. The veneer so cut is characteristically marked with irregularly concentric zones of growth increments, similar in appearance to those in the central zones of flat-sawn boards. Most rotary-cut veneer is employed in utility plywood, with a small part going into decorative faces, such as shown in Fig. 11-4*b*.

Half-round veneer is cut by mounting a half log off center in a veneer lathe, with the bark side against the knife. The cut passes across the growth increments at an angle; this increases the amount of figure in the veneer face. Most stump veneers are cut by the half-round method (Fig. 11-5). An alternate method is to mount the section of log or stump as above, except that the pith side is against the knife. Veneers produced in this way are called ***back-cut*** veneers.

There are a few other methods of special veneer cutting that are little used. A patented method known as ***cone cutting*** yields a radially oriented figure in a circular sheet. The veneer is peeled from the conical end of a cylindrical bolt of figured wood in much the same manner as a pencil is sharpened. The diameter of the

FIG. 11-4 Black walnut (*Juglans nigra* L.) veneers.
(*a*) Two-piece book-matched flat-cut veneers with a partial curly figure.
(*b*) Rotary-cut veneer.
(*Courtesy of American Walnut Mfg. Association.*)

FIG. 11-5 Black walnut (*Juglans nigra* L.) stumpwood veneers, cut half-round and matched four ways. (*Courtesy of Fine Hardwoods Association.*)

bolt and the angle of the point of the cone, which is formed by the knife at the end of the bolt, control the diameter of the circular sheet and determine the number of times a figure is repeated (Fig. 11-6). Another method which is rarely used is one employed to make artificial curly grain from straight-grained wood by cutting with a wavy-edged knife. This results in corrugated sheets of veneer, which after pressing and sanding yield a surface resembling true curly-grained wood.

Some special effects are obtained from figured veneers by the method in which the sheets of veneer are matched as they are made up into plywood or panels. When two successively cut sheets of veneer are turned over, as one would open the pages of a book, the veneers are *book-matched*. This type of edge matching yields a bilaterally symmetrical pattern from unsymmetrical figure in the veneer sheets (Figs. 11-4, 11-9*a*, 11-12*a*). Four-piece book- and end-matched veneers, as shown in Fig. 11-5, result from book matching of two consecutive pairs of veneer sheets, first along an edge or the end, followed by book matching of each separate pair. Book matching of veneers with a strong diagonal stripe yields a *V-matched* or *herringbone* figure. Book and endmatching diagonally striped veneers forms a diamond pattern.

B. Figure Caused by Tissue Orientation in Wood

As stated before, the general direction of the cells in wood, as well as the planes of cleavage traceable to their orientation, is known as the *grain* of wood. If the direction of the grain is parallel to the axis of the trunk or the axis of the board, the wood is *straight-grained*. Boards which are cut parallel to the pith in a tapered log will have diagonal grain orientation, as seen from the edge of the board; such pieces are *cross-grained* or *diagonal-grained*. On the other hand, if the grain of the wood

FIG. 11-6 Table top of black walnut (*Juglans nigra* L.) made from cone-cut veneer with six repetitions of the figure. (*Mersman Bros. Corporation.*)

in a tree stem is slanted either right or left in a helical orientation about the axis of the tree, the wood has *spiral grain.*

Truly straight-grained wood is more of a rarity than most people realize. Though cross grain can be eliminated by correct sawing of the lumber, the spiral grain present in most trees, to some extent, cannot be corrected in manufacture.

1. RIBBON OR STRIPE FIGURE IN WOOD

In some trees the direction of spiral grain may be periodically reversed; i.e., the angle of the grain inclination changes from a right- or left-hand helix to the opposite slope at intervals in the life of the tree. When this reversal of grain direction occurs it continues for the life of the tree. Wood with grain of this type is said to be *interlocked.* Wood with pronounced interlocked grain is present in sycamore, blackgum, and American elm. It is present to a lesser extent in a number of other native woods, such as basswood. Interlocked grain is the common condition in tropical timbers.

The quarter-sawn surface of interlocked-grained wood shows a characteristic *ribbon* or *striped* figure (Fig. 11-7). The finished surface of a board with interlocked grain exhibits dark and light streaks corresponding to the zones of different

FIG. 11-7 A block of Honduras mahogany (*Swietenia macrophylla* King), illustrating the relationship of interlocked grain at the split end to ribbon-stripe figure.

FIG. 11-8 Curly grain in a split block of white ash (*Fraxinus americana* L.). Natural size. (*Sample courtesy of J. A. Cope.*)

spiral-grain direction in the board. The color differences in the bands arise because light is reflected differently in each separate zone of grain orientation. This can be easily verified by reversing the ends of a board possessing ribbon figure. Each reversal causes an interchange of the light and dark bands on the face of the piece.

2. FIGURE CAUSED BY WAVINESS OR TWISTING OF THE GRAIN

The distinctive patterns in this classification fall into two general groups: figure resulting from localized waves that do not disturb the overall grain direction, and figure caused by grain which is twisted and lacks any consistent direction.

a. Curly or Wavy-grain Figure. This figure results from waves in the grain direction approximately at right angles to the longitudinal axis of the board. Split faces will show these waves on the radial face and usually on the tangential face as well (Fig. 11-8). Variable light reflection produces the curly appearance on flat surfaces in a manner previously described (Fig. 11-4). When the grain corrugations are close and abrupt, the resulting pattern is called *fiddleback figure,* because of the common use of maple so figured for violin backs (Fig. 11-9*c*).

Among domestic woods, wavy and curly figures are most frequent in maple and birch but occur sporadically in other woods as well. Curly-grained sugar maple wood is often known as *tiger maple,* because of its dark-golden, horizontal striping. Curly grain may be irregularly distributed within the stem and often occurs in the vicinity of knots, even though it is found nowhere else in the tree, because of the tissue distortion in these regions.

b. Broken Stripe. A combination of interlocked and wavy grain results in a ribbon figure which does not extend the full length of the piece. If the ribbon figure is broken into stripes of a foot or more in length and appears twisted, the pattern is called a *broken stripe* (Fig. 11-9). When a broken-stripe figure is interrupted by irregular horizontal wrinkles or waves in the grain, a *mottled* figure is produced. A very fine mottled figure results in the relatively rare *bee's-wing figure* of mahogany.

c. Blister and Quilted Figure. The tangential surface of the logs in some trees of maple, birch, and mahogany shows a pattern of interlacing grooves. When flat-sawn boards or rotary-cut veneers are produced from such logs, the surface appears to be marked with depressions or mounds, separated by narrow ridges or grooves. The seeming depression or elevation of the flat surface is the result of differences in the light reflection from the varying grain directions of the figure. When the interlacing grooves enclose a series of irregularly rounded areas, a *blister figure* is produced. If the enclosed areas are longer in the direction across the grain than parallel to the grain, the wood has a *quilted figure* (Fig. 11-10).

Neither of these figures is common, and even in those trees in which they occur, usually only a part of the stem will be well figured, while the other parts will show straight grain.

d. Bird's-eye and Dimples. This kind of figure is due to local distortions in fiber alignment that are occasioned by conical indentations in the growth increments.

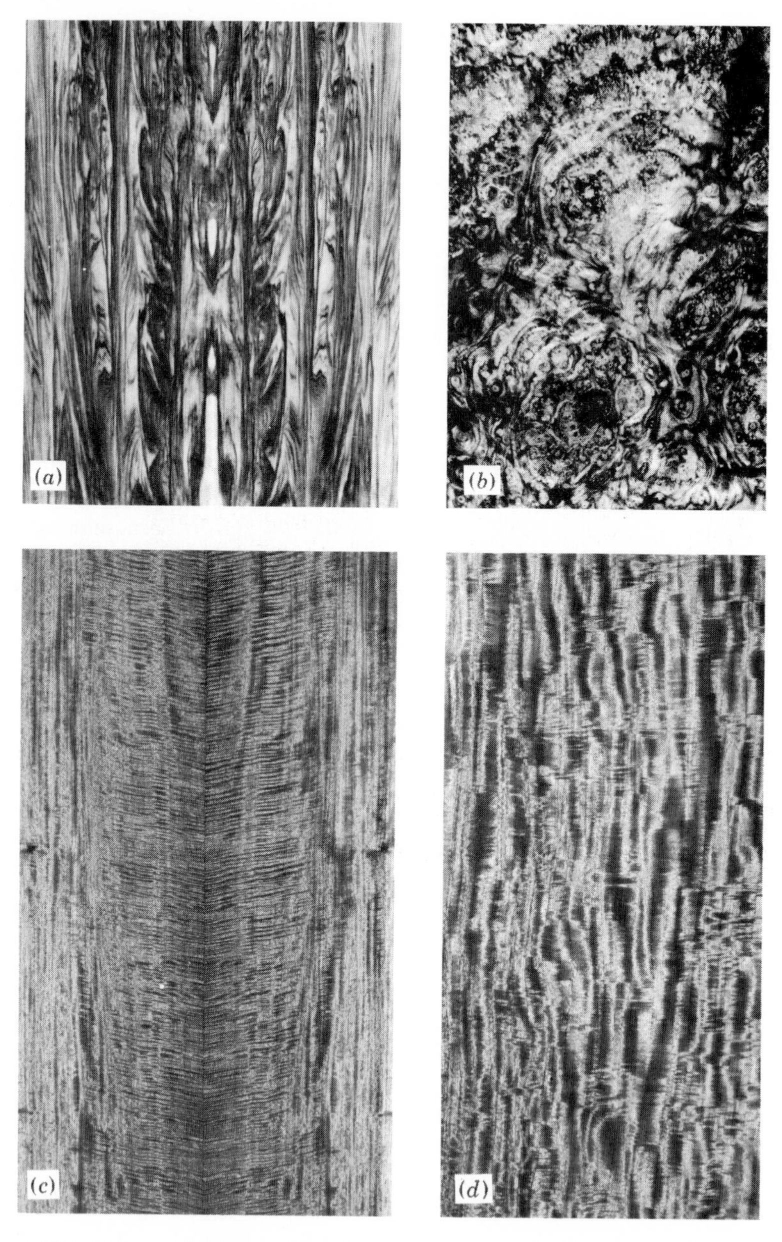

FIG. 11-9 (*a*) Pigment-figured sweetgum (*Liquidambar styraciflua* L.). Two-piece book-matched. (*Courtesy of American Walnut Mfg. Association.*)

(*b*) Burl figure in black walnut (*Juglans nigra* L.). (*Courtesy of American Walnut Mfg. Association.*)

(*c*) Cuban mahogany (*Swietenia mahagoni* Jacq.), book-matched veneer. The central portion shows a fiddleback figure and the edges a broken stripe. (*Courtesy of Mahogany Association.*)

(*d*) Strong block mottle figure in African mahogany (*Khaya* sp.). (*Courtesy of Mahogany Association.*)

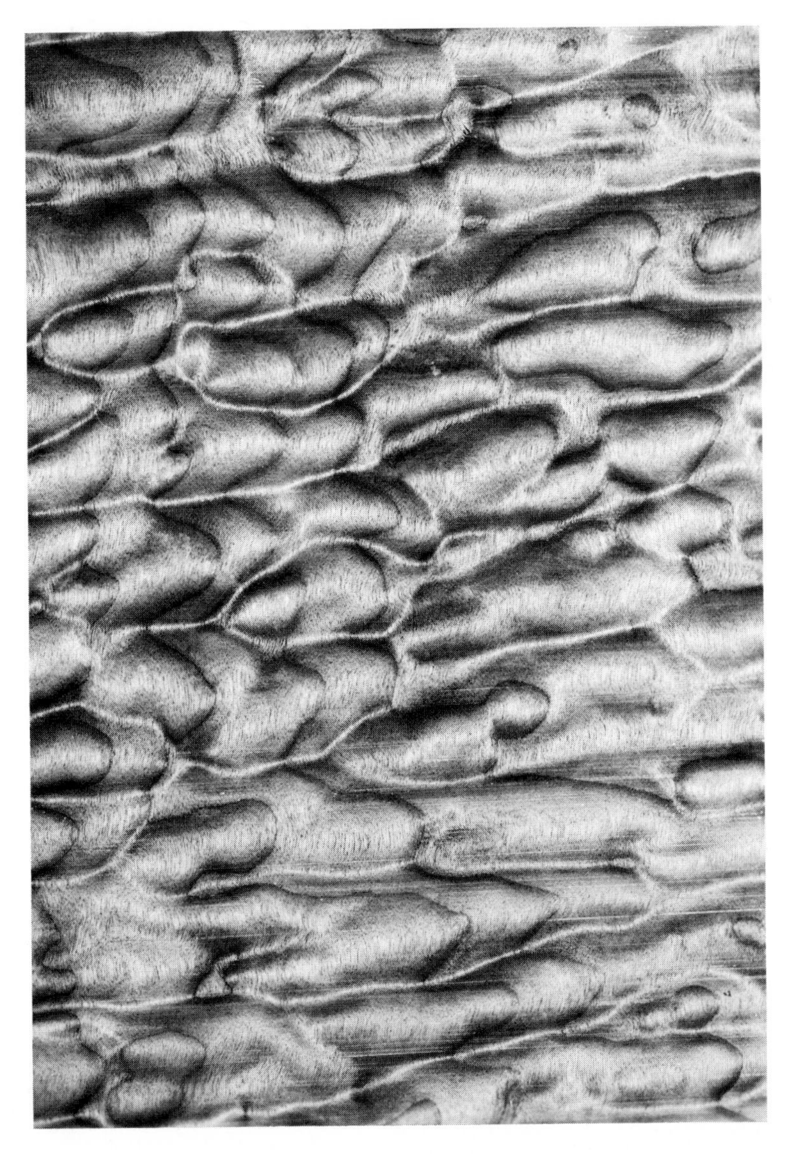

FIG. 11-10 Quilted maple veneer obtained from selected stock of *Acer macrophyllum* Pursh. Material supplied by P. J. Landry, Kelso Veneer Co., Kelso, Wash. (Four-fifths natural size.)

These indentations extend from the surface of the bark inward toward the pith and, once started, continue in successive growth layers for many years, frequently throughout the life of the tree. Conical elevations are usually present on the surface of the inner bark, projecting into the indentations in the wood. When logs with bird's-eye grain are plain-sawn or rotary-cut, the area on the surface of the wood in which the tissue is distorted resembles bird's eyes, hence the name.

The bird's-eye formation may extend throughout the length of the tree and even into the branches. More frequently, however, the figure is confined to one side of the tree or is restricted to irregular patches, scattered over the trunk, that are separated by normal wood. Considerable variation is also manifested in the size and in the proximity of the eyes on the surface.

Bird's-eye figure is most common and most characteristically developed in hard maple, but a similar figure is also found in soft maple, birch, and white ash. No explanation of the cause of bird's-eye figure that is wholly acceptable has yet been given.

Similar localized conical depressions, formed by the cambium in a radial direction, result in *dimples* on the tangential surface as in lodgepole pine (*Pinus contorta* Dougl., Fig. 11-11).

e. Crotch and Stump Figures. The twisted grain in large crotches or forks, and stump swells, especially in trees that produce ornamental timber, yields highly figured and valuable veneers. Usually the veneers obtained from crotches and stumps of trees are identified by the name of the part from which they come, with a descriptive adjective and the name of the kind of wood, e.g., walnut feather crotch.

A crotch is the portion of a tree trunk at the junction of a major fork. In commercial practice the crotch is trimmed to include mostly the wood forming an acute angle between the branches, since it is in this zone that the growth of the woody tissue is crowded and twisted. The typical pattern obtained by cutting through the central portion of the crotch, as shown in Fig. 11-12, is called a *feather-crotch* figure. When the twisted grain is rather open, a *moonshine-crotch* figure is formed.

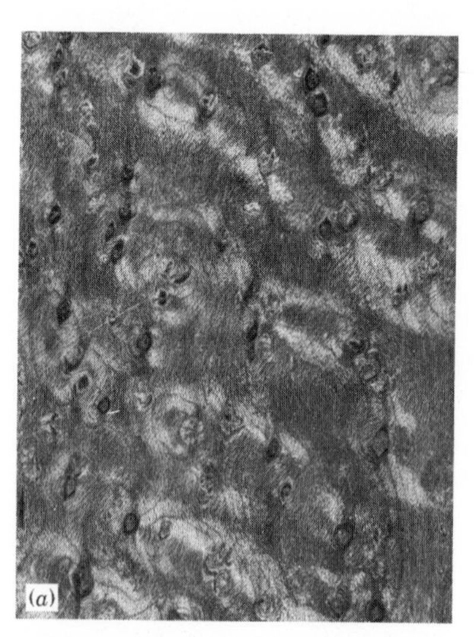

FIG. 11-11 (*a*) Bird's-eye figure in veneer of sugar maple (*Acer saccharum* Marsh.). Half natural size.

(*b*) Dimples on a split tangential face of lodgepole pine (*Pinus contorta* Dougl.). Natural size.

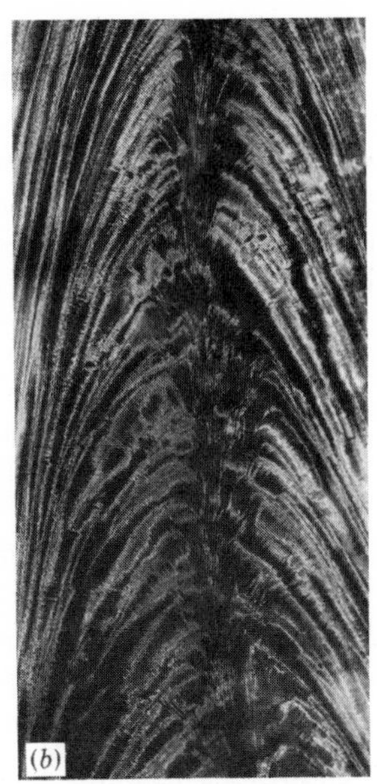

FIG. 11-12 (*a*) Feather-crotch figure in black walnut (*Juglans nigra* L.). Two-piece book-matched veneers. (*Courtesy of the American Walnut Mfg. Association.*)
(*b*) Crotch figure in African mahogany (*Khaya* sp.). (*Courtesy of the Mahogany Association.*)

Veneers cut toward the outside of the crotch show much-less-twisted grain and have *swirl* figures. Feather-crotch veneers are usually used with the figure inverted from the direction in which it grew. Ornamental crotch figures of North American woods come almost entirely from black walnut.

Stump wood comes from the enlarged portion of the base of the stem, just above the roots. To be valuable for veneers, stumps must possess irregular grain, which is usually indicated on the outside by fluting, ridges, or other irregularities. A great variety of highly figured veneer is cut from stumps, but no single pattern is characteristic. Most of the stump veneers cut in this country are obtained from black walnut and are used in matched patterns for ornamental purposes.

f. Burl Figures. These figures are obtained from large, abnormal bulges or excrescences that form on the trunk or limbs of nearly any kind of tree. The surface of these burls, or burrs, as they are sometimes called, is usually smooth in the hardwoods and may be corrugated in conifers. Burl figure is caused by the extremely irregular grain orientation that is often combined with small, partly developed buds and streaks of darker color (Fig. 11-9*b*). Among our native trees, burls from black walnut, maple, Oregon-myrtle, and redwood are well-known sources of burl veneer. The last two are widely used also for making turned articles. Burls are often quite small, but they can be larger than the stem or branch to which they are attached, and the basal burls of redwood may be more than 5 feet across.

A closely related type of veneer is produced from the swollen and distorted top of a tree which has been pollarded, i.e., one whose crown has been repeatedly trimmed back to the trunk every few years. Pollard veneers show a twisted figure with many small branch traces. This type of veneer is obtained from oak grown in England and the European continent.

C. Figure Caused by Color Distribution

Pigment figure is traceable to irregular streaks and patches of color, darker than the background and bearing no relationship to the growth increments.

Pigment figures are reasonably common in our native hardwoods and appear in such decorative woods as black walnut, often in combination with grain figures. Some heartwood of sweetgum is strongly marked with irregular pigment streaks; it is termed ***figured redgum*** in this country and ***satin walnut*** abroad (Fig. 11-9*a*). Among the imported timbers, pigment figure is common in Circassian walnut (***Juglans regia*** L.) and in the sharply striped zebrano and zebrawood (***Microberlinia brazzavillensis*** A. Chev.). The latter wood has deep-brown to yellowish background, and possesses probably the most strikingly regular pigment figure found in wood.

Completely irregular pigment streaks have no definite patterns on either flat-sawn or quartered boards. However, pigment which is associated with early or late wood will produce figures similar to those from the growth increments (Fig. 11-4). A stripe figure is obtained from walnut, especially ***Juglans regia*** L., by quartering logs with a regular zonation of dark pigment.

Color differences between heartwood and sapwood can result in irregular figures. For example, much of the black cherry lumber has some sapwood pattern showing, and the same is true for black walnut. Among the conifers, eastern redcedar (***Juniperus virginiana*** L.) is one of the few in which the contrast between the light-colored sapwood and the dark heartwood frequently results in an ornamental figure (Fig. 11-2).

V. WOOD IDENTIFICATION BY CHEMICAL MEANS

Although the composition of the cell wall substances, in terms of cellulose, hemicelluloses, and lignin, varies in different woods, these variations are not of a magnitude to allow identification of the individual species, based either on quantitative or qualitative analysis of these substances. The chemical differences of timber are due mainly to the kinds and the amounts of extractives present in the wood, mainly in the heartwood portion of it. Therefore identification of a wood by chemical means depends on the possibility of determining the presence of one or more extraneous materials, specific to a given wood, either by color reaction produced by a reagent applied to the wood surface, or by one of the several analytical methods employed in chemistry.

Spot tests with reagents are seldom sufficiently specific, or constant enough, to be eliable. The analytical methods, requiring qualitative determinations or the use of hromatographic methods, require laboratory techniques and equipment too omplicated to be used for routine identification purposes. Comprehensive discus- ion of such analytical methods is outside the scope of this text. Therefore only a few xamples will be offered by way of illustrating the different approaches to the lentification of woods by chemical means that have been tried.

A. Identification by the Use of Chemical Reagents

hese tests consist of the application of a chemical reagent to the wood surface and bservation of the resulting color reaction. For instance, heartwood of red oak can be ifferentiated from that of white oak by application of a solution consisting of 5 rams of benzidine, dissolved in 23 cubic centimeters of 25 percent hydrochloric cid and 970 cubic centimeters of water, and mixed just before application with an qual amount of 10 percent solution of sodium nitrate in water. When this mixture is pplied to a wood surface of red oak, it develops a reddish-orange color, while that f white oak becomes dark greenish brown.

Similar color tests have been employed for differentiating sapwood from heart- vood, especially in the species in which there is little color difference between the wo. Information on the various reagents that can be used for this purpose, and their elative effectiveness when applied to the softwood species, is discussed by Kutscha nd Sachs.[4]

B. Identification by Extraction

hese tests are based on chemical analysis of extractives, obtained usually from the eartwood either by fractional distillation or by extraction with water or some inor- anic or organic solvent. Identification of the substances and determination of their elative abundance are made either by the standard analytical procedures or by the olorimetric and chromatographic methods, or by a combination of these tests. A ew examples of the use of these techniques follow.

Slippery elm can be identified by the yellow and orange materials extracted with enzene. The yellow compound isolated in the form of brown-yellow needles was hown to be $C_{15}H_{16}O_2$, and the orange compound either a 6- or 7-methoxy derivative f it. The yellow material turned intense green with ferric chloride and a dark purple vith nitrous acid.[2]

The aqueous extracts from the heartwood of Douglas-fir and western larch levelop a reddish-purple color when first reduced with zinc and acetic acid and then reated with hydrochloric acid.[1]

Heartwood of jack pine can be distinguished from that of loblolly pine by frac- ional distillation or by the gas-liquid chromatographic analysis of their terpenes. Those of jack pine consist of large amounts of α- and β-pinene, while those of lob- olly pine contain β-phellandrene.

Jeffrey pine wood can be separated from ponderosa pine by the composition of the essential oils, which consist of 90 to 95 percent normal heptane in Jeffrey pine and o a mixture of α- and β-pinene and limonene in ponderosa pine.

Heartwood of eastern white pine can be distinguished from that of western white pine by subjecting their acetone extractives to simple paper chromatography. When treated with chromagenic spray, eastern white pine develops yellowish-orange spots resulting from a high percentage of cryptostrobin and strobobanksin, while western white pine shows pinkish-red spots indicative of a preponderance o pinocembrin.[6]

Many attempts have been made to develop a simple surface treatment to distinguish between western hemlock and true firs. After trying more than 500 reagents the Timber Research and Development Association of Great Britain published a method consisting of spraying ethyl glycol in concentrated hydrochloric acid (9:1) on wood samples or pulp chips; when these were heated to 100°C for several minutes hemlock was presumed to become red and fir brown. This test proved to be only partially successful because it was based on the assumption that hemlock contain leucoanthocyanins and that fir does not, when, in fact, sapwood of both species contains these materials.

A more reliable method for differentiation between hemlock and true firs has been suggested by Swan.[9] This method consists of extracting the finely ground wood mea with cold carbon tetrachloride and subjecting the extract, after vacuum separation to thin-layer chromatography, using methylene dichloride-acetic acid (100:1 solvent. The separation is based on the identification of wax mixture in hemlock and of manoyl oxide in the fir.[9]

VI. EXOTIC TIMBER SPECIES

Consideration was given to including information on the exotic timber species imported to the United States. The idea was rejected because the number of such woods is rapidly increasing and there is no logical way of deciding which species should be included and which left out, without either further expanding the scope o this text or arbitrarily limiting the treatments to only a few best-known kinds, such as true and Philippine mahoganies, teak, and rosewoods. This latter approach would not serve the needs of those seriously interested in acquiring information about the exotic timbers and might frustrate those wishing to identify a given piece of foreign wood which, by chance, was not included in the text.

Information on woods imported to this country is available from the U.S. Forest Products Laboratory, Madison, Wisconsin, and from a manual entitled "Commercial Foreign Woods on the American Market—A Manual to Their Structure, Identification, Uses and Distribution," by D. A. Kribs, Dover Publications, Inc. New York, 1968.

A comprehensive list of the principal standard references on foreign woods, by geographical regions, is included in "Identification of Hardwoods—A Microscopic ey," by J. D. Brazier and G. L. Franklin, Forest Prod. Res. Bull. 46, Dept. Sci. & Ind. es., Charles House, 5–11 Regent St., London, S.W.I. This publication, in itself, rovides detailed information on the anatomical features of some 380 commercial pecies from the temperate and tropical regions.

Extensive coverage of the tropical woods is also given by the Timber Research and)evelopment Association, High Wycombe, Bucks, England. This organization prints eriodically leaflets dealing with the characteristics, properties, and uses of ndividual species.

Wood, a periodical published by Nema Press, London, reproduces in each issue a elected timber in natural color, with a brief statement of its characteristics.

ELECTED REFERENCES

1. Barton, G. M., and J. A. T. Gardner: Determination of Dihydroquercitin in Douglas Fir and Western Larch, *Anal. Chem.,* **30**(2):279–281 (1958).
2. Fracheboud, M., et al.: New Sesquiterpenes from Yellow Wood of Slippery Elm, *Forest Prod. J.,* **18**(2):37–44 (1968).
3. Gray, V. R.: The Color of Wood and Its Change, *J. Inst. Wood Sci.,* **8**:35–57 (1961).
4. Kutscha, N. P., and I. B. Sachs: Color Tests for Differentiating Heartwood and Sapwood in Certain Softwood Tree Species, *U.S. Forest Prod. Lab. Rept.* 2246, 1962.
5. Nambiyar, V. N. P., and S. Patton: The Peroxide Theory of the Russel Effect, *J. Inst. Wood Sci.,* **5**:48–57 (1960).
6. Seikel, M. K., et al.: Chemotaxonomy as an Aid in Differentiating Wood of Eastern and Western White Pine, *Am. J. Botany,* **52**(1):1046–1049 (1965).
7. Sullivan, J. P.: Color Characterization of Wood: Spectrophotometry and Wood Color, *Forest Prod. J.,* **17**(7):43–48 (1967).
8. ———: Color Characterization of Wood: Color Parameters of Individual Species, *Forest Prod. J.,* **17**(8):25–29 (1967).
9. Swan, E. P.: Chemical Methods of Differentiating the Wood of Several Western Conifers, *Forest Prod. J.,* **16**(1):51–54 (1966).

12

Keys for Identification and Descriptions of Woods by Species

This chapter contains five keys for identification, descriptions of woods by species, and comparative physical data for the more important woods native to the United States and Canada. Softwoods and hardwoods are dealt with in separate sections in order to bring the pertinent material closer together.

This material is arranged as follows:

The key to coniferous woods, I, and the key to hardwoods, IV, are based on features visible with a 10X hand lens or without magnification (gross features). The keys to coniferous woods, II, and to hardwoods, V, are based on minute features requiring a microscope for their determination. Key III to conifers is based on a combination of gross and minute features. Each key has a particular value, depending on the size and nature of the sample to be identified and the accessibility of a compound microscope.

In order to determine the presence or absence of many of the gross features with Keys I and IV it is necessary to cut a smooth surface with a *sharp* knife or a safety razor blade. Unless a clean cut is made with a sharp cutting tool, details to be observed on the surface will be obscured.

A presumed identification is made when the point is reached in the key at which the name of the particular wood or genus is given. However this identification should not be accepted as final until additional verification is made with the low-power (5X) cross-section photographs, accompanying Keys I and IV, and the descriptions by species, following the keys.

As was stated in Chap. 11, the identification of a wood by means of Keys II, III, and V should not be attempted without first studying the information in Chaps. 4 and 5, dealing with the minute structure of coniferous woods and hardwoods.

For purposes of study of the microscopic features, permanent mounts of wood cut with a sliding microtome are desirable. However, identification of wood on the basis of minute features can be done very well with temporary sections made freehand with a safety razor blade. To make such slides, cut a clean surface in the area to be observed, moisten the surface with water or alcohol, and slice off a small piece using a sharp, new safety razor blade. The important points to bear in mind when making freehand sections are to use a sharp edge, to slice across the surface with a shearing motion, and to cut the section as thin as possible. Temporary sections of this sort mounted in water, alcohol, glycerin, or white corn syrup under a cover glass can demonstrate all the details evident in a permanent mount—however, usually not over so large an area.

As in the case of keys based on the gross features, the final identification should be verified by reference to the accompanying high-power photographs and by studying the more complete information on the minute anatomy of a given species presented in the description of woods by species section, which follows the keys.

When the available pieces of wood are too small to be sectioned, the material can be macerated to produce fibrous material that can be identified with the fiber key, i.e., Key VI, Chap. 13.

Users of this textbook should be cautioned about placing too much reliance on the measurements of the cell dimensions and the cell wall thicknesses, or on the information on frequency of cell distribution and volume occupied by the different kinds of cells. This is so because of the great variability in size and distribution of cells that exists not only among individual trees of a species, but also in different parts of the same tree and even in the same growth ring. Considerable discrepancies may also arise depending on the method used in selecting and

preparing the sample, the number of measurements made, and the techniqu
employed in making the measurements. The average values given in this text ar
useful as an aid to identification only when they are proved to be reasonabl
constant within the limits given and when the spread in the range of the averag
values between the species is appreciable. For comprehensive treatment of thi
subject, see article by D. M. Smith, Microscopic Methods for Determining Cros
sectional Cell Dimensions, *U.S. Forest Service Res. Paper* FPL 79, 1969.

Descriptions of Woods by Species, pages 448 and 537, present a summary c
the general characteristics and minute anatomy of the more important woods c
the United States and Canada. Several less-important species are also included t
illustrate the range of anatomical features found in the native woods. Wher
several closely related species, e.g., southern pines, cannot be separated wit
any degree of certainty on the basis of either gross features or minute anatom
of the wood, they are discussed under a single heading.

Each description of a wood, or a group of closely related woods, is divided int
two sections: General Description and Minute Anatomy. The first deals with woo
characteristics that can be recognized either without magnification or with th
aid of a 10X hand lens; the section entitled Minute Anatomy presents wood feature
which require a compound microscope for their identification.

These two sections are followed by a paragraph summarizing the most im
portant uses of that wood. This information is not necessarily arranged in th
order of importance of these uses but is rather grouped, when appropriate, unde
major categories, such as round-wood uses, lumber uses, veneer and plywooc
and miscellaneous. In some instances, when information of special interest or c
value in separating allied species is available, this is included under Remarks

For convenience of comparing the various physical, mechanical, and nor
mechanical properties of the native woods, these properties are presented i
tabulated form in Tables 12-1 and 12-2, immediately following descriptions b
species sections.

The reader should be reminded that summations of the anatomical features c
the coniferous woods and hardwoods are presented in Table 4-7, page 146, an
Table 5-10, pages 190 to 197, respectively.

Coniferous Woods—Keys and Descriptions by Species

KEY TO CONIFEROUS WOODS—GROSS FEATURES*

Woods nonporous (without vessels); cross sections consisting of distinct radial rows of tracheids; rays indistinct to the naked eye.

1. Resin canals present; longitudinal (*x*)† canals appear as small openings mostly in the outer part of the late wood; transverse canals included in some rays, which then appear larger (Figs. 12-7, 12-8, 12-10 to 12-22).. **2**

1. Resin canals normally absent; longitudinal wound (traumatic) canals occasionally present in a tangential row (*x*).. **5**

2. Resin canals numerous, evenly distributed in outer portion of every ring (*x*); generally visible to the naked eye‡ as dark- or light-colored dots, easily seen with a hand lens (Figs. 12-12 through 12-19)......... **3**

2. Resin canals sparse, unevenly distributed, and sometimes absent in some rings (occasionally appear as 2–many in a tangential row), barely visible to the naked eye as light or dark flecks, inconspicuous without a hand lens (Figs. 12-7, 12-8, 12-10, 12-11, 12-20 to 12-22).... **4**

3. Transition from early to late wood gradual; wood soft, light, quite even in density (Figs. 12-12, 12-14, 12-15, 12-19).

A. Wood relatively coarse-textured; resin canals appearing as minute openings to the naked eye (*x*).
Sugar pine—*Pinus lambertina* Dougl. Fig. 12-14. Desc. p. 448

B. Wood medium to relatively fine-textured, with resinous odor; heartwood creamy white to light brown or reddish brown, turning darker with age and exposure, split tangential surface not dimpled; resin canals appearing as light- or dark-colored dots (*x*).
Eastern white pine—*Pinus strobus* L. Fig. 12-19. Desc. p. 452
Western white pine—*Pinus monticola* Dougl. Fig. 12-15. Desc. p. 450

* This key is designed for the separation of the more important temperate North American woods based on features visible with a 10X hand lens or without magnification.
A series of photographs (Figs. 21-1 to 12-30, pp. 429 to 433) at low magnification (5×) accompanies this key, to illustrate the normal appearance of the various kinds of woods under a hand lens.

† (*x*) indicates that the feature is observed on the cross section; (*r*) and (*t*) refer to the radial and tangential surfaces respectively.

‡ Numerous but relatively inconspicuous in *Pinus resinosa* Ait., *Pinus banksiana* Lamb., *Pinus contorta* Dougl., and sometimes in *Pinus ponderosa* Laws.

C. Wood medium- to relatively fine-textured, without resinous odor; heartwood light pinkish yellow to pale brown with purplish tinge, darkening to silver-brown on exposure; split tangential surface often noticeably dimpled; resin canals appearing as light- or dark-colored dots.

Sitka spruce*—*Picea sitchensis* (Bong.) Carr. Fig. 12-12. Desc. p. 472

3. Transition from early to late wood quite abrupt; wood medium-hard to hard; uneven in density (Figs. 12-13, 12-16 to 12-18).

A. Split tangential surface with numerous depressions which give it a dimpled effect. This feature may be present in:

Lodgepole pine—*Pinus contorta* Dougl. Desc. p. 454

Jack pine—*Pinus banksiana* Dougl. Fig. 12-13. Desc. p. 463

Ponderosa pine†—*Pinus ponderosa* Laws. Fig. 12-17. Desc. p. 459

B. Split tangential surface not dimpled.

Hard pines—*Pinus* spp. Fig. 12-16. Desc. p. 456

4. Transition from early to late wood gradual, late wood distinct but not pronounced.

A. Heartwood light pinkish yellow to pale brown with purplish tinge, darkening to silvery brown on exposure, dull to somewhat lustrous, medium-textured; split tangential surface often noticeably dimpled.

Sitka spruce—*Picea sitchensis* (Bong.) Carr. Fig. 12-12. Desc. p. 472

B. Heartwood yellowish to pale reddish yellow to orange-red or deep red; wood with characteristic odor on freshly cut surface, medium-textured, contours of growth rings frequently wavy.

Douglas-fir (fast growth)—*Pseudotsuga menziesii* (Mirb.) Franco. Fig. 12-21. Desc. p. 474

C. Heartwood not evident; wood nearly white to pale yellow or light yellow-brown, lustrous, relatively fine-textured; split tangential surface only occasionally dimpled.

Spruce—*Picea* spp. Figs. 12-10, 12-11. Desc. p. 469

4. Transition from early to late wood abrupt; wood medium hard to hard, uneven in density.

A. Heartwood yellowish or pale reddish yellow to orange-red or deep red; wood with characteristic odor on freshly cut surface,

* Occasional samples of *Picea sitchensis* (Bong.) Carr. are about as coarse-textured as many pieces of *Pinus lambertiana* Dougl.

† Typical ponderosa and lodgepole pine wood is rather soft, with the late-wood zones appearing as narrow lines, indistinct without magnification, the wood then resembling that of the soft pines.

not oily; contours of growth rings often wavy; resin canals (*x*) frequently in short tangential lines.*

Douglas-fir—*Pseudotsuga menziesii* (Mirb.) Franco. Fig. 12-20, 12-22. Desc. p. 474

B. Heartwood with distinct brownish cast, wood lacking characteristic odor on freshly cut surface, somewhat oily; contours of growth rings usually smooth; resin canals (*x*) sometimes in tangential groups of 2–5, or single and sporadic, not in tangential lines.

Tamarack and western larch—*Larix* spp. Figs. 12-7, 12-8. Desc. p. 465

5. Heartwood with distinct odor, pleasantly fragrant or ill-scented **6**

5. Wood without distinct odor.. **9**

6. Wood with fragrant, "cedar-like" or sweetish odor....................... **7**

6. Wood ill-scented, or with rancid odor **8**

7. Heartwood dull red, or rose-red to purplish, reddish brown, or dull brown.

A. Wood fine-textured; heartwood rose-red or purplish when first exposed, aging to dull red or reddish brown.

Eastern redcedar—*Juniperus virginiana* L. Fig. 12-6. Desc. p. 499

B. Wood medium- to coarse-textured, firm; texture within growth rings even; wood frequently pecky; heartwood reddish brown to dull brown, sometimes with a lavender tinge, with a distinct acrid taste.†

Incense-cedar—*Libocedrus decurrens* Torr. Fig. 12-9. Desc. p. 489

C. Wood medium- to coarse-textured, soft; texture within growth rings uneven, with the late wood considerably harder than the early wood; wood never pecky; heartwood reddish, pinkish brown to dull brown, with faint sweetish odor.

Western redcedar—*Thuja plicata* Donn. Fig. 12-27. Desc. p. 493

7. Heartwood yellowish white to clear yellow, straw-brown or light brown tinged with red or pink.

A. Heartwood clear yellow, darkening on exposure; wood with odor resembling raw potatoes on freshly cut surfaces.

Alaska-cedar—*Chamaecyparis nootkatensis* (D. Don) Spach. Fig. 12-4. Desc. p. 496

B. Heartwood yellowish white to pale brown, with pungent gingery odor on freshly cut surfaces.

Port-Orford-cedar—*Chamaecyparis lawsoniana* (A. Murr.) Parl. Fig. 12-3. Desc. p. 494

* Color and characteristic odor of heartwood are usually sufficient to separate *Pseudotsuga* from *Larix*. However, the most positive separation is the presence of spiral thickenings in all early-wood tracheids in Douglas-fir, as seen with a microscope.

† This test is made by placing a thin shaving of wood on the tongue; a stinging sensation (not a bitter taste) after a short time indicates the acrid taste.

C. Heartwood light brown, tinged with red or pink (roseate); zonate parenchyma visible with a hand lens and sometimes with a naked eye as darker lines (*x*).

Atlantic white-cedar—*Chamaecyparis thyoides* (L.) B.S.P. Fig. 12-5. Desc. p. 498

D. Heartwood uniformly straw-brown; parenchyma not visible (*x*).

Northern white-cedar—*Thuja occidentalis* L. Fig. 12-26. Desc. p. 491

8. Heartwood clear yellow.

A. Heartwood with odor resembling raw potatoes on freshly cut surfaces.

Alaska-cedar—*Chamaecyparis nootkatensis* (D. Don) Spach. Fig. 12-4. Desc. p. 496

B. Heartwood with strong unpleasant odor and mild characteristic bitterish taste.

California torreya—*Torreya californica* Torr. Fig. 12-28. Desc. p. 502

8. Heartwood variable in color, ranging from yellowish to light or dark brown, reddish brown to almost black, with rancid odor on freshly cut surfaces; wood often greasy or oily and frequently pecky (see p. 347); strand parenchyma visible along the grain as dark lines.

Baldcypress—*Taxodium distichum* (L.) Rich. Fig. 12-24. Desc. p. 487

9. Heartwood distinct, bright orange to rose-red, light cherry-red, or deep reddish brown; sapwood nearly white, grayish white, or light yellow.

A. Heartwood light cherry-red to deep reddish brown; sapwood nearly white or grayish white, wood coarse-textured, light to moderately light, soft to moderately hard.

Redwood—*Sequoia sempervirens* (D. Don) Endl. Fig. 12-23 Desc. p. 485

B. Heartwood light orange to rose-red; sapwood light yellow; wood fine-textured, dense, fairly heavy.

Pacific yew—*Taxus brevifolia* Nutt. Fig. 12-25. Desc. p. 501

9. Heartwood not distinct; wood creamy white to buff, yellow, or yellowish brown to brown, with or without a lavender, roseate, or reddish-brown tinge.

A. Wood light buff to light brown, sometimes with a reddish-brown to pinkish tinge; early wood usually with a reddish tinge; transition from early to late wood semiabrupt to abrupt.

Hemlock—*Tsuga* spp. Figs. 12-29, 12-30. Desc. p. 477

B. Wood whitish to creamy white or pale brown, early wood whitish, changing gradually to a darker lavender late wood (color contrast more pronounced in wide rings); transition from early to late wood gradual to semiabrupt.

True fir—*Abies* spp. Figs. 12-1, 12-2. Desc. p. 480

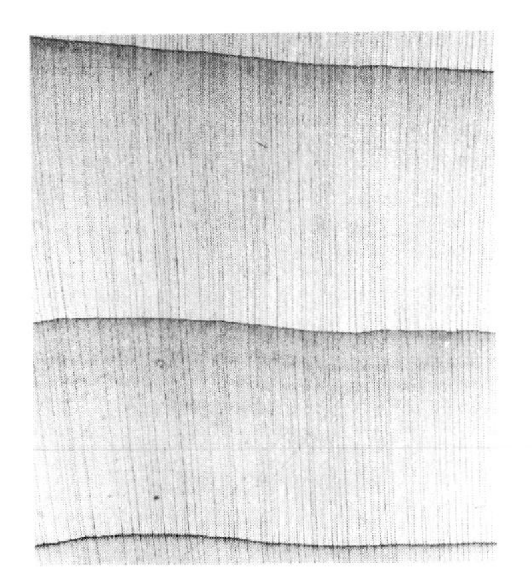

FIG. 12-1 *Abies balsamea* (L.) Mill. (x—5×.)

FIG. 12-2 *Abies concolor* (Gord. & Glend.) Lindl. (x—5×.)

FIG. 12-3 *Chamaecyparis lawsoniana* (A. Murr.) Parl. (x—5×.)

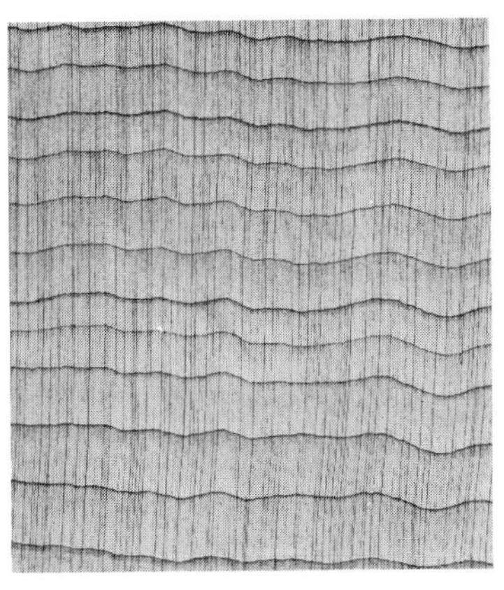

FIG. 12-4 *Chamaecyparis nootkatensis* (D. Don) Spach. (x—5×.)

FIG. 12-5 *Chamaecyparis thyoides* (L.) B.S.P. (x—5×.)

FIG. 12-6 *Juniperus virginiana* L. (x—5×.)

FIG. 12-7 *Larix laricina* (DuRoi) K. Koch. (x—5×.)

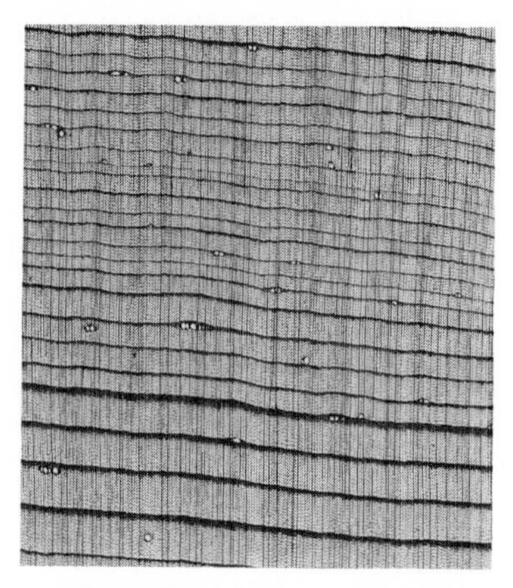

FIG. 12-8 *Larix occidentalis* Nutt. (x—5×.)

FIG. 12-9 *Libocedrus decurrens* Torr. (x—5×.)

FIG. 12-10 *Picea engelmannii* Parry. (x—5×.)

FIG. 12-11 *Picea rubens* Sarg. (x—5×.)

FIG. 12-12 *Picea sitchensis* (Bong.) Carr. (x—5×.)

FIG. 12-13 *Pinus banksiana* Lamb. (x—5×.)

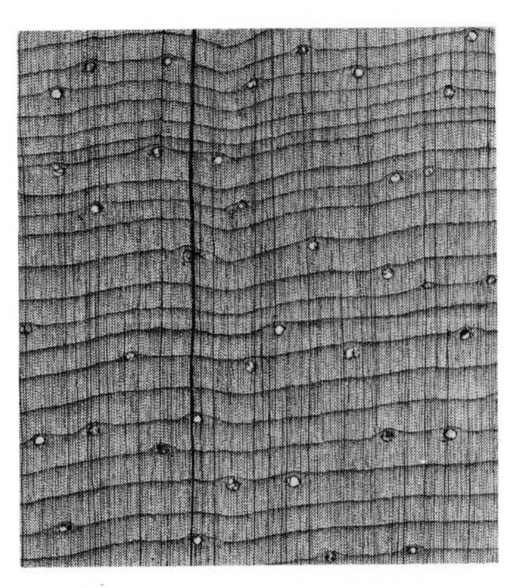

FIG. 12-14 *Pinus lambertiana* Dougl. (x—5×.)

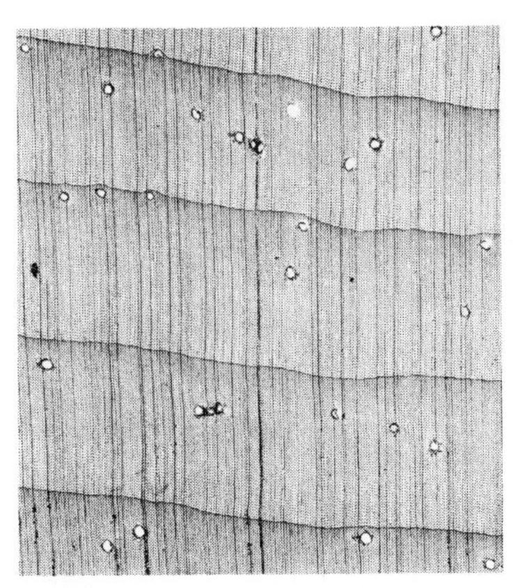

FIG. 12-15 *Pinus monticola* Dougl. (x—5×.)

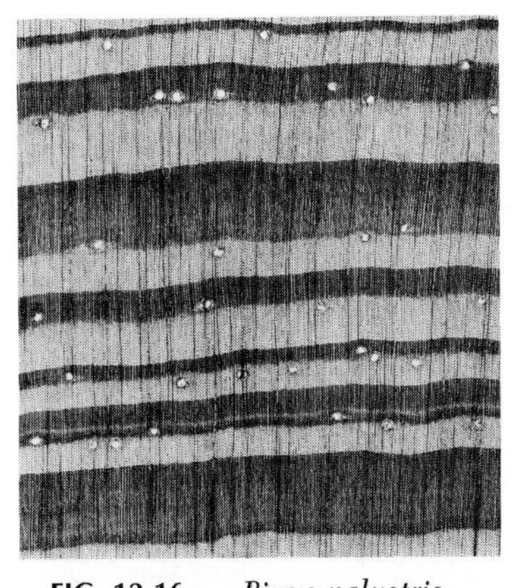

FIG. 12-16 *Pinus palustris* Mill. (x—5×.)

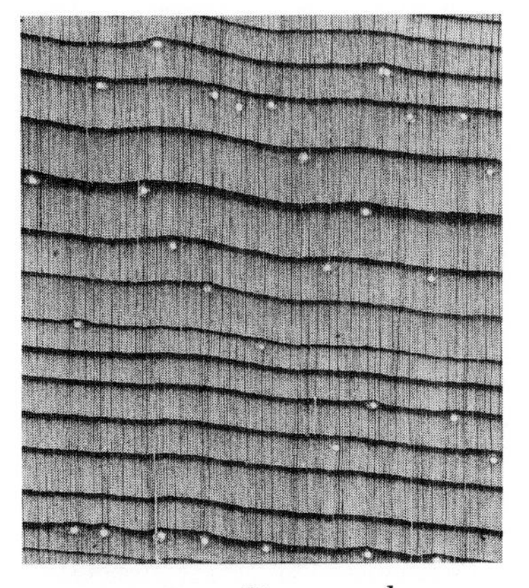

FIG. 12-17 *Pinus ponderosa* Laws. (x—5×.)

FIG. 12-18 *Pinus resinosa* Ait. (x—5×.)

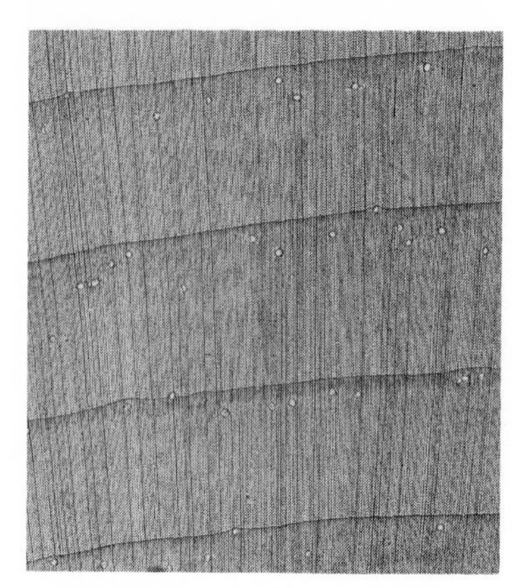

FIG. 12-19 *Pinus strobus* L. (x—5×.)

FIG. 12-20 *Pseudotsuga menziesii* (Mirb.) Franco (slow growth, x—5×).

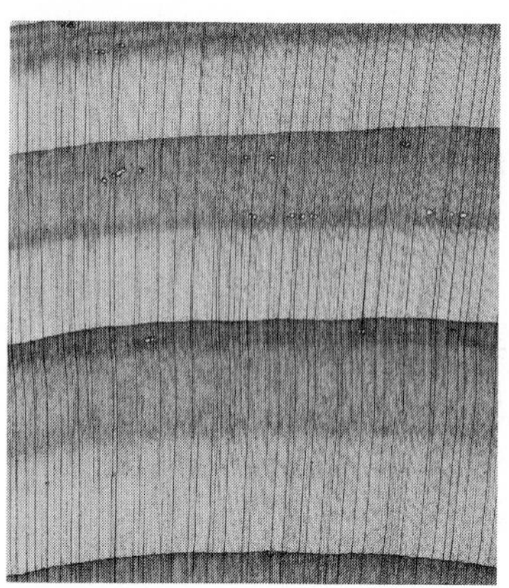

FIG. 12-21 *Pseudotsuga menziesii* (Mirb.) Franco (fast growth, x—5×).

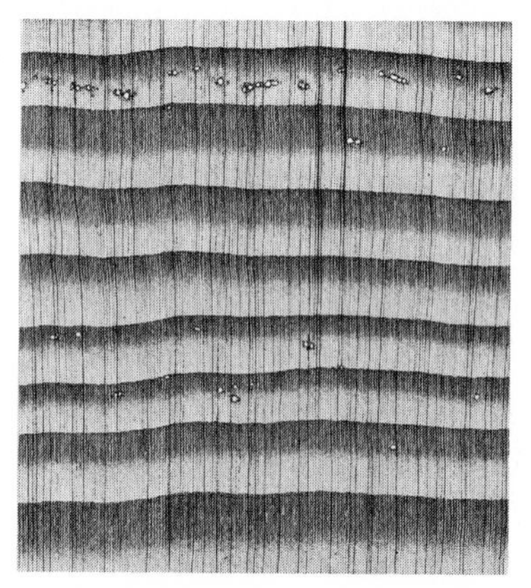

FIG. 12-22 *Pseudotsuga menziesii* (Mirb.) Franco (average growth, x—5×).

FIG. 12-23 *Sequoia sempervirens* (D. Don) Endl. (x—5×.)

FIG. 12-24 *Taxodium distichum* (L.) Rich. (x—5×.)

FIG. 12-25 *Taxus brevifolia* Nutt. (x—5×.)

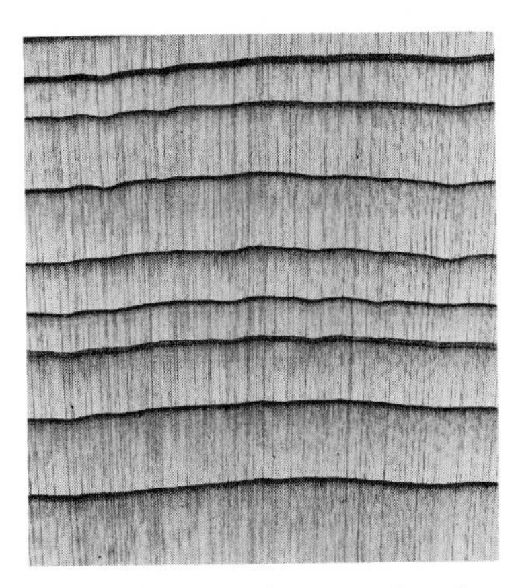

FIG. 12-26 *Thuja occidentalis* L. (x—5×.)

FIG. 12-27 *Thuja plicata* Donn. (x—5×.)

FIG. 12-28 *Torreya californica* Torr. (x—5×.)

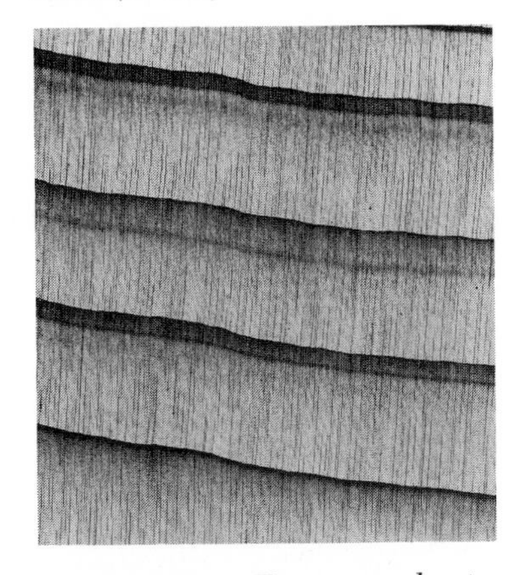

FIG. 12-29 *Tsuga canadensis* (L.) Carr. (x—5×.)

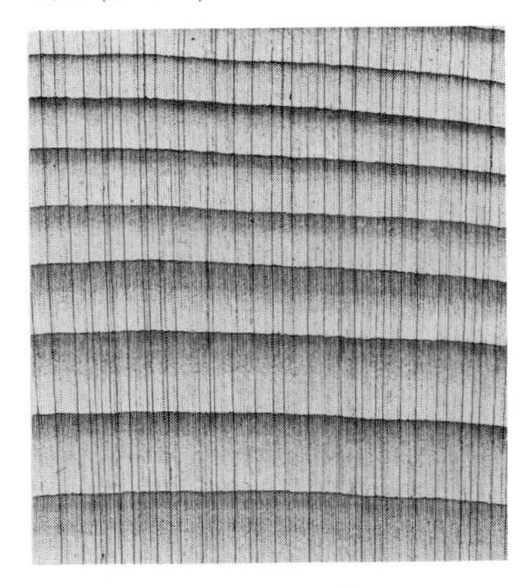

FIG. 12-30 *Tsuga heterophylla* (Raf.) Sarg. (x—5×.)

II. KEY TO CONIFEROUS WOODS—MINUTE FEATURES*

1. Longitudinal and transverse resin canals both present........................ 2

1. Longitudinal and transverse canals normally absent; traumatic canals† occasionally present, aligned in a tangential row (*x*)......................... 1

2. Ray parenchyma cross-field pitting window-like, lenticular, or pinoid (see Figs. 12-31 to 12-35, inc.); epithelial cells thin-walled (Fig. 4-14*a* and *c*)... 3

2. Ray parenchyma cross-field pitting in form of small pits resembling in appearance normal bordered pits in the ray tracheids (see Figs. 12-36 to 12-42, inc.); epithelial cells thick-walled (Fig. 4-14*b* and *e*)... 11

3. Ray tracheids dentate (*r*) (Figs. 12-31 to 12-33, inc.)......................... 4

3. Ray tracheids not dentate (*r*) (Fig. 12-34).. 9

4. Ray parenchyma cross-field pits large, rounded window-like (fenestriform), usually 1 or 2 per cross field (Fig. 4-8*a*).
Red pine—*Pinus resinosa* Ait. Desc. p. 461
Scotch pine—*Pinus sylvestris* L.

4. Ray parenchyma cross-field pits pinoid to lenticular, generally variable in shape and size, usually 1–6 per cross field (Figs. 4-8*b*, *c*)..... 5

5. Wall thickness (radial direction) of late-wood tracheids less than 6 μ; late-wood band (*x*) usually 3–10 cells in width................................ 6

5. Wall thickness (radial direction) of late-wood tracheids 6 μ or more; late-wood band (*x*) usually 10 or more cells in width.............................. 8

6. Average number of contiguous rows of ray parenchyma cells per ray in the direction of the tracheid axis 3.5 or more.
Ponderosa pine—*Pinus ponderosa* Laws. Desc. p. 459
Jeffrey pine—*Pinus jeffreyi* Grev. & Balf. Desc. p. 459

6. Average number of contiguous rows of ray parenchyma cells per ray in the direction of the tracheid axis less than 3.5......................... 7

7. Ray tracheids prominently dentate; average horizontal diameter of resin canal in fusiform rays (*t*) 45 μ; average number of rows of ray parenchyma cells in vertical direction per ray approximately 3.
Lodgepole pine—*Pinus contorta* Dougl. Desc. p. 454

7. Ray tracheids shallowly dentate (seldom extending to the center of the cell cavity); average horizontal diameter of resin canals in fusiform rays (*t*) 30–35 μ; average number of rows of ray parenchyma cells in vertical direction per ray, approximately 2.
Jack pine—*Pinus banksiana* Lamb. Desc. p. 463

* This key is designed to separate the more important North American coniferous woods on features requiring the use of a compound microscope.

The key is accompanied by a series of 14 figures (Figs. 12-31 to 12-44), at 300X, illustrating the cross-field pitting in conifers.

† Canals supposedly formed as a result of wounding; resembling normal canals when viewed on transverse section, but arranged in a row along the growth ring and often extending for some little distance.

8. Radial diameter of the lumen of the last-formed late-wood tracheids (x) 1–3 times the radial thickness of the secondary wall; average number pits per ray parenchyma cross field in the early wood less than 2.8; less than 20 percent of pits in the ray parenchyma cross fields paired vertically.

Ponderosa pine—*Pinus ponderosa* Laws. Desc. p. 459

Jeffrey pine—*Pinus jeffreyi* Grev. & Balf. Desc. p. 459

8. Radial diameter of the lumen of the last-formed late-wood tracheids (x) about equal to the radial thickness of the secondary wall; average number of pits per ray parenchyma cross field more than 2.8; more than 20 percent of the pits in the ray parenchyma cross fields paired vertically.

Southern hard pines—*Pinus* spp. Desc. p. 456

9. Ray parenchyma cross-field pits broad-oval to fenestriform, in transverse rows of pits of 2 or commonly 3, paired pits separated laterally by a distance of 8 μ or more; longitudinal resin canals with tangential diameter* up to 300 μ (avg. between 175 and 225).

Sugar pine—*Pinus lambertiana* Dougl. Fig. 12-35. Desc. p. 448

9. Ray parenchyma cross-field pits fenestriform, transverse rows of pits of 3 per cross field absent or rare, lateral separation between paired pits less than 8 μ; longitudinal resin canals with maximum tangential diameter up to 200 μ **10**

10. Paired fenestriform pits in the early-wood tracheids occur less than 60 percent of the time; ray parenchyma cross-field pits rarely lenticular, occasionally superimposed (margins overlapping laterally).

Eastern white pine—*Pinus strobus.* L. Desc. p. 452

10. Paired fenestriform pits in early-wood tracheids occur more than 60 percent of the time; lenticular ray parenchyma cross-field pits sometimes present, rarely superimposed.

Western white pine—*Pinus monticola* Dougl. Desc. p. 450

11. Longitudinal tracheids with spiral thickening.

Douglas-fir—*Pseudotsuga menziesii* (Mirb.) Franco. Desc. p. 474

11. Longitudinal tracheids lacking spiral thickening (rarely evident in the late-wood tracheids of tamarack)........ **12**

12. Late wood pronounced, with abrupt transition; bordered pits on the longitudinal tracheids frequently biseriate (r).

Larch—*Larix* spp. Fig. 12-36. Desc. pp. 465 to 468

12. Late wood not pronounced, transition usually gradual; bordered pits on the longitudinal tracheids uniseriate or biseriate (r)........ **13**

13. Ray cells (t) oblong to oval in shape; bordered pits on the longitudinal tracheids rarely biseriate.

White, Black, Red and Englemann spruce—*Picea* spp. Desc. p. 469

* The diameter of the canal inclusive of the epithelium.

13. Ray cells (*t*) orbicular to oval in shape; abundant yellowish-brown deposits in the ray cells; bordered pitting on the longitudinal tracheids frequently biseriate; occasional crystals in ray cells.
Sitka spruce—*Picea sitchensis* (Bong.) Carr. Desc. p. 472

14. Longitudinal tracheids with spiral thickening............................ **15**
14. Longitudinal tracheids without spiral thickening........................ **16**

15. Angle of spiral thickening about 80° from the tracheid axis; spirals unevenly spaced; ray parenchyma cross-field pitting commonly in transverse rows of 3 or more; widest early wood tracheids 50 μ or more.
California torreya—*Torreya californica* Torr. Desc. p. 502

15. Angle of spiral thickening about 60° from the tracheid axis; spirals quite evenly spaced; ray parenchyma cross-field pitting never or rarely in transverse rows of 3 or more; widest early-wood tracheids less than 50 μ wide.
Pacific yew—*Taxus brevifolia* Nutt. Desc. p. 501

16. Parenchyma (longitudinal) fairly abundant to abundant, present in every growth ring, excluding marginal parenchyma................... **17**
16. Parenchyma wanting or very sparse and sporadic (sometimes abundant in a given growth ring but wanting in adjacent rings), marginal parenchyma present or absent... **24**

17. End walls of ray parenchyma cells nodular (*r*)................................ **18**
17. End walls of ray parenchyma cells smooth (*r*)................................. **20**

18. Ray tracheids (*r*) abundant in low rays, often constituting the entire ray.
Alaska-cedar—*Chamaecyparis nootkatensis* (D. Don) Spach. Figs. 12-41, 12-42. Desc. p. 496

18. Ray tracheids (*r*) absent or extremely sparse............................. **19**

19. Intercellular spaces at rounded corners of the longitudinal tracheids (*x*) frequent and conspicuous; end walls of ray parenchyma cells with indentures similar to those in *Thuja plicata* (Fig. 12-44); ray tracheids absent.
Redcedar—*Juniperus* spp. Desc. p. 499

19. Intercellular spaces at corners of the longitudinal tracheids (*x*) lacking or rarely present; end walls of ray parenchyma cells lacking indentures; ray tracheids occasionally present.
Incense-cedar—*Libocedrus decurrens* Torr. Desc. p. 489

20. Transverse walls of longitudinal parenchyma cells nodular (*t*)....... **21**
20. Transverse walls of longitudinal parenchyma cells not nodular or inconspicuously nodular (*t*).. **23**

21. End walls of ray parenchyma cells with indentures (Fig. 12-44).
Western redcedar—*Thuja plicata* Donn. Desc. p. 493

21. End walls of ray parenchyma cells lacking indentures....................... **22**

22. Intertracheid pitting in 1–4 longitudinal rows (frequently in 3); angle of ray parenchyma cross-field pit apertures 80° from the tracheid axis.

Baldcypress—*Taxodium distichum* (L.) Rich. Fig. 12-40. Desc. p. 487

22. Intertracheid pitting in one longitudinal row or occasionally in part biseriate; angle of ray parenchyma cross-field pit apertures 40° from the tracheid axis.

Port-Orford-cedar—*Chamaecyparis lawsoniana* (A. Murr.) Parl. Desc. p. 494

23. Intertracheid pitting in 1–3 longitudinal rows (usually 2); ray parenchyma cross-field pitting taxodioid.

Redwood—*Sequoia sempervirens* (D. Don) Endl. Fig. 12-39. Desc. p. 485

23. Intertracheid pitting in 1 longitudinal row or rarely paired; ray parenchyma cross-field pitting cupressoid.

Atlantic white-cedar—*Chamaecyparis thyoides* (L.) B.S.P. Desc. p. 498

24. Ray tracheids present (*r*).. **25**

24. Ray tracheids absent (*r*).. **27**

25. Ray tracheids fairly abundant, at least in the low rays; rays generally less than 25 cells in height.. **26**

25. Ray tracheids sporadic, wanting from many rays; rays frequently 25 or more cells in height.

Balsam fir—*Abies balsamea* (L.) Mill.* Fig. 12-38. Desc. p. 480

26. Ray tracheids usually restricted to one row on the upper and lower margins of the rays, low along the grain.

Hemlock—*Tsuga* spp.† Desc. pp. 477 to 480

26. Ray tracheids usually wanting in the high rays, abundant in the low rays and frequently constituting the entire ray, high along the grain.

Alaska-cedar—*Chamaecyparis nootkatensis* (D. Don) Spach. Figs. 12-41 and 12-42. Desc. p. 496

27. Ray parenchyma cells with end walls nodular and horizontal walls conspicuously pitted.. **28**

27. Ray parenchyma cells with end walls smooth and horizontal walls inconspicuously pitted.. **29**

28. Ray cells (*t*) oblong, contents colorless.

Eastern fir—*Abies* spp. Desc. p. 480

28. Ray cells (*t*) orbicular to oval, contents reddish brown.

Western fir—*Abies* spp. Desc. p. 482

* Ray tracheids occasionally occur sporadically on the margins of rays in western redcedar, *Thuja plicata* Donn. The separation can then be made on the basis of ray heights (see descriptions).

† For separation of eastern and western hemlock see remarks under the description of western hemlock, p. 480.

29. Bordered pitting in one longitudinal row or very rarely biseriate on the radial walls; tracheids fine, up to 35 (avg 20–30) μ in diameter.
Northern white-cedar—***Thuja occidentalis*** L. Desc. p. 491

29. Bordered pitting in 1–2 longitudinal rows on the radial walls; tracheids medium coarse, up to 50 (avg 32–38) μ in diameter.
Western redcedar—***Thuja plicata*** Donn. Desc. p. 493

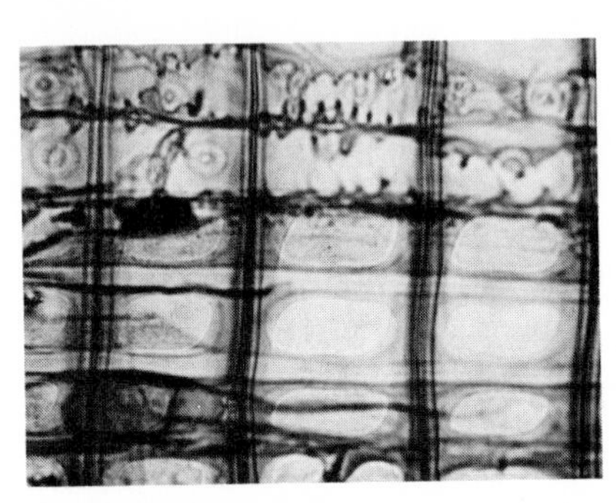

FIG. 12-31 Radial section of the wood of red pine, ***Pinus resinosa*** Ait., showing a portion of a ray composed of two rows of dentate ray tracheids at the top, and three complete rows of ray parenchyma cells, the latter possessing angular, window-like (fenestriform) pits in the back walls of their cross fields. (300×.)

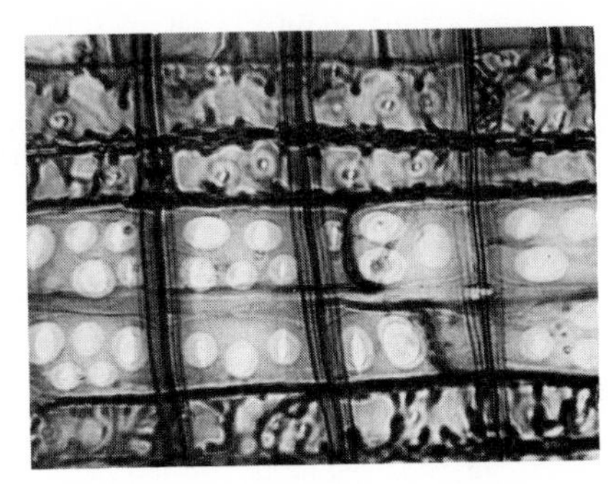

FIG. 12-32 Radial section of the wood of ponderosa pine, ***Pinus ponderosa*** Laws., showing a portion of a ray composed of three rows of dentate ray tracheids (two at top and one below), and two rows of ray parenchyma cells with rounded pinoid pits in the back walls of their cross fields. (300×.)

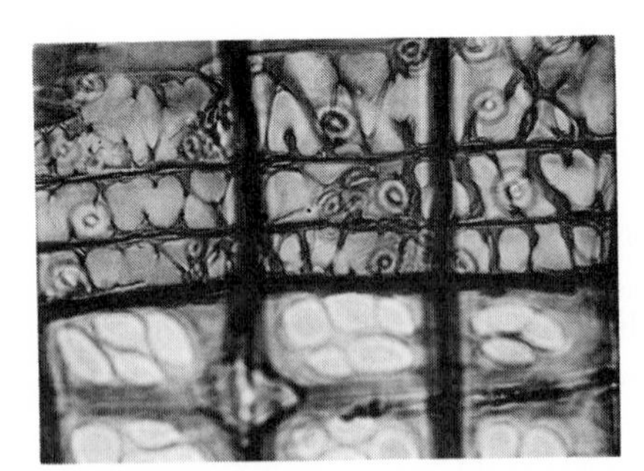

FIG. 12-33 Radial section of the wood of longleaf pine, ***Pinus palustris*** Mill., showing a portion of a ray composed of three rows of dentate ray tracheids at the top and one complete row of ray parenchyma cells. The latter cells possess cross fields with several pinoid pits having acute apices. (300×.)

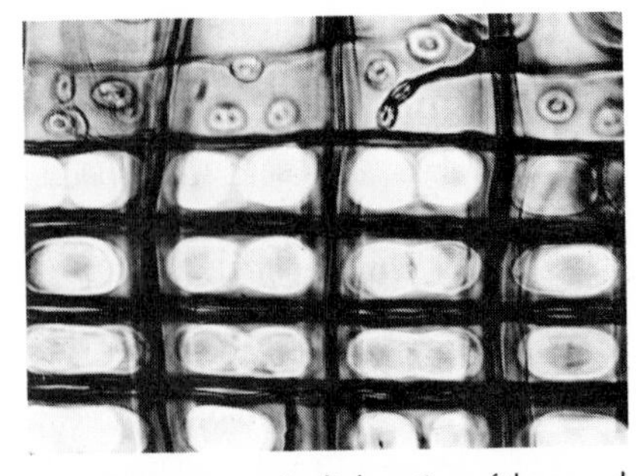

FIG. 12-34 Radial section of the wood of eastern white pine, ***Pinus strobus*** L., showing portion of a ray composed of one row of nondentate ray tracheids at the top, and three complete rows of ray parenchyma cells, the ray parenchyma cross fields exhibiting one to two window-like (fenestriform) pits in each back wall. (300×.)

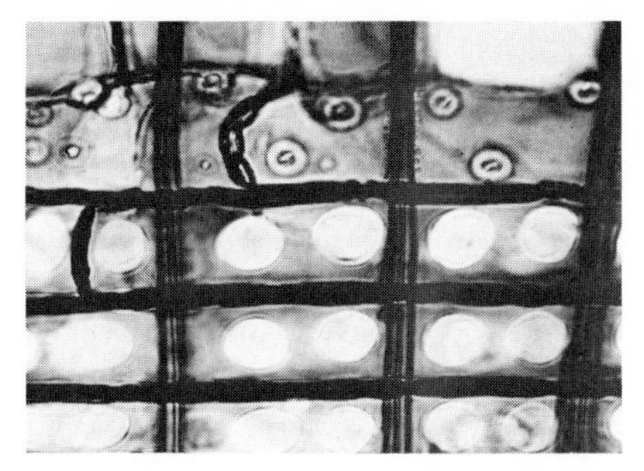

FIG. 12-35 Radial section of the wood of sugar pine, ***Pinus lambertiana*** Dougl., showing portion of a ray composed of one row of nondentate ray tracheids at the top and two complete rows of ray parenchyma cells, the ray parenchyma cross fields having paired, rounded, window-like pits in the back walls. (300×.)

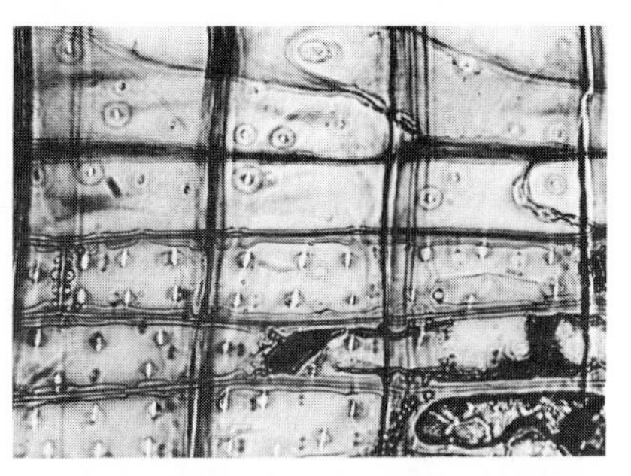

FIG. 12-36 Radial section of the wood of western larch, ***Larix occidentalis*** Nutt., showing portion of a ray composed of two rows of nondentate ray tracheids at the top and two complete rows of ray parenchyma cells, the ray parenchyma cross fields with small, piceoid pits arranged in part in two horizontal rows. (300×.)

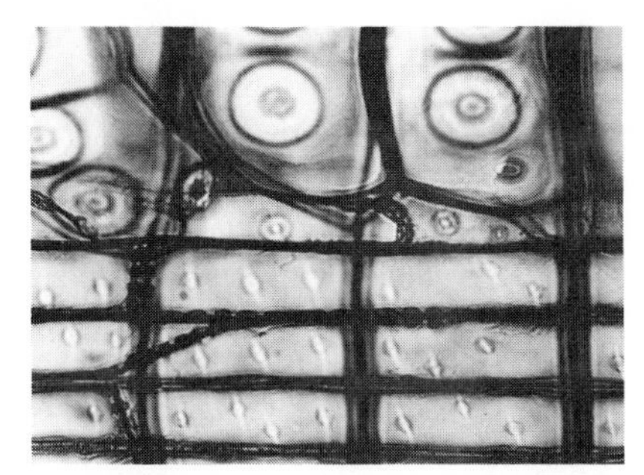

FIG. 12-37 Radial section of the wood of Sitka spruce, ***Picea sitchensis*** (Bong.) Carr., showing portion of a ray composed of one row of nondentate ray tracheids at the top and three rows of ray parenchyma cells, the ray parenchyma cross fields exhibiting small piceoid pits in their back walls arranged in single transverse rows, for the most part. (300×.)

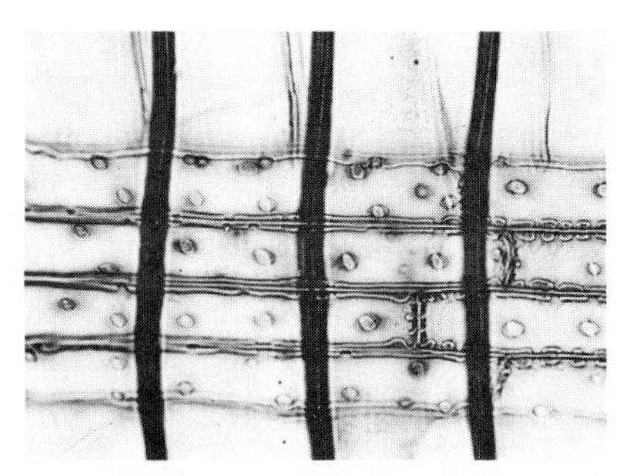

FIG. 12-38 Radial section of the wood of balsam fir, ***Abies balsamea*** (L.) Mill., showing portion of a ray entirely composed of ray parenchyma cells whose cross-field pits are small and taxodioid. (300×.)

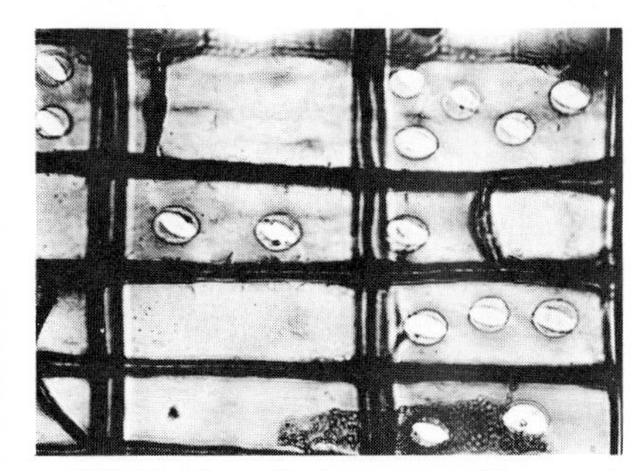

FIG. 12-39 Radial section of the wood of redwood, ***Sequoia sempervirens*** (D. Don) Endl., showing portion of a ray composed entirely of ray parenchyma cells whose cross-field pits are large, oval, and taxodioid. (300×.)

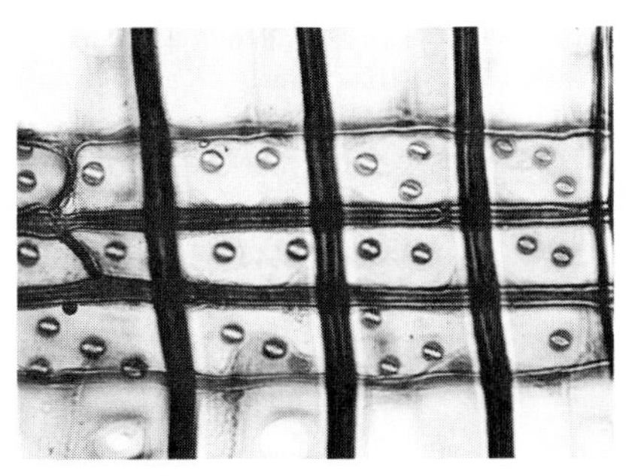

FIG. 12-40 Radial section of the wood of baldcypress, ***Taxodium distichum*** (L.) Rich., showing portion of a ray composed entirely of ray parenchyma cells whose cross fields are equipped with small, oval-orbicular, taxodioid-cupressoid pits. (300×.)

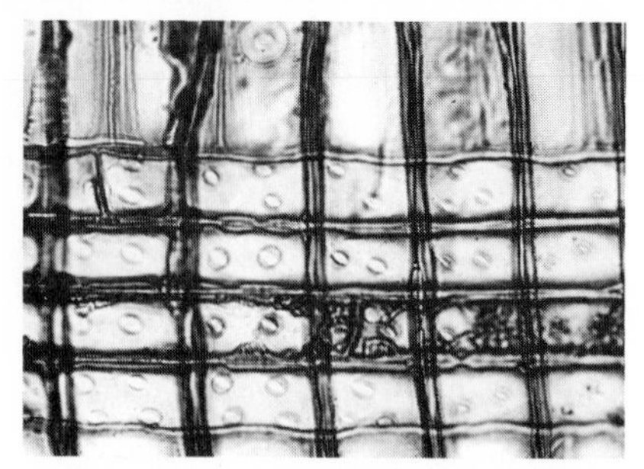

FIG. 12-41 Radial section of the wood of Alaska-cedar, ***Chamaecyparis nootkatensis*** (D. Don) Spach, showing portion of a ray composed entirely of ray parenchyma cells whose cross-field pits are small, orbicular, and cupressoid. (300×.)

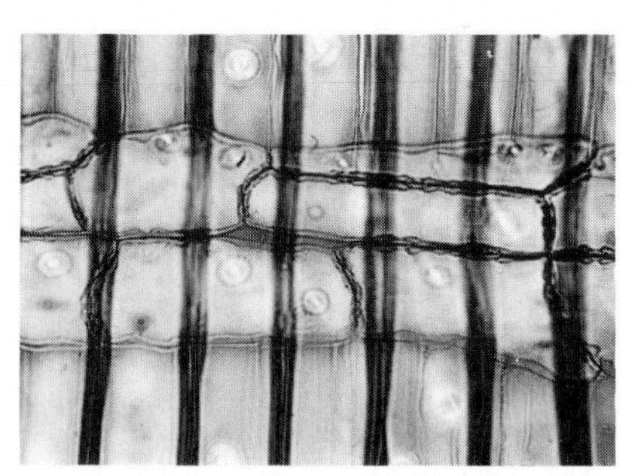

FIG. 12-42 Same as in Fig. 12-41 except that the low ray shown consists entirely of ray-tracheid cells. (300×.)

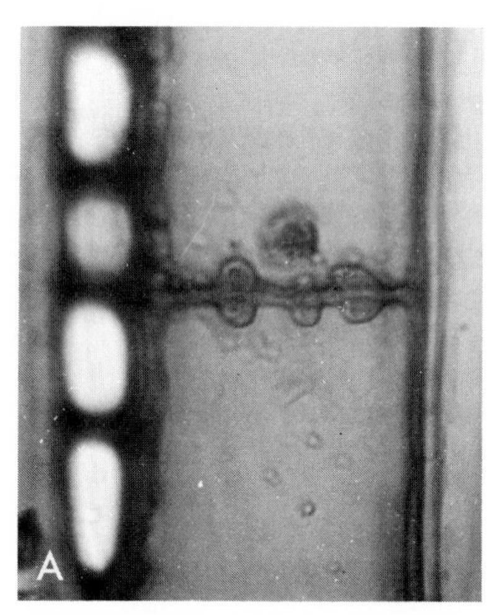

FIG. 12-43 Nodular transverse (end) walls in the longitudinal parenchyma (*t*) of baldcypress, ***Taxodium distichum*** (L.) Rich. (680×.)

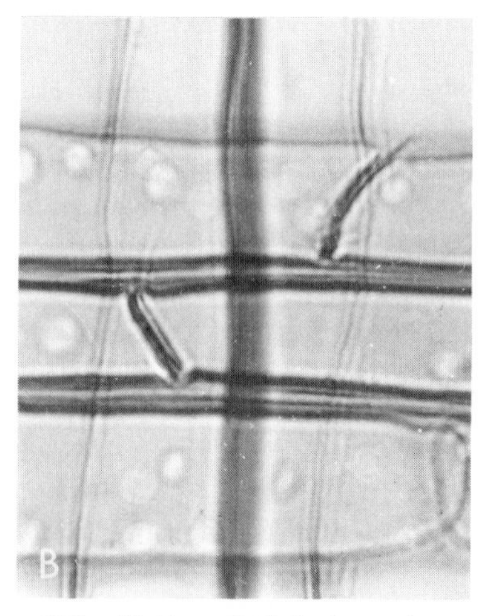

FIG. 12-44 Radial view of ray parenchyma cells in a ray of western redcedar, ***Thuja plicata*** Donn, showing indentures (pitlike depressions) where the end walls of the ray parenchyma cells abut on the horizontal walls. (710×.)

III. KEY TO CONIFEROUS WOODS BASED ON GROSS AND MINUTE FEATURES

1. Longitudinal and transverse resin canals present; longitudinal canals appearing as small openings or flecks (x) confined for the most part to the outer portion of the growth rings; transverse canals present in some of the rays, forming radial streaks on the transverse surface (Figs. 12-7, 12-8, and 12-10 to 12-22, inc.).. **2**

1. Longitudinal and transverse resin canals normally absent; longitudinal wound canals (traumatic) occasionally present, aligned in a tangential row (x) (Figs. 12-1 to 12-6, 12-9, 12-23 to 12-30, inc.)..................... **18**

2. Longitudinal resin canals numerous, quite evenly distributed in the outer portion of every growth ring, generally visible to the naked eye* as dark- or light-colored dots or as small openings, conspicuous or relatively conspicuous with a hand lens (Figs. 12-12 to 12-19, inc.); epithelial cells thin-walled except in *Picea sitchensis* (Bong.) Carr. (see Fig. 4-10).. **3**

2. Longitudinal resin canals generally sparse, unevenly distributed, sometimes absent in portions of some growth rings (if numerous, 2–many in a tangential row), invisible or barely visible as small dark or white flecks to the naked eye, not conspicuous with a hand lens (Figs. 12-7, 12-8, 12-10, 12-11, 12-20 to 12-22, inc.); epithelial cells thick-walled.. **14**

3. Wood soft, light, quite even-grained; transition from early to late wood gradual (Figs. 12-12, 12-14, 12-15, 12-19); ray tracheids not dentate... **4**

3. Wood soft to medium-hard or hard, more or less uneven-grained; transition from early to late wood more or less abrupt (Figs. 12-16 to 12-18, inc.); ray tracheids dentate.. **7**

4. Wood medium-coarse to relatively coarse-textured; longitudinal tracheids with average tangential diameter of 35–45 μ................. **5**

4. Wood medium-textured; longitudinal tracheids with average tangential diameter of 25–35 μ.
Eastern white pine—*Pinus strobus* L. Desc. p. 452

5. Wood with a resinous odor; heartwood cream-colored to light or deep reddish brown; tangential surface (split) not noticeably dimpled; ray parenchyma cross-field pits large and window-like (fenestriform, Fig. 12-35), 1–2 per cross field.. **6**

5. Wood without resinous odor; heartwood light pinkish yellow to pale brown with a purplish cast, darkening on exposure to silvery brown with a faint tinge of red; tangential surface (split) frequently noticeably

* Numerous but relatively inconspicuous in red pine—*Pinus resinosa* Ait., jack pine—*Pinus banksiana* Lamb., lodgepole pine—*Pinus contorta* Dougl., and sometimes in ponderosa pine—*Pinus ponderosa* Laws.

dimpled; ray parenchyma cross-field pits small, piceoid (Fig. 12-37), quite uniform in size, 1–4 (generally 2–3) per cross field.

Sitka spruce—*Picea sitchensis* (Bong.) Carr. Desc. p. 472

6. Wood with unusually prominent dark streaks along the grain (resin canals), frequently exuding a sugary substance when green; seasoned heartwood a pale reddish brown; resin canals appearing to the naked eye as minute openings on the transverse surface; window-like pits in the ray parenchyma cross fields of the early-wood tracheids with rounded or lemon-shaped apertures, not occupying most of the cross field, frequently in transverse rows of 3 or 4; longitudinal resin canals with maximum tangential diameter* of 300 μ (avg between 175–225).

Sugar pine—*Pinus lambertiana* Dougl. Desc. p. 448

6. Wood without unusually prominent dark streaks along the grain (resin canals), lacking sugary exudations when green; seasoned heartwood dark reddish brown; resin canals appearing to the naked eye as dark- or light-colored dots; window-like pits in the ray parenchyma cross fields of the early-wood tracheids with angular orifices, occupying most of the cross field, rarely in transverse rows of 3; longitudinal resin canals with maximum tangential diameter of 200 μ (avg between 135–150).

Western white pine—*Pinus monticola* Dougl. Desc. p. 450

7. Tangential surface (split) with numerous depressions which give it a dimpled appearance (Fig. 11-10*b*); outer margin of the growth increment (*r*) with depressions or sinuses.. **8**

7. Tangential surface (split) not dimpled; outer margin of the growth increment (*r*) not obviously distorted.. **9**

8. Longitudinal resin canals (*x*) small, inconspicuous or not visible with the naked eye, with maximum tangential diameter of 100 μ (avg between 80–90).

Lodgepole pine—*Pinus contorta* Dougl. Desc. p. 454

8. Longitudinal resin canals comparatively large (*x*), plainly visible to the naked eye, with maximum tangential diameter of 230 μ (avg between 160–185).

Ponderosa pine—*Pinus ponderosa* Laws. Desc. p. 459
Jeffrey pine—*Pinus jeffreyi* Grev. & Balf. Desc. p. 459

9. Wood hard, heavy, strong, generally highly resinous; bands of late wood broad; late-wood tracheids forming a band 10–many cells in width (*x*). **10**

9. Wood approaching that of the soft pines, soft to moderately hard, light to moderately heavy; bands of late wood usually narrow; late-wood tracheids forming a band 3–10 cells in width (*x*).............................. **12**

10. Longitudinal resin canals small (*x*), inconspicuous or not visible to the naked eye, the majority with a tangential diameter of less than

* Diameter inclusive of the epithelium.

125 μ; dentations in the ray tracheids shallow (not prominent), seldom extending across the cell (*r*)... **11**

10. Longitudinal resin canals comparatively large, plainly visible to the naked eye, the majority with a tangential diameter of more than 125 μ; dentations in the ray tracheids prominent, frequently extending across the cell (*r*) forming a reticulate pattern.
Southern hard pines—*Pinus* spp. Desc. p. 456
Ponderosa pine—*Pinus ponderosa* Laws. Desc. p. 459
Jeffrey pine—*Pinus jeffreyi* Grev. & Balf. Desc. p. 459

11. Pits in the ray parenchyma cross fields in the early wood large, window-like, rounded (Fig. 12-31), 1–2 per cross field.
Red pine—*Pinus resinosa* Ait. Desc. p. 461
Scotch pine—*Pinus sylvestris* L.

11. Pits in the ray parenchyma cross fields in the early wood small, pinoid, similar to those in *Pinus palustris* (ex., Fig. 12-33) and variable in shape, 1–6 (generally 4–6), in 2 transverse rows per cross field.
Jack pine—*Pinus banksiana* Lamb. Desc. p. 463

12. Heartwood light red to light brown; resin canals small (*x*), inconspicuous or not visible to the naked eye, the majority with a tangential diameter of less than 125 μ; dentations in the ray tracheids shallow and not prominent, seldom extending across the cell....... **13**

12. Heartwood yellowish, not oily; resin canals large (*x*), visible to the naked eye, the majority with a tangential diameter of more than 125 μ; dentations in the ray tracheids prominent, frequently extending across the cell (*r*), forming a reticulate pattern.
Ponderosa pine—*Pinus ponderosa* Laws. Desc. p. 459
Jeffrey pine—*Pinus jeffreyi* Grev. & Balf. Desc. p. 459

13. Ray parenchyma cross-field pits in the early wood, large, oval, window-like (Fig. 12-31), 1–2 per cross field.
Red pine—*Pinus resinosa* Ait. Desc. p. 461

13. Ray parenchyma cross-field pits in the early wood, small pinoid, and variable in shape, 1–6 (generally 4–6 in 2 transverse rows) per cross field.
Jack pine—*Pinus banksiana* Lamb. Desc. p. 463

14. Late wood not pronounced; transition from early to late wood gradual.. **15**

14. Late wood pronounced (the band of late wood frequently narrow); transition from early to late wood abrupt................................ **16**

15. Heartwood light pinkish yellow to pale brown with a purplish cast, darkening on exposure to silvery brown with a faint tinge of red, dull to somewhat lustrous, medium-textured; tangential surface (split) frequently noticeably dimpled; longitudinal tracheids with average tangential diameter between 35 and 45 μ; ray cells (*t*) orbicular to oval.
Sitka spruce—*Picea sitchensis* (Bong.) Carr. Desc. p. 472

15. Heartwood not distinct; wood nearly white to pale yellowish white or light yellowish brown, lustrous, relatively fine-textured; tangential surface (split) not noticeably dimpled, longitudinal tracheids with average tangential diameter between 25 and 30 μ; ray cells (*t*) oblong to oval.
White, Black, Red and Engelmann spruce—*Picea* spp. Desc. p. 469

16. Wood with characteristic odor on fresh-cut surface, not oily; contour of growth rings frequently wavy; heartwood yellowish or pale reddish yellow to orange-red or deep red (without brownish cast); early-wood longitudinal tracheids with obvious spiral thickenings; longitudinal resin canals (*x*) in short tangential lines; transverse resin canals (*t*) usually with 6 epithelial cells.
Douglas-fir—*Pseudotsuga menziesii* (Mirb.) Franco. Desc. p. 474

16. Wood without characteristic odor on fresh-cut surface, more or less oily; contour of growth rings seldom wavy; heartwood with brownish cast; longitudinal tracheids without spiral thickenings or the spirals sporadic and confined to the late wood (*Larix laricina*); longitudinal resin canals (*x*) not in tangential lines (sometimes in tangential groups of 2–5); transverse resin canals (*t*) with up to 12 or more epithelial cells.......... **17**

17. Wood medium-textured; heartwood yellowish brown (occasionally reddish brown); growth rings generally wide (5–20 per in.), variable in width; bands of late wood usually wide; longitudinal tracheids with average diameter between 30 and 40 μ.*
Eastern larch—*Larix laricina* (Du Roi) K. Koch. Desc. p. 465

17. Wood coarse-textured; heartwood russet or reddish brown; growth rings generally narrow (15—30+ per in.), quite uniform in width; bands of late wood usually narrow; longitudinal tracheids with average tangential diameter between 40 and 50 μ.
Western larch—*Larix occidentalis* Nutt. Desc. p. 467

18. Wood fragrant when freshly cut.......... **19**

18. Wood not fragrant, sometimes ill-scented when freshly cut.......... **25**

19. Heartwood purplish or rose-red to dull red, reddish or pinkish brown, or dull brown.......... **20**

19. Heartwood yellowish white to clear yellow, straw-brown, or light brown with a tinge of pink.......... **22**

20. Wood medium- to coarse-textured; heartwood reddish to pinkish brown or dull brown, sometimes with a lavender tinge; longitudinal tracheids with average tangential diameter between 30 and 40 μ; intercellular spaces few or wanting at the corners of the tracheids (*x*), if present, inconspicuous.......... **21**

* Computed by counting the number of tracheids falling under the calibrated scale of an eyepiece micrometer, using a 10X eyepiece and a 16-mm objective, *making no allowance* for the rays. To obtain an approximate average, 5 random positions on the section are recommended. A similar procedure should be followed whenever data of this general type are required.

20. Wood fine-textured; heartwood purplish or rose-red when first exposed, aging to dull brown or reddish brown; longitudinal tracheids with average tangential diameter between 20 and 30 μ; intercellular spaces frequent at the rounded corners of the tracheids (x).
Redcedar—*Juniperus* spp. Desc. p. 499

21. Wood firm when cut across the grain with a knife, with pungent, acrid taste (spicy),* often pecky;† early and late wood of about the same hardness; ray parenchyma cross-field pits cupressoid (see Fig. 4-8*f*); rays occasionally in part biseriate, the ray cells in the heartwood with deposits of dark gum.
Incense-cedar—*Libocedrus decurrens* Torr. Desc. p. 489

21. Wood crumbly when cut across the grain with a knife, without pungent, acrid (spicy) taste; late wood considerably harder than the early wood; ray parenchyma cross-field pits taxodioid; rays strictly uniseriate, the ray cells in the heartwood with scanty, gummy infiltration.
Western redcedar—*Thuja plicata* Donn. Desc. p. 493

22. Heartwood yellowish white to bright clear yellow or yellowish brown **23**

22. Heartwood light brown tinged with red or pink (roseate), or straw-brown **24**

23. Wood with a pungent, ginger-like odor on fresh-cut surfaces; heartwood yellowish white to pale yellowish brown; annual rings usually averaging 6–15 per in.; rays consisting entirely of ray parenchyma or very rarely with ray tracheids.
Port-Orford-cedar—*Chamaecyparis lawsoniana* (A. Murr.) Parl. Desc. p. 494

23. Wood with an odor resembling that of raw potatoes on fresh-cut surfaces; heartwood bright clear yellow, darkening somewhat upon exposure; growth rings usually very narrow, generally averaging 30—50+ per in.; low rays frequently consisting entirely of ray tracheids, the high rays usually composed entirely of ray parenchyma.
Alaska-cedar—*Chamaecyparis nootkatensis* (D. Don) Spach. Desc. p. 496

24. Wood somewhat oily, rubbery when cut across the grain with a knife; heartwood light brown, tinged with red or pink; parenchyma fairly abundant to abundant, present in every growth ring; ray parenchyma cross-field pits cupressoid.
Atlantic white-cedar—*Chamaecyparis thyoides* (L.) B.S.P. Desc. p. 498

24. Wood dry, crumbly when cut across the grain with a knife; heartwood uniformly straw-brown; parenchyma lacking, or very sparse and sporadic (sometimes abundant in a growth ring but then wanting in adjacent rings); ray parenchyma cross-field pits taxodioid.
Northern white-cedar—*Thuja occidentalis* L. Desc. p. 491

* See second footnote (†) in Key I, p. 427.

† See Chap. 10 for description of pecky condition.

25. Wood very coarse-textured; longitudinal tracheids with average tangential diameter between 45 and 65 μ; strands of longitudinal parenchyma readily visible along the grain with a hand lens as dark lines (dark gummy infiltration in the cells).. **26**

25. Wood fine- to medium-coarse-textured; longitudinal tracheids with tangential diameter between 15 and 45 μ; strands of longitudinal parenchyma wanting or, if present, not readily visible along the grain with a hand lens.. **27**

26. Wood more or less oily (with a greasy feel), often with a rancid odor on fresh-cut surface, frequently pecky; heartwood very variable in color, ranging from yellowish white to light or dark brown, reddish brown, or almost black; contour of individual rings frequently irregular; pits in the ray parenchyma cross fields in the early wood orbicular to oval-orbicular, the long axis of the aperture of the inconspicuously taxodioid pit less than 8 μ; end walls of longitudinal parenchyma conspicuously nodular.

Baldcypress—*Taxodium distichum* (L.) Rich. Desc. p. 487

26. Wood dry and brittle, not pecky; heartwood clear light red to deep reddish brown; contour of individual growth rings usually quite regular; pits in the ray parenchyma cross fields in the early wood oval for the most part, the long axis of the aperture of the conspicuously taxodioid pit more than 8 μ; end walls of longitudinal parenchyma inconspicuously or sporadically nodular.

Redwood—*Sequoia sempervirens* (D. Don) Endl. Desc. p. 485

27. Spiral thickening present in longitudinal tracheids........................... **28**

27. Spiral thickening absent in longitudinal tracheids............................ **29**

28. Heartwood bright orange to rose-red; sapwood nearly white to light yellow; wood heavy; longitudinal tracheids with average tangential diameter between 15 and 20 μ; spiral thickenings evenly spaced.

Pacific yew—*Taxus brevifolia* Nutt. Desc. p. 501

28. Heartwood clear yellow; sapwood yellowish white not much different from heartwood; wood moderately heavy; longitudinal tracheids with average tangential diameter between 25 and 35 μ; spiral thickenings fine, in widely spaced bands.

California torreya—*Torreya californica* Torr. Desc. p. 502

29. Wood usually without characteristic odor; heartwood generally wanting, the color whitish or creamy white through shades of pale straw-buff and light brown; early and late wood generally differentiated through differences in color... **30**

29. Wood with a characteristic odor resembling that of raw potatoes on a fresh-cut surface; heartwood distinct, bright clear yellow with no conspicuous distinction between color of early and late wood; ray tracheids abundant in the low rays (frequently composing the entire ray), usually lacking in the high rays.

Alaska-cedar—*Chamaecyparis nootkatensis* (D. Don) Spach. Desc. p. 496

30. Wood light buff to light brown; early wood usually with a reddish tinge; late wood darker, with a reddish-brown to purplish tinge; pits in the ray parenchyma cross fields cupressoid to piceoid; ray tracheids present, usually restricted to 1 row on the upper and lower margins of the ray, lower than the cells of the ray parenchyma...... **31**

30. Wood whitish to creamy white or pale brown; early wood whitish passing gradually into darker, usually lavender late wood (color contrasts of early and late wood more pronounced in wider rings); pits in the ray parenchyma cross fields taxodioid; ray tracheids absent or, if present, sporadic and wanting from many rays............ **32**

31. Wood dry, brittle, harsh under tools, uneven-grained; transition from early to late wood frequently rather abrupt; longitudinal strands of resin cells and wound canals (termed ***bird peck***) rarely present.
Eastern hemlock—*Tsuga canadensis* (L.) Carr. Desc. p. 477

31. Wood not very brittle, hard but not harsh under tools, quite even-grained; transition from early to late wood gradual; longitudinal strands of resin cells and wound canals (termed ***bird peck***) frequently present.
Western hemlock—*Tsuga heterophylla* (Raf.) Sarg. Desc. p. 478

32. Ray cells (*t*) oblong.
Eastern fir—*Abies* spp. Desc. p. 480

32. Ray cells (*t*) orbicular to oval.
Western fir—*Abies* spp. Desc. p. 482

Description of Coniferous Woods by Species

In spite of repeated warnings against placing too much reliance on the measurements of the various wood cell features, in this section the numerical values of the anatomical structures have been largely retained. This is because the tangential dimensions of the anatomical structures in the coniferous woods tend to remain fairly constant within relatively narrow limits; i.e., they are not excessively affected by the location of a cell within the growth rings or the various parts of the stem. Furthermore, the softwoods possess only a few different kinds of cells, by far the most abundant of which are the longitudinal tracheids, which comprise upward of 90 percent of the total volume of a softwood. Because of these factors the numerical values for wood cells in the coniferous woods are considered to be quite reliable and are given in the descriptions that follow. *Measurements of the diameter of the resin canals include the epithelium.* Further numerical information on the anatomical features of the coniferous woods is found in Tables 4-4, 4-5, and 4-6 in Chap. 4.

SUGAR PINE

Pinus lambertiana Dougl.

GENERAL CHARACTERISTICS

Sapwood nearly white to pale yellowish white, narrow to medium wide, *frequently discolored by blue stain*; *heartwood* light brown to pale reddish brown (never deep reddish brown as in the eastern and western white pines), frequently discolored with brown stain; *wood* with a faint, noncharacteristic odor, often exuding a sugary substance when green but without characteristic taste when dry, straight- and even-grained, relatively coarse-textured, light (sp gr approx 0.35 green, 0.38 ovendry), moderately soft (some samples spongy, others quite hard). *Growth rings* distinct, delineated by a rather narrow band of darker late wood at the outer margin, narrow to medium wide. Early-wood zone usually wide, appearing to occupy most of the ring; transition from early to late wood gradual; late wood as described above, not appreciably more resistant to tools than the early wood. *Parenchyma* not visible. *Rays* very fine (x), not visible to the naked eye except where they include a transverse resin canal, forming a fine, close, inconspicuous fleck on the quarter surface. *Resin canals* present, longitudinal and transverse; (a) longitudinal canals conspicuous (the orifice visible to the naked eye), numerous, confined largely to the central and outer portions of the

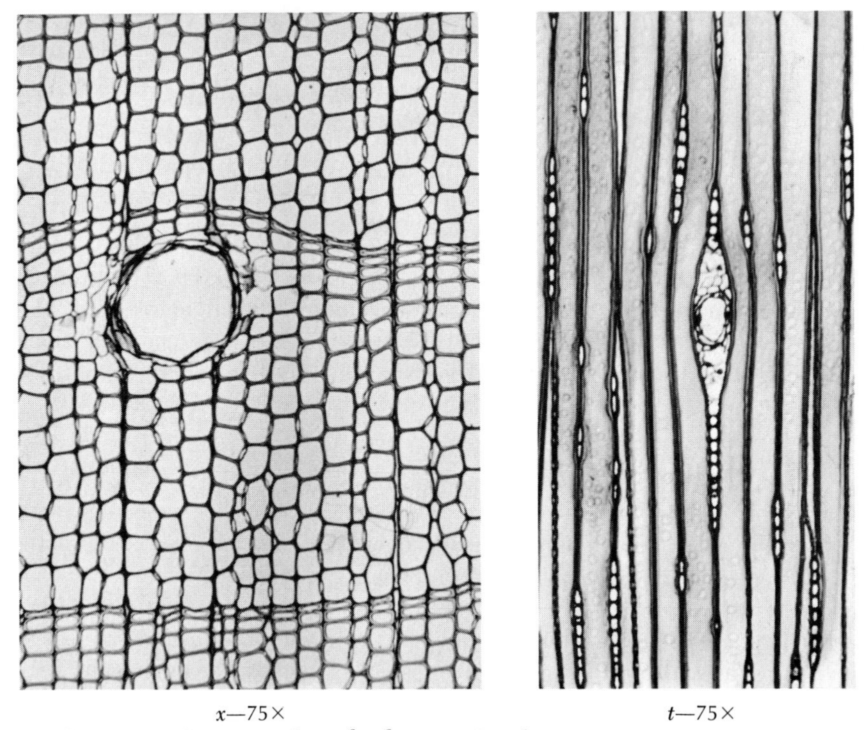

x—75× *t*—75×

FIG. 12-45 Sugar pine, *Pinus lambertiana* Dougl.

ring, solitary or rarely 2–3 contiguous in the tangential plane, appearing as prominent dark streaks along the grain, especially in heartwood, where dark streaks are caused by dust in resin canals (*t*); (*b*) transverse canals less conspicuous than the longitudinal canals, appearing as whitish, rather prominent wood rays spaced at irregular intervals on the transverse surface, visible with a hand lens as brownish specks on the tangential surface.

MINUTE ANATOMY

Tracheids up to 65 (avg 40–50) μ in diameter; bordered pits in 1–2 rows on the radial walls; tangential pitting present in the last few rows of late-wood tracheids; pits leading to ray parenchyma large (window-like), 1–4 (generally 2 and not infrequently 3) per cross field, those in the early wood rounded to elliptical (lemon-shaped) and more or less widely spaced (Fig. 4-8, p. 134), occasionally superimposed. ***Longitudinal parenchyma*** wanting. ***Rays*** of two types, uniseriate and fusiform; (*a*) uniseriate rays numerous (*t*), 1–12 plus cells in height; (*b*) fusiform rays scattered, with a transverse resin canal, 2–4-seriate through the central thickened portion, tapering above and below to uniseriate margins similar to the *a* rays, up to 30 plus cells in height; end walls nodular; ray tracheids present in both types of rays, marginal and interspersed, nondentate. ***Resin canals*** with thin-walled epithelium, frequently occluded with tylosoids in the heartwood; longitudinal canals with maximum diameter of 300 (avg 175–225) μ, the transverse canals much smaller (usually less than 80 μ).

REMARKS

Sometimes confused with eastern white pine (***Pinus strobus*** L.) and with western white pine (***Pinus monticola*** Dougl.), but the heartwood never aging to deep reddish brown as in these species; coarser-textured, with more rounded (lemon-shaped) and widely spaced pits in the early-wood ray crossings, and with larger longitudinal resin canals resulting in more prominent streaks along the grain.

USES

Boxes and ***crates*** (because of light weight and color, nailing properties, and freedom from odor and taste); ***millwork*** (door, sash, interior and exterior trim, siding, panels, etc.; especially suited for such uses because of ease of working and ability to stay in place and to take and hold paint);* lower grades for ***building construction*** (sheathing, subflooring, roofing, etc); ***foundry patterns*** for which it is in increasing demand because of the scarcity of high-grade eastern white pine (considered as standard for wood patterns), its ability to meet the exacting requirements of this use, and its availability in defect-free, wide and thick pieces; ***signs***; ***piano keys*** and ***organ pipes***; ***plywood***; ***pulpwood***.

WESTERN WHITE PINE, IDAHO WHITE PINE

Pinus monticola Dougl.

GENERAL CHARACTERISTICS

Sapwood nearly white to pale yellowish white, narrow to medium wide; ***heartwood*** cream colored to light brown or reddish brown, turning darker on exposure; ***wood*** with a slightly resinous, noncharacteristic odor, without characteristic taste, straight- and even-grained, medium-coarse to rather coarse-textured, moderately light (sp gr approx 0.36 green, 0.42 ovendry), moderately soft. ***Growth rings*** distinct, delineated by a band of darker late wood, generally narrow (mature trees). Early-wood zone usually wide; transition from early to late wood gradual; late-wood zone usually narrow, not appreciably more resistant to tools than the early wood. ***Parenchyma*** not visible. ***Rays*** very fine (x), not visible to the naked eye except where they include a transverse resin canal, forming a fine, inconspicuous fleck on the quarter surface. ***Resin canals*** present, longitudinal and

* The resin in the heartwood tends to discolor paint.

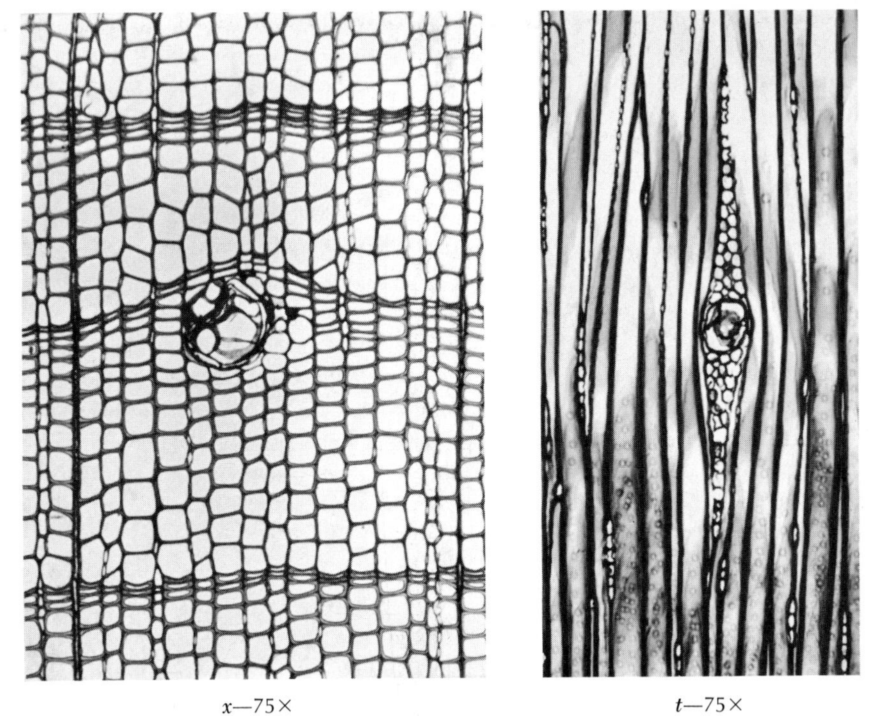

x—75× t—75×

FIG. 12-46 Western white pine, *Pinus monticola* Dougl.

transverse; (*a*) longitudinal canals appearing as whitish flecks to the naked eye, numerous, confined largely to the central and outer portions of the ring, solitary or rarely 2–3 contiguous on the tangential plane, forming more or less prominent streaks along the grain (*t*); (*b*) transverse canals less conspicuous than the longitudinal canals, appearing as whitish, rather prominent wood rays spaced at irregular intervals on the transverse surface, scarcely visible with a hand lens on the tangential surface.

MINUTE ANATOMY

Tracheids up to 60 (avg 35–45) μ in diameter; bordered pits in one row (occasionally in two) on the radial walls; tangential pitting present in the last few rows of late-wood tracheids; pits leading to ray parenchyma large (window-like), 1–4 (mostly 1–2) per cross field, occasionally superimposed, those in the early wood more or less angled and occupying most of the back wall. ***Longitudinal parenchyma*** wanting. ***Rays*** of two types, uniseriate and fusiform; (*a*) uniseriate rays numerous (*t*) 1–12 plus cells in height; (*b*) fusiform rays scattered, with a transverse resin canal, 2–4-seriate through the central thickened portion, tapering above and below to uniseriate margins similar to the *a* rays, up to 20 plus cells in height; end walls nodular; ray tracheids present in both types of rays, marginal and interspersed, nondentate. ***Resin canals*** with thin-walled epithelium, frequently occluded with tylosoids in the heartwood; longitudinal canals with maximum diameter of 200 (avg 135–150), the transverse canals much smaller (usually less than 80 μ).

REMARKS

Very similar to eastern white pine (***Pinus strobus*** L.) but usually somewhat coarser-textured, with narrower, more uniform rings and with smaller and more angular pits in the early-wood ray crossings and larger longitudinal resin canals.

These two species cannot be separated positively on the basis of minute wood anatomy. Their heartwood, however, can be distinguished by subjecting the heartwood acetone extractives to simple paper chromatography (ref. 6, Chap 11). When sprayed with chromagenic spray, eastern white pine develops yellowish-orange spots because of the presence of a high percentage of cryptostrobin and strobobanksin, while western white pine shows pinkish-red spots indicative of a preponderance of pinocembrin.

USES

Matches; *boxes* and *crates* (very satisfactory because of light weight and color, ability to take nails and screws without splitting, and freedom from odor and taste); *building construction* (lower grades for sheathing, subflooring, roof boards; higher grades for siding, exterior and interior trim, partitions, paneling, etc.); *millwork* (such as sash, frames, doors) for which it is very well suited because of the ease of working and its ability to stay in place and hold paint; *patterns*; *car construction*; *fixtures*; *caskets*; *core stock* for plywood (especially table tops); *pulpwood*.

EASTERN WHITE PINE, NORTHERN WHITE PINE

Pinus strobus L.

GENERAL CHARACTERISTICS

Sapwood nearly white to pale yellowish white, narrow to medium wide; *heartwood* cream colored to light brown or reddish brown, turning much darker on exposure; *wood* with a slightly resinous, noncharacteristic odor, without characteristic taste, generally straight- and even-grained,medium-textured, light (sp gr approx 0.34 green, 0.37 ovendry), moderately soft. *Growth rings* distinct, delineated by a band of darker late wood, narrow to wide according to the age and vigor of the tree. Early-wood zone usually wide; transition from early to late wood gradual; late-wood zone narrow (wood from mature trees) to wide (second-growth stock), not appreciably more resistant to tools than the early wood. *Paren-*

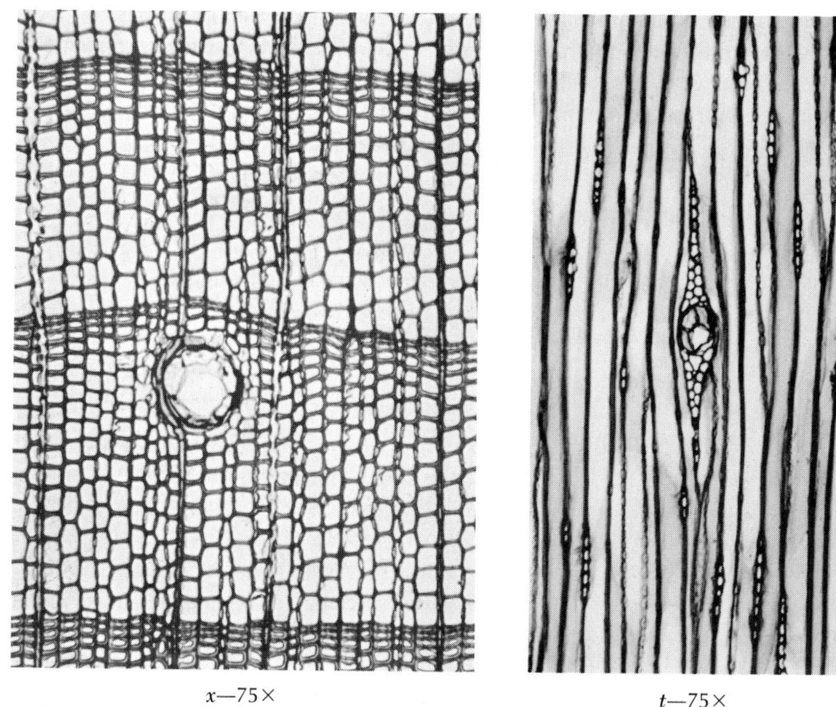

x—75× *t*—75×

FIG. 12-47 Eastern white pine, *Pinus strobus* L.

chyma not visible. *Rays* very fine (*x*), not visible to the naked eye except where they include a transverse resin canal, forming a fine, close, inconspicuous fleck on the quarter surface. *Resin canals* present, longitudinal and transverse; (*a*) longitudinal canals appearing as whitish flecks to the naked eye, numerous, confined largely to the central and outer portions of the ring, solitary or rarely 2–3 contiguous on the tangential plane, forming more or less prominent streaks along the grain (*t*); (*b*) transverse canals less conspicuous than the longitudinal canals, appearing as whitish, rather prominent wood rays spaced at irregular intervals, on the transverse surface, barely visible with a hand lens on the tangential surface.

MINUTE ANATOMY

Tracheids up to 45 (avg 25–35) μ in diameter; bordered pits in one row or occasionally paired on the radial walls; tangential pitting present in the last few rows of late-wood tracheids; pits leading to ray parenchyma large (window-like), 1–2 (mostly 1) per cross field, those in the early wood mostly oblong and occupying most of the back wall (Fig. 12-34, p. 438). *Longitudinal parenchyma* wanting. *Rays* of two types, uniseriate and fusiform; (*a*) uniseriate rays numerous (*t*), 1–8 plus cells in height; (*b*) fusiform rays scattered, with a transverse resin canal, 2–3 seriate through the central thickened portion, tapering above and below to uniseriate margins similar to the *a* rays, up to 30 plus cells in height; end walls nodular; ray tracheids present in both types of rays, marginal and interspersed, nondentate. ***Resin canals*** with thin-walled epithelium, frequently occluded with tylosoids in the heartwood; longitudinal canals with maximum diameter of 150 (avg 90–120) μ, the transverse canals much smaller (usually less than 60 μ).

REMARKS

See Remarks under western white pine (***Pinus monticola*** Dougl., p. 452).

USES

Boxes and ***crates*** (one of the principal species, second-growth stock being used for the most part), for which it is specially suited because of its light weight, good color for stenciling, and lack of objectionable odor and taste; ***patterns*** (formerly the standard pattern wood, owing to its uniform texture, ease of cutting in any direction, minimal shrinkage and swelling, ability to stay in place, ease of gluing, freedom from resin, and strength to withstand rough handling; replaced largely by sugar and western white pines, because of scarcity of high-grade stock); ***mill-work*** (principally sash and doors); ***toys, woodenware*** and ***novelties***; ***signs***; ***caskets***; ***building construction*** (formerly widely used for practically every part of a house, but now infrequently because of the scarcity of suitable stock); ***matches*** (formerly the leading wood, now largely replaced by western white pine and aspen); ***shade*** and ***map rollers***; ***venetian blinds***; ***dairy*** and ***poultry supplies***; ***boot*** and ***shoe findings***.

LODGEPOLE PINE

Pinus contorta Dougl.

GENERAL CHARACTERISTICS

Sapwood nearly white to pale yellow, narrow; ***heartwood*** light yellow to pale yellowish brown, often scarcely darker than the sapwood and not clearly distinct; ***wood*** with a distinct, noncharacteristic, resinous odor (especially when green), without characteristic taste, generally straight- but somewhat uneven-grained, medium fine-textured, frequently prominently dimpled on the tangential surface (split), moderately light (sp gr approx 0.38 green, 0.43 ovendry), moderately soft. ***Growth rings*** distinct, not as conspicuous as in many of the other hard pines, delineated by a band of darker late wood. Early-wood zone wide or narrow (outer rings of mature trees); transition from early to late wood more or less abrupt; late-wood zone narrow but distinct, not appreciably more resistant to tools than the early wood. ***Parenchyma*** not visible. ***Rays*** very fine (*x*), not visible to the naked eye, appearing whitish or brownish with a lens where they contain a transverse

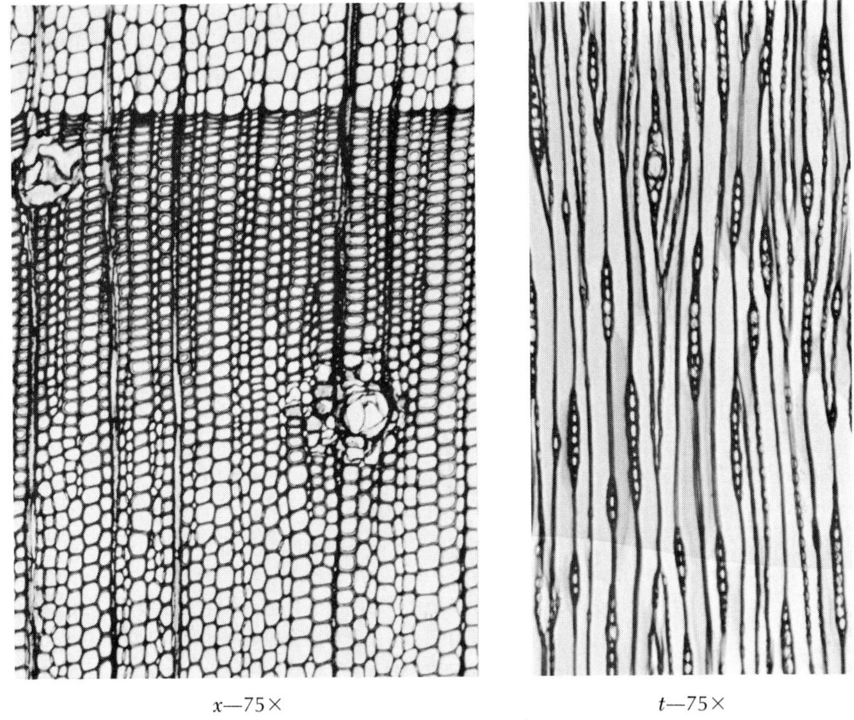

FIG. 12-48 Lodgepole pine, *Pinus contorta* Dougl.

resin canal, forming a fine, close, inconspicuous fleck on the quarter surface. *Resin canals* present, longitudinal and transverse; (*a*) longitudinal canals relatively inconspicuous or not visible to the naked eye, numerous, confined largely to the central and outer portions of the ring, solitary for the most part, forming fairly conspicuous, brownish streaks along the grain (*t*); (*b*) transverse canals less conspicuous than the longitudinal canals, appearing as brownish radial lines spaced at irregular intervals on the transverse surface, barely visible with a hand lens on the tangential surface.

MINUTE ANATOMY

Tracheids up to 55 (avg 35–45) μ in diameter, those in the dimpled areas swirled as viewed in the tangential section; bordered pits in one row or occasionally paired on the radial walls; tangential pitting wanting in the last few rows of late-wood tracheids (sometimes appearing to be present owing to a twist in the tracheids); pits leading to ray parenchyma pinoid, variable in size and shape, 1–6 (generally 2–4) per cross field. *Longitudinal parenchyma* wanting. *Rays* of two types, uniseriate or in part biseriate, and fusiform; (*a*) uniseriate rays numerous (*t*), 1–15 plus cells in height; biseriate rays frequent in the areas of swirled tissue; (*b*) fusiform rays scattered or rarely 2–3 confluent along the grain in the swirled areas, with a transverse resin canal, 2–3-seriate through the central thickened portion, tapering above and below to uniseriate margins similar to the *a* rays, up to 15 plus cells in height; ray tracheids present in both types of rays, marginal and interspersed, prominently dentate [with teeth frequently extending across the cell forming a reticulate pattern (*r*)]; marginal tracheids often in several rows; low rays frequently consisting entirely of ray tracheids; ray parenchyma thin-walled. *Resin canals* with thin-walled epithelium, frequently occluded with tylosoids in the heartwood; longitudinal canals with maximum diameter of 110 (avg 80–90) μ, the transverse canals much smaller (usually less than 50 μ).

REMARKS

Lodgepole pine can usually be separated from the other commercial pines, except ponderosa and jack pines, by the dimpling on the split tangential surface. Ponderosa pine has much larger resin canals (see Remarks, p. 461). Resin canals, especially horizontal, tend to be smaller in jack pine than in lodgepole,* but this feature does not provide a reliable means of separation. Black† reports that ray parenchyma in lodgepole pine have thin walls, lacking in localized thickenings, while those of jack pine are slightly and irregularly thickened. Positive identification of these two species is possible on the basis of the analysis of terpenes; jack pine has large amounts of α- and β-terpenes, while lodgepole pine contains β-phellandrene.

USES

Mine timbers; *poles*; *posts*; *railroad ties*; *pulpwood*; suitable for *hardboard* and *particle board*; *lumber* for rough construction, frequently mixed with spruce; *planing mill products*, such as siding, knotty pine paneling, and flooring; *railroad car decking*; *veneer* for decorative faces for prefinished plywood; *structural plywood*; locally for *corral rails*; *orchard props*; *rustic furniture*; and *shingles*.

SOUTHERN PINE

Longleaf pine (*Pinus palustris* Mill.)
Shortleaf pine (*Pinus echinata* Mill.)
Loblolly pine (*Pinus taeda* L.)
Slash pine (*Pinus elliottii* Engelm.)
Pitch pine (*Pinus rigida* Mill.)
Pond pine (*Pinus serotina* Michx.)
Etc.

GENERAL CHARACTERISTICS

Sapwood nearly white to yellowish or orange-white or pale yellow, thin to very thick; *heartwood* distinct, ranging through shades of yellow and orange to reddish

* According to the Ottawa Laboratory the average diameter of the horizontal resin canals in lodgepole pine is 45 μ, and in jack pine 33μ. (The Annual Report, Forest Prod. Res. Branch, Ottawa Laboratory, 1961–62.)

† T. M. Black, Some Features of the Timber Anatomy of *Pinus contorta* Loud. and *Pinus banksiana* Lamb., *J. Inst. Wood Sci.*, **11**:57–65 (1963).

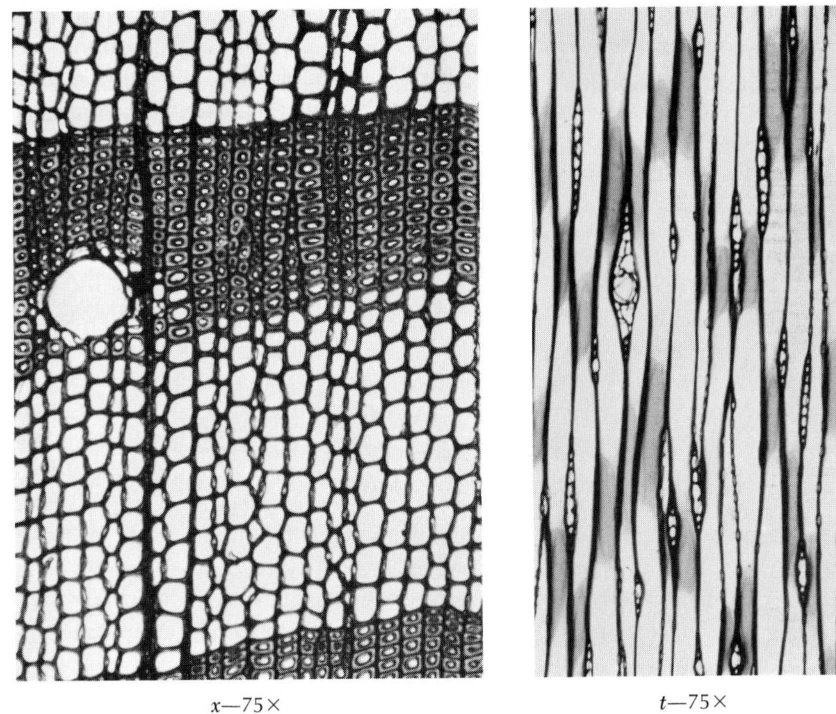

x—75× *t*—75×

FIG. 12-49 Shortleaf pine, *Pinus echinata* Mill.

brown or light brown, resinous;* *wood* with a distinct, noncharacteristic, resinous odor, without characteristic taste, generally straight- but uneven-grained, medium-textured, moderately heavy to very heavy (avg sp gr 0.45–0.56 green, 0.52–0.66 ovendry),† moderately hard to hard. *Growth rings* distinct, delineated by a pronounced band of darker late wood. Early-wood zone varying from wide or very wide (slash pine, loblolly pine) to narrow (slow-growth shortleaf pine, etc.); transition from early to late wood abrupt, the contrast frequently very striking; late-wood zone ranging from broad to narrow, varying greatly in width and density according to age of the tree, conditions of growth, and within general limits according to species. *Parenchyma* not visible. *Rays* very fine (*x*), not visible to the naked eye except where they include a transverse resin canal, forming a fine, close, inconspicuous fleck on the quarter surface. *Resin canals* present, longitudinal and transverse; (*a*) longitudinal canals appearing as whitish or brownish flecks which are conspicuous or relatively conspicuous to the naked eye, plainly

* In longleaf pine the sapwood contains about 2 percent resin, while the average for the heartwood is 7 to 10 percent. The average resin content for the heartwood in butt logs may be 15 percent and that in the stump 25 percent.

† In order to assist in selection of wood suitable for structural purposes, the grading rules for southern yellow pine contain the *density rule* provision. To qualify under this provision, structural timber must show on one end of the piece an average of not fewer than 6 rings per inch and one-third summerwood.

distinct with a hand lens, numerous, confined largely to the central and outer portions of the ring, solitary or rarely 2–3 contiguous in the tangential plane, generally visible as relatively inconspicuous streaks along the grain (*t*); (*b*) transverse canals less conspicuous than the longitudinal canals, appearing as whitish, relatively inconspicuous wood rays spaced at irregular intervals on the transverse surface, not visible or barely visible with a hand lens on the tangential surface.

MINUTE ANATOMY

Tracheids up to 60 (avg 35–45) μ in diameter; bordered pits in one row or not infrequently paired on the radial walls; tangential pitting wanting on the last few rows of late-wood tracheids; pits leading to ray parenchyma pinoid, variable in size and shape, 1–6 (generally 2–5) per cross field (Fig. 12-33, p. 438). ***Longitudinal parenchyma*** wanting. ***Rays*** of two types, uniseriate and fusiform; (*a*) uniseriate rays numerous (*t*), 1–10 plus cells in height; (*b*) fusiform rays scattered, with a transverse resin canal, 2–4-seriate through the central thickened portion, tapering above and below to uniseriate margins similar to the *a* rays, up to 12 plus cells in height; ray tracheids present in both types of rays, marginal and interspersed, prominently dentate [the teeth frequently extending across the cell, forming a reticulate pattern (*r*)]; marginal and interspersed tracheids often in several rows; low rays frequently consisting entirely of ray tracheids; ray parenchyma thin-walled. ***Resin canals*** with thin-walled epithelium, frequently occluded with tylosoids in the heartwood; longitudinal canals with maximum diameter of 180 (avg 90–150) μ, the transverse canals much smaller (usually less than 70 μ).

REMARKS

The yellow or hard pines of Southeastern and Eastern United States cannot be separated on the basis of wood structure; in the trade, southern pine is usually marketed according to density. Kukachka* reports that growth rings of longleaf and slash pines frequently exhibit multiple late-wood bands and that the pith of longleaf pine is 0.2 in. in diameter as against 0.1 in. or less for other southern pines.

USES

Pulpwood (converted largely by the sulfate process for use in the manufacture of kraft paper and insulation and other types of fiberboard); ***hardboard, particle board; poles; mine timbers; piling; railroad ties; slack cooperage*** (shortleaf and loblolly); ***veneer*** and ***plywood; excelsior; structural timbers*** (such as stringers, beams, and joists in bridge, trestle, warehouse, and factory construction, for which it is highly suited because of its strength, stiffness, and hard-

* B. F. Kukachka, Identification of Coniferous Woods, *Tappi*, **43:**887–896 (1960).

ness); *building construction* (joists, rafters, studdings, and building lumber);* *boxes; baskets; crates* (boards and veneer), and *pallets; planing-mill products* and *millwork* (because of hardness and good wearing qualities); *railroad-car construction; agricultural implements; ship-* and *boatbuilding; paving blocks; tanks* and *silos; woodenware* and *novelties; destructive distillation* (mostly heartwood from the stumps and tops of longleaf pine).

PONDEROSA PINE AND JEFFREY PINE

Ponderosa pine, western yellow pine (*Pinus ponderosa* Laws.)

Jeffrey pine (*Pinus jeffreyi* Grev. & Balf.)

GENERAL CHARACTERISTICS

Sapwood nearly white to pale yellowish, wide (often composed of 80 or more rings); *heartwood* of ponderosa pine yellowish to light reddish or orange-brown, that of Jeffrey pine sometimes with a pinkish cast; *wood* of ponderosa pine with distinct, noncharacteristic, resinous odor, that of Jeffrey pine sweet (applelike) when first exposed; woods of both species without characteristic taste, generally straight- and quite even- to very uneven-grained, medium coarse-textured, frequently dimpled on the tangential surface (split) but the dimples less conspicuous than in lodgepole pine, moderately light (sp gr approx 0.38 green, 0.42 ovendry), moderately soft. *Growth rings* distinct, inconspicuous to conspicuous, delineated by a band of darker late wood. Early-wood zone narrow to wide (wide-ringed stock); transition from early to late wood abrupt; late-wood zone broad and conspicuous in wide rings, in narrow rings much reduced and often appearing as a narrow line which is not visible without magnification, the wood then resembling soft pine. *Parenchyma* not visible. *Rays* very fine (*x*), not visible to the naked eye except where they include a transverse resin canal, forming a fine, close, inconspicuous fleck on the quarter surface. *Resin canals* present, longitudinal and transverse; (*a*) longitudinal canals conspicuous (the orifice generally visible to the naked eye), numerous, confined largely to the central and outer portions of the ring, solitary or rarely 2–3 contiguous in the tangential plane, appearing as relatively prominent dark streaks along the grain (*t*); (*b*) transverse canals less conspicuous than the longitudinal canals, appearing as whitish, relatively inconspicuous wood rays spaced at irregular intervals on the transverse surface, barely visible with a hand lens as brownish specks on the tangential surface.

MINUTE ANATOMY

Tracheids up to 60 (avg 35–45) μ in diameter; bordered pits in one row (occasionally in two) on the radial walls; tangential pitting wanting in the last few rows

* Longleaf and slash pines are used for heavy construction such as for factories, warehouses, bridges, and docks, whereas loblolly and shortleaf pines are used principally for such building materials as interior finish, frame and sash, wainscoting, joists, and subflooring.

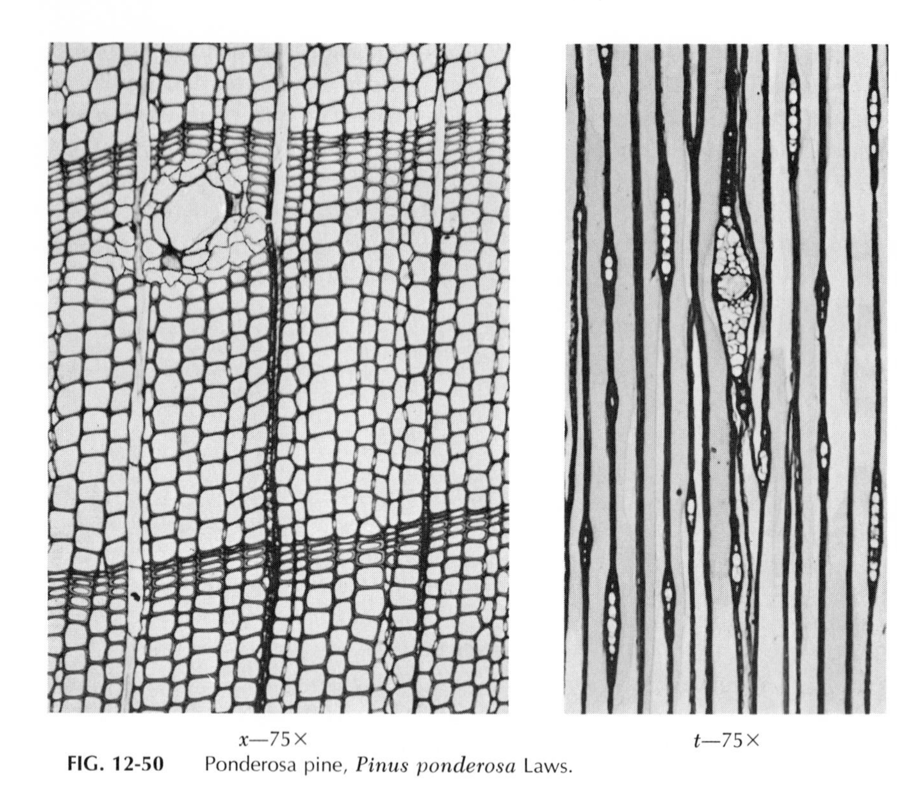

FIG. 12-50 Ponderosa pine, *Pinus ponderosa* Laws.

of late-wood tracheids; pits leading to ray parenchyma pinoid, variable in shape and size, 1–7 (generally 4–5) per cross field (Fig. 12-32). ***Longitudinal parenchyma*** wanting. ***Rays*** of two types, uniseriate and fusiform; (*a*) uniseriate rays numerous (*t*), 1–12 plus cells in height; (*b*) fusiform rays scattered, with a transverse resin canal, 3–5-seriate through the central thickened portion, tapering above and below to uniseriate margins similar to the *a* rays, up to 20 plus cells in height; ray tracheids present in both types of rays, marginal and occasionally interspersed, prominently dentate [the teeth frequently extending across the cell, forming a reticulate pattern (*r*)]; marginal tracheids often in several rows; low rays frequently consisting entirely of ray tracheids; ray parenchyma thin-walled. ***Resin canals*** with thin-walled epithelium, frequently occluded with tylosoids in the heartwood; longitudinal canals with maximum diameter of 230 (avg 160–185) μ, the transverse canals much smaller (usually less than 70 μ).

REMARKS

Ponderosa and Jeffrey pines cannot be separated on the basis of wood structure. The heartwood of Jeffrey pine is generally softer than that of ponderosa pine and frequently has a pinkish cast which is lacking in ponderosa pine; the odor of fresh-cut Jeffrey pine is sweet (hence the term *apple pine*), in contrast to the resinous odor of ponderosa pine. The two pines can be positively separated on the basis of the composition of their essential oils; Jeffrey pine contains 90 to 95 percent heptane, while ponderosa pine oils are a mixture of α-pinene, β-pinene, and limonene.

Dimpled ponderosa pine is sometimes confused with lodgepole pine, which also contains this feature. Ponderosa pine differs from the latter by possessing darker heartwood and larger longitudinal resin canals [ponderosa pine—230 (avg 160–185) μ; lodgepole pine—110 (avg 80–90) μ].

USES

Boxes and *crates*, usually lower grades (for which it is well suited because of its light weight and color, ability to take nails and screws without splitting, strength to withstand rough handling, and freedom from objectionable odors and taste); *millwork* (the softer grades), especially for sash, doors, and screens (because of its softness and uniformity of grain which permit of accurate machining, and its moderately low shrinkage, ability to stay in place, resistance to abrasion, and attractive clear color); *planing-mill products* [interior finish, trim, siding, paneling, including knotty grade (the difference between the color of the heart- and sapwood is accentuated upon exposure, necessitating care in matching the wood for interiors)]; *building construction* (the heavier, wider-ringed, more resinous wood), as joists, rafters, studdings, sills, and sheathing; *turned work* (porch columns, posts, balusters, stair rails); *caskets* and *coffins*; *furniture*; *shade* and *map rollers*; *patterns* (not considered equal to the white pines); *trunks*; *toys*; *piling; poles; posts; mine timbers; plywood; pulpwood.*

RED PINE, NORWAY PINE

Pinus resinosa Ait.

GENERAL CHARACTERISTICS

Sapwood nearly white to yellowish, narrow to medium wide; *heartwood* light red to orange-brown or reddish brown; *wood* somewhat oily, with a fairly strong, resinous, noncharacteristic odor, without characteristic taste, usually straight- and quite even-grained, medium-textured, moderately heavy (sp gr approx 0.41 green, 0.44 ovendry), moderately soft. *Growth rings* distinct, delineated by a band of darker late wood, narrow to wide. Early-wood zone generally wide; transition from early to late wood more or less abrupt; late-wood zone narrow to fairly wide (in wide rings), darker and appreciably denser than the early-wood zone. *Parenchyma* not visible. *Rays* very fine (x), not visible with the naked eye, appearing whitish with a lens where they contain a transverse resin canal, forming a fine, close, inconspicuous fleck on the quarter surface. *Resin canals* present, longitudinal and transverse; (*a*) longitudinal canals relatively inconspicuous to the naked eye, appearing as minute, brownish flecks, relatively conspicuous with a hand lens, numerous, confined largely to the central and outer portions of the ring, solitary or rarely 2–3 contiguous in the tangential plane, not visible or forming relatively inconspicuous streaks along the grain (*t*); (*b*) transverse canals less conspicuous than longitudinal canals, appearing as whitish, radial lines spaced at irregular intervals on the transverse surface, not visible with hand lens on the tangential surface.

MINUTE ANATOMY

Tracheids up to 45 (avg 30–40) μ in diameter; bordered pits in one row or occasionally paired on the radial walls; tangential pitting wanting or very sporadic in the last few rows of late-wood tracheids; pits leading to ray parenchyma large (window-like), 1–2 (mostly 1) per cross field (Fig. 12-31). ***Longitudinal parenchyma*** wanting. ***Rays*** of two types, uniseriate and fusiform; (*a*) uniseriate rays numerous (*t*), 1–10 plus cells in height; (*b*) fusiform rays scattered, with a transverse resin canal, 2–3-seriate through the central thickened portion, tapering more or less abruptly above and below to uniseriate margins, up to 10 plus cells in

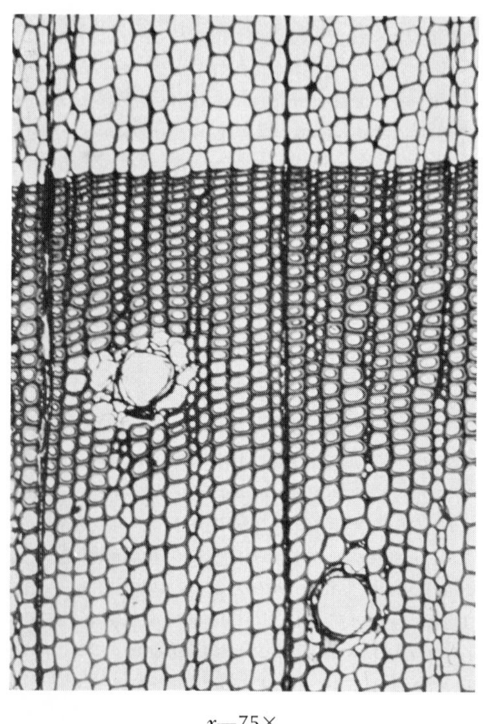

x—75×

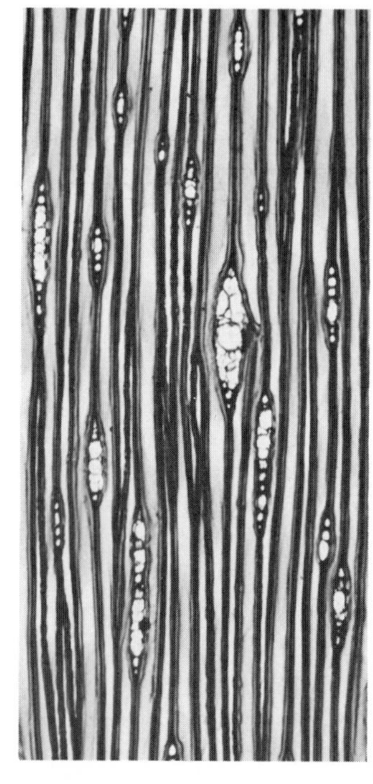

t—75×

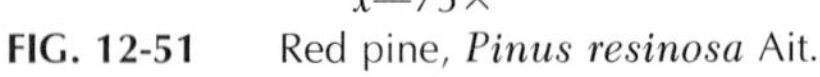

FIG. 12-51 Red pine, *Pinus resinosa* Ait.

height; ray tracheids present in both types of rays, marginal and interspersed, shallowly dentate (the teeth seldom extending to the center of the cell cavity). ***Resin canals*** smaller than in most pines, with thin-walled epithelium, frequently occluded with tylosoids in the heartwood; longitudinal canals with maximum diameter of 120 (avg 80–110) μ, the transverse canals much smaller (usually less than 50 μ).

REMARKS

Norway pine is sometimes confused with the soft pines but can readily be separated by the more or less abrupt transition between early and late wood and the presence of dentate ray tracheids (nondentate in the soft pines). The window-like pits that feature the ray crossings are similar to those of the soft pines and serve to distinguish Norway pine from all other domestic hard pines.

USES

Pulpwood (converted by sulfate process); ***railroad ties***; ***poles***; ***posts***; ***building construction*** (locally); ***boxes***; ***crates***; ***pallets***; ***planing-mill products*** (sash, doors, blinds, interior and exterior finish).

JACK PINE

Pinus banksiana Lamb.

GENERAL CHARACTERISTICS

Sapwood nearly white, wide; ***heartwood*** formation frequently delayed until trees reach the age of 40 to 50 years, light orange to light brown, somewhat resinous; ***wood*** with a distinct, noncharacteristic, resinous odor, without characteristic taste, generally straight- but somewhat uneven-grained, medium-textured, moderately heavy (sp gr approx 0.40 green, 0.45 ovendry), moderately soft; split tangential surface sometimes dimpled. ***Growth rings*** distinct, delineated by a band of darker late wood. Early-wood zone variable in width; transition from early to late wood abrupt; late-wood zone narrow to fairly wide, darker and appreciably denser than the early-wood zone. ***Parenchyma*** not visible. ***Rays*** very fine (x), not visible to the naked eye, appearing whitish with a lens where they contain a transverse resin canal, forming a fine, close, inconspicuous fleck on the quarter surface. ***Resin canals*** present, longitudinal and transverse; (a) longitudinal canals relatively inconspicuous to the naked eye, numerous, confined for the most part to the central and outer portions of the ring, mostly solitary, forming inconspicuous, brownish streaks along the grain (t); (b) transverse canals inconspicuous, appearing with a hand lens as whitish, radial lines spaced at irregular intervals on the transverse surface, barely visible with a hand lens on the tangential surface.

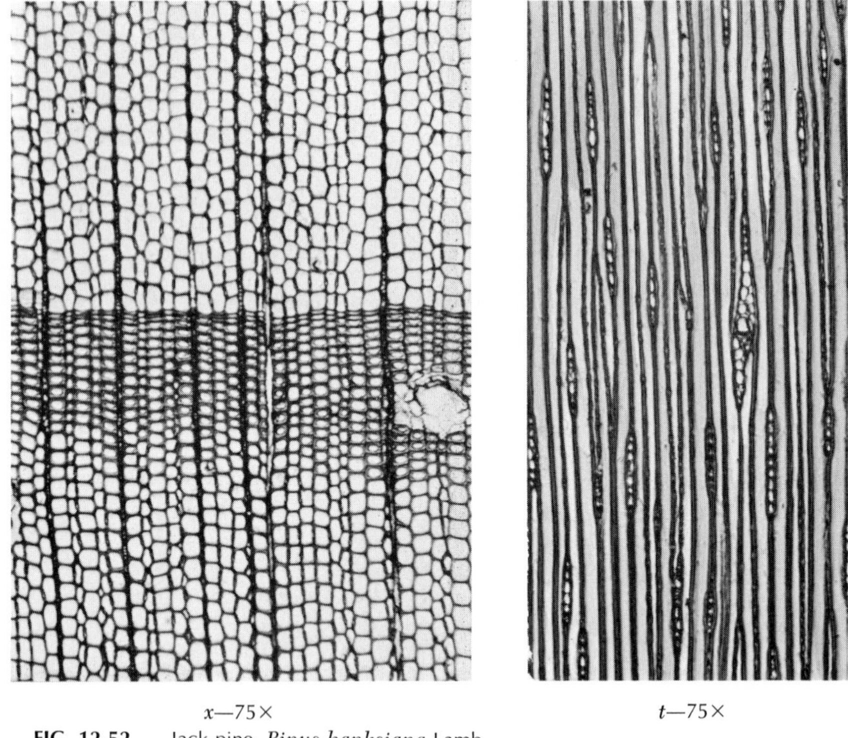

FIG. 12-52 Jack pine, *Pinus banksiana* Lamb.

MINUTE ANATOMY

Tracheids up to 45 (avg 27–37) μ in diameter; bordered pits in one row or occasionally paired on the radial walls; tangential pitting wanting in the last few rows of late-wood tracheids; pits leading to ray parenchyma pinoid, variable in size and shape, 1–6 (generally 4–6 in two rows) per cross field. ***Longitudinal parenchyma*** wanting. ***Rays*** of two types, uniseriate and fusiform; (*a*) uniseriate rays numerous (*t*), 1–10 plus cells in height; (*b*) fusiform rays scattered, with a transverse resin canal, 2–3-seriate through the central thickened portion, tapering more or less abruptly to uniseriate margins similar to the *a* rays, up to 10 plus cells in height; ray tracheids present in both types of rays, shallowly dentate [the teeth seldom extending to the center of the cell cavity (*r*)]; marginal tracheids often in several rows; low rays frequently consisting entirely of ray tracheids; ray parenchyma thin-walled. ***Resin canals*** smaller than in most pines (comparable in size to those of lodgepole and red pine), with thin-walled epithelium, frequently occluded with tylosoids in the heartwood; longitudinal canals with maximum diameter of 100 (avg 75–90) μ, the transverse canals much smaller (usually less than 45 μ). See footnote p. 456.

REMARKS

For separation of jack pine from ponderosa pine and lodgepole pine, see Remarks, p. 456.

USES

Pulpwood (converted usually by sulfate process); ***poles***; ***posts***; ***mine timbers***. Lumber generally knotty and considered less desirable than that of red pine, used principally for ***boxes*** and ***crates***; ***rough construction***; better grades generally marketed as red pine and used for ***planing-mill products*** (sash, doors, interior and exterior finish).

TAMARACK, EASTERN LARCH
Larix laricina (Du Roi) K. Koch

GENERAL CHARACTERISTICS

Sapwood whitish, narrow; ***heartwood*** yellowish to russet brown or occasionally reddish brown (fast-grown stock); ***wood*** generally without characteristic odor or taste, more or less oily and with somewhat greasy feel, frequently spiral-grained, medium fine-textured, moderately heavy (sp gr approx 0.49 green, 0.57 ovendry), moderately hard. ***Growth rings*** distinct, delineated by a pronounced band of darker late wood, generally moderately wide to wide (8–20 per in.) and variable in width. Early-wood zone usually occupying three-fourths or more of the ring; transition from early to late wood abrupt; late-wood zone narrow to wide, conspicuous to the naked eye. ***Parenchyma*** not visible. ***Rays*** very fine (*x*), not distinct to the naked eye, forming a fine, close, inconspicuous fleck on the quarter surface. ***Resin canals*** present, longitudinal and transverse; (*a*) longitudinal canals small, inconspicuous, not visible to the naked eye or appearing as whitish or dark flecks, sparse, confined largely to the central and outer portions of the ring, solitary or 2–several contiguous in the tangential plane, occasionally in a tangential line extending for some distance along the ring; (*b*) transverse canals smaller than the longitudinal canals, appearing with a hand lens as somewhat broader, whitish rays spaced at irregular intervals on the transverse surface, not visible or barely visible with a hand lens on the tangential surface.

MINUTE ANATOMY

Tracheids up to 45 (avg 28–35) μ in diameter, those in the late wood occasionally with spiral thickening; bordered pits in 1–2 rows on the radial walls; tangential pits present in the last few rows of late-wood tracheids; pits leading to ray parenchyma piceoid, small, quite uniform in size, with distinct border, 1–12 (generally

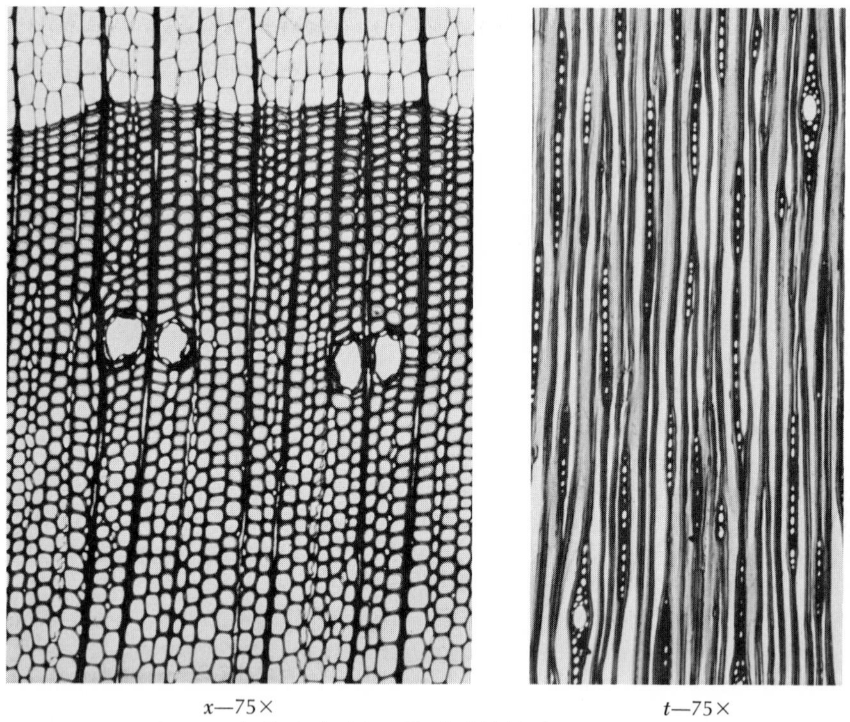

FIG. 12-53 Tamarack, *Larix laricina* (Du Roi) K. Koch

4–6) per cross field, often in a double horizontal row. ***Longitudinal parenchyma*** marginal and very sparse, or wanting. ***Rays*** of two types, uniseriate or rarely in part biseriate, and fusiform; (*a*) uniseriate rays numerous (*t*), 1–20 plus cells in height; biseriate rays very sparse and scattered, or wanting; (*b*) fusiform rays scattered, with a transverse resin canal, 2–3-seriate through the central thickened portion, tapering above and below to uniseriate margins similar to the *a* rays, up to 20 plus cells in height; end walls nodular, with indentures; ray tracheids in both types of rays, marginal and rarely interspersed, nondentate; marginal tracheids usually in one row. ***Resin canals*** with thick-walled epithelium, occasionally with tylosoids in the heartwood; longitudinal canals with maximum diameter of 110 (avg 60–90) μ; transverse canals much smaller (usually less than 25 μ).

REMARKS

Tamarack or eastern larch is sometimes confused with western larch; for distinguishing characters, see Remarks under Western Larch, p. 468.

USES

Locally for *rough construction*; *posts*; *poles*; *ties*; *novelties*; *boxes, crates,* and *pallets*; *pulpwood.*

WESTERN LARCH
Larix occidentalis Nutt.

GENERAL CHARACTERISTICS

Sapwood whitish to pale straw-brown, narrow (rarely over 1 in. in width); ***heartwood*** russet or reddish-brown; ***wood*** without characteristic odor or taste; with characteristic oily appearance, straight-grained, coarse-textured, moderately heavy (sp gr approx 0.48 green, 0.55 ovendry), moderately hard. ***Growth rings*** distinct, delineated by a pronounced band of darker late wood, generally very narrow (30–60 per in.) and quite uniform in width. Early-wood zone usually occupying two-thirds or more of the ring, appearing porous with a hand lens (large tracheids); transition from early to late wood very abrupt; late-wood zone generally very narrow, sharply delineated and conspicuous to the naked eye. ***Parenchyma*** not visible. ***Rays*** very fine (*x*), not distinct to the naked eye, forming a fine, close, inconspicuous fleck on the quarter surface. ***Resin canals*** present, longitudinal and transverse; (*a*) longitudinal canals small, inconspicuous, not visible to the naked eye or appearing as whitish or dark flecks, sparse, confined for the most part to the narrow bands of late wood, solitary or 2–several contiguous in the tangential plane, not visible or forming inconspicuous streaks along the grain; (*b*) transverse canals smaller than the longitudinal canals, appearing with a

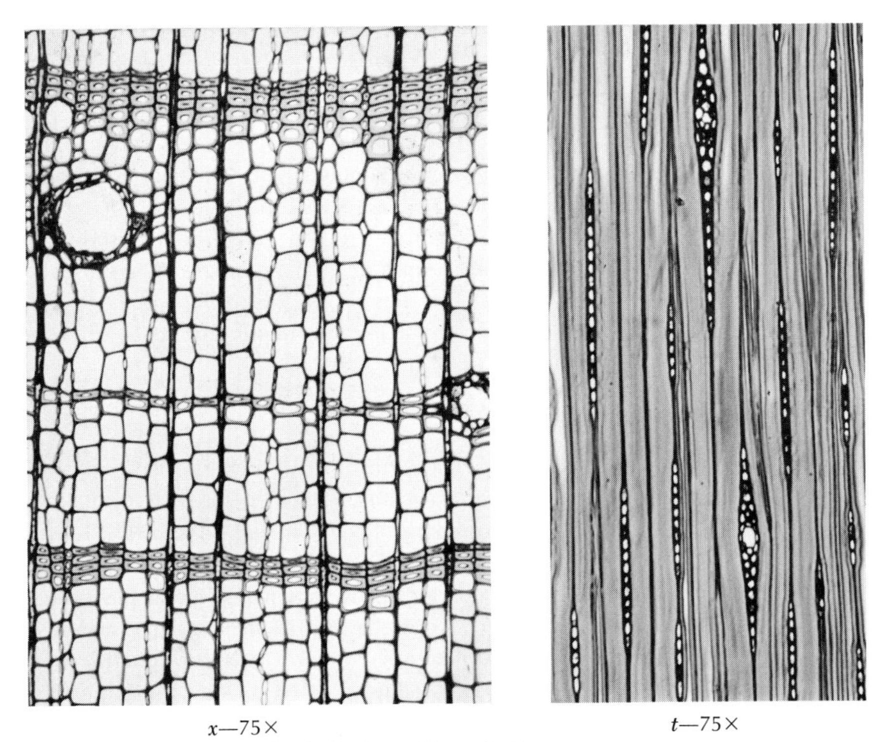

FIG. 12-54 Western larch, *Larix occidentalis* Nutt.

hand lens as somewhat broader, whitish rays spaced at irregular intervals on the transverse surface, not visible or barely visible with a hand lens on the tangential surface.

MINUTE ANATOMY

Tracheids up to 60 (avg 38–50) μ in diameter, those in the late wood occasionally with spiral thickening; bordered pits 1–2 rows on the radial walls; tangential pitting present on the last few rows of late-wood tracheids; pits leading to ray parenchyma piceoid, small, quite uniform in size, with distinct border, 1–10 (generally 4–6) per cross field, often in a double horizontal row (Fig. 12-36). *Longitudinal parenchyma* marginal and very sparse, or wanting. *Rays* of two types, uniseriate or rarely in part biseriate, and fusiform; (*a*) uniseriate rays numerous (*t*), 1–20 plus cells in height; biseriate rays very sparse and scattered, or wanting; (*b*) fusiform rays scattered, with one or very rarely two transverse resin canals, 2–3-seriate through the central thickened portion, tapering above and below to uniseriate margins similar to the *a* rays, up to 20 plus cells in height; end walls nodular, with indentures; ray tracheids in both types of rays, marginal and rarely interspersed, nondentate; marginal tracheids usually in one row. *Resin canals* with thick-walled epithelium, occasionally with tylosoids in the heartwood; longitudinal canals with maximum diameter of 135 (avg 60–90) μ; transverse canals much smaller (usually less than 25 μ).

REMARKS

For differences between western larch and Douglas-fir, see Remarks under Douglas-fir, p. 476. Western larch is sometimes confused with tamarack or eastern larch but can be separated by the following characters: the heartwood of eastern larch is generally yellowish brown in contrast to the russet- or reddish-brown heartwood of western larch; the annual rings in the eastern species are usually much wider and less uniform in width, the grain is not so straight, and the texture is finer [tracheids with maximum diameter of 45 (avg 30–40) μ].

USES

Building materials (principal use) in the form of small timbers, planks, boards, rough-dimension stock; *glue-laminated beams*, esteemed for their coloration; *planing-mill products*, especially for edge-grained *flooring* (for which it is very satisfactory because of hardness resulting from narrow rings with a large percentage of late wood; flat-grained flooring tends to separate along the annual increments), also for *interior finish* (because of the distinctive figure and ease with which it can be finished in the natural color); *railroad car construction*; *boxes*; *crates* (shows a tendency to split if large, pointed nails are used); *pallets*; *veneer*, rotary converted mostly into structural plywood, sliced veneer for faces in ***prefinished plywood; ties; poles; fuel wood*** (rated high in fuel value compared with other softwoods of the Inland Empire); future growing use for ***pulpwood*** and ***particle board*** is indicated. Western larch, as well as tamarack, contains considerable quantities of a water-soluble gum (arabogalactan), especially in butt logs; it can be easily extracted and oxidized into mucic acid, used in making baking soda.

TRANSCONTINENTAL, RED, ENGELMANN SPRUCE

Transcontinental spruce
White spruce [*Picea glauca* (Moench) Voss]
Black spruce [*Picea mariana* (Mill.) B.S.P.]
Red spruce (*Picea rubens* Sarg.)
Engelmann spruce (*Picea engelmannii* Parry)

GENERAL CHARACTERISTICS

Wood nearly white to pale yellowish brown (***heartwood*** not distinct), lustrous, without characteristic odor or taste, even- and usually straight-grained, medium- to fine-textured, light (Engelmann spruce, avg sp gr approx 0.31 green, 0.35 ovendry) to moderately light (other spruces, avg sp gr approx 0.37 green, 0.41–0.45 ovendry), soft to moderately soft. ***Growth rings*** distinct, delineated by the contrast between the late wood and the early wood of the succeeding ring, narrow to wide. Early-wood zone usually a number of times wider than the late wood, grading into the late wood; late-wood zone distinct to the naked eye but usually not pronounced, somewhat darker than the early wood, generally narrow. ***Parenchyma*** not visible. ***Rays*** very fine (*x*), not distinct to the naked eye or barely visible where they include a transverse resin canal, forming a fine, close, inconspicuous fleck on the quarter surface. ***Resin canals*** present, longitudinal and transverse; (*a*) longitudinal canals small, generally not visible to the naked eye, appearing as white flecks with a hand lens, solitary or 2–several contiguous in the tangential plane or not infrequently grouped in a tangential line extending for some distance along the ring, often irregularly distributed and wanting entirely from portions of some rings, not visible or barely visible along the grain; (*b*) transverse canals smaller than the longitudinal canals, visible with a hand lens and occasionally to the naked eye as somewhat broader, whitish rays spaced at irregular intervals on the transverse surface, not visible or indistinct with a hand lens on the tangential surface.

MINUTE ANATOMY

Tracheids up to 35 (avg 25–30) μ in diameter; bordered pits in one row or very rarely paired on the radial walls; tangential pitting present in the last few rows of late-wood tracheids; pits leading to ray parenchyma piceoid, small, quite uniform in size, with distinct border, 1–6 (generally 2–4) per cross field, generally in a

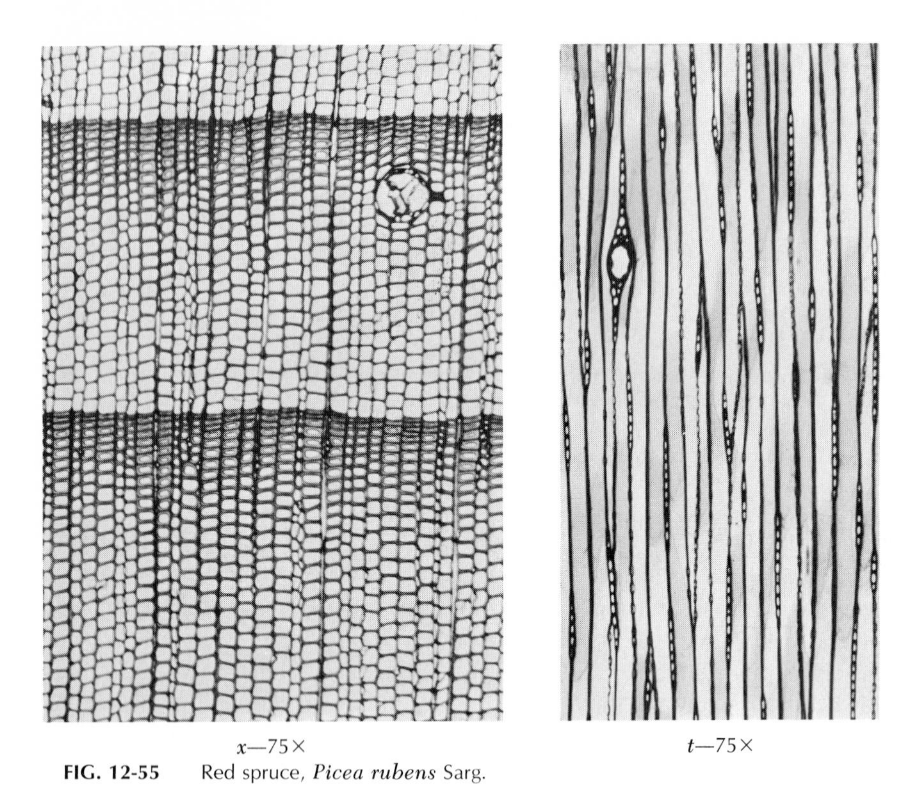

x—75× *t*—75×

FIG. 12-55 Red spruce, *Picea rubens* Sarg.

single horizontal row. ***Longitudinal parenchyma*** wanting. ***Rays*** of two types, uniseriate or rarely in part biseriate, and fusiform; (*a*) uniseriate rays numerous (*t*), 1–20 plus cells in height, biseriate rays very sparse and scattered, or wanting; (*b*) fusiform rays scattered, with one or rarely two transverse resin canals, 2–3-seriate through the central thickened portion, tapering above and below to uniseriate margins like the *a* rays, up to 20 plus cells in height; end walls nodular, with indentures; ray tracheids in both types of rays, usually restricted to one row on the upper and lower margins, nondentate. ***Resin canals*** with thick-walled epithelium, occasionally with tylosoids in the heartwood; longitudinal canals with maximum diameter of 135 (avg 50–90) μ; transverse canals much smaller (usually less than 30 μ).

REMARKS

The woods of white, red, black, and Engelmann spruce cannot be separated with certainty by either gross characteristics or minute anatomy. Engelmann spruce is usually slightly lighter and weaker than other spruces; in some samples of Engelmann spruce the transition from early to late wood is more abrupt than in other spruces, and the late wood is then usually appreciably denser than the early wood.

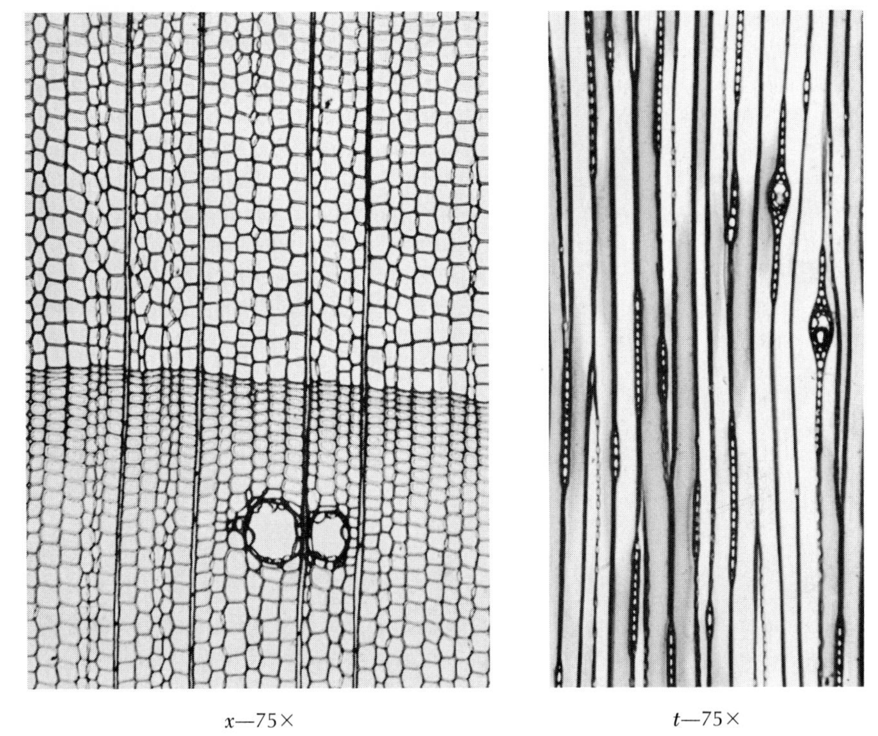

x—75× *t*—75×

FIG. 12-56 Engelmann spruce, *Picea engelmannii* Parry

Using a statistical approach, Denyer et al.* have found that on the average white spruce has tracheids and resin canals of larger diameter (28.3 μ versus 25.3 μ and 71.0 μ versus 55.0 μ respectively) than black spruce. White spruce has less late wood (10.6 percent versus 16.2 percent) than black spruce. All these differences, however, amount to only one standard deviation, or less.

Spruce wood is sometimes confused with that of the true firs (*Abies* spp.) but can be readily separated from it by resin canals of the normal type which do not occur in the latter.

USES

Pulpwood (principal use), reduced mainly by the sulfite and mechanical processes, and converted into a wide range of papers (from newsprint to high-grade writing paper); *fiber molded products*; *insulating boards*; *hardboard*; *particle board*; *round timbers* (poles, mine props); *railroad ties* (especially Engelmann spruce); *cooperage* (staves and headings); lumber used for *boxes* and

* W. B. G. Denyer, D. J. Gerrard, and R. W. Kennedy, A Statistical Approach to the Separation of White and Black Spruce on the Basis of Xylem Anatomy, *Forest Sci.*, **12**(2):177–184 (1966).

crates (very satisfactory in every way except when exceptionally high nail-holding capacity is required; *pallets*; *general building purposes* (boards, planks, joists, dimension stock); *planing-mill products* (doors, sash, casing, interior finish, roof decking, paneling); *laminated timbers*; *railroad car construction* and *repairs*; *boatbuilding*; *musical instruments*, especially for sounding boards (long a favorite because of its high resonant qualities); *furniture* and *kitchen cabinets*; *woodenware* and *novelties*; *ladder rails*; *paddles* and *oars* (because of its lightness combined with the required stiffness and strength); *veneer for containers*; *structural* and *prefinished interior plywood* on a limited scale (especially Engelmann spruce; because of the scarcity of peeler logs it seldom meets the requirements for rotary-cut veneer, but is considered to be excellent for production of cants for sliced veneer).

SITKA SPRUCE

Picea sitchensis (Bong.) Carr.

GENERAL CHARACTERISTICS

Sapwood creamy white to light yellow, grading into the darker heartwood, wide; *heartwood* light pinkish yellow to pale brown with purplish cast, darkening on exposure to silvery brown with a faint tinge of red; *wood* more or less lustrous, without characteristic odor or taste, generally straight- and even-grained and frequently dimpled on the tangential surface, medium-textured, moderately light (sp gr approx 0.37 green, 0.42 ovendry), moderately soft. *Growth rings* distinct, delineated by a band of darker late wood, narrow to medium wide. Early-wood zone usually occupying one-half to two-thirds of the ring; transition from early to late wood gradual; late-wood zone somewhat darker and denser than the band of early wood. *Parenchyma* not visible. *Rays* very fine (*x*), not visible to the naked eye except where they include a transverse resin canal, darker than the background (infiltration) and forming a fine, rather conspicuous fleck on the quarter surface. *Resin canals* present, longitudinal and transverse; (*a*) longitudinal canals fairly large, appearing as white flecks in the dark heartwood to the naked eye, sparse to fairly numerous, solitary or 2–several contiguous in the tangential plane or rarely in longer tangential lines, more or less regularly distributed and often wanting entirely from portions of some rings, appearing as scattered streaks along the grain; (*b*) transverse canals smaller than the longitudinal canals, visible against the dark background of heartwood on the transverse surface as somewhat broader whitish rays spaced at irregular intervals, usually distinct with a hand lens on the tangential surface.

MINUTE ANATOMY

Tracheids up to 55 (avg 35–45) μ in diameter; bordered pits in one row or rarely paired on the radial walls; tangential pitting present in the last few rows of late-

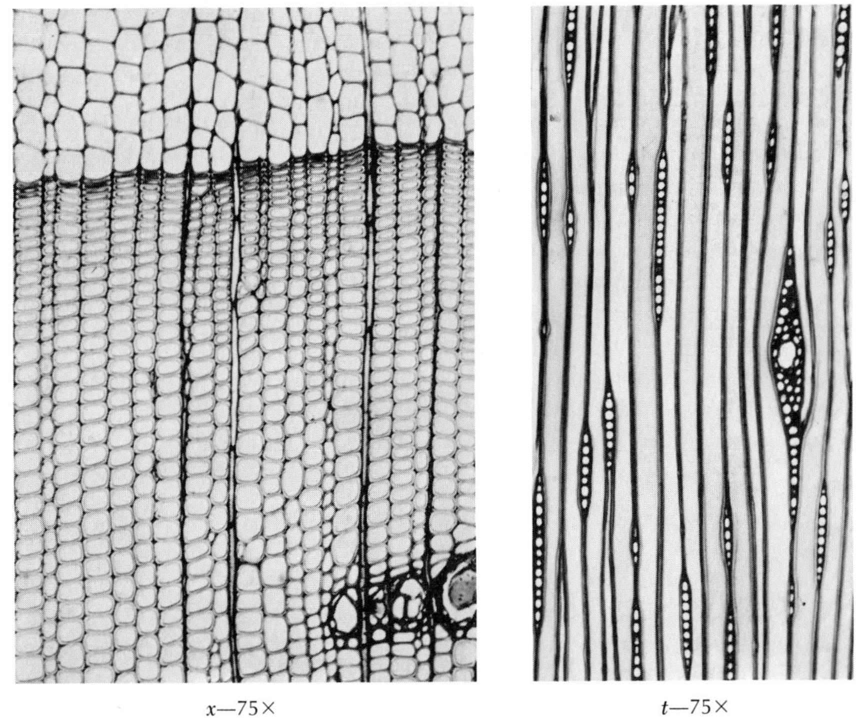

FIG. 12-57 Sitka spruce, *Picea sitchensis* (Bong.) Carr.

wood tracheids; pits leading to ray parenchyma piceoid, small, quite uniform in size, with distinct border, 1–6 (commonly with 3 or more) per cross field, generally in a single horizontal row (Fig. 12-37). *Longitudinal parenchyma* wanting. *Rays* of two types, uniseriate and fusiform; (*a*) uniseriate rays numerous (*t*), 1–20 plus cells in height; (*b*) fusiform rays scattered, with a transverse resin canal, 3–5-seriate through the central portion, tapering above and below to uniseriate margins like the *a* rays, up to 16 plus cells in height, end walls nodular, with indentures; ray tracheids in both types of rays, usually restricted to one row on the upper and lower margins, nondentate; ray parenchyma in the heartwood generally with gummy infiltration. *Resin canals* with thick-walled epithelium, occasionally with tylosoids in the heartwood; longitudinal canals with maximum diameter of 135 (avg 60–90) μ; transverse canals much smaller (usually less than 35 μ).

REMARKS

Distinct from white, red, black and Engelmann spruces in color and texture (coarser-textured and more or less woolly-fibered). Sometimes confused with eastern white pine (*Pinus strobus* L.) because of its dark-colored heartwood and unusually large resin canals but readily separated through the absence of resinous odor and window-like pits in the cross fields, and by its thick-walled epithelium.

USES

Pulpwood (reduced by the sulfite and mechanical processes); *poles*; *sailing-boat masts*. *Lumber* used for *boxes* and *crates*; *pallets*; *general construction*; *planing-mill products* and *millwork* (flooring, siding, ceiling, paneling, sash, doors, blinds); *laminated beams*; *musical instruments*, especially sounding boards (slow-growing material preferred because of its high resonant qualities); *furniture* (mostly hidden parts); *kitchen cabinets*; *boatbuilding*; *railroad car construction* and *repairs*; formerly extensively used for *aircraft construction* because of its high strength properties on the basis of weight, availability in clear pieces of large size, uniform texture, and freedom from hidden defects); *portable bleachers* and *gymnasium seats*; *overhead garage* and *warehouse doors*; *cooperage* (mostly staves and headings for slack barrels); *ladder rails*; *woodenware* and *novelties*. *Veneer* for plywood core stock.

DOUGLAS-FIR

Pseudotsuga menziesii (Mirb.) Franco

GENERAL CHARACTERISTICS

Sapwood whitish to pale yellowish or reddish white, narrow (Rocky Mountain type) to several inches in width (Pacific Coast type); *heartwood* ranging from yellowish or pale reddish yellow (slow-grown stock) to orange-red or deep red (fast-grown stock), the color varying greatly in different samples; *wood* with a characteristic resinous odor when fresh (different from that of pine), without characteristic taste, usually straight- and even- or uneven-grained, medium- to fairly coarse-textured, moderately light (sp gr approx 0.43 green, 0.48 ovendry—Rocky Mountain type) to moderately heavy (sp gr approx 0.45 green, 0.51 ovendry—Pacific Coast type), moderately hard. *Growth rings* very distinct, frequently wavy, delineated by a pronounced band of darker late wood, narrow (yellow fir) to very wide (red fir); growth increments (striae) usually less conspicuous than in pine on the radial surface. Early-wood zone usually several times wider than the band of late wood; transition from early to late wood generally abrupt (in wide rings sometimes more or less gradual); late-wood zone pronounced, very narrow in slow-grown stock to very wide and dense in wide rings. *Parenchyma* not visible. *Rays* very fine (*x*), not visible to the naked eye, forming a fine, close, inconspicuous fleck on the quarter surface. *Resin canals* present, longitudinal and transverse; (*a*) longitudinal canals small, barely visible or indistinct to the naked eye, plainly visible with a hand lens as dark spots or openings, confined largely to the outer half of the ring, sparse and scattered or numerous and exhibiting more or less a tendency (especially in certain rings) toward alignment in tangential rows of 2–30 plus, not visible or forming relatively inconspicuous streaks along the grain (*t*); (*b*) transverse canals smaller than the longitudinal canals, appearing with a hand lens as somewhat broader rays spaced at irregular intervals on the transverse surface, not visible or barely visible with a hand lens on the tangential surface.

MINUTE ANATOMY

Tracheids up to 55 (avg 35–45) μ in diameter, characterized (especially in early wood) by fine, close bands of spiral thickening; bordered pits in one row or occasionally paired on the radial walls; tangential pitting present in the last few rows of late-wood tracheids; pits leading to ray parenchyma piceoid, small, quite uniform in size, with distinct border, 1–6 (generally 4) per cross field. ***Longitudinal parenchyma*** marginal and very sparse, or wanting. ***Rays*** of two types, uniseriate or rarely in part biseriate, and fusiform; (*a*) uniseriate rays numerous (*t*), 1–25 plus cells in height; biseriate rays very sparse and scattered, or wanting; (*b*) fusiform rays scattered, with one or very rarely two transverse resin canals, 3–5-seriate through the central thickened portion, tapering above and below to uniseriate margins similar to the *a* rays, up to 16 plus cells in height; end walls nodular; ray tracheids present in both types of rays, marginal and very rarely interspersed, nondentate, occasionally with spiral thickening; marginal tracheids usually in one row. ***Resin canals*** with thick-walled epithelium, constricted at intervals and occasionally with tylosoids in

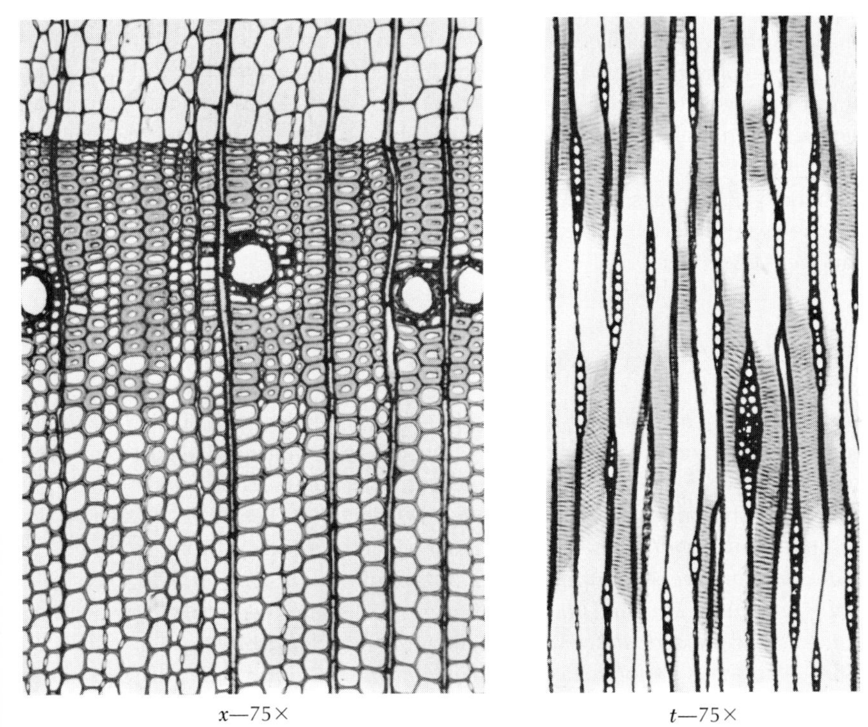

x—75× *t*—75×

FIG. 12-58 Douglas-fir, *Pseudotsuga menziesii* (Mirb.) Franco

the heartwood; longitudinal canals with maximum diameter of 150 (avg 60–90) μ transverse canals much smaller (usually less than 25 μ).

REMARKS

Douglas-fir wood is very variable in color, ring width, strength, and working qualities. Narrow-ringed Douglas-fir is quite uniform-textured, moderately soft, and easily worked. It tends to be yellow or pale reddish yellow in color, and because of this it was formerly known in the trade as "yellow fir." In contrast, Douglas-fir with wide rings is characterized by wide bands of reddish late wood, which is responsible for its overall orange-red or deep-red color; such wood is coarser-grained and more unevenly textured than the narrow-ringed stock and also stronger and more refractory under tools. Because of its prevailingly reddish cast the wide-ringed Douglas-fir is sometimes sold as "red fir."

Douglas-fir grown in the Rocky Mountains, especially in the southern parts of the range, tends to contain smaller percentages of late wood, at any rate of growth, and to be lower in specific gravity and strength than Douglas-fir grown in the coastal areas, the Cascades, and the Sierras. Likewise, wide-ringed, second-growth Douglas-fir from the coastal areas is frequently lighter in weight and lower in strength properties than virgin-growth wood from the same regions.

Douglas-fir wood is sometimes confused with southern pine and western larch. It can usually be separated from the former on the basis of color, distinctive odor, and the smaller resin canals, which in Douglas-fir show a tendency toward alignment in tangential rows of 2–20 plus. The heartwood of western larch, in contrast to that of Douglas-fir, has a distinct brownish cast. Presence of spiral thickening in both the early- and the late-wood tracheids in Douglas-fir permits of certain identification of this species; spiral thickening is found only occasionally in the late-wood tracheids in western larch, and only rarely in southern pines, in wood immediately adjacent to the pith.

USES

Building and *construction* in the form of lumber, timbers, and piling; large *laminated beams* and *arches* for churches, schools, and commercial buildings; *veneer*, converted largely into *structural plywood*, extensively used in building and construction. The principal uses of *lumber*, in addition to general construction, are *planing-mill products* (sash, doors, flooring, and general millwork); *railroad car construction*; *boxes*, *crates* and *pallets*; *containers* for corrosive chemicals; *silos* and *tanks*; *ship* and *boatbuilding*, especially ship knees; *furniture*. *Motion-picture* and *theatrical scenery* (mainly plywood). *Pulpwood* (leading pulpwood species in the west for sulfate pulp; *insulating* and *other types* of *fiberboard*; *particle* and *hardboard*. *Poles*; *piling*; *mine timbers*; *railroad ties*; *cooperage*.

EASTERN HEMLOCK, HEMLOCK, CANADA HEMLOCK

Tsuga canadensis (L.) Carr.

GENERAL CHARACTERISTICS

Wood buff to light brown, the late-wood portion of the ring frequently with a roseate or reddish-brown tinge (*heartwood* not distinct, but the last few rings near the bark usually somewhat lighter in color), odorless or with a sour odor when fresh, without characteristic taste, uneven- and frequently spiral-grained, coarse- to medium-textured, moderately light (sp gr approx 0.38 green, 0.42 ovendry), moderately hard. *Growth rings* distinct, delineated by a pronounced band of darker late wood, narrow to wide (proximate rings not infrequently very variable in thickness; individual rings often variable in thickness and hence more or less sinuate in contour). Early-wood zone usually occupying two-thirds or more of the ring; transition from early to late wood gradual to abrupt; late-wood zone distinct to the naked eye, decidedly darker (roseate or reddish brown) and denser than the early wood, variable in thickness in different rings and often in a given ring. *Parenchyma* not visible. *Rays* very fine (x), not distinct to the naked eye, forming a fine, close, inconspicuous

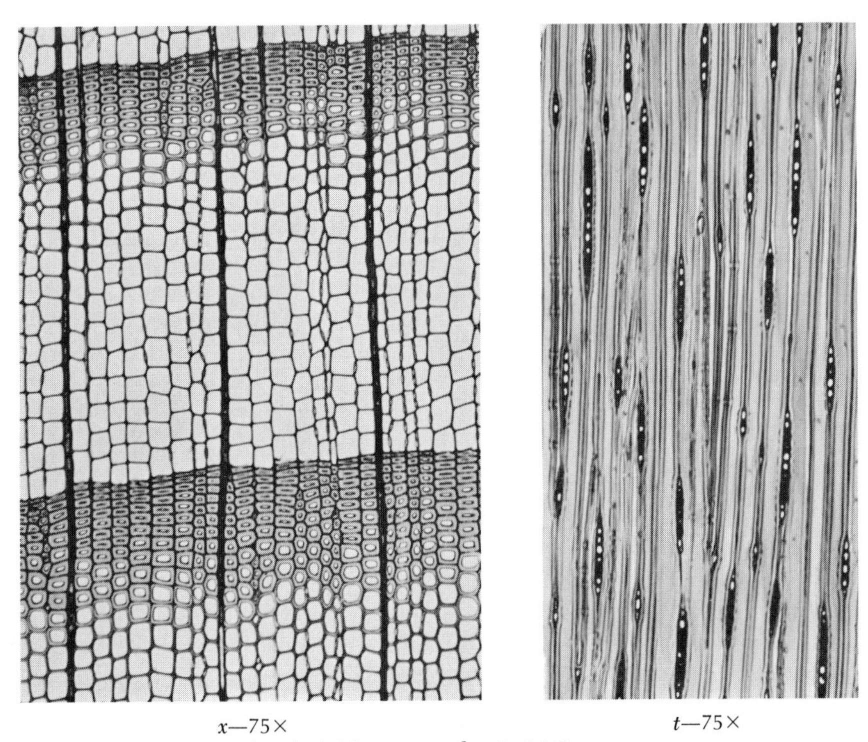

x—75× *t*—75×

FIG. 12-59 Eastern hemlock, *Tsuga canadensis* (L.) Carr.

fleck on the quarter surface. Normal ***resin canals*** wanting; traumatic canals very rare.

MINUTE ANATOMY

Tracheids up to 45 (avg 28–40) μ in diameter; bordered pits in 1–2 (mostly 1) rows on the radial walls; tangential pitting present in the last few rows of late-wood tracheids; pits leading to ray parenchyma piceoid to cupressoid, small, quite uniform in size, with distinct border, 1–5 (generally 3–4) per cross field. ***Longitudinal parenchyma*** marginal and very sparse, or wanting. ***Rays*** uniseriate, 1–12 plus cells in height; ray tracheids present, usually restricted to one row on the upper and lower margins of the ray. End walls nodular, with indentures.

REMARKS

For differences between eastern and western hemlock, see Remarks under Western Hemlock. Eastern hemlock differs from the spruces (***Picea*** spp.) in the absence of normal resin canals and from the firs (***Abies*** spp.) in possessing ray tracheids [ray tracheids sporadic in ***A. balsamea*** (L.) Mill., wanting in other North American species of ***Abies***].

USES

Pulpwood (converted largely by the sulfite process); ***general construction*** (framing, sheathing, roofing, subflooring); ***boxes*** and ***crates*** (principally for heavy shipping containers); ***pallets; railroad car construction.*** The bark contains 10–13 percent of tannin and has been an important domestic source of this product.

WESTERN HEMLOCK, WEST COAST HEMLOCK

Western hemlock [*Tsuga heterophylla* (Raf.) Sarg.]
Mountain hemlock [*Tsuga mertensiana* (Bong.) Carr.]

GENERAL CHARACTERISTICS

Wood whitish to light yellowish brown, the late-wood portion of the ring frequently with a roseate, purplish, or reddish-brown tinge (***heartwood*** not distinct, but the last few rings near the bark almost white), frequently with dark streaks (termed ***bird pecks*** in the trade) caused by the maggots of a small black fly (***Chilosia alaskensis*** Hunter),* occasionally with whitish spots (floccosoids),† odorless or with a sour odor when fresh, without characteristic taste, generally straight- and quite even-grained,

* See p. 371.

† The floccosoids are produced by whitish deposits of α-conidendrin-4′-glucoside within the wood cells (see p. 364).

medium- to fine-textured, moderately light (sp gr approx 0.42 green, 0.47 ovendry), moderately hard. *Growth rings* distinct, delineated by a band of darker late wood, narrow to wide but generally narrower and more uniform in width than in the eastern hemlock; individual rings frequently variable in thickness and hence more or less sinuate in contour. Early-wood zone usually occupying two-thirds or more of the ring; transition from early to late wood more or less gradual; late-wood zone distinct to the naked eye, darker and denser than the early wood (the contrast between early and late wood less accentuated than in the eastern hemlock). *Parenchyma* not visible. *Rays* very fine (x), not distinct to the naked eye, forming a fine, close, inconspicuous fleck on the quarter surface. Normal *resin canals* wanting; longitudinal strands of traumatic resinous cells or wound canals sometimes present, the latter sporadic and often in widely separated rings, arranged in a tangential row (x), appearing as dark streaks along the grain.

MINUTE ANATOMY

Tracheids up to 50 (avg 30–40) μ in diameter; bordered pits in 1–2 (mostly 1) rows on the radial walls; tangential pitting present in the last few rows of late-wood tracheids; pits leading to ray parenchyma piceoid to cupressoid, small, quite uni-

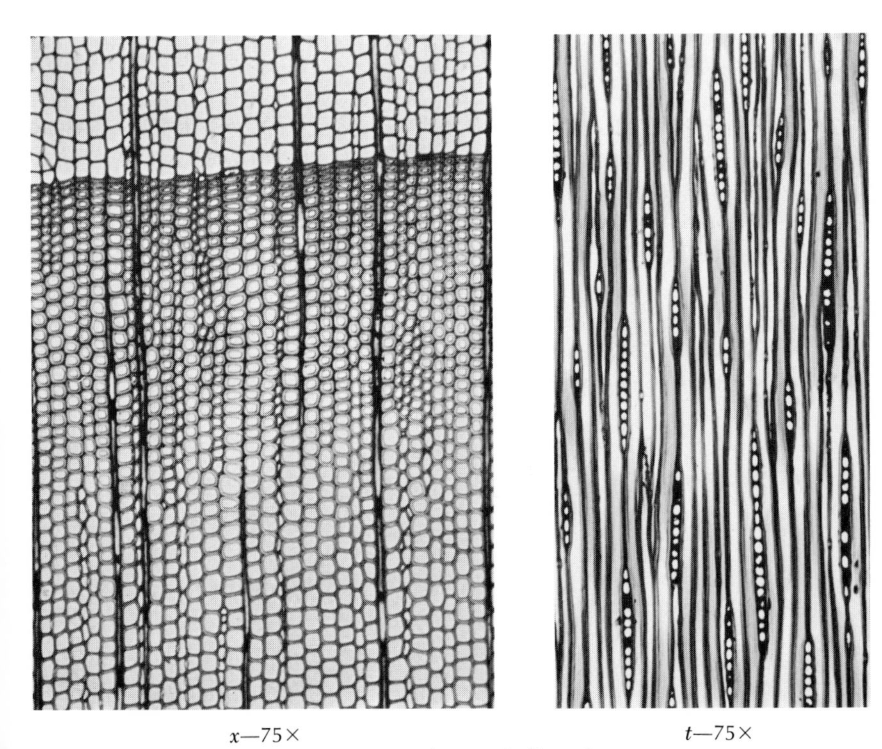

x—75× t—75×

FIG. 12-60 Western hemlock, *Tsuga heterophylla* (Raf.) Sarg.

form in size, with distinct border, 1–4 (generally 2–3) per cross field. ***Longitudinal parenchyma*** very sparse and terminal, or wanting. ***Rays*** uniseriate or very rarely with paired cells, 1–16 plus cells in height; end walls nodular with indentures; ray tracheids present, usually restricted to one row on the upper and lower margins of the ray.

REMARKS

Wood of the two species of western hemlock is indistinguishable and is considered to be comparable in strength properties, with the possible exception of the modulus of elasticity, which is lower on the average for mountain hemlock. Western hemlock cannot always be separated with certainty from eastern hemlock, but the following characters will serve to distinguish these woods in most instances: roseate cast less pronounced than in the eastern species; straighter-grained and less harsh under tools; early wood somewhat finer-textured; rings generally narrower and more uniform in width; transition from early to late wood more gradual and the latter less accentuated; longitudinal strands of resin cells and wound canals, termed ***bird pecks*** in the trade, not infrequent (rare in eastern hemlock).

Western hemlock is sometimes confused with western fir; for distinguishing characteristics, see Remarks under Western Fir, p. 484.

USES

Pulpwood (important source in the Pacific Northwest; converted by the sulfite and mechanical processes); ***general construction*** (all but the heaviest type); ***planing-mill products*** (such as siding, ceiling, flooring, shiplap, and finish); ***paneling***, ***boxes*** and ***crates*** (generally in the type requiring rather thick boards because of the tendency to split in nailing); ***railroad car construction***; ***slack cooperage***, formerly used for sugar and flour barrels (because of freedom from resinous materials and the absence of taste, and its clear color); ***ladder rails***, for which purpose it competes with spruce; ***plywood***. The bark contains 12–22 percent of tannin and may become an important domestic source of this product.

BALSAM AND FRASER FIR

Balsam fir [*Abies balsamea* (L.) Mill.]

Fraser fir, southern balsam fir [*Abies fraseri* (Pursh) Poir.]

GENERAL CHARACTERISTICS

Wood whitish to creamy white or pale brown (especially the early wood), the late-wood portion of the ring frequently with a lavender tinge (***heartwood*** not distinct), without characteristic odor, tasteless or with a slight salty tang, straight- and even-grained, medium-textured, light (sp gr approx 0.34 green, 0.38 ovendry), soft.

Growth rings distinct, delineated by the contrast between the somewhat denser late wood and the early wood of the succeeding ring, medium wide to wide or narrow in the outer portion of mature trees. Early-wood zone usually occupying two-thirds or more of the ring; transition from early to late wood very gradual; late-wood zone distinct to the naked eye, somewhat darker than the early wood, generally narrow. *Parenchyma* not visible. *Rays* very fine (x), not distinct to the naked eye, forming a fine, close, inconspicuous fleck on the quarter surface. Normal *resin canals* wanting; longitudinal wound (traumatic) canals sometimes present, sporadic and often in widely separated rings, arranged in a tangential row (x) which frequently extends for some distance along the ring, appearing as dark streaks along the grain.

MINUTE ANATOMY

Tracheids up to 50 (avg 30–40) μ in diameter; bordered pits in one row or very rarely paired on the radial walls; tangential pitting present in the last few rows of late-wood tracheids; pits leading to ray parenchyma taxodioid, small, quite uniform in size, with distinct border, 1–3 (generally 2–3) per cross field (Fig. 12-38); end walls nodular. *Longitudinal parenchyma* very sparse, marginal, or wanting.

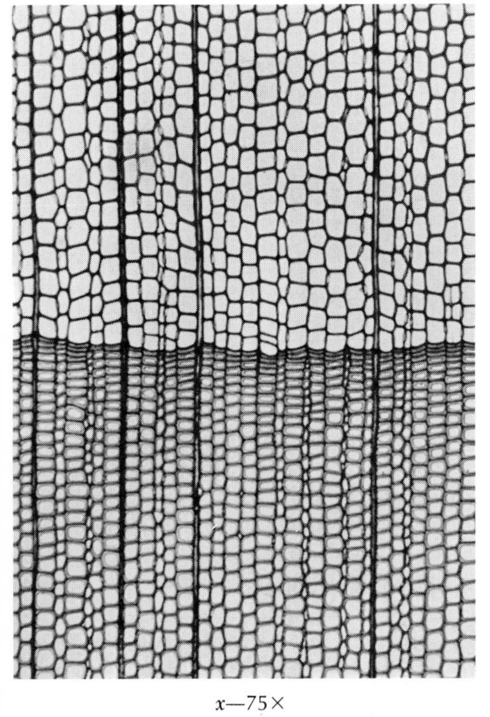

x—75×

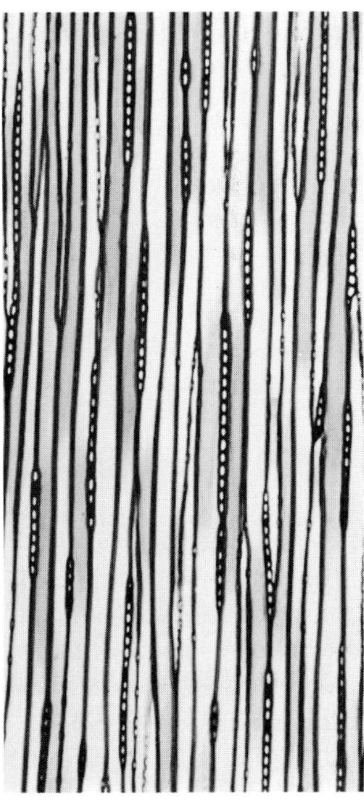

t—75×

FIG. 12-61 Balsam fir, *Abies balsamea* (L.) Mill.

Rays uniseriate, very variable in height (1–30 plus cells), consisting wholly of ray parenchyma or rarely with a row of ray tracheids on the upper and lower margins [*Abies balsamea* (L.) Mill.]; end walls nodular.

REMARKS

Lighter, softer, and weaker than spruce. Frequently confused with eastern spruce but readily separated in that ray tracheids [*A. balsamea* (L.) Mill. excepted] and normal resin canals do not occur in the true firs (*Abies* spp.). The longitudinal wound (traumatic) canals, which are occasionally present, sometimes resemble normal canals; in case of doubt, examine a tangential section of the wood and note the absence (true fir) or presence (spruce) of transverse resin canals in the wood rays.

USES

Pulp (in admixture with eastern spruce); *cooperage* (sugar and butter tubs, because of the absence of objectionable taste and resinous materials that might taint). Lumber used for *boxes* and *crates* (especially cheeseboxes); *pallets*; *millwork* (siding, molding); *general construction*; *novelties* and *woodenware*; bark on young trees and branches contains blisters filled with an *oleoresin* (Canada balsam), used in cementing lenses and mounting specimens for observation with a microscope.

WESTERN FIRS

White fir [*Abies concolor* (Gord. & Glend.) Lindl.]
Grand fir [*Abies grandis* (Dougl.) Lindl.]
Subalpine fir [*Abies lasiocarpa* (Hook.) Nutt.]
California red fir (*Abies magnifica* A. Murr.)
Noble fir (*Abies procera* Rehd.)
Pacific silver fir [*Abies amabilis* (Dougl.) Forbes]

GENERAL CHARACTERISTICS

Wood whitish or light buff to yellowish brown or light brown, the late-wood portion of the ring frequently with a roseate, reddish-brown, or lavender tinge (*heartwood* not distinct), without characteristic odor (sometimes with a slight disagreeable odor when green), tasteless, generally straight- and quite even-grained, medium- to somewhat coarse-textured, light (sp gr approx 0.35–037 green, 0.30–0.45 ovendry; occasionally as high as 0.56 in noble fir), moderately soft. *Growth rings* distinct, delineated by a band of darker late wood, medium wide to wide (3–4 per in.) or narrow in the outer portion of mature trees. Early-wood zone usually occupying one-half or more of the ring; transition from early to late wood gradual; late-wood zone distinct to the naked eye, variable in depth of color and density according to conditions of growth, ranging from broad in wide rings to very narrow. *Parenchyma* not visible. *Rays* very fine (*x*), not distinct to the naked

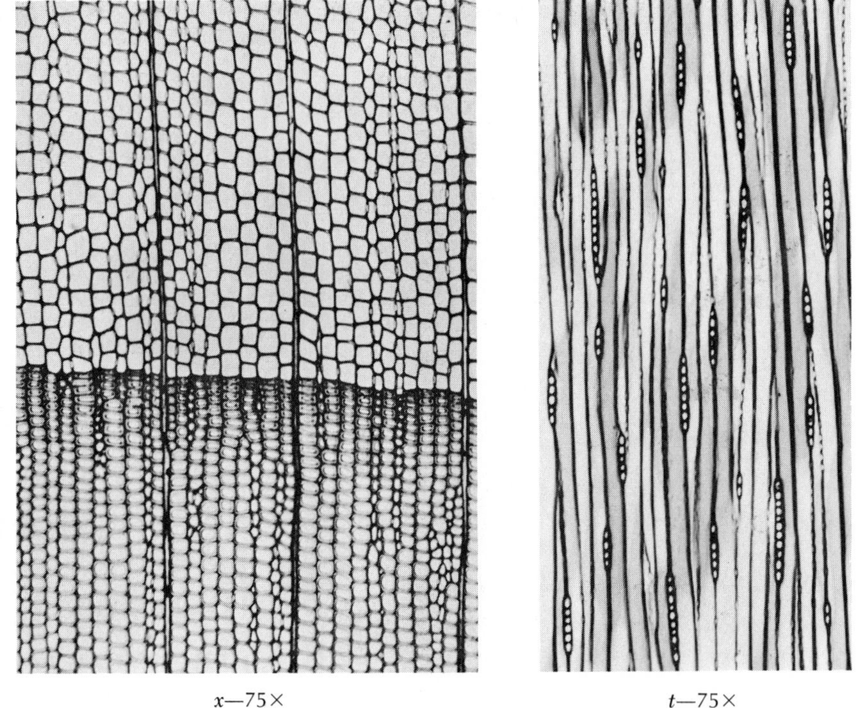

FIG. 12-62 White fir, *Abies concolor* (Gord. & Glend.) Lindl.

eye, forming a fine, close, inconspicuous fleck on the quarter surface. Normal *resin canals* wanting; longitudinal wound (traumatic) canals sometimes present, sporadic and often in widely separated rings, arranged in a tangential row (*x*) which frequently extends some distance along the ring, appearing as dark streaks along the grain.

MINUTE ANATOMY

Tracheids up to 60 (avg 35–45) μ in diameter; bordered pits in one row or occasionally biseriate on the radial walls; tangential pitting present in the last few rows of late-wood tracheids; pits leading to ray parenchyma taxodioid, small, quite uniform in size, with distinct border, 1–4 (generally 2–4) per cross field. *Longitudinal parenchyma* marginal and very sparse, or wanting.* *Rays* uniseriate (or occasionally in part biseriate in *Abies magnifica* A. Murr.), very variable in height (1–30 plus cells), consisting wholly of ray parenchyma; end walls nodular. Crystals are regularly found in the marginal parenchyma cells of *A. procera*, *A. grandis*, *A. magnifica*, and *A. concolor*; they are absent in *A. amabilis* and *A. lasiocarpa*, as well as in *A. balsamea*. (See reference cited in footnote, p. 145.)

* The strands of traumatic resinous cells, forming black streaks and termed *bird pecks*, that are sometimes found in the body of the ring as a transitional stage in the formation of longitudinal wound canals, are not interpreted here as true longitudinal parenchyma.

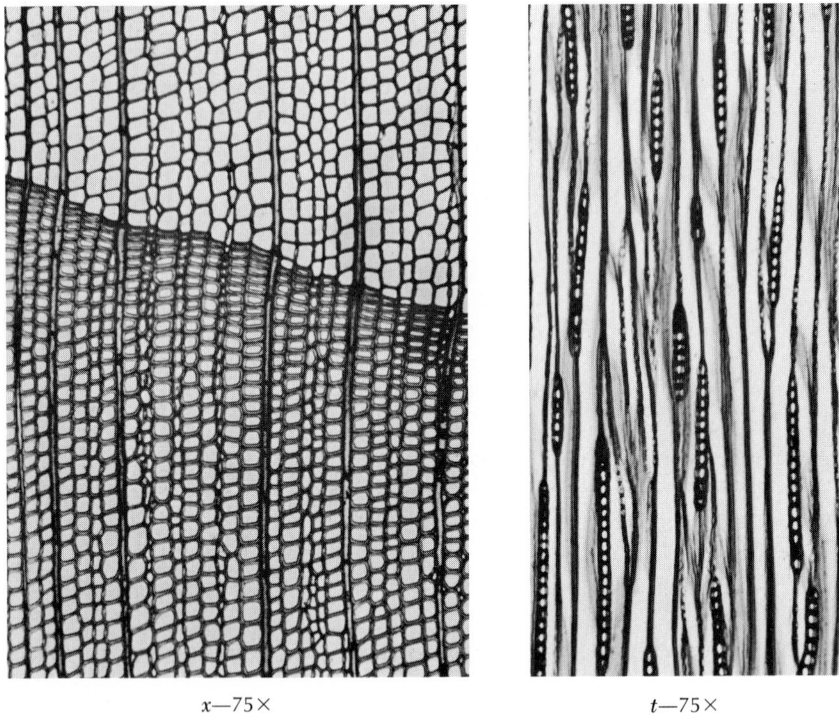

FIG. 12-63 California red fir, *Abies magnifica* A. Murr.

REMARKS

The timber of noble fir (*A. procera* Rehd.) resembles that of western hemlock [*Tsuga heterophylla* (Raf.) Sarg.] but can be separated as follows: the wood of western hemlock is more brittle, and harsher under tools; the rings are generally narrower (tolerant tree) and more uniform in width than those of the noble fir, and the early wood is finer and more even-textured (early wood coarse and somewhat variable in texture in noble fir); marginal-ray tracheids are invariably found in the wood rays of hemlock and are generally wanting in the western firs. The woods of the other western balsam firs resemble balsam fir [*Abies balsamea* (L.) Mill.] in color, ranging from white and creamy white to pale brown with lavender, reddish, or reddish-brown bands of late wood. The true firs (*Abies* spp.) cannot be separated on the basis of wood anatomy.

For chemical and chromatographic methods of separating true firs from western hemlock see p. 420.

USES

General construction (rough and dressed boards, dimension timbers); *boxes* and *crates* (because of its light weight, clean appearance, freedom from odors and stains, and relatively low cost); *planing-mill products* (especially sash, doors, and trim); *pulpwood* (used in manufacture of printing paper and high-grade wrapping paper); *plywood*.

REDWOOD, COAST REDWOOD

Sequoia sempervirens (D. Don) Endl.

GENERAL CHARACTERISTICS

Sapwood nearly white, narrow; *heartwood* clear light red to deep reddish brown; *wood* without characteristic odor or taste, generally straight- and even- (slow growth) or uneven- (second growth) grained (rarely wavy-grained), coarse-textured, light to moderately light (sp gr 0.38 green, 0.41 ovendry), soft to moderately hard. *Growth rings* distinct, delineated by a band of darker late wood, very narrow (sometimes only 2–3 tracheids in width and not infrequently discontinuous in old trees) to very wide (coppice-grown stock). Early-wood zone porous (the openings of the tracheids readily visible with a hand lens), narrow to wide; transition from early to late wood generally abrupt; late-wood zone distinct to the naked eye, darker and denser than the early wood, narrow or wide in broad rings. *Parenchyma* abundant, present in every growth ring; cells scattered, readily visible in the sapwood with a lens and sometimes to the naked eye because of their dark resinous contents, inconspicuous in the dark-colored heartwood; strands of parenchyma visible along the grain (split) as dark lines. *Rays* coarse (*x*) for a coniferous wood, lighter than the background in the heartwood and hence generally visible to the naked eye, forming a fine, close, relatively conspicuous fleck on the quarter surface. Normal *resin canals* wanting; longitudinal wound (traumatic) canals sometimes present, sporadic and often in widely separated rings, arranged in a tangential row (*x*), appearing as dark streaks along the grain.

MINUTE ANATOMY

Tracheids up to 80 (avg 50–65) μ in diameter; bordered pits in 1–3 (generally 2) rows on the radial walls; tangential pitting present in the last few rows of late-wood tracheids; pits leading to ray parenchyma taxodioid, fairly large, quite uniform in size, oval for the most part (the long axis of the pit orifice more than 8 μ), 1–4 (generally 2–3) per cross field (Fig. 12-39). *Longitudinal parenchyma* apotracheal-diffuse to diffuse-in-aggregates (the cells solitary or occasionally 2–several contiguous in the tangential plane, conspicuous because of their dark resinous contents); end walls smooth. *Rays* uniseriate or not infrequently in part biseriate, consisting entirely of ray parenchyma or rarely with marginal or isolated ray tracheids,* the tallest up to 40 (mostly 10–15) cells in height.

* M. Gordon, Ray Tracheids in *Sequoia sempervirens*, *New Phytol.*, **11**:1–7 (1912).

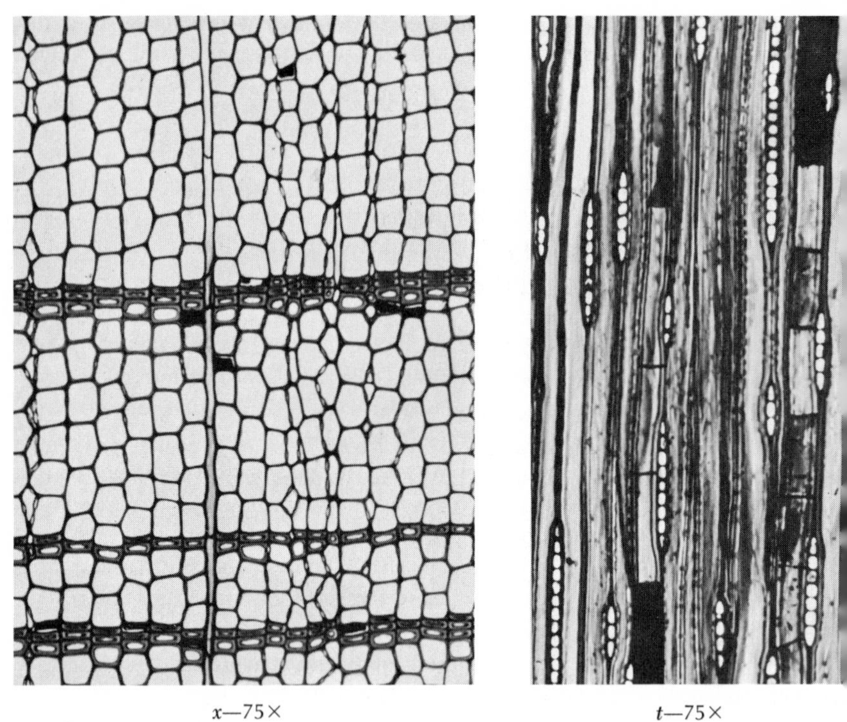

FIG. 12-64 Redwood, *Sequoia sempervirens* (D. Don) Endl.

REMARKS

Unique among domestic coniferous woods because of its distinctive characteristics (straightness of grain, coarseness of texture, freedom from oily materials that characterize some coniferous woods, etc.). Resembles some of the cedars in color but much coarser-textured, without characteristic odor, and with more conspicuous parenchyma. Readily separated from baldcypress by its color, lack of greasiness, and the nature of the pits in the cross fields (compare Figs. 12-39 and 12-40). Some samples of redwood are peculiar in exhibiting even-width zones of compression wood composed of a number of annual increments which alternate at regular intervals with bands of normal wood.

Sierra redwood is the product of an allied tree, the bigtree or giant sequoia of the high Sierra, *Sequoia gigantea* (Lindl.) Decne. [*Sequoiadendron giganteum* (Lindl.) Buchholz]; it is occasionally used for fence posts, vineyard stakes, shakes, shingles, and for lumber, but the supply is becoming more limited as the groves of this species are incorporated into parks. Formerly employed for siding but generally too light and too soft for this purpose.

The wood of Sierra redwood closely resembles that of coast redwood. The following characters will usually suffice to separate these two timbers: Sierra redwood generally softer and more brittle; heartwood darker, often with a purplish cast;

bands of late wood generally narrower (1–4 tracheids in width); rays less conspicuous and less abundant, and biseriate rays less frequent.

USES

General and *building construction* (timbers, sills, joists, building lumber), because of its durability combined with strength and availability in large sizes; *planing-mill products* and *millwork* (especially sash, doors, and blinds, also siding, ceiling, and general millwork): *fences* and *patios*; *boxes* and *crates*; *caskets* and *coffins* (because of high natural durability); *tanks, vats* and *cooling towers* (because of durability, high resistance to action of chemicals, low shrinkage, and freedom from twisting and warping); *cigar-, tobacco-,* and *candy-boxes*; *conduits* such as pipes and flumes (because of durability, low cost, smoothness of surface, and ability to withstand extremes of temperature); *garden furniture*; *apiary supplies*; *signs*; *stadium seats*; *ship-* and *boatbuilding*; *greenhouses*; *plywood* for exterior and interior novelty application; *woodenware* and *novelties*; *shingles* and *shakes* (sawed and hand-split); wood from *burls* used for turned and carved articles and occasionally for table tops; *shredded bark* converted into wool-like insulating sorptive material; *pulpwood*; *particle board.*

BALDCYPRESS, RED CYPRESS, YELLOW CYPRESS, SOUTHERN CYPRESS

Taxodium distichum (L.) Rich.

GENERAL CHARACTERISTICS

Sapwood pale yellowish white, merging more or less gradually into the heartwood, usually 1 in. or more in width; *heartwood* very variable in color, ranging from yellowish to light or dark brown, reddish brown, or almost black (in the Gulf Coast and South Atlantic regions, the color averages darker than farther north); *wood* with greasy feel (especially along the grain), often with a rancid odor (light-colored stock, sometimes odorless),* generally straight- and even- or uneven-grained, coarse-textured, moderately heavy (sp gr approx 0.42 green, 0.47 ovendry), moderately hard. *Growth rings* distinct, delineated by a band of darker late wood, narrow and occasionally discontinuous in old, overmature trees, or wide (proximate rings not infrequently variable in thickness; individual rings often variable in thickness and the rings then sinuate in contour). Early-wood zone narrow to wide, usually several times wider than the late-wood; transition from early to late wood more or less abrupt; late-wood zone conspicuous (in the darker grades) or inconspicuous (lighter stock), narrow or broad in wide rings. *Parenchyma* abundant,

* In spite of the presence of an oil that is responsible for the greasy feel and rancid odor, cypress wood does not impart taste, odor, or color to food products that come into contact with it.

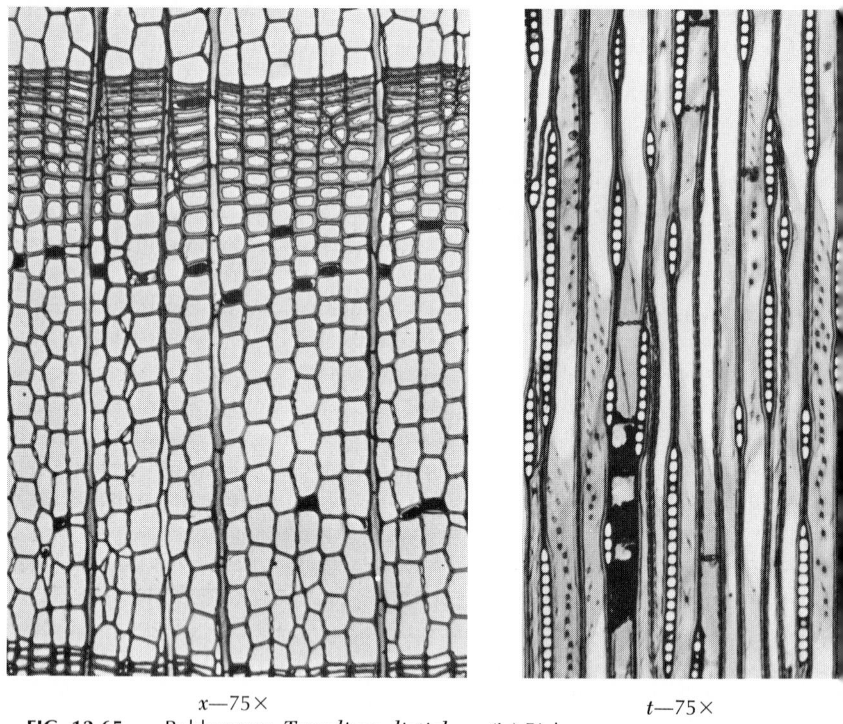

x—75× *t*—75×

FIG. 12-65 Baldcypress, *Taxodium distichum* (L.) Rich.

present in every growth ring; cells scattered, readily visible in the lighter-colored grades with a hand lens and sometimes to the naked eye because of their dark resinous contents; strands of parenchyma visible along the grain as dark lines. ***Rays*** rather coarse (*x*) for a coniferous wood, forming a fine relatively conspicuous fleck on the quarter surface. ***Resin canals*** wanting. Finger-shaped pockets extending along the grain and containing a brown friable mass of decayed wood, frequently present; this defect, known as brown rot or peckiness, is caused by the fungus *Stereum* spp.

MINUTE ANATOMY

Tracheids up to 70 (avg 45–60) μ in diameter; bordered pits in 1–4 (frequently 3) rows on the radial walls; tangential pitting present in the last few rows of late-wood tracheids; pits leading to ray parenchyma taxodioid or more rarely cupressoid, fairly large, quite uniform in size, orbicular for the most part (the long axis of the pit orifice less than 8 μ), 1–6 (generally 4) per cross field (Fig. 12-40). ***Longitudinal parenchyma*** apotracheal-diffuse to diffuse-in-aggregates (the cells solitary or occasionally 2–several contiguous in the tangential plane, conspicuous because of their dark resinous contents); end walls nodular. ***Rays*** uniseriate or rarely in part biseriate, consisting entirely of ray parenchyma, the tallest more than 15 (up to 60) cells in height; end walls occasionally nodular.

REMARKS

Baldcypress is one of the most variable woods in the United States in color, weight, and durability, and various grades such as red cypress, tidewater red cypress, yellow cypress, and white cypress are recognized by the trade; red cypress is usually slightly heavier and more durable than the yellow or white grades. The term *yellow cypress* is also applied to the wood of Alaska-cedar [*Chamaecyparis nootkatensis* (D. Don) Spach]. Pondcypress, *T. distichum* var. *nutans* (Ait.) Sweet, occurs on the coastal plain in Southeastern United States, but the wood is not distinguished from that of baldcypress.

USES

Most of the major uses of this wood depend on its reputed durability when exposed to conditions favorable to decay: *construction*, especially parts exposed to the weather (such as siding, beams, posts, timbers in docks and bridges);* *caskets* and *coffins*; *millwork* (especially doors, sash, and blinds, also interior trim and paneling): *kitchen cabinets*; *tanks*, *vats*, and *silos*; *containers* for corrosive chemicals; *laundry appliances*; *greenhouses*; *machine parts*; *ship-* and *boat-building*; *patterns*; *fixtures*; *stadium seats*; *boxes* and *crates* (considerable quantities, but mostly low-grade stock). In forms other than lumber, cypress is used extensively for *railroad ties* and in limited quantities for *poles*, *piling*, *fence posts*, and *cooperage*.

INCENSE-CEDAR, PENCIL CEDAR

Libocedrus decurrens Torr.

GENERAL CHARACTERISTICS

Sapwood nearly white, thin; *heartwood* reddish brown to dull brown, sometimes with a lavender tinge; *wood* with characteristic pungent odor and spicy acrid taste, straight- and even-grained, medium-textured, light (sp gr approx 0.35 green, 0.38 ovendry), moderately soft. *Growth rings* distinct, delineated by a band of darker late wood, quite uniform in width, medium broad to broad (avg 9–12 per in.). Early-wood zone usually broad, occupying most of the ring; transition from early to late wood gradual; late-wood zone fairly conspicuous, narrow. *Parenchyma* very abundant, (*a*) apotracheal-diffuse and not visible or barely

* Second growth stock, regardless of its color or source, is only moderately decay resistant and should not be used in contact with the ground, unless treated. R. N. Campbell and J. W. Clark, Decay Resistance of Baldcypress Heartwood, *Forest Prod. J.*, **10**(5):250–253 (1960).

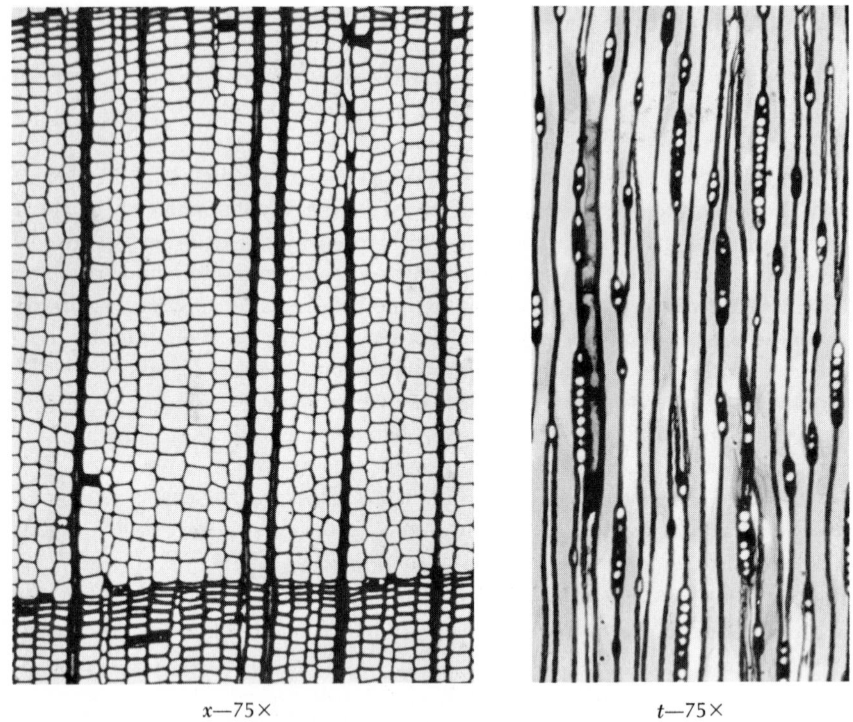

FIG. 12-66 Incense-cedar, *Libocedrus decurrens* Torr.

visible with a hand lens in the early wood, (*b*) not infrequently banded in 1–2 dark bands in the late wood which are visible with a hand lens and often to the naked eye. ***Rays*** fine (*x*), plainly visible with a hand lens, forming a fine, close fleck on the quarter surface. ***Resin canals*** wanting. Finger-shaped pockets extending along the grain and containing a brown mass of decayed wood frequently present; this defect, known as *peckiness*, is caused by the fungus *Polyporus amarus* Hedg.

MINUTE ANATOMY

Tracheids up to 50 (avg 35–40) μ in diameter; bordered pits in 1–2 rows on the radial walls; tangential pitting present in the last few rows of late-wood tracheids; pits leading to ray parenchyma cupressoid, small, oval, quite uniform in size, with distinct border and lenticular orifice, 1–4 (generally 1–2) per cross field. Cells of *longitudinal parenchyma* solitary, 2–several contiguous in the tangential plane, and in interrupted bands in the outer late wood, with dark gummy contents; end walls nodular. *Rays* uniseriate or not infrequently in part biseriate, mostly low (tallest rays 15 plus cells in height), consisting entirely of ray parenchyma or with occasional ray tracheids; broadest rays (uniseriate) 15–25 μ wide; ray cells in the heartwood with deposits of dark gum; end walls nodular.

REMARKS

Sometimes confused with western redcedar (*Thuja plicata* Donn). For distinguishing characteristics of these two woods, see Remarks under Western redcedar, p. 494.

USES

Fence posts (because of its durability and ease of splitting); *ties* (very satisfactory as to durability but so soft that they require tie plates). Lumber used for *pencil slats* (principal species), for which purpose it is especially suited because of the ease with which it may be whittled owing to the evenness of the grain (early and late wood of approximately the same hardness), its uniformity of texture, and its relative firmness; *woodenware* and *novelties*; *millwork* (interior finish, outside trim, doors, and sash); *mothproof chests* and *closets*; *furniture*; *veneer*, sliced for decorative interior grades of plywood.

NORTHERN WHITE-CEDAR, EASTERN ARBORVITAE
Thuja occidentalis L.

GENERAL CHARACTERISTICS

Sapwood nearly white, narrow; *heartwood* uniformly straw-brown; *wood* with characteristic cedary odor [distinct from that of Atlantic white-cedar (*Chamaecyparis thyoides* (L.) B.S.P.)], with faint bitter taste, usually straight- and even-grained, fine-textured, very light (sp gr approx 0.29 green, 0.31 ovendry), soft. *Growth rings* distinct, delineated by a darker band of late wood, usually narrow. Late-wood zone occupying most of the ring; transition from early to late wood more or less gradual; late-wood zone somewhat denser than the early wood, narrow. *Parenchyma* not visible. *Rays* very fine (*x*), forming a fine, close, inconspicuous fleck on the quarter surface. *Resin canals* wanting.

MINUTE ANATOMY

Tracheids fine [up to 35 (avg 20–30) μ in diameter]; bordered pits in one row or very rarely paired on the radial walls; tangential pitting present in the last few rows of late-wood tracheids; pits leading to ray parenchyma taxodioid, small, orbicular or nearly so, quite uniform in size, with distinct border and lenticular orifice, 1–4 per cross field. *Longitudinal parenchyma* variable in distribution, (*a*) sparse (apotracheal-diffuse) and often apparently wanting, or (*b*) abundant and banded in a given growth ring and then often wanting in neighboring rings; end walls nodular. *Rays* uniseriate, 1–8 plus cells in height, consisting entirely of ray parenchyma or sometimes with an occasional ray tracheid; ray cells empty or with scanty infiltration; end walls with indentures.

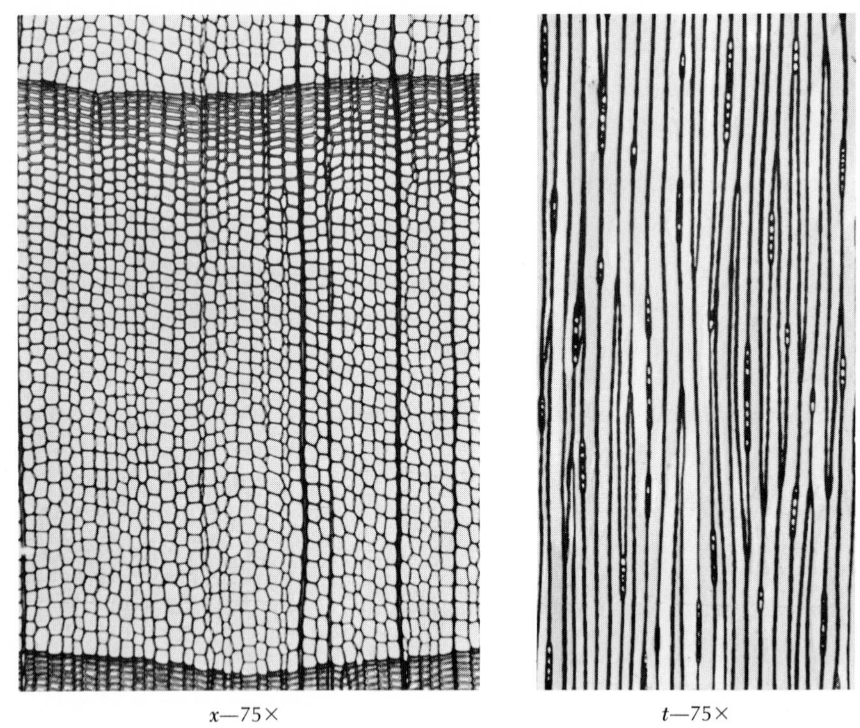

FIG. 12-67 Northern white-cedar, *Thuja occidentalis* L.

REMARKS

Comparable to western redcedar (*Thuja plicata* Donn), but lighter in color, finer-textured, and generally with less-prominent growth rings. Sometimes confused with Atlantic white-cedar [*Chamaecyparis thyoides* (L.) B.S.P.], and most readily separated by gross features: drier (less oily) and more brittle; heartwood straw-brown in contrast to the roseate-tinted heartwood of Atlantic white-cedar; with a characteristic odor that is different from that of the last-named species. Structurally, these two woods are very similar, but the parenchyma is much more abundant and more evenly distributed in Atlantic white-cedar (see description of parenchyma under each species).

USES

Principally for *poles* and *posts* (because of its durability when in contact with the ground); *cabins*; *ties* (so soft that it requires tie plates); *pails* and *tubs*; *rustic furniture*; *shingles*. Lumber used for *tanks*; *boatbuilding* (especially canoe ribs); *fish-net floats* and *imitation minnows* (because of the extreme lightness of the wood); *woodenware* and *novelties*.

WESTERN REDCEDAR, GIANT ARBORVITAE

Thuja plicata Donn

GENERAL CHARACTERISTICS

Sapwood nearly white, narrow; ***heartwood*** reddish or pinkish brown to dull brown; ***wood*** with a characteristic sweet, fragrant (cedary) odor and faint bitter taste, straight- and quite even-grained, medium- to somewhat coarse-textured, light (sp gr approx 0.31 green, 0.33 ovendry), moderately soft. ***Growth rings*** distinct and generally quite conspicuous, delineated by a darker band of late wood, narrow to fairly wide. Early-wood zone occupying most of the ring, soft under tools; transition from early to late wood more or less abrupt; late-wood zone narrow, hard. ***Parenchyma*** not visible or barely distinct with a hand lens as a narrow line in the late wood. ***Rays*** fine (x), forming a fine, close, inconspicuous fleck on the quarter surface. ***Resin canals*** wanting.

MINUTE ANATOMY

Tracheids medium coarse [up to 45 (avg 30–40) μ in diameter]; bordered pits in 1–2 rows on the radial walls; tangential pitting present in the last few rows of

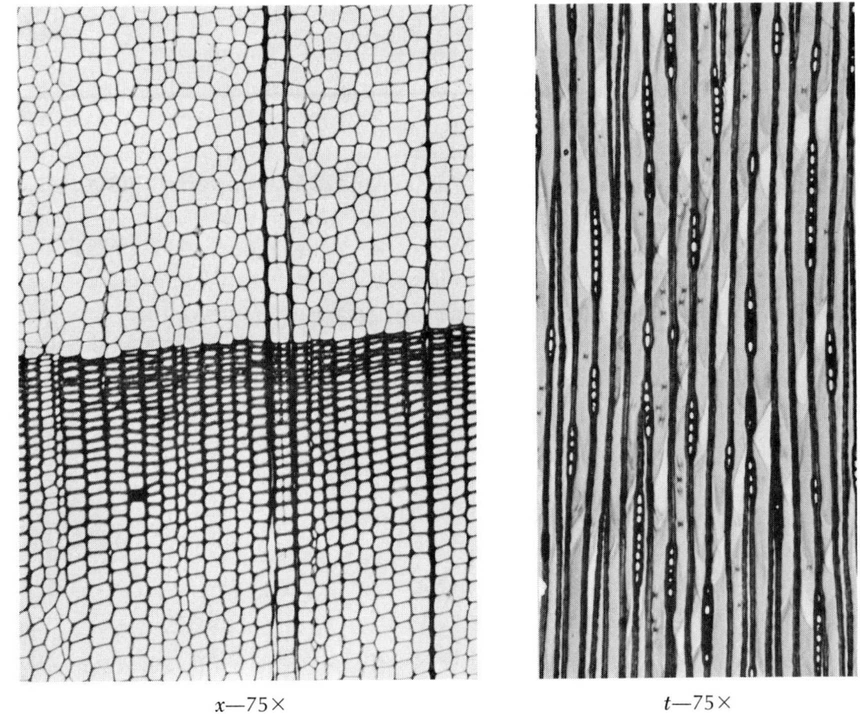

x—75× t—75×

FIG. 12-68 Western redcedar, *Thuja plicata* Donn

late-wood tracheids; pits leading to ray parenchyma taxodioid, small, orbicular or nearly so, quite uniform in size, with distinct border and lenticular orifice, 1–4 per cross field. Lines of banded *parenchyma* very variable in distribution, (*a*) present in every ring in some samples, (*b*) in other instances sporadic in distant rings, the wood then often apparently without parenchyma; end walls nodular. *Rays* uniseriate, 1–12 plus cells in height, consisting entirely of ray parenchyma or rarely with a more or less continuous or interrupted row of ray tracheids on the upper and low margins; ray cells with scanty gummy infiltration; end walls with indentures.

REMARKS

Similar in color to, and hence often confused with, incense-cedar (*Libocedrus decurrens* Torr.) and sometimes with redwood; softer, more brittle (drier), and less firm under tools than incense-cedar, without the relatively high infiltration content and acrid taste that are features of this wood; parenchyma generally much less abundant; rays strictly uniseriate (those of incense-cedar not infrequently in part biseriate). Can be easily separated from redwood, which lacks the cedar-like odor characteristic of western redcedar. For differences between *Thuja plicata* Donn and *T. occidentalis* L., see Remarks under the latter, p. 492.

USES

Shingles (more than 95 percent of wood shingles manufactured in the United States), for which purpose its durability, ease of working, and lightness make this the premier wood; *poles* (because of the good form of the tree, large sizes available, and its durability; shipped to all parts of the United States); *posts*; *piling* (has a tendency to crush in driving); *boxes* and *crates*. The *lumber* is used for all purposes where durability and ease of working are of first importance [especially for *caskets* and *coffins*, *siding* (most important western species for this purpose), *tank stock*, *porch columns*, *hothouse construction*, *boatbuilding* (planking of racing shells)], also for *interior finish*; *veneer*, used for faces and backs, especially in the manufacture of exterior siding, for which purpose it is more desirable than Douglas-fir.

PORT-ORFORD-CEDAR, PORT-ORFORD WHITE-CEDAR

Chamaecyparis lawsoniana (A. Murr.) Parl.

GENERAL CHARACTERISTICS

Sapwood nearly white to pale yellowish white, frequently not clearly distinguishable from the heartwood, narrow to fairly wide (1–3 in.); *heartwood* yellowish white to pale yellowish brown; *wood* with a characteristic pungent, ginger-like odor when freshly cut,* with bitter, somewhat spicy taste, straight- and even-grained, medium- to somewhat coarse-textured, moderately light (sp gr approx 0.40 green, 0.44 ovendry), moderately soft. ***Growth rings*** distinct but not conspicuous, delineated by a band of somewhat darker late wood, narrow to fairly wide (usually of medium width). Early-wood zone generally broad, occupying most of the ring; transition from early to late wood more or less gradual; late-wood zone only slightly denser than the early wood, narrow. ***Parenchyma*** abundant, not distinct or barely visible with a hand lens. ***Rays*** narrow (*x*), appearing fine with a hand lens, forming a close, low, inconspicuous fleck on the quarter surface. ***Resin canals*** wanting.

MINUTE ANATOMY

Tracheids up to 50 (avg 35–45) μ in diameter; bordered pits in one row or occasionally in part biseriate on the radial walls; tangential pitting present on the last few rows of late-wood tracheids, pits leading to ray parenchyma cupressoid, small, orbicular or nearly so, quite uniform in size, with distinct border and len-

* This odor is due to a volatile oil, which, if inhaled continuously for a long time, may cause kidney complications. Men working at the sawmills cutting Port-Orford-cedar use masks as a protection against the oil and the fine dust resulting from sawing this wood.

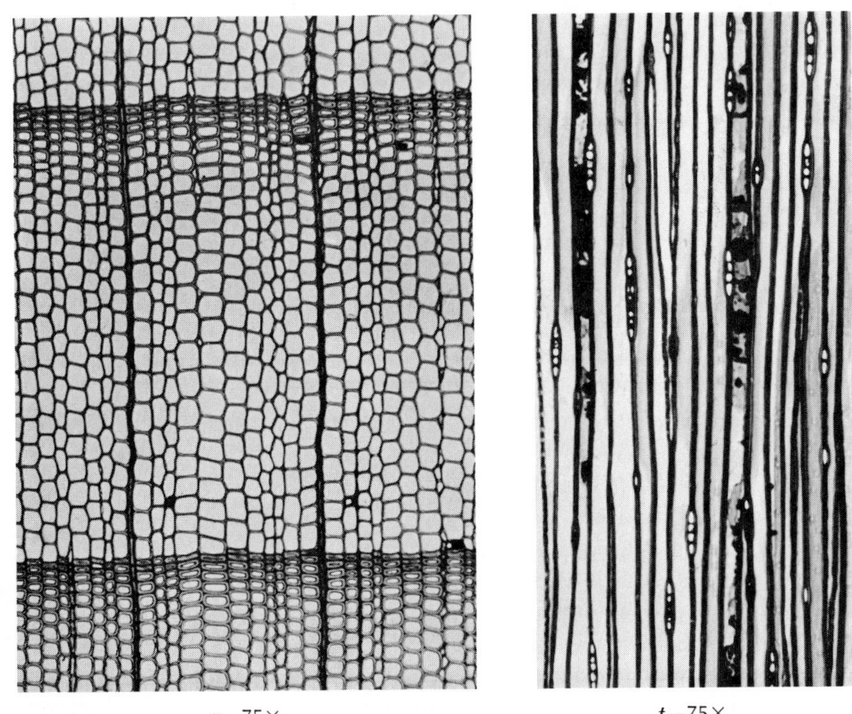

x—75× *t*—75×

FIG. 12-69 Port-Orford-cedar, *Chamaecyparis lawsoniana* (A. Murr.) Parl.

ticular orifice, 1–4 per cross field. ***Longitudinal parenchyma*** apotracheal-diffuse or occasionally exhibiting more or less a tendency toward zonation; cells usually conspicuous on account of their gummy contents, solitary or 2–several contiguous in the tangential plane; end walls nodular. ***Rays*** uniseriate or rarely with paired cells, low (1–6 plus cells in height), narrow (7–12 μ), consisting entirely of ray parenchyma or very rarely with ray tracheids; gummy infiltration fairly abundant in the ray cells in the heartwood.

REMARKS

Distinctive among North American coniferous woods because of its pungent, ginger-like odor. Comparable to Alaska-cedar [*Chamaecyparis nootkatensis* (D. Don) Spach] but without the pronounced yellow cast that is a feature of this wood, with a different odor, with broader rings, somewhat coarser-textured, and seldom with ray tracheids.

USES

Formerly the principal wood for ***storage-battery separators*** because of its electrical resistance and high resistance to the action of acids; ***millwork*** (sash, doors, interior finish); ***lining*** for ***mothproof boxes*** and ***closets*** (because of its reputed moth-repellent qualities; closets should be aired when new, as otherwise the volatile oil from the wood condenses on hardware, buttons, and glass); ***woodenware*** and ***novelties***; ***arrowshafts***; ***boat construction***, especially ***decking***; ***general construction***; ***tanks***; ***planking***; ***mine timbers*** (lower grades).

ALASKA-CEDAR, YELLOW CYPRESS

Chamaecyparis nootkatensis (D. Don) Spach

GENERAL CHARACTERISTICS

Sapwood nearly white to yellowish white, very narrow; ***heartwood*** bright clear yellow, darkening upon exposure; ***wood*** with a characteristic odor which in some samples resembles that of raw potatoes, with a faint bitter, somewhat spicy taste, straight- and even-grained, fine- to medium-textured, moderately heavy (sp gr approx 0.42 green, 0.46 ovendry), moderately hard. ***Growth rings*** not visible or barely visible to the naked eye, generally plainly visible with a hand lens, delineated by a darker band of late wood, narrow to very narrow (50–90 per in.), quite uniform in width. Early-wood zone occupying most of the ring; transition from early to late wood more or less abrupt; late-wood zone appreciably denser than the early wood, narrow, not conspicuously differentiated from the early wood through discrepancy in color. ***Parenchyma*** not distinct at low magnifications. ***Rays*** fine (x), appearing fine with a hand lens, forming a fine, low, inconspicuous fleck on the quarter surface. ***Resin canals*** wanting.

MINUTE ANATOMY

Tracheids up to 40 (avg 25–35) μ in diameter; bordered pits in one row or rarely paired on the radial walls; tangential pitting present on the last few rows of latewood tracheids; pits leading to ray parenchyma cupressoid, small, orbicular or nearly so, quite uniform in size, with distinct border and lenticular orifice, 1–3 per cross field (Fig. 12-41). ***Longitudinal parenchyma*** sparse, or not infrequently wanting altogether, to abundant; cells apotracheal-diffuse, usually distinguishable on account of their gummy contents, solitary or 2–several contiguous in the tangential plane; end walls nodular. ***Rays*** uniseriate or rarely with paired cells, 1–20 plus (the majority less than 12) cells in height; end walls nodular with indentures; low rays with ray tracheids, these frequently constituting the entire ray (Fig. 12-42); ray tracheids (*r*) short and high (along the grain); gummy infiltration fairly abundant in some of the ray cells.

REMARKS

Comparable to Port-Orford-cedar [***Chamaecyparis lawsoniana*** (A. Murr.) Parl.]; for distinguishing characters, see Remarks under this species, p. 496. Not to be confused with yellow cypress produced by ***Taxodium distichum*** (L.) Rich.

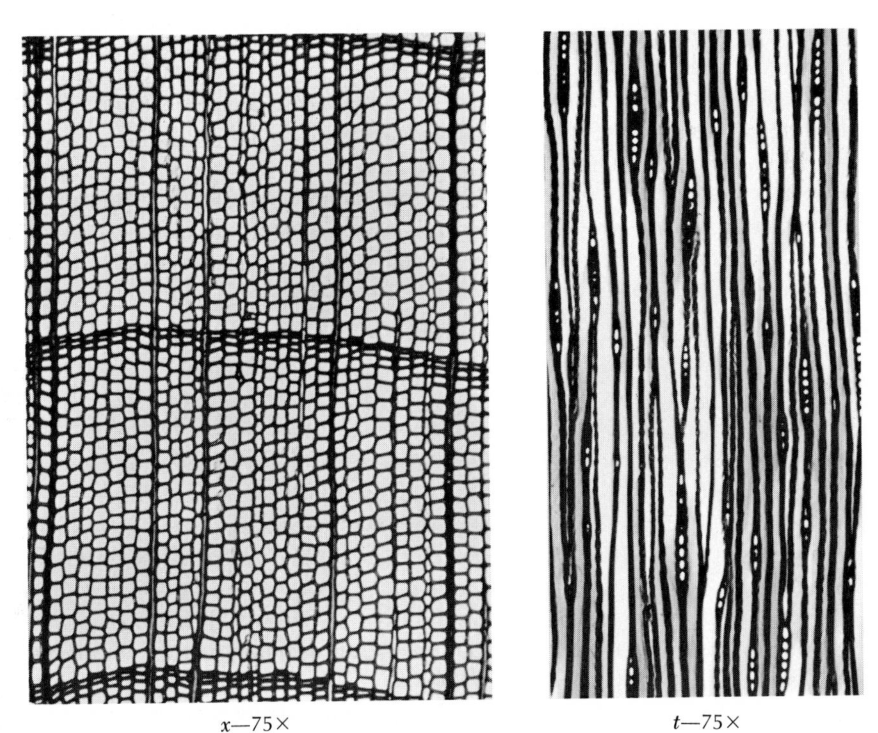

x—75× *t*—75×

FIG. 12-70 Alaska-cedar, ***Chamaecyparis nootkatensis*** (D. Don) Spach

USES

Locally for *poles, interior finish, furniture, cabinetwork, novelties, caskets,* and *hulls* of small boats; valuable for *patterns* (because of its good working qualities, low shrinkage, and ability to stay in place); *canoe paddles* (long preferred by the Alaska Indians); *oars; piling* and *marine buoys* (because of its resistance to marine borers); *greenhouse* and *conservatory* construction (because of its resistance to decay); *acid tanks* and *chemical containers* (because of its acid resistance); *sounding boards* of musical instruments (because of its uniform density, straightness and evenness of grain, low shrinkage, and ability to stay in place). A few logs are shipped to Japan, where the wood is used in temple buildings and other important structures, because of its purported resistance to termite attack.

ATLANTIC WHITE-CEDAR, SOUTHERN WHITE-CEDAR

Chamaecyparis thyoides (L.) B.S.P.

GENERAL CHARACTERISTICS

Sapwood whitish, narrow; *heartwood* light brown tinged with red or pink (roseate); *wood* somewhat oily, with a characteristic cedary odor [distinct from that of northern white-cedar (*Thuja occidentalis* L.)] and faint bitter taste, usually straight- and even-grained, fine-textured, light (sp gr 0.31 green, 0.35 ovendry), moderately soft. *Growth rings* distinct but not conspicuous, delineated by a darker band of late wood, usually fairly wide. Early-wood zone occupying most of the ring; transition from early to late wood more or less gradual; late-wood zone somewhat denser than the early wood, narrow. *Parenchyma* fairly abundant to abundant (present in every growth ring), banded and apotracheal-diffuse, in the former instance appearing as darker concentric lines which are visible with a hand lens and not infrequently to the naked eye. *Rays* very fine (x), forming a fine, close, inconspicuous fleck on the quarter surface. *Resin canals* wanting.

MINUTE ANATOMY

Tracheids fine [up to 40 (avg 25–35) μ in diameter], without or with few intercellular spaces at the corners; bordered pits in one row or very rarely paired on the radial walls; tangential pitting present in the last few rows of late-wood tracheids; pits leading to ray parenchyma cupressoid, small, orbicular or nearly so, quite uniform in size, with distinct border and lenticular orifice, 1–4 per cross field. Cells of *longitudinal parenchyma* solitary and often scattered, or 2–several contiguous in the tangential plane and banded, generally distinguishable in the transverse section by their contents. *Rays* uniseriate or rarely with paired cells, 1–12 plus cells in height, consisting entirely of ray parenchyma or with an occasional ray tracheid; ray cells empty or with scanty infiltration.

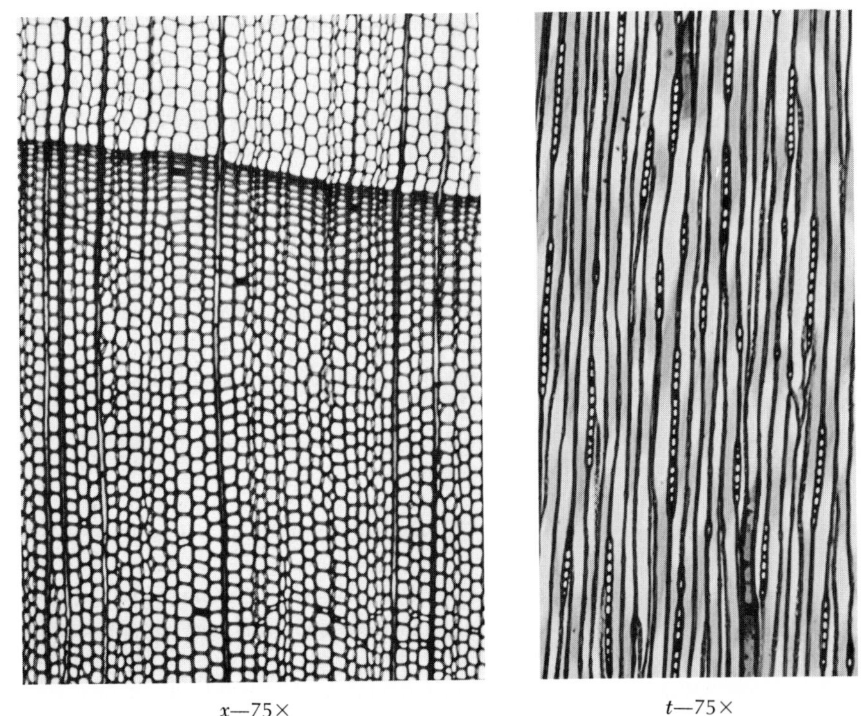

x—75×　　　*t*—75×

FIG. 12-71 Atlantic white-cedar, *Chamaecyparis thyoides* (L.) B.S.P.

REMARKS

Similar in structure to northern white-cedar (*T. occidentalis* L.); for distinguishing characters see Remarks under Northern white-cedar, p. 492.

USES

Principally for *poles, shingles, woodenware* (tubs and pails), and *lumber* (with many defects, particularly knots and wind shakes), *water tanks* (because of its lightness and durability); *boat construction.* Lumber used for *planing-mill* products (such as siding, porch lumber); *boxes* and *crates* (especially for vegetables and small fruits); *fencing.*

REDCEDAR

Eastern redcedar, Tennessee redcedar (*Juniperus virginiana* L.)

Southern redcedar [*Juniperus silicicola* (Small) Bailey]

GENERAL CHARACTERISTICS

Usually available only in small sizes and generally knotty; *sapwood* nearly white, thin; *heartwood* purplish or rose-red when first exposed, aging to dull red or reddish brown, sometimes with lighter streaks of included sapwood;* *wood* with mild characteristic "pencil-cedar" odor and taste, straight- (except near knots) and even-grained, fine-textured, moderately heavy (sp gr approx 0.44 green, 0.49 ovendry), hard. ***Growth rings*** distinct, delineated by a band of darker late wood, narrow to wide (a given ring often variable in width and the rings hence sinuate in contour), not infrequently eccentric; false rings occasioned by 2–3 bands of late wood sometimes present. Early-wood zone usually broad, occupying most of the ring; transition from early to late wood gradual to rather abrupt; late-wood zone conspicuous, much darker than the early wood, narrow. ***Parenchyma*** very abundant, banded, in 1–several (usually 1–2) dark bands which are visible with a hand lens and often to the naked eye (especially in the sapwood). ***Rays*** very fine (x), darker than the background and forming a fine, low fleck on the quarter surface. ***Resin canals*** wanting.

MINUTE ANATOMY

Tracheids fine [up to 35 (avg 20–30) μ in diameter], often with intercellular spaces at the corners (x); bordered pits in one row or very rarely paired on the radial walls; tangential pitting present in the last few rows of late-wood tracheids; pits leading to ray parenchyma cupressoid, small, orbicular or nearly so, quite uniform in size, with distinct border and lenticular orifice, 1–4 per cross field. Cells of ***longitudinal parenchyma*** solitary or 2–3 contiguous in the tangential plane, and banded, with dark gummy contents; end walls nodular. ***Rays*** uniseriate, low (1–6 plus cells and less than 250 μ in height), consisting entirely of ray parenchyma; ray cells in the heartwood with copious deposits of dark gum; end walls nodular with indentures.

REMARKS

Distinctive among North American coniferous woods because of the color of the heartwood. The two species producing this timber cannot be separated on the basis of wood structure.

USES

Fence posts (principal use from the standpoint of quantity). Lumber used for ***chests, wardrobes,*** and ***closet linings*** (because of its color, excellent working qualities, fragrance, and reputed moth-repellent properties); ***millwork*** (sash, doors, interior finish); ***pencil slats*** (formerly the most important use; today, because of the scarcity of suitable material, composing less than 10 percent of the output); ***woodenware*** and ***novelties.***

* McGinnes et al. have found that included sapwood zones in redcedar were more numerous at the base of trees, decreasing in frequency with the tree height. Longer patches were usually associated with some injury, e.g., frost damage and bird pecks, and were characterized by abnormal tissue formation; smaller patches of included sapwood appeared to consist of normal wood tissue. E. A. McGinnes, Jr., S. A. Kandel, and P. S. Szopa, Frequency and Selected Anatomical Features of Included Sapwood in Eastern Redcedar, *Wood Science*, **2**(2):100–106(1969).

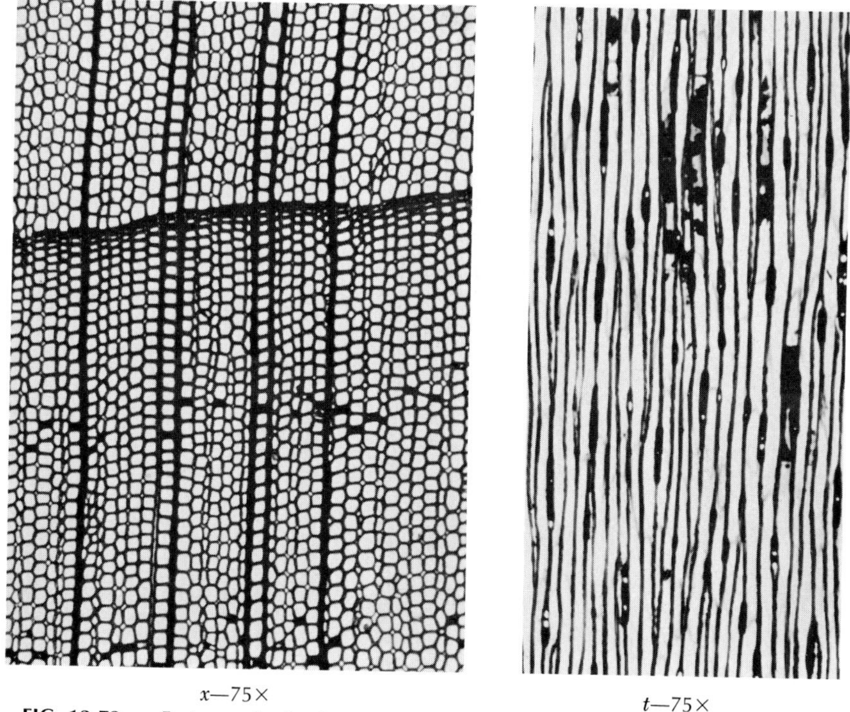

x—75× *t*—75×

FIG. 12-72 Eastern redcedar, *Juniperus virginiana* L.

PACIFIC YEW

Taxus brevifolia Nutt.

GENERAL CHARACTERISTICS

Sapwood light yellow, thin; *heartwood* bright orange to rose-red; *wood* without characteristic odor or taste, even-grained, very fine-textured, heavy (sp gr approx 0.60 green, 0.67 ovendry), very hard. *Growth rings* distinct, delineated by a band of darker late wood, narrow to medium wide or broad. Early-wood zone usually occupying about one-half of the ring; transition from early to late wood very gradual; late wood very dense. *Parenchyma* not visible. *Rays* very fine (*x*), not distinct to the naked eye, forming a fine, close, inconspicuous fleck on the quarter surface. Normal and wound (traumatic) *resin canals* wanting.

MINUTE ANATOMY

Tracheids fine [up to 25 (avg 15–20) μ in diameter], more or less rounded, featured by fine, close bands of spiral thickening; bordered pits in one row on the radial walls; tangential pitting present in the last few rows of late-wood tracheids; pits leading to ray parenchyma cupressoid, small, orbicular or nearly so, quite uniform in size, with distinct border and lenticular orifice, 1–4 (generally 1–2) per

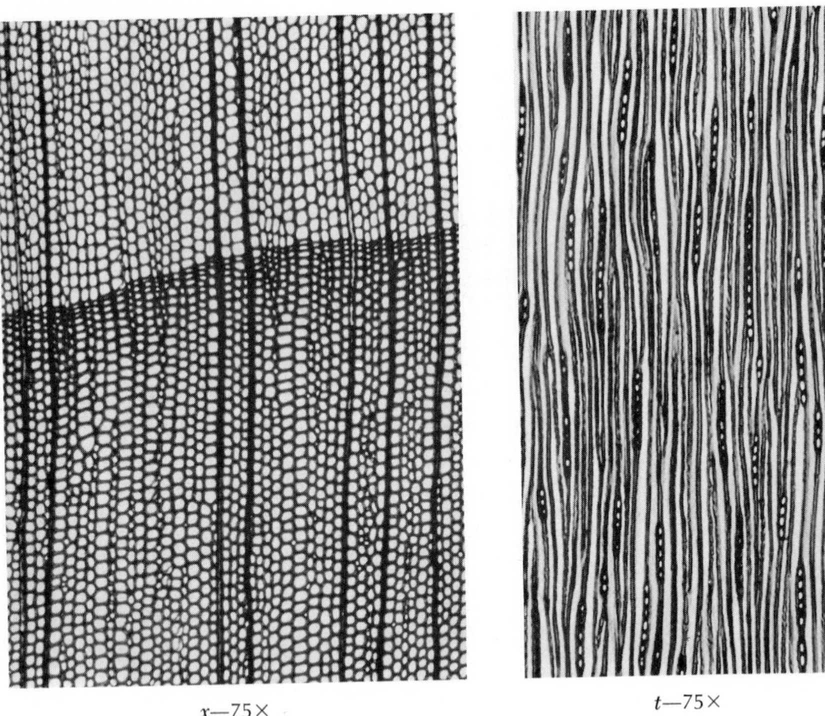

FIG. 12-73 Pacific yew, *Taxus brevifolia* Nutt.

cross field. *Longitudinal parenchyma* wanting. *Rays* uniseriate or very rarely with paired cells, 1–25 cells in height, consisting entirely of ray parenchyma.

REMARKS

Of limited commercial importance because of its scarcity and small size.

USES

Poles, bows, canoe paddles, small turned articles, and *carvings;* locally for *paneling* and *furniture.*

CALIFORNIA TORREYA AND FLORIDA TORREYA

California torreya (*Torreya californica* Torr.)

Florida torreya (*Torreya taxifolia* Arn.)

GENERAL CHARACTERISTICS

Sapwood whitish to yellowish white, frequently difficult to distinguish from the heartwood; *heartwood* clear yellow; *wood* with a characteristic odor (strong and unpleasant in *T. taxifolia*), with a mild, characteristic, bitterish taste, even-grained, medium- to somewhat coarse-textured, moderately light to moderately heavy (avg sp gr approx 0.48 green), soft to moderately hard. ***Growth rings*** fairly distinct but not prominent, delineated by a band of darker late wood. Early wood occupying most of the ring; transition from early to late wood very gradual; late wood dense. ***Parenchyma*** not visible. ***Rays*** very fine (*x*), not distinct to the naked eye, forming a fine, close, inconspicuous fleck on the quarter surface. Normal and wound (traumatic) ***resin canals*** wanting.

MINUTE ANATOMY

Tracheids fine [up to 55 (avg 25–35) μ in diameter], featured by fine, widely spaced bands of spiral thickening; bordered pits in 1–2 rows on the radial walls; tangential pitting sparse, usually confined to the late-wood tracheids; pits leading to ray parenchyma cupressoid, small, orbicular or nearly so, 2–7 (generally 2–4) per cross field. ***Longitudinal parenchyma*** wanting. ***Rays*** uniseriate, 1–12 plus cells in height, consisting entirely of ray parenchyma.

USES

Of no commercial importance; used locally for *fence posts, cabinets, models, novelties*, and *patterns*.

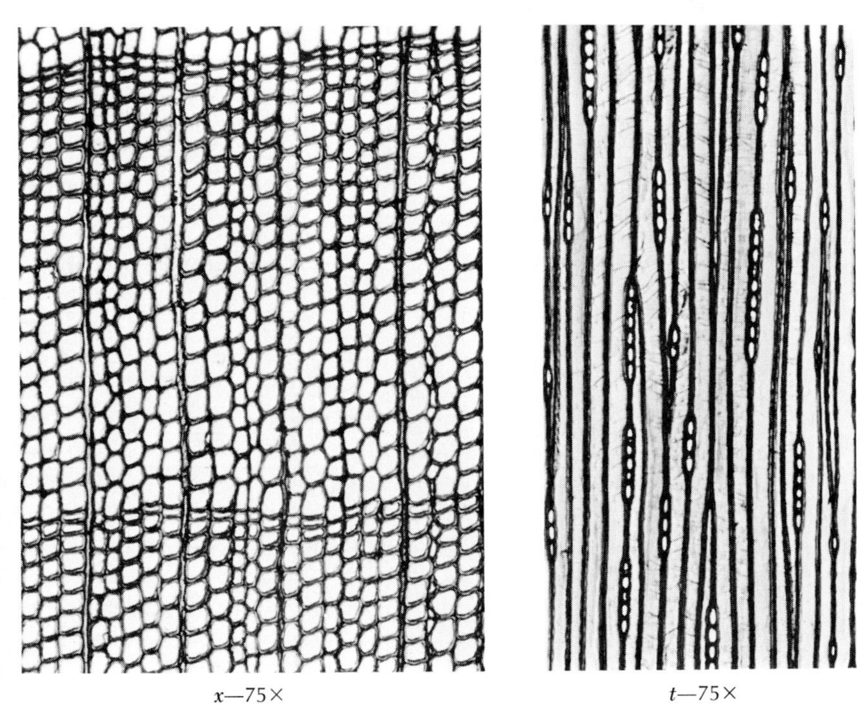

x—75× *t*—75×

FIG. 12-74 California torreya, *Torreya californica* Torr.

TABLE 12-1 Comparative Physical Data for Coniferous Woods Native to the United States[a]

Common name of wood[b]	Sp gr, avg		Weight, lb/cu ft		Shrinkage based on green dimension, %[c]				Index classes to mechanical properties[d]											
	Green	12%	Green	12%	Vol.	T	R	T/R	Bending strength	Compression as a post	Hardness	Stiffness	Shock resistance	Splitting	Screw holding	Ability to stay in place	Drying defects	Machining qualities	Gluing qualities	Durability rating
(1)	(2)	(3)	(4)	(5)	(6)	(7)	(8)	(9)	(10)	(11)	(12)	(13)	(14)	(15)	(16)	(17)[e]	(18)[f]	(19)[g]	(20)[h]	(21)[i]
Alaska-cedar	0.42	0.44	36	31	9.2	6.0	2.8	2.1	4	3	4	3	4	4	3	1	1	*G*	3	*R*
Baldcypress	0.42	0.46	51	32	10.5	6.2	3.8	1.6	4	3	5	3	4	4	3	2	1	*A*	3	*R,M*
Douglas-fir (coast type)	0.45	0.48	38	34	11.8	7.8	5.0	1.6	3	2	4	1	4	4	3	3	2	*A*	2	*M*
Douglas-fir (inland)	0.45	0.48	38	36	10.9	7.6	4.1	1.9	3	2	4	1	4	4	3	3	2	*A*	2	*M*
Fir, balsam	0.34	0.36	45	25	11.2	6.9	2.9	2.4	5	4	5	4	5	5	4	3	2	*P*	1	*S*
Fir, white	0.35	0.37	47	26	9.8	7.1	3.2	2.2	5	4	5	3	5	4	4	3	2	*A*	1	*S*
Hemlock, eastern	0.38	0.40	50	28	9.7	6.8	3.0	2.3	4	3	4	3	5	4	4	3	2	*P*	2	*S*
Hemlock, western	0.42	0.44	41	31	11.9	7.9	4.3	1.8	4	3	4	3	4	4	4	3	2	*A*	2	*S*
Incense-cedar	0.35	0.37	45	26	7.6	5.2	3.3	1.6	4	3	5	4	5	4	5	1	1	*G*	1	*R*
Larch, western	0.48	0.51	50	36	13.2	8.1	4.2	1.9	2	2	4	1	4	4	2	3	2	*A*	1	*M*
Pine, eastern white	0.34	0.35	36	25	8.2	6.0	2.3	2.6	5	4	5	4	5	5	5	1	2, 4	*G*	2	*M*
Pine, jack	0.40	0.43	40	30	10.4	6.5	3.4	1.9	4	4	4	3	4	4	3	3	2	*A*	2	*S+*
Pine, loblolly	0.47	0.51	53	36	12.3	7.4	4.8	1.5	3	3	4	2	4	4	3	3	2	*A*	2	*S+*
Pine, lodgepole	0.38	0.41	39	29	11.5	6.7	4.5	1.5	5	4	5	3	5	4	4	2	2	*A*	2	*S+*
Pine, longleaf	0.54	0.58	55	41	12.2	7.5	5.1	1.5	2	2	3	1	3	4	2	3	2	*A*	2	*M*
Pine, ponderosa	0.38	0.40	45	28	9.6	6.3	3.9	1.6	5	4	5	4	5	4	4	2	2	*A*	2	*S+*
Pine, red	0.41	0.44	49	30	11.5	7.2	4.6	1.6	5	4	5	3	4	4	3	3	2	*A*	2	*S+*
Pine, shortleaf	0.46	0.51	52	36	12.3	7.7	4.4	1.8	3	3	4	2	4	4	3	3	2	*A*	2	*S+*
Pine, slash	0.56	0.61	58	43	12.2	7.8	5.5	1.4	2	2	3	1	3	4	2	3	2	*A*	2	*M*
Pine, sugar	0.35	0.36	52	25	7.9	5.6	2.9	1.9	5	4	5	4	5	4	5	1	2	*G*	2	*S+*
Pine, western white	0.36	0.38	35	27	11.8	7.4	4.1	1.8	5	4	5	3	5	4	4	2	2	*G*	2	*S+*
Port-Orford-cedar	0.40	0.42	36	29	10.1	6.9	4.6	1.5	4	3	4	2	5	5	4	2	1	*G*	3	*R*
Redcedar, eastern	0.44	0.47	37	33	7.8	4.7	3.1	1.5	3	3	3	5	3	4	3	1	1	*G*	2	*R*
Redcedar, western	0.31	0.33	27	23	6.8	5.0	2.4	2.1	5	4	5	4	5	5	5	1	1, 5	*A*	1	*R*
Redwood	0.38	0.40	50	28	6.8	4.4	2.6	1.7	3	2	4	3	5	4	4	2	1, 5	*A*	1	*R*

(1)	(2)	(3)	(4)	(5)	(6)	(7)	(8)	(9)	(10)	(11)	(12)	(13)	(14)	(15)	(16)	(17)	(18)	(19)	(20)	(21)
Spruce, eastern and Engelmann	0.36	0.38	36	27	12.9	7.4	4.0	1.8	5	4	5	3	5	5	4	2	2	*A*	1	*S*
Spruce, Sitka	0.37	0.40	33	28	11.5	7.5	4.3	1.7	5	4	5	3	5	4	4	2	2	*A*	1	*S*
Tamarack	0.49	0.53	47	37	13.6	7.4	3.7	2.0	3	3	5	3	4	4	2	3	2	*P*	2	*M*
White-cedar, Atlantic	0.31	0.32	26	23	8.8	5.4	2.9	1.9	5	4	5	5	5	5	5	2	1	*A*	2	*R*
White-cedar, northern	0.29	0.31	28	22	7.2	4.9	2.2	2.2	5	5	5	5	5	5	5	1	1	*A*	2	*R*
Yew, Pacific	0.60	0.62	54	44	9.7	5.4	4.0	1.4	2	1	2	4	3	3	2	2	2	*G*	2	*R*

[a] Based on information from "Wood Handbook," U.S. Department of Agriculture Handbook 72, 1955, and ***U.S. Dept. Agr. Bull.*** 479, 1935, except for columns 9 and 18.

[b] Common names according to "Check List of Native and Naturalized Trees of the United States," U.S. Department of Agriculture Handbook 41, 1953. See text for scientific names.

[c] Shrinkage values are for change from green to ovendry.

[d] Coding of mechanical properties according to index classes:

Column No.	*Mechanical property*	*Index classes*				
		5	4	3	2	1
10	Modulus of rupture in bending, green wood, psi	<6000	6000 to	7000 to	8000 to	11,000+
11	Maximum crushing strength parallel, green wood, psi	<2200	2200 to	3000 to	3600 to	4500+
12	Side hardness, green wood, lb	< 400	400 to	600 to	900 to	1200+
13	Modulus of elasticity in bending, psi × 1000	< 800	800 to	1000 to	1300 to	1500+
14	Height of drop with 50-lb hammer, in.	15 to	25 to	35 to	50 to	70+
15	Load, lb/in. of width to split green wood	< 150	150 to	250 to	350 to	450+
16	Nail and screw withdrawal as function of sp gr OD	< 0.40	0.40 to	0.45 to	0.55 to	0.70+

[e] Dimensional changes with moisture in normal use ranges from class 1 with little change, to class 5 with greatest dimensional change.

[f] Drying defects classed as follows: 1. Slight warping and little dimensional change. 2. Distinct warping. 3. Pronounced warping. 4. Tendency to check and split. 5. Subject to collapse and honeycombing.

[g] Machining qualities: *G* = good to excellent, smooth surface. *A* = average. *P* = poor, requiring care to be acceptable.

[h] Gluing qualities: 1. Glues very easily. 2. Glues well. 3. Requires control. 4. Difficult to glue, requires close control.

[i] Heartwood durability ratings: *R* = resistant to decay. *M* = moderate resistance to decay. *S* = low decay resistance. + indicates high rating within the group.

Hardwoods—Keys and Descriptions by Species

IV. KEY TO HARDWOODS—GROSS FEATURES*

Woods porous (with vessels); cross section consisting of pores (vessels) embedded in a mass of fibers and parenchyma tissue; rays distinct or indistinct to the naked eye (Figs. 12-75 to 12-140, inc., pp 514–525).

1. Early-wood pores **conspicuously larger** than the late-wood pores, distinct with the naked eye† (ex., Figs. 12-101, 12-108, 12-130); (*a*) transition from large early-wood pores to small late-wood pores abrupt (**wood ring-porous**, ex., Figs. 12-101, 12-130); (*b*) transition from large early-wood pores to small late-wood pores somewhat gradual (**wood semi-ring-porous**, ex., Fig. 12-108) **2**

1. Early-wood pores **not conspicuously larger** than the late-wood pores and indistinct to the naked eye (ex., Fig. 12-87); early-wood zones not sharply defined (**wood diffuse-porous**) **13**

2. Transition in size of pores from early to late wood abrupt, wood ring-porous **3**

2. Transition in size of pores from early to late wood somewhat gradual, wood semi-ring-porous **12**

3. Broad rays present, conspicuous (*x*; Figs. 12-127 to 12-130, inc.) often 1 in. or more in height along the grain (*t*; Fig. 12-126), forming a broad ray fleck on the radial surface.

A. Late-wood pores distinct with a hand lens, not numerous, thick-walled, the orifices plainly visible, rounded; tyloses usually absent or sparse in the early-wood pores; heartwood usually pinkish or pale brown.

Red oak—*Quercus* spp. Figs. 12-128, 12-130. Desc. pp. 566–568

B. Late-wood pores indistinct with a hand lens, numerous, thin-walled, the orifices scarcely visible, angular; tyloses generally present in the early-wood pores (heartwood); heartwood rich light to dark brown, usually without flesh-colored cast.

* This key is designed for the separation of the more important temperate North American hardwoods based on features visible with a 10X hand lens or without magnification.

A series of photographs (Figs. 12-75 to 12-140, inc., pp. 514 to 525) at low magnification (5X) accompanies this key to illustrate the normal appearance of the various kinds of woods under a hand lens.

† In osage-orange and black locust (Figs. 12-112, 12-133) early-wood pores are large but poorly defined because of occlusion with tyloses.

White oak—*Quercus* spp. Figs. 12-127, 12-129. Desc. pp. 569–572.

3. Broad rays absent (ex., Fig. 12-77) **4**

4. Late wood figured with wavy concentric (tangential) bands of pores which are mostly continuous and separated by bands of mechanical tissue (*x*; ex., Fig. 12-137) **5**

4. Late wood not figured with wavy, concentric (tangential) bands of pores (ex., Fig. 12-103) **7**

5. Early-wood pores in a single line.

A. Early-wood pores plainly visible without a hand lens, approximately equal in size and quite evenly spaced in a more or less continuous row; tyloses sparse.

American elm—*Ulmus americana* L. Fig. 12-137. Desc. p. 572

B. Early-wood pores scarcely visible without a hand lens, the larger spaced at intervals in an interrupted row and separated by smaller pores; tyloses abundant.

Hard elm—*Ulmus* spp. Fig. 12-139. Desc. p. 575

5. Early-wood pores in several rows (ex., Fig. 12-138) **6**

6. Early-wood pores in the heartwood completely occluded with tyloses, their contours poorly defined.

A. Heartwood golden yellow to bright orange, darkening upon exposure, often with reddish streaks along the grain; coloring matter readily soluble in water.

Osage-orange—*Maclura pomifera* (Raf.) Schneid. Fig. 12-112. Desc. p. 580

B. Heartwood greenish yellow to dark yellowish or golden brown; coloring matter not readily soluble in water.

Black locust—*Robinia pseudoacacia* L. Fig. 12-133. Desc. p. 598

6. Early-wood pores in the heartwood not completely occluded with tyloses, their contours distinct.

A. Wood light brown to dark reddish brown; rays usually indistinct without a lens.

Slippery elm—*Ulmus rubra* Mühl. Fig. 12-138. Desc. p. 574

B. Sapwood pale yellow to grayish or greenish yellow; imperfectly developed heartwood yellowish gray or light brown streaked with yellow; rays distinctly visible to the naked eye (*x*).

Hackberry—*Celtis* spp. Fig. 12-95. Desc. p. 577

C. Sapwood yellowish, narrow; heartwood orange-yellow to golden brown, turning dull dark brown on exposure; ray fleck conspicuous on the radial surface.

Red mulberry—*Morus rubra* L. Fig. 12-115. Desc. p. 578

7. Late-wood parenchyma appearing under the lens as fine, numerous, continuous, or broken, light-colored, tangential lines (ex., Fig. 12-91), or closely and evenly punctate (ex., Fig. 12-98).

A. Sapwood creamy white when freshly cut, darkening on exposure to light yellow-brown (sometimes with grayish-brown stain); heartwood rarely present, then only in small bands, blackish brown to black; wood with storied rays, forming ripple marks on the tangential surface; parenchyma (*x*) closely and evenly punctate, relatively inconspicuous.

Common persimmon—*Diospyros virginiana* L. Fig. 12-98. Desc. p. 620

B. Sapwood whitish to pale brown; heartwood pale brown to brown or reddish brown (that of pecan rich brown, sometimes with streaks of a darker hue); wood without storied rays; parenchyma conspicuous, in concentric tangential lines in the late wood.

Hickory—*Carya* spp. Figs. 12-89 to 12-91, inc. Desc. p. 541

7. Late-wood parenchyma not evident, or if evident associated with pores (paratracheal) and occasionally connecting the pores in the outer late-wood zone.. **8**

8. Early-wood pores in the heartwood completely occluded with tyloses, their contours poorly defined.

A. Heartwood golden yellow to bright orange, darkening upon exposure, often with reddish streaks along the grain; coloring matter readily soluble in water.

Osage-orange—*Maclura pomifera* (Raf.) Schneid. Fig. 12-112. Desc. p. 580

B. Heartwood greenish yellow to dark yellowish or golden brown; coloring matter not readily soluble in water.

Black locust—*Robinia pseudoacacia* L. Fig. 12-133. Desc. p. 598

8. Early-wood pores in the heartwood open or partly occluded with tyloses, their contours distinct.. **9**

9. Late-wood pores small, thin-walled, grouped in patches of porous tissue, which are obliquely radial (flame-shaped); rays very fine, barely visible with a hand lens.

A. Early-wood pores in one somewhat interrupted row.

Golden chinkapin—*Castanopsis chrysophylla* (Dougl.) A.DC. Fig. 12-93. Desc. p. 561

B. Early-wood pores very large, in several rows.

American chestnut—*Castanea dentata* (Marsh.) Borkh. Fig. 12-92. Desc. p. 559

9. Late-wood pores solitary or in radial rows of 2–3, in nestlike groups or in interrupted tangential bands; rays plainly visible with the naked eye or a hand lens (*x*)... **10**

10. Pores in the late wood solitary or in radial rows of 2–3, frequently united laterally by parenchyma toward the outer margin of the ring.

A. Wood lustrous, strong; sapwood nearly white, wide; heartwood light brown to pale yellow streaked with brown.

White ash—*Fraxinus* spp. Figs. 12-101, 12-102. Desc. p. 621

B. Wood dull, weak; sapwood whitish to light brown, narrow; heartwood grayish brown to brown.
Black ash—*Fraxinus nigra* Marsh. Fig. 12-103. Desc. p. 624

C. Sapwood light yellow; heartwood dull grayish brown to orange-brown or dark brown; freshly cut surface with aromatic odor.
Sassafras—*Sassafras albidum* (Nutt.) Nees. Fig. 12-135. Desc. p. 587

10. Pores in the late wood in nestlike groups or in interrupted tangential bands **11**

11. Heartwood light red to reddish brown.

A. Pores in the outer late wood solitary, in short radial rows or in small groups, embedded in short tangential bands of parenchyma; rays conspicuous to the naked eye (*x*); reddish gum deposits in the pores are frequent in the heartwood.
Honeylocust—*Gleditsia triacanthos* L. Fig. 12-104. Desc. p. 595

B. Pores in the outer late wood in nestlike groups which occasionally coalesce laterally; short bands of parenchyma not evident; rays not conspicuous to the naked eye; gum deposits in the pores infrequent in the heartwood.
Kentucky coffeetree—*Gymnocladus dioicus* (L.) K. Koch. Fig. 12-105. Desc. p. 594

11. Heartwood grayish brown or orange-yellow to golden brown.

A. Heartwood grayish brown to brown, with a faint odor resembling that of kerosene; rays relatively inconspicuous to the naked eye; the ray fleck on the radial surface not pronounced.
Catalpa—*Catalpa* spp. Fig. 12-94. Desc. p. 625

B. Heartwood orange-yellow to golden brown, turning dull dark brown on exposure, without characteristic odor; rays conspicuous to the naked eye (*x*), forming a pronounced fleck on the radial surface.
Red mulberry—*Morus rubra* L. Fig. 12-115. Desc. p. 578

12. Heartwood rich chocolate, purplish, or gray-brown, or brown; late-wood parenchyma frequently appearing under a hand lens as fine, broken or continuous lines, or closely and evenly punctate, or in bands toward the outer margin of the growth rings (*x*); rays distinct with a hand lens.

A. Heartwood rich chocolate or purplish brown, dull, with characteristic odor; wood relatively hard, heavy; late-wood parenchyma frequently appearing under a hand lens as fine, broken or continuous lines, or closely and evenly punctate (*x*).
Black Walnut—*Juglans nigra* L. Fig. 12-108. Desc. p. 539

B. Heartwood light chestnut-brown, lustrous, without characteristic odor; wood soft (readily dented with the thumbnail), light; late-wood parenchyma frequently appearing as fine, numerous broken or continuous lines, or closely and evenly punctate (*x*).
Butternut—*Juglans cinerea* L. Fig. 12-107. Desc. p. 538

C. Heartwood grayish brown, occasionally with a lavender tinge, with a faint odor resembling that of kerosene; wood soft (readily dented with the thumbnail), light; parenchyma in bands toward the outer margin of the growth ring, frequently associated with groups of small late-wood pores (*x*).

Catalpa—*Catalpa* spp. Fig. 12-94. Desc. p. 625

12. Heartwood grayish white to light grayish or pale reddish brown; wood soft; parenchyma not evident; rays barely distinct even with a hand lens.

Cottonwood—*Populus* spp. Fig. 12-122. Desc. p. 544
Willow—*Salix* spp. Fig. 12-134. Desc. p. 548

13. Rays wholly or in part broad (*x*), the broadest fully twice as wide as the largest pores (ex., Figs. 12-99, 12-120).. **14**

13. Rays narrow (*x*), the broadest less than twice the width of the largest pores (ex., Figs. 12-78, 12-87, 12-110).. **17**

14. Rays nearly uniform in width, close, appearing to the naked eye on the tangential surface as closely packed, broken lines (Fig. 12-121); grain generally interlocked.

American sycamore—*Platanus occidentalis* L. Fig. 12-120. Desc. p. 590

14. Rays obviously of two widths, broad and very narrow;
(*a*) broad rays separated by several of the narrow type, conspicuous (*x*; ex., Fig. 12-99), or
(*b*) broad rays relatively inconspicuous because of color, sporadic and widely spaced (ex., Fig. 12-81).................................. **15**

15. Broad rays sporadic, rather inconspicuous because of color (*x*), up to $\frac{4}{5}$+ in. in height (*t*) along the grain (Fig. 12-82); pores numerous, quite evenly spaced except in the outer late wood where they may be crowded; wood light and soft.

Red alder—*Alnus rubra* Bong. Fig. 12-81. Desc. p. 556

15. Broad rays common, conspicuous; wood hard and heavy................. **16**

16. Pores irregularly spaced, arranged in radial fan-shaped or stream-like groups extending across the rings; broad rays frequently extending 1 in. or more along the grain.

A. Sapwood reddish brown; heartwood brown tinged with red; banded parenchyma plainly visible to the naked eye, in ragged tangential lines (*x*).

Tanoak—*Lithocarpus densiflorus* (Hook. & Arn.) Rehd. Fig. 12-111. Desc. p. 562

B. Sapwood whitish to grayish brown; heartwood dull brown to gray-brown; banded parenchyma barely visible with a hand lens, appearing closely and evenly punctate (*x*).

Live oak—*Quercus virginiana* Mill. Fig. 12-131. Desc. p. 564

16. Pores quite evenly spaced except in the outer late wood where they may be crowded, but not in radial or fan-shaped groups; rays plainly visible to the naked eye on the tangential surface as short, staggered lines (Fig. 12-100).

American beech—*Fagus grandifolia* Ehrh. Fig. 12-99. Desc. p. 558

17. Wood ivory-white, creamy to grayish white, grayish brown, or shades of yellow or green **18**

17. Wood pinkish, light to dark brown or reddish brown **23**

18. Rays distinct to the naked eye (*x*) **19**

18. Rays not distinct to the naked eye (*x*) **21**

19. Wood shades of yellow or green or grayish green to greenish brown, or grayish brown.

A. Heartwood variable in color, ranging from clear yellow to dark yellowish green, green, or greenish brown to greenish black (in magnolia); growth rings delineated by a whitish line of marginal parenchyma; rays normally spaced, distinct with a naked eye.

Yellow-poplar—*Liriodendron tulipifera* L. Fig. 12-110. Desc. p. 584

Magnolia—*Magnolia* spp. Figs. 12-113, 12-114. Desc. p. 581

B. Heartwood greenish or brownish gray; growth rings not delineated by marginal parenchyma (indistinct); rays very close together, seemingly occupying half the area on the cross section (*x*, impression gained by casual observation).

Black tupelo, blackgum—*Nyssa* spp. Figs. 12-116, 12-117. Desc. p. 610

C. Heartwood grayish brown, frequently with darker streaks of pigment figure, aromatic on the freshly cut surface; pores encircled by a whitish sheath of parenchyma.

California laurel—*Umbellularia californica* (Hook. & Arn.) Nutt. Fig. 12-140. Desc. p. 585

19. Wood ivory-white, creamy, or creamy brown **20**

20. Wood ivory-white, frequently with bluish cast; hard; pores small, grouped in radial strings (*x*); rays distinct with a naked eye.

American holly—*Ilex opaca* Ait. Fig. 12-106. Desc. p. 600

20. Wood creamy white to pale creamy brown; often appearing lacelike under a lens (*x*); rays quite uniformly spaced.

Basswood—*Tilia* spp. Fig. 12-136. Desc. p. 610

21. Rays plainly visible with a hand lens (*x*) **22**

21. Rays barely visible or indistinct with a hand lens (*x*).

A. Sapwood whitish to grayish white, gradually merging into creamy white to pale yellowish-white heartwood, frequently with grayish streaks caused by oxidative sap stain; ripple marks sometimes evident on the tangential surface; pores mostly constant in size, minute, quite evenly distributed throughout the growth rings.

Buckeye—*Aesculus* spp. Fig. 12-80. Desc. p. 607

B. Heartwood uniformly grayish or pale reddish brown; pores gradually decreasing in size from early to late wood, the largest in the early wood barely visible to the naked eye.
Cottonwood—*Populus* spp. Fig. 12-122. Desc. p. 544
Willow—*Salix* spp. Fig. 12-134. Desc. p. 548

C. Heartwood pale creamy white to light grayish brown; pores small, not visible with the naked eye, decreasing gradually in size from early to late wood.
Aspen—*Populus* spp. Figs. 12-123, 12-124. Desc. p. 546

22. Wood creamy white to pale creamy brown, soft and light, often appearing lacelike under the lens (*x*).
Basswood—*Tilia* spp. Fig. 12-136. Desc. p. 610

22. Heartwood greenish or brownish gray or grayish brown, moderately heavy to heavy, moderately hard, not appearing lacelike under a hand lens (*x*).

A. Wood aromatic on the freshly cut surface; heartwood grayish brown, frequently with darker streaks of pigment figure; pores encircled by a whitish sheath of parenchyma; rays normally spaced.
California-laurel—*Umbellularia californica* (Hook. & Arn.) Nutt. Fig. 12-140. Desc. p. 585

B. Wood not aromatic; heartwood greenish or brownish gray; pores not encircled by a whitish sheath of parenchyma; rays very close together, seemingly occupying half the area on the cross section (*x*) (impression gained by casual observation).
Black tupelo, blackgum—*Nyssa* spp. Figs. 12-116, 12-117. Desc. p. 610

23. Rays distinct to the naked eye (*x*)........................ **24**
23. Rays not distinct to the naked eye (*x*)........................ **28**

24. Rays variable in width, the broadest separated by several narrow rays which are scarcely visible with the naked eye.

A. Sapwood whitish with a reddish tinge, narrow; heartwood uniformly light brown to reddish brown; outer margin of the growth ring usually sharply marked by a narrow darker line of denser fibrous tissue; wider rings usually quite sharply delineated against the background of pores and fibrous tissue.
Hard maple—*Acer* spp. Fig. 12-78. Desc. p. 602

B. Sapwood carneous to light pinkish brown, wide; heartwood, when present, dark brown, frequently variegated; outer margin of the growth ring and wider rays are not sharply delineated against the background of pores and fibrous tissue.
Dogwood—*Cornus* spp. Fig. 12-97. Desc. p. 615

24. Rays nearly uniform in width (ex., Fig. 12-125)........................ **25**

25. Rays very close, seemingly occupying half the area of the transverse surface (*x*, impression gained by casual examination); sapwood pinkish; heartwood carneous gray to varying shades of reddish brown, frequently with darker shades of reddish-brown pigment figure.
Sweetgum—*Liquidambar styraciflua* L. Fig. 12-109. Desc. p. 588

25. Rays normally spaced, not seemingly occupying half the area on the transverse surface of the wood.. **26**

26. Wood aromatic on the freshly cut surface; heartwood light brown to grayish brown, frequently with darker streaks of pigment figure; pores encircled by a whitish sheath of parenchyma.
California-laurel—*Umbellularia californica* (Hook. & Arn.) Nutt. Fig. 12-140. Desc. p. 585

26. Wood not aromatic; pores not encircled with parenchyma........... **27**

27. Pores in the early wood in a distinct uniseriate row; heartwood distinctly light to dark red-brown.
Black cherry—*Prunus serotina* Ehrh. Fig. 12-125. Desc. p. 592

27. Pores in the early wood not in a distinct uniseriate row.

A. Wood soft, light brown, sometimes with a pinkish cast, often appearing lacelike under a lens (*x*); rays uniformly spaced, of the same color as the background; ray flecks (*r*) high, rather distant.
Basswood—*Tilia* spp. Fig. 12-136. Desc. p. 610

B. Wood hard; heartwood pale brown, frequently with grayish or greenish cast; rays appearing on the tangential surface as short crowded lines, visible without magnification.
Soft maple—*Acer* spp. Figs. 12-75 to 12-77, inc. Desc. p. 605

C. Wood hard; heartwood light to dark brown or reddish brown; largest pores obviously wider than the broadest rays; ray fleck (*t*) inconspicuous.
Birch—*Betula* spp. Figs. 12-84 to 12-87, inc. Desc. p. 552

28. Rays barely visible with a hand lens.

A. Wood whitish, aging to flesh colored or light brown with a reddish tinge, frequently stained with gray oxidative stain; rays very close together, seemingly occupying half the area (*x*, impression gained by casual examination).
Red alder—*Alnus rubra* Bong. Fig. 12-81. Desc. p. 556

B. Heartwood light brown to pale reddish brown; rays normally spaced.
Cottonwood—*Populus* spp. Fig. 12-122. Desc. p. 544
Willow—*Salix* spp. Fig. 12-134. Desc. p. 548

28. Rays plainly visible with a hand lens....................................... **29**

29. Wood aromatic on the freshly cut surface; heartwood light brown to grayish brown, frequently with darker streaks of pigment figure; pores encircled by a whitish sheath of parenchyma.
California-laurel—*Umbellularia californica* (Hook. & Arn.) Nutt. Fig. 12-140. Desc. p. 585

29. Wood not aromatic; heartwood light brown, with or without grayish or greenish cast, brown or reddish brown; pores not encircled by white sheaths of parenchyma.. **30**

30. Rays very close together, seemingly occupying half the area on the transverse surface (*x*), (impression gained by casual examination); sapwood pinkish; heartwood carneous gray to varying shades of brown, frequently with darker streaks of pigment figure.
Sweetgum—*Liquidambar styraciflua* L. Fig. 12-109. Desc. p. 588

30. Rays normally spaced, not seemingly occupying half the area on the transverse surface.. **31**

31. Late wood figured with wavy, concentric bands of pores which are mostly continuous and separated by bands of mechanical tissue.
Hard elm—*Ulmus* spp. Fig. 12-139. Desc. p. 575

31. Late wood not figured with wavy, concentric bands of pores............. **32**

32. Wood soft, light brown, sometimes with a pinkish cast; often appearing lacelike under a lens (*x*); rays uniformly spaced, of the same color as background (*x*); ray flecks (*r*) high, rather distant.
Basswood—*Tilia* spp. Fig. 12-136. Desc. p. 610

32. Wood hard, light to dark brown or reddish brown, not lacelike under a lens (*x*).

A. Heartwood pale brown, frequently with grayish or greenish cast; rays on tangential surface appearing as short, crowded lines, visible without magnification.
Soft maple—*Acer* spp. Figs. 12-75 to 12-77, inc. Desc. p. 605

B. Heartwood light to dark brown or reddish brown; largest pores wider than the broadest rays; ray fleck (*t*) inconspicuous.
Birch—*Betula* spp. Figs. 12-84 to 12-87, inc. Desc. p. 552

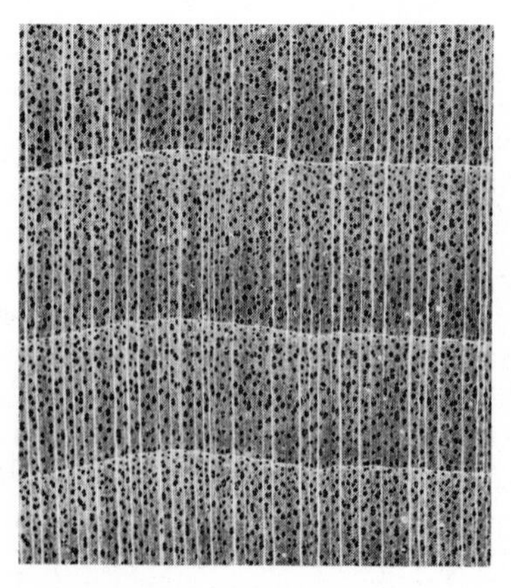

FIG. 12-75 *Acer macrophyllum* Pursh. (*x*—5×.)

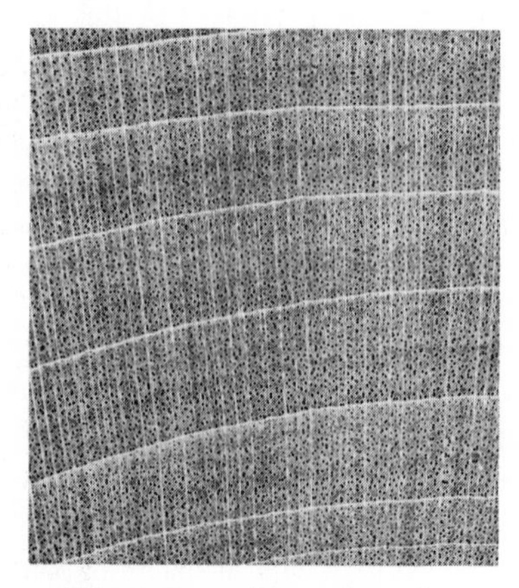

FIG. 12-76 *Acer rubrum* L. (*x*—5×.)

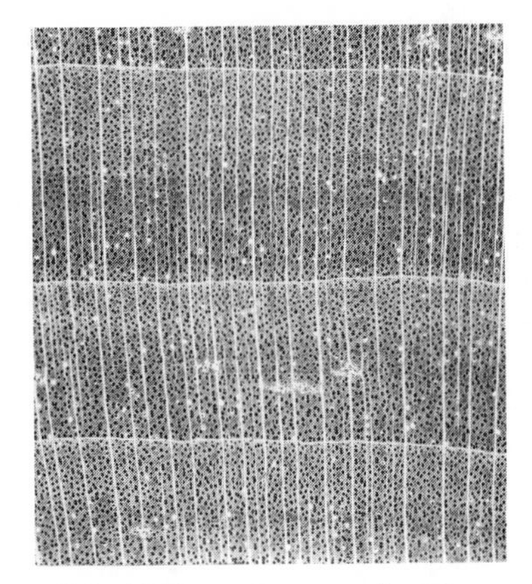

FIG. 12-77 *Acer saccharinum* L. (*x*—5×.)

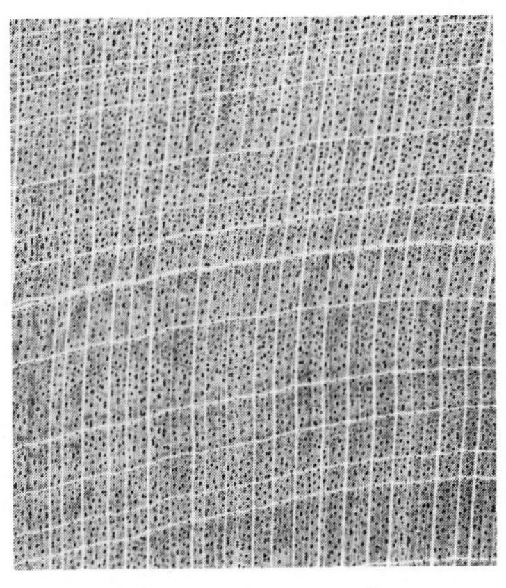

FIG. 12-78 *Acer saccharum* Marsh. (*x*—5×.)

FIG. 12-79 *Acer saccharum* Marsh. (*t*—natural size.)

FIG. 12-80 *Aesculus octandra* Marsh. (*x*—5×.)

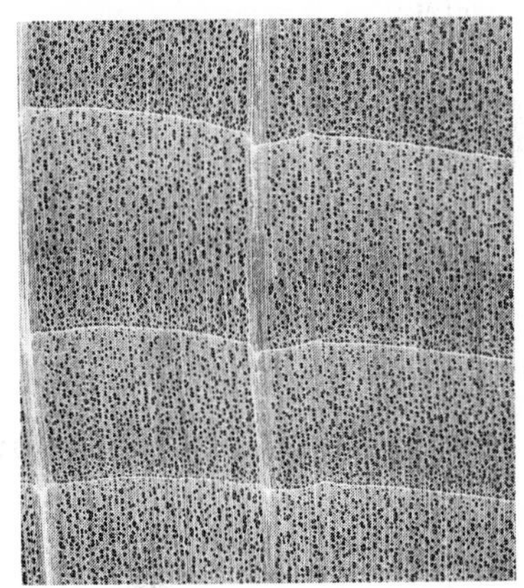

FIG. 12-81 *Alnus rubra* Bong. (*x*—5×.)

FIG. 12-82 *Alnus rubra* Bong. (*t*—natural size.)

FIG. 12-83 *Arbutus menziesii* Pursh. (x—5×.)

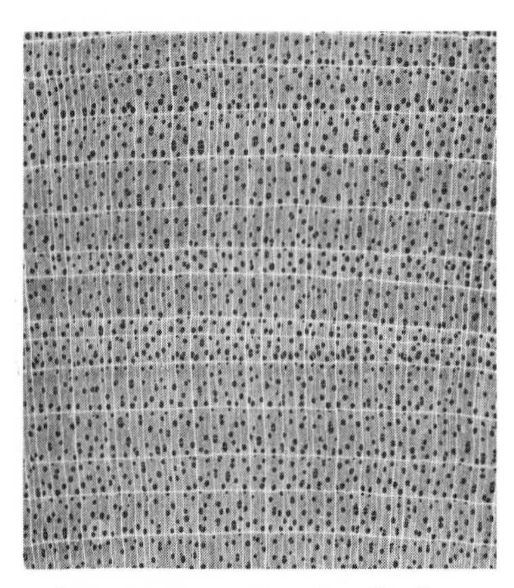

FIG. 12-84 *Betula alleghaniensis* Britton. (x—5×.)

FIG. 12-85 *Betula alleghaniensis* Britton. (t—natural size.)

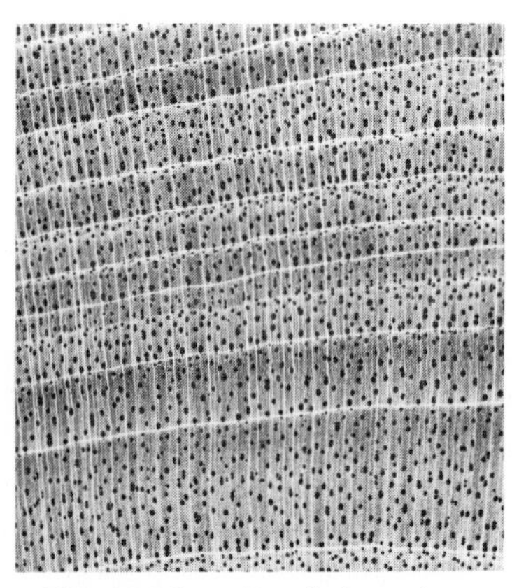

FIG. 12-86 *Betula lenta* L. (x—5×.)

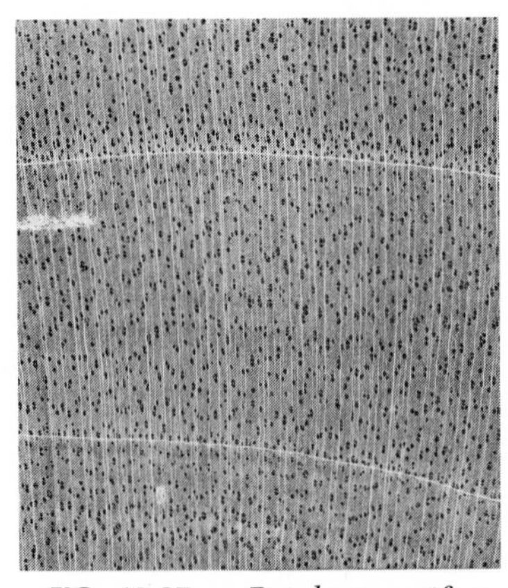

FIG. 12-87 *Betula papyrifera* Marsh. (x—5×.)

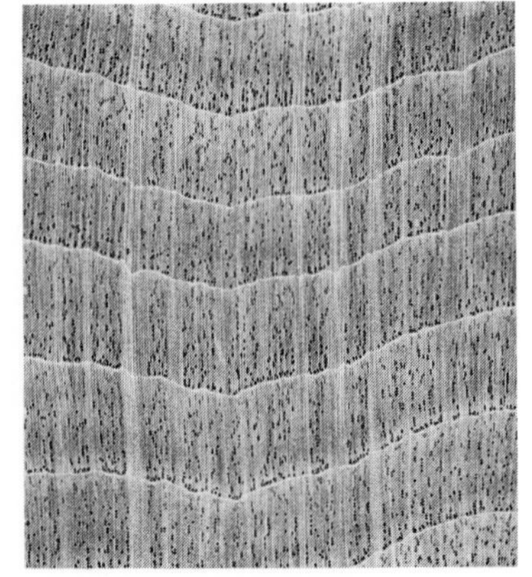

FIG. 12-88 *Carpinus caroliniana* Walt. (x—5×.)

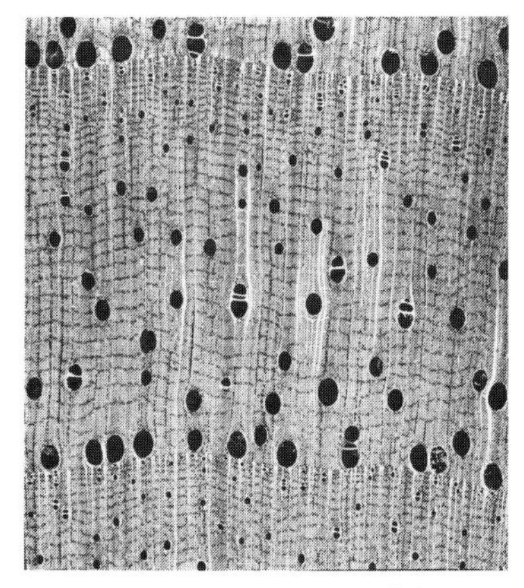

FIG. 12-89 *Carya cordiformis* (Wangenh.) K. Koch. (*x*—5×.)

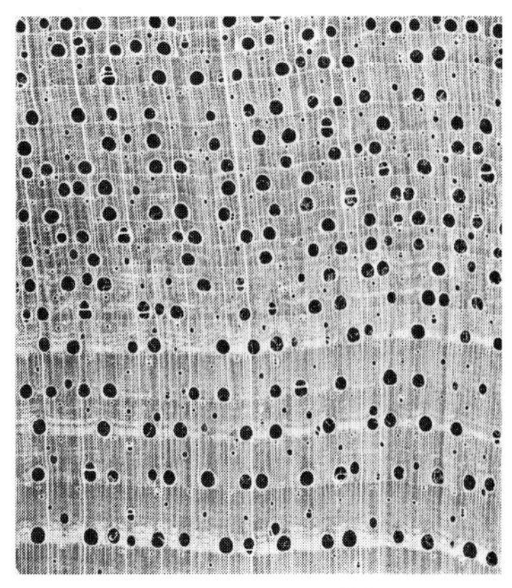

FIG. 12-90 *Carya glabra* (Mill.) Sweet. (*x*—5×.)

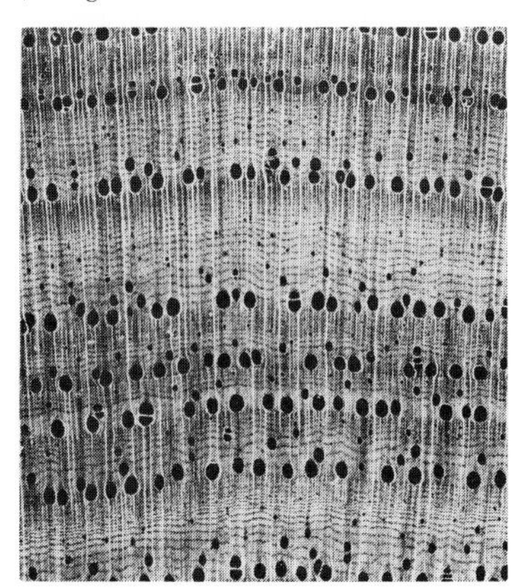

FIG. 12-91 *Carya ovata* (Mill.) K. Koch. (*x*—5×.)

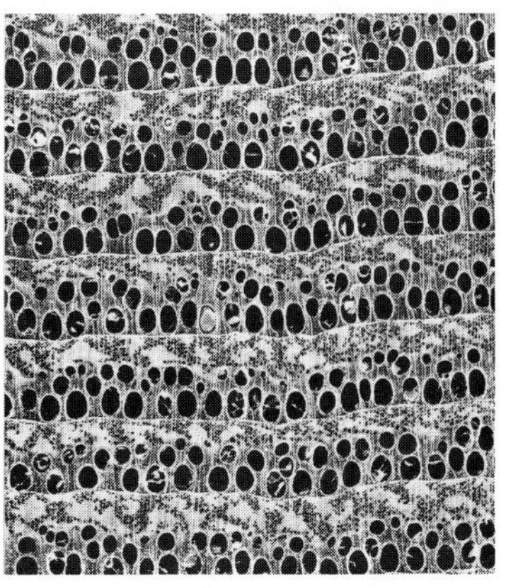

FIG. 12-92 *Castanea dentata* (Marsh.) Borkh. (*x*—5×.)

FIG. 12-93 *Castanopsis chrysophylla* (Dougl.) A.DC. (*x*—5×.)

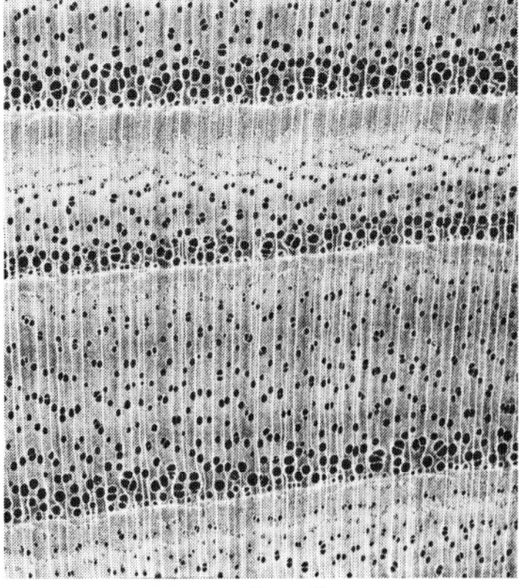

FIG. 12-94 *Catalpa speciosa* Warder. (*x*—5×.)

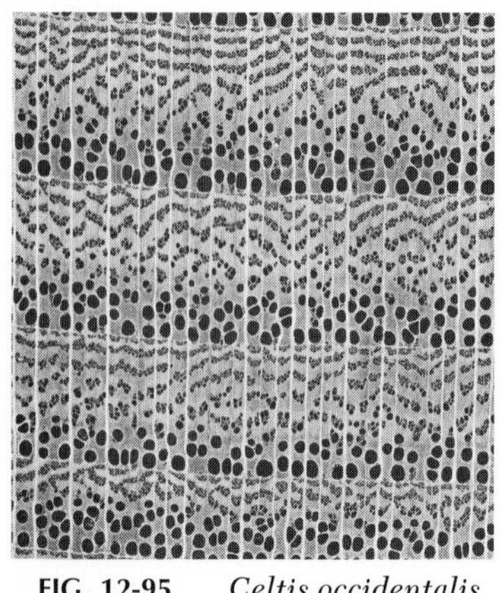

FIG. 12-95 *Celtis occidentalis* L. (*x*—5×.)

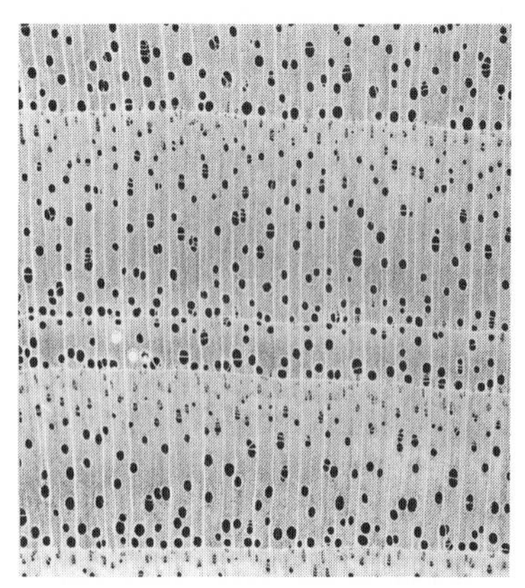

FIG. 12-96 *Cladrastris lutea* (Michx.f.) K. Koch. (*x*—5×.)

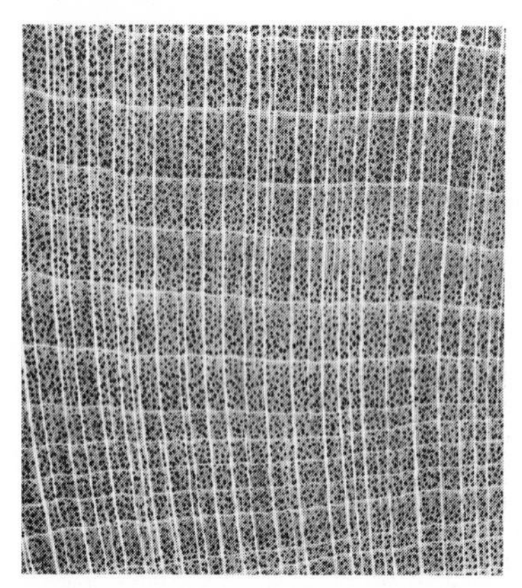

FIG. 12-97 *Cornus florida* L. (*x*—5×.)

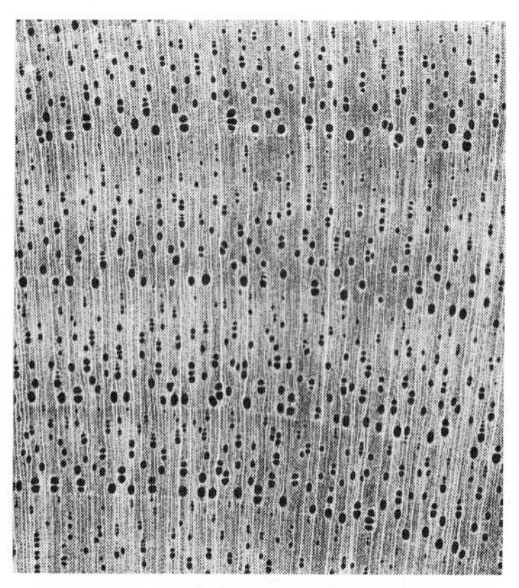

FIG. 12-98 *Diospyros virginiana* L. (*x*—5×.)

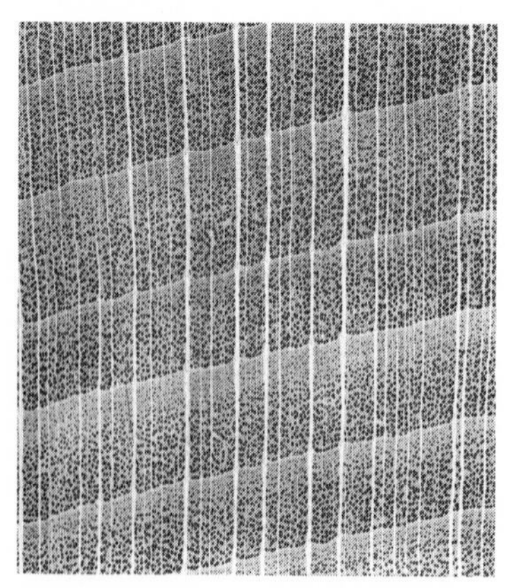

FIG. 12-99 *Fagus grandifolia* Ehrh. (*x*—5×.)

FIG. 12-100 *Fagus grandifolia* Ehrh. (*t*—natural size.)

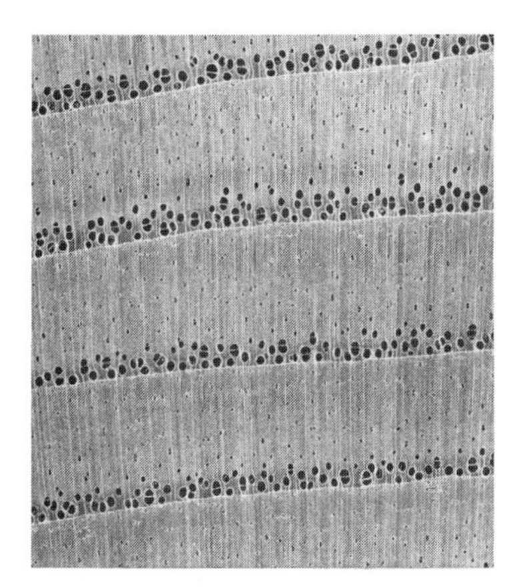

FIG. 12-101 *Fraxinus americana* L. (x—5×.)

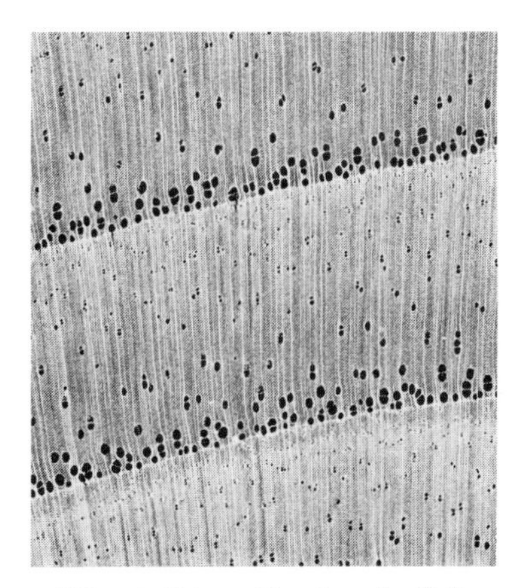

FIG. 12-102 *Fraxinus latifolia* Benth. (x—5×.)

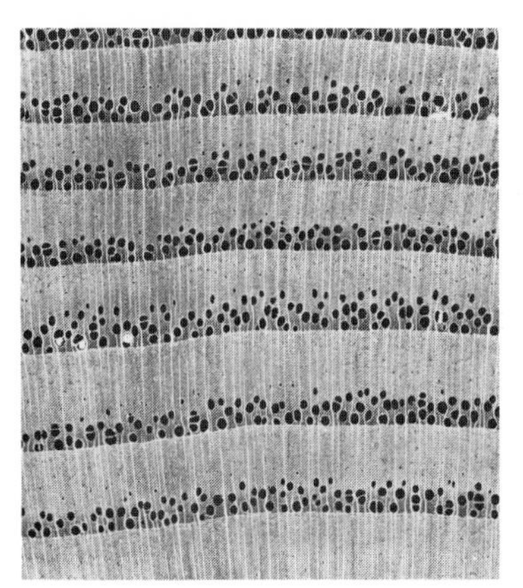

FIG. 12-103 *Fraxinus nigra* Marsh. (x—5×.)

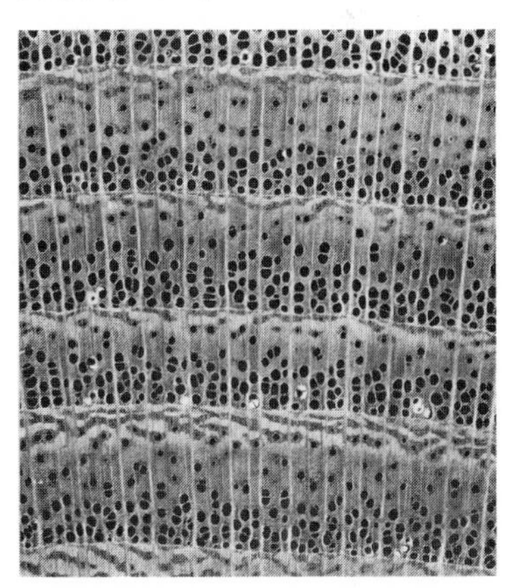

FIG. 12-104 *Gleditsia triacanthos* L. (x—5×.)

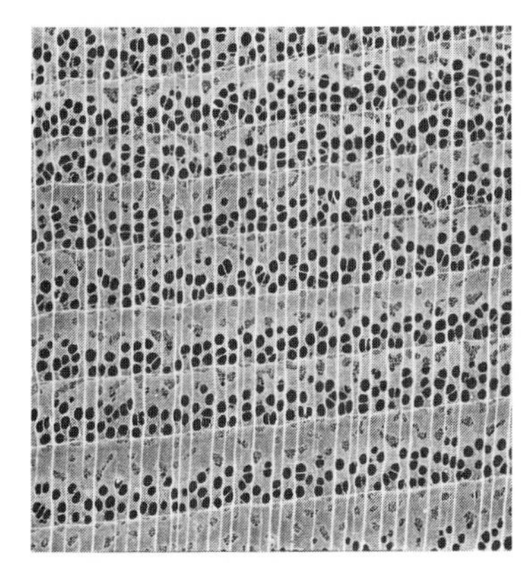

FIG. 12-105 *Gymnocladus dioicus* (L.) K. Koch. (x—5×.)

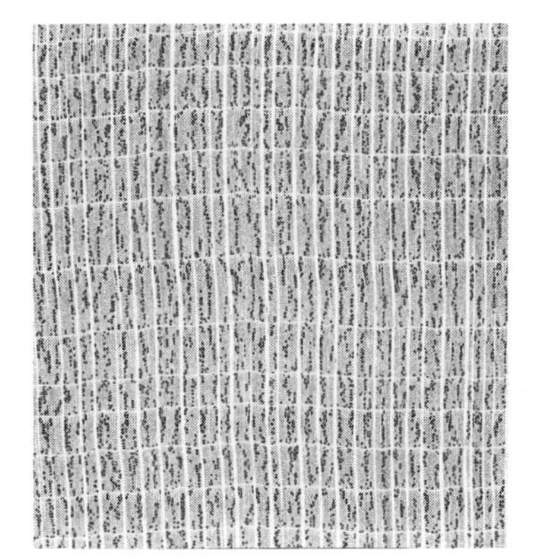

FIG. 12-106 *Ilex opaca* Ait. (x—5×.)

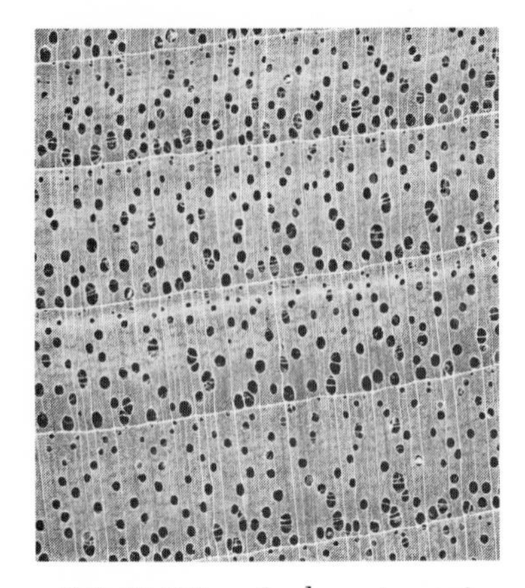

FIG. 12-107 *Juglans cinerea* L. (*x*—5×.)

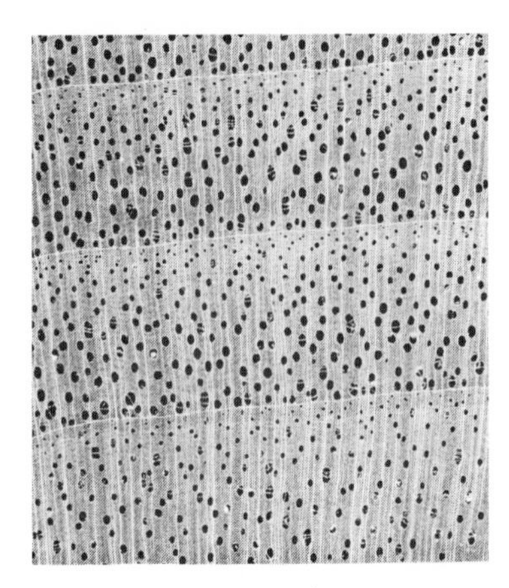

FIG. 12-108 *Juglans nigra* L. (*x*—5×.)

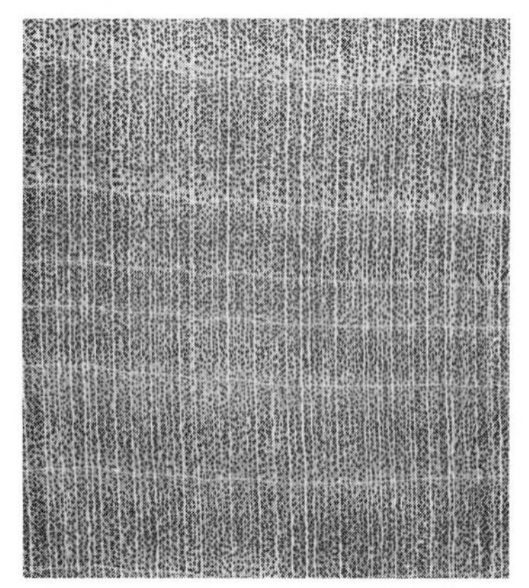

FIG. 12-109 *Liquidambar styraciflua* L. (*x*—5×.)

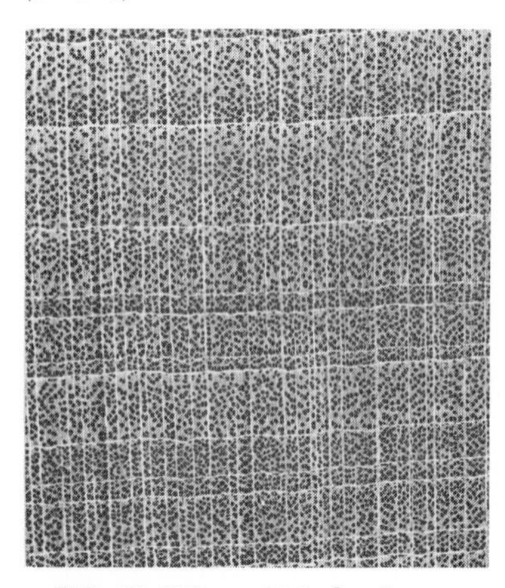

FIG. 12-110 *Liriodendron tulipifera* L. (*x*—5×.)

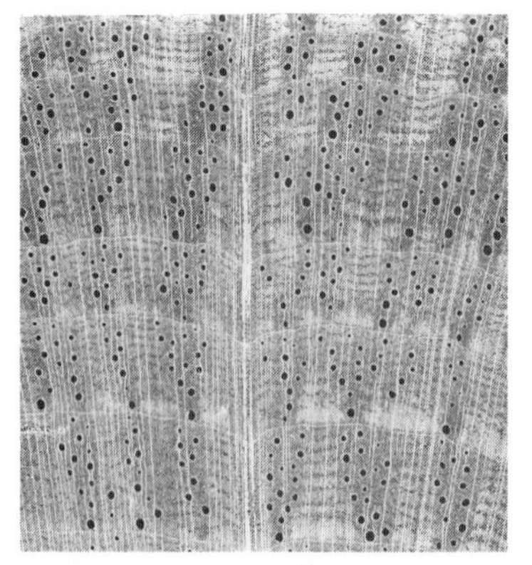

FIG. 12-111 *Lithocarpus densiflorus* (Hook. & Arn.) Rehd. (*x*—5×.)

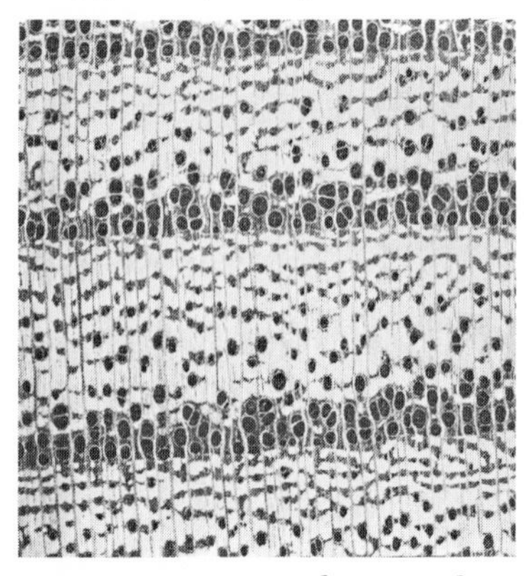

FIG. 12-112 *Maclura pomifera* (Raf.) Schneid. (*x*—5×.)

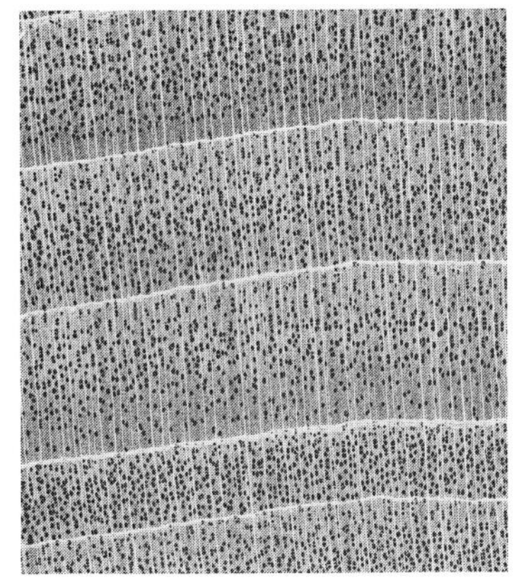

FIG. 12-113 *Magnolia acuminata* L. (*x*—5×.)

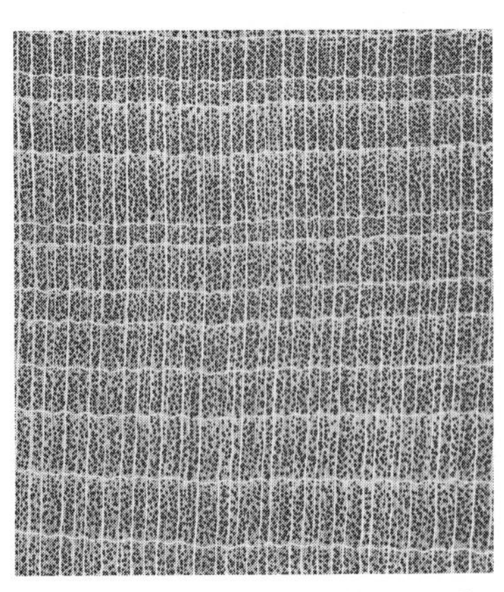

FIG. 12-114 *Magnolia grandiflora* L. (*x*—5×.)

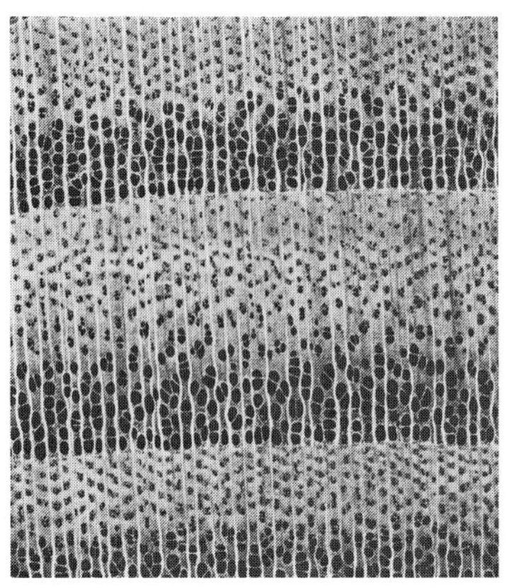

FIG. 12-115 *Morus rubra* L. (*x*—5×.)

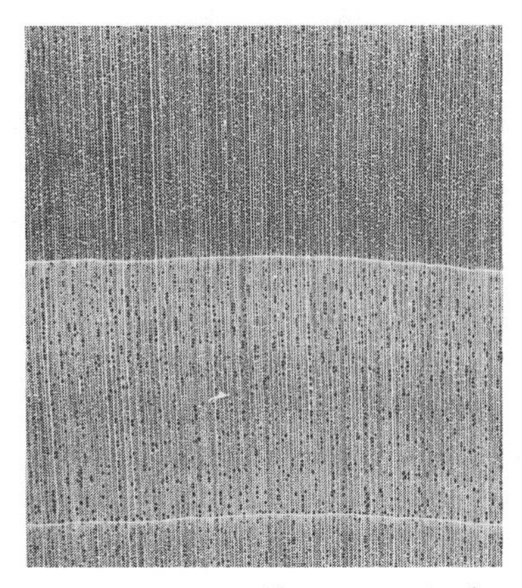

FIG. 12-116 *Nyssa aquatica* L. (*x*—5×.)

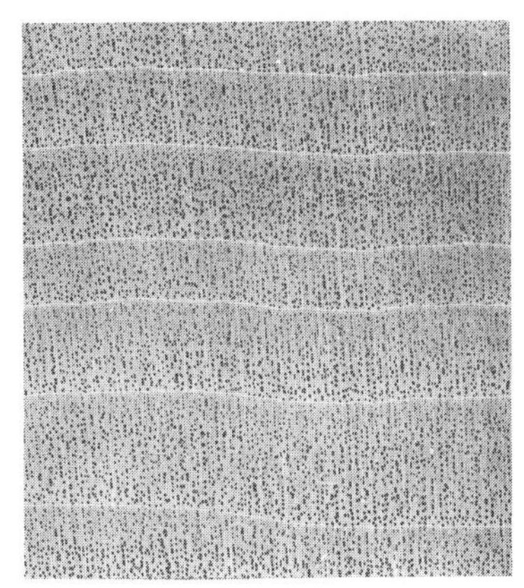

FIG. 12-117 *Nyssa sylvatica* Marsh. (*x*—5×.)

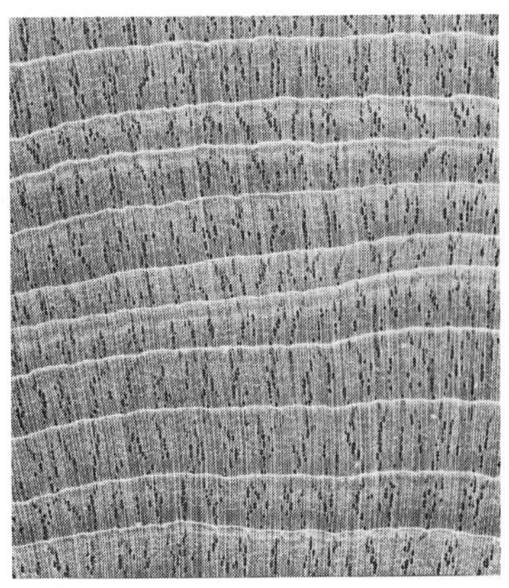

FIG. 12-118 *Ostrya virginiana* (Mill.) K. Koch. (*x*—5×.)

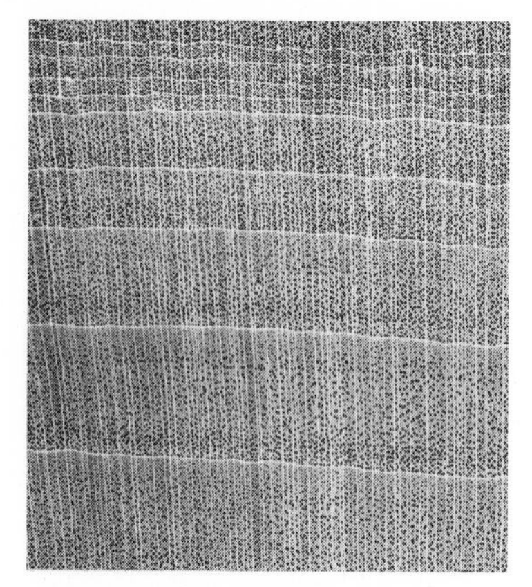

FIG. 12-119 *Oxydendrum arboreum* (L.) DC. (*x*—5×.)

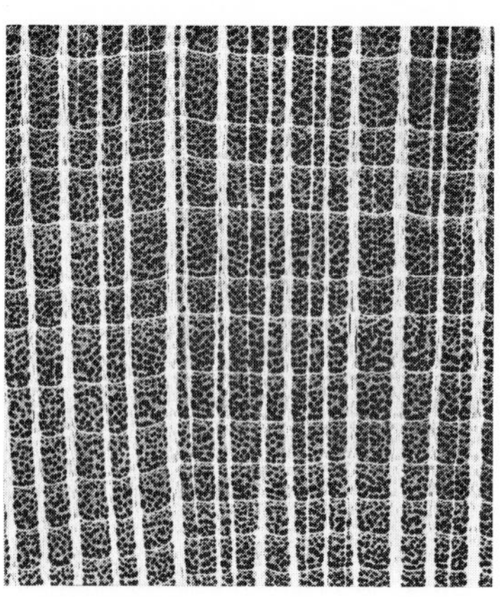

FIG. 12-120 *Platanus occidentalis* L. (*x*—5×.)

FIG. 12-121 *Platanus occidentalis* L. (*t*—natural size.)

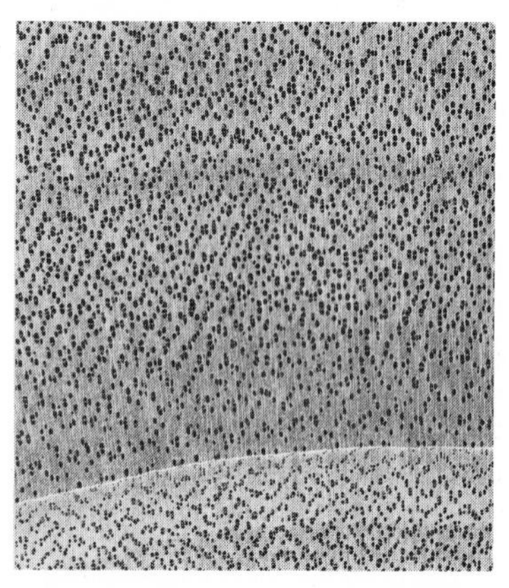

FIG. 12-122 *Populus deltoides* Bartr. (*x*—5×.)

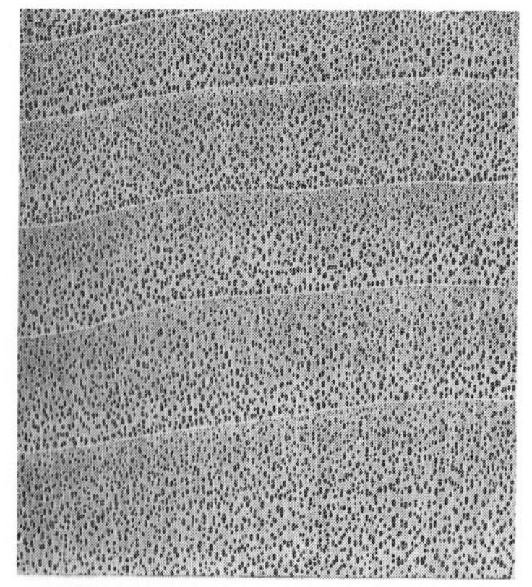

FIG. 12-123 *Populus grandidentata* Michx. (*x*—5×.)

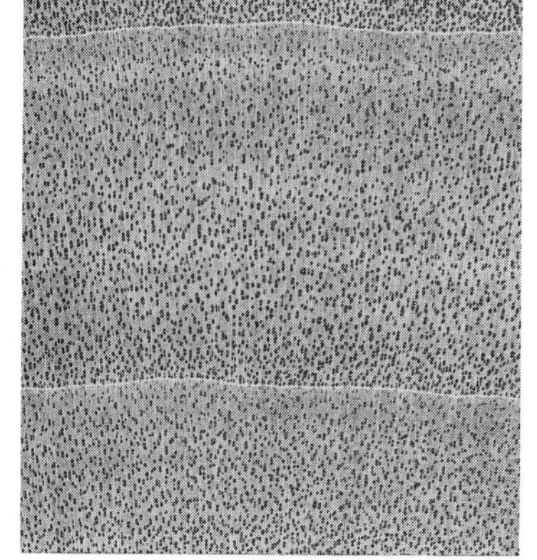

FIG. 12-124 *Populus tremuloides* Michx. (*x*—5×.)

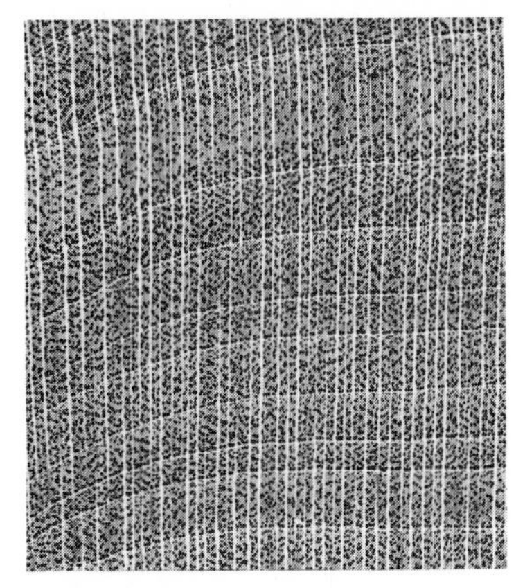

FIG. 12-125 *Prunus serotina* Ehrh. (x—5×.)

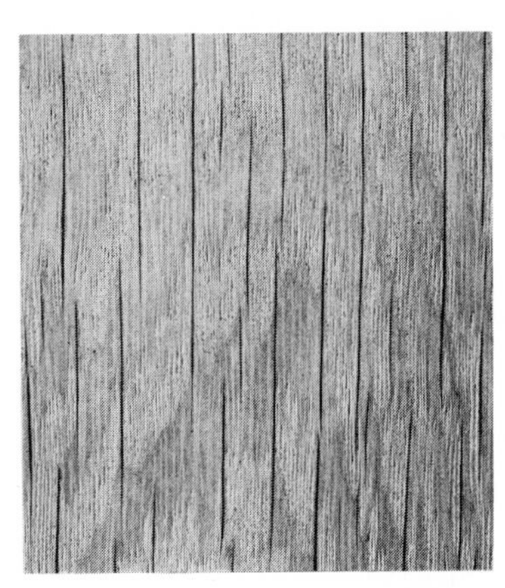

FIG. 12-126 *Quercus alba* L. (t—natural size.)

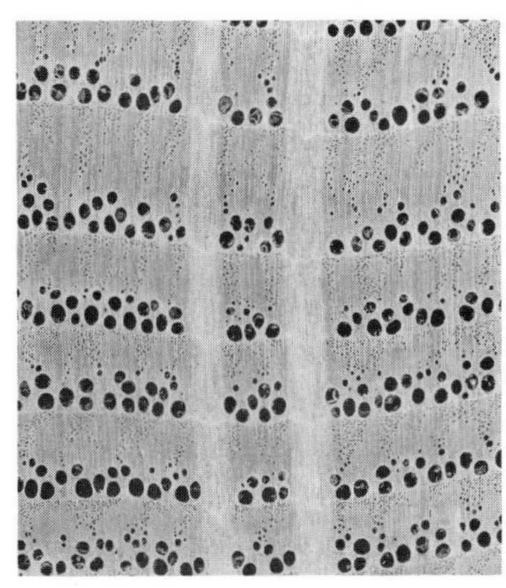

FIG. 12-127 *Quercus bicolor* Willd. (x—5×.)

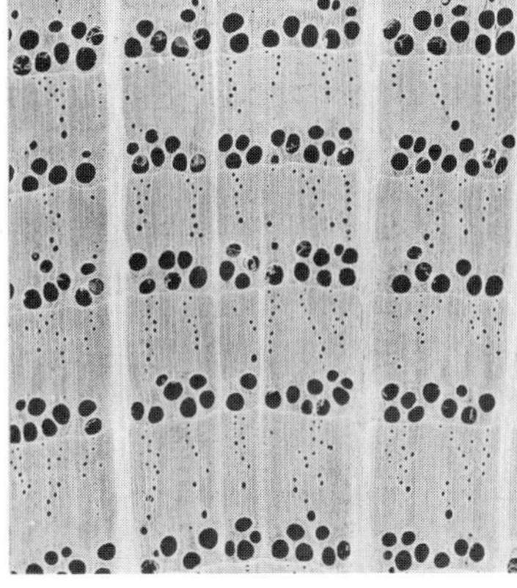

FIG. 12-128 *Quercus coccinea* Muenchh. (x—5×.)

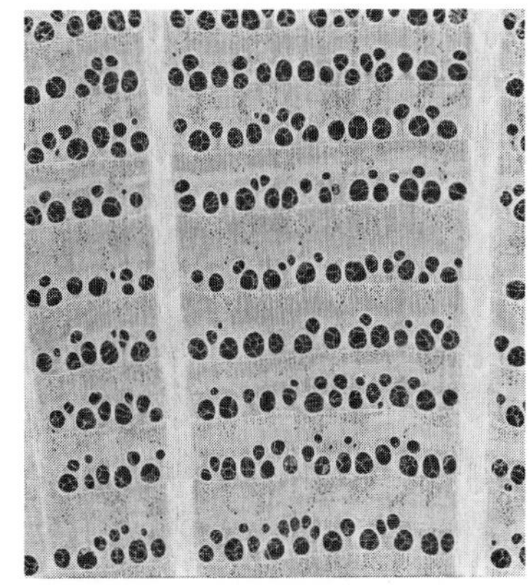

FIG. 12-129 *Quercus stellata* Wangenh. (x—5×.)

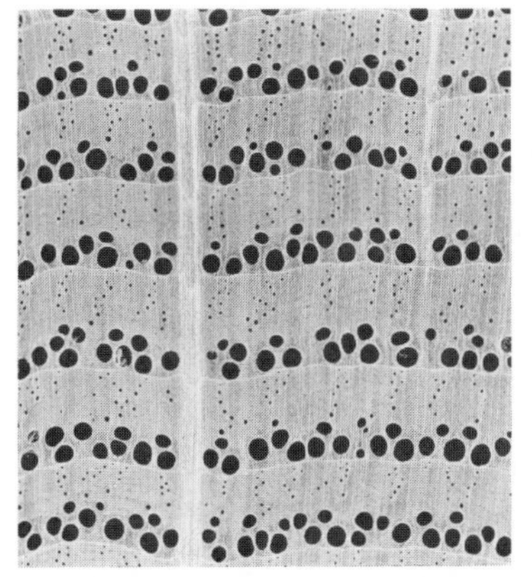

FIG. 12-130 *Quercus velutina* Lam. (x—5×.)

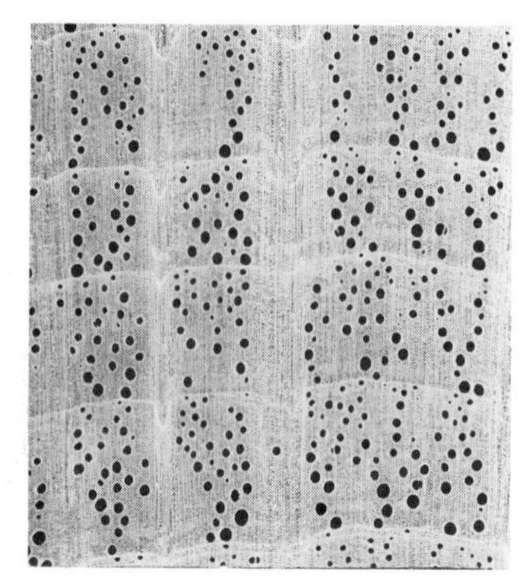

FIG. 12-131 *Quercus virginiana* Mill. (x—5×.)

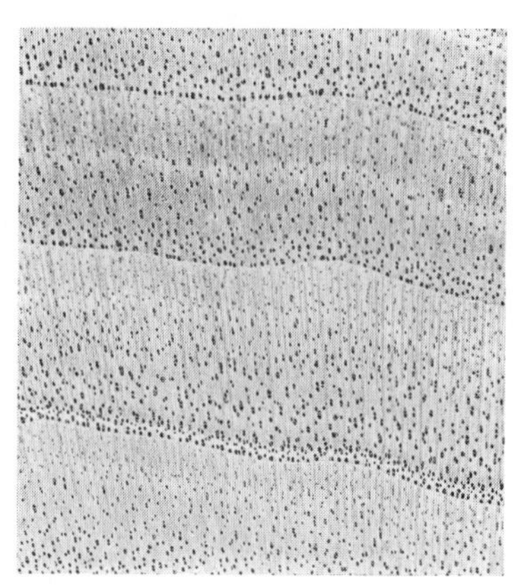

FIG. 12-132 *Rhamnus purshiana* DC. (x—5×.)

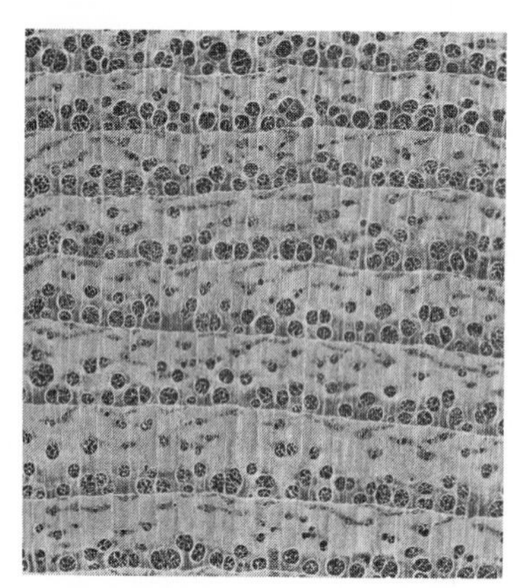

FIG. 12-133 *Robinia pseudoacacia* L. (x—5×.)

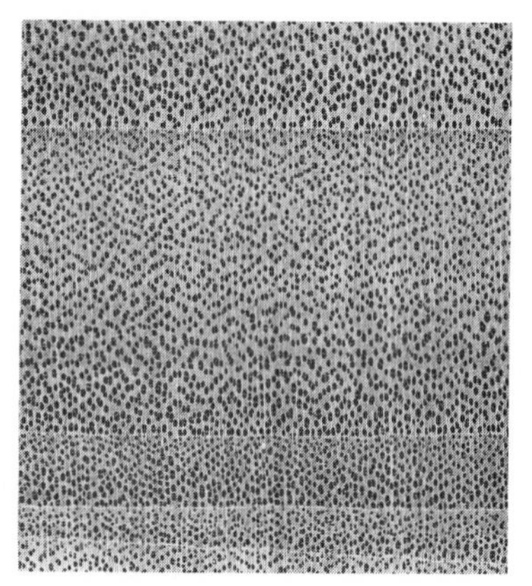

FIG. 12-134 *Salix nigra* Marsh. (x—5×.)

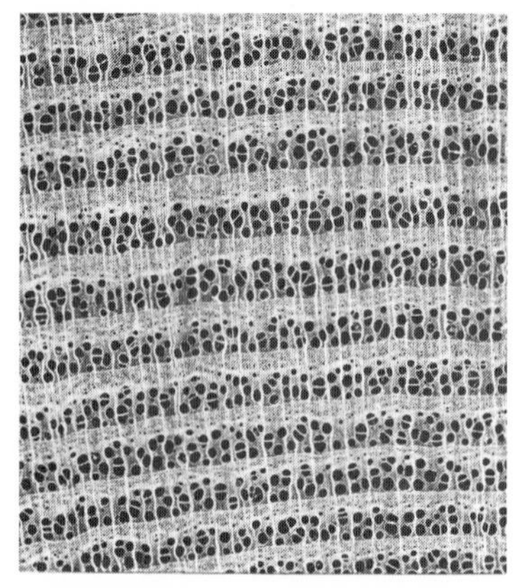

FIG. 12-135 *Sassafras albidum* (Nutt.) Nees. (x—5×.)

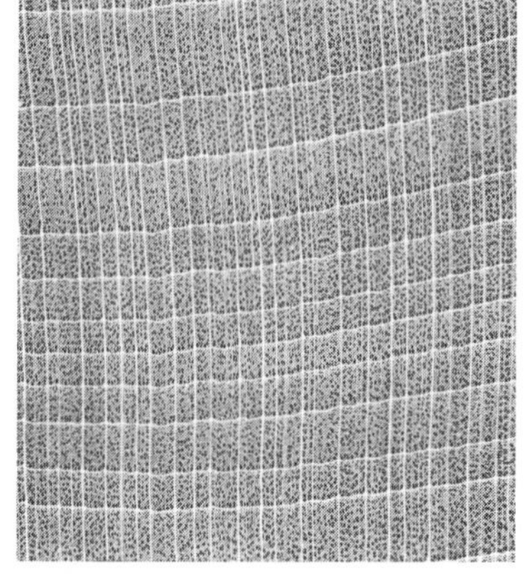

FIG. 12-136 *Tilia americana* L. (x—5×.)

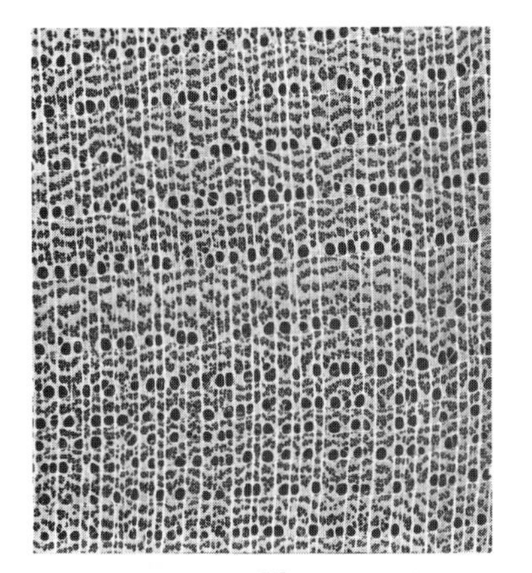

FIG. 12-137 *Ulmus americana* L. (x—5×.)

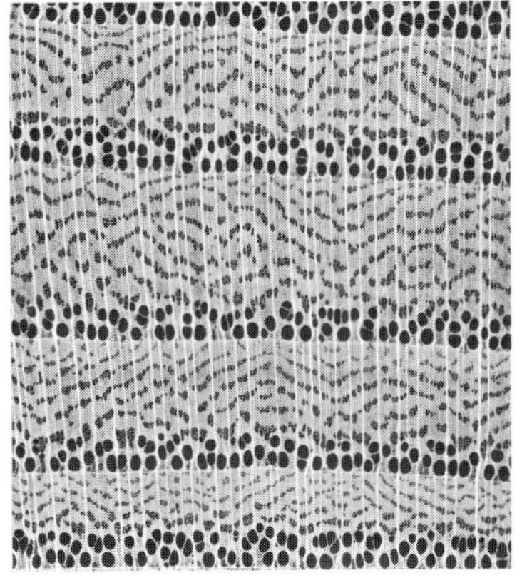

FIG. 12-138 *Ulmus rubra* Mühl. (x—5×.)

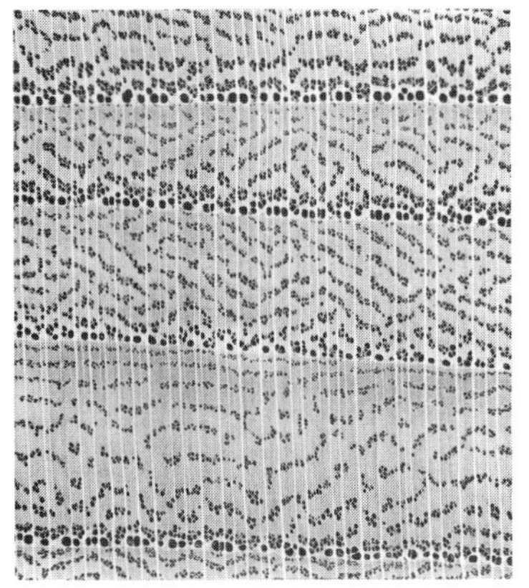

FIG. 12-139 *Ulmus thomasii* Sarg. (x—5×.)

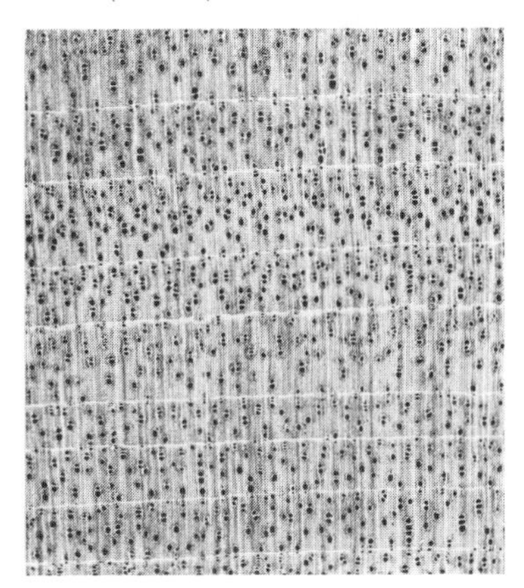

FIG. 12-140 *Umbellularia californica* (Hook. & Arn.) Nutt. (x—5×.)

V. KEY TO HARDWOODS—MINUTE FEATURES*

1. Early-wood vessels obviously larger than those in the late wood (*x*); wood ring-porous **2**

1. Early-wood vessels not larger or but slightly larger than those in the late wood (*x*); wood diffuse-porous **34**

2. Broad rays of the oak type present **3**

2. Broad rays of the oak type absent **4**

3. Late-wood vessels not numerous, thick-walled, rounded (thicker-walled than the surrounding cells); tyloses sparse or absent in the early-wood vessels in the heartwood.
Red oak—*Quercus* spp. Desc. p. 566

3. Late-wood vessels numerous, thin-walled, angular (as thin-walled as the surrounding cells, or thinner-walled); tyloses abundant in the early-wood vessels in the heartwood.
White oak—*Quercus* spp. Desc. p. 569

4. Early-wood vessels grading in size into those of the late wood; early-wood zone not sharply defined; the transition from early to late wood more or less gradual; wood semi-ring-porous **5**

4. Early-wood vessels not grading in size into those of the late wood (or if so, the first-formed early-wood vessels much larger than those in the outer portions of the ring); early-wood zone sharply defined, the transition from early to late wood abrupt; wood typically ring-porous **14**

5. Rays 1–6 seriate; parenchyma included in the body of the growth ring, or wanting (*x*) **6**

5. Rays uniseriate; parenchyma marginal (*x*) **12**

6. Parenchyma relatively abundant; vessels without spiral thickening or the spirals restricted to those of the late wood; largest vessels more than 100 μ in diameter† (ranging from 120–300) **7**

6. Parenchyma extremely sparse or wanting; vessels with spiral thickening (Fig. 12-141); largest vessels less than 100 μ in diameter (ranging from 60–90).
Black cherry—*Prunus serotina* Ehrh. Desc. p. 592

7. Intervessel pits vestured, similar to those in *Robinia pseudoacacia* (see Fig. 12-144).
Yellowwood—*Cladrastis lutea* (Michx. f.) K. Koch. Desc. p. 597

7. Intervessel pits not vestured **8**

8. Late-wood vessels with spiral thickening, those in the outer portion of the growth ring associated with parenchyma forming tangential,

* This key is designed to separate the more important North American hardwoods on features requiring the use of a compound microscope.

The key is accompanied by a series of 20 photomicrographs at 400X (Figs. 12-141 to 12-160, inc.) illustrating the intervessel pitting in hardwoods.

† The diameter stated is always the tangential diameter.

several seriate, more or less continuous bands, with one or more bands in the outer portion of the ring consisting entirely of parenchyma.
Catalpa—*Catalpa* spp. Desc. p. 625

8. Late-wood vessels without spiral thickening, solitary or in radial groups of 2–4; apotracheal parenchyma present throughout the growth ring.......... **9**

9. Rays storied (*t*); intervessel pits minute, 2–several frequently joined laterally (Fig. 12-142).
Common persimmon—*Diospyros virginiana* L. Desc. p. 620

9. Rays unstoried (*t*); intervessel pits of normal size, rounded or angular, generally not joined laterally (see Fig. 12-143).......... **10**

10. Lines of banded parenchyma conspicuous, 1–4-seriate; fibers with maximum diameter of less than 25 μ.
Pecan hickory group—*Carya* spp. Desc. p. 541

10. Lines of banded parenchyma inconspicuous, usually uniseriate; fibers with maximum diameter of more than 25 μ.......... **11**

11. Parenchyma frequently crystalliferous; ray cells mostly round (*t*).
Black walnut—*Juglans nigra* L. Desc. p. 539

11. Parenchyma not crystalliferous; ray cells mostly elliptical (*t*).
Butternut—*Juglans cinerea* L. Desc. p. 538

12. Rays essentially homocellular (see Fig. 5-11*a*).......... **13**

12. Rays essentially heterocellular (see Fig. 5-11*b*).
Willow—*Salix* spp. Desc. p. 548

13. Vessels 30–145 per sq mm, the largest 75–150 μ in diameter.
Cottonwood—*Populus* spp. Desc. p. 544

13. Vessels 85–180 per sq mm, the largest 50–100 μ in diameter.
Aspen—*Populus* spp. Desc. p. 546

14. Late wood figured with concentric, wavy, more or less continuous bands of porous tissue composed largely of vessels (*x*).......... **15**

14. Late wood without concentric bands of vessels (*x*) or occasionally with interrupted bands in the outer portion of the growth ring consisting of vessels and appreciable amounts of parenchyma; apotracheal parenchyma sometimes present.......... **19**

15. Large early-wood vessels in one row.......... **16**

15. Large early-wood vessels in several rows.......... **17**

16. Large early-wood vessels with maximum diameter of over 200 μ.
American elm—*Ulmus americana* L. Desc. p. 572

16. Large early-wood vessels with maximum diameter of less than 200 μ.
Hard elms—*Ulmus* spp. Desc. p. 575

17. Rays 1–6-seriate, essentially homocellular.......... **18**

17. Rays 1–13-seriate, essentially heterocellular.
Hackberry—*Celtis* spp. Desc. p. 577

18. Early-wood zone consisting essentially of large pores.
Slippery elm—*Ulmus rubra* Mühl. Desc. p. 574

18. Early-wood zone consisting of large and small pores.
American elm—*Ulmus americana* L. Desc. p. 572

19. Rays uniseriate or rarely in part biseriate.. **20**

19. Rays 1–many-seriate.. **21**

20. Early-wood vessels usually in one row, the largest 150–200 μ in diameter.
Golden chinkapin—*Castanopsis chrysophylla* (Dougl.) A. DC. Desc. p. 561

20. Early-wood vessels in several rows, the largest 240–360 μ in diameter.
American chestnut—*Castanea dentata* (Marsh.) Borkh. Desc. p. 559

21. Rays 1–4-seriate.. **22**

21. Rays 1–5- or more-seriate.. **27**

22. Parenchyma banded (*x*), in concentric rows distributed throughout the body of the growth ring.
Hickory—*Carya* spp. Desc. p. 541

22. Parenchyma not banded or, if banded, the rows restricted to short broken lines in the outer portion of the ring............................ **23**

23. Late-wood vessels in part with scalariform perforation plates; rays with oil cells.
Sassafras—*Sassafras albidum* (Nutt.) Nees. Desc. p. 587

23. Vessels in early and late wood with simple perforation plates; rays without oil cells.. **24**

24. Early-wood vessels completely occluded with tyloses; tyloses small, appearing cellular.. **25**

24. Early-wood vessels open or only partly occluded with tyloses; tyloses large, not appearing cellular.. **26**

25. Intervessel pits vestured (Fig. 12-144).
Black locust—*Robinia pseudoacacia* L. Desc. p. 598

25. Intervessel pits not vestured (Fig. 12-145).
Osage-orange—*Maclura pomifera* (Raf.) Schneid. Desc. p. 580

26. Late-wood vessels with spiral thickening; vessels in the outer part of the growth ring associated with parenchyma forming tangential, several-seriate, more or less continuous bands (one or more bands in the outer portion of the ring occasionally consisting entirely of parenchyma); fibers with maximum diameter of more than 25 μ (range from 16–32).
Catalpa—*Catalpa* spp. Desc. p. 625

26. Late-wood vessels without spiral thickening, solitary or in radial rows of 2–several; fibers with maximum diameter of less than 25 μ (range from 12–22).
Ash—*Fraxinus* spp. Desc. pp. 621 to 625

27. Parenchyma banded (*x*), in concentric, 1–4-seriate bands distributed throughout the body of the growth ring (*x*); late-wood vessels solitary or in radial rows of 2–3, without spiral thickening.
Hickory—*Carya* spp. Desc. p. 541

27. Parenchyma not banded (*x*); late-wood vessels solitary, in radial rows, in nests, or in interrupted bands consisting largely of vessels or of vessels and parenchyma, with spiral thickenings....................................... **28**

28. Early-wood vessels in one row; intervessel pits vestured.
Yellowwood—*Cladrastis lutea* (Michx. f.) K. Koch. Desc. p. 597

28. Early-wood vessels in several rows; intervessel pits vestured or not vestured.. **29**

29. Early-wood vessels in the heartwood partly or wholly occluded with tyloses, with or without gummy deposits....................................... **30**

29. Early-wood vessels without tyloses or tyloses only occasionally present, sometimes with gummy deposits... **33**

30. Early-wood vessels partly occluded with tyloses; tyloses large, not appearing cellular... **31**

30. Early-wood vessels completely occluded with tyloses; tyloses small, appearing cellular... **32**

31. Fibers 20–52 μ in diameter, not gelatinous; rays 1–6-seriate (mostly 2–3).
Catalpa—*Catalpa* spp. Desc. p. 625

31. Fibers 18–26 μ in diameter, frequently gelatinous; rays 1–8-seriate (mostly 5–7).
Red mulberry—*Morus rubra* L. Desc. p. 578

32. Intervessel pits vestured.
Black locust—*Robinia pseudoacacia* L. Desc. p. 598

32. Intervessel pits not vestured.
Osage-orange—*Maclura pomifera* (Raf.) Schneid. Desc. p. 580

33. Rays 1–14-seriate, the tallest more than 1200 μ in height; porous tissue toward the outer margin of the growth ring consisting of small vessels embedded in short, tangential bands of parenchyma.
Honeylocust—*Gleditsia triacanthos* L. Desc. p. 595

33. Rays 1–7-seriate, the tallest less than 1200 μ in height; porous tissue toward the outer portion of the growth ring consisting mostly of vessels.
Kentucky coffeetree—*Gymnocladus dioicus* (L.) K. Koch. Desc. p. 594

34. Early-wood vessels somewhat larger than those in the late wood; wood semi-diffuse-porous.. **35**

34. Vessels exhibiting little or no variation in size (except in extreme outer portion of certain species); wood typically diffuse-porous..... **46**

35. Rays 1–many-seriate; parenchyma present in the body of the ring, or wanting... **36**

35. Rays uniseriate; parenchyma confined to the outer margin of the ring.. **44**

36. Longitudinal parenchyma relatively abundant; vessels without spiral thickening or the spirals restricted to those of the late wood; largest vessels more than 100 μ in diameter (100–320).............. **37**

36. Longitudinal parenchyma absent or extremely sparse; vessels throughout the ring with spiral thickening (Fig. 12-141); largest vessels less than 100 μ in diameter (70–90).
Black cherry—*Prunus serotina* Ehrh. Desc. p. 592

37. Intervessel pits vestured.
Yellowwood—*Cladrastis lutea* (Michx. f.) K. Koch. Desc. p. 597

37. Intervessel pits not vestured... **38**

38. Late-wood vessels with spiral thickening, those toward the outer margin of the ring associated with parenchyma forming tangential, several-seriate, more or less continuous bands (one or more bands in the outer portion of the ring occasionally consisting entirely of parenchyma).
Catalpa—*Catalpa* spp. Desc. p. 625

38. Late-wood vessels without spiral thickening, solitary or in radial groups of 2–4; banded parenchyma present, the lines closely spaced throughout the growth ring.. **39**

39. Rays storied (*t*); intervessel pits minute, 2–several frequently joined (coalescent apertures, Fig. 12-142).
Common persimmon—*Diospyros virginiana* L. Desc. p. 620

39. Rays unstoried (*t*); pits on vessel walls of normal size, generally not joined laterally (Fig. 12-143).. **40**

40. Lines of banded parenchyma conspicuous, 1–4-seriate............... **41**

40. Lines of banded parenchyma inconspicuous, usually uniseriate..... **43**

41. Rays of one type, 1–5-seriate; vessels solitary and in short radial rows (*x*).
Water hickory—*Carya aquatica* (Michx. f.) Nutt. Desc. p. 541

41. Rays of two types, uniseriate and broad, the latter sometimes aggregate, up to 50+ cells in width; vessels in streamlike clusters which extend for some distance radially across several rings when they are evident....... **42**

42. Largest vessels 100 to 160 μ in diameter; banded parenchyma in 1–4-seriate lines; most of broad rays distinctly aggregate.
Tanoak—*Lithocarpus densiflorus* (Hook. & Arn.) Rehd. Desc. p. 562

42. Largest vessels 200–280 μ in diameter; banded parenchyma in 1–2 (mostly 1)-seriate lines; broad rays distinctly aggregate.
Live oak—*Quercus virginiana* Mill. Desc. p. 564

43. Parenchyma frequently crystalliferous; ray cells mostly round (*t*).
Black walnut—*Juglans nigra* L. Desc. p. 539

43. Parenchyma not crystalliferous; ray cells mostly elliptical (*t*).
Butternut—*Juglans cinerea* L. Desc. p. 538

44. Rays essentially homocellular (Fig. 5-11*a*)................................ **45**

44. Rays essentially heterocellular (Fig. 5-11*b*).
Willow—*Salix* spp. Desc. p. 548

45. Vessels 30–145 per sq mm, the largest 75–150 μ in diameter.
Cottonwood—*Populus* spp. Desc. p. 544
45. Vessels 85–180 per sq mm, the largest 50–100 μ in diameter.
Aspen—*Populus* spp. Desc. p. 546
46. Widest rays more than 8-seriate.. **47**
46. Widest rays never more than 8-seriate (aggregate rays when present not considered)... **48**
47. Rays of one type, intergrading in size, 1–14-seriate (mostly 3–14); intervessel pits not crowded (Fig. 12-149).
American sycamore—*Platanus occidentalis* L. Desc. p. 590
47. Rays of two types; narrow rays 1–5-seriate; broad rays 15–25+-seriate; intervessel pits crowded (Fig. 12-150).
American beech—*Fagus grandifolia* Ehrh. Desc. p. 557
48. Rays of two types, simple and aggregate, aggregate rays often at wide intervals... **49**
48. Rays of one type, simple... **50**
49. Simple rays uniseriate or rarely in part biseriate; perforation plates scalariform with 15+ thin bars; intervessel pits not crowded (Fig. 12-147).
Red alder—*Alnus rubra* Bong. Desc. p. 556
49. Simple rays 1–4-seriate (mostly 1–2); perforation plates predominantly simple; scalariform perforation plates sporadic, with 3–6 thick bars; intervessel pits crowded (Fig. 12-148).
American hornbeam—*Carpinus caroliniana* Walt. Desc. p. 550
50. Perforation plates exclusively simple....................................... **51**
50. Perforation plates exclusively scalariform, or simple and scalariform and sometimes reticulate.. **65**
51. Rays 1–8-seriate.. **52**
51. Rays uniseriate... **62**
52. Parenchyma absent or extremely sparse................................... **53**
52. Parenchyma fairly abundant to abundant, sometimes restricted to the outer face of the ring... **55**
53. Vessels solitary and in more or less irregular groups of 2–several which are frequently oblique to the rays, generally more numerous and crowded in the early wood, 100–180 per sq mm.
Black cherry—*Prunus serotina* Ehrh. Desc. p. 592
53. Vessels solitary and in radial rows of 2–5, seldom contiguous in the tangential plane, quite evenly distributed as groups throughout the ring, 20–90 per sq mm.. **54**
54. Rays of two widths, the narrow uniseriate (rarely 2–3-seriate) and numerous, the broad 3–8-seriate (mostly 5–7).
Hard maple—*Acer* spp. Desc. p. 602
54. Rays intergrading in width, 1–5-seriate (mostly 3–4), the uniseriate rays sparse (especially in bigleaf maple).
Soft maple—*Acer* spp. Desc. p. 605
55. Parenchyma in the body of the ring at least in part banded apotracheal.... **56**
55. Parenchyma in the body of the ring never banded........................... **57**

56. Vessels unevenly distributed, frequently in radial rows of 2–6+ (x) which are often further aggregated into flame-shaped groups; rays 1–3-seriate, the tallest less than 1000 μ in height.
Eastern hophornbeam—*Ostrya virginiana* (Mill.) K. Koch. Desc. p. 551

56. Vessels quite evenly distributed, the radial grouping not pronounced (x); rays 1–6-seriate, the tallest more than 1000 μ in height.
Basswood—*Tilia* spp. Desc. p. 610

57. Intervessel pits vestured.
Yellowwood—*Cladrastis lutea* (Michx. f.) K. Koch. Desc. p. 597

57. Intervessel pits not vestured.. **58**

58. Paratracheal parenchyma abundant, forming a 1–3-seriate sheath; rays typically heterocellular, frequently with oil cells; fibers in part septate.
California-laurel—*Umbellularia californica* (Hook. & Arn.) Nutt. Desc. p. 585

58. Paratracheal parenchyma absent or sparse and then forming a broken uniseriate sheath; rays homocellular to heterocellular; fibers not septate.. **59**

59. Marginal parenchyma absent.
Cascara buckthorn—*Rhamnus purshiana* DC. Desc. p. 609

59. Marginal parenchyma present.. **60**

60. Marginal parenchyma conspicuous, forming a line 1–several cells in thickness; pitting on the vessel walls generally scalariform, the pits linear, 12–50 μ in diameter (Fig. 12-152).
Magnolia—*Magnolia* spp. Desc. p. 581

60. Marginal parenchyma inconspicuous and restricted to a more or less continuous uniseriate line, or sporadic; pits on the vessel walls orbicular to hexagonal, 4–10 μ in diameter (Fig. 12-153)............. **61**

61. Rays of two widths, the narrow uniseriate (rarely 2–3) and numerous, the broad 3–8-seriate (mostly 5–7).
Hard maple—*Acer* spp. Desc. p. 602

61. Rays intergrading in width, 1–5-seriate (mostly 3–4), the uniseriate rays sparse (especially in bigleaf maple).
Soft maple—*Acer* spp. Desc. p. 605

62. Largest vessels 40–60 μ in diameter, with or without spiral thickening; intervessel pits of medium size (Fig. 12-154); parenchyma marginal and paratracheal (restricted to occasional cells); rays and longitudinal elements storied or unstoried.
Buckeye—*Aesculus* spp. Desc. p. 607

62. Largest vessels 60–160 μ in diameter, without spiral thickening; intervessel pits large (Fig. 12-155); parenchyma strictly marginal; rays and longitudinal elements unstoried.................................. **63**

63. Rays essentially homocellular.. **64**

63. Rays essentially heterocellular.
Willow—*Salix* spp. Desc. p. 548

64. Vessels 30–145 per sq mm, the largest 75–150 μ in diameter.
Cottonwood—*Populus* spp. Desc. p. 544

64. Vessels 85–180 per sq mm, the largest 50–100 μ in diameter.
Aspen—*Populus* spp. Desc. p. 546

65. Spiral thickening in the vessels and at least in some of the fibers......... **66**

65. Spiral thickening restricted to the vessels or absent........................... **67**

66. Vessels arranged in radial strings or clusters which often extend across the rings; the wider rays are high, many 30 or more cells in height.
American holly—*Ilex opaca* Ait. Desc. p. 600

66. Vessels solitary or in multiples of 2 to several, frequently further aggregated into tangentially aligned clusters or bands, the zones of pores then alternating with bands of fibrous tissue; broader rays low, under 25 cells in height.
Pacific madrone—*Arbutus menziesii* Pursh. Desc. p. 618

67. Rays typically heterocellular, the upper and lower margins of the broad rays consisting of 1–5+ rows of upright cells; upright cells of the broad rays 30–100 μ in height along the grain...................................... **68**

67. Rays homocellular to heterocellular; when heterocellular the upper and lower margins of the broad rays generally consisting of 1–several (mostly 1) rows of upright cells; upright cells of the broad rays generally less than 60 μ in height along the grain... **70**

68. Rays 1–8-seriate; uniseriate rays usually consisting entirely of upright cells; apotracheal parenchyma relatively abundant.
Dogwood—*Cornus* spp. Desc. p. 615

68. Rays 1–4-seriate; uniseriate rays consisting of both upright and procumbent cells, apotracheal parenchyma relatively sparse............ **69**

69. Pits on the vessel walls (*t*) in transverse rows of 1–3, frequently linear through coalescence, the pit contour generally rounded (Fig. 12-157); interfiber pits conspicuously bordered, large (7–9 μ in diameter); longitudinal traumatic gum canals sometimes present.
Sweetgum—*Liquidambar styraciflua* L. Desc. p. 588

69. Pits on the vessel walls (*t*) in transverse rows of 1–5+, seldom linear, the pit contour frequently angular to rectangular (Fig. 12–159); interfiber pits inconspicuously bordered, small (3–4 μ in diameter); traumatic gum canals absent.
Tupelo—*Nyssa* spp. Desc. p. 612

70. Pits on the vessel walls orbicular to elliptical to rectangular or linear; medium-sized to large (4–50 μ in diameter)................... **71**

70. Pits on the vessel walls orbicular to hexagonal, minute (2–4 μ in diameter, Fig. 12-158).
Birch—*Betula* spp. Desc. p. 552

71. Parenchyma paratracheal or apotracheal diffuse, or both, the cells scattered; scalariform perforation plates with 10–50 bars.................. **72**

71. Parenchyma marginal, the band 1–several cells in thickness; scalariform perforation plates with 6–10 bars... **74**

72. Spiral thickening in the vessels restricted to the tapering ends of the members; perforation plates exclusively scalariform; rays 6–13 per mm (x) **73**

72. Spiral thickening in the vessels not restricted to the tapering ends of the elements; perforation plates simple and scalariform; rays 3–5 per mm (x).
Sourwood—*Oxydendrum arboreum* (L.) DC. Desc. p. 617

73. Pits on the vessel walls (t) in transverse rows of 1–3, frequently linear through coalescence, the pit contour generally rounded (Fig. 12-157); interfiber pits conspicuously bordered, large (7–9 μ in diameter); longitudinal traumatic gum canals sometimes present.
Sweetgum—*Liquidambar styraciflua* L. Desc. p. 588

73. Pits on the vessel walls (t) in transverse rows of 1–5+, seldom linear, the pit contour frequently angular to rectangular (Fig. 12-159); interfiber pits inconspicuously bordered, small (3–4 μ in diameter); longitudinal traumatic gum canals absent.
Tupelo—*Nyssa* spp. Desc. p. 612

74. Bars of scalariform perforation plates thick (more than 1 μ); pits on the vessel walls linear, frequently extending across the vessel, 12–50 μ in diameter (similar to Fig. 12-152).
Southern magnolia—*Magnolia grandiflora* L. Desc. p. 582

74. Bars of scalariform perforation plates thin (less than 1 μ); pits on the vessel walls orbicular to elliptical or linear, seldom extending across the vessel, 6–40 (usually 6–20) μ in diameter (Fig. 12-160).
Yellow-poplar—*Liriodendron tulipifera* L. Desc. p. 584

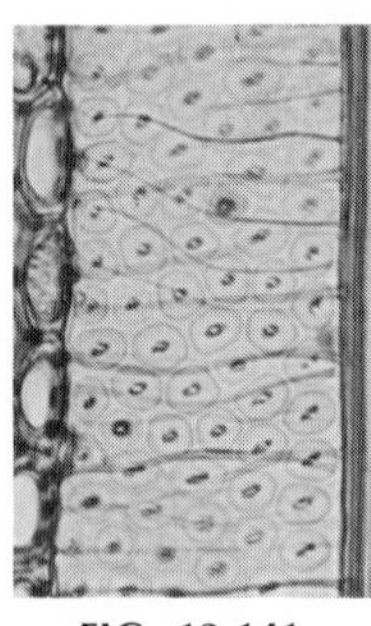
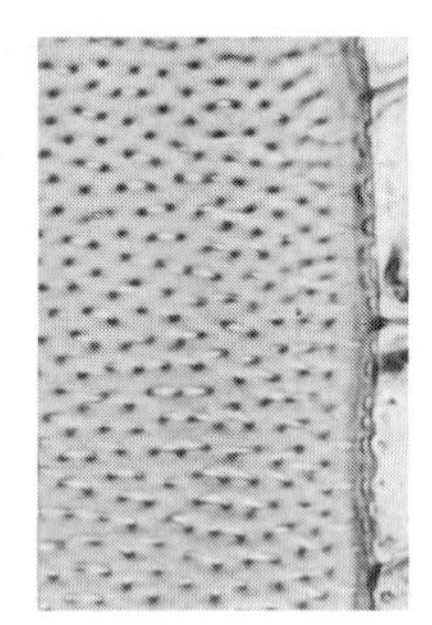
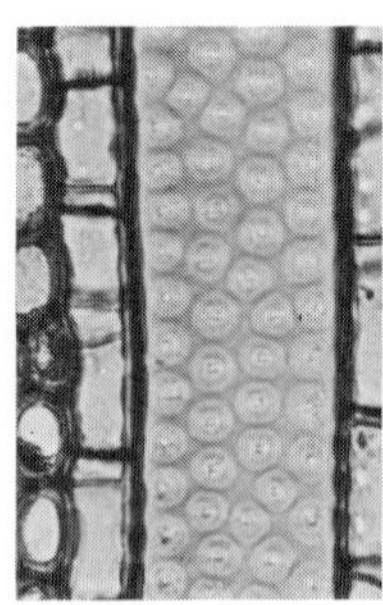
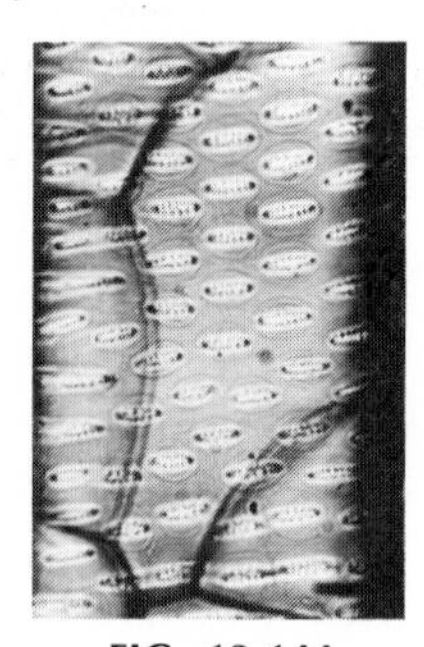

FIG. 12-141 **FIG. 12-142** **FIG. 12-143** **FIG. 12-144**

FIG. 12-141 Portion of the wall of a vessel element (t) from the wood of black cherry, *Prunus serotina* Ehrh., showing intervessel pitting and spiral thickening. (400×.)

FIG. 12-142 Portion of the wall of a vessel element (t) from the wood of common persimmon, *Diospyros virginiana* L., showing minute intervessel pits with coalescent apertures. (400×.)

FIG. 12-143 Portion of the wall of a vessel element (t) from the wood of shagbark hickory, *Carya ovata* (Mill.) K. Koch, showing intervessel pitting. The punctate appearance of the pits is the result of artifacts and is not vesturing. (400×.)

FIG. 12-144 Portion of the wall of a vessel element (t) from the wood of black locust, *Robinia pseudoacacia* L., showing vestured intervessel pits. (400×.)

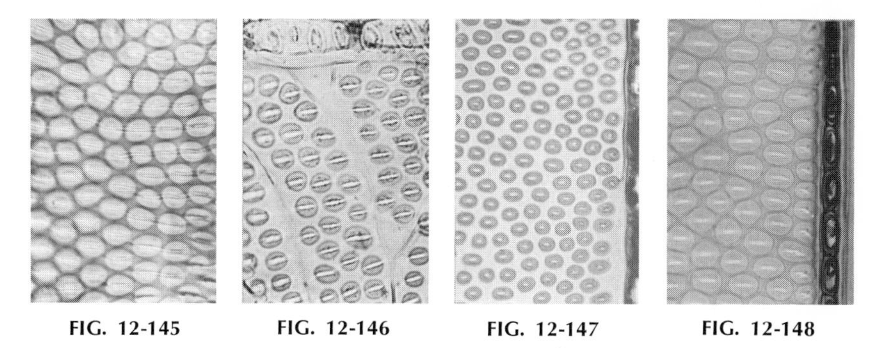

FIG. 12-145 FIG. 12-146 FIG. 12-147 FIG. 12-148

FIG. 12-145 Portion of the wall of a vessel element (*t*) from the wood of Osage-orange, *Maclura pomifera* (Raf.) Schneid., showing intervessel pits with coalescent apertures. (400×.)

FIG. 12-146 Portions of the walls of three vasicentric tracheids and a wood ray (*r*) from the wood of Shumard oak, *Quercus shumardii* Buckl., showing intertracheary pitting. The pits have a punctate appearance because of artifacts; they are not vestured. (400×.)

FIG. 12-147 Portion of the wall of a vessel element (*t*) from the wood of red alder, *Alnus rubra* Bong., showing widely spaced intervessel pitting. (400×.)

FIG. 12-148 Portion of the wall of a vessel element (*t*) from the wood of American hornbeam, *Carpinus caroliniana* Walt., showing crowded, angular intervessel pitting. (400×.)

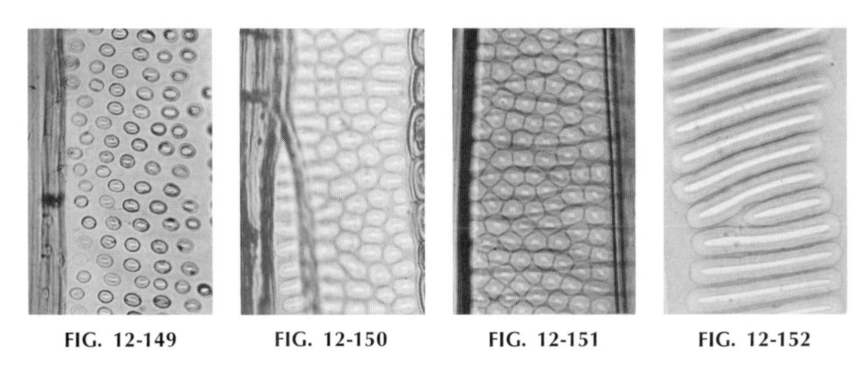

FIG. 12-149 FIG. 12-150 FIG. 12-151 FIG. 12-152

FIG. 12-149 Portion of the wall of a vessel element (*t*) from the wood of American sycamore, *Platanus occidentalis* L., showing widely spaced intervessel pitting. (400×.)

FIG. 12-150 Portions of the walls of two vessel elements (*t*) from the wood of American beech, *Fagus grandifolia* Ehrh., showing crowded intervessel pits. (400×.)

FIG. 12-151 Portions of the wall of a vessel element (*t*) from the wood of basswood, *Tilia americana* L., showing crowded intervessel pits and spiral thickening. (400×.)

FIG. 12-152 Portion of the wall of a vessel element (*t*) from the wood of cucumbertree, *Magnolia acuminata* L., showing scalariform intervessel pitting. (400×.)

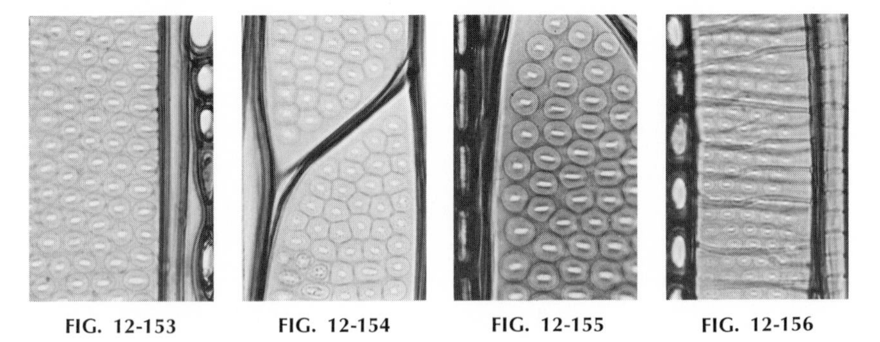

FIG. 12-153 FIG. 12-154 FIG. 12-155 FIG. 12-156

FIG. 12-153 Portion of the wall of a vessel element (*t*) from the wood of silver maple, *Acer saccharinum* L., showing intervessel pitting. (400×.)

FIG. 12-154 Portions of the walls of two vessel elements (*t*) from the wood of yellow buckeye, *Aesculus octandra* Marsh., showing medium-sized intervessel pits. The pits in the upper right corner are punctate because of artifacts and are not vestured. (400×.)

FIG. 12-155 Portion of the wall of a vessel element (*t*) from the wood of eastern cottonwood, *Populus deltoides* Bartr., showing large intervessel pits. (400×.)

FIG. 12-156 Portion of the wall of a vessel element (*t*) from the wood of American holly, *Ilex opaca* Ait., showing intervessel pitting and spiral thickening. The portion of a fiber at the right of the picture also shows spiral thickening. (400×.)

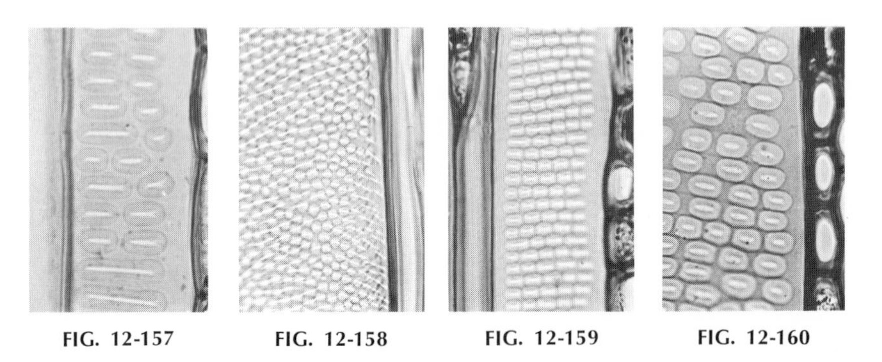

FIG. 12-157 FIG. 12-158 FIG. 12-159 FIG. 12-160

FIG. 12-157 Portion of the wall of a vessel element (*t*) from the wood of sweetgum, *Liquidambar styraciflua* L., showing oval and linear intervessel pits. (400×.)

FIG. 12-158 Portion of the wall of a vessel element (*t*) from the wood of yellow birch, *Betula alleghaniensis* Britton, showing small intervessel pits with coalescent apertures. (400×.)

FIG. 12-159 Portion of the wall of a vessel element (*t*) from the wood of black tupelo, *Nyssa sylvatica* Marsh., showing opposite intervessel pitting. (400×.)

FIG. 12-160 Portion of the wall of a vessel element (*t*) from the wood of yellow-poplar, *Liriodendron tulipifera* L., showing opposite intervessel pitting. (400×.)

Descriptions of Hardwoods by Species

In the case of the hardwoods, the great complexity of kinds and sizes of the cells, coupled with considerable variation in their dimensions in the same growth ring, as well as in different parts of the stem, makes definite numerical values not very meaningful. Values are presented, therefore, by the general descriptive terms which can be related to a range of numerical values by reference to the tables which follow.

Distribution of the Pores as Seen in Cross Section

Class	*Number per sq mm*
Very few	Less than 16
Few	16–25
Moderately few	26–50
Moderately numerous	51–75
Numerous	76–100
Very numerous	More than 100

Tangential Diameter of the Pores in Microns, Including the Thickness of the Walls

Class	*Tangential diameter, μ*
Very small	Less than 50
Small	51–100
Medium	101–150
Large	151–200
Very large	201–300
Extremely large	More than 300

Tangential Diameter of the Fibers in Microns, Including the Thickness of the Walls

Class	*Tangential diameter, μ*
Fine	Up to 16
Medium	16–25
Coarse	26–30
Very coarse	More than 30

Further numerical information on the anatomical features of hardwoods is presented in Tables 5-5, 5-6, 5-9, and 5-10 in Chap. 5.

BUTTERNUT, WHITE WALNUT

Juglans cinera L.

GENERAL CHARACTERISTICS

Sapwood white to light grayish brown, rarely more than 1 in. wide; *heartwood* light chestnut-brown, frequently variegated with pigment figure, lustrous; *wood* without characteristic odor or taste, straight-grained, moderately light (sp gr approx 0.36 green, 0.46 ovendry), moderately soft. *Growth rings* distinct, delineated through an abrupt difference in size between the pores of the outer late wood and those in the early wood of the succeeding ring. *Pores* scattered (never approximating half the area between the rays), those in the early wood readily visible to the naked eye, decreasing gradually in size toward the outer margin of the ring (wood semi-ring-porous), solitary and in multiples of 2–several; tyloses fairly abundant. *Parenchyma* visible with a hand lens (especially in the outer portion of the ring), arranged in fine, numerous, more or less continuous, tangential lines. *Rays* fine, indistinct without a hand lens.

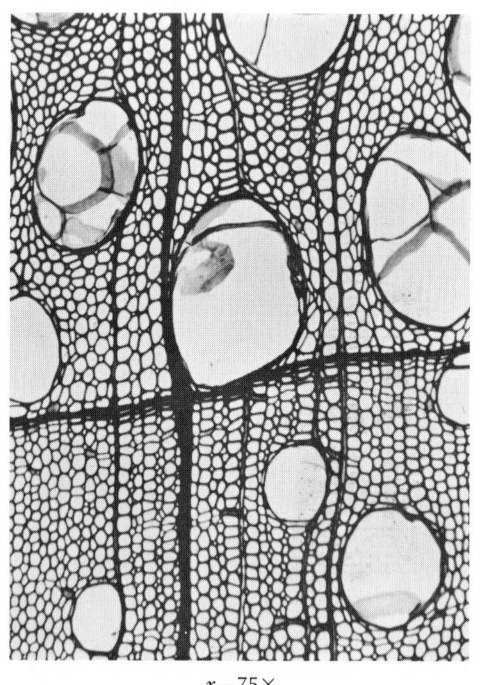

x—75×

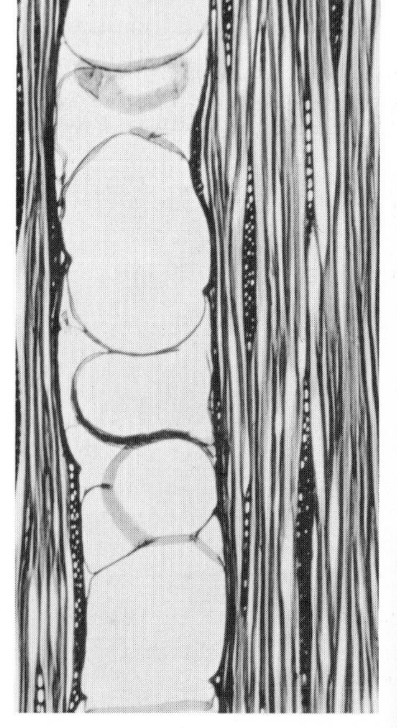

t—75×

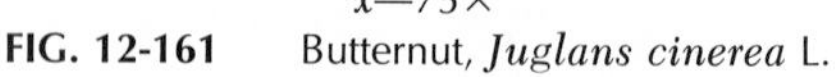

FIG. 12-161 Butternut, *Juglans cinerea* L.

MINUTE ANATOMY

Vessels very few, the largest large to very large; perforation plates simple; intervessel pits orbicular to oval or angular through crowding, large (8–16 μ in diameter). *Parenchyma* apotracheal-diffuse to banded, and scanty paratracheal to vasicentric, noncrystalliferous; lines of zonate parenchyma mostly uniseriate. *Fiber tracheids* thin-walled, medium to very coarse. *Rays* unstoried, 1–4-seriate, homocellular to heterocellular; ray-vessel pitting similar to intervessel type.

USES

Furniture; *cabinetwork*; *instrument cases*; *boxes* and *crates*; *millwork* (such as interior trim, paneling, sash, and doors); *woodenware, toys,* and *novelties*; *face veneer* for interior-grade plywood.

BLACK WALNUT, EASTERN BLACK WALNUT

Juglans nigra L.

GENERAL CHARACTERISTICS

Sapwood whitish to yellowish brown (in the trade commonly darkened by steaming or staining to match the heartwood); *heartwood* light brown to rich chocolate or purplish brown (the lighter shades from trees grown in the open), dull; *wood* with mild characteristic odor when worked, tasteless, straight- or irregular-grained (the wavy, curly, and mottled figures for which this wood is famous are obtained from burls, crotches, and stumpwood; quarter-sawn wood frequently shows alternate stripes due to uneven pigmentation), heavy (sp gr approx 0.51 green, 0.56 ovendry), hard. *Growth rings* distinct, delineated through an abrupt difference in size between the pores of the outer late wood and those in the early wood of the succeeding ring. *Pores* scattered (never approximating half the area between the rays), those in the early wood readily visible to the naked eye, decreasing gradually in size toward the outer margin of the ring (wood semi-ring-porous), solitary and in multiples of 2–several; tyloses fairly abundant. *Parenchyma* barely visible with a hand lens, arranged in fine, numerous, continuous, or broken tangential lines. *Rays* fine, indistinct without a hand lens.

MINUTE ANATOMY

Vessels very few, the largest large to very large; perforation plates simple; intervessel pits orbicular to oval or angular through crowding, large (10–16 μ in diameter). *Parenchyma* apotracheal-diffuse to banded, and scanty paratracheal to vasicentric, frequently crystalliferous; lines of banded parenchyma mostly uniseriate. *Fiber tracheids* thin- to moderately thick-walled, medium to very coarse.

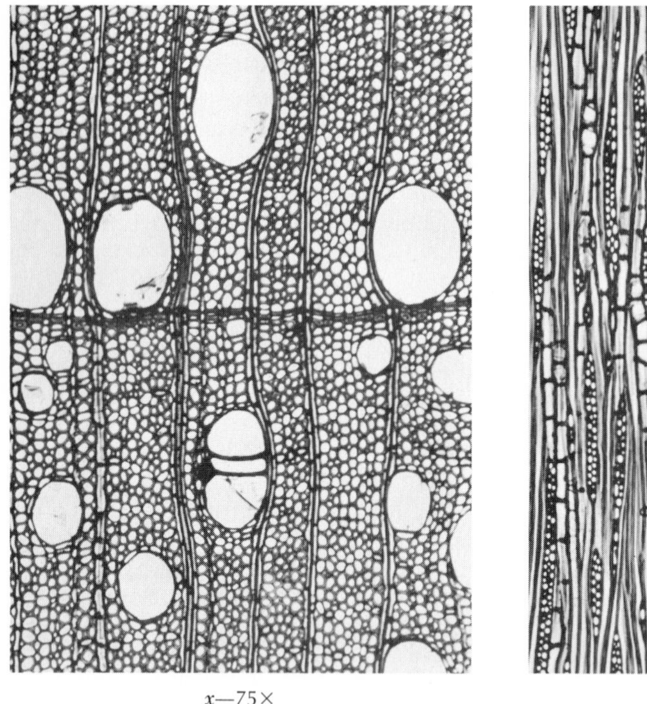

x—75× *t*—75×

FIG. 12-162 Black walnut, *Juglans nigra* L.

Rays 1–5-seriate, homocellular to heterocellular; ray-vessel pitting similar to intervessel type.

REMARKS

Unique among the woods of the United States in the color of its heartwood, and hence easily identified. Unquestionably our finest domestic cabinet wood. The Claro walnut of the trade is obtained from ***Juglans hindsii*** Jepson, a tree native to California; the heartwood is tan-brown, with prominent dark stripes and spots.

USES

Veneer (largely sliced, rotary-cut to a limited extent), used extensively for plywood faces in the manufacture of furniture, in cabinetwork, and in interior paneling; *gunstocks* (leading wood because of sufficient strength, shock resistance, and hardness, ability to stay in place, and good machining and finishing properties combined with sufficient coarseness of texture to permit of a good grip on the gun). Lumber used principally for ***furniture***, especially for dining-room tables (because of its ability to stand hard usage), chairs, bedroom, and living-room suites, office furniture; *fixtures*; *caskets* and *coffins* (because of good working qualities and pleasing appearance, combined with durability); *radio*, *television*, and *phonograph cabinets*; *piano cases*; *millwork* (such as doors, sash, frames, and interior finish); *sewing machines*; *woodenware* and *novelties*.

HICKORY

True Hickories

Shagbark hickory [*Carya ovata* (Mill.) K. Koch]
Shellbark hickory [*Carya laciniosa* (Michx. f.) Loud.]
Pignut hickory [*Carya glabra* (Mill.) Sweet]
Mockernut hickory (*Carya tomentosa* Nutt.)

Pecan Hickories

Bitternut hickory [*Carya cordiformis* (Wangenh.) K. Koch]
Pecan [*Carya illinoensis* (Wangenh.) K. Koch]
Nutmeg hickory [*Carya myristicaeformis* (Michx. f.) Nutt.]
Water hickory [*Carya aquatica* (Michx. f.) Nutt.]

GENERAL CHARACTERISTICS

Sapwood whitish to pale brown; *heartwood* pale brown to brown, or reddish brown (heartwood of pecan rich reddish brown, sometimes containing streaks of

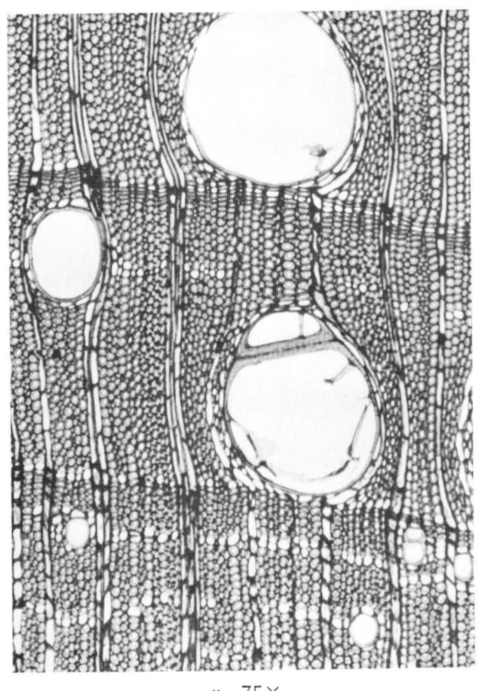

x—75×

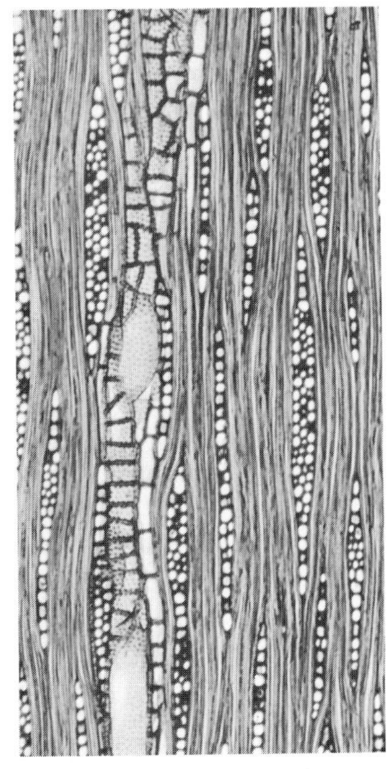

t—75×

FIG. 12-163 Mockernut hickory, *Carya tomentosa* Nutt

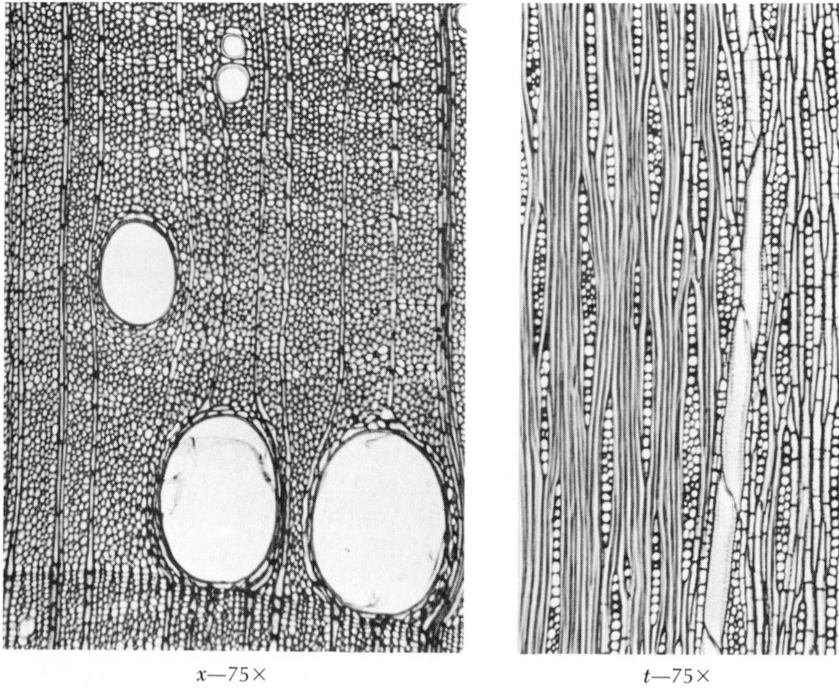

x—75× *t*—75×

FIG. 12-164 Shagbark hickory, *Carya ovata* (Mill.) K. Koch

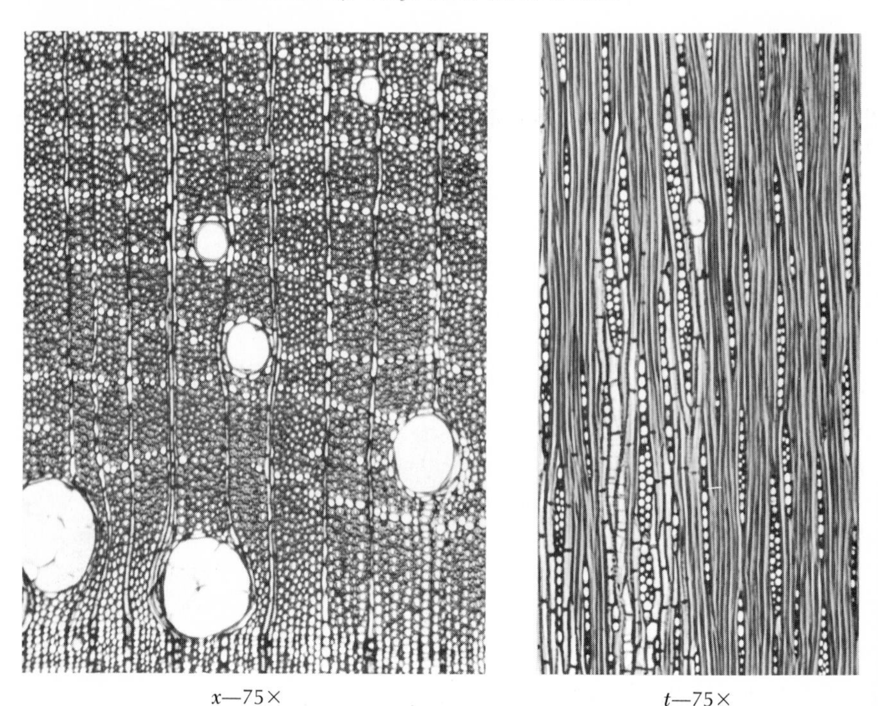

x—75× *t*—75×

FIG. 12-165 Bitternut hickory, *Carya cordiformis* (Wangenh.) K. Koch

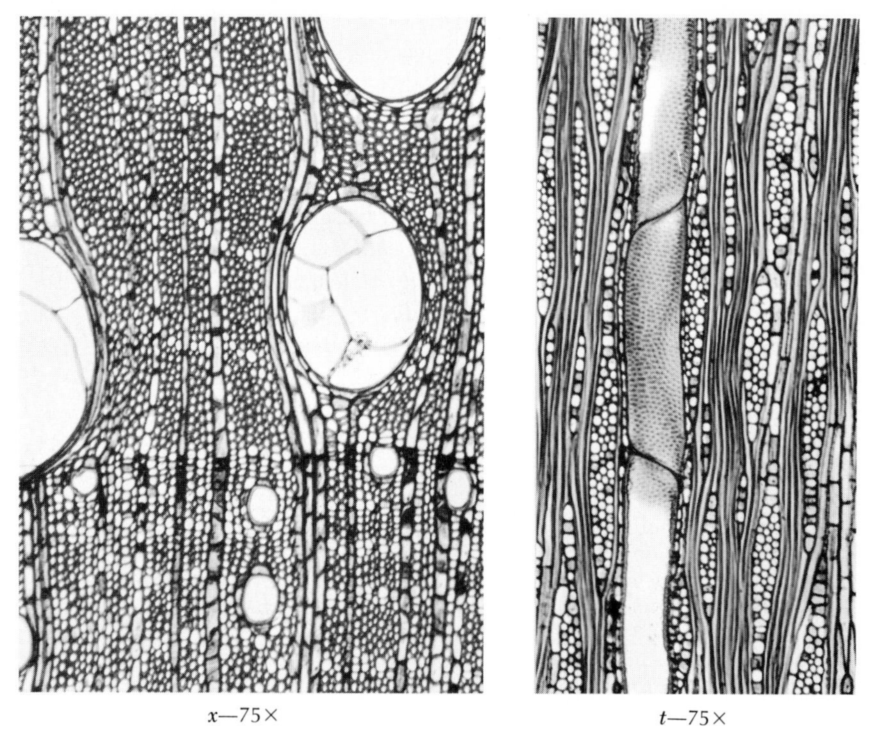

FIG. 12-166 Pecan, *Carya illinoensis* (Wangenh.) K. Koch

a slightly darker hue); *wood* without characteristic odor or taste, straight-grained, heavy to very heavy (sp gr 0.56–0.66 green, 0.62–0.78 ovendry), very hard. *Growth rings* distinct (wood ring-porous or semi-ring-porous). *Early-wood pores* large, visible to the naked eye; transition from early to late wood abrupt (true hickories) or more or less gradual (pecan hickories); *late-wood pores* small, visible with a hand lens, solitary and in multiples of 2–3. *Parenchyma* conspicuous with a hand lens, in fine, continuous, tangential lines which are arranged irrespective of the pores. *Rays* indistinct without a hand lens.

MINUTE ANATOMY

Vessels very few; largest late-wood vessels large to extremely large; perforation plates simple; intervessel pits orbicular to oval or angular through crowding (Fig. 12-143, p. 534), 6–8 μ in diameter. *Parenchyma* apotracheal-diffuse and in aggregates, and marginal; lines of banded parenchyma 1–4-seriate. *Fiber tracheids* thin- to thick-walled, frequently gelatinous, fine to medium. *Rays* 1–5-seriate, homocellular to heterocellular; ray-vessel pitting similar to intervessel type.

REMARKS

The woods of the pecan hickories (except *Carya cordiformis*) exhibit apotracheal banded parenchyma in the early-wood zones (*x*), while the true hickories differ in not showing parenchyma bands in the early-wood.* Pecan also displays a gradation in size of pores from early- to late-wood (wood semi-ring-porous).

USES

Tool handles (especially for "impact" tools such as hammers, axes, picks, and sledges) for which use hickory is unsurpassed because of its inherent qualities of hardness, strength, toughness, and resiliency; *ladders*; *furniture*; *sporting goods* (such as skis, gymnastic bars); *agricultural implements*; *flooring* (only pecan); *woodenware* and *novelties*; *special products* requiring a strong, tough, elastic wood (such as sucker rods—long straight-grained strips used in boring wells, picker sticks in cotton and silk mills, dowel pins, skewers); for *smoking meats*; *fuel wood* because of its high caloric value.

COTTONWOOD

Eastern cottonwood, eastern poplar (*Populus deltoides* Bartr.)
Balsam poplar, tacamahac poplar (*Populus balsamifera* L.)
Swamp cottonwood, swamp poplar (*Populus heterophylla* L.)
Black cottonwood (*Populus trichocarpa* Torr. & Gray)
Etc.

GENERAL CHARACTERISTICS

Sapwood whitish, frequently merging into the heartwood and hence not clearly defined, thin or thick; *heartwood* grayish white to light grayish brown (sometimes dull brown in *P. heterophylla* L.); *wood* odorless or with a characteristic disagreeable odor when moist, without characteristic taste, usually straight-grained, medium light to light (sp gr 0.32–0.37 green, 0.37–0.43 ovendry), moderately soft to soft. *Growth rings* distinct but inconspicuous, narrow to very wide. *Pores* numerous, small, the largest barely visible to the naked eye and more crowded in the first-formed early wood, decreasing gradually in size through the late wood (wood semi-ring to diffuse-porous), solitary and in multiples of 2–several. *Parenchyma* marginal, the narrow, light-colored line more or less distinct. *Rays* very fine, scarcely visible with a hand lens.

MINUTE ANATOMY

Vessels moderately few to very numerous, the largest small to medium, perforation plates simple; intervessel pits orbicular to oval or angular through crowding, large

* M. A. Taras and B. F. Kukachka, Separating Pecan and Hickory Lumber, *Forest Prod. J.*, **40**(4):58–59 (1970).

Some progress has been made in the separation of the woods of *Carya* spp. by color reactions. For further information, see I. H. Isenberg and M. A. Buchanan, A Color Reaction of Wood with Methanol-hydrochloric Acid, *J. Forestry*, **43:**888–890 (1945).

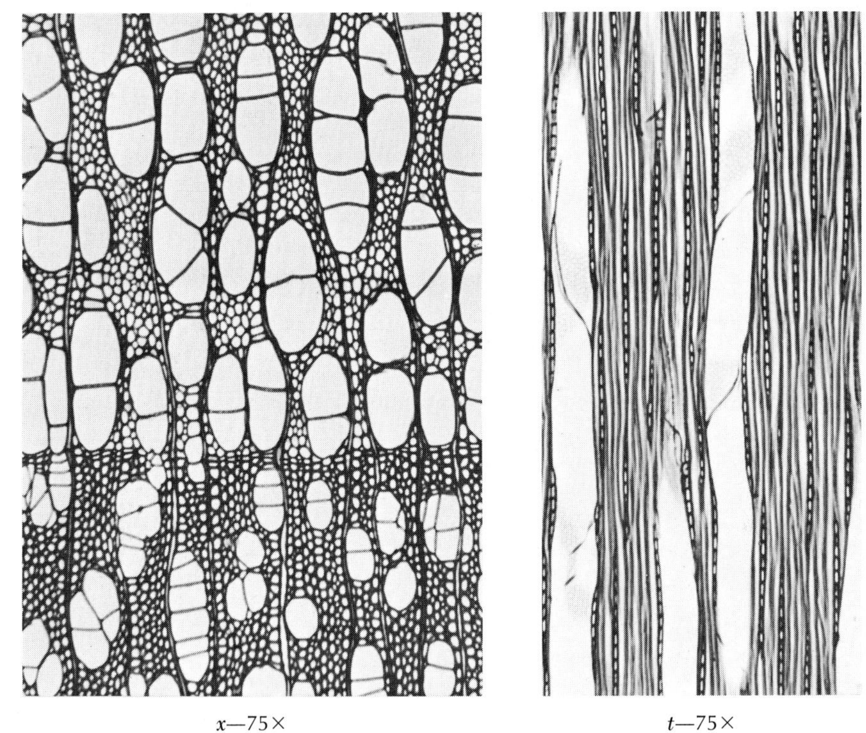

x—75× *t*—75×

FIG. 12-167 Eastern cottonwood, *Populus deltoides* Bartr

(9–13 μ in diameter). *Parenchyma* marginal, forming a narrow, continuous, or interrupted line. *Libriform fibers* coarse to very coarse, thin- to medium thick-walled, occasionally gelatinous, medium to very coarse. *Rays* unstoried, uniseriate, essentially homocellular; pits leading to vessels confined to the marginal cells or occurring in occasional rows in the body of the ray as well, simple to bordered.

REMARKS

The heartwood of *Populus heterophylla* L. is somewhat darker than that of the other species. The aspens are similar in structure to the cottonwoods but finer textured. For differences between the woods of *Salix* and *Populus*, see Remarks under Willow, p. 549.

USES

Pulp (manufactured by an alkaline process) for high-grade book and magazine paper; *excelsior* (the aspens and cottonwood are the principal excelsior woods, for which purpose they are especially well suited because of freedom from staining materials, light color, light weight, and uniformity of texture and straightness of

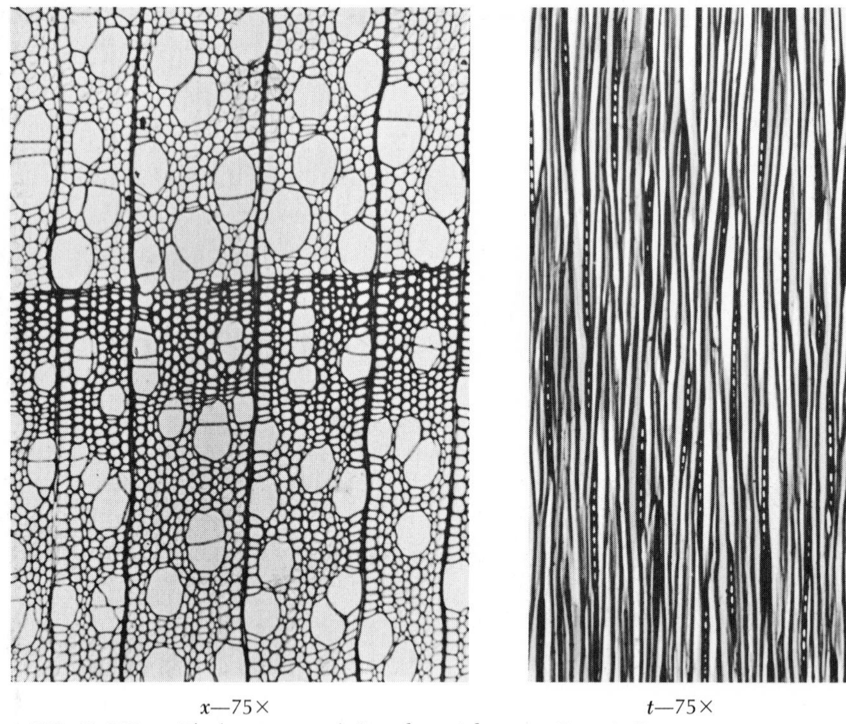

x—75× *t*—75×

FIG. 12-168 Black cottonwood, *Populus trichocarpa* Torr. & Gray

grain which permit of easy shredding into soft but strong and resilient strands); *veneer* for the manufacture of plywood for furniture (mostly as core and cross-banding stock), *musical instruments, containers* (such as berry boxes). *Lumber* (it is estimated that 85 percent of the lumber classified as "cottonwood" is cottonwood and balsam poplar, and the remainder aspen) used for *pallets, boxes,* and *crates* (about two-thirds to three-fourths of all "cottonwood" lumber) for which it is especially well suited because of light weight, ease of nailing without splitting, and good color for stenciling; *furniture* (concealed parts); *poultry* and *apiary supplies* (especially poultry coops and brooders); *laundry appliances* (such as ironing boards); *tubs* and *pails* for butter, lard, jelly, and other food products.

ASPEN

Quaking aspen, trembling aspen (*Populus tremuloides* Michx.)

Bigtooth aspen (*Populus grandidentata* Michx.)

GENERAL CHARACTERISTICS

Sapwood whitish to creamy colored, generally merging gradually into heartwood and hence not clearly defined, wide; *heartwood* whitish, creamy to light grayish brown; *wood* with a characteristic disagreeable odor when wet, odorless when dry, without characteristic taste, with a pronounced silky luster, usually straight-grained, medium light (sp gr approx 0.36 green, 0.40 ovendry), soft. *Growth rings* distinct because of darker late wood but not conspicuous, wide. *Pores* numerous, small, not visible without a hand lens, more crowded in the early wood, decreasing gradually in size through the late wood, solitary or in multiples of 2–several. *Parenchyma* marginal, indistinct. *Rays* very fine, scarcely visible with a hand lens.

MINUTE ANATOMY

Vessels numerous to very numerous, the largest very small to small in diameter; perforation plates simple; intervessel pits orbicular to oval or angular through crowding, large (8–12 μ in diameter). *Parenchyma* terminal, forming a narrow, continuous, or interrupted line. *Libriform fibers* thin- to medium thick-walled, occasionally gelatinous, 20–30 μ, medium to coarse. *Rays* unstoried, uniseriate, essentially homocellular; pits leading to vessels confined to the marginal cells or occurring in occasional rows in the body of the ray as well, simple to bordered.

REMARKS

The woods of the two aspens are quite similar in appearance and properties and cannot be separated from each other. The woods of the cottonwoods are also quite

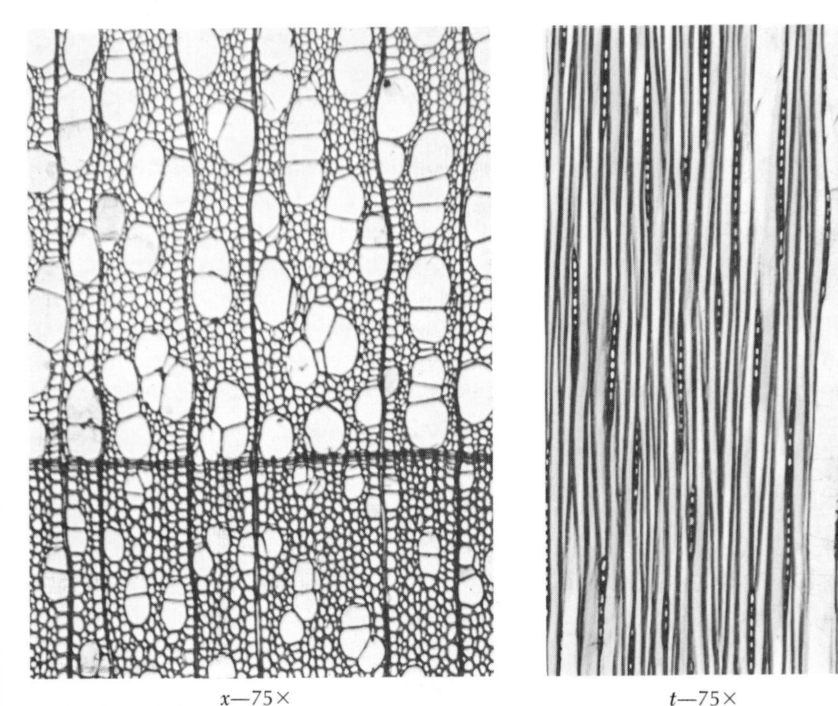

x—75× *t*—75×

FIG. 12-169 Bigtooth aspen, *Populus grandidentata* Michx.

similar to aspen, but coarser in texture, somewhat darker in color (never creamy), and devoid of luster.

USES

Pulp (manufactured by an alkaline process for high-grade book and magazine paper, or in admixture with spruce and balsam for sulfite pulp); *excelsior* (aspen and the cottonwoods are the principal excelsior woods, but the use of wood excelsior is rapidly declining) for which purpose they are especially well suited because of freedom from staining materials, light color, light weight, and uniformity of texture and straightness of grain which permit of easy shredding into soft but strong and resilent strands; logs are used for *cabin construction* and, when kept away from the ground, give satisfactory service; *matches* (veneer-cut by the rotary method). Lumber used for *boxes* and *crates*; *pallets*; *core stock* in plywood panels; *laundry appliances*; *poultry* and *apiary supplies*; *tubs* and *pails* for food products; *clothespins*; *rough construction* (locally); *veneer* for containers; research in Canada and the United States is directed toward assessing suitability of aspen for production of structural and general-utility plywood. Increasing quantities of pulpwood used for *particle board.*

BLACK WILLOW

Salix nigra Marsh.

GENERAL CHARACTERISTICS

Sapwood whitish, thin or thick; *heartwood* light brown to pale reddish or grayish brown, frequently with darker streaks along the grain; *wood* without characteristic odor or taste, usually straight-grained, moderately light to light (sp gr approx 0.34 green, 0.41 ovendry), moderately soft. *Growth rings* inconspicuous, narrow to wide. *Pores* numerous, small, the largest barely visible to the naked eye in the early wood, decreasing gradually in size through the late wood (wood semi-ring- to diffuse-porous), solitary and in multiples of 2–several. *Parenchyma* marginal, generally not visible at low magnifications. *Rays* very fine, scarcely visible with a hand lens.

MINUTE ANATOMY

Vessels moderately few to very numerous, the largest small to large; perforation plates simple; intervessel pits orbicular to oval or angular through crowding, large (6–10 μ in diameter). *Parenchyma* marginal, forming a narrow, continuous or interrupted, 1–2-seriate line. *Libriform fibers* thin- to moderately thick-walled, medium to coarse. *Rays* unstoried, uniseriate, heterocellular; upright cells in 1–several (mostly 1) marginal rows and not infrequently also in the body of the ray;

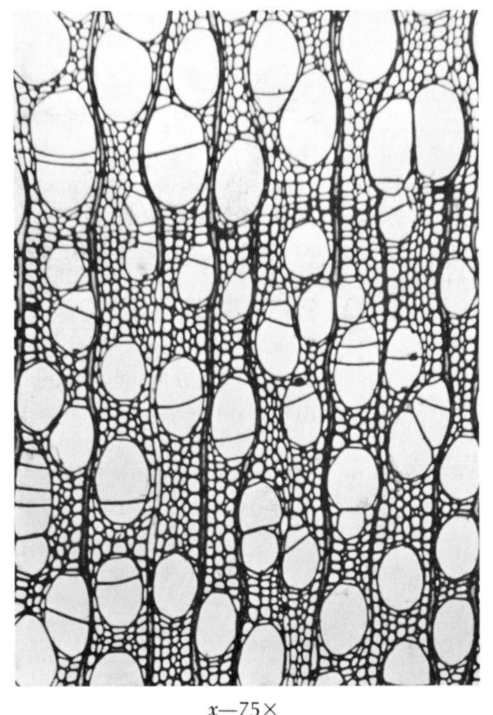

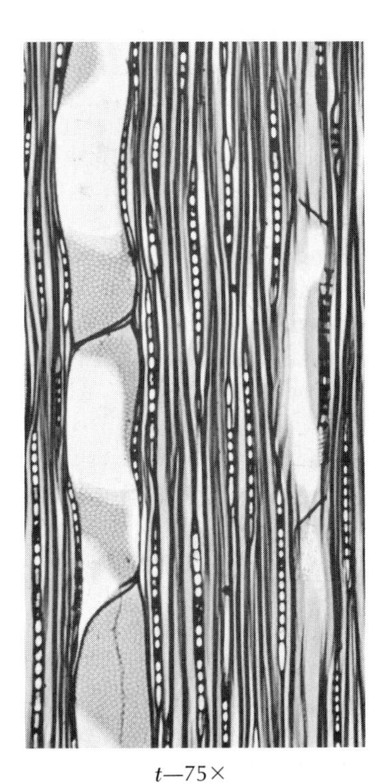

x—75×　　*t*—75×

FIG. 12-170 Black willow, *Salix nigra* Marsh.

pits leading to vessels restricted to the upright cells, fairly numerous and forming a more or less reticulate pattern, similar to intervessel type.

REMARKS

The woods of *Salix* and *Populus* are very similar but can usually be separated through color; the willows exhibit a decided brown or reddish-brown cast in contrast to the grayish-white or light grayish-brown shades that characterize *Populus* spp. Accurate identification as to genus is assured by the fact that the rays are always heterocellular in *Salix* and essentially homocellular in *Populus*.

USES

Boxes and *crates*; *furniture* (core stock, turned pieces, table tops); *slack cooperage*; *woodenware* and *novelties*; frequently sold in admixture with, and used for the same purposes as, cottonwood lumber (see p. 545); *caskets*; *veneer*; *charcoal*, resulting from destructive distillation is especially suitable for black-powder manufacture.

AMERICAN HORNBEAM, BLUE BEECH

Carpinus caroliniana Walt.

GENERAL CHARACTERISTICS

Sapwood nearly white, thick; *heartwood* pale yellowish or brownish white; *wood* without characteristic odor or taste, heavy (sp gr approx 0.58 green, 0.72 ovendry), hard. *Growth rings* usually distinct (wood diffuse-porous), delineated by a narrow, whitish band at the outer margin, sinuate (trunk of tree fluted). *Pores* small, indistinct without a hand lens, arranged for the most part in multiples of 2–several which are usually further aggregated into radial strings. *Parenchyma* distinct or indistinct with a hand lens, when visible appearing as fine light-colored lines spanning the intervals between the rays. *Rays* of two types, broad (aggregate) and narrow (simple); (*a*) aggregate rays visible to the naked eye (*x*), numerous, more or less irregularly distributed, separated by several–many narrow rays; (*b*) narrow rays indistinct without a hand lens.

MINUTE ANATOMY

Vessels moderately numerous, the largest small; perforation plates simple or rarely scalariform with 3–6 thick bars; spiral thickening present; intervessel pits

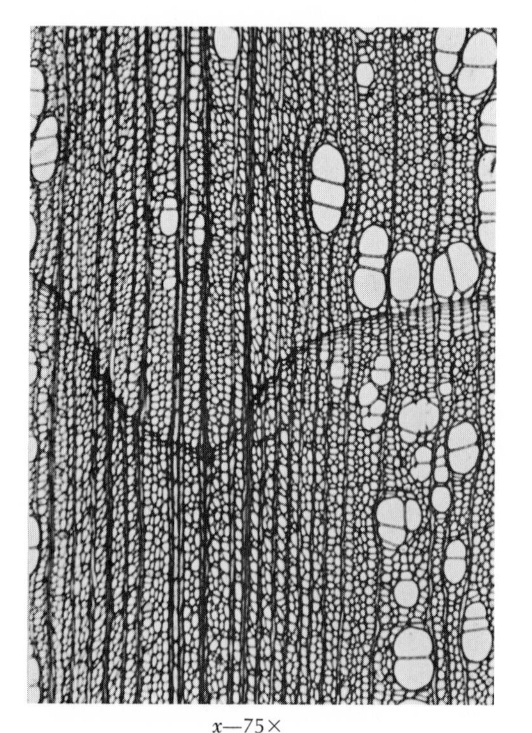

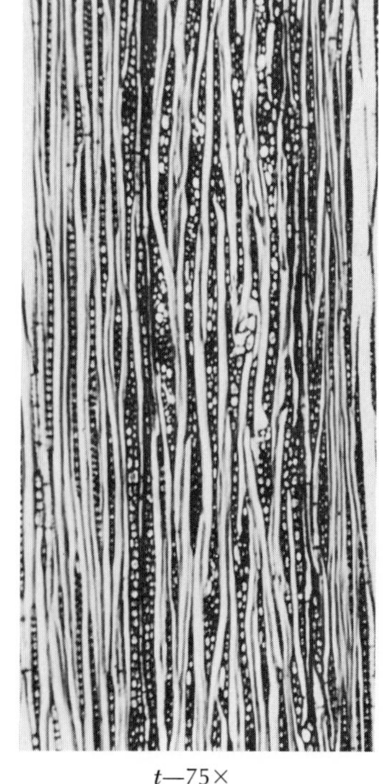

x—75×　　*t*—75×

FIG. 12-171 American hornbeam, *Carpinus caroliniana* Walt.

orbicular to oval or angular through crowding, large (6–15 μ in diameter). *Parenchyma* apotracheal-diffuse to banded and marginal; banded parenchyma very abundant, the lines uniseriate; diffuse and marginal parenchyma relatively sparse. *Fiber tracheids* thin- to moderately thick-walled, fine to medium. Narrow *rays* 1–4 (mostly 1–2)-seriate, homocellular to heterocellular; aggregate rays consisting (1) of units similar to the narrow rays and (2) of included fibers and vessels; ray-vessel pitting simple to bordered.

USES

Locally for *handles, farm vehicle parts, fuel wood.*

EASTERN HOPHORNBEAM

Ostrya virginiana (Mill.) K. Koch

GENERAL CHARACTERISTICS

Sapwood whitish, wide; *heartwood* whitish to light brown tinged with red; *wood* without characteristic odor or taste, very heavy (sp gr approx 0.63 green, 0.78 ovendry), very hard. *Growth rings* inconspicuous (wood diffuse-porous), with ragged (erose) contours, the outer late wood often specked with white. *Pores* small, indistinct or barely visible to the naked eye, solitary and in multiples of 2–several which are more or less unevenly distributed throughout the growth ring and are often further aggregated into flamelike groups. Banded *parenchyma* generally visible with a lens in the outer portion of the ring. *Rays* fine, indistinct to the naked eye, close and often seemingly occupying half of the area on the transverse surface of the wood.

MINUTE ANATOMY

Vessels moderately numerous to numerous, small to medium; perforation plates simple or rarely scalariform with 3–6 bars; spiral thickening present; intervessel pits orbicular to oval or angular through crowding, large (6–10 μ in diameter). *Parenchyma* apotracheal-diffuse to banded, and marginal; lines of banded parenchyma more evident in the outer portion of the ring, uniseriate; marginal parenchyma sparse. *Libriform fibers* medium thick-walled, fine to medium. *Rays* unstoried, 1–3-seriate, homocellular to heterocellular, the tallest less than 1000 μ in height along the grain; ray-vessel pitting simple to bordered.

REMARKS

Sometimes confused with birch, from which it differs in possessing ragged growth rings which are often specked with white near the outer margin, vessels which exhibit more or less a tendency toward aggregation into flamelike groups, vessel

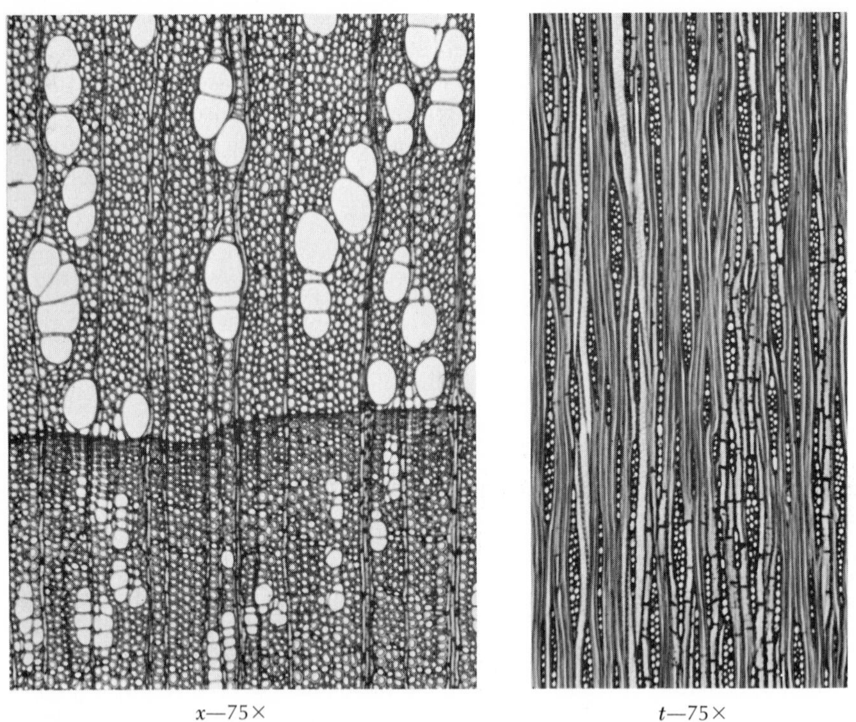

x—75× *t*—75×

FIG. 12-172 Eastern hophornbeam, *Ostrya virginiana* (Mill.) K. Koch

elements with predominantly simple perforations and spiral thickening, and zonate parenchyma in the outer portion of the ring.

USES

Locally for *furniture*; *farm vehicles* (such parts as axles); *handles*; *levers*; *mallets*; *canes*; *woodenware* and *novelties*; *fuel wood*.

BIRCH

Yellow birch (*Betula alleghaniensis* Britton)
Sweet birch, black birch, cherry birch (*Betula lenta* L.)
River birch, red birch (*Betula nigra* L.)
Paper birch, white birch (*Betula papyrifera* Marsh.)
Gray birch (*Betula populifolia* Marsh.)

GENERAL CHARACTERISTICS

Sapwood whitish, pale yellow, or light reddish brown; *heartwood* light to dark brown or reddish brown; *wood* without characteristic odor or taste, straight-

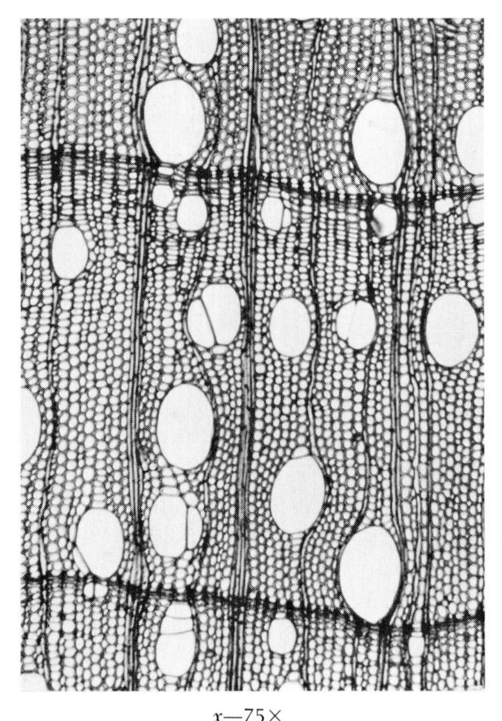

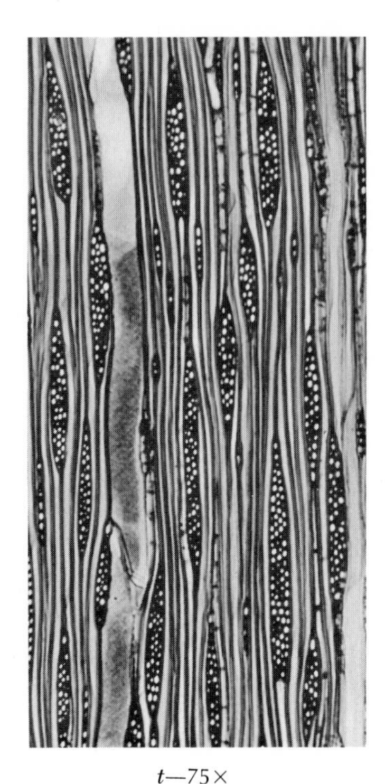

x—75×　　　　*t*—75×

FIG. 12-173 Sweet birch, *Betula lenta* L.

grained, moderately heavy to very heavy (sp gr 0.45–0.60 green, 0.55–0.71 oven-dry), moderately hard to hard. *Growth rings* frequently not very distinct without a lens (wood diffuse-porous), delineated by a fine line of denser fibrous tissue at the outer margin and usually by smaller pores in the late-wood portion of the ring. *Pores* appearing as whitish dots to the naked eye, the larger obviously wider than the widest rays, nearly uniform in size and evenly distributed throughout the growth ring, solitary and in multiples of 2–several. *Parenchyma* not visible. *Rays* fine, generally not distinct to the naked eye but plainly visible with a hand lens, narrower than the largest pores.

MINUTE ANATOMY

Vessels few to moderately numerous, the largest small to medium; perforation plates scalariform; intervessel pits orbicular to broad-oval or angular through crowding (Fig. 12-158, p. 536), minute (2–4 μ in diameter), the orifices frequently confluent. *Parenchyma* apotracheal-diffuse, and in aggregates, paratracheal, and marginal. *Fiber tracheids* thin- to moderately thick-walled, medium to coarse. *Rays* unstoried, 1–5-seriate, homocellular; ray-vessel pitting similar to intervessel type.

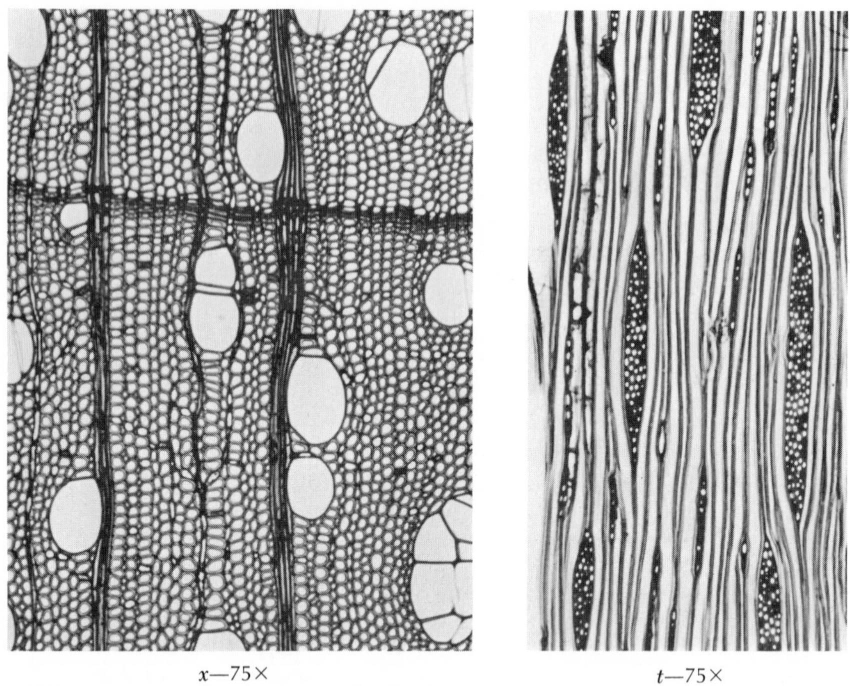

FIG. 12-174 Yellow birch, *Betula alleghaniensis* Britton

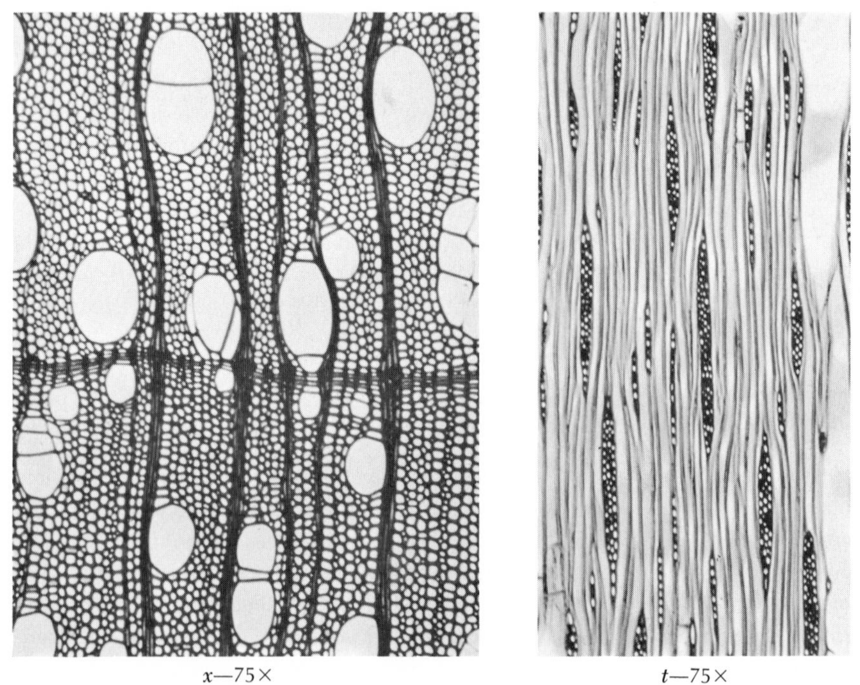

FIG. 12-175 River birch, *Betula nigra* L.

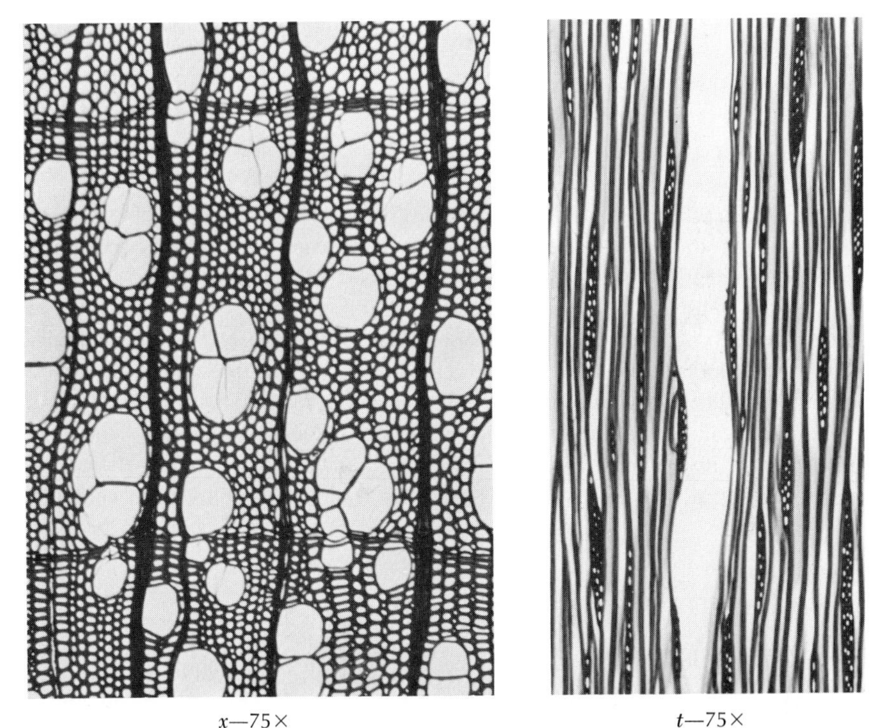

FIG. 12-176 Paper birch, *Betula papyrifera* Marsh.

REMARKS

The woods of the different species of *Betula* cannot be separated with certainty on the basis of either gross structure or minute anatomy; black birch and yellow birch are harder, heavier, and stronger than the other native species.

USES

Veneer (the largest and the best logs of yellow and black birch, used largely in furniture and interior paneling); *hardwood distillation*; *cooperage* (slack); *railroad ties*. Lumber used for *furniture* (one of the principal furniture woods of the United States, for which purposes its good working and finishing qualities, hardness, pleasing figure, and attractive color recommend it very highly); *radio*, *television*, and *stereo cabinets*; *kitchen cabinets*; *boxes* and *crates*; *woodenware* and *novelties*; *toys*; *planing-mill products* (especially interior trim, flooring, sash, and doors); *shuttles*, *spools*, *bobbins*, and a variety of other turned articles; *butcher blocks*; *agricultural implements*; *musical* and *scientific instruments*; *toothpicks* and *shoe pegs* (paper birch). *Pulpwood*.

RED ALDER
Alnus rubra Bong.

GENERAL CHARACTERISTICS

Wood whitish when first sawed, aging to flesh color or light brown with a reddish tinge, subject to oxidative sap stain and sticker stain;* *heartwood* indistinct; without characteristic odor or taste, straight-grained, moderately light (sp gr approx 0.37 green, 0.43 ovendry), soft to moderately soft. *Growth rings* distinct (wood diffuse porous), delineated by a whitish or brownish line at the outer margin. *Pores* small, indistinct without a hand lens, solitary, in multiples of 2–several and in small clusters. *Parenchyma* indistinct. *Rays* of two types, narrow (simple) and broad (aggregate); (*a*) narrow rays closely spaced, not visible without a hand lens; (*b*) aggregate rays at irregular and often at wide intervals, not sharply delineated and relatively inconspicuous to the naked eye (*x*), up to 4/5 plus in. in height (*t*) along the grain.

MINUTE ANATOMY

Vessels numerous, the largest small; perforation plates scalariform with 15 plus thin bars; intervessel pits orbicular to oval, quite widely spaced, fairly small (4–8 μ in diameter). *Parenchyma* scanty paratracheal, apotracheal-diffuse, and in aggregates, and occasionally marginal; (*a*) paratracheal parenchyma sparse, restricted to occasional cells (not forming a sheath); (*b*) apotracheal parenchyma sparse to fairly abundant, the cells solitary or in short tangential rows of 2–several; (*c*) marginal parenchyma present or wanting, when present forming an interrupted uniseriate line. *Fiber tracheids* thin- to moderately thick-walled, medium to very coarse. *Rays* unstoried, homocellular; (*a*) narrow rays uniseriate or rarely in part biseriate; (*b*) aggregate rays consisting of units similar to the narrow rays, and of included fibers and vessels; ray-vessel pitting similar to intervessel type.

REMARKS

Sometimes confused with birch (*Betula* spp.) but softer (readily dented with the thumbnail) and lighter, with aggregate rays (at least at wide intervals). Readily separated from cottonwood (*Populus* spp.) and willow (*Salix* spp.) through the presence of scalariform perforation plates with many bars (perforation plates simple in *Populus* and *Salix*), parenchyma in the body of the ring (parenchyma strictly marginal in *Populus* and *Salix*), and aggregate rays (at least at wide intervals).

* These stains can be prevented by steaming lumber at 212°F and 100 percent relative humidity for at least 4 hours before air or kiln drying. The color of steamed alder wood varies in shade from nearly white to reddish brown. (C. J. Kozlik, Establishing Color in Red Alder Lumber, *Forest Res. Lab., Oregon State Univ.*, Report D-8, 1967.)

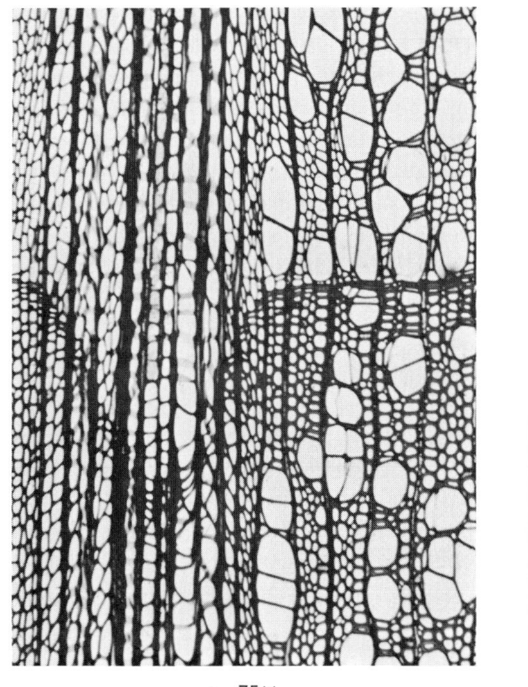
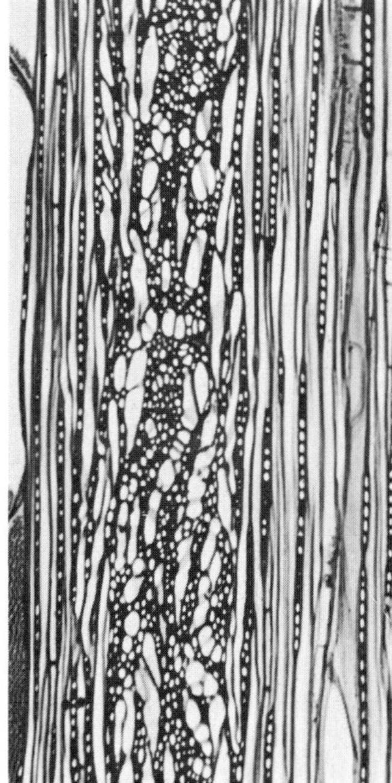

x—75× *t*—75×

FIG. 12-177 Red alder, *Alnus rubra* Bong.

USES

Red alder is the leading hardwood of the Pacific Northwest. The principal uses include *furniture* (mainly for turned and other exposed parts in stained and enameled furniture and in chairs, for which it is especially suited because of its workability, its pleasing grain, the ease with which it takes glue, its finishing qualities, and its ability to take and hold nails and screws); *core stock* and *cross-bands* in plywood, for which it is considered particularly suitable because of moderate shrinkage and good gluing qualities; *woodenware* and *novelties* (such as floor lamps, candlesticks, umbrella stands, ferneries, hatracks); *sash*, *doors*, *paneling*, and other *millwork*; *fixtures*; *handles*; *charcoal*; important source of *pulpwood*.

AMERICAN BEECH

Fagus grandifolia Ehrh.

GENERAL CHARACTERISTICS

Sapwood whitish; *heartwood* whitish with a reddish tinge to reddish brown; *wood* without characteristic odor or taste, straight- to interlocked-grained, heavy (sp gr approx 0.56 green, 0.67 ovendry), hard. *Growth rings* distinct (wood diffuse-porous), delineated by a dark line or band of denser late wood. *Pores* small, indistinct without a hand lens, usually crowded and largest in the early wood, decreasing in number and size through the central portion of the ring, scattered and very small in the late wood. *Parenchyma* not visible with a hand lens or zonate in the outer late wood, the lines appearing very finely punctate. *Rays* of two types, broad (oak-type) and narrow; (*a*) broad rays plainly visible to the naked eye, separated by several narrow rays, appearing on the tangential surface as short, rather widely spaced, staggered lines which are visible without magnification; (*b*) narrow rays fine, not visible without magnification.

MINUTE ANATOMY

Vessels moderately few to very numerous, the largest small; tyloses present in the heartwood, in uniseriate longitudinal rows for the most part (the upper and lower walls in contact and appearing as nearly horizontal transverse partitions arranged in a ladder-like series); perforation plates simple or those in the smaller vessels occasionally scalariform; intervessel pits oval to long-elliptical (Fig. 12-150, p. 535), with horizontal or nearly horizontal orifices, 6–20 μ in diameter. *Paren-*

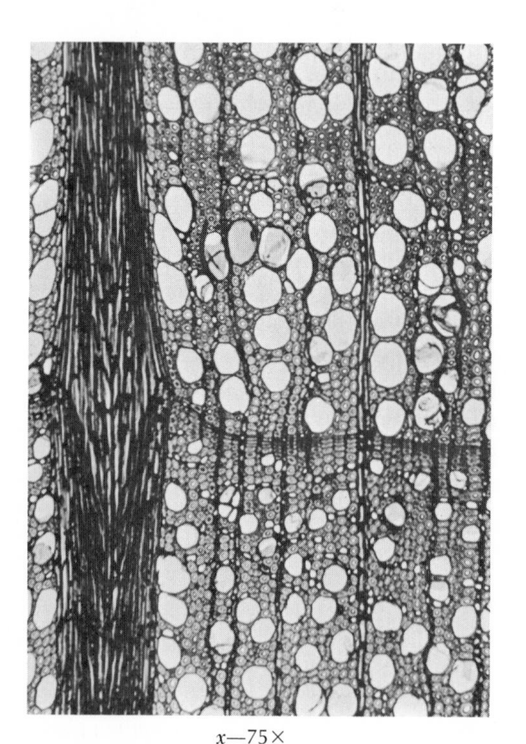

x—75×

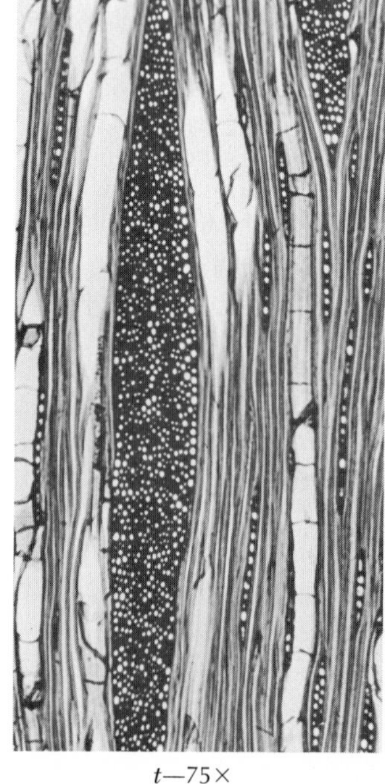

t—75×

FIG. 12-178 American beech, *Fagus grandifolia* Ehrh.

chyma abundant, apotracheal-diffuse to banded, the lines of the latter more evident toward the outer margin of the ring. *Fiber tracheids* thick-walled, medium. *Rays* unstoried, homocellular or with marginal upright cells; (*a*) broad (oak-type) rays 15–25 plus seriate, 1–several mm in height along the grain; (*b*) narrow rays much more numerous than the broad oak-type rays, 1–5-seriate, up to 500 plus μ in height; ray-vessel pitting simple to bordered.

REMARKS

Comparable to the oaks (*Quercus* spp.) in possessing two types of rays, but diffuse-porous.

USES

Charcoal production; *railroad ties* (treated with creosote by pressure methods); *pulp* (by soda method); *slack cooperage* (largely for vegetable and fruit barrels); *veneer*, mainly for the manufacture of crates, baskets, and fruit containers and to some extent for furniture and interior plywood; *fuel wood* (ranks high in fuel value). Lumber used for *boxes* and *crates*; *pallets*; *furniture* (especially for curved and turned parts of chairs); *handles* and brush backs; *woodenware* and *novelties* (including toys, spools, clothespins, and a variety of other small turned articles); *planing-mill products* (especially flooring).

AMERICAN CHESTNUT

Castanea dentata (Marsh.) Borkh.

GENERAL CHARACTERISTICS

Sapwood narrow, whitish to light brown; *heartwood* grayish brown to brown, turning darker with age; *wood* without characteristic odor, with mild astringent taste due to its tannin content, straight-grained, moderately light (sp gr approx 0.40 green, 0.45 ovendry), moderately hard. *Growth rings* conspicuous except in very narrow ringed stock (wood ring-porous). *Early-wood pores* very large, plainly visible to the naked eye, forming a broad, conspicuous band several pores in width; transition from early to late wood abrupt; *late-wood pores* small, arranged in obliquely radial (flame-shaped) patches of light tissue which are less obvious in narrow rings. *Parenchyma* indistinct. *Rays* very fine, barely visible with a hand lens.

MINUTE ANATOMY

Vessels in the late wood numerous to very numerous; largest early-wood vessels very large to extremely large; perforation plates simple, or those in the late-wood

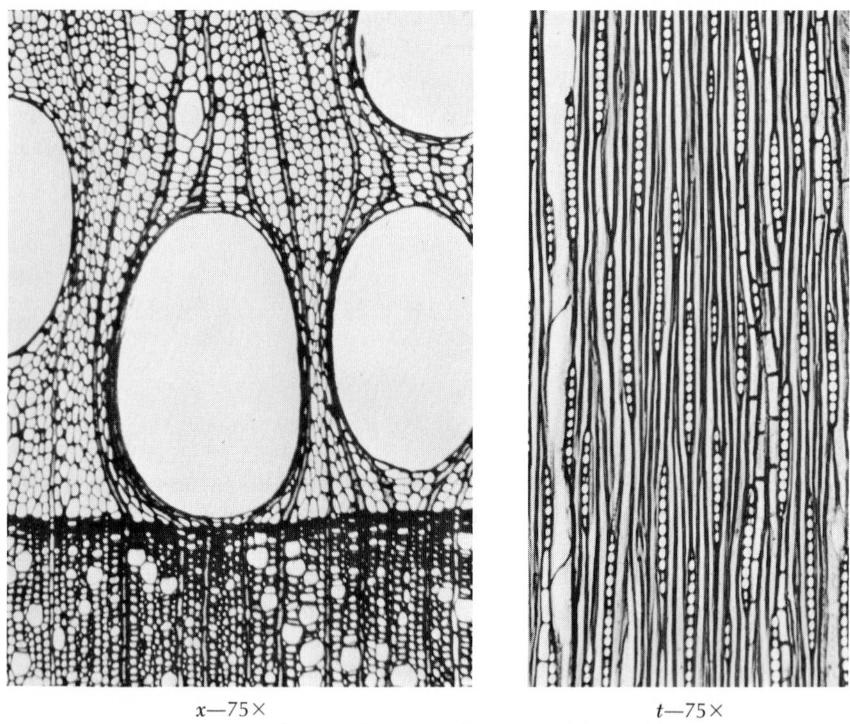

x—75× *t*—75×

FIG. 12-179 American chestnut, *Castanea dentata* (Marsh.) Borkh.

vessels occasionally scalariform; pits leading to contiguous tracheary cells orbicular to long-elliptical, with horizontal or nearly horizontal orifice, 8–18 μ in diameter. *Vasicentric tracheids* present, confined to the vicinity of the early-wood vessels. *Parenchyma* paratracheal-scanty and apotracheal-diffuse; (*a*) paratracheal parenchyma sparse, restricted to occasional cells (not forming a sheath); (*b*) apotracheal-diffuse parenchyma abundant, especially in the late wood. *Fibers* range from *libriform* to *fiber tracheids*, very coarse. *Rays* unstoried, uniseriate or rarely in part biseriate, homocellular; ray-vessel pitting simple to bordered.

REMARKS

Distinctive among North American woods. Occasionally confused with oak (produced by *Quercus* spp.) but lighter in weight and softer, without the broad rays that are a feature of that timber.

USES

Most of the chestnut has been killed by the blight [*Endothia parasitica* (Murr.) A. & A.]; the uses listed below are indicative of the important rank *formerly held* by this wood. *Tannin* (principal domestic source of tannin; obtained by soaking the wood chips in hot water and evaporating the resulting liquor to the desired concentration); *semichemical pulp* (manufactured from the extracted chips as well as from fresh wood), for fiberboard; *poles, fence posts,* and *railroad ties,* for which it is very suitable because of its natural durability combined with sufficient strength and hardness; *slack cooperage.* Lumber used for *furniture; caskets* and *coffins* (a leading wood because of its outstanding durability and good working qualities); *boxes* and *crates; millwork* (especially sash, doors, and paneling); *plywood* (as a core stock and cross-banding material), for which its freedom from warping, low density, moderate shrinkage, and good gluing qualities recommend it very highly.

GOLDEN CHINKAPIN

Castanopsis chrysophylla (Dougl.) A.DC.

GENERAL CHARACTERISTICS

Sapwood narrow, light brown with a pinkish tinge, hardly distinguishable from the heartwood; *heartwood* light brown tinged or striped with pink; *wood* without

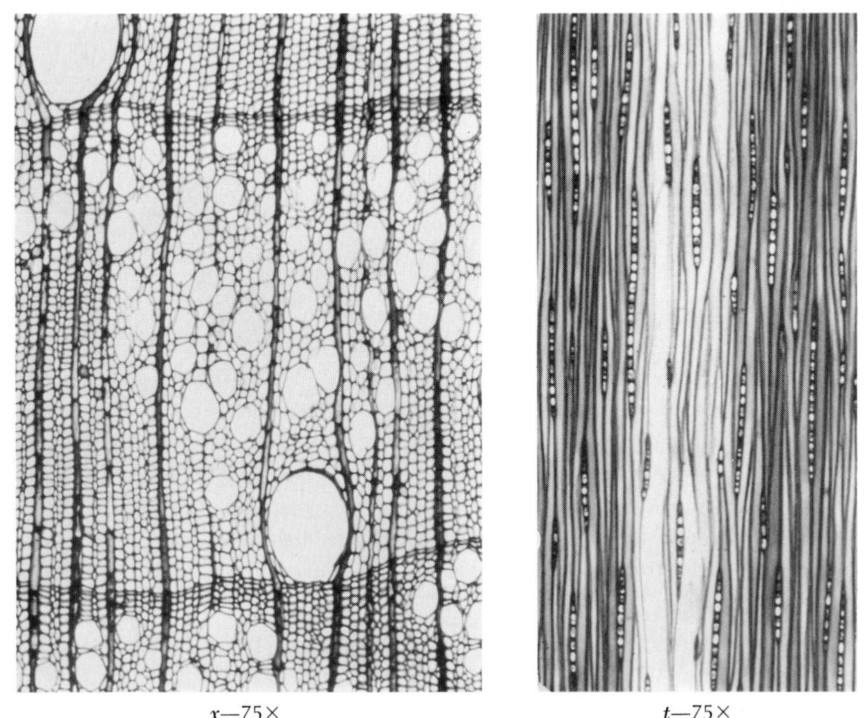

x—75× *t*—75×

FIG. 12-180 Golden chinkapin, *Castanopsis chrysophylla* (Dougl.) A.DC.

characteristic odor or taste, moderately heavy (sp gr approx 0.42 green, 0.48 oven-dry), fairly hard. *Growth rings* conspicuous (wood ring-porous). *Early-wood pores* plainly visible to the naked eye, in a uniseriate (very rarely 2- or more-seriate) interrupted row; transition from early to late wood abrupt; *late-wood pores* small, indistinct without a hand lens, arranged in obliquely radial (flame-shaped) patches of light tissue which are less obvious in narrow rings. *Parenchyma* indistinct. *Rays* very fine, barely visible with a hand lens.

MINUTE ANATOMY

Vessels in the late wood numerous to very numerous; largest early-wood vessels large; perforation plates simple, or those in the smaller early-wood vessels occasionally scalariform; pits leading to contiguous tracheary cells orbicular to long-elliptical, 6–18 μ in diameter. *Vasicentric tracheids* present, confined to the vicinity of the early-wood vessels. *Parenchyma* paratracheal-scanty, apotracheal-diffuse, and banded; (*a*) paratracheal parenchyma sparse, restricted to occasional cells (not forming a sheath); (*b*) apotracheal-diffuse parenchyma abundant, the cells sometimes banded together in more or less interrupted, inconspicuous, 1–3 (mostly 1)-seriate lines of (*c*) banded parenchyma. *Fibers* range from *libriform* to *fiber tracheids*, fine to coarse. *Rays* unstoried, uniseriate or rarely in part biseriate, homocellular; ray-vessel pitting simple to bordered.

USES

Locally for *paneling*; *furniture* and *novelties*; *fuel wood.*

TANOAK

Lithocarpus densiflorus (Hook. & Arn.) Rehd.

GENERAL CHARACTERISTICS

Sapwood light reddish brown when first exposed, turning darker with age and then difficult to distinguish from the heartwood, wide; *heartwood** light brown tinged with red, aging to dark reddish brown; *wood* without characteristic odor or taste, heavy (sp gr approx 0.58 green), hard. *Growth rings* scarcely distinct or wanting, when visible, delineated by a faint narrow line of darker (denser) fibrous tissue at the outer margin. *Pores* barely visible to the naked eye, unevenly distributed, inserted in light-colored tissue in streamlike clusters which extend for

* Prestemon states that no normal heartwood is present in tanoak but that trees frequently contain an irregularly shaped, darker-colored core, ranging from brown to almost black. The stained portions of the stem, called "heart stain," generally contain several colored zones resembling those caused by white-rot fungi; but no evidence of fungi has been found. "Heart stain" is more prevalent among large, old trees that have grown slowly. [D. R. Prestemon, Variations in Heart Stain and Density in Tanoak, *Forest Prod. J.*, **17**(7):33–41 (1967).]

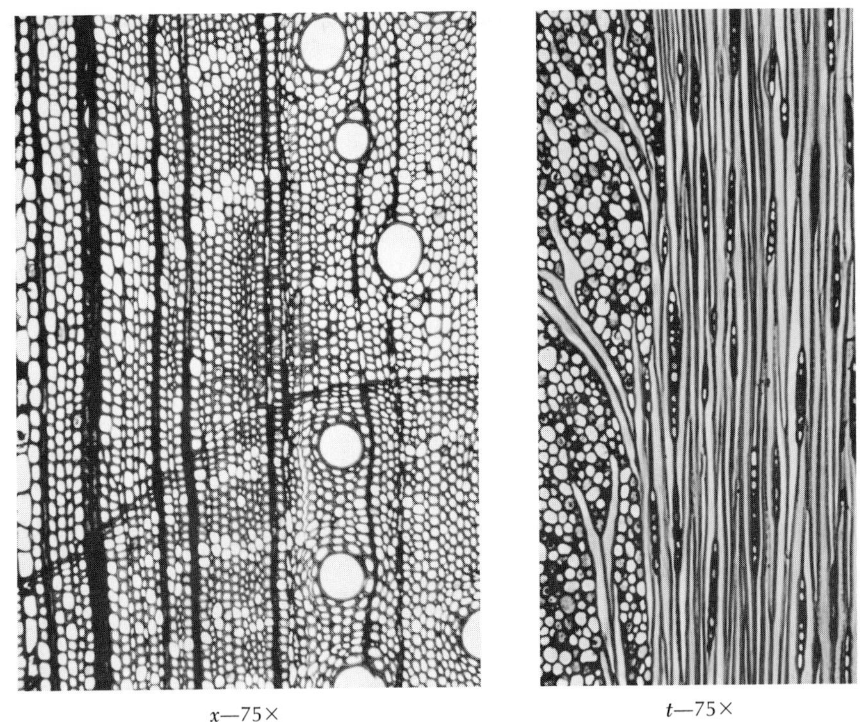

x—75× *t*—75×

FIG. 12-181 Tanoak, *Lithocarpus densiflorus* (Hook. & Arn.) Rehd.

some distance radially (across several–many rings when these are evident), those in the early wood usually somewhat larger but the transition in pore size from early to late wood gradual. ***Parenchyma*** visible with a hand lens, in tangential, rather wide, ragged lines which usually do not contrast sharply with the background of fibrous tissue and are hence poorly defined. ***Rays*** of two types (*x*), broad (aggregate and oak-type) and narrow; (*a*) broad rays present in some samples and wanting in others, when present, distinct and frequently conspicuous to the naked eye on the transverse section, separated by several–many narrow rays, of the same color as the background on the tangential surface and hence indistinct, forming a high fleck on the quarter; (*b*) narrow rays much more numerous than the broad rays, indistinct without magnification.

MINUTE ANATOMY

Vessels rather thick-walled, orbicular or oval for the most part, very few, the largest medium to large; perforation plates simple; pits leading to contiguous tracheary cells orbicular, 6–10 μ in diameter. ***Vasicentric tracheids*** present, forming a 1–several (mostly 1–2)-seriate sheath about the vessels which is rarely interrupted by parenchyma, passing over into wide-lumened fibers. ***Parenchyma***

paratracheal-scanty, apotracheal-diffuse to banded; (*a*) paratracheal parenchyma restricted to occasional cells contiguous to the vessels; (*b*) apotracheal-banded parenchyma very abundant, conspicuous, in numerous, concentric, more or less continuous, 1–3 (mostly 1–2)-seriate, ragged lines; (*c*) apotracheal-diffuse parenchyma sparse, appearing as occasional cells in the tracts of fibrous tissue. ***Fibers libriform*** and ***fiber tracheids***, thick-walled, frequently gelatinous, fine to medium. ***Rays*** unstoried, homocellular or nearly so; (*a*) broad (aggregate and oak-type) rays several–many-seriate, up to 400 μ in width, many cells (into the hundreds) in height along the grain (*t*); aggregate rays composed of smaller rays separated by strands of fibrous tissue, otherwise comparable to the broad rays; (*b*) narrow rays very numerous, uniseriate, variable in height (1–12 plus cells) along the grain; ray-to-vessel pitting simple to bordered.

USES

Locally for ***fuel wood***; occasionally for ***furniture***; ***mine timbers***; ***baseball bats***; ***face veneer***; ***pulpwood***. ***Tannin*** in commercial quantities is obtainable from the bark; hence the common name of the tree, "tanoak."

LIVE OAK

Quercus virginiana Mill.

GENERAL CHARACTERISTICS

Sapwood whitish to grayish brown; ***heartwood*** dull brown to gray-brown; ***wood*** without characteristic odor or taste, irregular grained, exceedingly heavy (sp gr approx 0.81 green, 0.98 ovendry), exceedingly hard. ***Growth rings*** scarcely distinct, delineated by a faint narrow line of darker (denser) fibrous tissue at the outer margin. ***Pores*** barely visible to the naked eye, unevenly distributed in stream-like clusters which extend for some distance radially (across several–many rings when these are evident), those in the early wood usually somewhat larger but the transition in pore size from early to late wood gradual. ***Parenchyma*** barely visible with a hand lens, in closely spaced, punctate lines which are poorly differentiated against the background of fibrous tissue. ***Rays*** of two types (*x*), broad (aggregate and oak-type) and narrow; (*a*) broad rays usually conspicuous to the naked eye on the transverse section, separated by several–many narrow rays, appearing on the tangential surface as rather widely spaced, staggered lines of varying length, forming a high fleck on the quarter; (*b*) narrow rays much more numerous than the broad rays, indistinct at low magnification.

MINUTE ANATOMY

Vessels rather thick-walled, orbicular or oval for the most part, very few, the largest large to very large; perforation plates simple; pits leading to contiguous

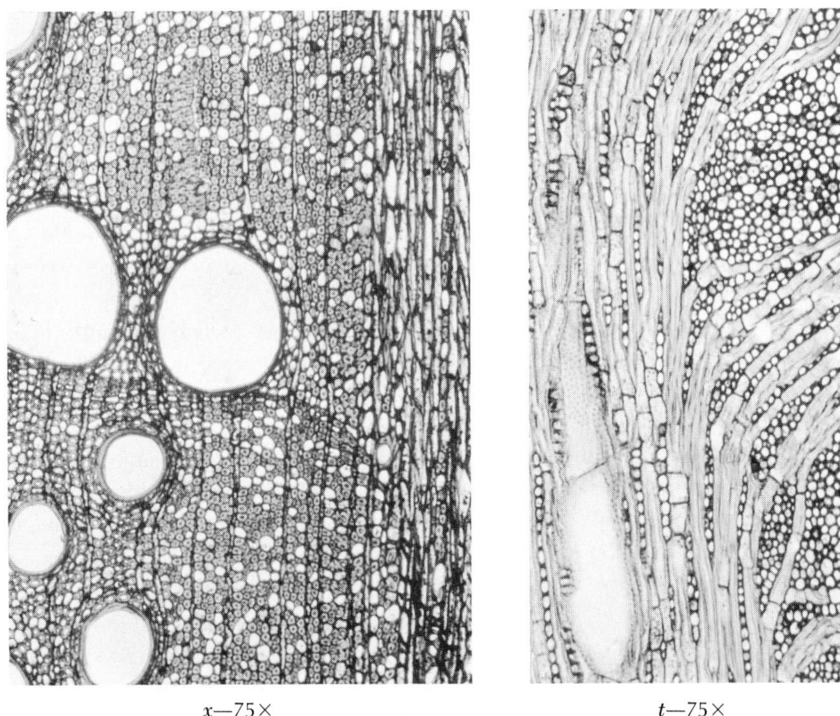

x—75× *t*—75×

FIG. 12-182 Live oak, *Quercus virginiana* Mill.

tracheary cells orbicular to oval, 6–10 μ in diameter. ***Vasicentric tracheids*** present, forming a 1–several-seriate sheath about the vessels which is interrupted by parenchyma. ***Parenchyma*** paratracheal-scanty, apotracheal-diffuse, and banded; (*a*) paratracheal parenchyma restricted to occasional cells contiguous to the vessels; (*b*) apotracheal-diffuse parenchyma abundant, scattered among the fibers as single cells or in small cell clusters, passing over into numerous, conspicuous, 1–2 (mostly 1)-seriate, ragged, anastomosing lines of (*c*) banded parenchyma. ***Fibers libriform*** and ***fiber tracheids*** thick-walled, sometimes gelatinous, fine to medium. ***Rays*** unstoried, homocellular or nearly so; (*a*) broad (oak-type and aggregate) rays many-seriate, up to 600 μ in width through the central portion, many cells (into the hundreds) in height along the grain (*t*); aggregate rays composed of smaller rays separated by strands of fibrous tissue, otherwise comparable to the other broad rays; (*b*) narrow rays very numerous, 1–3 (mostly 1)-seriate, variable in height (1–20 plus cells) along the grain; ray-vessel pitting simple to bordered.

USES

Valued for shipbuilding before the advent of steel vessels; used locally for articles requiring exceptional strength and toughness.

RED OAK

Northern red oak (*Quercus rubra* L.)
Black oak (*Quercus velutina* Lam.)
Shumard oak (*Quercus shumardii* Buckl.)
Scarlet oak (*Quercus coccinea* Muenchh.)
Pin oak (*Quercus palustris* Muenchh.)
Willow oak (*Quercus phellos* L.)
Other species of the *Erythrobalanus* group

GENERAL CHARACTERISTICS

Sapwood whitish to grayish or pale reddish brown; ***heartwood*** pinkish to light reddish brown, the flesh-colored cast generally pronounced, occasionally light brown; ***wood*** without characteristic odor or taste, generally straight-grained, heavy to very heavy (sp gr 0.52–0.61 green, 0.62–0.76 ovendry), hard to very hard. ***Growth rings*** very distinct (wood ring-porous). ***Early-wood pores*** large, distinctly visible to the naked eye, forming a conspicuous band 1–4 pores in width, with few or no tyloses in the heartwood; transition from early to late wood gradual to more or less abrupt; ***late-wood pores*** abundant (less numerous than in white oak), small, indistinct or barely visible to the naked eye but distinct with a hand lens, associated with light-colored tissue in radial, mostly uniseriate, occasionally forking rows or

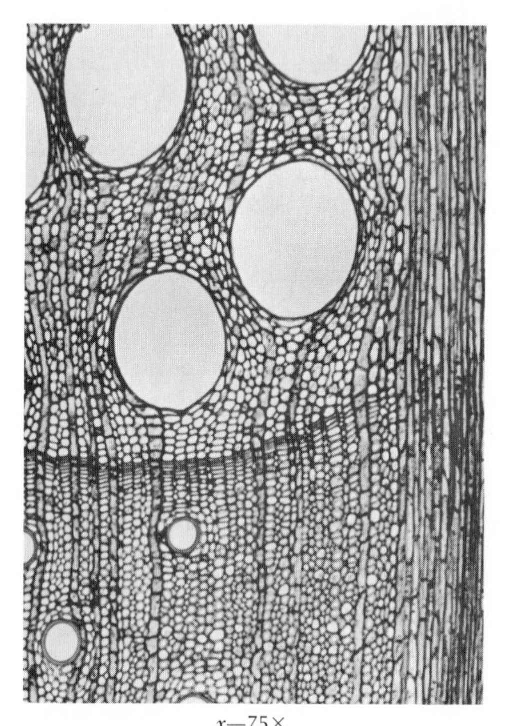

x—75×

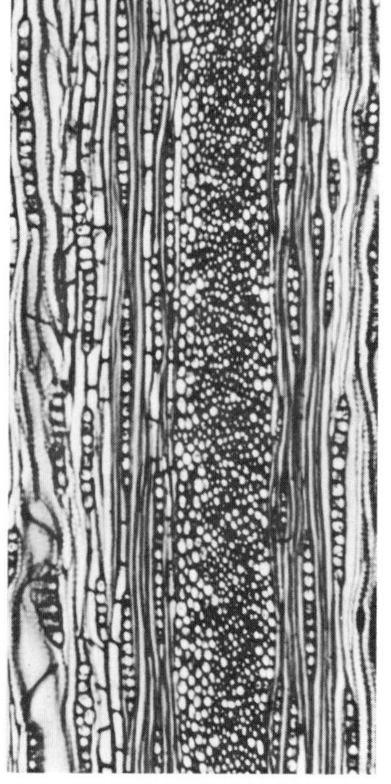

t—75×

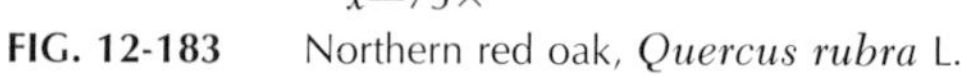

FIG. 12-183 Northern red oak, *Quercus rubra* L.

scattered in radially aligned, flame-shaped tracts of nonfibrous tissue, thick-walled. *Parenchyma* visible with a hand lens, (1) forming part of the conjunctive tissue between the early-wood pores and the rays, (2) composing most of the light-colored tissue in which the late-wood pores are inserted, (3) usually zonate in fine, more or less regular tangential lines in the outer portion of the ring. *Rays* of two types (*x*), broad (oak-type) and narrow (simple); (*a*) broad rays very conspicuous to the naked eye, separated by several–many narrow rays, appearing on the tangential surface as rather widely spaced, staggered lines of varying length which sometimes extend ½ in. or more along the grain, forming a handsome high fleck on the quarter; (*b*) narrow rays much more numerous than the broad rays, indistinct without magnification.

MINUTE ANATOMY

Vessels in the late wood very few to few; largest early-wood vessels large to extremely large; perforation plates simple; pits leading to contiguous tracheary cells orbicular to oval (Fig. 12-146, p. 535), 6–10 μ in diameter. *Vasicentric tracheids* present, intermingled with parenchyma, (1) forming most of the conjunctive tissue between the early-wood vessels and the rays and (2) composing some of the light-colored tissue in which the late-wood pores are inserted. *Paren-*

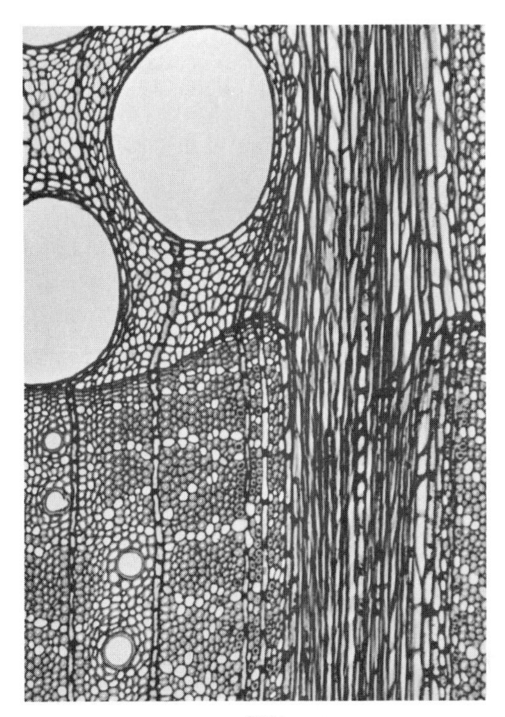

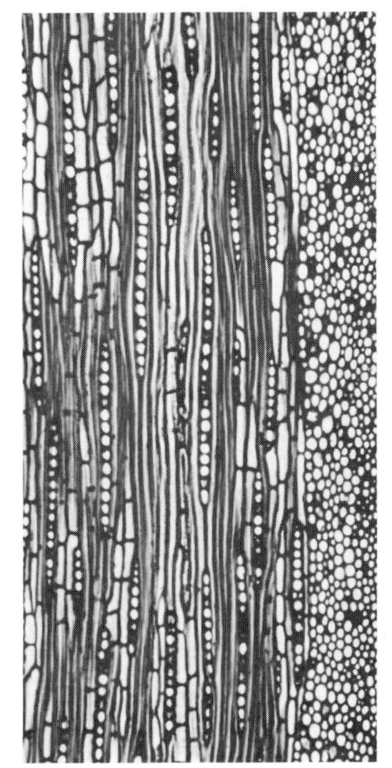

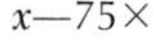

x—75× *t*—75×

FIG. 12-184 Black oak, *Quercus velutina* Lam.

chyma abundant, paratracheal, apotracheal-diffuse, and usually banded; (*a*) paratracheal parenchyma intermingled with tracheids and distributed as described above; (*b*) apotracheal-diffuse parenchyma restricted to the fibrous tracts and toward the outer margin of the ring (particularly in wide rings) exhibiting more or less a tendency toward aggregation into concentric lines of (*c*) banded parenchyma. ***Fibers libriform*** and ***fiber tracheids*** medium thick- to thick-walled, frequently gelatinous, fine to medium. ***Rays*** unstoried, homocellular; (*a*) broad (oak-type) rays 12–30 plus seriate and 150–400 plus μ wide through the central portion, many cells (into the hundreds) in height along the grain (*t*); (*b*) narrow rays very numerous, uniseriate or occasionally in part biseriate through the central portion, very variable in height (1–20 plus cells) along the grain; ray-vessel pitting simple to bordered.

REMARKS

The woods of the various red oaks belonging to the red oak group (***Erythrobalanus***) cannot be separated with certainty.

Red oaks (***Erythrobalanus*** group) can be distinguished from white oaks (***Leucobalanus*** group) by the following characters:

White oaks	*Red oaks*
1. Heartwood rich light brown to dark brown, without flesh-colored cast	1. Heartwood pinkish or pale reddish brown
2. Transition from early to late wood generally abrupt	2. Transition from early to late wood gradual to more or less abrupt
3. Early-wood pores in the heartwood usually occluded with tyloses	3. Early-wood pores in the heartwood usually open
4. Late-wood pores thin-walled, more or less angular, not sharply defined with a hand lens	4. Late-wood pores plainly visible with a hand lens, thick-walled, rounded
5. Large rays averaging ½–1¼ in. in height, frequently taller than 1½ in.	5. Large rays averaging ¼–½ in. in height, rarely taller than 1½ in.

NOTE: The fourth criterion is by far the most reliable.

Heartwood of red oak can be distinguished from that of white oak by the benzidine-sodium nitrate test (see p. 419).

USES

Same as white oak (see p. 572) except where exposed to decay (railroad ties, fence posts, mine timbers, etc.), under which conditions it requires preservative treatment; unsuited for tight cooperage without treatment with paraffin, silicate of soda, or a similar substance because of the unoccluded early-wood vessels.

WHITE OAK

White oak (*Quercus alba* L.)
Bur oak (*Quercus macrocarpa* Michx.)
Overcup oak (*Quercus lyrata* Walt.)
Post oak (*Quercus stellata* Wangenh.)
Swamp chestnut oak, basket oak (*Quercus michauxii* Nutt.)
Chestnut oak, rock oak (*Quercus prinus* L.)
Swamp white oak (*Quercus bicolor* Willd.)
Other species of the *Leucobalanus* group

GENERAL CHARACTERISTICS

Sapwood whitish to light brown, thin or thick; *heartwood* rich light brown to dark brown; *wood* without characteristic odor or taste, usually straight-grained, heavy to very heavy (sp gr 0.55–0.64 green, 0.66–0.79 ovendry), hard to very hard. *Growth rings* very distinct except in slow-grown stock (wood ring-porous). *Early-wood pores* large, distinctly visible to the naked eye, forming a conspicuous band 1–3 pores in width, often occluded with tyloses in the heartwood; transition from early to late wood abrupt or somewhat gradual; *late-wood pores* numerous, small, not sharply defined with a hand lens, scattered in radially aligned, flame-

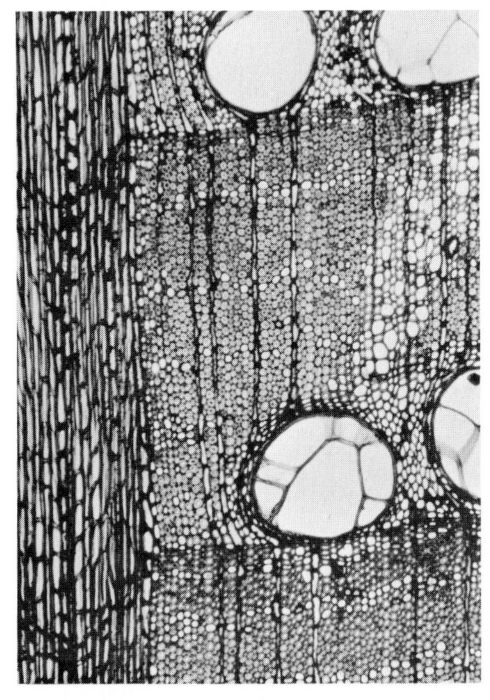

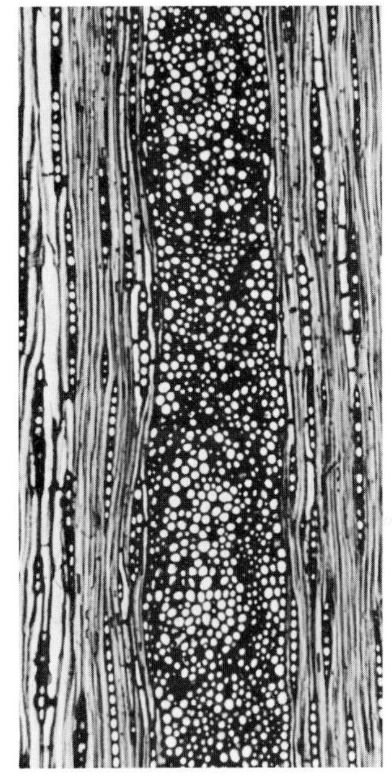

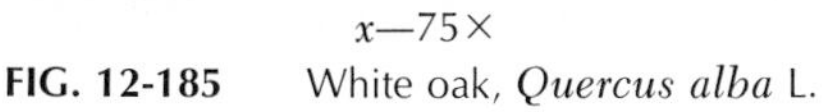

x—75×   *t*—75×

FIG. 12-185 White oak, *Quercus alba* L.

shaped tracts of light-colored tissue, thin-walled. *Parenchyma* visible with a hand lens, (1) forming part of the conjunctive tissue between the early-wood pores and the rays, (2) composing most of the tissue in the flame-shaped tracts in which the late-wood pores are inserted, (3) usually banded in fine, more or less regular, tangential lines in the outer portion of the ring. *Rays* of two types (x), broad (oak-type) and narrow (simple); (a) broad rays very conspicuous to the naked eye, separated by several–many narrow rays, appearing on the tangential surface as rather widely spaced, staggered lines of varying length which frequently extend 1 in. or more along the grain, forming a handsome, high fleck on the quarter; (b) narrow rays much more numerous than the broad rays, indistinct without magnification.

MINUTE ANATOMY

Vessels in the late wood few to numerous; largest early-wood vessels large to extremely large; perforation plates simple; pits leading to contiguous tracheary cells orbicular to oval, 6–10 μ in diameter. *Vasicentric tracheids* present, intermingled with parenchyma, (1) forming most of the conjunctive tissue between the early-wood vessels and the rays and (2) composing part of the flame-shaped tracts

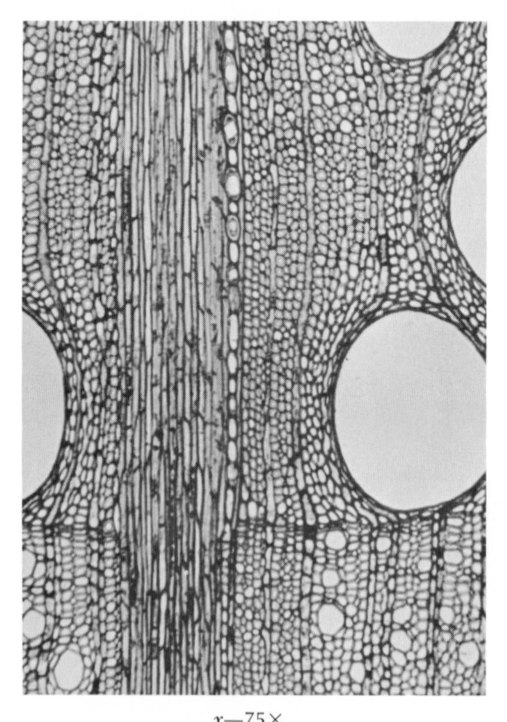

x—75×

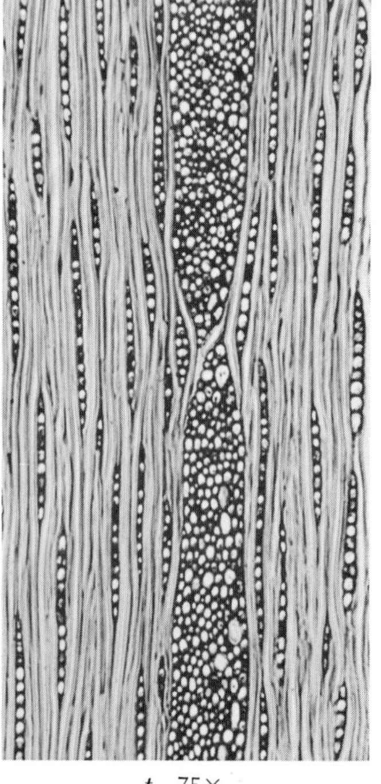

t—75×

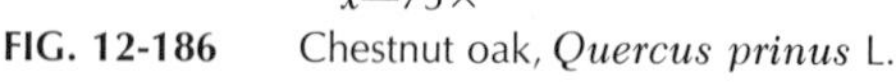

FIG. 12-186 Chestnut oak, *Quercus prinus* L.

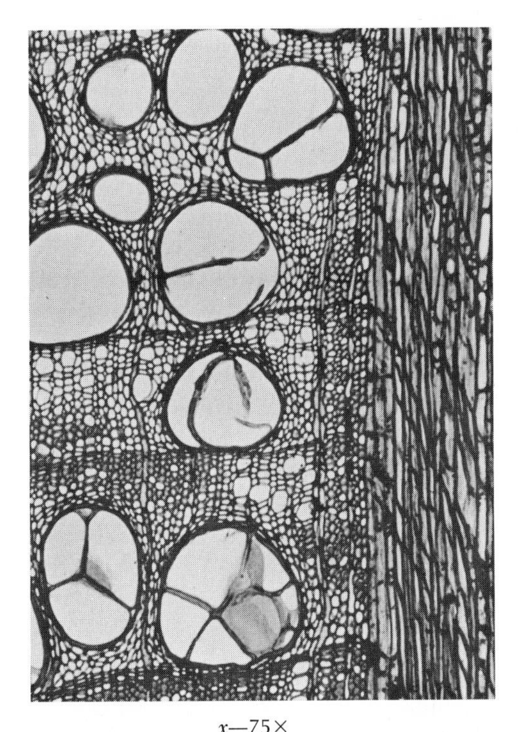

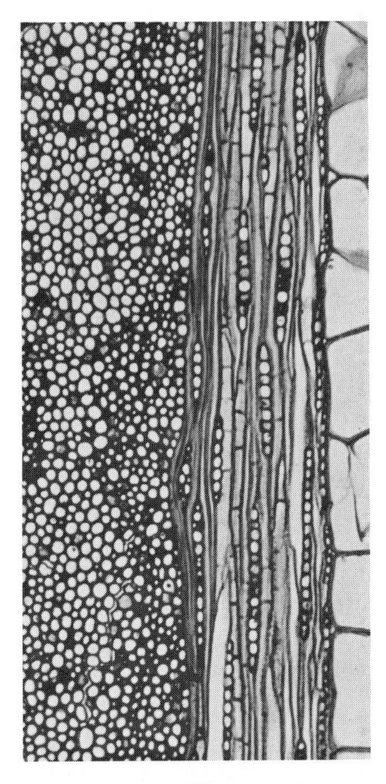

x—75× *t*—75×

FIG. 12-187 Post oak, *Quercus stellata* Wangenh.

in which the late-wood vessels are inserted. ***Parenchyma*** abundant, paratracheal, apotracheal-diffuse, and usually banded; (*a*) paratracheal parenchyma intermingled with tracheids and distributed as described above; (*b*) apotracheal-diffuse parenchyma restricted to the fibrous tracts and toward the outer portion of the ring (particularly in wide rings) exhibiting more or less a tendency toward aggregation into concentric lines of (*c*) banded parenchyma. ***Fibers libriform*** and ***fiber tracheids*** medium thick- to thick-walled, frequently gelatinous, fine to medium. ***Rays*** unstoried, homocellular; (*a*) broad (compound) rays 12–30 plus seriate and 150–400 plus μ wide through the central portion, many cells (into the hundreds) in height along the grain (*t*); (*b*) narrow rays very numerous, uniseriate or occasionally in part biseriate through the central portion, very variable in height (1–20 plus cells) along the grain; ray-vessel pitting simple to bordered.

REMARKS

The woods of the various oaks belonging to the white oak group (*Leucobalanus*) cannot be separated with certainty.

For differences between white oak and red oak, see Remarks under Red Oak, p. 568.

USES

Railroad ties (favorite tie wood because of its hardness, resiliency, and natural durability); *tight cooperage* (owing to its impermeability to liquids, strength, and durability); *slack cooperage; fence posts; mine timbers; poles; piling; export logs* and *timbers; veneer* (sliced), used largely in furniture and interior paneling; *firewood* (very high in fuel value). Lumber used for *flooring* (principal species, widely used because of its hardness, high resistance to abrasion, ability to finish smoothly, and attractive figure); *farm (nonmotor) vehicles; planing-mill* products (sash, doors, trim, wainscoting, general millwork); *furniture* (especially desks and tables, chairs, frames for upholstered furniture); *kitchen cabinets; fixtures; railroad cars; boxes, crates* and *pallets* (good nail-holding ability, but heavy and tends to split in nailing), where strength is important; *skids; ship-* and *boatbuilding; agricultural implements; caskets* and *coffins; handles. Pulpwood.*

AMERICAN ELM, WHITE ELM

Ulmus americana L.

GENERAL CHARACTERISTICS

Sapwood grayish white to light brown, thick; *heartwood* light brown to brown, frequently with a reddish tinge; *wood* generally without characteristic odor or taste, straight- or sometimes interlocked-grained, moderately heavy (sp gr approx 0.46 green, 0.55 ovendry), moderately hard. *Growth rings* distinct (wood ring-porous), usually fairly wide. *Early-wood pores* large, distinctly visible to the naked eye, arranged in a 1 (rarely 2–3)-seriate, more or less continuous row, those at the beginning of the ring approximately equal in size; transition from early to late wood abrupt; *late-wood pores* small, numerous, arranged in more or less continuous, wavy, concentric bands. *Parenchyma* not visible. *Rays* not distinct to the naked eye.

MINUTE ANATOMY

Vessels in the late wood moderately numerous to very numerous; *largest early-wood vessels* large to very large and usually in a single row; perforation plates simple; spiral thickening present in the smaller vessels; intervessel pits orbicular or angular through crowding, 10–12 μ in diameter. *Vascular tracheids* present both in the early-wood porous zone and in the wavy bands of late-wood vessels, grading into small vessels, with spiral thickening. *Parenchyma* paratracheal-scanty to vasicentric and apotracheal-diffuse; (*a*) paratracheal parenchyma abundant, associated with vascular tracheids, (1) contiguous to but never forming a continuous sheath about the large early-wood pores, (2) marginal to and included in the clusters of smaller early-wood vessels and vascular tracheids, and (3) marginal

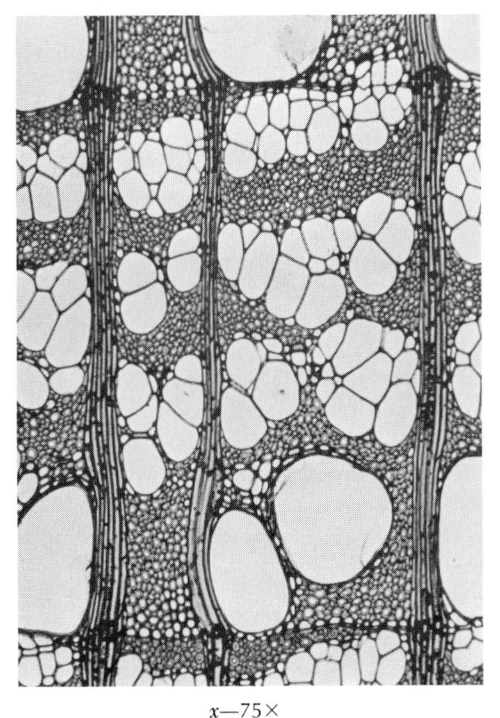

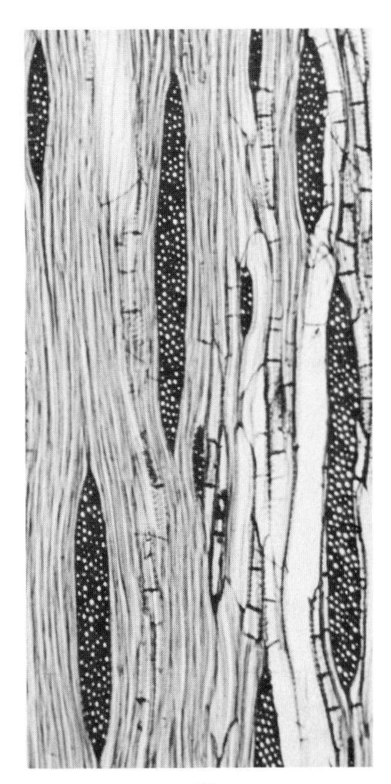

x—75×　　　　*t*—75×

FIG. 12-188 American elm, *Ulmus americana* L.

to and included in the wavy concentric bands of late-wood vessels and vascular tracheids; (*b*) apotracheal-diffuse parenchyma sparse, scattered in the fibrous tracts. *Libriform fibers* medium thick- to thick-walled, frequently gelatinous, fine to coarse. *Rays* unstoried, 1–7 (mostly 4–6)-seriate, essentially homocellular; ray-vessel pitting simple to bordered.

USES

Slack cooperage for staves and hoops (although elm* is still considered superior to other native woods for slack staves, it ranks below redgum, pine, ash, and tupelo gum in the quantity used; elm hoops have been largely displaced by metal hoops); *boxes* and *crates*, and other containers required to withstand rough handling, for which strength, toughness, and superior bending qualities make it very serviceable; *pallets*; *veneer* for fruit, vegetable containers, cheeseboxes,

* In the trade, the term *elm* is applied indiscriminately to the timbers of American (white) elm, slippery (red) elm, and hard elm (see p. 575). American elm outranks the other species in the quantity consumed for the uses here stated. Commercial stands of *elm* have been seriously depleted by the Dutch elm disease, to the extent that in many areas elm timber is no longer available in commercial quantities.

and plywood; ***furniture*** (especially bent parts such as rockers and arms, also for upholstery frames and dinettes, a recent favorite for "Danish-type" furniture); ***vehicles***; ***dairy***, ***poultry***, and ***apiary supplies***; ***interior trim***; ***agricultural implements***.

SLIPPERY ELM, RED ELM

Ulmus rubra Mühl.

GENERAL CHARACTERISTICS

Sapwood grayish white to light brown, narrow, with a faint characteristic odor resembling that of the inner bark; ***heartwood*** brown to dark brown, frequently with shades of red, usually odorless; sometimes contains yellow and orange compounds, readily extractable with benzene (see p. 419); ***wood*** without characteristic taste, straight- or sometimes interlocked-grained, moderately heavy (sp gr approx 0.48 green, 0.57 ovendry), moderately hard. ***Growth rings*** distinct (wood ring-porous). ***Early-wood pores*** large, distinctly visible to the naked eye, forming a conspicuous band 2–4 pores in width; transition from early to late wood abrupt; ***late-wood pores*** very small, numerous, arranged in more or less continuous, wavy, concentric bands. ***Parenchyma*** not visible. ***Rays*** not distinct to the naked eye.

MINUTE ANATOMY

Vessels in the late wood numerous to very numerous; ***largest early-wood vessels*** very large and in several rows; perforation plates simple; spiral thickening present in the smaller vessels; intervessel pits orbicular or angular through crowding, 8–12 μ in diameter. ***Vascular tracheids*** present in both the early-wood porous zone and in the wavy bands of late-wood vessels, grading into small vessels, with spiral thickening. ***Parenchyma*** paratracheal-scanty to vasicentric and apotracheal-diffuse; (*a*) paratracheal parenchyma abundant, (1) associated with vascular tracheids and forming tracts of conjunctive tissue between the early-wood vessels, (2) marginal to and occasionally included in the wavy concentric bands of late-wood vessels and vascular tracheids; (*b*) apotracheal-diffuse parenchyma sparse, scattered in the fibrous tracts. ***Libriform fibers*** medium thick- to thick-walled, frequently gelatinous, fine to medium. ***Rays*** unstoried, 1–5 (mostly 3–4)-seriate, essentially homocellular; ray-vessel pitting simple to bordered.

USES

Commonly marketed with American elm as "elm," or sometimes as "soft elm" and used for the same purposes.

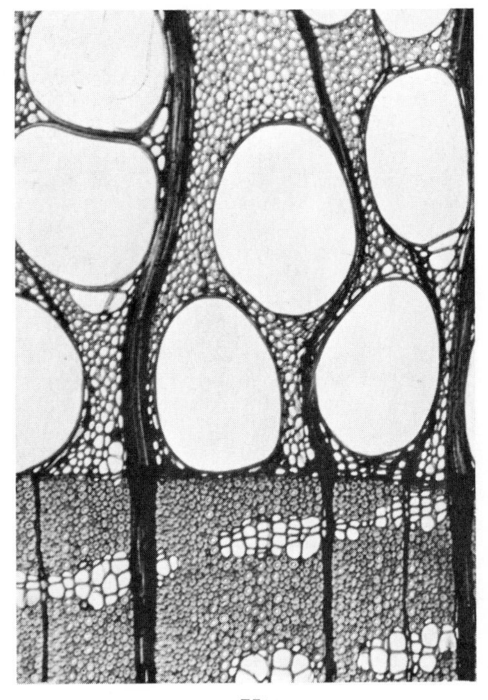

x—75× *t*—75×

FIG. 12-189 Slippery elm, *Ulmus rubra* Mühl.

HARD ELM

Rock elm, cork elm (*Ulmus thomasii* Sarg.)
Winged elm (*Ulmus alata* Michx.)
Cedar elm (*Ulmus crassifolia* Nutt.)

GENERAL CHARACTERISTICS

Sapwood light brown to brown, narrow; *heartwood* light brown to brown, frequently with a reddish tinge; *wood* without characteristic odor or taste, straight- or sometimes interlocked-grained, heavy (sp gr approx 0.57 green, 0.66 ovendry), hard. *Growth rings* fairly distinct (wood ring-porous). *Early-wood pores* variable in size; larger pores scarcely visible to the naked eye, spaced at more or less irregular intervals in an interrupted row and separated by smaller pores; transition from early to late wood more or less gradual; *late-wood pores* small, numerous, arranged in more or less continuous, wavy, concentric bands. *Parenchyma* not visible. *Rays* not distinct to the naked eye.

MINUTE ANATOMY

Vessels in the late wood numerous to very numerous; largest early-wood vessels medium to large and usually in an interrupted row; perforation plates simple;

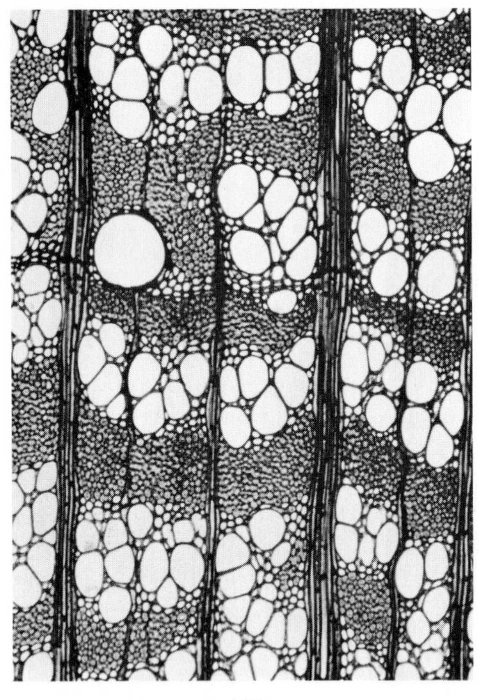

x—75×

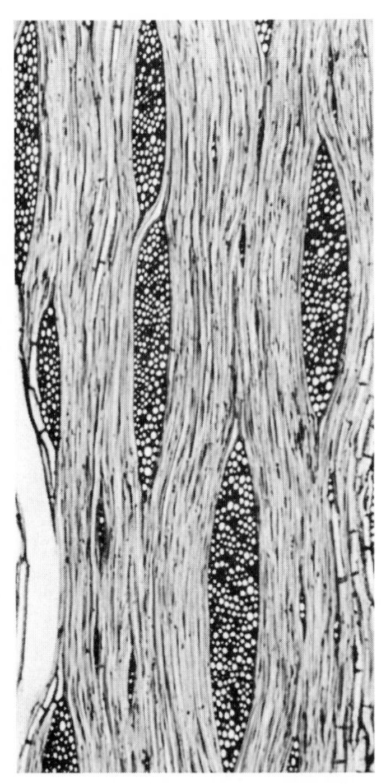

t—75×

FIG. 12-190 Rock elm, *Ulmus thomasii* Sarg.

spiral thickening present in the smaller vessels; intervessel pits orbicular or angular through crowding 8–12 μ in diameter. ***Vascular tracheids*** present both in the early-wood porous zone and in the wavy bands of late-wood vessels, grading into small vessels, with spiral thickening. ***Parenchyma*** paratracheal-scanty to vasicentric and apotracheal-diffuse; (*a*) paratracheal parenchyma abundant, associated with vascular tracheids, (1) contiguous to but never forming a continuous sheath about the larger early-wood pores, (2) marginal to and included in the clusters of smaller early-wood vessels and vascular tracheids, and (3) marginal to and included in the wavy concentric bands of late-wood vessels and vascular tracheids; (*b*) apotracheal-diffuse parenchyma sparse, scattered in the fibrous tracts. ***Libriform fibers*** medium thick- to thick-walled, frequently gelatinous, fine to medium. ***Rays*** unstoried, 1–10 (mostly 4–6)-seriate, essentially homocellular; ray-vessel pitting simple to bordered.

USES

Hard elm is used for the manufacture of the same products as American and slippery elms; it is preferred, however, where hardness and ability to resist shock are of primary importance.

HACKBERRY

Hackberry, common hackberry (*Celtis occidentalis* L.)

Sugarberry (*Celtis laevigata* Willd.)

GENERAL CHARACTERISTICS

Sapwood pale yellow to grayish or greenish yellow, frequently discolored with blue sap stain, wide; ***heartwood,*** when present, yellowish gray to light brown streaked with yellow; ***wood*** without characteristic odor or taste, straight- or sometimes interlocked-grained, moderately heavy (sp gr approx 0.49 green, 0.59 ovendry), moderately hard. ***Growth rings*** distinct (wood ring-porous). ***Early-wood pores*** large, distinctly visible to the naked eye, forming a conspicuous band 2–5 pores in width; transition from early to late wood more or less abrupt; ***late-wood pores*** small, numerous, arranged in more or less continuous, wavy, concentric bands. ***Parenchyma*** not visible. ***Rays*** distinctly visible to the naked eye.

MINUTE ANATOMY

Vessels in the late wood moderately to very numerous; ***largest early-wood vessels*** very large to extremely large, in several rows; perforation plates simple;

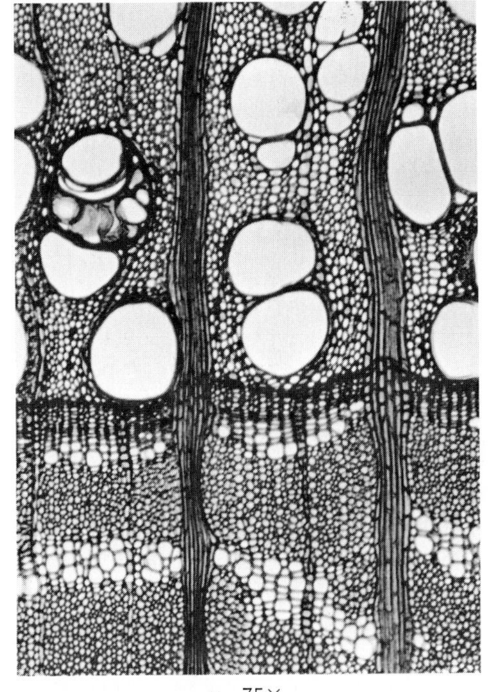

x—75×

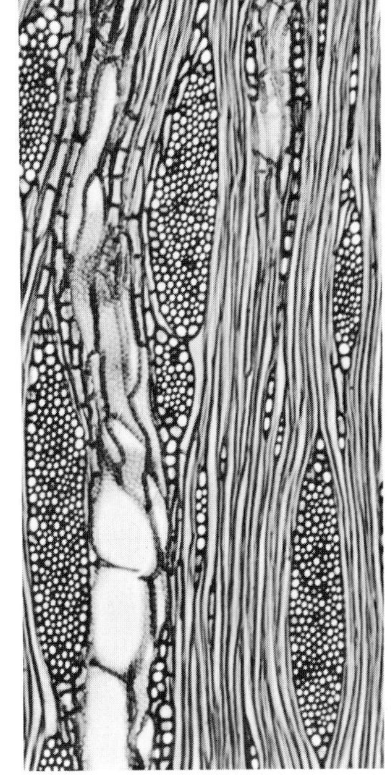

t—75×

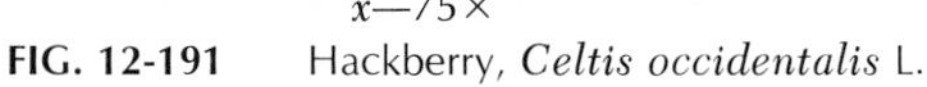

FIG. 12-191 Hackberry, *Celtis occidentalis* L.

spiral thickening present in the smaller vessels; intervessel pits orbicular or angular through crowding, 8–12 μ in diameter. ***Vascular tracheids*** present in both the early-wood porous zone and in the wavy bands of late-wood vessels, grading into small vessels, with spiral thickening. ***Parenchyma*** paratracheal-scanty to vasicentric and apotracheal-diffuse; (*a*) paratracheal parenchyma abundant, (1) associated with vascular tracheids and forming tracts of conjunctive tissue between the early-wood vessels, (2) marginal to and occasionally included in the wavy concentric bands of late-wood vessels and vascular tracheids; (*b*) apotracheal-diffuse parenchyma very sparse, scattered in the fibrous tracts. ***Libriform fibers*** moderately thick- to thick-walled, frequently gelatinous, fine to medium. ***Rays*** unstoried, 1–13 (mostly 5–8)-seriate, heterocellular for the most part; ray-vessel pitting simple.

REMARKS

The woods of common hackberry and sugarberry are indistinguishable. Hackberry is sometimes confused with elm (*Ulmus* spp.) but has a wider sapwood, with a distinct yellowish tinge (whitish to light brown in elm), and wider and heterocellular rays.

USES

At many mills, not distinguished from and used for the same purposes as American and slippery elms, and white ash. The better grades are used principally for *furniture*, and to a lesser extent for *millwork*, and *sporting* and *athletic* goods; the low-grade lumber is made up largely into *boxes* and *crates*; *veneer* for containers and interior plywood faces.

RED MULBERRY

Morus rubra L.

GENERAL CHARACTERISTICS

Sapwood yellowish, narrow; ***heartwood*** orange-yellow to golden brown, turning russet-brown on exposure; ***wood*** without characteristic odor or taste, straight-grained, heavy (sp gr approx 0.59 green), hard. ***Growth rings*** distinct (wood ring-porous). ***Early-wood pores*** large, plainly visible to the naked eye, forming a band 2–8 pores in width, occasionally with white deposits; transition from early to late wood abrupt; ***late-wood pores*** small, arranged in nestlike groups showing more or less of a tendency, especially in the outer late wood, toward aggregation into concentric, wavy, interrupted bands, sometimes solitary. ***Parenchyma*** not visible. ***Rays*** plainly visible to the naked eye, forming a pronounced fleck on the radial surface.

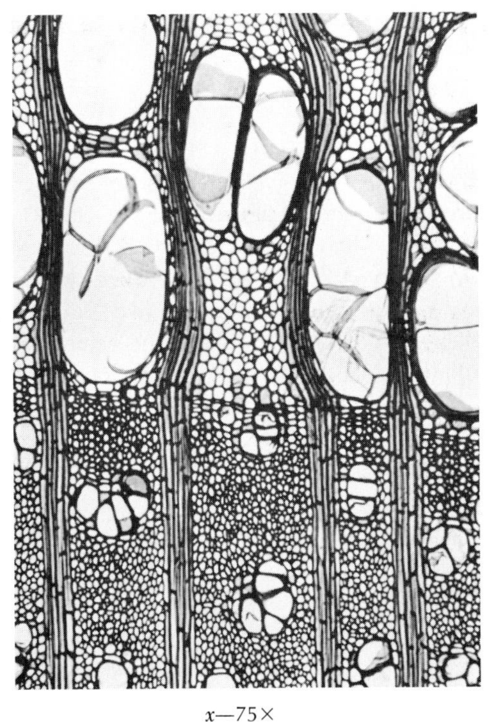

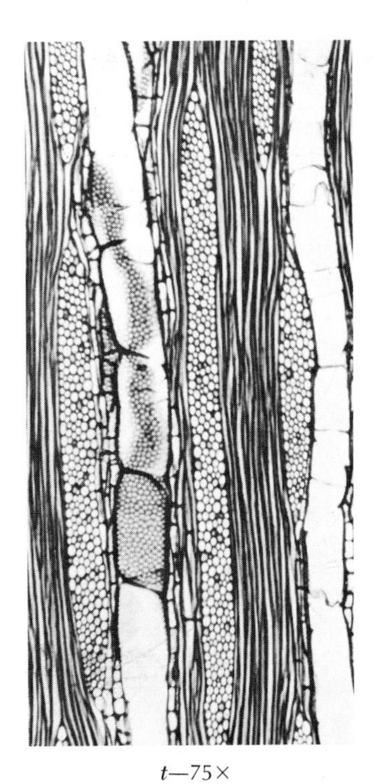

x—75×　　*t*—75×

FIG. 12-192 Red mulberry, *Morus rubra* L.

MINUTE ANATOMY

Vessels few to numerous, the largest medium to very large; perforation plates simple; spiral thickening present in the small vessels in the late wood; intervessel pits orbicular to oval or angular crowding, 8–10 μ in diameter; early-wood vessels in the heartwood partly occluded with tyloses, sometimes with white deposits. ***Parenchyma*** paratracheal-vasicentric and apotracheal-diffuse; (*a*) paratracheal parenchyma (1) abundant in the porous early-wood zone and composing part of the conjunctive tissue between the vessels and the rays, (2) less abundant elsewhere in the ring, forming a 1–several (mostly 1)-seriate, often interrupted sheath about the vessels or vessel groups; (*b*) apotracheal-diffuse parenchyma sparse, confined to the fibrous tracts. ***Libriform fibers*** thin- to thick-walled, frequently gelatinous, medium to coarse. ***Rays*** unstoried, 1–8 (mostly 5–7)-seriate, homocellular to heterocellular; ray-vessel pitting simple to bordered.

USES

Locally for *fence posts; furniture; interior finish; caskets; agricultural implements; wagon stock; cooperage.*

OSAGE–ORANGE

Maclura pomifera (Raf.) Schneid.

GENERAL CHARACTERISTICS

Sapwood light yellow, narrow; ***heartwood*** golden-yellow to bright orange, darkening upon exposure, often with reddish streaks along the grain; coloring matter readily soluble in tepid water; ***wood*** without characteristic odor or taste, straight-grained, exceedingly heavy (sp gr approx 0.76 green, 0.84 ovendry), very hard. ***Growth rings*** distinct (wood ring-porous). ***Early-wood pores*** large, forming a band 2–3 pores in width, completely occluded with tyloses in the heartwood and the contours of the individual pores hence indistinct, the early-wood zone appearing as a light-colored band (*x*) to the naked eye; transition from early to late wood abrupt; ***late-wood pores*** small, arranged in nestlike groups which coalesce in the outer late wood, forming interrupted concentric bands. ***Parenchyma*** not distinct as such. ***Rays*** barely visible to the naked eye.

MINUTE ANATOMY

Vessels very few to moderately numerous, the largest large to very large; perforation plates simple; spiral thickening present in the small vessels in the late wood; tyloses in the early-wood vessels small, appearing cellular *en masse*; intervessel pits orbicular to oval or angular through crowding, 8–12 μ in diameter, the orifices frequently confluent (Fig. 12-145, p. 535). ***Parenchyma*** abundant, paratracheal-vasicentric, paratracheal-confluent, and marginal; (*a*) paratracheal-vasicentric parenchyma (1) intermingled with some fusiform parenchyma cells (substitute fibers) and fibers and composing an appreciable portion of the conjunctive tissue between the vessels and the rays in the porous early-wood zone, (2) forming an irregular, 1–several-seriate, continuous or interrupted sheath about the vessels or groups of vessels farther out in the ring, (3) toward the outer margin of the ring passing over into (*b*) paratracheal-confluent parenchyma and frequently serving as conjunctive tissue, uniting proximate groups of vessels laterally (across the rays) or in the extreme outer portion of the ring forming interrupted tangential bands devoid of vessels; (*c*) marginal parenchyma abundant but not forming a distinct line, passing over into the paratracheal parenchyma of the succeeding ring. ***Fusiform parenchyma cells*** (substitute fibers) occasional, intermingled with parenchyma. ***Libriform fibers*** moderately thick- to thick-walled, fine. ***Rays*** unstoried, 1–6 (mostly 2–4)-seriate, homocellular for the most part; ray-vessel pits simple or bordered.

REMARKS

The wood of Osage-orange is frequently confused with that of black locust (***Robinia pseudoacacia*** L.); for differences between these timbers, see Remarks under Black Locust, p. 600.

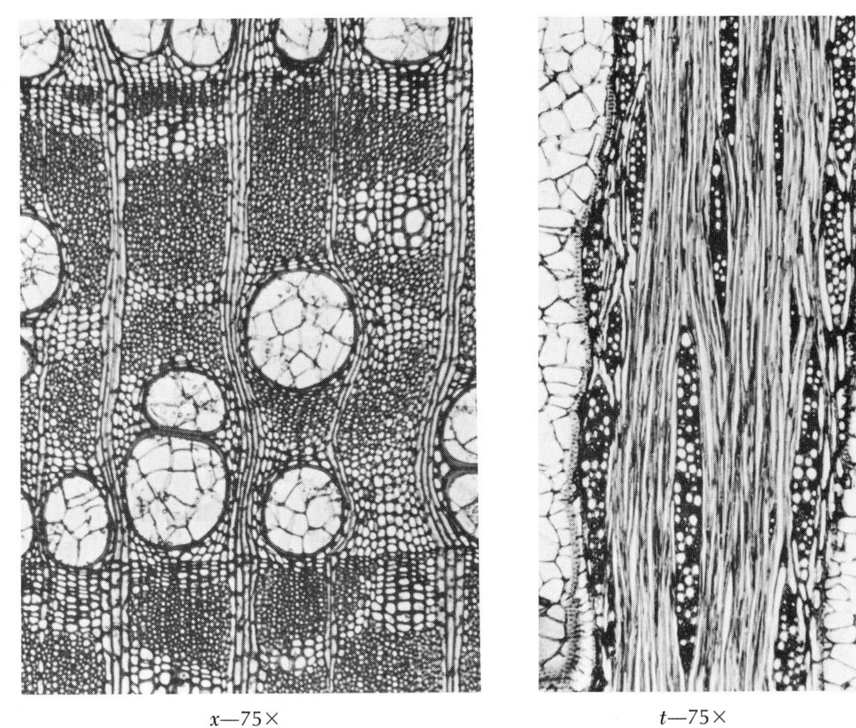

x—75× *t*—75×

FIG. 12-193 Osage-orange, *Maclura pomifera* (Raf.) Schneid.

USES

Fence posts (very durable); *insulator pins* (because of durability, strength, and minimal shrinking and swelling which tend to prevent the pins from loosening in the crossarms); locally for *nonmotor vehicle parts* (especially for hubs), *treenails*, *machinery parts*; *archery* (bows); *dyewood* (contains yellow, green, and brown coloring principles; now seldom used and only locally because of the advent of synthetic dyes).

MAGNOLIA

Cucumbertree, cucumber magnolia. (*Magnolia acuminata* L.)

Southern magnolia, evergreen magnolia. (*Magnolia grandiflora* L.)

GENERAL CHARACTERISTICS

Sapwood whitish, narrow or wide; *heartwood* yellow, greenish yellow to brown, or greenish black (the latter is said to be common in southern magnolia); *wood* without characteristic odor or taste, straight-grained, fairly heavy (sp gr 0.44–0.46 green, 0.52–0.53 ovendry), moderately hard to hard. *Growth rings* distinct, delineated by a whitish line of marginal parenchyma. *Pores* small, indistinct without

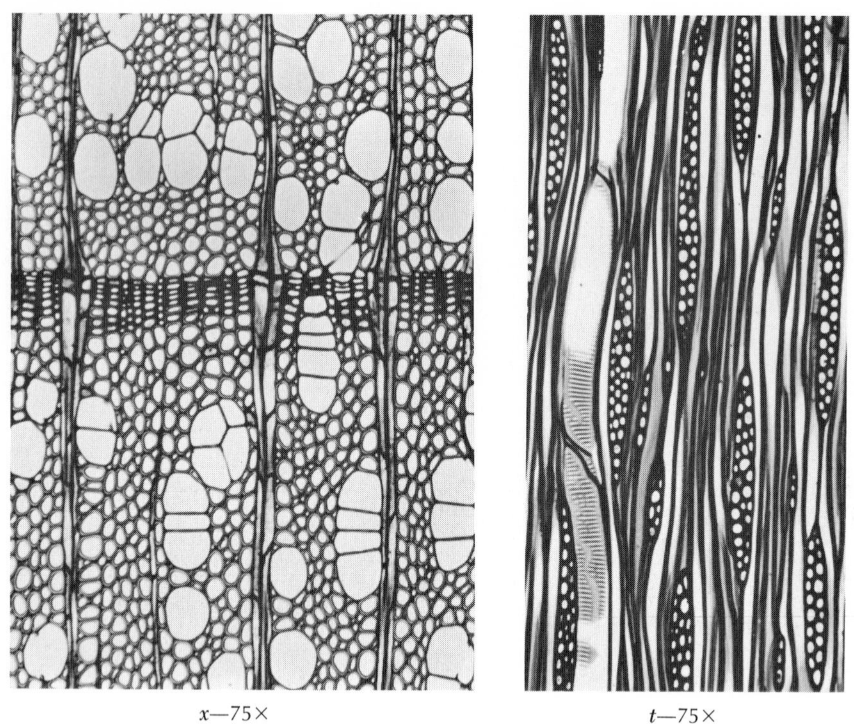

x—75× *t*—75×

FIG. 12-194 Cucumbertree, *Magnolia acuminata* L.

a hand lens, quite uniform in size, fairly evenly distributed throughout the ring (wood diffuse-porous), solitary or in multiples of 2–several. ***Parenchyma*** marginal, the line plainly visible to the naked eye. ***Rays*** distinct to the naked eye (*x*), nearly uniform in width.

MINUTE ANATOMY

Vessels moderately numerous to numerous, the largest small; perforation plates simple or occasionally scalariform in the first few annual rings in *Magnolia acuminata* L., mostly scalariform with 6–10 stout bars in *M. grandiflora* L.; spiral thickening present in the vessels of *M. grandiflora* L.; intervessel pitting scalariform (Fig. 12-152, p. 535), the pits linear or rarely elliptical, 12-60 μ in diameter. Line of marginal *parenchyma* 1–several-seriate. ***Fiber tracheids*** thin- to moderately thick-walled, coarse to very coarse. ***Rays*** 4–7 per mm (*x*), unstoried, 1–5 (mostly 1–2 in *M. acuminata*)-seriate, homocellular to heterocellular, where heterocellular the upper and lower margins generally consisting of 1 row of upright cells; upright cells less than 60 μ in height along the grain; ray-vessel pits simple to bordered.

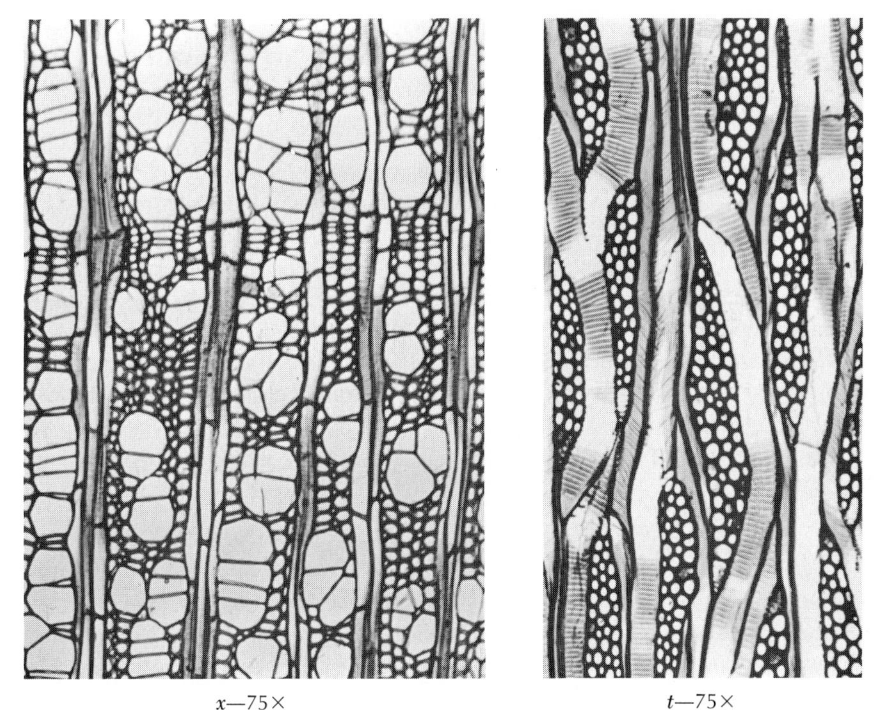

FIG. 12-195 Southern magnolia, *Magnolia grandiflora* L.

REMARKS

Cucumbertree and southern magnolia cannot always be separated with certainty; the latter can frequently be distinguished by its darker (greenish-black) heartwood, by the presence of scalariform perforation plates in addition to those of the simple type in the mature wood, and by the spiral thickenings in the vessels; the rays of *M. grandiflora* L. are also wider, ranging from 1–5 cells in breadth. For separation of magnolia timbers from yellow-poplar, see Remarks under the latter, p. 584.

USES

Frequently mixed with yellow-poplar and sold as such or under its own name (the dark-colored heartwood of southern magnolia commands a higher price; preferred for *furniture* and other uses where hardness is a factor). Other important uses for magnolia are for *fixtures*; *venetian blinds* (because of good finishing qualities and freedom from twisting and warping); *interior finish*; *sash*; *doors*; *general millwork*; *boxes* and *crates*; *pulpwood*.

YELLOW-POPLAR, TULIP-POPLAR, WHITEWOOD
Liriodendron tulipifera L.

GENERAL CHARACTERISTICS

Sapwood whitish, often variegated or striped, narrow; ***heartwood*** variable in color, ranging from clear yellow to tan or greenish brown, frequently marked with shades of purple, dark green, blue, and black (these colorations have no apparent effect on the physical properties of this wood, and affect its value only where a natural finish is desired);* ***wood*** without characteristic odor or taste, straight-grained, moderately light (sp gr approx 0.38 green, 0.43 ovendry), moderately soft. ***Growth rings*** distinct, delineated by a whitish line of marginal parenchyma. ***Pores*** small, not visible without a lens, fairly uniformly distributed throughout the ring (wood diffuse-porous), solitary and in multiples of 2–several. ***Parenchyma*** marginal, the line plainly visible to the naked eye. ***Rays*** distinct to the naked eye (*x*), nearly uniform in width.

MINUTE ANATOMY

Vessels moderately numerous to numerous, the largest small to medium; perforation plates scalariform, with 2–16 thin bars; intervessel pits oval or oval-angular for the most part (Fig. 12-160), arranged in transverse rows (opposite pitting), 6–15 μ in diameter, rarely linear and then up to 20 μ in diameter. Line of marginal ***parenchyma*** 1–several-seriate. ***Fiber tracheids*** thin- to moderately thick-walled, medium to coarse. ***Rays*** 4–7 per mm (*x*), unstoried or rarely somewhat storied, 1–5 (mostly 2–3)-seriate, homocellular to heterocellular, where heterocellular the upper and lower margins generally consisting of 1 row of upright cells; upright cells less than 60 μ in height along the grain; ray-vessel pits simple to bordered.

REMARKS

Yellow-poplar resembles magnolia, but the figure of the flat-sawn stock is usually less prominent because the zones of terminal parenchyma are narrower and hence less accentuated. Accurate identification is assured with a compound microscope. In yellow-poplar the pits are oval or oval-angular and are arranged in transverse rows (see Fig. 12-160, p. 536); the intervessel pits of magnolia, in contrast, are predominantly linear and extend nearly across the vessel wall (see Fig. 12-152, p. 535). The vessels of yellow-poplar are invariably devoid of spiral thickening; the same holds for the vessels of ***Magnolia acuminata*** L., but those of ***M. grandiflora*** L. occasionally possess spirals. The perforation plates of ***M. acuminata*** L. are simple in the mature wood, those of ***M. grandiflora*** L. mostly scalariform; the vessels of this last-named species also occasionally exhibit spiral thickening, a feature that is never present in yellow-poplar.

* Areas of plain brown, untinged by yellow or green, however, indicate tissue affected by fungous growth.

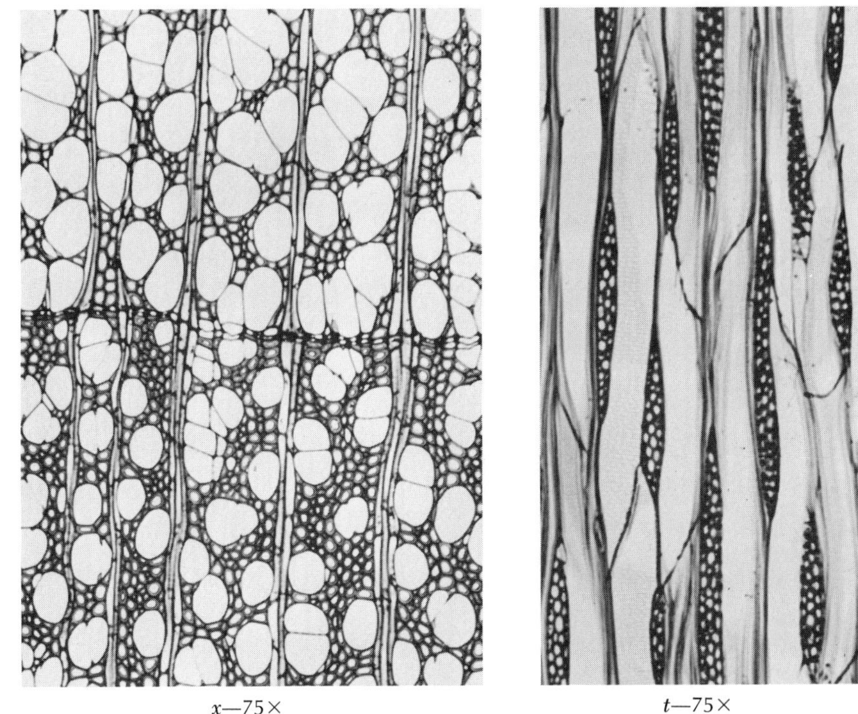

x—75× *t*—75×

FIG. 12-196 Yellow-poplar, *Liriodendron tulipifera* L.

USES

Veneer (large, clear logs are well suited to the manufacture of rotary-cut veneer) used extensively for berry and fruit boxes, also in the form of plywood for interior finish, furniture, cabinetwork, and piano cases, and as corestock; *pulp* (by the soda process); *hat blocks*. Lumber used for *furniture, cabinetwork*, and *finish* that is to be painted or enameled; *boxes, crates* and *pallets* (lower grades); *millwork* (sash, doors, and blinds); *musical instruments* (especially television, radio, and phonograph cabinets); *fixtures*; *coffins* and *caskets*; *miscellaneous uses*, such as kitchen utensils, toys, novelties, patterns, cigar boxes.

CALIFORNIA-LAUREL, OREGON-MYRTLE

Umbellularia californica (Hook. & Arn.) Nutt.

GENERAL CHARACTERISTICS

Sapwood whitish to light brown, thick; *heartwood* light brown to grayish brown, frequently with darker streaks of pigment figure; *wood* with a characteristic spicy odor (strong to very mild), without characteristic taste, straight- or interlocked-

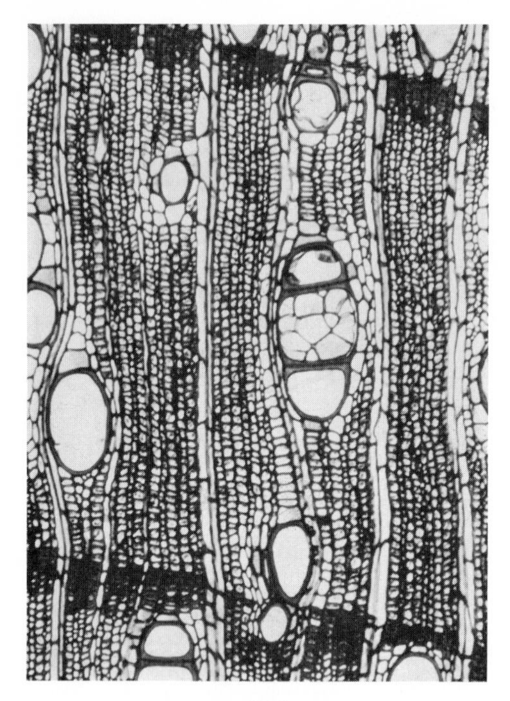

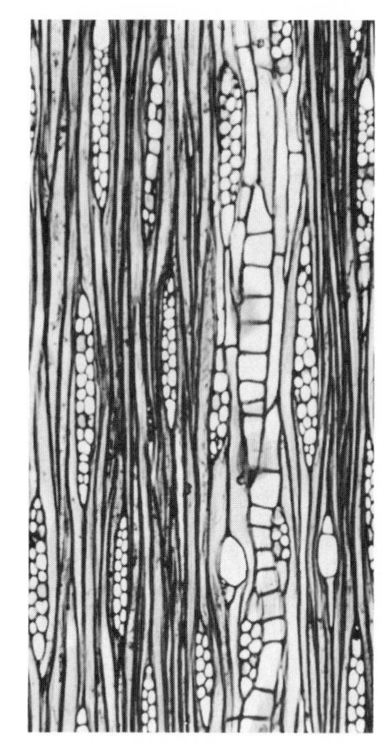

x—75× t—75×

FIG. 12-197 California-laurel, *Umbellularia californica* (Hook. & Arn.) Nutt.

grained, heavy (sp gr approx 0.51 green, 0.58 ovendry), moderately hard, subject to attack by powder-post beetles. *Growth rings* distinct, delineated by a darker band of denser late wood. *Pores* small, barely visible to the naked eye, rather distant, evenly distributed throughout the growth ring (wood diffuse-porous), solitary and in multiples of 2–several, encircled by a whitish sheath of vasicentric *parenchyma* which is about as wide as the pore. *Rays* fine, not distinct to the naked eye, plainly visible with a hand lens.

MINUTE ANATOMY

Vessels very few to moderately few, the largest medium to large; perforation plates simple; spiral thickening occasional; intervessel pits orbicular to angular, 6–10 μ in diameter. *Parenchyma* paratracheal-vasicentric, abundant, forming a 1–3-seriate sheath. *Libriform fibers* thin-walled, medium to very coarse, frequently gelatinous, in part septate. *Rays* unstoried, 1–4 (mostly 2)-seriate, typically heterocellular, frequently with oil cells; ray-vessel pitting simple.

REMARKS

This wood darkens appreciably when soaked in water, a treatment that is often followed prior to its utilization.

USES

Veneer, used in the manufacture of furniture and for paneling (handsome burl stock is obtained from this species); ***novelties*** and ***woodenware; turnery; furniture squares; cabinetwork; interior trim.*** Used under the keel in launching ships; appears to resist crushing better and to have more "slip" than any other local species.

SASSAFRAS

Sassafras albidum (Nutt.) Nees

GENERAL CHARACTERISTICS

Sapwood light yellow; ***heartwood*** dull grayish brown to orange-brown or dark brown; ***wood*** with odor of sassafras on fresh-cut surface, with spicy taste, straight-grained, moderately heavy (sp gr approx 0.42 green, 0.47 ovendry), moderately hard. ***Growth rings*** very distinct (wood ring-porous). ***Early-wood pores*** large, plainly visible to the naked eye, forming a band 3–8 pores in width; transition from early to late wood abrupt; ***late-wood pores*** small, solitary, and in multiples of 2–3. ***Parenchyma*** visible with a hand lens, forming a narrow sheath around the late-wood pores and extending laterally from their flanks. ***Rays*** barely visible to the naked eye.

MINUTE ANATOMY

Vessels very few to few, the largest medium to large; perforation plates simple or occasionally scalariform in the late-wood vessels; intervessel pits orbicular to oval, generally not crowded, 6–10 μ in diameter. ***Parenchyma*** paratracheal-vasicentric, rarely paratracheal-confluent in the late wood, and apotracheal-diffuse; sheath of paratracheal parenchyma around the late-wood vessels 1–several-seriate, extending from the flanks (aliform); apotracheal-diffuse parenchyma very sparse. ***Fiber tracheids*** thin-walled, medium to very coarse. ***Rays*** unstoried, 1–4-seriate, homocellular to heterocellular, frequently containing oil cells; ray-vessel pits simple to bordered.

REMARKS

Frequently confused with black ash (***Fraxinus nigra*** Marsh.), which it resembles in color, grain, and texture. Distinct from this species in possessing a characteristic odor on freshly exposed surfaces traceable to oil cells in the wood rays, and in the presence of aliform parenchyma extending from the flanks of the late-wood pores.

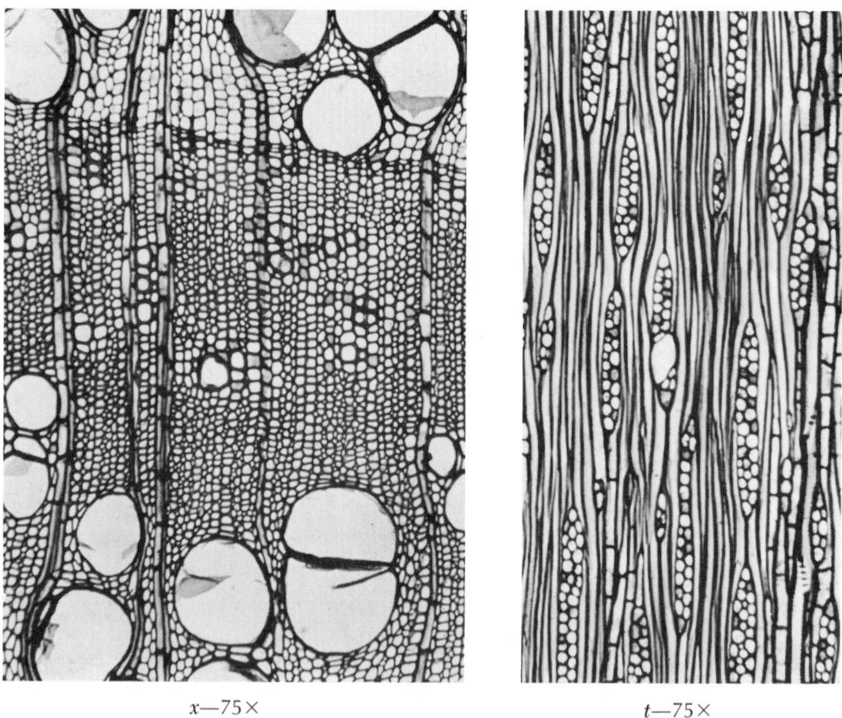

x—75× *t*—75×

FIG. 12-198 Sassafras, *Sassafras albidum* (Nutt.) Nees

USES

Locally for *fence posts* and *general millwork; rails; foundation posts; wooden pails; small boat construction*; larger logs are occasionally sawed into lumber which is often sold as "black ash."

SWEETGUM, REDGUM

Liquidambar styraciflua L.

GENERAL CHARACTERISTICS

Sapwood (called "sap-gum" in the trade) white, frequently with a pinkish tinge, often discolored with blue sap stain; *heartwood* (called "redgum" in the trade) carneous gray to varying shades of reddish brown, the darker grades frequently with darker streaks of pigment figure (called "figured redgum" in the trade);* *wood* without characteristic odor or taste, frequently interlocked-grained, moderately heavy (sp gr approx 0.46 green, 0.54 ovendry), fairly hard. *Growth rings*

* Dull yellowish to brown discolorations in redgum indicate early stages of decay; the figure thus resulting should not be confused with that of figured redgum.

inconspicuous (wood diffuse-porous). ***Pores*** small, not visible to the naked eye, quite uniform in size throughout the growth ring, numerous and frequently crowded, solitary, in multiples of 2–3 (mostly 2), or paired laterally (overlapping vessel segments). ***Parenchyma*** not visible. ***Rays*** not distinct to the naked eye, very close, seemingly occupying half the area on the transverse surface of the wood. Longitudinal wound (traumatic) ***gum canals*** sometimes present (Fig. 12-199*x*), in tangential rows that usually appear at wide intervals, frequently occluded with white deposits.

MINUTE ANATOMY

Vessels very numerous, the largest small to medium; perforation plates scalariform, with many bars (15 plus) 1 μ or more in thickness, or reticulate; spiral thickening present, restricted to the tapering ends of the vessel elements; intervessel pits in transverse rows of 1–3 (Fig. 12-157, p. 536), orbicular to oval or linear through fusion, 6–30 μ in diameter, oppositely arranged. ***Parenchyma*** paratracheal-scanty and apotracheal-diffuse, sparse. ***Fiber tracheids*** moderately thick-walled, medium to very coarse, with conspicuous bordered pits 7–9 μ in diameter. ***Rays*** 6–9 per mm (*x*), unstoried, of two types; (*a*) narrow rays uniseriate,

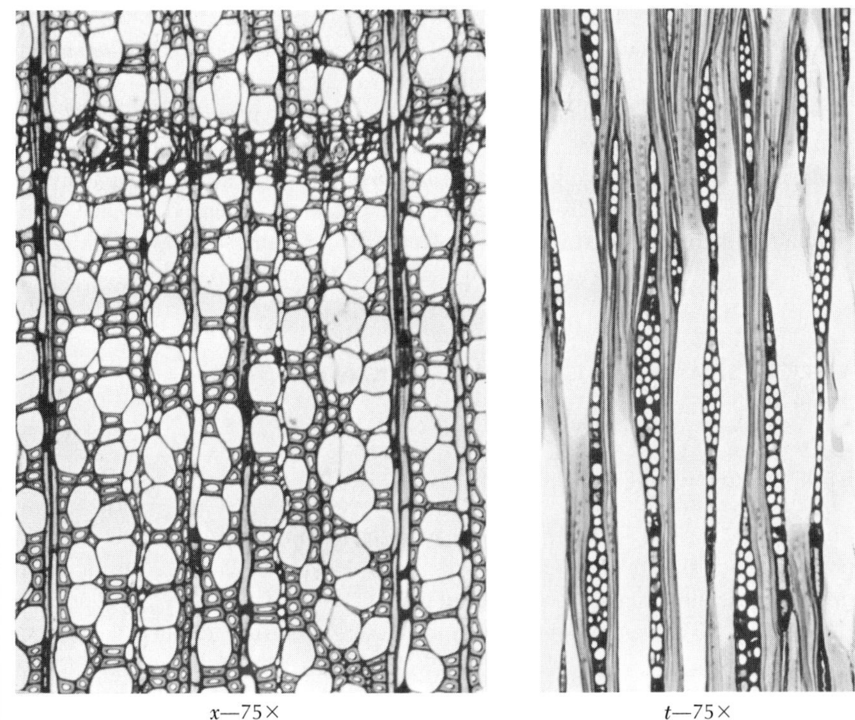

x—75× *t*—75×

FIG. 12-199 Sweetgum, ***Liquidambar styraciflua*** L.

homocellular and consisting entirely of upright cells or heterocellular with both upright and procumbent cells; (*b*) wider rays 2–3 (mostly 2)-seriate through the central portion, which consists of procumbent cells, with uniseriate extensions (*t*) above and below consisting wholly or mainly of upright cells; several rays of the "*b*" type not infrequently confluent along the grain; upright cells in 1–5 plus rows (*r*), 30–100 μ in height along the grain; ray-vessel pitting simple to bordered. *Gum canals,* when present, arranged in a uniseriate tangential row (*x*), with angled orifice.

REMARKS

Unique among North American hardwoods. Sweetgum, sometimes called "hazelwood" in the trade, is cut from the heartwood; the sapwood is marketed as "sapgum" in the United States and as "hazel pine" in England. Figured sweetgum (pigment figure) resembles Circassian walnut (*Juglans regia* L.) and is often marketed abroad as "satin walnut." The pigment figure of sweetgum is best shown in flat-sawn timber and in rotary-cut veneer; a handsome ribbon stripe is obtained by quartering interlocked-grained stock.

USES

*Veneer** used extensively in furniture, in panels, and in various cheap containers such as fruit baskets, boxes, crates, and cigar boxes; *slack cooperage*; *mine props*; *railroad ties*; *pulp* (alkaline process). Lumber used for *planing-mill products* (especially trim), *furniture, radio, television,* and *kitchen cabinets, boxes, crates,* and *pallets* (lower grades); *turned articles.* A considerable quantity of redgum lumber is normally shipped abroad, mainly to England, France, and Germany, where it is used in the manufacture of furniture, some of which is re-exported to the United States as "satin walnut" (see under Remarks).

AMERICAN SYCAMORE, BUTTONWOOD, AMERICAN PLANETREE

Platanus occidentalis L.

GENERAL CHARACTERISTICS

Sapwood whitish to light yellowish or reddish brown; *heartwood* light to dark brown or reddish brown when distinguishable; *wood* without characteristic odor or taste, generally irregularly interlocked-grained, moderately heavy (sp gr approx 0.46 green, 0.54 ovendry), moderately hard. *Growth rings* distinct, delineated

* The most important species in the United States for hardwood veneer production.

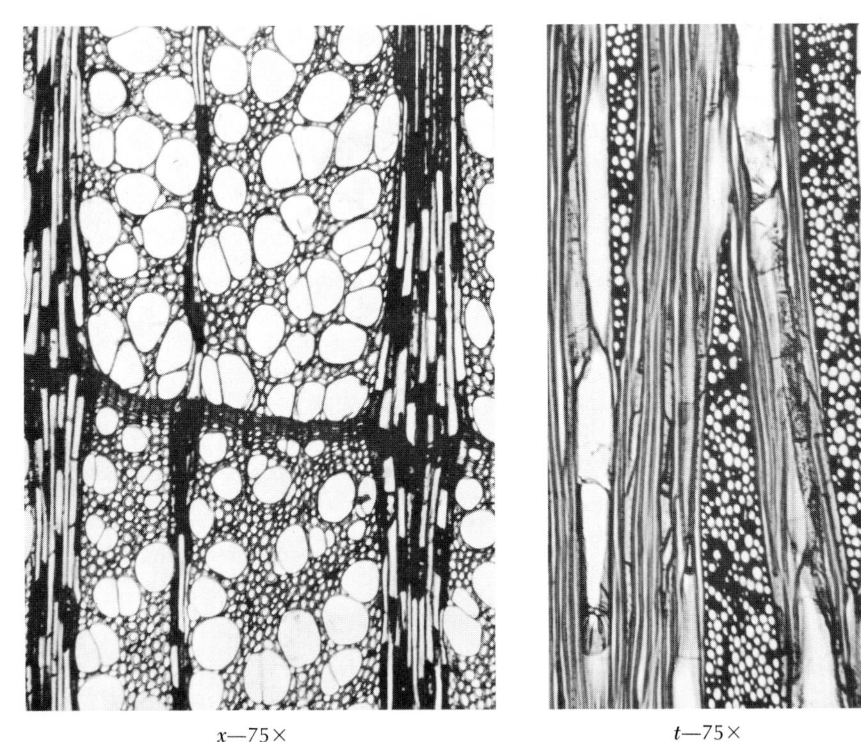

x—75× *t*—75×

FIG. 12-200 American sycamore, *Platanus occidentalis* L.

by a narrow band of lighter tissue at the outer margin. *Pores* small, indistinct, or barely visible to the naked eye, numerous and frequently crowded. *Parenchyma* not visible. *Rays* comparatively wide, conspicuous to the naked eye (*x*), nearly uniform in width, close, appearing as short, closely packed lines on the tangential surface (the pattern is characteristic), forming a high, reddish brown or silvery fleck on the radial surface.

MINUTE ANATOMY

Vessels very numerous, the largest small; perforation plates simple for the most part, occasionally scalariform with a few bars; intervessel pits oval to orbicular (Fig. 12-149, p. 535), widely spaced, small (4–6 μ in diameter). *Parenchyma* paratracheal-scanty and apotracheal-diffuse, and in aggregates; (*a*) paratracheal parenchyma restricted to occasional cells, never forming a sheath; (*b*) apotracheal-diffuse parenchyma abundant, the cells scattered and (*c*) in short lines which exhibit no regularity. *Fiber tracheids* moderately thick-walled, medium to very coarse. *Rays* unstoried, 1–14 (mostly 3–14)-seriate, up to 3 plus mm in height along the grain, homocellular; ray-vessel pitting similar to intervessel type.

REMARKS

Distinct among the North American woods because of its characteristic figure, either quartered or flat-sawn; the former is highly ornamental because of the large crowded rays.

USES

Veneer, for fruit and vegetable baskets, for cigar boxes, and occasionally for paneling (quarter-sawn); *slack cooperage* (especially sugar and flour barrels because it does not impart taste, odor, or stain). Lumber used for *boxes, crates* (especially for plug-tobacco boxes), and *pallets*; *furniture* (largely concealed parts, such as drawer sides and bottoms, backs for cheaper chairs); *millwork* (interior trim and paneling; the beauty of quarter-sawn stock has not been sufficiently recognized).

BLACK CHERRY, CHERRY

Prunus serotina Ehrh.

GENERAL CHARACTERISTICS

Sapwood whitish to light reddish brown, narrow; *heartwood* light to dark reddish brown, dull; *wood* without characteristic odor or taste, straight-grained, moderately heavy (sp gr approx 0.47 green, 0.53 ovendry), moderately hard. *Growth rings* fairly distinct, delineated by a narrow, inconspicuous band of porous tissue in the early wood. *Pores* small, indistinct to the naked eye, those at the beginning of the ring somewhat larger, closer, and aligned in a more or less definite uniseriate row (wood semi-ring-porous), elsewhere in the ring quite uniform in size, fairly evenly distributed, and arranged solitary, in multiples, and in nests which are frequently oblique to the rays. *Parenchyma* not visible. *Rays* plainly visible to the naked eye. Longitudinal traumatic, lysigenous *gum canals* sometimes present (Fig. 5-15*b*).

MINUTE ANATOMY

Vessels moderately numerous to very numerous, the largest small; perforation plates simple; spiral thickening present (Fig. 12-141, p. 534); intervessel pits broad-oval to orbicular or somewhat angular through crowding, 7–10 μ in diameter; gummy infiltration occasionally present. *Parenchyma* very sparse, confined to occasional cells, or usually absent. *Fiber tracheids* thin- to thick-walled, medium to coarse. *Rays* unstoried, 1–6 (mostly 3–4)-seriate, homocellular to heterocellular; ray-vessel pitting similar to intervessel type.

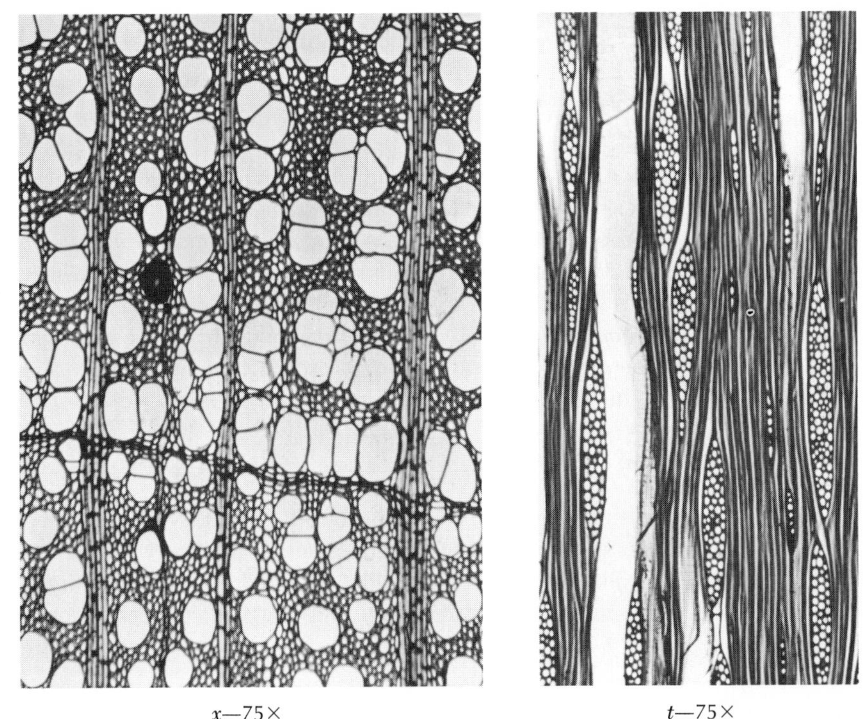

x—75× *t*—75×

FIG. 12-201 Black cherry, *Prunus serotina* Ehrh.

REMARKS

One of the most handsome of our domestic woods because of its reddish-brown color and luster when properly finished. An excellent cabinet wood for the reasons mentioned above, coupled with low shrinkage and freedom from warping and checking. Not used so extensively as formerly owing to its relative scarcity.

USES

Printers' blocks for mounting electrotypes (for which it is especially suited because of its strength, moderate hardness, moderate shrinkage, and ability to stay in place); *furniture* (valued for its beautiful natural color and its good working and finishing qualities); *patterns; professional* and *scientific instruments; piano actions; cores* for high-class panels; *interior trim* (in building, cars, and boats); *handles; woodenware, novelties* and *toys, face veneer* for furniture and interior grades of plywood.

KENTUCKY COFFEETREE, COFFEETREE
Gymnocladus dioicus (L.) K. Koch

GENERAL CHARACTERISTICS

Sapwood yellowish white, narrow; *heartwood* light red to red or reddish brown; *wood* without characteristic odor or taste, straight-grained, heavy (sp gr approx 0.50 green), hard. *Growth rings* conspicuous (wood ring-porous). *Early-wood pores* large, plainly visible to the naked eye, forming a band 3–6 pores in width, open, occasionally with deposits of reddish gum; transition from early to late wood abrupt; *late-wood pores* small, visible with a hand lens, in the outer late wood in nestlike groups, which occasionally coalesce. *Parenchyma* not visible as such. *Rays* visible but not very conspicuous to the naked eye.

MINUTE ANATOMY

Vessels very few to moderately numerous, the largest large to very large; perforation plates simple; spiral thickening present in the small vessels in the late wood; intervessel pits orbicular to oval, vestured, 5–9 μ in diameter. *Parenchyma* paratracheal-vasicentric, paratracheal-confluent, and marginal; (*a*) paratracheal-

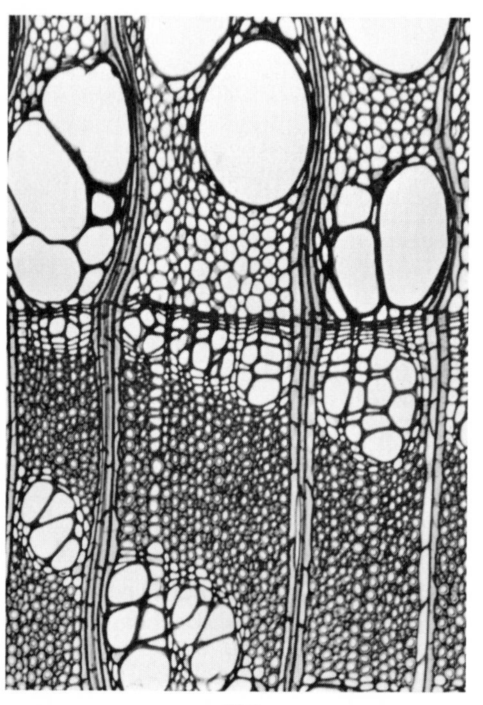

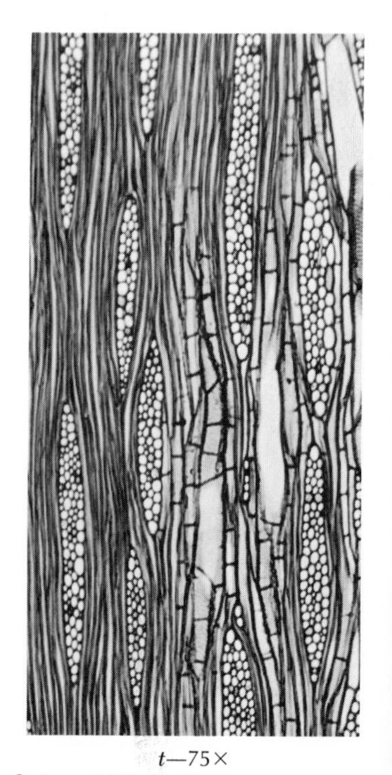

x—75×　　*t*—75×

FIG. 12-202 Kentucky coffeetree, *Gymnocladus dioicus* (L.) K. Koch

vasicentric parenchyma fairly abundant, forming a 1–several (mostly 1)-seriate, occasionally interrupted sheath about the vessels or vessel groups, in the outer portion of the ring, not infrequently extending tangentially as (*b*) paratracheal-confluent parenchyma and joining proximate vessel groups; (*c*) marginal parenchyma abundant but not forming a distinct line, passing over into paratracheal-confluent parenchyma. ***Libriform fibers*** thin- to thick-walled, medium. ***Rays*** unstoried, 1–7 (mostly 3–5)-seriate, homocellular, the tallest less than 1200 μ in height along the grain; ray-vessel pitting similar to intervessel type.

REMARKS

This wood is frequently confused with that of honeylocust (***Gleditsia triacanthos*** L.), but can be readily distinguished with a hand lens. The vessels in the late wood of coffeetree are readily visible at low magnification (10X) and are arranged in nestlike groups which appear to be unaccompanied by parenchyma; in honeylocust, in contrast, the vessels toward the outer margin of the ring are barely visible with a hand lens, are solitary, in short radial rows or in small groups, and are embedded in short tangential bands of parenchyma. Honeylocust is also characterized by broader rays and frequent deposits of reddish gum in the vessels.

USES

Locally for *fence posts* and *rails*; *crossties*; *rough construction*; *furniture*; *cabinetwork*; *interior finish*; *fuel*.

HONEYLOCUST

Gleditsia triacanthos L.

GENERAL CHARACTERISTICS

Sapwood yellowish, wide; ***heartwood*** light red to reddish brown; ***wood*** without characteristic odor or taste, very heavy (sp gr approx 0.60 green, 0.67 ovendry), very hard. ***Growth rings*** conspicuous (wood ring-porous). ***Early-wood pores*** large, plainly visible to the naked eye, forming a band 3–5 pores in width, frequently with deposits of reddish-brown gum; transition from early to late wood abrupt; ***late-wood pores*** small, barely visible with a hand lens, solitary, in short radial rows, or in small groups, embedded in short, tangential bands of parenchyma in the late wood. ***Parenchyma*** as above, the whitish bands visible with a hand lens in the outer late wood. ***Rays*** conspicuous to the naked eye.

MINUTE ANATOMY

Vessels very few to moderately few, the largest large to very large; perforation plates simple; spiral thickening confined to the small vessels in the late wood; intervessel pits orbicular to oval or somewhat angular through crowding, vestured,

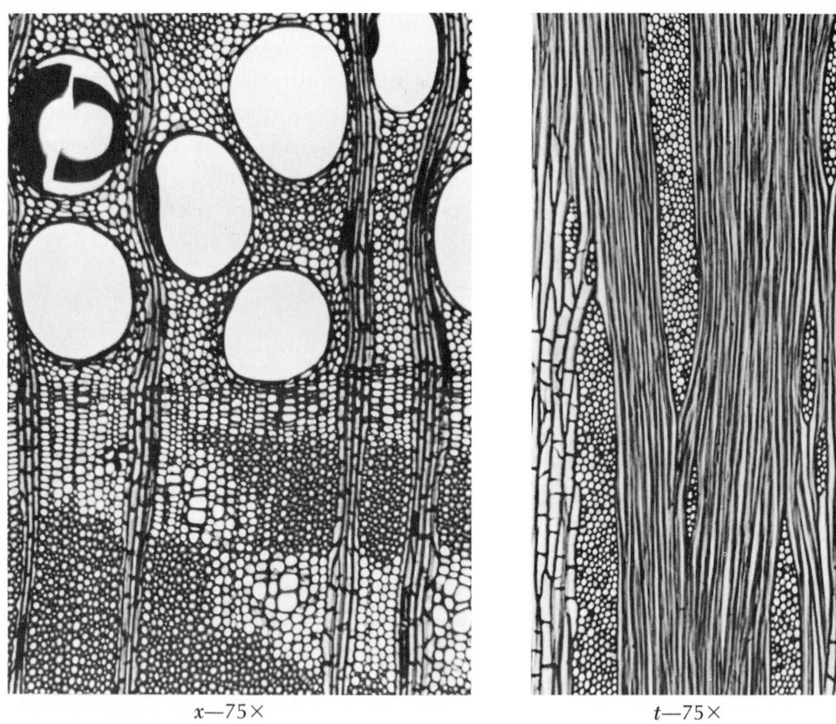

x—75× *t*—75×

FIG. 12-203 Honeylocust, *Gleditsia triacanthos* L.

5–10 μ in diameter, the apertures frequently confluent. *Parenchyma* abundant, paratracheal-vasicentric, paratracheal-confluent, and marginal; (*a*) paratracheal-vasicentric parenchyma (1) composing an appreciable portion of the conjunctive tissue between the vessels and the rays in the porous early wood zone, (2) forming fairly extensive tracts about the vessels farther out in the ring, (3) in the outer late wood extending from the flanks of the vessels and frequently uniting with parenchyma from proximate vessels and forming short, tangential, 1–8 plus seriate bands of (*b*) paratracheal-confluent parenchyma in which the vessels or vessel groups are included; (*c*) marginal parenchyma abundant but not forming a distinct line, passing over into the paratracheal parenchyma of the succeeding ring. *Fibers* moderately thick- to thick-walled, medium. *Rays* unstoried, 1–14 (mostly 6–9)-seriate, essentially homocellular, the tallest more than 1200 μ in height along the grain; ray-vessel pitting similar to intervessel type.

REMARKS

Honeylocust wood is frequently confused with that of coffeetree [*Gymnocladus dioicus* (L.) K. Koch]; for differences between these timbers see Remarks under Coffeetree, p. 595.

USES

Locally for *fence posts* and *rails*; *generally construction*; *furniture* (the wood possesses many desirable qualities such as attractive figure and color, strength, and hardness but is little used because of its scarcity); *interior trim*; *vehicles* (especially for wagon wheels).

YELLOWWOOD

Cladrastis lutea (Michx. f.) K. Koch

GENERAL CHARACTERISTICS

Sapwood narrow, nearly white; *heartwood* clear yellow, changing on exposure to light brown or yellow streaked with brown; *wood* without characteristic odor or taste, heavy (sp gr approx 0.52 green), hard. *Growth rings* distinct but not prominent (wood semi-ring-porous to diffuse-porous), delineated by a whitish line of marginal parenchyma and also through a fairly abrupt difference in size between the pores of the late wood and those in the early wood of the succeeding ring. *Pores* scattered except in the early wood, where they are often arranged in

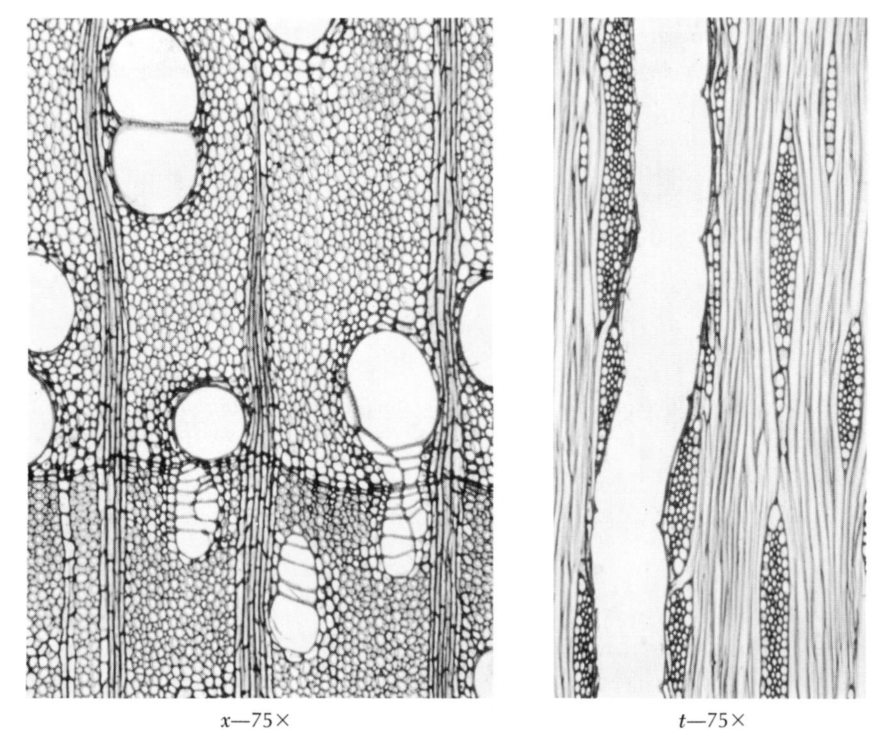

x—75× *t*—75×

FIG. 12-204 Yellowwood, *Cladrastis lutea* (Michx. f.) K. Koch

a uniseriate, more or less continuous row, decreasing gradually in size toward the outer margin of the ring, showing as white specks to the naked eye, appearing thick-walled (because of encircling parenchyma) with a hand lens, solitary and in multiples of 2–several (sometimes in nests in the late wood); tyloses sparse; brownish gummy deposits fairly abundant. *Parenchyma* as above and marginal, the whitish line usually distinct with a hand lens. *Rays* visible to the naked eye.

MINUTE ANATOMY

Vessels very few to few, the largest medium to large; perforation plates simple; spiral thickening occasionally present in the small vessels in the late wood, inconspicuous; intervessel pits mostly orbicular to angular, occasionally elliptical to linear, 6–18 μ in diameter, vestured. *Parenchyma* abundant, paratracheal, apotracheal-diffuse, and marginal; (*a*) paratracheal parenchyma forming an irregular, 1–several-seriate continuous sheath about the vessels or groups of vessels; (*b*) apotracheal-diffuse parenchyma sparse, restricted to occasional cells; (*c*) terminal parenchyma abundant, forming a distinct 1–several-seriate line. *Libriform fibers* moderately thick- to thick-walled, frequently gelatinous (especially in the outer late wood), fine to very coarse. *Rays* unstoried, heterocellular; (*a*) broader rays mostly 4–6-seriate, the upright cells frequently flanking the core of the rays as sheath cells, also in 1–several rows on the upper and lower margins of the rays; (*b*) narrow rays scarce, mostly 1–2-seriate; uniseriate rays consisting entirely of upright cells; biseriate rays with the upright cells confined to the upper and lower margins; ray-vessel pitting similar to intervessel type.

USES

Veneer (largely logs cut from trees used for ornamental plantings); *gunstocks* (occasionally); *fuel wood.*

BLACK LOCUST

Black locust (*Robinia pseudoacacia* L.)

Shipmast locust (*Robinia pseudoacacia* var. *rectissima* Raber)

GENERAL CHARACTERISTICS

Sapwood yellowish, narrow; *heartwood* greenish yellow to dark yellowish, greenish, or golden-brown (that of shipmast locust darker, ranging from deep yellow to golden or rich reddish brown); *wood* without characteristic odor or taste, very heavy (sp gr approx 0.66 green, 0.73 ovendry), very hard. *Growth rings* distinct (wood ring-porous). *Early-wood pores* large, forming a band 2–3 pores in width, completely occluded with tyloses in the heartwood and the contours of the individual pores hence indistinct, the early-wood zone appearing as a light-colored band (*x*) to the naked eye; transition from early to late wood abrupt; ***late-wood pores*** small, arranged in nestlike groups which coalesce laterally in the outer late wood, forming interrupted concentric bands. ***Parenchyma*** not distinct as such. ***Rays*** generally visible to the naked eye.

MINUTE ANATOMY

Vessels very few to moderately few, the largest large to very large; perforation plates simple; spiral thickening present in the small vessels in the late wood; tyloses in the early-wood vessels small, appearing cellular *en masse*; intervessel pits orbicular to oval or angular through crowding, vestured, 5–12 μ in diameter, the orifices frequently confluent. ***Parenchyma*** abundant, paratracheal-vasicentric, paratracheal-confluent, and marginal; (*a*) paratracheal-vasicentric parenchyma (1) intermingled with substitute fibers and some fibers and composing an appreciable portion of the conjunctive tissue between the vessels and the rays in the porous early-wood zone, (2) forming an irregular, 1–several-seriate, continuous or interrupted sheath about the vessels or groups of vessels farther out in the ring, (3) toward the outer margin of the ring passing over into (*b*) paratracheal-confluent

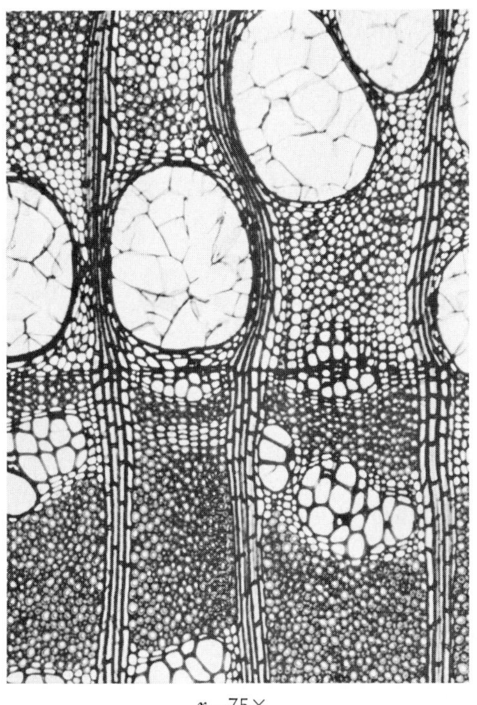

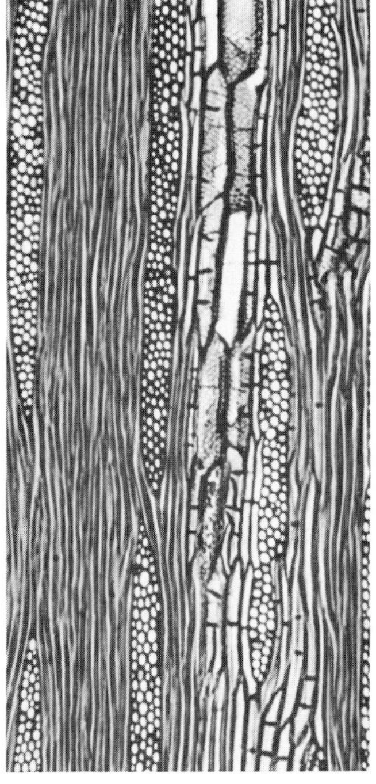

x—75× *t*—75×

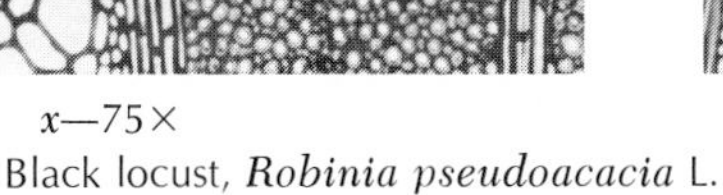

FIG. 12-205 Black locust, *Robinia pseudoacacia* L.

parenchyma and frequently serving as conjunctive tissue uniting proximate groups of vessels laterally (across the rays); (*c*) marginal parenchyma abundant but not forming a distinct line, passing over into the paratracheal parenchyma of the succeeding ring. ***Fusiform parenchyma cells*** (substitute fibers) present, intermingled with parenchyma. ***Libriform fibers*** moderately thick- to thick-walled, occasionally gelatinous, medium. ***Rays*** unstoried, 1–7 (mostly 3–5)-seriate, homocellular to somewhat heterocellular; ray-vessel pitting simple to bordered.

REMARKS

The wood of black locust is frequently confused with that of Osage-orange [*Maclura pomifera* (Raf.) Schneid.], but these woods can be distinguished by the following characters. The fresh-cut surface of Osage-orange is usually of a deeper shade of yellow or orange-brown, and the yellow coloring matter readily dissolves when shavings are placed in tepid water; very little coloring matter can be extracted by this method from black locust. Positive identification of black locust is ensured through the presence of vestured intervessel pits, a feature that is wanting in Osage-orange.

USES

Fence posts, mine timbers, poles, railroad ties, stakes (especially well suited for the above uses because of its durability, hardness, and strength); ***insulator pins*** (formerly the principal product made from black locust; well adapted for this purpose because of its strength, durability, and moderate shrinkage and swelling which tend to prevent the pins from loosening in the crossarms); *machine parts; woodenware* and *novelties; boxes* and *crates* (requiring exceptional strength); *planing-mill products; treenails; mine equipment.*

AMERICAN HOLLY

Ilex opaca Ait.

GENERAL CHARACTERISTICS

Sapwood white; *heartwood* ivory-white, frequently with bluish streaks or a bluish cast; *wood* without characteristic odor or taste, heavy (sp gr approx 0.50 green, 0.60 ovendry), hard. *Growth rings* barely distinct (wood diffuse-porous), delineated by a faint line of denser fibrous tissue at the outer margin. *Pores* very small, barely visible with a hand lens, grouped in radial strings (pore clusters) which not infrequently extend across the rings and cross the ring boundaries. *Parenchyma* indistinct. *Rays* of two widths; (*a*) broader rays visible to the naked eye, wider than the largest pores; (*b*) narrow rays barely visible with a hand lens, several between the broad rays.

MINUTE ANATOMY

Vessels very numerous, the largest small; perforation plates scalariform with many bars (15 plus); spiral thickening present; intervessel pits generally orbicular to oval and mostly opposite in a low spiral, 4–7 μ in diameter, occasionally linear and then up to 25 μ in diameter (Fig. 12-156, p. 536). ***Parenchyma*** paratracheal-scanty and apotracheal-diffuse; (*a*) paratracheal parenchyma very sparse, restricted to occasional cells; (*b*) apotracheal-diffuse parenchyma sparse to fairly abundant, the cells scattered. ***Fiber tracheids*** thin- to moderately thick-walled, with spiral thickenings, medium. ***Rays*** unstoried, heterocellular; (*a*) broader rays 3–5 (mostly 5)-seriate, many cells in height, with 1–several (mostly 1) rows of upright cells on the upper and lower margins; (*b*) narrow rays uniseriate, few–many cells in height, consisting entirely of upright cells; ray-vessel pitting similar to intervessel type.

REMARKS

Distinct among North American woods because of its ivory-white heartwood, small pores in radial strings (pore clusters) which frequently extend across the

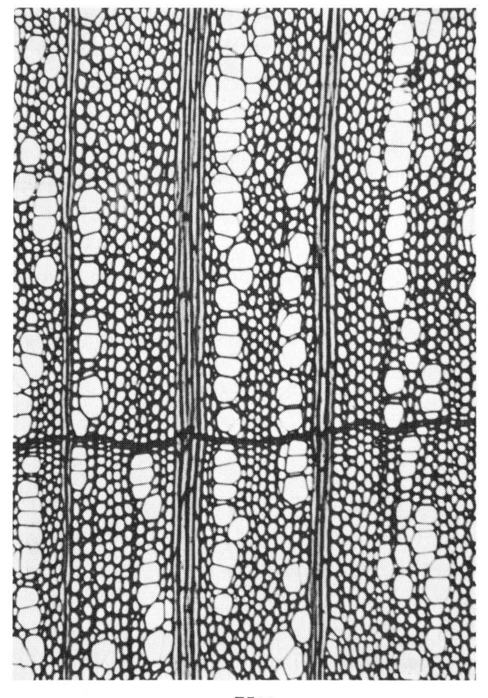

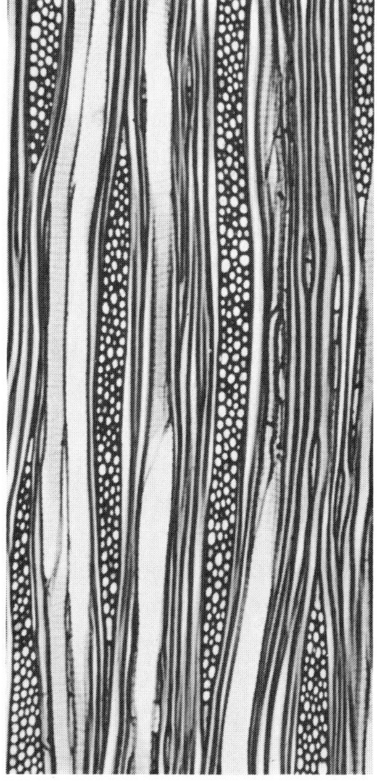

x—75×　　　　*t*—75×

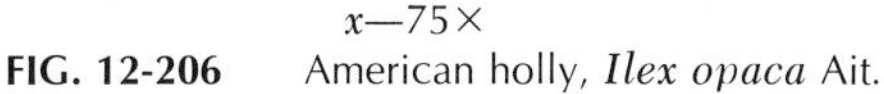

FIG. 12-206 American holly, *Ilex opaca* Ait.

rings and occasionally across the ring boundaries, latticed vessel perforations with many bars (15 plus), and spiral thickening in the vessel segments and fibers. The wood of American hornbeam (*Carpinus caroliniana* Walt.) approaches holly in color but is quite distinct in its structural features (aggregate rays, vessel perforations simple or scalariform with 3–6 bars, and with spiral thickening confined to the vessels).

USES

Furniture (largely inlay work); *fixtures*; *brush backs*; *handles*; *turnery*; *novelties*; *engravings*, *scrollwork*; *carvings*; *scientific instruments* (measuring scales and rules). Occasionally stained black to imitate ebony produced by *Diospyros* and *Maba* spp., especially for piano keys.

HARD MAPLE

Sugar maple (*Acer saccharum* Marsh.)

Black maple (*Acer nigrum* Michx. f.)

GENERAL CHARACTERISTICS

Sapwood white with a reddish tinge, narrow; *heartwood* uniform light reddish brown; *wood* without characteristic odor or taste, straight-grained (occasionally curly- or wavy-grained or with bird's-eye figure), heavy (sp gr 0.52–0.56 green, 0.62–0.68 ovendry), hard. *Growth rings* usually fairly distinct, delineated by a narrow, darker line of denser fibrous tissue. *Pores* small, indistinct without a hand lens, quite uniform in size, evenly distributed throughout the growth ring (wood diffuse-porous), solitary and occasionally in multiples of 2–several. *Parenchyma* not visible. *Rays* of two widths; (*a*) broader rays visible to the naked eye (*x*), fully as wide as the largest pores, separated by several narrow rays, forming a pronounced close ray fleck on the quarter surface, appearing on the tangential surface as short crowded lines which are visible without magnification; (*b*) narrow rays scarcely visible with a hand lens.

MINUTE ANATOMY

Vessels few to moderately numerous, the largest small; perforation plates simple; spiral thickening present; intervessel pits orbicular or angular through crowding, 6–10 μ in diameter. *Parenchyma* sparse, restricted to occasional cells, marginal, paratracheal-scanty, and apotracheal-diffuse. *Fibers libriform* and *fiber tracheids* thin- to moderately thick-walled, fine to coarse. *Rays* unstoried, essentially homocellular; (*a*) broader rays 3–8 (mostly 5–7)-seriate, up to 800 plus μ in height along the grain; (*b*) narrow rays 1–3 (mostly 1)-seriate, much lower than the broader rays (majority less than 200 μ in height); ray-vessel pits similar to intervessel type.

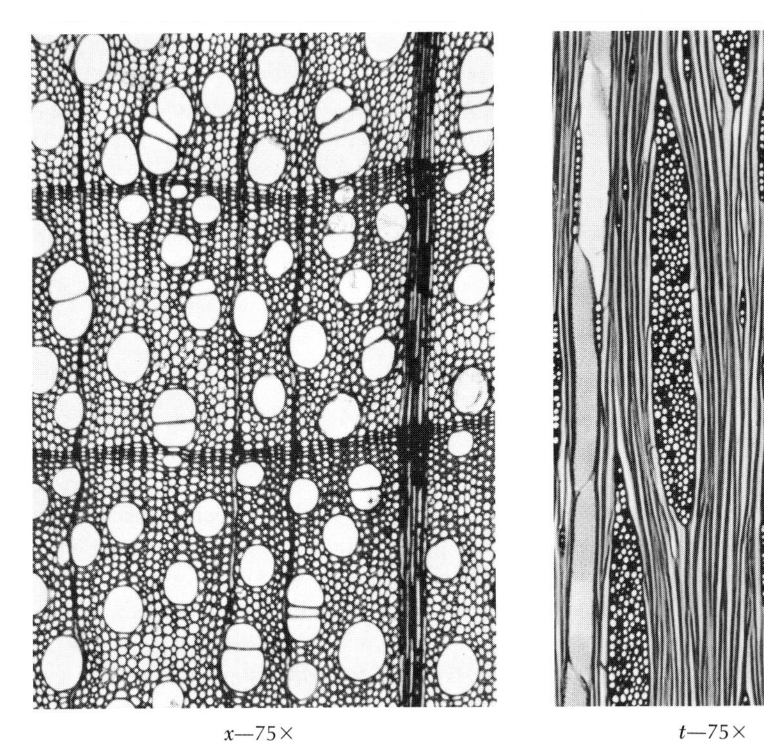

x—75× *t*—75×

FIG. 12-207 Sugar maple, *Acer saccharum* Marsh.

REMARKS

Sugar maple and black maple are sold indiscriminately as hard maple. Hard maple can be distinguished from the soft maples (*A. rubrum* L., *A. saccharinum* L.) and from bigleaf maple (*A. macrophyllum* Pursh) in that the rays intergrade in size (*x*) in these last-named species and the widest rays are not so broad. Sometimes confused with dogwood (*Cornus* spp.); for distinguishing features, see Remarks under Dogwood, p. 615.

USES

Charcoal production (standard wood along with beech and birch); *railroad ties* (very suitable as to hardness and resiliency but must be pressure-treated against decay); *veneer* (used in the manufacture of panels for furniture and musical instruments, and to some extent for interior grades of plywood); *shoe lasts*; *bowling pins*; *fuel wood*. Lumber used for *furniture* (one of the leading native woods because of its hardness, strength, and good working and finishing qualities combined with pleasing color and grain); *flooring* (long a favorite because of its uniform texture and hardness resulting in high resistance to abrasion, even under such severe conditions of use as in bowling alleys, and dance and factory floors); *boxes, crates,* and *pallets*; *boot* and *shoe findings*; *farm vehicles*; *handles*; *woodenware* and *novelties*; *shuttles, spools, bobbins* and other turned products; *toys*; *butchers' blocks* and *skewers*; *general millwork* (such as sash and doors); *sporting* and *athletic goods* (bowling pins, billiard cues, croquet mallets and balls, dumbbells); *musical instruments* [especially piano frames and the backs (curly stock) of violins]. The standard wood in the United States for *shear-test blocks*. *Pulpwood.*

BIGLEAF MAPLE, OREGON MAPLE

Acer macrophyllum Pursh

GENERAL CHARACTERISTICS

Sapwood reddish white, sometimes with a grayish cast; *heartwood* pinkish brown; *wood* without characteristic odor or taste, generally straight- but occasionally wavy-grained, moderately heavy (sp gr approx 0.44 green, 0.51 ovendry), moderately hard. *Growth rings* not very distinct, delineated by a narrow light line of fibrous tissue. *Pores* moderately small to medium-sized, indistinct without a hand lens, evenly distributed in the growth ring or somewhat more numerous in the

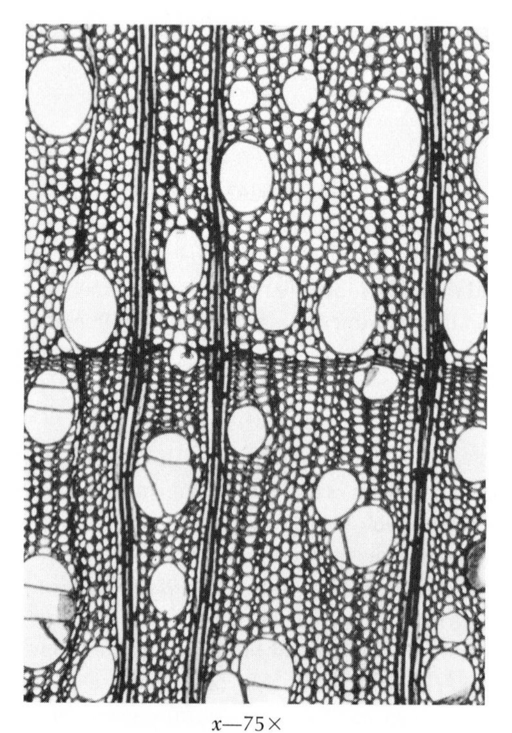

x—75×

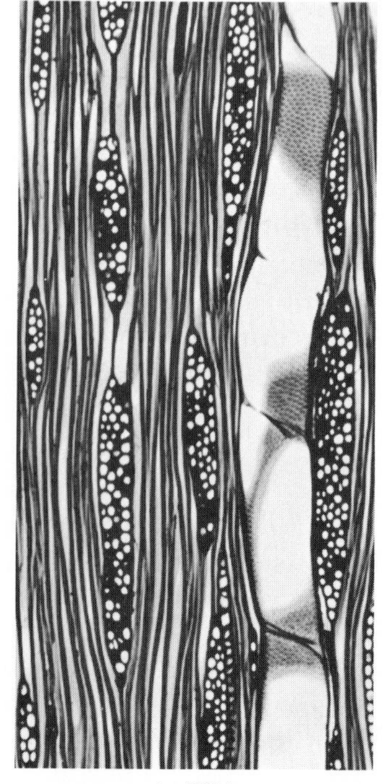

t—75×

FIG. 12-208 Bigleaf maple, *Acer macrophyllum* Pursh

first-formed early wood, solitary and occasionally in multiples of 2–several. *Parenchyma* not visible. *Rays* visible to the naked eye, intergrading in width, the broadest about as wide as the largest pores, forming a pronounced close ray fleck on the quarter surface, appearing on the tangential surface as short crowded lines which are visible without magnification.

MINUTE ANATOMY

Vessels few to moderately numerous, 30–80 per sq mm, the largest small to medium; perforation plates simple; spiral thickening present; intervessel pits orbicular or angled through crowding, 4–10 μ in diameter; gummy deposits not infrequent. *Parenchyma* sparse, marginal, paratracheal-scanty, and apotracheal-diffuse. *Fibers libriform* and *fiber tracheids*, thin- to moderately thick-walled, fine to coarse, *Rays* unstoried, 1–5 (mostly 3–5)-seriate, essentially homocellular; ray-vessel pitting similar to intervessel type.

REMARKS

Bigleaf maple differs from the soft maples in the color of the heartwood, which has a decidedly pinkish cast, and in possessing larger pores (60–120 μ in diameter in bigleaf maple; 60–80 μ in the soft maples) and fewer uniseriate rays. For differences between bigleaf maple and the hard maples, see Remarks under Hard Maple, p. 603.

USES

Veneer (plain and figured); the plain veneer is largely made into backing and cross-banding in the furniture industry; the figured consists of burl, blister, curly, and quilted patterns* and is used for veneer faces and inlay work for expensive furniture; *furniture* including living-room, dining-room, and bedroom sets [finished as walnut or mahogany, enameled or natural (if of fancy veneer)], also overstuffed and upholstered articles and chairs; *handles, fixtures*; *woodenware* and *novelties*; lower grades of lumber for *boxes, crates*, and *pallets*.

SOFT MAPLE

Red maple (*Acer rubrum* L.) Silver maple (*Acer saccharinum* L.)

GENERAL CHARACTERISTICS

Sapwood white, wide; *heartwood* light brown, sometimes with a grayish or greenish tinge or with a faint purplish cast; *wood* without characteristic odor or

* Quilted maple (Fig. 11-10) was so named by P. J. Landry of the Kelso Veneer Company, Kelso, Wash., and is traceable to blister-like elevations in the wood which form under the bark. The quilted structure does not extend around the trunk but is confined to the concave side of leaning trees and to the wider side of trunks with eccentric heartwood; it may continue lengthwise along the trunk from the root crown into the smaller limbs.

taste, straight-grained (sometimes curly-grained), moderately heavy (sp gr 0.44–0.49 green, 0.51–0.55 ovendry), moderately hard to hard. *Growth rings* not very distinct, delineated by a narrow, darker line of denser fibrous tissue. *Pores* small, indistinct without a hand lens, evenly distributed throughout the growth ring, solitary and occasionally in multiples of 2–several. *Parenchyma* not visible. *Rays* visible to the naked eye, intergrading in width, the broadest about as wide as the largest pores, forming a pronounced close ray fleck on the quarter surface, appearing on the tangential surface as short crowded lines which are visible without magnification.

MINUTE ANATOMY

Vessels moderately few to moderately numerous, the largest small; perforation plates simple; spiral thickening present; intervessel pits orbicular to broad-oval or somewhat angled through crowding, 5–10 μ in diameter; gummy deposits not infrequent. *Parenchyma* sparse, restricted to occasional cells, marginal, paratracheal-scanty, and apotracheal-diffuse. *Fibers libriform* and *fiber tracheids* thin- to moderately thick-walled, medium to coarse. *Rays* unstoried, 1–5-seriate (uniseriate rays usually sparse in *A. saccharinum*, common in *A. rubrum*), essentially homocellular; ray-vessel pitting similar to intervessel type.

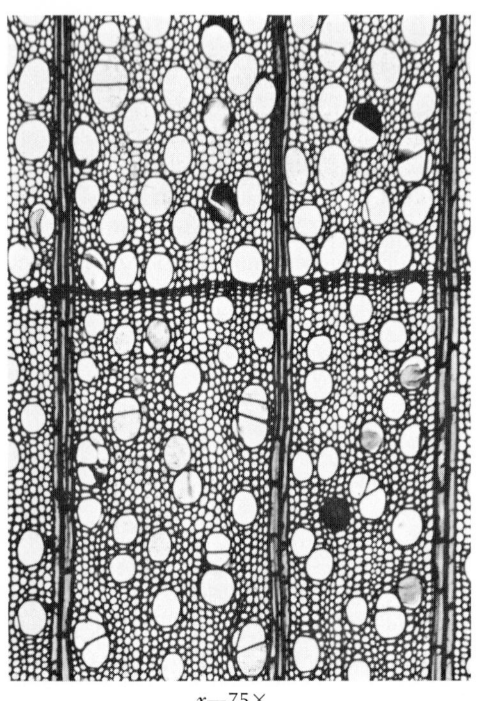

x—75×

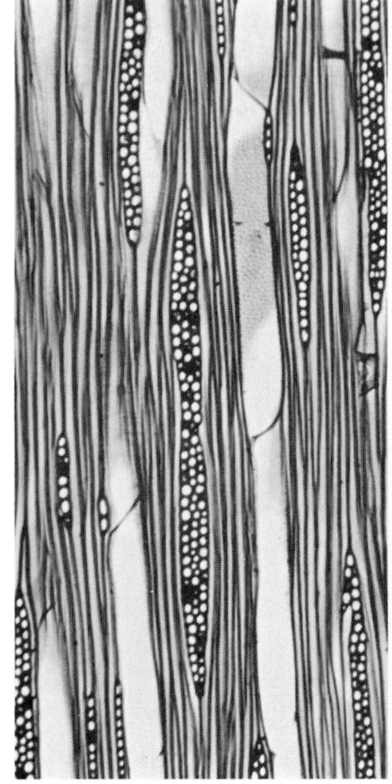

t—75×

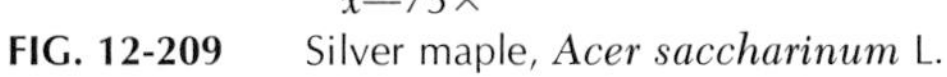

FIG. 12-209 Silver maple, *Acer saccharinum* L.

REMARKS

As the common name indicates, the wood of the soft maples is not so hard as that of hard maple produced by *A. saccharum* Marsh. and *A. nigrum* Michx. f.; heartwood frequently with a grayish or greenish tinge. For other differences between soft and hard maple, see Remarks under Hard Maple, p. 603.

USES

Used largely for the same purposes as hard maple (except where strength and hardness are a primary requisite), principally for *furniture* (upholstered frames); *boxes, crates,* and *pallets; food containers* (especially butter tubs); truck bodies; *wall paneling; planing-mill products;* preferred to hard maple for *core stock; veneer* for furniture and interior grade plywood. *Pulpwood.*

BUCKEYE

Yellow buckeye (*Aesculus octandra* Marsh.)

Ohio buckeye (*Aesculus glabra* Willd.)

GENERAL CHARACTERISTICS

Sapwood white to grayish white, gradually merging into the heartwood; *heartwood* creamy white to pale yellowish white, frequently with darker (grayish) streaks of oxidative sap stain; *wood* odorless or with a mild characteristic odor (especially when moist) resembling that of basswood (*Tilia* spp.), without characteristic taste, light (sp gr approx 0.33 green, 0.38 ovendry), soft. *Growth rings* not visible or barely distinct and delineated by a light-colored line, generally narrow. *Pores* numerous, minute, not visible without a hand lens, nearly constant in size and quite evenly distributed throughout the growth ring (wood diffuse-porous), solitary and in multiples of 2–several. *Parenchyma* not visible or barely visible as a light line terminating the growth ring. *Rays* very fine, scarcely visible with a hand lens, close and seemingly forming half of the area on the transverse surface, storied or unstoried, in the first instance forming fine ripple marks on the tangential surface.

MINUTE ANATOMY

Vessels very numerous, the largest small, vessel members storied with the other elements, or unstoried; perforation plates simple; spiral thickening present or wanting; when present, occasionally restricted to the vessels in the late wood; intervessel pits orbicular to oval or angular through crowding (Fig. 12-154, p. 536), 5–10 μ in diameter. *Parenchyma* marginal and paratracheal-scanty, the cambiform rows along the grain storied with the other elements, or unstoried; (*a*) terminal parenchyma in a more or less continuous, 1–several-seriate line; (*b*) paratracheal

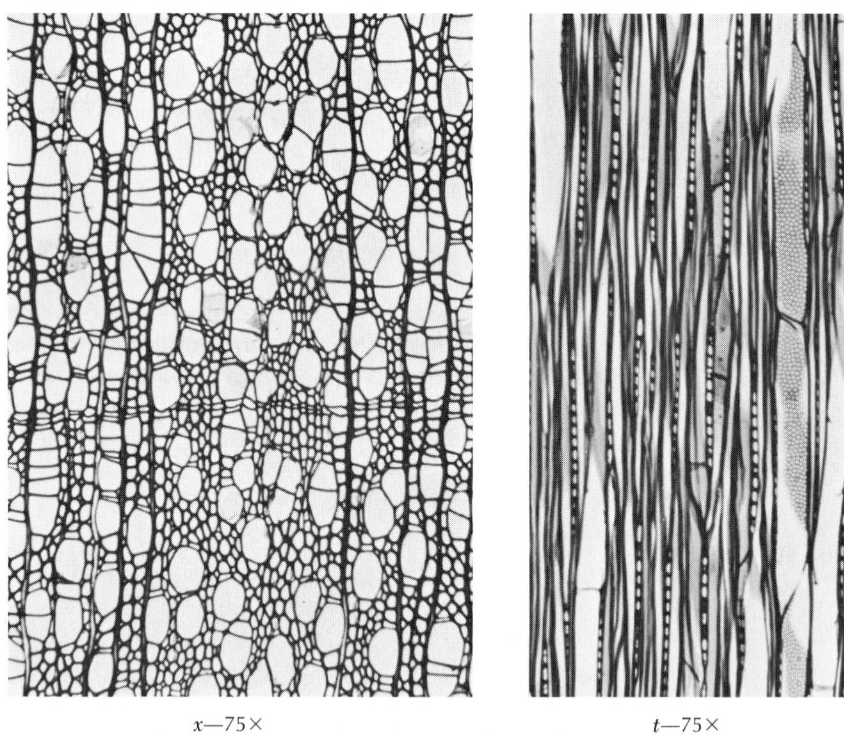

x—75×   *t*—75×

FIG. 12-210 Yellow buckeye, *Aesculus octandra* Marsh.

parenchyma sparse, restricted to occasional cells (not forming a sheath). ***Fibers libriform*** and ***fiber tracheids*** thin-walled, medium to coarse, the central portion conforming to the storied structure of the wood when this situation prevails. ***Rays*** storied with the other elements, or unstoried, uniseriate, homocellular to heterocellular; ray-vessel pitting simple.

REMARKS

Yellow buckeye can usually be distinguished from Ohio buckeye in that it possesses distinct ripple marks traceable to storied rays and longitudinal elements. Ripple marks are wanting or sporadic in Ohio buckeye since the elements are never regularly storied. Buckeye is sometimes confused with aspen (*Populus* spp.) which is also fine-textured.

USES

Boxes, *crates*, and other shipping containers; *cigar* and *tobacco boxes*; formerly for *artificial limbs*; *woodenware* and *novelties*; *toys*; *furniture* (unexposed parts); *trunks* and *valises* (usually as plywood); excellent for *drawing boards*; *plaques* for pyrography.

CASCARA BUCKTHORN
Rhamnus purshiana DC.

GENERAL CHARACTERISTICS

Sapwood yellowish white; *heartwood* yellowish brown tinged with red; *wood* without characteristic odor or taste, hard (sp gr approx 0.50 green, 0.54 ovendry), heavy. *Growth rings* fairly distinct, delineated by a difference in the size and number of the pores in the late wood and in the early wood of succeeding rings. *Pores* small, indistinct without a hand lens, more numerous and somewhat larger in the early wood (wood diffuse-porous) and sometimes forming a more or less continuous uniseriate row at the beginning of the ring, solitary and in multiples of 2–several. *Parenchyma* not visible. *Rays* visible to the naked eye.

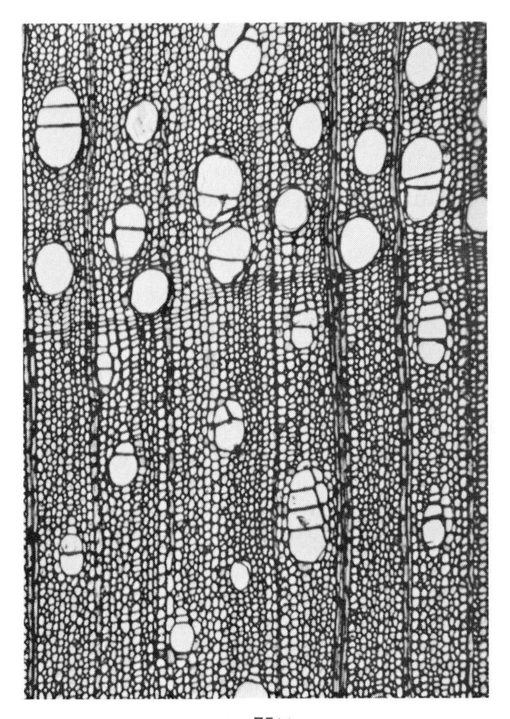

x—75×

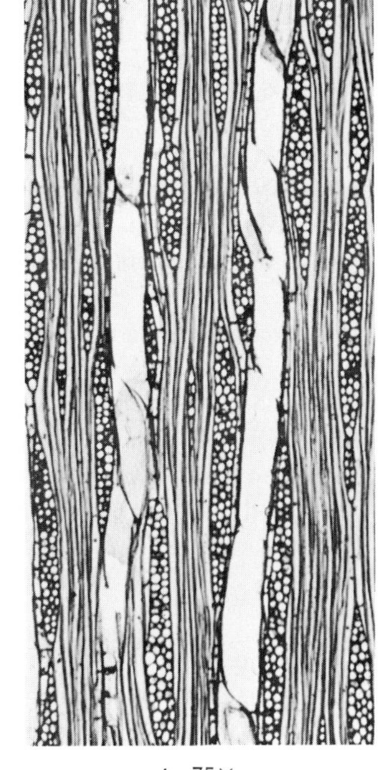

t—75×

FIG. 12-211 Cascara buckthorn, *Rhamnus purshiana* DC.

MINUTE ANATOMY

Vessels few to moderately few, the largest small; perforation plates simple; spiral thickening present; intervessel pits not crowded, orbicular to oval, 6–8 μ in diameter, occasionally oval-linear and then up to 12 μ in diameter. ***Parenchyma*** paratracheal-scanty to vasicentric and apotracheal-diffuse; (*a*) paratracheal parenchyma forming an interrupted uniseriate sheath around the vessels or groups of vessels; (*b*) apotracheal-diffuse parenchyma sparse, confined to occasional cells. ***Libriform fibers*** thin-walled, fine to medium. ***Rays*** unstoried, 1–5 (mostly 3–4)-seriate, essentially homocellular; ray-vessel pitting similar to intervessel type.

USES

Posts, built-up turned articles, and other local uses. The chief value of this species is in its bark; an extract prepared from it is widely used for laxative purposes in the United States and abroad. The finely ground wood also yields an extract up to 75 percent of efficiency.

BASSWOOD

American basswood (*Tilia americana* L.)

White basswood (*Tilia heterophylla* Vent.)

GENERAL CHARACTERISTICS

Sapwood whitish to creamy white or pale brown, merging more or less gradually into the darker heartwood; ***heartwood*** pale brown, sometimes with a reddish tinge; ***wood*** with a faint characteristic odor on fresh-cut surface (especially when wet), tasteless, straight-grained, light (sp gr approx 0.32 green, 0.40 ovendry), soft. ***Growth rings*** fairly distinct, delineated by a difference in size of the pores in the late wood and in the early wood of the succeeding ring. ***Pores*** numerous, small, distinctly visible with a hand lens, quite evenly distributed (wood diffuse-porous), solitary and in multiples or tangential groups of 2–several. ***Parenchyma*** not distinct or barely visible with a hand lens, banded. ***Rays*** variable in width, the broader not distinct on the transverse surface without a hand lens, forming high, scattered ray flecks on the quarter surface.

MINUTE ANATOMY

Vessels moderately few to very numerous, the largest small to medium; perforation plates simple; spiral thickening present; intervessel pits polygonal through crowding, 5–8 μ in diameter, rarely orbicular to broad-oval. ***Parenchyma*** abundant, marginal and apotracheal-banded, the lines of the latter numerous and uniseriate. ***Fiber tracheids*** thin-walled, medium to very coarse. ***Rays*** unstoried or rarely somewhat

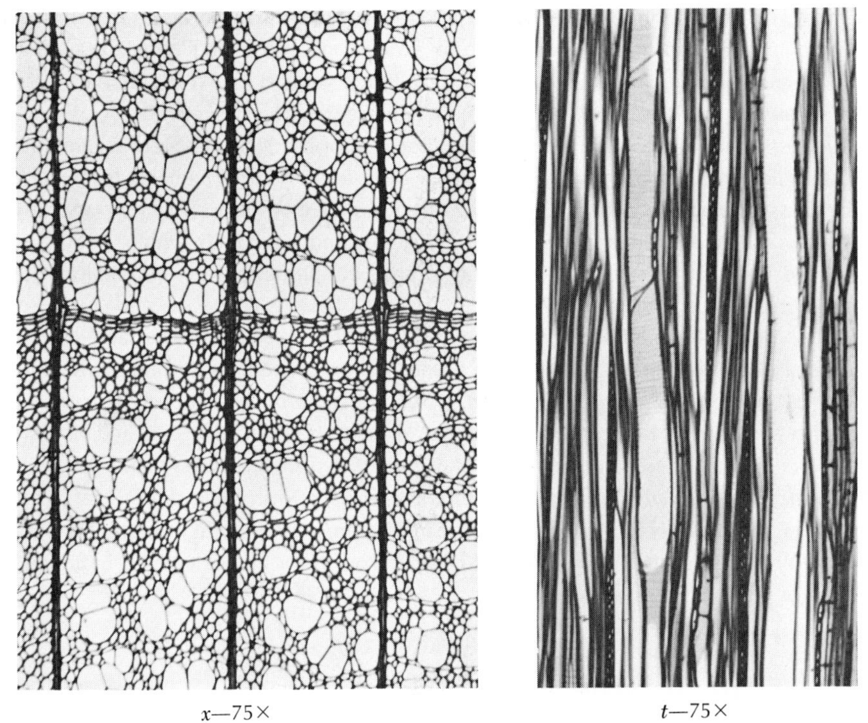

FIG. 12-212 American basswood, *Tilia americana* L.

storied, of two widths; (*a*) broader rays 1–6-seriate, up to 1.2 plus mm in height along the grain (*t*), essentially homocellular; (*b*) narrow rays uniseriate for the most part, much lower than the broad rays (the majority less than 300 μ in height), the cells nearly uniform in size but higher than those in the broad rays; ray-vessel pitting simple to bordered.

REMARKS

Devoid of figure. Readily recognized by its faint but characteristic odor, the smoothness with which it cuts, the ease of indentation with the thumbnail, and its characteristic, high, widely spaced ray fleck. Sometimes confused with buckeye (*Aesculus* spp.) but distinct in color (without the yellowish cast of buckeye) and in possessing larger pores, banded parenchyma, and wider rays (see description of buckeye, p. 607).

USES

Veneer (cut almost entirely by the rotary method) used in the manufacture of *plywood*, for drawer panels, mirror backings, and other concealed parts of furniture (because of its freedom from checking and warping, the light color of the wood, and ease of gluing) and for trunk panels and valises (because it does not split readily under impact); as *core stock* (thick veneer) on which thin veneers of expensive cabinet woods are glued; as *veneer stock* for baskets. *Slack cooperage*, especially heading for flour barrels (because of its clear appearance, freedom from checking and warping, and the lack of taste and color pigments that might be transmitted to the flour). *Excelsior* (a good native wood for this purpose, because of light weight and color, straight grain, and ease with which it can be shaved into thin but tough and resilient strands but prone to develop a malodor if used under moist conditions; not used at present so extensively as formerly because of the steady demand for other purposes). *Lumber* used for *boxes* and *crates* (because of its lightness, light color, good nailing properties, and freedom from objectionable odor and taste); *dairy, poultry*, and *apiary supplies*; *furniture* (concealed parts); *core stock* for panels; *millwork* (especially sash and doors); *trunks* and *valises*; *woodenware* and *novelties*; *toys*; *caskets* and *coffins*; *laundry appliances*; *novelties*; *kitchen cabinets*; *handles*; *shade* and *map rollers*; *fixtures*; *venetian-blind slats*; *piano keys* (lumber cut from frozen logs; carefully air-seasoned in diffused light to preserve color and selected for straightness of grain).

BLACK TUPELO, BLACKGUM, TUPELO, AND WATER TUPELO

Black tupelo, blackgum, pepperidge (*Nyssa sylvatica* Marsh.)
Swamp tupelo, swamp blackgum [*Nyssa sylvatica* var. *biflora* (Walt.) Sarg.]
Water tupelo, tupelo-gum (*Nyssa aquatica* L.)

GENERAL CHARACTERISTICS

Sapwood white to grayish white, gradually merging into the darker heartwood; *heartwood* greenish or brownish gray; *wood* without characteristic odor or taste, usually interlocked-grained and hence showing a distinct ribbon figure when quarter-sawn (especially blackgum), moderately heavy (sp gr approx 0.46 green, 0.52–0.55 ovendry), moderately hard. *Growth rings* generally indistinct, even under a lens (wood diffuse-porous). *Pores* small, not visible to the naked eye, quite uniform in size, numerous and fairly evenly distributed, solitary or occasionally in short radial groups and in multiples. *Parenchyma* not visible. *Rays* fine, not distinct on the transverse surface without a hand lens, very close and seemingly occupying one-half the area on the transverse surface of the wood.

MINUTE ANATOMY

Vessels numerous to very numerous, the largest small; perforation plates exclusively scalariform with numerous thin bars, less than 1 μ in thickness; spiral

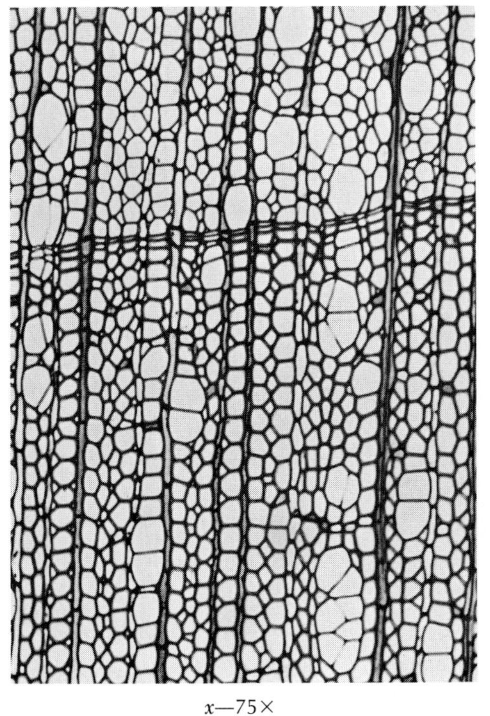

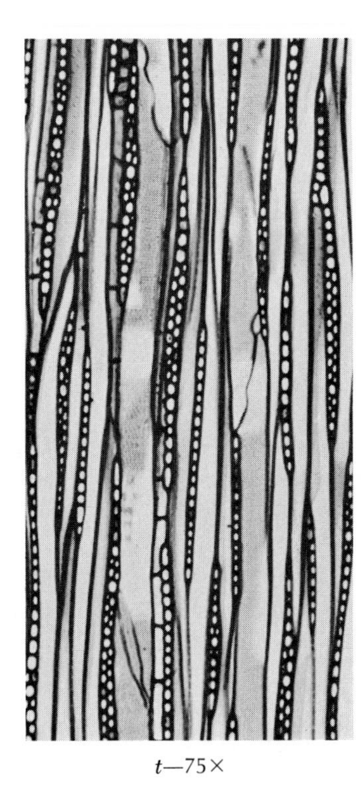

x—75×　　*t*—75×

FIG. 12-213 Water tupelo, *Nyssa aquatica* L.

thickening occasionally present, restricted to the tapering ends of the vessel elements; intervessel pits in transverse rows of 1–5 plus (opposite pitting), for the most part oval-rectangular and 5–12 μ in diameter, occasionally linear through fusion and then up to 20 μ in diameter. ***Parenchyma*** paratracheal-scanty and apotracheal-diffuse, the cells scattered. ***Fiber tracheids*** moderately thick- to thick-walled, medium to very coarse. ***Rays*** 8–13 per mm (*x*), unstoried, 1–4-seriate, heterocellular; upright cells generally restricted to one on the upper and lower margins, less than 60 μ in height along the grain; ray-vessel pits similar to intervessel type.

REMARKS

Black tupelo, swamp tupelo, and water tupelo are difficult to separate with certainty, even at high magnifications, but the last two woods are usually somewhat softer, lighter, and more porous, with more crowded, slightly larger vessels. The vessels are frequently greatly restricted in number in wood from the swollen butt log of water tupelo; since the fibers are also thinner-walled and radially aligned, the tissue, as viewed in the transverse section at high magnifications, strikingly resembles that of a coniferous wood.

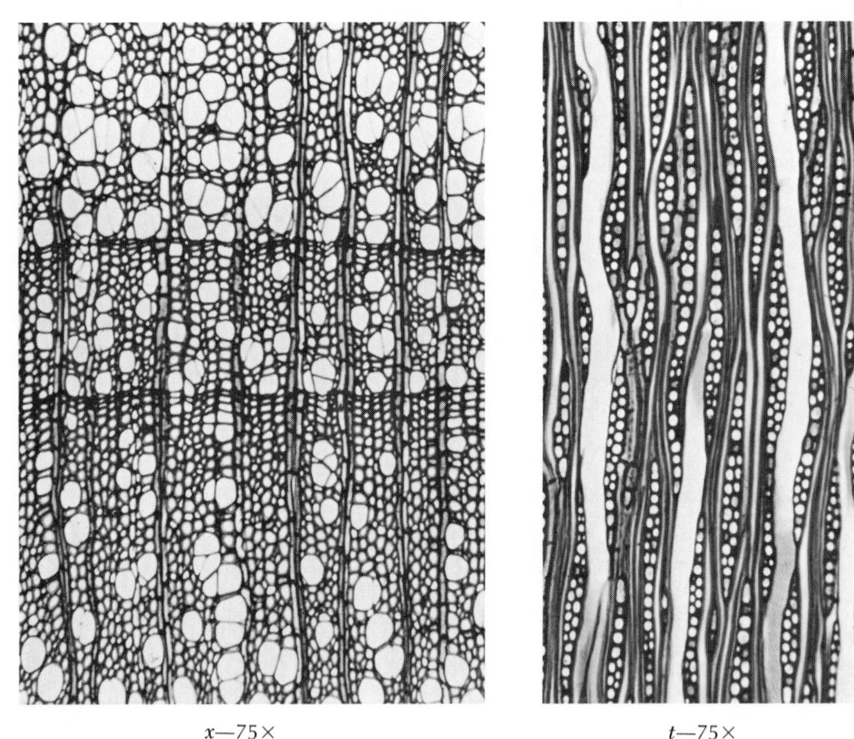

FIG. 12-214 Black tupelo, *Nyssa sylvatica* Marsh.

USES

Veneer, most of which is converted into fruit and berry boxes and similar containers; *plywood*, for panels (not too satisfactory as it tends to warp); *pulp* (alkaline process; these woods can also be pulped satisfactorily by the sulfite and semichemical processes); *cooperage* (mostly slack staves and heading); *railroad ties* (heartwood takes preservative readily, and interlocked grain obviates use of antisplitting irons). *Lumber* used for *boxes, crates*, and *pallets*, for which its light color and toughness recommend it very highly (such boxes are extensively used in the export trade because the wood shows stenciling well); *furniture*, especially in concealed parts (such as drawer sides and bottoms); *kitchen cabinets; factory floors* and *platform planking* required to withstand heavy wear; *planing-mill products* (such as trim, molding, and built-in cabinets); *farm vehicles; cigar boxes* [as core stock covered with a veneer of Spanish cedar (*Cedrela* spp.) or with paper bearing the image (along the grain) of this species]; *woodenware, novelties*, and *fixtures; handles*.

FLOWERING DOGWOOD, DOGWOOD

Flowering dogwood (*Cornus florida* L.)

Pacific dogwood (*Cornus nuttallii* Audubon)

GENERAL CHARACTERISTICS

Sapwood carneous to light pinkish brown, very wide; *heartwood*, when present, dark brown, frequently variegated; *wood* without characteristic odor or taste, heavy to very heavy (sp gr 0.58–0.64 green, 0.70–0.80 ovendry), hard to very hard. *Growth rings* distinct but not sharply delineated (wood diffuse-porous). *Pores* small, not visible without a hand lens, solitary and in multiples of 2–several. *Parenchyma* not distinct or barely distinct with a hand lens, banded. *Rays* of two widths; (*a*) broader rays visible to the naked eye but not sharply delineated against the background of pores and fibrous tissue; (*b*) narrow rays scarcely distinct with a lens.

MINUTE ANATOMY

Vessels medium few to very numerous, the largest small to medium; perforation plates scalariform, with many (20 plus) fine bars; intervessel pits oval to linear

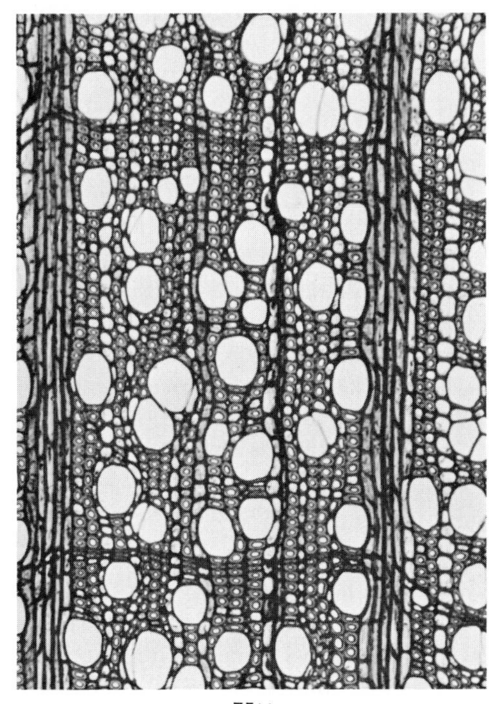

x—75×

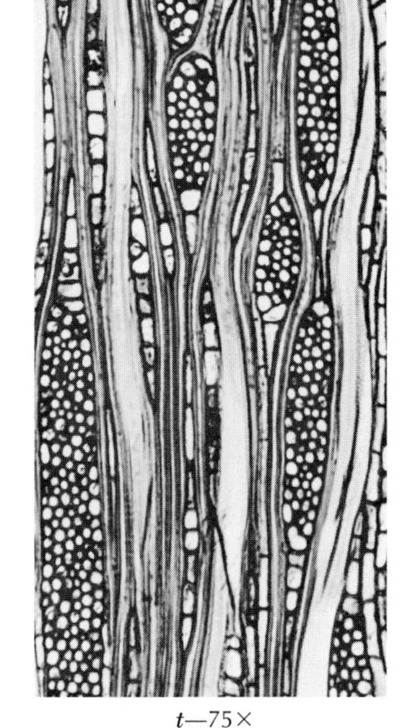

t—75×

FIG. 12-215 Flowering dogwood, *Cornus florida* L.

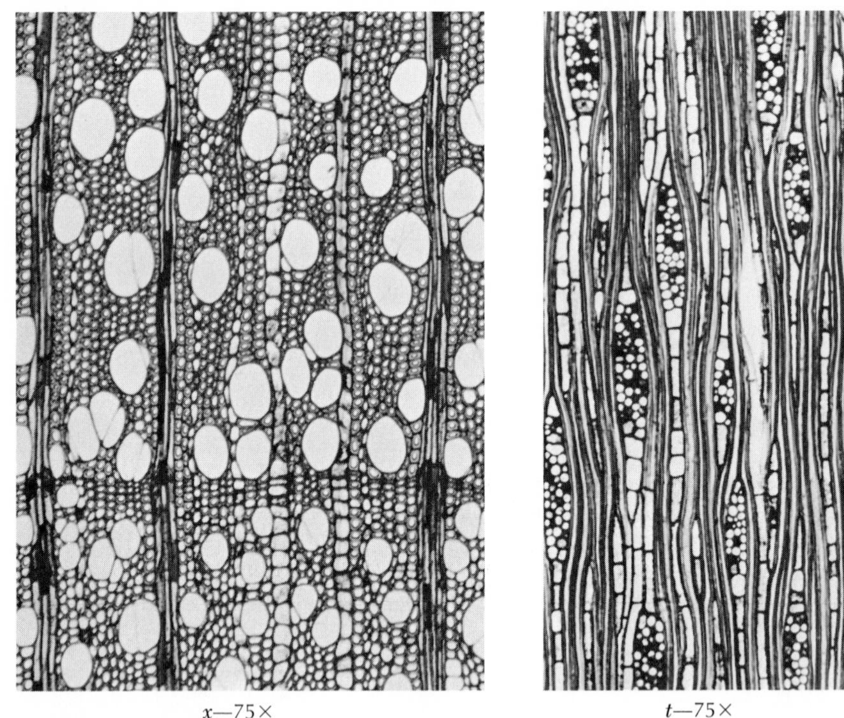

FIG. 12-216 Pacific dogwood, *Cornus nutallii* Audubon

(the scalariform pitting common), 8–20 plus μ in diameter. *Parenchyma* paratracheal-scanty, and apotracheal-diffuse to banded; (*a*) paratracheal parenchyma restricted to occasional cells, never forming a sheath; (*b*) banded and (*c*) apotracheal-diffuse parenchyma abundant, the former in broken tangential, mostly uniseriate lines which form no definite pattern. *Fiber tracheids* moderately thick- to thick-walled, medium to very coarse. *Rays* unstoried, heterocellular; (*a*) broader rays 3–8-seriate and composed of horizontal cells through the central portion, with uniseriate margins of 1–5 plus (occasionally many) rows of upright cells; narrow rays 1–2-seriate, composed wholly or largely of upright cells; ray-vessel pitting similar to intervessel type.

REMARKS

Pacific dogwood resembles the eastern species very closely, and samples cannot always be identified with certainty. The broad rays in *C. florida* are 3–8-seriate through the central portion and up to 80 cells in height (*t*); the western species, in contrast, is featured by broad rays which are 3–5-seriate for the most part and seldom over 40 cells in height. Dogwood is sometimes confused with hard maple produced by *Acer saccharum* Marsh., and *A. nigrum* Michx. f., respectively. Dogwood usually has a decided flesh-colored cast (sapwood), the annual rings are less distinct than those of hard maple, and the wood rays do not stand out so sharply on the transverse surface. Positive identification of dogwood is assured through the presence of banded parenchyma, scalariform plates with many bars, linear intervessel pits, and heterocellular rays.

USES

Shuttles for textile weaving (approx 90 percent of the "cut"); *spools*; *bobbin heads*; *small pulleys*; *mallet heads*; *golf-club heads* (occasionally); *turnpins* for shaping the ends of lead pipes; *jewelers' blocks*; *machinery bearings*. The major uses of this wood are contingent on its hardness and close texture which cause it to work and stay smooth under continuous wear.

SOURWOOD

Oxydendrum arboreum (L.) DC.

GENERAL CHARACTERISTICS

Sapwood yellowish brown to light pinkish brown, wide (up to 80 plus growth rings); *heartwood* brown tinged with red when first exposed, becoming duller with age; *wood* without characteristic odor or taste, heavy (sp gr approx 0.50 green, 0.59 ovendry), hard. *Growth rings* distinct but not prominent (wood diffuse-porous), delineated by a narrow band of denser fibrous tissue at the outer margin, narrow to medium wide. *Pores* numerous, very small (indistinct to the naked

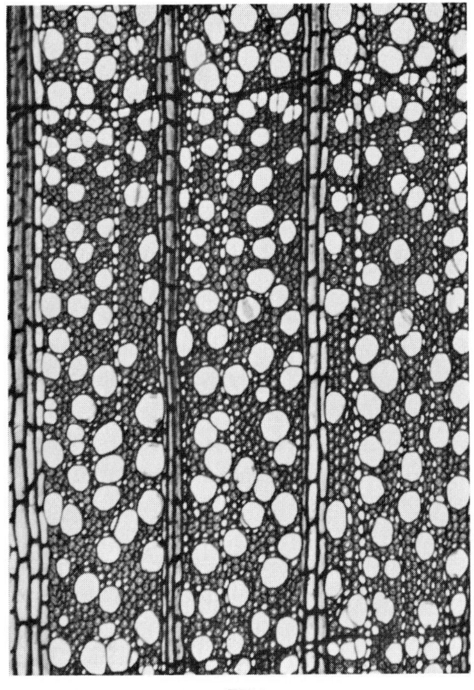

x—75×

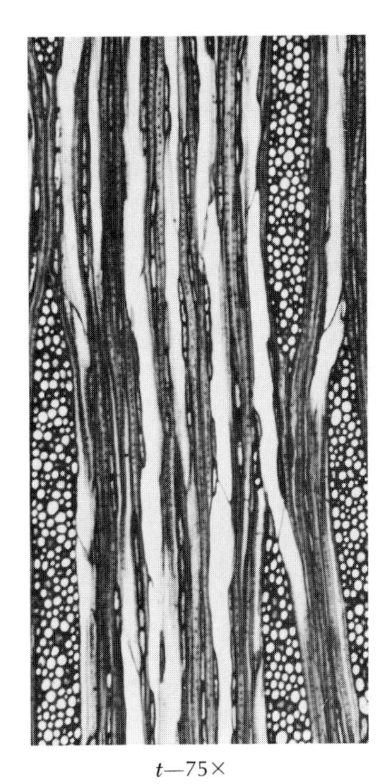

t—75×

FIG. 12-217 Sourwood, *Oxydendrum arboreum* (L.) DC.

eye) and nearly uniform in size except in the outer late wood, solitary for the most part, quite evenly distributed throughout the growth ring. *Parenchyma* not distinct. *Rays* not distinct on the transverse surface without a hand lens, appearing to be quite uniform in size, nearly as wide as the largest pores.

MINUTE ANATOMY

Vessels very numerous, the largest very small to small; spiral thickening present, often inconspicuous; perforation plates simple, scalariform or occasionally reticulate (foraminate), the first type the most frequent; tyloses wanting; pits leading to contiguous tracheary cells orbicular to oval or rarely linear, 4–11 μ in diameter. *Parenchyma* sparse, apotracheal-diffuse and very rarely paratracheal-scanty, in the latter instance restricted to occasional cells. *Fiber tracheids* thick-walled, with bordered pits, medium. *Rays* 3–5 per mm, unstoried, heterocellular; (*a*) broader rays 5–7-seriate, many cells in height, the upright cells confined to 1–several rows on the upper and lower margins; (*b*) narrow rays uniseriate, 1–6 plus cells in height, consisting entirely of upright cells; ray-vessel pitting similar to intervessel type.

USES

Locally for *tool handles*; *bearings of machinery*; *sled runners*.

PACIFIC MADRONE, MADROÑO

Arbutus menziesii Pursh

GENERAL CHARACTERISTICS

Sapwood white or cream colored, frequently with a pinkish tinge; *heartwood* light reddish brown; *wood* without characteristic odor or taste, heavy (sp gr approx 0.58 green, 0.69 ovendry), hard. *Growth rings* barely visible with a hand lens, delineated by a continuous, uniseriate row of pores in the first-formed early wood. *Pores* numerous, minute, nearly uniform in size except for a row of large pores in the first-formed early wood (wood diffuse-porous), arranged as described above in the early wood, elsewhere solitary, in multiples, or in short radial rows and frequently further clustered or in bands, the zones of pores then usually alternating with bands of dense (darker) fibrous tissue. *Parenchyma* not evident. *Rays* barely to readily visible with a hand lens.

MINUTE ANATOMY

Vessels very numerous, the largest very small to small; perforation plates scalariform or occasionally reticulate (foraminate); spiral thickening present; intervessel

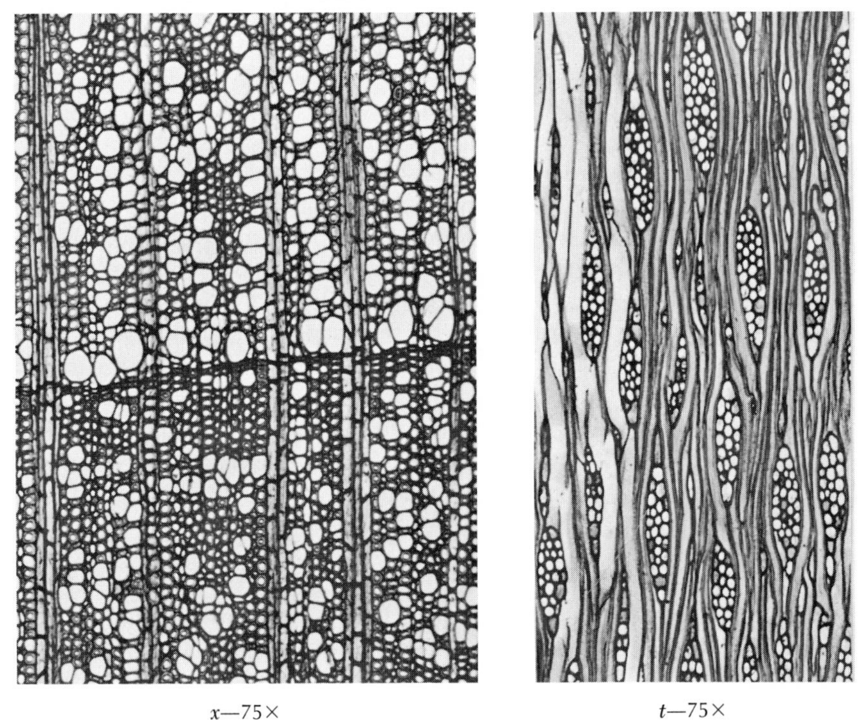

x—75× *t*—75×

FIG. 12-218 Pacific madrone, *Arbutus menziesii* Pursh

pits orbicular, 3–5 μ diameter. *Parenchyma* paratracheal-scanty, very sparse, restricted to occasional cells. *Fiber tracheids* moderately thick- to thick-walled, (*a*) in the proximity of vessels with conspicuous bordered pits and frequently with spiral thickening, (*b*) elsewhere with small, inconspicuous bordered pits and devoid of spiral thickening, frequently septate, medium. *Rays* unstoried, heterocellular; (*a*) broader rays 3–5-seriate, low (usually under 25 cells in height), with 1–several (mostly 1) rows of upright cells on the upper and lower margins; (*b*) narrow rays 1–2-seriate, less than 15 cells in height; uniseriate rays of the *b* type consisting entirely of upright cells; biseriate rays of the *b* type consisting of horizontal cells through the central portion, with 1–several rows of marginal upright cells; ray-vessel pitting simple to bordered, about the same size as intervessel type.

USES

Locally for *fuel*; as a substitute for dogwood (*Cornus florida* L. and *C. nuttallii* Audubon), for *shuttles*; as *rollers*, 6 to 8 in. in diameter, in shifting heavy cargoes aboard ships; *furniture*; *sliced veneer* for paneling. Formerly converted into *charcoal*.

COMMON PERSIMMON

Diospyros virginiana L.

GENERAL CHARACTERISTICS

Sapwood white to creamy white when freshly cut, darkening on exposure to light yellowish brown or grayish brown, wide (practically all the stock used for commercial purposes is sapwood); ***heartwood*** usually very small, blackish brown to black, often streaked, irregular in outline (*x*); ***wood*** without characteristic odor or taste, very heavy (sp gr approx 0.64 green, 0.79 ovendry), very hard. ***Growth rings*** distinct but not conspicuous (wood semi-ring-porous). ***Early-wood pores*** visible to the naked eye, decreasing gradually or somewhat abruptly in size toward the outer margin of the ring, solitary and in multiples of 2–3. ***Parenchyma*** visible with a hand lens, appearing closely and evenly punctate, the lines quite distinct. ***Rays*** indistinct without a hand lens, storied with the longitudinal elements and forming ripple marks on the tangential surface.

MINUTE ANATOMY

Vessels very few to few, the largest medium to large, vessel members storied with the other elements; perforation plates simple; intervessel pits orbicular to broad-

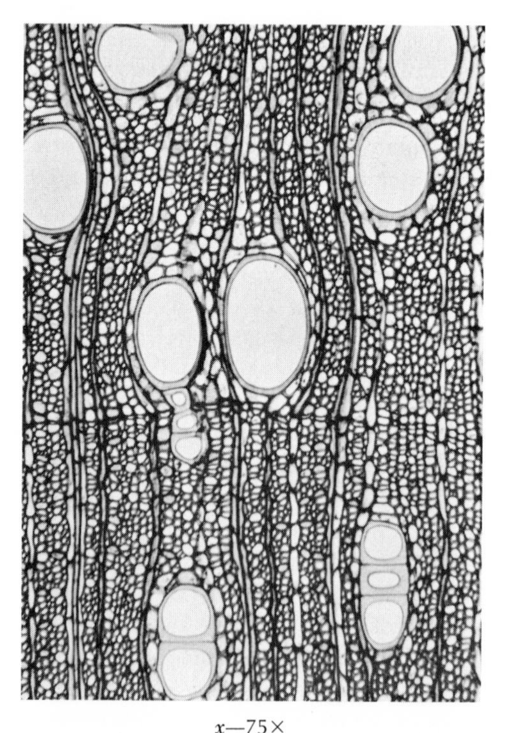

x—75×

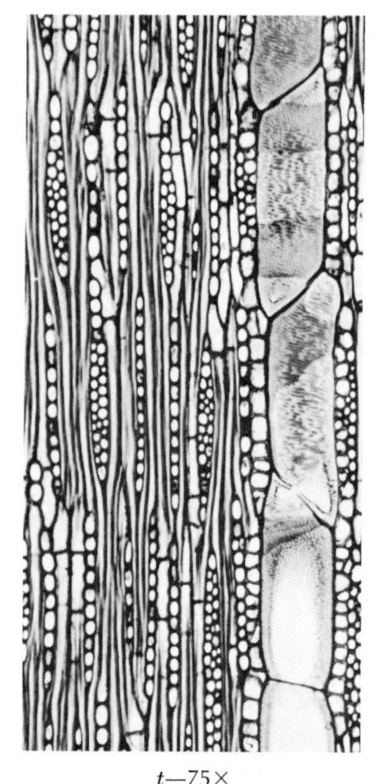

t—75×

FIG. 12-219 Common persimmon, *Diospyros virginiana* L.

oval, minute (2–4 μ in diameter), the apertures of several frequently confluent. *Parenchyma* paratracheal-vasicentric, apotracheal-banded, and marginal, in cambiform rows along the grain which are storied with the other elements; (*a*) paratracheal-vasicentric parenchyma confined to the immediate vicinity of the vessels, the sheath narrow and mostly 1–2-seriate; (*b*) apotracheal-banded parenchyma very abundant, the lines close, 1–2-seriate; (*c*) marginal parenchyma forming a 1–2-seriate line, the cells somewhat larger than those in the body of the ring. *Libriform fibers* relatively thin- to thick-walled, fine to coarse, the central portion conforming to the storied structure of the wood. *Rays* storied, 1–3-seriate, homocellular to heterocellular; ray-vessel pits similar to intervessel type.

REMARKS

Characterized by unusually thick-walled vessels, minute intervessel pits the apertures of which are frequently confluent, close narrow lines of apotracheal-banded parenchyma [cells appearing punctate (*x*) at low magnification], and ripple marks (*t*) traceable to storied wood rays and longitudinal elements. Sometimes confused with hickory (produced by *Carya* spp.) but distinct in possessing the minute intervessel pits (those of hickory 6–8 μ diameter) and ripple marks (*t*).

USES

Shuttles for textile weaving; *spools*; *bobbins*; *golf-club heads* (all these uses are based on the inherent hardness, strength, and toughness of this wood, and its ability to stay smooth under friction); *boxes* and *crates*; *shoe lasts* (formerly widely used but now largely replaced by less-expensive woods, such as maple); *handles*.

WHITE ASH AND OREGON ASH

White ash (*Fraxinus americana* L.)
Green ash (*Fraxinus pennsylvanica* Marsh.)
Oregon ash (*Fraxinus latifolia* Benth.)

GENERAL CHARACTERISTICS

Sapwood nearly white, wide; *heartwood* grayish brown, light brown, or pale yellow streaked with brown; *wood* somewhat lustrous, without characteristic odor or taste, straight-grained, heavy (sp gr 0.50–0.56 green, 0.58–0.64 ovendry), hard. *Growth rings* distinct (wood ring-porous). *Early-wood pores* large, distinctly visible to the naked eye, forming a band 2–4 pores in width; transition from early to late wood abrupt; *late-wood pores* small, barely visible to the naked eye, solitary and in multiples of 2–3. *Parenchyma* visible with a hand lens in the late wood, forming a narrow sheath about the pores and frequently uniting them laterally

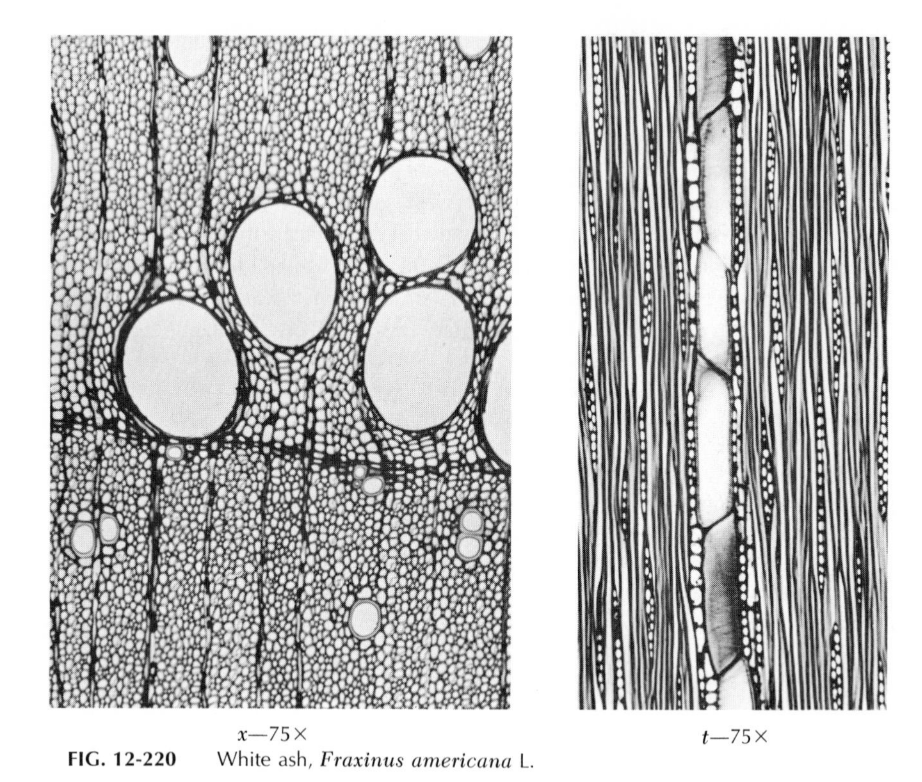

x—75× t—75×

FIG. 12-220 White ash, *Fraxinus americana* L.

toward the outer margin of the ring. *Rays* not distinct or barely visible to the naked eye.

MINUTE ANATOMY

Vessels very few; largest early-wood vessels large to very large; perforation plates simple; intervessel pits orbicular to short-oval or occasionally somewhat angular through crowding, 3–5 μ in diameter. *Vasicentric tracheids* present, confined to the immediate vicinity of the early-wood vessels. *Parenchyma* paratracheal-vasicentric, paratracheal-aliform to confluent in the outer late wood, and marginal. *Libriform fibers* thin- to medium thick-walled, fine to medium. *Rays* unstoried, 1–3-seriate, homocellular; ray-vessel pitting similar to intervessel type.

REMARKS

The woods of red ash (*F. pennsylvanica* Marsh.) and Oregon ash (*F. latifolia* Benth.) are very similar to that of white ash (*F. americana* L.) and are not distinguished from it in the trade. For separation of white ash and Oregon ash from black ash, see Remarks under the latter, p. 625.

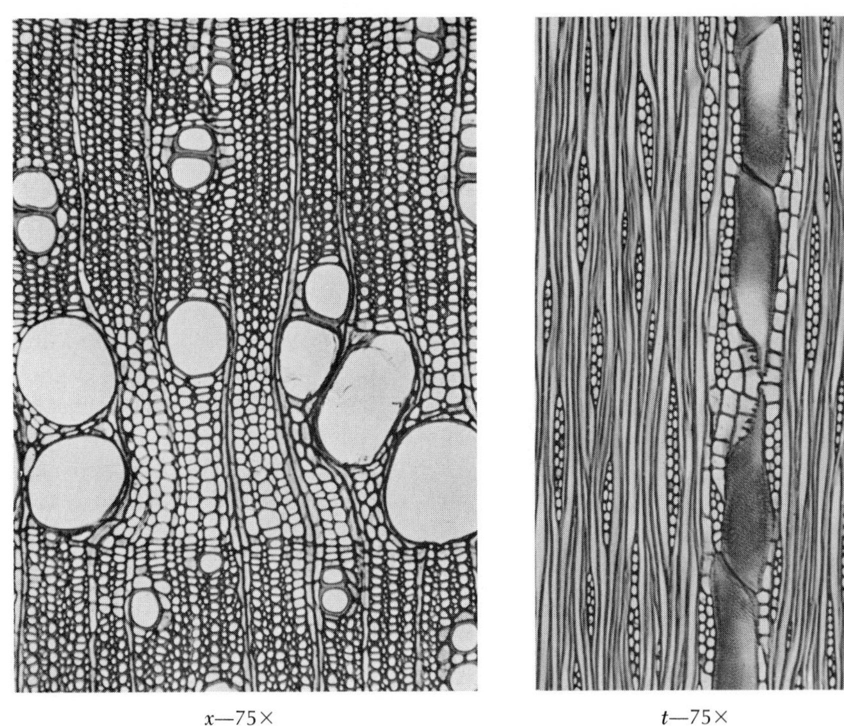

x—75× *t*—75×

FIG. 12-221 Oregon ash, *Fraxinus latifolia* Benth.

USES

Handles, for which it is second only to hickory in importance (the standard wood for shovel, spade, rake, and other long handles, because of its straightness of grain, stiffness, hardness, strength, moderate weight, good bending qualities, and capacity for wearing smooth in use): *furniture*, especially for bent parts and chair bottoms (it is especially well adapted for the latter use because it can be easily split into thin but tough and elastic strips); *vehicle parts*, such as poles, shafts, trees, and braces for wagons, and bottom boards; formerly used for frames in automobiles and airplanes; *railroad cars*; *sporting* and *athletic goods* (practically all baseball bats and long oars, a considerable number of short oars and paddles, also tennis-racket frames, snowshoes, ski, polo, and hockey sticks, and other sporting goods); *boxes, baskets*, and *crates*; *pallets*; *kitchen cabinets*; *agricultural implements*; *planing-mill products* (such as trim, for which old-growth ash is preferred); *ship-* and *boatbuilding*; *dairy, poultry*, and *apiary supplies*; *wood pipe*; *toys*; *woodenware* and *novelties*; *cooperage*, principally for slack staves and headings, formerly for tight cooperage used for butter tubs, and oil and pork barrels (for which its freedom from odor and taste and its good working qualities recommend it very highly); *veneer*.

BLACK ASH, BROWN ASH
Fraxinus nigra Marsh.

GENERAL CHARACTERISTICS

Sapwood whitish to light brown, narrow; *heartwood* grayish brown to brown [darker than that of the white ash (*F. americana* L.)], dull; *wood* without characteristic odor or taste, straight-grained, medium heavy (sp gr approx 0.45 green, 0.53 ovendry), medium hard. *Growth rings* distinct (wood ring-porous), frequently narrow. *Early-wood pores* large, distinctly visible to the naked eye, forming a band 2–4 pores in width; transition from early to late wood abrupt; *late-wood pores* small, barely visible to the naked eye, solitary and in multiples of 2–3, rarely joined laterally by parenchyma in the outer late wood (usually connected by parenchyma in white ash). *Parenchyma* visible with a hand lens, forming a narrow sheath around the pores in the late wood, rarely uniting them laterally as described above. *Rays* indistinct or barely visible to the naked eye.

MINUTE ANATOMY

Vessels very few to moderately few; largest early-wood vessels large to very large; perforation plates simple; intervessel pits orbicular to short-oval or occa-

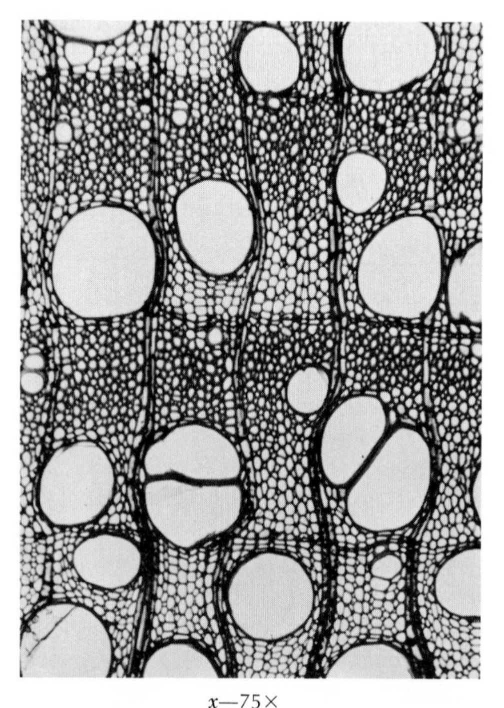

x—75×

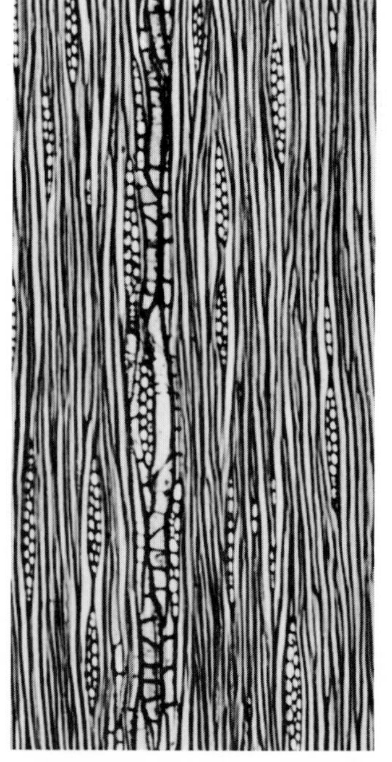

t—75×

FIG. 12-222 Black ash, *Fraxinus nigra* Marsh.

sionally somewhat angular through crowding, 3–6 μ in diameter. ***Vasicentric tracheids*** present, confined to the vicinity of the early-wood vessels. ***Parenchyma*** paratracheal-vasicentric, rarely paratracheal-confluent in the outer late wood, and marginal; sheath of paratracheal-vasicentric parenchyma around the late-wood vessels uniseriate for the most part; marginal parenchyma fairly abundant, grading into the tissue of the succeeding ring, not forming a distinct line. ***Libriform fibers*** thin- to fairly thick-walled, fine to medium. ***Rays*** unstoried, 1–3-seriate, homocellular; ray-vessel pits similar to intervessel type.

REMARKS

Black ash intergrades with white ash (various species) and is frequently confused with this timber. Black ash is generally lighter in weight and weaker, and the heartwood is usually of a more decided brown and less lustrous; hence the term "brown ash," which is sometimes used in the trade. The rings are generally narrower in black ash but, in contrast, the porous zone of early wood is usually wider (linear measurement, not in number of pores) and occupies more space in the ring. In black ash, the pores in the outer part of the ring are seldom united by paratracheal-confluent parenchyma, a feature that generally characterizes the white ashes and Oregon ash.

USES

The better grades of lumber are used for the same purposes as white ash. "Old-growth" black ash is utilized for ***planing-mill*** products (such as interior trim) and for ***cabinetwork*** (frequently preferred to white ash for these purposes because it exhibits a more handsome figure and retains its shape better); ***basketmaking*** because the wood splits readily through the early-wood zone of the annual increments when it is pounded. Stock for pack baskets, etc., is made in this way by local craftsmen on a number of the Indian reservations.

CATALPA

Northern catalpa (*Catalpa speciosa* Warder)

Southern catalpa (*Catalpa bignonioides* Walt.)

GENERAL CHARACTERISTICS

Sapwood pale gray, narrow; ***heartwood*** grayish brown, occasionally with a lavender tinge; ***wood*** with a faint aromatic, noncharacteristic odor, without characteristic taste, straight-grained, moderately light (sp gr approx 0.38 green, 0.42 ovendry), moderately soft. ***Growth rings*** distinct (wood ring-porous), generally wide; individual rings often variable in width and the rings then sinuate in contour. ***Early-wood pores*** large, distinctly visible to the naked eye, arranged in a band 3–5 plus pores in width, somewhat lighter in color than the denser late wood; transition from early to late wood abrupt, or more or less gradual; *late-wood pores* small, arranged in small groups which are further aggregated into interrupted or continuous concentric bands toward the outer margin of the ring. *Parenchyma* not distinct, or associated with pores and then distinct and zonate toward the outer margin of the ring. *Rays* usually indistinct to the naked eye, plainly visible with a hand lens.

MINUTE ANATOMY

Vessels very few to few, the largest large to very large; perforation plates simple; spiral thickening rather inconspicuous, present in some of the small vessels in the late wood; intervessel pits orbicular or nearly so, 6–8 μ in diameter. *Parenchyma* paratracheal-vasicentric to aliform and paratracheal-banded, the banded parenchyma more frequent near the outer margin of the ring and the outermost band not infrequently devoid of vessels. *Libriform fibers* thin-walled, medium to very coarse. *Rays* unstoried, 1–6 (mostly 2–3)-seriate, homocellular to heterocellular; ray-vessel pitting similar to intervessel type.

USES

Locally for *fence posts; rails; general construction work; interior finish; cabinetwork; fuel.*

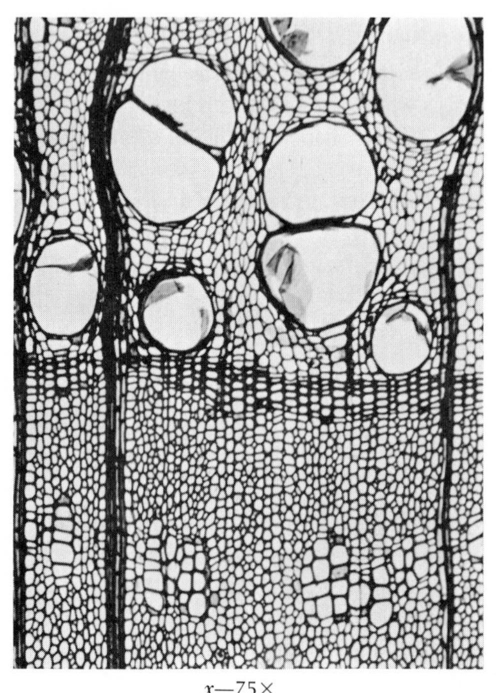

x—75×

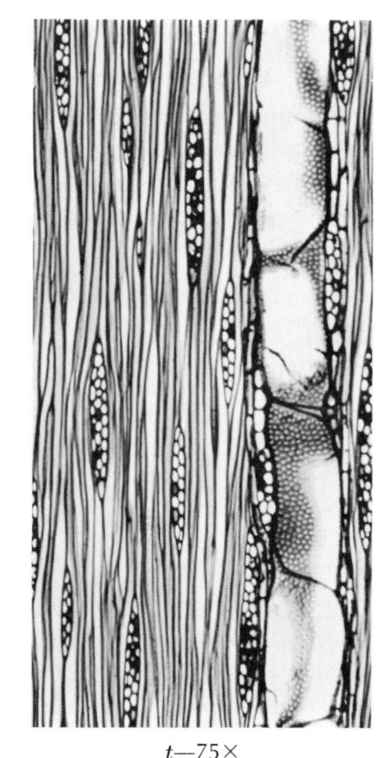

t—75×

FIG. 12-223 Northern catalpa, *Catalpa speciosa* Warder

TABLE 12-2 Comparative Physical Data for Hardwoods Native to the United States[a]

Common name of wood[b]	Sp gr, avg		Weight, lb/cu ft		Shrinkage based on green dimension, %[c]				Index classes to mechanical properties[d]											
	Green	12%	Green	12%	Vol.	T	R	T/R	Bending strength	Compression as a post	Hardness	Stiffness	Shock resistance	Splitting	Screw holding	Ability to stay in place	Drying defects	Machining qualities	Gluing qualities	Durability rating
(1)	(2)	(3)	(4)	(5)	(6)	(7)	(8)	(9)	(10)	(11)	(12)	(13)	(14)	(15)	(16)	(17)[e]	(18)[f]	(19)[g]	(20)[h]	(21)[i]
Alder, red	0.37	0.41	46	28	12.6	7.3	4.4	1.7	4	2	4	3	5	4	4	3	2	*G*	2	*S*
Ash, black	0.45	0.49	52	34	15.2	7.8	5.0	1.6	4	4	4	3	3	3	3	4	3	*G*	3	*S*
Ash, white	0.55	0.60	48	42	13.4	7.8	4.8	1.6	2	2	2	2	3	2	2	3	3	*G*	3	*S*
Aspen, quaking	0.35	0.38	43	26	11.5	6.7	3.5	1.9	4	5	5	4	5	5	4	3	2	*G*	1	*S*
Basswood, American	0.32	0.37	42	26	15.8	9.3	6.6	1.4	4	4	5	3	5	4	4	3	2	*A*	2	*S*
Beech, American	0.56	0.64	54	45	16.3	11.0	5.1	2.2	2	3	3	2	3	2	2	4	3, 4, 5	*G*	4	*S*
Birch, paper	0.48	0.55	50	38	16.2	8.6	6.3	1.4	4	4	4	3	3	4	2	4	2	*G*	2	*S*
Birch, yellow	0.55	0.62	57	43	16.7	9.2	7.2	1.3	2	3	3	1	3	3	2	4	2	*A*	4	*S*
Buckeye, yellow	0.33	0.36	49	25	12.5	8.1	3.6	2.2	5	5	5	4	5	4	5	3	2	*A*	2	*S*+
Buckthorn, cascara	0.50	0.52	50	36	7.6	4.6	3.2	1.4	4	3	3	5	2	3	2	4	3	*A*	4	*S*
Butternut	0.36	0.38	46	27	10.6	6.4	3.4	1.9	4	4	5	4	5	4	4	2	2	*A*	2	*S*
California-laurel	0.51	0.55	54	39	12.4	8.6	2.9	2.9	4	3	2	5	2	2	2	3	3, 4	*A*	4	*S*
Catalpa, northern	0.38	0.41	41	29	7.3	4.9	2.5	2.0	4	4	4	4	3	4	4	3	2	*A*	3	*R*
Cherry, black	0.47	0.50	45	35	11.5	7.1	3.7	1.9	3	3	3	2	4	3	3	3	2	*A*	3	*R*
Chestnut, American	0.40	0.43	55	30	11.6	6.7	3.4	2.0	4	4	4	4	4	4	3	2	2	*G*	1	*R*
Chinkapin, golden	0.42	0.46	61	32	13.2	7.4	4.6	1.6	3	3	4	3	4	4	3	3	3	*A*	3	*S*
Cottonwood, eastern	0.37	0.40	49	28	14.1	9.2	3.9	2.4	4	4	5	3	5	4	4	4	3, 5	*P*	1	*S*
Dogwood, flowering	0.64	0.73	64	51	20.8	11.8	7.4	1.6	2	2	1	3	2	3	1	5	2	*A*	3	*S*
Elm, American	0.46	0.50	54	35	14.6	9.5	4.2	2.3	3	4	4	3	3	2	2	5	3	*P*	2	*S*
Elm, rock	0.57	0.63	53	44	14.1	8.1	4.8	1.7	2	2	2	3	2	2	2	5	3	*P*	2	*S*
Elm, slippery	0.48	0.53	56	37	13.8	8.9	4.9	1.8	3	3	3	3	2	2	2	5	3	*P*	2	*S*
Hackberry	0.49	0.53	50	37	16.9	8.9	4.8	1.9	4	4	3	4	3	2	2	4	3	*G*	2	*S*

TABLE 12-2 Comparative Physical Data for Hardwoods Native to the United States (*Continued*)

Common name of wood[b]	Sp gr, avg		Weight, lb/cu ft		Shrinkage based on green dimension, %[c]				Index classes to mechanical properties[d]											
	Green	12%	Green	12%	Vol.	T	R	T/R	Bending strength	Compression as a post	Hardness	Stiffness	Shock resistance	Splitting	Screw holding	Ability to stay in place	Drying defects	Machining qualities	Gluing qualities	Durability rating
(1)	(2)	(3)	(4)	(5)	(6)	(7)	(8)	(9)	(10)	(11)	(12)	(13)	(14)	(15)	(16)	(17)[e]	(18)[f]	(19)[g]	(20)[h]	(21)[i]
Hickory, shagbark	0.64	0.72	64	50	16.7	10.0	7.0	1.4	1	1	1	1	1	2	2	5	3	*G*	4	*S*
Holly, American	0.50	0.57	57	40	16.9	9.9	4.8	2.1	4	4	3	4	2	2	2	4	2, 4	*G*	3	*S*
Honeylocust	0.60	0.67	61	44	10.8	6.6	4.2	1.6	2	2	1	3	3	1	2	2	3, 4	*A*	3	*M*+
Hophornbeam, eastern	0.63	0.70	60	49	19.4	10.0	8.5	1.2	2	3	2	3	1	3	1	5	3	*A*	4	*S*
Hornbeam, American	0.58	0.70	53	49	19.1	11.4	5.7	2.0	4	4	2	4	1	3	1	5	3	*A*	4	*S*
Locust, black	0.66	0.69	58	48	10.2	7.2	4.6	1.6	1	1	1	1	3	2	1	2	2	*A*	4	*R*+
Madrone, Pacific	0.58	0.65	60	45	18.1	12.4	5.6	2.2	3	3	2	4	3	2	2	5	3	*P*	4	*S*
Magnolia, southern	0.46	0.50	59	35	12.3	6.6	5.4	1.2	3	4	3	3	2	3	3	4	2	*A*	2	*S*
Maple, red	0.49	0.54	50	38	13.1	8.2	4.0	2.0	3	3	3	2	4	3	2	3	3,4	*A*	3	*S*
Maple, sugar	0.56	0.63	56	44	14.9	9.5	4.9	1.9	2	2	2	1	3	1	2	4	3,4	*G*	4	*S*
Oak, live	0.81	0.89	76	62	14.7	9.5	6.6	1.4	1	1	1	1	2	1	1	4	3, 4, 5	*P*	4	*R*
Oak, red[j]	0.57	0.62	64	43	14.7	8.9	4.2	2.1	2	3	2	2	3	2	2	4	3, 4, 5	*G*	3	*S*+
Oak, white[k]	0.59	0.67	63	45	15.9	9.2	5.3	1.7	2	2	2	3	3	2	1	4	3, 4, 5	*G*	3	*R*
Osage-orange	0.76	0.80	62	56	9.2	....	...	...	1	1	1	1	1	1	1	3	3	*A*	4	*R*+
Persimmon, common	0.64	0.74	63	52	19.1	11.2	7.9	1.4	2	2	1	2	3	2	1	4	3	*A*	4	*S*
Sassafras	0.42	0.45	44	31	10.3	6.2	4.0	1.6	4	4	4	4	3	3	3	3	2	*A*	3	*R*
Sourwood	0.50	0.55	53	38	15.2	8.9	6.3	1.4	3	3	3	2	3	2	2	4	2	*A*	3	*S*
Sweetgum	0.46	0.52	55	35	15.0	9.9	5.2	1.9	3	3	4	3	3	3	3	4	3	*A*	2	*S*+
Sycamore, American	0.46	0.49	52	34	14.2	7.6	5.1	1.5	4	4	4	3	4	3	3	4	3, 4	*P*	3	*S*
Tupelo, black	0.46	0.50	45	35	13.9	7.7	4.4	1.8	3	3	3	4	4	3	2	5	3	*P*	3	*S*

(1)	(2)	(3)	(4)	(5)	(6)	(7)	(8)	(9)	(10)	(11)	(12)	(13)	(14)	(15)	(16)	(17)[e]	(18)[f]	(19)[g]	(20)[h]	(21)[i]
Walnut, black	0.51	0.55	58	38	12.8	7.8	5.5	1.4	2	2	3	2	3	2	2	2	2, 5	*G*	3	*R*
Willow, black	0.34	0.37	50	26	14.4	8.1	2.6	3.1	5	5	5	5	3	4	4	3	3	*P*	1	*S*
Yellow-poplar	0.40	0.42	46	30	12.3	7.1	4.0	1.8	4	4	4	3	4	4	4	2	2	*A*	1	*S*

[a] Based on information from "Wood Handbook," U.S. Department of Agriculture Handbook 72, 1955, and *U.S. Dept. Agr. Bull.* 479, 1935, except for columns 9 and 18.

[b] Common names according to "Check List of Native and Naturalized Trees of the United States," U.S. Department of Agriculture Handbook 41, 1953. See text for scientific names.

[c] Shrinkage values are for change from green to ovendry condition.

[d] Coding of mechanical properties according to index classes:

		Index classes				
Column No.	*Mechanical property*	5	4	3	2	1
10	Modulus of rupture in bending, green wood, psi	<6000	6000 to	7000 to	8000 to	11,000+
11	Maximum crushing strength parallel, green wood, psi	<2200	2200 to	3000 to	3600 to	4500+
12	Side hardness, green wood, lb	< 400	400 to	600 to	900 to	1200+
13	Modulus of elasticity in bending, psi × 1000	< 800	800 to	1000 to	1300 to	1500+
14	Height of drop with 50-lb hammer, in.	15 to	25 to	35 to	50 to	70+
15	Load, lb/in. of width to split green wood	< 150	150 to	250 to	350 to	450+
16	Nail and screw withdrawal as function of sp gr OD	< 0.40	0.40 to	0.45 to	0.55 to	0.70+

[e] Dimensional changes with moisture in normal use ranges from class 1 with little change, to class 5 with greatest dimensional change.

[f] Drying defects classed as follows: 1. Slight warping and little dimensional change. 2. Distinct warping. 3. Pronounced warping. 4. Tendency to check and split. 5. Subject to collapse and honeycombing.

[g] Machining qualities: *G* = good to excellent, smooth surface. *A* = average. *P* = poor, requiring care to be acceptable.

[h] Gluing qualities: 1. Glues very easily. 2. Glues well. 3. Requires control. 4. Difficult to glue, requires close control.

[i] Heartwood durability ratings: *R* = resistant to decay. *M* = moderate resistance to decay. *S* = low decay resistance. + indicates high rating within the group.

[j] Indicates commercial red oak as weighted averages for black, laurel, pin, northern red, scarlet, water, and willow oaks.

[k] Indicates commercial white oak as weighted averages for bur, chestnut, post, swamp chestnut, white, and swamp white oaks.

13

Wood Fibers and Their Identification

The term *fiber*, as here used, designates any type of wood cell that is retained in the wood pulp. The majority of these cells are tracheids in the case of softwoods, and the fibrous cells, i.e., libriform fibers, fiber tracheids, tracheids, and some of the longer vessel elements in the case of hardwoods.

One of the principal uses of wood, in other than its solid form, is as a source of fiber, principally in pulp production. It has been shown by numerous investigators that the characteristics of wood pulp and products made of it are determined by the properties of wood, used as a raw material, and by the morphology and chemistry of the wood cells, as well as by the processes employed in separating wood into its component cells and in converting the resulting pulp-mass into finished products. In this section attention will be centered on the properties of wood and the morphological and chemical characteristics of wood cells, which influence the qualities of wood pulp and paper, since discussion of the manufacturing processes involved in conversion of wood into pulp and paper products is outside the scope of the text.

In considering wood as a source of fiber, two factors must be taken into account: yield of fiber per given volume or weight of wood, and the quality of the resulting fiber. The former depends on the characteristics of wood prior to pulping and th-processes employed in its conversion into pulp, while the latter is mainly a result of the morphological features of the individual fibers and the modification brought about in them by the methods of conversion. Fiber quality is also a variable quantity in the sense that interpretation of the quality aspects of fibers depends on the specific requirements of the final product to be made from the wood pulp. For example, quality requirements for wood fibers in production of strong kraft paper are quite different from those for high-grade writing paper, and they are quite unlike those for filter paper. The question of wood-fiber quality is still further complicated by the lack of agreement among the technical people and the producers of pulp products on the interpretation of the qualitative features of fibers, and by the difficulties encountered in determining these features in a practical way. A feeling exists among some converters that because of great variance in the properties of wood, in terms of species and the variability of wood characteristics within the species and even within a given tree, it is frequently easier to modify the process of conversion, or the specifications for the final product, to suit the available raw material, than to segregate wood on the basis of the anatomical characteristics of its cells. This thinking implies that the availability of the raw material at a favorable price is more important than the characteristics of wood or its cells.

I. EFFECT OF WOOD PROPERTIES ON PULP PRODUCTION

As previously stated, the two principal considerations in dealing with wood as a source of fiber are yield of fiber per given volume or weight of wood, and the characteristics of the resulting pulp.

The fiber yield may be expressed in terms of (1) the percentage of dry weight of pulp produced from a given weight of wood, e.g., so many pounds of ovendry pulp per hundredweight of ovendry wood, or (2) as a percentage of the ovendry pulp obtained from a given volume of wood at a known moisture content, e.g., so many pounds of ovendry pulp per cubic foot of wood at a stated moisture content. The quantity of fiber produced per given measure of wood is dependent on the characteristics of the raw material, i.e., species, and such growth factors as density of wood, the relationship between the early and late wood, age of the tree, position of the sample in the tree, and defects.

The fiber yield is also affected by the pulping process employed, ranging from better than 90 percent for groundwood to less than 50 percent in the case of some chemical processes.

The quality of the resulting fibers depends on the wood structure, i.e., the type of cells present in a given wood, the morphological characteristics of the individual cells, and to a lesser degree on the chemical composition of the cell wall material. The final fiber quality is also a result of modifications in the original fiber brought about by the pulping process employed and the subsequent treatment accorded to the pulp in its conversion to the finished product.

A. Species

With improvement in the methods of conversion, the specific characteristics of a kind of wood are often less important than the availability of the raw material and the market requirements for the pulp products. Wood of different species can be successfully blended to produce pulp of desired quality. Nevertheless pulping characteristics of a species of wood and the quality of the resulting fiber must be known in order to design proper pulping procedures and to assure the desired quality of pulp. For instance, the southern hard pines will require a different method of pulping than spruce, because of the presence of resin in the former. The fiber obtained from the southern pines is longer and stiffer than that of spruce, and it will produce bulkier and coarser paper, with high resistance to tear. Spruce fiber, on the other hand, is more pliable and will convert into denser and smoother paper, with higher tensile and bursting strength.

Hardwoods require different treatment, and the resulting pulp consists of relatively short fibers. Such pulp is used for special products but also can be made to serve many of the same purposes as the long-fibered pulp from the coniferous woods by blending the species and by modifying the process of conversion. Lower-density hardwoods such as aspen are preferred, but high-density hardwoods, e.g., maple or oak, are also used in ever-increasing amounts.

B. Wood Density

Variations in wood density within a species affect both the yield and the quality of pulp. When pulp yield is expressed in terms of the ovendry weight of wood fiber per unit volume of wood, measured green, i.e., above the fiber saturation point, then the density of wood, or its specific gravity, is the most important single factor in determining the pulp yield.[3,6,14] For instance, studies made on yields of kraft pulp from the southern hard pines found yield to be practically a straight-line relationship, amounting to an increase of 1 lb of pulp for every 2-lb increase in wood density, on a cubic foot basis (Fig. 13-1). A similar general relationship between wood density and fiber yield was found to hold for other species, though the exact amount of fiber increase varies with the given species of wood and the process of pulping employed.

Variations in wood density within a species may also affect the quality of pulp. These effects are related to the differences in cell size and "mix" of the various types of cells, and variations in the cell wall thickness, especially in the wood with pronounced early and late wood, as in southern hard pines or Douglas-fir. The effect of density on the quality of pulp is discussed in the subsequent paragraphs.

C. Early and Late Wood

In the species with prominent difference between early and late wood, ratios between these two components of the growth ring affect the yield, as well as the quality of pulp.

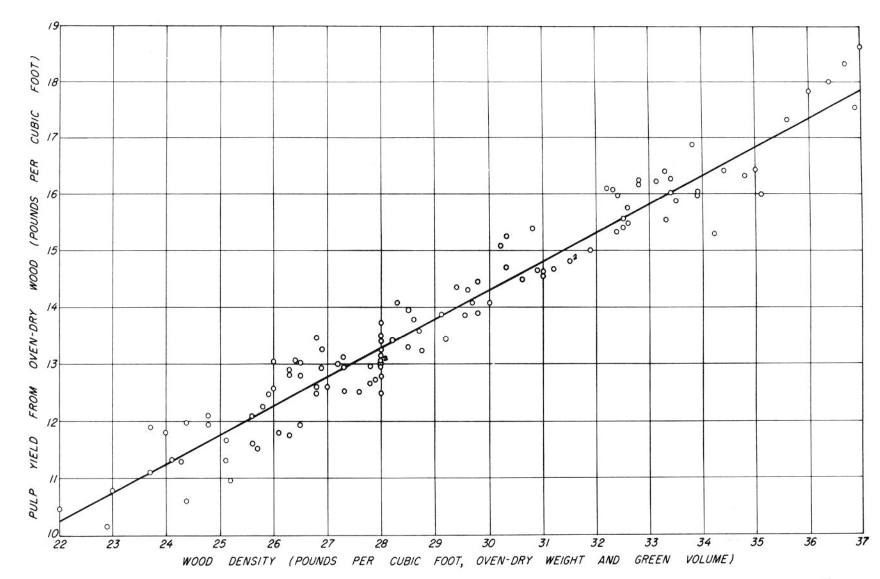

FIG. 13-1 Relationship between wood density and kraft pulp yields of southern yellow pine pulpwood. (*Courtesy of H. L. Mitchell, Forest Products Lab., U.S. Forest Service.*)

Pulp yield is affected because increase or decrease in the late-wood content is usually associated with a corresponding variation in wood density and hence with variation in the yield of fiber per unit volume. Fluctuations in the amount of early wood, of course, have the opposite effect.

Pulp quality is affected by variations in the early- and late-wood ratios because differences in the cell wall thickness in the two zones of the growth ring, and the corresponding differences in the ratios of the cell wall thickness to the cell wall diameter, affect such properties of pulp and paper as its strength, bulk, and surface qualities. These variables also influence the response of the wood fibers to pulp purification and bleaching, and paper sheet formation.

D. Rate of Growth and Age

In the past, rate of growth, expressed in terms of the number of growth rings per inch of radius of the cross section of wood, was considered an indication of wood density, especially in the coniferous woods. It was assumed that in the softwoods, at least, the specific gravity of wood tended to decrease with increase in the ring width. This assumption did not take into consideration the fluctuating ratios of the early and late wood in a given growth ring, the position of the sample in the tree, or the variations in the thickness of the cell walls, whether caused by environment or heredity.

More thorough studies in this country and abroad, based on sound statistical analysis, have shown conclusively that changes in the rate of growth, as expressed by width of the growth rings, are not a reliable indicator of variations in the wood density.[4,5,6] These studies further revealed that much of the variation in wood density among trees of the same species, as well as within a given stem, is due to the age of the tree at the time a given piece of wood was formed, as well as to the location of the sample in terms of stem height in the tree.[2,6,14] Of particular importance are differences in the fiber morphology, wood density, and chemical constituents of the cell wall in the juvenile (core) wood, as compared to the mature (outer) wood. (For more detailed discussion of these subjects, see Chap. 7.)

E. Reaction Wood

Reaction wood, i.e., compression wood in conifers and tension wood in hardwoods, is less desirable for papermaking than the normal wood, because of the generally adverse effect of such wood on strength properties of the chemical pulp. Because of the higher density of reaction wood, the yield of pulp, when determined on a volume basis, may be greater than that from the normal wood. However, the chemical pulps and the groundwood produced from compression wood are inferior in strength. Furthermore, the chemical pulp from compression wood requires a longer cooking and bleaching time, because of the greater amount of lignin present in such wood.

The tension wood of hardwoods possesses higher cellulose and lower lignin and pentosan content. Because of this, tension wood delignifies and bleaches more easily than the normal wood. It produces good dissolving pulp* and superior groundwood, but the chemical pulp for papermaking from the tension wood has exceptionally low strength.

II. EFFECT OF CHEMICAL COMPOSITION OF WOOD ON PULP PRODUCTION

There is a scarcity of information to indicate the exact amounts of the cell wall constituents that would produce pulp with the most desirable characteristics for a given use.[8] Of the three principal cell wall constituents, lignin is considered undesirable in the chemically produced pulp. Since large quantities of lignin in the wood are found in the compound middle lamella which is destroyed in pulping, a considerable amount of the lignin is removed in the process. Further amounts of lignin, remaining in the secondary cell wall of the fibers, are removed in the bleaching operations. The amount of lignin remaining in the pulp after bleaching affects the papermaking properties of the pulp by making the fibers stiff and by decreasing the interfiber surface bonding activities of the hemicellulose, resulting in paper of low strength and high opacity. On the other hand, lignin-rich fibers do not shrink much when dried, and therefore paper made of such fibers has good dimensional stability. In the groundwood all of the inter- and intracellular lignin is retained and only some lignin is removed by bleaching. Paper made of, or containing large

* A highly purified pulp, with the alpha cellulose content ranging from 89 to 99 percent, used in the manufacture of rayon, cellulose acetate, cellophane, nitrocellulose, and other cellulose derivatives.

quantities of, ground wood has low strength and tends to turn brown and crumble with age.

In chemical pulping, cellulose is the principal remaining constituent of the wood fibers, and as such it determines most of the properties of pulp and paper, because of its effect on the ultimate fiber strength, fiber bonding, and paper sheet characteristics. In combination with the residual hemicelluloses, it accounts for the variations in pulp yield.

The hemicellulosic material of the fiber retained after delignification forms a gel on the surface of fibers and in the interfibrillar spaces.[10] The hemicellulosic gel on the surface of the fibers provides a solid film bond between the surfaces of adjacent cellulosic fibers. Its effect on the fiber strength is to allow internal redistribution of stress to occur in a pulp or paper sheet, when external load is applied. Reduction in the hemicellulosic content results in replacing the relatively flexible bond between the hemicellulosic gel and the two adjacent fiber walls, consisting of cellulose, by a more rigid cellulose-to-cellulose bond. When this occurs the stress distribution is inhibited and the tensile strength properties of paper are lowered.

The interfibrillar hemicellulosic material makes fibers flexible. This fact plays an important role in beating the pulp, since the hemicelluloses act as plasticizers, by providing internal lubrication. The amount of the hemicellulosic material retained in pulp also increases the bulk yield of fiber.

Because of these favorable qualities of the hemicellulosic material, contrasted with the generally undesirable effects of lignin, most chemical processes are designed to favor delignification reaction, i.e., as complete removal of lignin as possible, and to minimize as much as practical the chemical reactions that would cause elimination of the hemicelluloses.

The effect of the various extraneous materials present in the wood, principally in the heartwood, is only partially understood. The presence of appreciable amounts of resinous materials in some conifers, such as hard pines and Douglas-fir, requires a process (sulfate) that can cope successfully with such products. Many other kinds of extraneous materials are destroyed in pulping. The most vexing problem is that of the extraneous materials retained in the pulp, which, because they affect the color or brightness of pulp, require expensive bleaching for their removal. The precise effect on the properties of paper of the residual extraneous materials that might still remain in the pulp after bleaching is difficult to evaluate because of the variety of such materials that might be present. However, it is believed that the amounts of such residual products are small and that their effect is insignificant in relation to other factors that influence the characteristics of finished paper.

Considerable evidence exists that the variations in the chemical composition of wood are less important in determining the quality of pulp and paper products than the morphological characteristics of the fibers. This is especially true of chemical pulp produced from the coniferous woods, because of the low degree of variations in the chemical composition of the cell wall components in the softwoods. In the hardwood chemical or semichemical pulps, the higher percentage of hemicelluloses, especially pentosans, has some beneficial effect on paper strength, but again this effect is less significant than the morphological characteristics of the hardwood fibers.

III. EFFECT OF THE CELL WALL MORPHOLOGY ON PULP AND PAPER PROPERTIES

Paper can be defined as a flat and relatively thin mat consisting of fibers deposited at various angles, but essentially in the plane of the sheet.[7] The fibers provide numerous crossover points at which they may be bonded together. The principal factors which control the characteristics of paper, such as its strength, density, porosity, surface qualities, etc., are (1) properties of the individual fibers, (2) the distribution and the form of fibers, i.e., whether the fibers are straight or curved within the sheet of paper, and (3) the ability of the adjacent fiber surfaces to form a bond with one another.

A. Properties of the Individual Fibers

Three fiber variables are considered to be of importance in determining the physical characteristics of pulp and paper. These are fiber density, fiber length, and fiber strength.

1. FIBER DENSITY

Fiber density has a positive effect on the yield of pulp.[4,11,12,13] When expressed either as thickness of the cell wall or as the ratio of wall thickness to cell diameter it is one of the most important factors influencing the characteristics of the resulting pulp and paper. Thick-walled cells, such as found in the late wood of the southern hard pines, resist the compacting forces and tend to maintain their original cross-sectional shape. This results in highly opaque, coarse, and bulky papers, with high resistance to tear. However, because of the reduced number of potential contact areas between the uncollapsed fibers, other strength properties associated with fiber bonding, such as bursting and tensile strength, and folding resistance, are appreciably reduced. Thin-walled fibers, on the other hand, collapse readily and form dense well-bonded papers, low in tear but high in other strength properties. It has been suggested that the optimum strength values in softwood pulp are reached with 20 percent late wood and that the desirable range for the coniferous pulpwood is 15 to 50 percent late wood. Hardwoods with a below-average wood density are considered to be more desirable for pulping than those with higher specific gravity. One rule of thumb for establishing the suitability of a hardwood species for pulpwood is that the ratio of twice the wall thickness to lumen diameter should be less than unity.[4,5]

2. FIBER LENGTH

In the past it was assumed that fiber length was the most important single feature of paper-making fibers in determining the properties of paper, especially its strength.

More recently it has been increasingly recognized that other fiber characteristics, such as the ratio between cell lumen width and cell wall thickness, and particularly the ratio of fiber length to its width, affect paper properties more than the fiber length alone.[4]

Fiber length is important insofar as a minimum length is required to provide sufficient bonding surface to spread the stresses evenly over the entire area of the sheet. Also, paper consisting of short fibers will have more free ends and therefore will offer less resistance to pulling apart. But, as pointed out by Barefoot et al. for loblolly pine,[1] though the longer fiber lengths are an important plus factor, their significance is affected by the fact that any portion of a tree will contain a range of fiber lengths, with many short fibers included. Therefore, they conclude: "These short fibers negate to an unknown extent the very benefits desired when making pulp from trees with high average fiber length. A dramatic reduction in the variation of the fiber lengths within trees will probably be necessary to achieve any great benefits to pulp and paper properties from merely an increased average length." It may be assumed that the same argument would hold for other species, since in all of them, cell length variations occur from the center of the tree (juvenile wood) to the periphery (mature wood) and from the butt to the crown of the stem, as well as within each growth ring, especially when pronounced difference exists between the early and late wood.

Finally it must be recognized that the length of fibers in the paper is not the original length of cells in a tree but the length remaining after completion of the various chemical and physical treatments to which wood and its fibers (pulp) are subjected in the course of pulp- and papermaking. Nevertheless it is recognized that when similar manufacturing processes are employed the lower strength values of hardwood pulps, as compared to that obtained from softwoods, are due to the considerably shorter fibers found in the former.

3. FIBER STRENGTH

It has been known for some time that the tensile strength in the direction of the fiber axis and the strength in shear in flexure of the individual fibers contribute significantly to the properties of pulp and paper. Van den Akker et al.[13] found that at least 40 percent of the fibers were broken in the tensile failure of unbeaten sheets and that the percentage increased considerably with beating.* However, these workers felt that fiber-to-fiber bonding may be more important in developing the tensile strength of papers than the strength of the individual fibers. Giertz[8] suggests that interfiber bonding is the determining factor where beating is slight. But with continuing beating the total bonded area will reach a critical point beyond which rupture in a sheet of paper will occur across the fibers rather than between, i.e., in the bond. In

* Beating is an operation in paper manufacture in which pulp suspended in water is given mechanical treatment resulting in cutting, splitting, and crushing of fibers. The beaten stock acquires a characteristic slimy feel and is transformed into a mass of pulp which can be spread into sheets.

this view the maximum strength of paper is dependent on the average strength of the individual fibers in the sheet of paper.

A compromise suggestion is that ultimate failure of paper depends on the relationship between tensile strength of the individual fibers and the shear strength of the fiber-to-fiber bonds. The failure of a sheet of paper may be visualized as a chain reaction initiated by the break of the weakest fiber or a bond. This would impose additional stress on the adjacent fibers and bonds, causing further failure, until a complete rupture of the sheet occurs.

Whatever is the real role of the individual fibers in determining the strength and other properties of paper, the main problem in evaluating their contribution quantitatively has been due to the technical difficulties encountered in designing meaningful strength tests of single wood cells. Even with improvements in experimental techniques, the necessity of including a large number of fibers in order to account for a wide variation in the fiber characteristics means that the testing of single cells not only is extremely tedious but requires a great deal of ingenuity in order to be valid.

On the basis of such tests Tamolang et al.[12] determined that as the density of the paper sheet increased the strength of individual fibers played an increasingly important role in developing the tensile and bursting strength of paper. It has been further shown that fiber strength is highly related to the cell wall area, i.e., cell wall thickness, and to the fibrillar angle. The fibrillar angle is believed to be the major factor in governing fiber strength, when fiber strength is expressed in terms of strength per *unit area* of the cell wall; increase in the fibrillar angle results in decrease of strength properties of the fiber and vice versa. It should be pointed out, however, that the average fibrillar angle of the secondary cell wall is a composite of the fibrillar angles of numerous lamellae, of which the *S*1, *S*2, and *S*3 layers of the secondary wall are composed. But since the *S*2 layer is generally by far the thickest of the three layers, presumably the overall fibrillar angle orientation in the lamellae of the *S*2 layer only was considered in this study.

IV. SUMMARY

The final characteristics of pulp or of a sheet of paper are a result of the combined effect of all the properties of wood and its individual cells, and of modifications in the properties of wood fibers brought about by the conversion processes. Frequently the latter, i.e., the manufacturing processes, exercise a greater influence over the ultimate quality of the final fiber product than the intrinsic properties of the raw material. In any case, no single property of wood or its fiber can be said to play a predominant role in determining the characteristics of pulp and paper.

Yield of pulp, when expressed in terms of weight per unit volume of wood, is influenced by wood density more than by any other property of the raw material.

The quality of pulp and paper is affected by the characteristics of fibers considered *en masse* and individually. For instance, the ratio between the early and late wood, in such wood as southern hard pines, determines the bulk of paper, its strength properties, and its surface characteristics.

The most important of the morphological characteristics of the individual fibers in papermaking are fiber density (i.e., cell wall thickness), fiber length, and fiber strength. The last appears to be strongly influenced by the fibrillar angle.

The chemical composition of the cell wall material is of some importance, especially in hardwoods, but its effect on the quality of pulp and paper is believed to be overshadowed by the anatomical characteristics of the fibers.

However, since all the intrinsic properties of wood and wood fiber are affected by the manufacturing processes, it may be concluded that the characteristics of the final wood-fiber product for a specific purpose are at best a compromise attained through interplay of the various properties of wood fibers, as modified by the various treatments to which they were subjected in the course of manufacture.

V. WOOD-FIBER IDENTIFICATION

Wood material to be identified with the fiber key must first be macerated, i.e., converted into a fibrous mass. Maceration procedures are simple and begin by reducing the wood to slivers somewhat thinner than a match stick. Two simple methods of maceration give good results. Franklin's method uses equal volumes of glacial acetic acid and hydrogen peroxide in sufficient quantity to cover the splinters in a test tube. The wood is kept in the corked tube at 60°C for about 2 days and then washed free of acid. Schultze's method employs concentrated nitric acid and a few crystals of potassium chlorate at room temperatures until the splinters appear white. After washing, the material which has been treated by either method is shaken vigorously in a little water or alcohol (glass beads may be added) until the splinters are reduced to pulp. A *little* of the pulp can then be transferred to a slide with a *clean* dissecting needle, carefully teased apart, and covered with a drop of water or alcohol and a cover glass. Such unstained preparations, with proper manipulation of the substage condenser, will show all necessary details of the fiber structure.

When fiber identification is to be made from commercial wood pulp or paper, a small quantity from these should be torn into small pieces, soaked in water, and vigorously shaken in a test tube containing water and glass beads. The pulpy mass is then handled as previously described. Frequently, a better resolution of minute structure is obtained by adding to the wash water a stain, such as light green, methyl green, or aqueous safranin.

VI. KEY TO IDENTIFICATION OF WOODY FIBERS WITH A MICROSCOPE*

1. Tracheids† of the coniferous type present; vessel elements lacking...... **2**
1. Fibers of the hardwood type present; vessel elements present............ **17**

* Designed to separate the more important temperate North American genera of trees producing papermaking fibers. This key is a modification of one prepared by C. C. Forsaith and revised by H. A. Core for use in teaching fiber identification at State University College of Forestry at Syracuse University. Reference is made throughout the key to pertinent figures

2. Early-wood tracheids with spiral thickening.
Douglas-fir—*Pseudotsuga menziesii* (Mirb.) Franco

2. Early-wood tracheids lacking spiral thickening............................ **3**

3. Ray parenchyma cross-field pits* window-like (fenestriform), or pinoid, lenticular or broad-oval, largest pit apertures more than 10 μ along their major axis, generally variable in shape and size (Figs. 12-31 to 12-35, inc.)... **4**

3. Ray parenchyma cross-field pits piceoid, cupressoid, or taxodioid; largest pit apertures less than 10 μ along the major axis, generally uniform in shape (Figs. 12-36 to 12-42, inc.)..................................... **6**

4. Ray parenchyma cross-field pits oval to lenticular (pinoid, Fig. 12-32, 12-33), variable in shape and size, averaging more than 2 per cross field; pits in cross fields commonly paired in the direction of the tracheid axis (more than 1 horizontal row of pits per cross field); all ray tracheids dentate.
Hard pines—*Pinus* spp.

4. Ray parenchyma cross-field pits window-like (fenestriform) to broad-oval (Figs. 12-31, 12-34, 12-35); more or less uniform in shape, averaging 2 or less than 2 per cross field (except sometimes in sugar pine); pits in cross fields seldom paired in the direction of the tracheid axis (single horizontal row of pits per cross field); ray tracheids dentate or nondentate.. **5**

5. Average number of fenestriform pits per ray parenchyma cross field more than 1.2; marginal ray tracheids nondentate (Figs. 12-34, 12-35).
Soft pines—*Pinus* spp.

5. Average number of fenestriform pits per ray parenchyma cross field less than 1.2; ray tracheids with small dentations (Fig. 12-31).
Red pine—*Pinus resinosa* Ait.
Scotch pine—*Pinus sylvestris* L.

6. Marginal ray tracheid pitting always present; ray parenchyma cross-field pits piceoid to cupressoid (Figs. 12-36, 12-37)..................... **7**

6. Marginal ray tracheid pitting absent or sporadic; ray parenchyma cross-field pits taxodioid or cupressoid (Figs. 12-38 to 12-42, inc.)..... **12**

7. Early-wood ray parenchyma cross fields commonly with more than 4 pits per transverse row (Figs. 12-36, 12-37).................................. **8**

* For this key the cross fields are subdivided into those formed by the ray parenchyma cells and the longitudinal tracheids, and the ones formed by the ray tracheids and the longitudinal tracheids, since the pitting is frequently different in these two groups of cross fields.

---

in the text, but for complete illustrations of the fibers of the various species of wood reference must be made to C. H. Carpenter, H. A. Core, L. Leney, W. A. Côté, Jr., and A. C. Day, Papermaking Fibers, *Tech. Publ.* 74, State University College of Forestry, Syracuse, N.Y., 1963.

† In this key, unless otherwise stated, the features being studied should be observed on the early-wood tracheids.

7. Early-wood ray parenchyma cross fields rarely with more than 4 pits per transverse row (generally less than 1 row in 20).............................. 9

8. Angle of ray parenchyma pit apertures 10° or smaller from the tracheid axis; late-wood tracheids relatively thin walled; average length of tracheids 4.5 mm or more.
Sitka spruce—*Picea sitchensis* (Bong.) Carr.

8. Angle of ray parenchyma pit apertures more than 10° from the tracheid axis; late-wood tracheids relatively thick walled; average length of tracheids less than 4.5 mm.
Larch—*Larix* spp.

9. Ray parenchyma cross fields mostly with 1–2 pits, more than 10 percent with only 1 pit.. 10

9. Ray parenchyma cross fields mostly with 2–3 pits, less than 10 percent with only 1 pit.. 11

10. Distance between pits in ray parenchyma cross fields measured along one edge of the tracheid and parallel to its axis approximately 15 μ; angle of ray parenchyma pits from tracheid axis 45° or less.
Western hemlock—*Tsuga heterophylla* (Raf.) Sarg.

10. Distance between pits in ray parenchyma cross fields measured along one edge of the tracheid and parallel to its axis approximately 11 μ; angle of cross-field pits from tracheid axis more than 45°.
Eastern hemlock—*Tsuga canadensis* (L.) Carr.

11. Average length of tracheids more than 4.5 mm.
Sitka spruce—*Picea sitchensis* (Bong.) Carr.

11. Average length of tracheids less than 4.0 mm.
Spruce—*Picea* spp.

12. Average length of tracheids more than 4.0 mm; intertracheid pitting commonly 3 or more in a row across the width of the tracheid; average number of pits per transverse row in ray parenchyma cross fields more than 2.. 13

12. Average length of tracheids less than 4.0 mm; intertracheid pitting less than 3 in a row across the width of the tracheid; average number of pits per transverse row in ray parenchyma cross fields less than 2.. 14

13. Ray parenchyma pit apertures more than 10 μ in length along the major axis; pits conspicuously taxodioid (Fig. 12-39); average length of tracheids 5.3 mm or more.
Redwood—*Sequoia sempervirens* (D. Don) Endl.

13. Ray parenchyma pit apertures less than 10 μ in length along the major axis, pits cupressoid to taxodioid (Fig. 12-40); average length of tracheids less than 5.3 mm.
Baldcypress—*Taxodium distichum* (L.) Rich.

14. Ray parenchyma pit apertures angular in outline; the largest ray contact areas along the tracheid length more than 250 μ; ray cells with both nodular end walls and heavily pitted transverse walls that appear nodular.
True firs—*Abies* spp.

14. Ray parenchyma pit apertures oval to slitlike; largest ray contact areas along the tracheid length less than 250 μ; ray cells with nodular thickening only on the end walls, if present; transverse walls smooth.............................. **15**

15. Ray contact areas along the tracheid length more than 200 μ.
Incense-cedar—*Libocedrus decurrens* Torr.

15. Ray contact areas along the grain less than 200 μ.............................. **16**

16. Angle of major axis of ray parenchyma pits more than 30° with the tracheid axis (Fig. 12-41).
Chamaecyparis spp.

16. Angle of major axis of ray parenchyma pits less than 30° with the tracheid axis.
Thuja spp.

17. Scalariform perforations in all vessel elements.............................. **18**

17. Scalariform perforations absent or found only occasionally in the small late-wood elements.............................. **24**

18. Vessel elements with spiral thickenings distributed over the entire length of the vessel elements; intervessel pitting scalariform.
Southern magnolia—*Magnolia grandiflora* L.

18. Vessel elements lacking spiral thickenings, or with spiral thickenings restricted to the ends of the vessel elements.............................. **19**

19. Scalariform perforations with more than 15 bars.............................. **20**

19. Scalariform perforations with less than 15 bars.............................. **23**

20. Spiral thickenings absent; pits alternately or spirally arranged, mostly less than 5 μ diameter (Fig. 12-147, 12-158).............................. **21**

20. Spiral thickenings restricted to the ends of the vessel elements; pits oppositely arranged, mostly considerably more than 5 μ in diameter (Figs. 12-157, 12-159).............................. **22**

21. Intervessel pits small, not crowded, pit apertures distinct, not confluent (Fig. 12-147); ligulate extensions of vessel elements rarely attenuate or mucronate.
Red alder—*Alnus rubra* Bong.

21. Intervessel pits crowded, extremely small, and generally confluent (Fig. 12-158); ligulate extensions of vessel elements generally attenuate or mucronate.
Birch—*Betula* spp.

22. Intervessel pits rectangular, not linear, 1–5 per transverse row (Fig. 12-159); ray-vessel pitting profuse over the entire ray contact area on the vessel elements; scalariform perforations never reticulate, bars less than 1 μ wide.

Tupelo—*Nyssa* spp.

22. Intervessel pits rectangular to linear, 1–3 per transverse row (Fig. 12-157); ray-vessel pitting more pronounced at the edges of the ray contact areas; scalariform perforations tending toward reticulate, bars 1 μ or more wide.

Sweetgum—*Liquidambar styraciflua* L.

23. Intervessel pits scalariform, the largest more than 30 μ wide (Fig. 12-152).

Cucumber—*Magnolia acuminata* L.

23. Intervessel pits opposite, largest pits less than 30 μ wide (Fig. 12-160).

Yellow-poplar—*Liriodendron tulipifera* L.

24. Scalariform perforations found only occasionally in the small vessels, with 1–3 thick bars **25**

24. Scalariform perforations absent.......... **28**

25. Vessel elements of two distinct size classes (wood ring-porous), the largest at least 200 μ wide.......... **26**

25. Vessel elements intergrading in size from large to small (wood diffuse-porous); the largest less than 200 μ wide.......... **27**

26. Vasicentric tracheids and fiber tracheids with conspicuous bordered pits present; late-wood vessel elements thin-walled; ray-vessel pitting uniformly oval and simple over the ray contact areas.

American chestnut—*Castanea dentata* (Marsh.) Borkh.

26. Vasicentric tracheids and fiber tracheids absent; late-wood vessel elements thick-walled; ray-vessel pitting not uniform over the ray contact areas, horizontally elongated in margins and rounded toward the central portion of the area.

Sassafras—*Sassafras albidum* (Nutt.) Nees

27. Intervessel pits predominantly oval, mostly less than 5 μ wide, not crowded (Fig. 12-149), ray-vessel pits similar in shape and size to intervessel pits and apparently lacking a border; vessel elements lacking slender ligulate extensions.

American sycamore—*Platanus occidentalis* L.

27. Intervessel pits predominantly hexagonal (angled), more than 5 μ wide, frequently crowded (Fig. 12-150); ray-vessel pits unlike the intervessel pits; vessel elements with slender ligulate extensions.

American beech—*Fagus grandifolia* Ehrh.

28. Vessel elements with spiral thickenings that may be present in all elements or restricted to the small vessel elements.......... **29**

28. Vessel elements without spiral thickenings.......... **34**

29. Vessel elements of two distinct size classes, the largest more than 170 μ wide.......... **30**

29. Vessel elements variable in size, the largest less than 170 μ wide.......... **31**

30. Vascular tracheids present;* libriform fibers medium-thick- to thin-walled; ray-vessel pitting elongated and not like intervessel pitting.
Elms—*Ulmus* spp.
Hackberry—*Celtis* spp.

30. Vascular tracheids absent; libriform fibers thin-walled; intervessel and ray-vessel pitting similar.
Catalpa—*Catalpa* spp.

31. Spiral thickenings generally more than 10 μ apart............................ **32**

31. Spiral thickenings generally less than 10 μ apart.............................. **33**

32. Ray-vessel cross fields with rows of 5–8 pits; pit apertures 7 μ long; ligulate extensions on vessel elements rare.
Basswood—*Tilia* spp.

32. Ray-vessel cross fields with rows of fewer than 5 pits; pit apertures about 4 μ; ligulate extensions on the vessel elements common (Fig. 12-141).
Black cherry—*Prunus serotina* Ehrh.

33. Spiral thickenings about 7 μ apart.
Soft maple—*Acer* spp.

33. Spiral thickenings 3 to 4 μ apart.
Hard maple—*Acer* spp.

34. Vessel elements with marked classification into large and small sizes, largest more than 170 μ wide (ring- and semi-ring-porous woods).. **35**

34. Vessel elements without marked distinction into small and large size classifications, largest less than 170 μ wide (diffuse and semi-diffuse porous woods).. **43**

35. Vasicentric tracheids with conspicuous bordered pits present............. **36**

35. Vasicentric tracheids absent.. **38**

36. Late-wood vessel elements relatively scarce, thick-walled.
Red oak—*Quercus* spp.

36. Late-wood vessel elements numerous, thin-walled...................... **37**

37. Thick-walled libriform fibers abundant; all late-wood vessel elements with simple perforations.
White oak—*Quercus* spp.

37. Fibers thin-walled; smallest late-wood vessel elements occasionally with scalariform perforations.
American chestnut—*Castanea dentata* (Marsh.) Borkh.

38. Cross fields conspicuous on vessel elements but not necessarily abundant; cross-field pit apertures large, oval to irregular in shape, and separated by a distance less than the minor axis of the pit...... **39**

38. Cross fields inconspicuous on vessel elements; ray-vessel pit apertures small, slitlike, and separated by a distance greater than the minor axis of the pit.. **40**

* This feature requires careful examination of the late-wood vessel elements since vascular tracheids are similar to them in appearance except that the end walls are imperforate.

39. Smaller vessel elements occasionally scalariform, thick-walled; vessel elements abundant.
Sassafras—*Sassafras albidum* (Nutt.) Nees

39. Smaller vessel elements thin-walled. Vessel elements often scarce.
Walnut and **Butternut**—*Juglans* spp.

40. Smaller vessel elements thin-walled; the larger cross-field and longitudinal pit apertures more than 2 μ wide.
Catalpa—*Catalpa* spp.

40. Smaller vessel elements thick-walled; all pit apertures 2 μ or less in width.. **41**

41. Vessel elements with average length less than 0.33 mm; larger-diameter vessel elements thin-walled; vasicentric tracheids occasionally present.
Ash—*Fraxinus* spp.

41. Vessel elements with average length greater than 0.33 mm; larger-diameter vessel elements thick-walled; relatively scarce.................. **42**

42. Intervessel pits more than 5 μ in diameter (Fig. 12-143), apertures not confluent; ray-vessel cross fields transversely elongated (procumbent), uncommon, 2 to 3 pits wide along the minor axis of the cross field.
Hickory—*Carya* spp.

42. Intervessel pits less than 5 μ in diameter (Fig. 12-142), apertures frequently confluent; ray-vessel cross fields in part longitudinally elongated (upright), common, often 15 cells high along the major axis of cross fields.
Common persimmon—*Diospyros virginiana* L.

43. Intervessel pitting scalariform, pits 12 to 60 μ wide (Fig. 12-152).
Cucumbertree—*Magnolia acuminata* L.

43. Intervessel pits oval to hexagonal, alternately arranged, less than 20 μ in width.. **44**

44. Cross fields inconspicuous on vessel elements, ray-vessel pit apertures small, slitlike, and separated by a distance greater than their minor axis.
American sycamore—*Platanus occidentalis* L.

44. Cross fields conspicuous on vessel elements but not necessarily abundant; ray-vessel pit apertures large, oval to irregular in shape, and separated by a distance smaller than the minor axis of the pit. **45**

45. Fiber contact areas on vessel elements with conspicuous uniseriate rows of pits; smaller late-wood vessel elements occasionally with scalariform perforations; fiber tracheids present.
American beech—*Fagus grandifolia* Ehrh.

45. Fiber contact areas on vessel elements without distinctive rows of pits; libriform fibers with inconspicuous pits.. **46**

46. Marginal portions of ray contact areas on vessel elements longitudinally elongated (upright); rays heterocellular; average length of vessel elements less than 0.5 mm.
Black willow—*Salix nigra* Marsh.

46. Marginal portions of ray contact areas on vessel elements transversely elongated (procumbent); rays homocellular; average length of vessel elements greater than 0.5 mm.
Aspen and Poplar—*Populus* spp.

SELECTED REFERENCES

1. Barefoot, A. C., R. G. Hitchings, and E. L. Ellwood: Wood Characteristics and Kraft Paper Properties of Four Selected Loblolly Pines, III, Effect of Fiber Morphology in Pulps Examined at a Constant Permanganate Number, *Tappi*, **49**(4):137–146 (1966).
2. Besley, L.: Relationship between Wood Fiber Properties and Paper Quality, Woodland Research Index 114, Pulp and Paper Research Institute of Canada, Montreal, 1959.
3. Bolam, F. (ed.): "The Formation and Structure of Paper," Tech. Sec., British Paper & Board Makers Assoc., 2 vols., London, 1962.
4. Dinwoodie, J. M.: The Relationship between Fiber Morphology and Paper Properties, A Review of Literature, *Tappi*, **48**(8):440–447 (1965).
5. ———: The Influence of Anatomic and Chemical Characteristics of Softwood Fibers on the Properties of Sulfate Pulp, *Tappi*, **49**(2):57–67 (1967).
6. Forest Biology Subcommittee No. 2, on Test and Quality Objectives: Pulpwood Properties: Response of Processing and of Paper Quality to Their Variation, *Tappi*, **43**(11):40A–60A (1960).
7. Gallay, W.: The Interdependence of Paper Properties, pp. 491–531 in F. Bolam (ed.) (ref. 3), vol. I, 1962.
8. Giertz, H. W.: Effect of Pulping Processes on Fiber Properties and Paper Structure, pp. 597–620 in F. Bolam (ed.) (ref. 3), vol. II, 1962.
9. Kellogg, R. M., and F. F. Wangaard: Influence of Fiber Strength and Sheet Properties of Hardwood Pulp, *Tappi*, **47**(6):361–367 (1964).
10. Spiegelberg, H. L.: The Effect of Hemicelluloses on the Mechanical Properties of Individual Pulp Fibers, *Tappi*, **49**(9):389–396 (1966).
11. Tamolang, F. N., and F. F. Wangaard: Relationship between Hardwood Fiber Characteristics and Pulp-sheet Properties, *Tappi*, **44**(3):201–216 (1961).
12. ———, ——— and R. M. Kellogg: Strength and Stiffness of Hardwood Fibers, *Tappi*, **50**(2):68–72 (1967).
13. Van den Akker, J. A., A. L. Lathrop, M. H. Voelker, and L. R. Dearth: Importance of Fiber Strength to Sheet Strength, *Tappi*, **41**(8):416–425 (1958).
14. Wangaard, F. F.: Research on Quality of Wood in Relation to Its Utilization as Fiber for Papermaking, *Saertrykk ar Norjk Skogindustri*, nr. 8, Oslo (1958).
15. ———: Contribution of Hardwood Fibers to the Properties of Kraft Pulps, *Tappi*, **45**(7): 548–556 (1962).

Glossary

adult wood: wood produced by cambial cells which have reached a period of maximal dimensions; characterized by relatively constant cell size, well-developed structural pattern, and stable physical behavior; also called *outer* or *mature wood;* see **juvenile wood.**

aggregate ray: composite structure, consisting of a number of small rays, fibers, and sometimes also vessels, which to the unaided eye or at low magnification appears as a single broad ray.

air-dry moisture content: the equilibrium moisture content of wood for conditions outdoors but undercover; the standard moisture content for this condition is taken at 12 percent.

aliform parenchyma: type of paratracheal parenchyma that extends out from the flanks of a pore, forming an eyelet with it.

alpha cellulose: the fraction of the total carbohydrate component of wood substance that is empirically defined as insoluble in 17.5 percent sodium hydroxide at 20°C.

alternate pitting: type of pitting in which bordered pits are arranged in diagonal rows across the cell; when crowded, the pits become polygonal in surface view.

amorphous regions: portions of the cell walls of plants in which the carbohydrate chains are not parallel or arranged in a crystal lattice; see also **crystallite.**

angstrom unit: one hundred-millionth of a centimeter; 10^{-8} cm.

anisotropic: exhibiting different properties when tested along axes in different directions.

annual growth: layer of wood laid down during a given year; same as annual or seasonal increment; see also **growth ring.**

annual increment: see **annual growth.**

annual ring: annual increment of wood as it appears on a transverse surface or in a transverse section; same as **growth ring.**

anticlinal cell division: division of the fusiform cambial initials in the radial-longitudinal plane, forming new xylem and phloem initials; also called **pseudotransverse cell division;** see **periclinal cell division.**

apical growing point: meristematic tissue at the apices of the tree, responsible for elongation of stems and roots.

apotracheal diffuse-in-aggregates parenchyma: apotracheal parenchyma cells that tend to be grouped in short tangential lines from ray to ray, as seen in cross section; same as *diffuse zonate* parenchyma.

apotracheal-diffuse parenchyma: single apotracheal parenchyma strands or cells distributed irregularly among fibers or tracheids, as seen in cross section; same as **metatracheal-diffuse** parenchyma.

apotracheal parenchyma: axial parenchyma independent of the pores or vessels; includes *marginal, diffuse, diffuse-in-aggregate,* and *banded* apotracheal parenchyma. Formerly known as *metatracheal parenchyma.*

apposition: the increase in thickness of the cell wall caused by the addition of cell wall material to the surface of the developing wall.

aspirated bordered pit pair: a bordered pit pair in which the torus is displaced against one or the other of the pit borders so that the aperture is blocked.

attenuate vessel-element extension: gradually tapering extension on one or both ends of a vessel element; see also **ligulate extension.**

axial parenchyma: parenchyma cells derived from fusiform cambial initials; also known as *longitudinal parenchyma.*

axial strand parenchyma: cells of axial parenchyma arranged in a row along the grain; such a row is formed through further (postcambial) division of a single axial (longitudinal) cell cut off from a fusiform initial in the cambium; same as *longitudinal strand parenchyma.*

banded parenchyma: axial parenchyma forming concentric lines or bands, as seen in cross section; termed *apotracheal banded,* if independent of the pores, and *paratracheal banded,* if definitely associated with the pores.

bark: the tissues in the cylindrical axis of a tree outside of the cambium; bark is composed of inner living bark and outer dead brown bark.

bark pockets: small patches of bark embedded in wood.

bars: remnants of the perforation plate between the openings in scalariform perforations.

bars of Sanio: same as *crassulae.*

bast fibers: fibers of the secondary phloem.

bending strength: the resistance to stress applied transversely in such a manner as to cause curvature of the member axis.

bird's-eye figure: figure on the plain-sawn and rotary-cut surface of wood exhibiting numerous rounded areas resembling a bird's eye; caused by local fiber distortions; most common in hard maple.

biseriate ray: ray consisting of two rows of cells, as viewed in the tangential section.

black check: bark pockets containing a certain amount of resin; common in western hemlock; also called *black streak.*

black streak: black streaks in western hemlock [*Tsuga heterophylla* (Raf.) Sarg.] caused by the maggots of a fly.
blind pit: pit leading into an intercellular space between two cells, i.e., a pit without a complementary pit in an adjacent cell.
blister figure: figure on smooth plain-sawn and rotary-cut surfaces that appears to consist of small, more or less widely spaced, elevated, or depressed areas of rounded contour.
bordered pit: pit with an overhanging margin, i.e., a pit in which the cavity becomes abruptly constricted during the thickening of the secondary wall (see Fig. 3-11).
bordered pit pair: two complementary bordered pits in adjacent cells.
boreholes: minute openings in cell walls, through which the hyphae of the fungus pass.
bound water: water in the wood that is associated with the cell wall material.
bowing: longitudinal warping, flatwise, from a straight line drawn end to end of the piece.
box-heart split: a split originating in the wood surrounding the pith during initial stages of drying; caused by stresses set up because of the differences in tangential and radial shrinkage of the wood near the pith.
brashness: a condition in wood which causes a sudden and complete break with very little splintering, usually under small loads and deformations.
brittle heart: a defect in hardwoods resulting from the presence of fibers with localized wrinkles (slip planes) that cause reduction in strength of the wood.
broken stripe: figure formed by tapering stripes, 1 ft or more in length, on the quartered surface of an interlocked grained wood.
brown rots: a type of wood-destroying fungi that decompose cellulose and the associated pentosans, leaving the lignin in a more or less unaltered state; the resultant mass of decayed wood is of a powdery consistency, of varying shades of brown.
burl: bulge or excrescence that forms on the trunk and branches of a tree.
cambial fusiform initial: cambial initial that, through repeated division, gives rise to a radially directed row of longitudinal elements of xylem and phloem; such a cell is fusiform in shape; same as longitudinal cambial initial; see **cambial ray initial.**
cambial initial: individual cell in the cambium; see also **cambial fusiform initial** and **cambial ray initial.**
cambial ray: that portion of a ray included in the cambium.
cambial ray initial: cambial initial that gives rise to ray cells through repeated division; unless the ray consists of a single row of cells (*t*), cambial ray initials are grouped in ray areas; the cells are often isodiametric, as seen in a tangential section.
cambial region: a term of convenience for the layer of varying width composed of cambial initials and their undifferentiated derivatives.
cambial zone: see **cambial region.**
cambium: growing (generative) layer between the xylem and phloem; a lateral meristem responsible for formation of xylem and phloem.
casehardening: a term applied to dry lumber with nearly uniform moisture content but characterized by the presence of residual stresses, tension in the interior of the piece (core), and compression in the outer layers of cells (shell).
cell wall: the wall that encloses the cell contents; in a mature cell it is compound; i.e., it consists of several layers.

cellulose: principal chemical constituent of the cell walls of the higher plants; complex carbohydrate occurring in the form of polymer chains and having the empirical formula $(C_6H_{10}O_5)n$; see also **alpha cellulose, holocellulose, true cellulose.**

checks: ruptures along the grain that develop during seasoning either because of a difference in radial and tangential shrinkage or because of uneven shrinkage of the tissue in adjacent portions of the wood.

chemical stain: stain that is caused by chemical changes in the materials present in the lumina of xylary cells.

chipped grain: wood surface from which pieces of wood have been scooped out or chipped by the action of cutting tools; also called *torn grain.*

closing membrane: see **pit membrane.**

coalescent pit apertures: slitlike inner-pit apertures of several pits, united to form spiral grooves.

coefficient of thermal expansion: the magnitude of change in a dimension per degree centigrade change in temperature; in wood there are three coefficients to correspond to the orthotropic axes, αL, αR, αT.

collapse: defect that sometimes develops *above the fiber saturation point* when very wet heartwood of certain species is dried; evidenced by abnormal and irregular shrinkage.

comb grain: see **edge grain.**

companion cells: special type of parenchyma cells associated structurally and functionally with sieve tubes in all angiosperms; lacking in pteridophytes and gymnosperms.

compound middle lamella: a term of convenience for the compound layer between the secondary walls of contiguous cells; this layer consists of intercellular substance (the true middle lamella) and the primary walls on each side of it.

compound ray: obsolete term formerly applied to the unusually large wood ray (accompanied by small rays) that is found in certain species of *Quercus,* in *Fagus grandifolia* Ehrh., etc.

compression failures: localized buckling of fibers and other longitudinal elements produced by the compression of wood along the grain beyond its proportional limit; compression failures sometimes develop in standing trees.

compression wood: wood formed on the lower side of branches and of curved stems of conifers; this tissue has abnormally high longitudinal shrinkage and physical properties that differ from those of normal wood; see also **reaction wood** and **tension wood.**

confluent parenchyma: see **paratracheal confluent parenchyma**.

coniferous wood: wood produced by coniferous trees; same as *softwood* or *nonporous wood.*

cork cambium: see **phellogen.**

cortex: that portion of the primary axis of a vascular plant that immediately surrounds the central cylinder (stele); on the outside, it is enveloped by a uniseriate layer—the epidermis (stems) or the exodermis (roots).

crassulae: thicker, generally arching portions of the intercellular layer above and below primary pit fields; formerly called *bars of Sanio.*

creep in wood: the deformation produced in wood under a constant load applied over a period of time.

crooking: longitudinal warping, edgewise, from a straight line drawn from end to end of the piece.

cross field: a term of convenience for the rectangle formed by the walls of a ray cell and a longitudinal tracheid, as seen in the radial section. Principal application is to conifers. Formerly known as *ray crossing.*

cross grain: in standing trees, grain in which the fiber alignment deviates from the axis of the stem; in wooden members, grain in which the fiber alignment deviates from the direction parallel to the long axis of the piece; see also **spiral grain.**

cross section: section cut at right angles to the grain; same as *transverse section.*

crotch: segment of a stem that forks.

crown-formed wood: wood formed in the part of the stem lying within the crown region of the tree.

crystalliferous: bearing crystals.

crystallites: regions in the cell wall of plants in which the cellulose is arranged in a highly ordered, crystal lattice of parallel chains; these regions are of limited size and are separated by regions of little crystalline order; see also **amorphous regions.**

cupping: warping of the face of a plank or board so that it assumes a trough-like shape, the edges remaining approximately parallel to each other.

cupressoid pit: a cross-field pit in early wood of conifers with an ovoid, included aperture that is rather narrower than the lateral space on either side between the aperture and the border, as in *Chamaecyparis, Libocedrus,* and *Juniperus.*

curly grain: grain that results from more or less abrupt and repeated right and left deviations from the vertical in fiber alignment; the radial split faces of such wood are corrugated, the split tangential faces smooth (see Fig. 11-8) or corrugated.

cutin: the layer of waxy, waterproofing material overlying the epidermis in plants.

defect: any abnormality or irregularity that lowers the commercial value of wood by decreasing its strength or affecting adversely its working or finishing qualities or its appearance.

deliquescent growth: growth in which the trunk divides rather abruptly into limbs.

density of wood: mass of wood per unit of volume, see also **weight density.**

dentate: toothed; with toothlike projections.

dermatogen: the region of incompletely differentiated tissue between the apical promeristem and the epidermis (stem) or exodermis (root) in the cylindrical axis of a vascular plant.

diagonal grain: type of cross grain resulting from failure to saw parallel to the growth increments.

diamonding: uneven shrinkage that causes "squares" to become diamond-shaped on drying; usually develops in pieces in which the growth increments extend diagonally, so that the faces of the piece are neither flat nor edge-grained.

dielectric constant (ϵ): a measure of the alternating-current-conducting capacity of a material.

diffuse parenchyma: apotracheal parenchyma the cells of which are scattered in the growth ring.

diffuse-porous wood: porous wood in which the pores exhibit little or no variation in size indicative of seasonal growth; see **ring-porous wood.**

discontinuous growth ring: a growth ring that is formed on only one side of the stem.

double ring: growth ring that appears to consist of two rings, one of which is a false ring; see **false ring.**

dry rot: a special type of brown rot, causing widespread damage in buildings; in the United States the causal organism is *Poria incrassata* (B. & C.) Curt.

early wood: that portion of a growth increment which is produced at the beginning of the growing season; *springwood;* see **late wood.**

edge grain: figure in lumber which has been sawn so that the face of the board is the radial plane of the log; commercially lumber is considered edge-grained when the angle between the surface and the annual rings lies between 45 and 90° with the wide surface of the piece; synonymous with *vertical grain, rift grain,* and *quarter-sawn.*

electrical conductivity: the reciprocal of resistivity.

elementary fibril: the smallest structural organization of cellulose molecules in the woody cell wall; a bundle of molecules with an aggregate diameter of approximately 35 Å.

encased knot: that portion of a branch which becomes embedded in the bole of a tree after the branch dies; also called *loose knot.*

end checks: seasoning checks that develop on the ends of a piece of wood.

end wall: the oblique or transverse wall between two cells, i.e., the wall normal to the longer cell axis.

endodermis: the innermost layer of cortex, one cell thick, without intercellular spaces and consisting of cells with suberized or cutinized walls.

enzymes: complex protein molecules produced by living cells; they act as organic catalysts making it possible for the biochemical reactions necessary for physiological processes to take place.

epidermis: the outermost, generally uniseriate, layer of primary tissue that is continuous over the younger portions of the aerial part of a plant except where interrupted by stomatal openings; in woody plants, the epidermis ceases to function after a periderm forms beneath it, and is subsequently cast.

epithelial: of the nature of, or pertaining to, epithelium; see **epithelium.**

epithelium: excreting parenchymatous tissue surrounding the cavity of resin and gum canals.

equilibrium moisture content: the balance of moisture content attained by wood at any given level of relative humidity and temperature of the surrounding atmosphere; abbreviated to E.M.C.

excurrent growth: growth in which the axis is prolonged, forming an undivided main trunk, as in pine.

exothermic reaction temperature of wood: the temperature level in heating wood at which the reaction itself begins to develop heat; usually accepted as approximately 273°C.

extended pit aperture: an inner pit aperture whose outline, in surface view, extends beyond the outline of the pit border; see **included pit aperture.**

face veneer: veneer that is used for exposed surfaces in plywood.

false heartwood: pathological heartwood formed in species that do not possess normal heartwood (on the basis of color).

false ring: band of what appears to be late wood followed outwardly by tissue resembling early wood, which in turn is followed by true late wood, wholly included within the boundaries of a true ring; see **double ring.**

feather crotch: figure with a design resembling a cluster of feathers, found in crotch veneer.

fenestriform pit: a large, window-like pit in the cross field of ray parenchyma cells in soft pines; also in *Pinus resinosa* and *Pinus sylvestris*.

fiber: an elongated cell with pointed ends and a thick or not infrequently a thin wall;

includes (1) *fiber tracheids* with bordered pits and (2) *libriform fibers* with simple pits.

fiber saturation point: point when all water is evaporated from the cell cavities, but the cell walls are still fully saturated with moisture.

fiber tracheid: a fiber-like cell with pointed ends and bordered pits having lenticular to slitlike apertures. This term is generally applied to the fibers with bordered pits in woody angiosperms, but it is also applicable to the late-wood tracheids of gymnosperms.

fibrils: an obsolete term referring to threadlike structures visible in the cell wall under a light microscope.

figure: in a broad sense, any design or distinctive markings on the longitudinal surfaces of wood; in a restricted sense, such decorative designs in wood as are prized in the furniture and cabinetmaking industries.

flat grain: figure in lumber which has been sawn so that the wide face of the board is approximately perpendicular (tangent) to the radius of the log; commercially lumber is considered to be flat-grained if the wide surface of the board is less than 45° from a tangent to the annual rings; synonymous with *slash grain;* see also **plain-sawn** and **flat-sawn.**

flat-sawn: said of wood so sawed that the tangential face of the wood is exposed on the surfaces of boards; same as *plain-sawn.*

floccosoids: white spots that frequently are present in the wood of western hemlock [*Tsuga heterophylla* (Raf.) Sarg.].

free moisture: water in wood that is found in cell cavities, either in liquid form or as a gas.

frost cracks: radial, longitudinal splits near the base of a tree, formed during extremely cold weather.

frost ring: brownish line extending circumferentially within a growth ring, consisting of collapsed cells and abnormal zones of parenchyma cells; traceable to injury of the cambium or of young, unlignified wood cells by either early or late frost.

fusiform cambial initial: cambial initial that, through repeated division, gives rise to a radially directed row of longitudinal elements of xylem and phloem; such a cell is fusiform in shape; same as *longitudinal cambial initial;* see **cambial-ray initial.**

fusiform parenchyma cell: parenchymatous cell that arises from a longitudinal cambial initial without subdivision; i.e., it has the shape of a short fiber.

fusiform ray: spindle-shaped ray, as viewed in a tangential section of wood, containing a resin canal.

gelatinous fiber: a fiber having a more or less unlignified inner layer in the secondary wall; generally associated with tension wood, but found also in the normal wood of the angiosperms.

grain of wood: arrangement and direction of alignment of wood elements when considered *en masse.*

growth ring: ring of wood on a transverse surface or in a transverse section, resulting from periodic growth; if but one growth ring is formed during a year it is called an *annual ring.*

growth stresses: internal stresses in the wood of living trees that are caused by cell growth, these stresses lead to warping and twisting of boards as they are relieved when sawing the log.

grub holes: oval, circular, or irregular holes (⅜ to 1 in. in diameter) in wood, caused by larvae and adult insects.

half-bordered pit-pair: a pairing of a bordered pit and a simple pit in adjacent cells.

hardwood: wood produced by broad-leaved trees such as oak, elm, and ash; same as *porous wood.*

heart checks: wood separation formed in the radial plane; also called *heart shakes* and *rift cracks;* see **ring shake.**

heart shake: separation of wood across the ring and generally following the rays; also called *heart check* and *rift crack.*

heartwood: dead inner core of a woody stem (or a log), generally distinguishable from the outer portion (sapwood) by its darker color; see **sapwood.**

heat of combustion of wood (H): the amount of heat in British thermal units (Btu) developed in burning a pound of wood; averages about 8400 Btu/lb of dry wood.

hemicellulose: group of carbohydrates found in the cell wall in more or less intimate association with cellulose; sometimes defined as those less-resistant substances in the cell wall which though insoluble in hot water can be removed with either hot or cold dilute alkalies or readily hydrolyzed into sugars and constituent acids by means of hot dilute acids.

herringbone figure: figure that results when two quarter-sawn pieces are matched so that the rays meet at an angle.

heterocellular ray: in the hardwoods, a ray consisting of two kinds of cells, *procumbant* and *upright;* in the softwoods, a ray consisting of *ray parenchyma* and *ray tracheids;* same as *heterogeneous ray.*

heterogeneous ray: see **heterocellular ray.**

holocellulose: the total carbohydrate fraction of wood remaining after the removal of lignin and substances extractable with solvents; see also **cellulose, alpha cellulose,** and **true cellulose.**

homocellular ray: a ray consisting entirely of one kind of cells; same as *homogeneous rays.*

homogeneous ray: see **homocellular ray.**

honeycombing: internal splitting in wood that develops in drying; caused by internal stresses or by closing of surface checks.

hydrolysis: a chemical process of decomposition involving addition of the elements of water; the presence of dilute acids, enzymes, or other agents may be needed to induce the reaction.

hypha: one of the threadlike elements of the mycelium of a fungus.

hypodermis: the tissue immediately underlying the epidermis, especially if this is different structurally from tissues deeper in the plant.

incipient decay: initial stage of decay.

included pit aperture: an inner pit aperture whose outline, in surface view, is included within the outline of the pit border; see **extended pit aperture.**

included sapwood: streaks or irregularly shaped areas of light-colored wood with the general appearance of normal sapwood, found embedded in the darker-colored heartwood; common in western redcedar.

indenture: a term applied to pitlike depressions in the coniferous woods, found in the corners of the ray cells where horizontal and vertical walls meet.

initial parenchyma: see **marginal apotracheal parenchyma.**

inner pit aperture: opening of the pit canal into the cell lumen.

intercellular canal: see **resin canal.**

intercellular layer: the layer of isotropic substance between cells; this layer is largely lignin and lacks cellulose; see **true middle lamella.**

intercellular space: a space between cells, generally at the corners where several cells are joined together, as in *Juniperus* spp.
intergrown knot: that portion of a branch which is embedded in the tree trunk while this branch is alive; also called *tight knot.*
interior stain: a form of sap stain that is confined to the interior of an infected piece and is not evident on the surface; caused by the same fungi as the ordinary sap stain.
interlocked grain: a condition produced in wood by the alternate orientation of fibers in successive layers of growth increments; the quarter-sawn face of such wood produces a ribbon figure.
intervascular pitting: pitting between adjacent vessel elements; also called *intervessel pitting.*
intervessel pitting: see **intervascular pitting.**
intrusive growth: elongation of a growing cell which forces the cell tip between adjacent cells.
intussusception: the introduction of new wall material into the existing net of the primary wall to cause an increase in surface area.
isotropic: having identical properties in all directions.
juvenile wood: wood formed near the pith, characterized by progressive increase in dimensions and changes in the cell characteristics, and in the pattern of cell arrangement; also called *core wood;* see **adult wood.**
kiln brown stain: chemical stain that develops during kiln drying; see **chemical stain** and **yard brown stain.**
knot: branch base that is embedded in the wood of a tree trunk or of a larger branch.
lamella: a thin sheet of microfibrils; a sublayer in the cellulose aggregation of the secondary cell wall.
late wood: that portion of an annual increment which is produced during the latter part of the growing season (during the summer); *summerwood;* see **early wood.**
latticed perforation plate: perforation plate with multiple perforations elongated and parallel, with barlike remnants of the plate between the openings; see **scalariform perforation plate.**
libriform fiber: an elongated, commonly thick-walled cell with simple pits, usually distinctly longer than the cambial fusiform initial; found in the woody angiosperms.
lignification: the process of deposition of the chemical constituent of the woody plant cell wall, known as *lignin.*
lignin: one of the principal constituents of woody cell walls, whose exact chemical composition is still unknown; the residue after treatment of solvent-extracted wood with strong mineral acids.
ligulate vessel-element extensions: tail-like extensions at one or both ends of a vessel element; *mucronate,* if extensions terminate abruptly; *attenuate,* if tapering gradually.
linear pit: a pit with an aperture that is long, narrow, and of more or less uniform breadth, as seen in surface view.
longitudinal parenchyma: see **axial parenchyma.**
longitudinal resin canal: resin canal extending with the grain, appearing as an opening or fleck on the transverse surface with the naked eye or hand lens.
longitudinal strand parenchyma: see **axial strand parenchyma.**

loose knot: that portion of any branch which is incorporated into the bole of a tree after the death of that branch.

loosened grain: loosened small portions of the wood on the flat-grained surfaces of boards, usually of the tops and edges of the growth increments.

lumen: the cavity of a cell (pl. *lumina*).

machine burns: brown stains caused by excessive friction developed when progress of the piece is impeded in a woodcutting operation.

marginal apotracheal parenchyma: apotracheal parenchyma, the cells of which occur singly or form a more or less continuous layer of variable width at the close of a season's growth, in which case it may also be called *terminal* parenchyma, or at the beginning of a season's growth, when it may also be termed *initial* parenchyma.

marginal cell: cell on the upper or the lower margin of a wood ray, as viewed in the tangential or radial section.

margo: outer ring of the pit membrane lying between the torus and the pit border in bordered pits of conifers.

marine borers: mollusks and crustaceans that attack submerged wood in salt and brackish water.

maximum moisture content: the maximum amount of water which can be contained in the combined spaces in the cell wall, the lumens, and intercellular spaces.

meristem: tissue consisting of living, thin-walled cells that are capable of repeated division.

metatracheal-diffuse parenchyma: see **apotracheal-diffuse parenchyma.**

metatracheal parenchyma: see **apotracheal parenchyma.**

metaxylem: the primary xylem formed *after* the differentiation of the protoxylem; when completely differentiated, primary xylem consists of protoxylem and metaxylem.

micelles: an obsolete term, formerly applied to discrete ultramicroscopic particles in the cell walls of the higher plants, as postulated by Nageli.

microfibril: a bundle of cellulose polymer chains and associated polysaccharides of other types that are united at some regions in highly ordered crystalline lattices known as *crystallites* and are less highly ordered in the zones between the crystallites (*amorphous regions*); it is the smallest natural unit of cell wall structure that can be distinguished with an electron microscope.

micrometer: a unit of length, one-millionth of a meter; equivalent to the better-known unit *the micron.*

micron: a unit of length one-millionth of a meter, or one-thousandth of a millimeter; equivalent to *the micrometer.*

microscopic checks: minute checks in wood between the fibrils in the secondary walls that cannot be detected without a compound microscope.

middle lamella: see **true middle lamella** and **compound middle lamella.**

mineral stain: olive and greenish-black streaks believed to designate areas of abnormal concentration of mineral matter; common in hard maple, hickory, and basswood; also called *mineral streak.*

mineral streak: see **mineral stain.**

modulus of elasticity in bending (E): the modulus of elasticity calculated from bending tests.

modulus of rupture (R): the maximum bending load to failure in pounds per square inch.

moisture content of wood: the weight of the moisture in wood, expressed as a percentage of its ovendry weight.

monopodial growth: same as *excurrent growth.*

moonshine crotch: swirling figure found in crotch veneer.

mottled figure: a broken stripe figure interrupted by irregular, horizontal waves in the grain.

multiple perforation: a perforated end wall in a vessel element consisting of two or more openings in a perforation plate; see also **simple perforation** and **scalariform perforation plate.**

multiple ring: growth ring that contains within its boundaries several false rings; see **false ring.**

multiseriate ray: ray consisting of several to many rows of cells, as viewed in the tangential section.

mycelium: the mass of interwoven threadlike filaments forming the vegetative part of a fungal plant; see **hypha.**

nodular end walls: the end wall of a parenchyma cell with a beaded appearance in sectional view.

nonporous wood: wood devoid of pores (vessels); same as *softwood* or *coniferous wood.*

opposite pitting: type of pitting in which bordered pits are arranged in transverse rows extending across the cell; when crowded, the outlines of the pits become rectangular in surface view.

orifice: mouth or opening, as of a tube or pit; opening; hole.

outer pit aperture: opening of the pit canal into the pit chamber.

ovendry weight: weight of wood obtained by drying in an oven at 105°C until there is no further weight loss.

oxidative stain: nonpathological stain in sapwood caused by chemical changes in the materials contained in the cells of the wood.

packing density: a term used by some authors to denote the actual cell wall density.

packing fraction: the ratio of the actual cell wall density to the density of solid cell wall substance.

paratracheal aliform parenchyma: see **aliform parenchyma.**

paratracheal confluent parenchyma: coalescent aliform parenchyma forming irregular tangential or diagonal bands, as seen in cross section.

paratracheal parenchyma: parenchyma the cells of which are obviously associated with the pores (vessels).

paratracheal vasicentric parenchyma: paratracheal parenchyma forming a complete sheath around a vessel.

parenchyma: tissue consisting of short, relatively thin-walled cells, generally with simple pits; concerned primarily with storage and distribution of carbohydrates; used specifically as a synonym for *axial parenchyma* which occurs in strands along the grain; may be visible with a hand lens on the transverse surface of wood as dots, as sheaths about pores, or as broken or continuous lines or bands.

peckiness: see **pecky dry rot.**

pecky dry rot: characterized by finger-sized pockets of decay in the living trees of incense-cedar (caused by *Polyporus amarus* Hedg.) and baldcypress (caused by *Stereum* sp.); it is also known as *peckiness* and *pocket dry rot.*

perforation: the opening (or openings) between two vessel members.

perforation plate: a term of convenience for the area of the wall (originally imperforate) involved in the coalescence of two elements of a vessel.

perforation rim: the remnant of a perforation plate forming a border about a simple perforation.

periblem: the region of incompletely differentiated tissue located between the apical promeristem and the cortex in the cylindrical axis of a vascular plant.

periclinal cell division: division of the cambial fusiform initials in the tangential-longitudinal plane, resulting in the formation of new xylem and phloem cells in the same radial file.

pericycle: outermost layer of the stele consisting normally of parenchyma but sometimes also of fibers.

periderm: a protective layer that forms in the epidermis, just beneath it, or in deeper-lying tissues, after which the epidermis ceases to function.

permanent set: nonrecoverable permanent deformation.

phellem: outermost layer of periderm, composed of cork cells formed to the outside by the phellogen; see **phellogen.**

phelloderm: innermost layer of periderm composed of cells formed to the inside by the phellogen; see **phellogen.**

phellogen: median layer of periderm from which the phellem and the phelloderm originate by cell division; cork cambium; see **phellem** and **phelloderm.**

phloem: inner bark; principal tissue concerned with the distribution of elaborated foodstuffs, characterized by the presence of sieve tubes.

phloem ray: that portion of a ray included in the phloem.

piceoid pit: a cross-field pit in early wood of conifers with a narrow and often slightly extended aperture, as in *Picea.*

pigment figure: figure in wood occasioned by irregular infiltration, resulting in dark lines, bands, zones, streaks, etc.

pinholes: small, round holes (1/100 to ¼ in. in diameter) in wood that result from the mining of ambrosia and similar beetles.

pinoid pit: a term of convenience for the smaller types of early-wood cross-field pits found in several species of *Pinus;* characteristically these are simple or with narrow borders and often variable in size and shape, but exclude the *window-like, fenestriform* pits in the ray parenchyma cross fields of soft pines and *Pinus resinosa* and *Pinus sylvestris.*

pit: a recess in the secondary wall of a cell, together with its external closing membrane; open internally to the lumen.

pit aperture: opening of a pit into a cell lumen or into a pit chamber; see **inner** and **outer pit aperture.**

pit canal: passage from the cell lumen to the chamber in bordered pits.

pit cavity: entire space within a pit from the membrane (middle lamella) to the lumen.

pit chamber: in a bordered pit, the space between the pit membrane (middle lamella) and the overhanging border.

pit field: areas on the radial walls of longitudinal coniferous tracheids bounded above and below by crassulae, containing one or more pits or devoid of pits.

pit membrane: that portion of the compound middle lamella which closes a pit cavity externally.

pit orifice: opening or mouth of a pit; same as *pit aperture.*

pit pair: two complementary pits of adjacent cells.
pitch pocket: lens-shaped opening in the grain at the common boundary of two growth increments, or sometimes within a growth increment, empty, or containing solid or liquid resin; found in certain coniferous woods.
pitch streaks: localized accumulation of resin, which permeates the cells, forming resin-soaked patches or streaks in coniferous woods.
pith: primary tissue in the form of a central parenchymatous cylinder found in stems and sometimes in roots.
pith flecks: small areas of wound tissue darker or lighter than the surrounding tissue, produced in wood through injury to the cambium by the larvae of flies of the genus *Agromyza* and subsequent occlusion of the resulting tunnels with parenchymatous cells.
plain-sawn: same as *flat-sawn;* also see **flat grain.**
plasmalemma: the outer, limiting membrane of the cytoplasm of the cell.
plasmodesmata: strands of cytoplasm passing through the primary walls and middle lamella of adjacent cells, and joining the cytoplasm in an interconnected system.
pocket dry rot: see **pecky dry rot.**
pore: cross section of a vessel; a vessel as it appears on a transverse surface or in a transverse section of wood.
pore chain: several to many pores arranged in a radial line or series, the adjacent pores retaining their separate identities.
pore cluster: nested pores or an irregular aggregation of pores.
pore multiple: group of two or more pores contiguous radially and flattened along the lines of contact so as to appear as subdivisions of a single pore.
porous wood: wood containing pores (vessels); same as *hardwood,* i.e., wood produced by broad-leaved trees.
powder-post damage: small holes (1/16 to 1/12 in. in diameter) filled with dry, pulverized wood, resulting from the work of beetles (largely *Lyctus*) in seasoned and unseasoned wood.
primary growth: elongation of the main and secondary axes, in both stems and roots, traceable to the activities of apical growing points.
primary phloem: first-formed phloem derived from an apical meristem.
primary pit field: an area on the cell wall in which the wall is thinner and within the boundary of which one or more pits usually develop; the true middle lamella is thinner in a pit field.
primary tissue: tissue arising from the activities of apical growing points.
primary wall: initial layer of the cell wall; formed during or following cell division and later modified during the postcambial differentiation of the cell; see also **compound middle lamella.**
primary xylem: first-formed xylem which originates from an apical meristem.
procambium: tissue differentiating to the rear of the plerome that is the precursor of the primary vascular tissue around the pith; the procambium may be a continuous layer or may consist of longitudinal strands.
procumbent ray cell: narrow cell elongated in the direction of the ray, of the type that composes *homocellular rays* and the body of *heterocellular rays.*
promeristem: meristem in a region of a growing plant body in which the formation of new organs or parts of organs is in progress.

proportional limit: the maximum load marking the limit of direct proportionality between the applied load and deformation produced, when testing a material mechanically.

prosenchyma: cells whose functions are mainly conductive and mechanical, equipped with bordered pits or sometimes with simple pits (libriform fibers); see **parenchyma.**

protective layer: a special cell wall layer that is formed in parenchyma cells preceding tylosis formation.

protolignin: lignin in the chemical form it has when in place in the cell wall.

protophloem: first-formed primary phloem.

protoxylem: first-formed primary xylem, with tracheary elements characterized by annular or spiral thickenings.

pseudotransverse cell division: division of the fusiform cambial initials in the radial-longitudinal plane, forming new xylem and phloem initials; also called *anticlinal cell division;* see **periclinal cell division.**

quarter sawn: the wide face of the board is the radial face of the log; same as *edge grained.*

quarter section: section cut along the grain parallel to the wood rays.

quarter surface: surface that is exposed when a log is cut along the grain in a radial direction (parallel to the wood rays).

quartered veneer: veneer produced by radial slicing or sawing.

quilted figure: blister-like figure found sometimes in bigleaf maple (*Acer macrophyllum* Pursh).

(*r*): symbol indicating a radial section or surface; see **radial section.**

radial section: section cut along the grain parallel to the wood rays and usually at right angles to the growth rings; see also **tangential section.**

raised grain: roughened condition of the surface of dressed lumber on which the hard late wood is raised above the softer early wood but is not torn loose from it.

ramiform pits: simple pits with coalescent canal-like pit cavities, as in stone cells.

ray: ribbon-shaped strand of tissue extending in a radial direction across the grain, so oriented that the face of the ribbon is exposed as a fleck on the quarter surface; see **wood ray.**

ray crossing: same as *cross field.*

ray fleck: portion of a ray as it appears on the quarter surface.

ray parenchyma: parenchyma included in rays, in contrast to *longitudinal parenchyma* which extends along the grain; see **ray tracheid.**

ray parenchyma cross field: the rectangular area of common wall between a ray parenchyma cell and a longitudinal tracheid; see **cross field.**

ray tracheid: cell with bordered pits and devoid of living contents, found in the wood rays of certain *softwoods*.

ray tracheid cross field: the rectangular or irregular area of common wall between a ray tracheid cell and a longitudinal tracheid; see **cross field.**

reaction wood: wood with distinctive anatomical and physical characteristics, formed in parts of leaning or crooked stems and in branches. In dicotyledons reaction wood is known as *tension wood,* and in gymnosperms as *compression wood.*

resilience: a measure of the recoverable energy in the elastic deformation of wood.

resin canal: tubular, intercellular space sheathed by secreting cells (epithelium), bearing resin in the sapwood.
resinous tracheid: tracheid containing lumps or amorphous deposits of reddish-brown or black resinous materials.
resistivity: the resistance offered by a cubic centimeter of a material to a steady, direct-current flow of electricity between opposite faces; expressed in ohm-centimeters; also known as *specific resistance.*
reticulate scalariform perforation plate: a multiple perforation plate, with netlike appearance.
reverse casehardening: a condition that develops in lumber as a result of oversteaming in the final stages of drying, characterized by a reversal of stresses, with compression in the interior of the piece and tension in the outer layers; see **casehardening.**
ribbon figure: figure consisting of changeable (with light) darker and lighter bands, obtained by quarter-sewing or slicing interlocked grain wood; also called *stripe.*
rift crack: see **heart shake.**
rift grain: see **edge grain.**
ring-porous wood: porous wood in which the pores formed at the beginning of the growing season (in the early wood) are much larger than those farther out in the ring, particularly if the transition from one to the other type is more or less abrupt; see **diffuse-porous wood.**
ring shake: rupture in wood that occurs between increments or less frequently within an annual growth layer; sometimes called *wind shake.*
ripewood trees: trees in which the wood of mature stems is of uniform color but which has an inner core of functionally differentiated heartwood with lower permeability than the sapwood at the outside of the stem; see also **sapwood trees.**
ripple marks: striations across the grain on the tangential surface of a wood, occasioned by storied rays or by these and other storied elements.
roe figure: figure formed by short stripes (less than 1 ft in length) on the quartered surface of an interlocked-grained wood.
rotary-cut veneer: veneer obtained by rotating a log against a cutting knife in such a way that a continuous sheet of veneer is unrolled spirally from the log.
S layers: the three layers of the secondary wall, designated as $S1$ or the outer layer, $S2$ the central layer, and $S3$ the inner layer; the last is called the *tertiary wall* by some authors.
sap stains: stains in the sapwood caused by wood-staining fungi or by the oxidation of compounds present in the lumina of living cells.
sapwood: outer (younger) portion of a woody stem (or a log), usually distinguishable from the core (heartwood) by its lighter color; see **heartwood.**
sapwood trees: trees in which the wood of mature stems is of uniform color and appears to be entirely sapwood.
scalariform perforation plate: a perforation plate with multiple elongated and parallel perforations, with barlike remnants of the plate between the openings; see also **simple perforation, multiple perforation.**
scalariform pitting: type of pitting in which bordered linear pits are arranged in a ladder-like series.

seasonal increment: layer of wood laid down during a given year; see **annual growth.**

secondary cell wall: that portion of the cell wall formed after the cell enlargement has been completed; see **primary wall.**

secondary growth: growth traceable to the activities of a lateral cambium; also called *secondary thickening.*

secondary periderms: periderms that form subsequently under the first periderm in trees, resulting in rough bark; see **periderm.**

secondary phloem: part of phloem (inner bark) produced by cambium.

secondary thickening: a means of thickening tree stems by growth in diameter, not traceable to terminal growing points.

secondary-wall layers: see **S layers.**

secondary xylem: wood produced by cambium.

semibordered pit pair: see **half-bordered pit pair.**

semi-diffuse-porous wood: wood intermediate between diffuse-porous and ring-porous wood; see **semi-ring-porous wood.**

semi-ring-porous wood: same as *semi-diffuse-porous wood.*

septate fiber: fiber provided with cross walls (septa).

shake: rupture of cells or between cells resulting in the formation of an opening in the grain of the wood; the opening may develop at the common boundary of two rings or within a growth ring.

sheath cells: upright cells on the margins of, and tending to form a sheath around, the procumbent cells of a multiseriate ray as seen in tangential section.

shrinkage: for wood this is expressed as the percentage change in dimension with respect to the swollen size as a basis.

sieve pitting: sievelike clustered pits in the ends of the narrow tubular processes between disjunctive cells.

sieve plates: thin areas in sieve tubes provided with small openings through which the protoplasts of two adjacent sieve-tube elements are connected.

sieve tube: composite structure found in the phloem of all vascular plants; it is composed of a longitudinal series of sieve-tube elements whose protoplasts are connected by strands of protoplasm extending through small openings in sieve plates.

sieve-tube element: one of the cellular units composing a sieve tube.

silver grain: quarter-sawn wood with conspicuous, lustrous rays.

simple perforation: a single and usually large and more or less rounded opening in the perforation plate; see also **multiple perforation.**

simple pit: a pit in which the cavity becomes wider, or remains of constant width, or only gradually narrows toward the cell lumen, during the growth in thickness of the secondary wall.

simple pit pair: a pairing of two simple pits in adjacent cells.

simple ray: small wood ray of the type that accompanies *compound* (oak-type), or *aggregate* rays, as in oak and alder; see **compound ray** and **aggregate ray.**

slash grain: see **flat grain.**

soft rot: decay caused by the Ascomycetes and Fungi Imperfecti; the surface of the affected wood is typically softened.

softwood: wood produced by coniferous trees; same as *nonporous wood.*

specific gravity of wood: the decimal ratio of the ovendry weight of a piece of wood to the weight of the water displaced by the wood at a given moisture content; abbreviated as sp gr or G.

specific strength of wood: an index of the efficiency of wood in stress resistance; the ratio of a given strength property to the specific gravity of the wood; in compression and tension the specific gravity is used to the first power, in bending the specific gravity is taken to the 1.5 power, and in working with E (modulus of elasticity) the bending modulus is divided by specific gravity squared.

spiral grain: grain in which the fibers are aligned in a helical orientation around the axis of the stem.

spiral thickening: ridges on the inner face of the secondary cell wall in the form of single or multiple helices around the cell axis.

springwood: see **early wood.**

star shake: heart shake that radiates from the pith.

stele: central core of the cylindrical axis of a vascular plant arising from the further differentiation of the plerome; it consists of pith, primary xylem, cambium, primary phloem, and pericycle.

stem-formed wood: wood formed in the stem of the tree below the crown; see **crown-formed wood.**

sticker markings: a special oxidative stain which occurs during air seasoning and kiln drying on and beneath the surface of the board, where stickers come in contact with it.

storied rays: rays arranged in tiers or in echelons, as viewed on a tangential surface or in a tangential section; see **ripple marks.**

straight grain: wood in which the direction of the fiber alignment is straight; and also refers to the grain orientation in standing trees which is parallel to the stem axis.

strain: the unit deformation produced by an applied stress; has similar designations to those given for stress.

strand parenchyma: see **longitudinal strand parenchyma.**

strand tracheids: tracheids in coniferous wood that arise from the further division of a cell which otherwise would have developed into the longitudinal tracheid; differing from the latter in being shorter and having one or both end walls at right angles to the longitudinal walls.

stress: force per unit of volume or area; expressed as primary stresses, *compression* with the forces acting toward each other, *tension* with forces acting against each other, or *shear* with forces sliding on each other; a combination of primary forces produces *bending stresses.*

stump wood: bell-shaped base of the tree just above the roots.

substitute fiber: fibrous parenchymatous cell in wood; fibrous cell in wood that remains living while it is a part of the sapwood.

summerwood: see **late wood.**

surface checks: seasoning checks that develop on the surface and extend into the wood for varying distances.

swirl crotch: figure obtained from that section of the tree where the typical crotch figure fades into that of normal stem wood.

(t): symbol indicating a tangential section or surface; see **tangential section.**

tangential section: section cut along the grain at right angles to the wood rays; see **radial section.**

taxodioid pit: a cross-field pit in early wood of conifers with large, ovoid to circular, included aperture that is wider than the lateral space on either side between the aperture and the border, as in *Sequoia, Taxodium,* and *Abies.*

tension wood: reaction wood formed typically on the upper sides of branches and the upper, usually concave side of leaning or crooked stems of hardwoods; characterized anatomically by lack of cell wall lignification and often by the presence of a gelatinous layer in the fibers; see also **compression wood, reaction wood.**

terminal parenchyma: see **marginal apotracheal parenchyma.**

texture of wood: expression that refers to the size and the proportional amounts of woody elements; in coniferous woods, the average tangential diameter of the tracheids is the best indicator of texture; in the hardwoods, the tangential diameters and number of vessels and rays.

thermal conductivity of wood: in the English system, the amount of heat in British thermal units (Btu) that will flow in 1 hr through a homogeneous material 1 in. thick and 1 ft square per 1°F temperature difference between the surfaces.

thermal insulating value of wood (R): the reciprocal of the thermal conductivity of wood.

tiger grain: a term applied to curly grain in hard maple.

tight knot: that portion of a branch which is embedded in the tree trunk while the branch is alive; also called *intergrown knot.*

torn grain: see **chipped grain.**

torus: central thickened portion of the pit membrane of a bordered pit.

toughness: the ability to absorb energy without separation of the material as failure progresses; the opposite of brashness.

trabecula: cylindrical, barlike structure extending across the lumen of a tracheid from one tangential wall to the other; trabeculae usually occur in series.

tracheid: fibrous lignified cell with bordered pits and imperforate ends; in coniferous wood, the tracheids are very long (up to 7 plus mm) and are equipped with large, prominent bordered pits on their radial walls; tracheids in hardwoods are shorter fibrous cells (seldom over 1.5 mm), are as long as the vessel elements with which they are associated, and possess small bordered pits; see **vascular tracheids** and **vasicentric tracheids.**

transverse resin canals: resin canals extending across the grain that are included in fusiform wood rays.

transverse section: section cut at right angles to the grain: see **cross section.**

traumatic resin canal: resin canal supposedly arising as a result of injury.

true cellulose: a carbohydrate with the empirical formula $(C_6H_{10}O_5)n$ that occurs naturally as a chain polymer and which hydrolyzes entirely to glucose residues; see also **alpha cellulose, holocellulose.**

true middle lamella: same as *intercellular layer.*

twisting: warping in which one corner of a piece of wood twists out of the plane of the other three.

tyloses: saclike or cystlike structures that sometimes develop in a vessel and rarely in a fiber through the proliferation of the protoplast (living contents) of a parenchyma cell through a pit pair (sing. *tylosis*).

tylosoids: structures in resin canals resembling tyloses in hardwoods; they arise through the proliferation of thin-walled epithelial cells (sing. *tylosoid*).

uniseriate ray: ray consisting of one row of cells, as viewed in the tangential section.

upright ray cell: short, high cell (at least twice the height of an ordinary ray cell), occurring on the margins and frequently in addition on the flanks and in the body of heterocellular ray; see **procumbent ray cell.**

vascular cambium: see **cambium.**

vascular plant: plant possessing specialized conducting tissue consisting of xylem and phloem.

vascular tissue: specialized conducting tissue consisting of xylem and phloem.

vascular tracheids: specialized cells in certain hardwoods, similar in shape, size, and arrangement to the small vessel elements but differing from them in being imperforate at the ends; see **vasicentric tracheids.**

vasicentric parenchyma: see **paratracheal vasicentric parenchyma.**

vasicentric tracheids: short, irregularly shaped fibrous cells with conspicuous bordered pits; vasicentric tracheids abound in the proximity of the large early-wood vessels of certain ring-porous hardwoods; they differ from vascular tracheids not only in shape but in arrangement (they are not arranged in definite longitudinal rows like vascular tracheids); see **vascular tracheids.**

veneer: thin sheet of wood sliced, sawed, or rotary-cut from a log or a flitch.

vessel: articulated, tubelike structure of indeterminate length in porous woods; formed through the fusion of the cells in a longitudinal row and perforation of common walls in one of a number of ways; see **multiple perforation, scalariform perforation plate,** and **simple perforation.**

vessel element: one of the cellular components of a vessel; synonym *vessel member;* formerly also known as *vessel segment.*

vessel member: same as *vessel element.*

vessel perforation: an opening from one vessel element to another.

vessel segment: an obsolete term; see **vessel element.**

vestured pit: bordered pit with its cavity wholly or partly lined with projections from the overhanging secondary wall.

warping: any distortion in a piece of wood from its true plane that may occur in seasoning.

warty layer: an isotropic layer of material deposited on the inner surface of the secondary wall of many kinds of wood; this layer frequently contains encysted globules of a dissimilar material; these inclusions produce the warts which lend the name to the layer.

wavy grain: grain due to undulations in the direction of fiber alignment; when a wavy grained wood is split radially, the exposed surfaces are wavy.

weathering: the cumulative effect of surface deterioration in wood exposed to weather and unprotected by paint or other means.

weight density of wood: the weight of a unit volume of wood; the weight of the wood is the sum of dry-wood weight and the water; volume is taken at the same moisture condition; commonly expressed as "weight of wood."

wetwood: the heartwood, and sometimes the inner sapwood, with higher moisture content than the adjacent sapwood; the excessive wetness is associated with bacterial action, resulting in fermentation odors. Wetwood is subject to excessive checking and collapse and causes difficulties in gluing.

white rot: a type of wood-destroying fungi that attack both cellulose and lignin; the resultant mass is spongy or stringy, usually white, but may assume various shades of yellow, tan, and light brown.

window-like pit: see **fenestriform pit.**

wind shake: see **ring shake.**
wood: xylary portion of fibrovascular tissue.
wood ray: that portion of a ray included in the wood; see **ray.**
wound heartwood: patches of dead sapwood that develop in the vicinity of wounds, i.e., sapwood in which the parenchyma is no longer living; wound heartwood is similar to normal heartwood except in location.
(x): symbol indicating a transverse section or surface made by cutting across the grain at right angles.
xylary ray: that portion of a ray included in the xylem; wood ray; see **ray.**
xylem: principal strengthening and water-conducting tissue of the stems, roots, and leaves of vascular plants, characterized by the presence of tracheary elements; the woody portion of vascular tissue.
xylem daughter cell: a xylem cell formed by the periclinal division of the xylem mother cell.
xylem mother cell: a cell formed by the periclinal division of a fusiform cambial initial on the xylem side, which undergoes further periclinal division before differentiation into a mature xylem cell.
yard brown stain: chemical stain that develops during air seasoning or in storage; see **chemical stain** and **kiln brown stain.**
Young's modulus of elasticity (Y): the measure of elasticity expressed as the ratio of stress to strain where both stress and strain are in measures of compressive or tensile response.
zonate: arranged in concentric lines or bands, as viewed in the transverse section; said of banded *pores* or *parenchyma*.

Index

Page numbers in **boldface** refer to the sections entitled "Descriptions of Woods by Species," pp. **448–503** and **537–626**.